1932	James Chadwick discovers the neutron.
1932	John Cockcroft and Ernest Walton produce the first nuclear reaction using a high-voltage accelerator.
1932	Ernest Lawrence produces first cyclotron for studying nuclear reactions.
1934	Irène and Frédéric Joliot-Curie discover artificially induced radioactivity.
1935	Hideki Yukawa proposes existence of medium-mass particles (mesons).
1938	Otto Hahn, Fritz Strassmann, Lise Meitner, and Otto Frisch discover nuclear fission.
1938	Hans Bethe proposes thermonuclear fusion reactions as the source of energy in stars.
1940	Edwin McMillan, Glenn Seaborg, and colleagues produce first synthetic transuranic elements.
1942	Enrico Fermi and colleagues build first nuclear fission reactor.
1945	Detonation of first fission bomb in New Mexico desert.
1946	George Gamow proposes big-bang cosmology.
1948	John Bardeen, Walter Brattain, and William Shockley demonstrate first transistor.
1952	Detonation of first thermonuclear fusion bomb at Eniwetok atoll.
1956	Frederick Reines and Clyde Cowan demonstrate experimental evidence for existence of neutrino.
1958	Rudolf L. Mössbauer demonstrates recoilless emission of gamma rays.
1960	Theodore Maiman constructs first ruby laser; Ali Javan constructs first helium-neon laser.
1964	Allan R. Sandage discovers first quasar.
1964	Murray Gell-Mann and George Zweig independently introduce three-quark model of elementary particles.
1965	Arno Penzias and Robert Wilson discover cosmic microwave background radiation.
1967	Jocelyn Bell and Anthony Hewish discover first pulsar.
1967	Steven Weinberg and Abdus Salam independently propose a unified theory linking the weak and electromagnetic interactions.
1974	Burton Richter and Samuel Ting and co-workers independently discover first evidence of fourth quark (charm).
1974	Joseph Taylor and Russell Hulse discover first binary pulsar.
1977	Leon Lederman and colleagues discover new particle showing evidence for fifth quark (bottom).
1981	Gerd Binnig and Heinrich Rohrer invent scanning-tunneling electron microscope.
1983	Carlo Rubbia and co-workers at CERN discover W and Z particles.
1986	J. Georg Bednorz and Karl Alex Müller produce first high-temperature superconductors.
1994	Investigators at Fermilab discover evidence for sixth quark (top).
1995	Eric Cornell and Carl Wieman produce first Bose-Einstein condensation.
1998	Discovery of neutrino oscillations shows that neutrinos have small but nonzero mass.
2003	WMAP satellite data reveal age and composition of universe.

MODERN PHYSICS

MODERN PHYSICS

Third edition

Kenneth S. Krane

DEPARTMENT OF PHYSICS
OREGON STATE UNIVERSITY

JOHN WILEY & SONS, INC

VP AND EXECUTIVE PUBLISHER	Kaye Pace
EXECUTIVE EDITOR	Stuart Johnson
MARKETING MANAGER	Christine Kushner
DESIGN DIRECTOR	Jeof Vita
DESIGNER	Kristine Carney
PRODUCTION MANAGER	Janis Soo
ASSISTANT PRODUCTION EDITOR	Elaine S. Chew
PHOTO DEPARTMENT MANAGER	Hilary Newman
PHOTO EDITOR	Sheena Goldstein
COVER DESIGNER	Seng Ping Ngieng
COVER IMAGE	CERN/SCIENCE PHOTO LIBRARY/Photo Researchers, Inc.

This book was set in Times by Laserwords Private Limited and printed and bound by R. R. Donnelley and Sons Company, Von Hoffman. The cover was printed by R. R. Donnelley and Sons Company, Von Hoffman.

This book is printed on acid free paper. ∞

Founded in 1807, John Wiley & Sons, Inc. has been a valued source of knowledge and understanding for more than 200 years, helping people around the world meet their needs and fulfill their aspirations. Our company is built on a foundation of principles that include responsibility to the communities we serve and where we live and work. In 2008, we launched a Corporate Citizenship Initiative, a global effort to address the environmental, social, economic, and ethical challenges we face in our business. Among the issues we are addressing are carbon impact, paper specifications and procurement, ethical conduct within our business and among our vendors, and community and charitable support. For more information, please visit our website: www.wiley.com/go/citizenship.

Evaluation copies are provided to qualified academics and professionals for review purposes only, for use in their courses during the next academic year. These copies are licensed and may not be sold or transferred to a third party. Upon completion of the review period, please return the evaluation copy to Wiley. Return instructions and a free of charge return mailing label are available at www.wiley.com/go/returnlabel. If you have chosen to adopt this textbook for use in your course, please accept this book as your complimentary desk copy. Outside of the United States, please contact your local sales representative.

Library of Congress Cataloging-in-Publication Data

Krane, Kenneth S.
    Modern physics/Kenneth S. Krane. -- 3rd ed.
       p. cm.
    Includes bibliographical references and index.
   ISBN 978-1-118-06114-5 (hardback)
1. Physics. I. Title.
   QC21.2.K7 2012
   539--dc23

                 2011039948

Printed in the United States of America
10 9 8 7 6 5 4 3

PREFACE

This textbook is meant to serve a first course in modern physics, including relativity, quantum mechanics, and their applications. Such a course often follows the standard introductory course in calculus-based classical physics. The course addresses two different audiences: (1) Physics majors, who will later take a more rigorous course in quantum mechanics, find an introductory modern course helpful in providing background for the rigors of their imminent coursework in classical mechanics, thermodynamics, and electromagnetism. (2) Nonmajors, who may take no additional physics class, find an increasing need for concepts from modern physics in their disciplines—a classical introductory course is not sufficient background for chemists, computer scientists, nuclear and electrical engineers, or molecular biologists.

Necessary prerequisites for undertaking the text include any standard calculus-based course covering mechanics, electromagnetism, thermal physics, and optics. Calculus is used extensively, but no previous knowledge of differential equations, complex variables, or partial derivatives is assumed (although some familiarity with these topics would be helpful).

Chapters 1–8 constitute the core of the text. They cover special relativity and quantum theory through atomic structure. At that point the reader may continue with Chapters 9–11 (molecules, quantum statistics, and solids) or branch to Chapters 12–14 (nuclei and particles). The final chapter covers cosmology and can be considered the capstone of modern physics as it brings together topics from relativity (special and general) as well as from nearly all of the previous material covered in the text.

The unifying theme of the text is the empirical basis of modern physics. Experimental tests of derived properties are discussed throughout. These include the latest tests of special and general relativity as well as studies of wave-particle duality for photons and material particles. Applications of basic phenomena are extensively presented, and data from the literature are used not only to illustrate those phenomena but to offer insight into how "real" physics is done. Students using the text have the opportunity to study how laboratory results and the analysis based on quantum theory go hand-in-hand to illuminate such diverse topics as Bose-Einstein condensation, heat capacities of solids, paramagnetism, the cosmic microwave background radiation, X-ray spectra, dilute mixtures of ^{3}He in ^{4}He, and molecular spectroscopy of the interstellar medium.

This third edition offers many changes from the previous edition. Most of the chapters have undergone considerable or complete rewriting. New topics have been introduced and others have been rearranged. More experimental results are presented and recent discoveries are highlighted, such as the WMAP microwave background data and Bose-Einstein condensation. End-of-chapter problem sets now include problems organized according to chapter section, which offer the student an opportunity to gain familiarity with a particular topic, as well as general problems, which often require the student to apply a broader array of concepts or techniques. The number of worked examples in the chapters and the number of end-of-chapter questions and problems have each increased by about 15% from the previous edition. The range of abilities required to solve the problems has been broadened, so that this edition includes both more straightforward problems that build confidence as well as more difficult problems that will challenge students. Each chapter now includes a brief summary of the important points. Some of the end-of-chapter problems are available for assignment using the WebAssign program (www.webassign.net).

A new development in physics teaching since the appearance of the 2nd edition of this text has been the availability of a large and robust body of literature from physics education research (PER). My own teaching style has been profoundly influenced by PER findings, and in preparing this new edition I have tried to incorporate PER results wherever possible. One of the major themes that has emerged from PER in the past decade or two is that students can often learn successful algorithms for solving problems while lacking a fundamental understanding of the underlying concepts. Many approaches to addressing this problem are based on pre-class conceptual exercises and in-class individual or group activities that help students to reason through diverse problems that can't be resolved by plugging numbers into an equation. It is absolutely essential to devote class time to these exercises and to follow through with exam questions that require similar analysis and articulation of the conceptual reasoning. More details regarding the application of PER to the teaching of modern physics, including references to articles from the PER literature, are included in the Instructor's Manual for this text, which can be found at www.wiley.com/college/krane. The Instructor's Manual also includes examples of conceptual questions for in-class discussion or exams that have been developed and class tested through the support of a Course, Curriculum and Laboratory Improvement grant from the National Science Foundation.

Specific changes to the chapters include the following:

Chapter 1: The sections on Units and Dimensions and on Significant Figures have been removed. In their place, a more detailed review of applications of classical energy and momentum conservation is offered. The need for special relativity is briefly established with a discussion of the failures of the classical concepts of space and time, and the need for quantum theory is previewed in the failure of Maxwell-Boltzmann particle statistics to account for the heat capacities of diatomic gases.

Chapter 2: Spacetime diagrams have been introduced to help illustrate relationships in the twin paradox. The application of the relativistic conservation laws to decay and collisions processes is now given a separate section to help students learn to apply those laws. The section on tests of special relativity has been updated to include recent results.

Chapter 3: The section on thermal radiation has been rewritten, and more detailed derivations of the Rayleigh-Jeans and Planck formulas are now given.

Chapter 4: New experimental results for particle diffraction and interference are discussed. The sections on the classical uncertainty relationships and on wave packet construction and motion have been rewritten.

Chapter 5: To help students understand the processes involved in applying boundary conditions to solutions of the Schrödinger equation, a new section on wave boundary conditions has been added. A new introductory section on particle confinement introduces energy quantization and helps to build the connection between the wave function and the uncertainty relationships. Time dependence of the wave function is introduced more explicitly at an earlier stage in the formulism. Graphic illustrations for step and barrier problems now show the real and imaginary parts of the wave function as well as its squared magnitude.

Chapter 6: The derivation of the Thomson model scattering angle has been modified, and the section on deficiencies of the Bohr model has been rewritten.

Chapter 7: To ease the entry into the 3-dimensional Schrödinger analysis of the hydrogen atom in spherical coordinates, a new section on the one-dimensional hydrogen atom has been added. Angular momentum concepts relating to the hydrogen atom are now introduced before the full solutions to the wave equation.

Chapter 8: Much of the material has been reorganized for clarity and ease of presentation. The screening discussion has been made more explicit.

Chapter 9: More emphasis has been given to the use of bonding and antibonding orbitals to predict the relative stability of molecules. Sections on molecular vibrations and rotations have been rewritten.

Chapter 10: This chapter has been extensively rewritten. A new section on the density of states function allows statistical distributions for photons or particles to be discussed more rigorously. New applications of quantum statistics include Bose-Einstein condensation, white dwarf stars, and dilute mixtures of ^{3}He in ^{4}He.

Chapter 11: The chapter has been rewritten to broaden the applications of the quantum theory of solids to include not only electrical conductivity but also the heat capacity of solids and paramagnetism.

Chapter 12: To emphasize the unity of various topics within modern physics, this chapter now includes proton and neutron separation energies, a new section on quantum states in nuclei, and nuclear vibrational and rotational states, all of which have analogues in atomic or molecular structure.

Chapter 13: The discussion of the physics of fission has been expanded while that of the properties of nuclear reactors has been reduced somewhat. Because much current research in nuclear physics is related to astrophysics, this chapter now features a section on nucleosynthesis.

Chapter 14: New material on quarkonium and neutrino oscillations has been added.

Chapter 15: Chapters 15 and 16 of the 2nd edition have been collapsed into a single chapter on cosmology. New results from COBE and WMAP are included, along with discussions of the horizon and flatness problems (and their inflationary solution).

Many reviewers and class-testers of the manuscript of this edition have offered suggestions to improve both the physics and its presentation. I am particularly grateful to:

David Bannon, Oregon State University

Gerald Crawford, Fort Lewis College

Luther Frommhold, University of Texas-Austin

Gary Goldstein, Tufts University

Leon Gunther, Tufts University

Gary Ihas, University of Florida

Paul Lee, California State University, Northridge

Jeff Loats, Metropolitan State College of Denver

Jay Newman, Union College

Stephen Pate, New Mexico State University

David Roundy, Oregon State University

Rich Schelp, Erskine College

Weidian Shen, Eastern Michigan University

Hongtao Shi, Sonoma State University

Janet Tate, Oregon State University

Jeffrey L. Wragg, College of Charleston

Weldon Wilson, University of Central Oklahoma

I am also grateful for the many anonymous comments from students who used the manuscript at the test sites. I am indebted to all those reviewers and users for their contributions to the project.

Funding for the development and testing of the supplemental exercises in the Instructor's Manual was provided through a grant from the National Science Foundation. I am pleased to acknowledge their support. Two graduate students at Oregon State University helped to test and implement the curricular reforms: K. C. Walsh and Pornrat Wattasinawich. I appreciate their assistance in this project.

The staff at John Wiley & Sons have been especially helpful throughout the project. I am particularly grateful to: Executive Editor Stuart Johnson for his patience and support in bringing the new edition into reality; Assistant Production Editor Elaine Chew for handling a myriad of complicated composition and illustration details with efficiency and good humor; and Photo Editor Sheena Goldstein for helping me navigate the treacherous waters of new copyright and permission restrictions.

In my research and other professional activities, I occasionally meet physicists who used earlier editions of this text when they were students. Some report that their first exposure to modern physics kindled the spark that led them to careers in physics. For many students, this course offers their first insights into what physicists really do and what is exciting, perplexing, and challenging about our profession. I hope students who use this new edition will continue to find those inspirations.

Corvallis, Oregon   Kenneth S. Krane
August 2011   kranek@physics.oregonstate.edu

CONTENTS

Chapter 1

THE FAILURES OF CLASSICAL PHYSICS

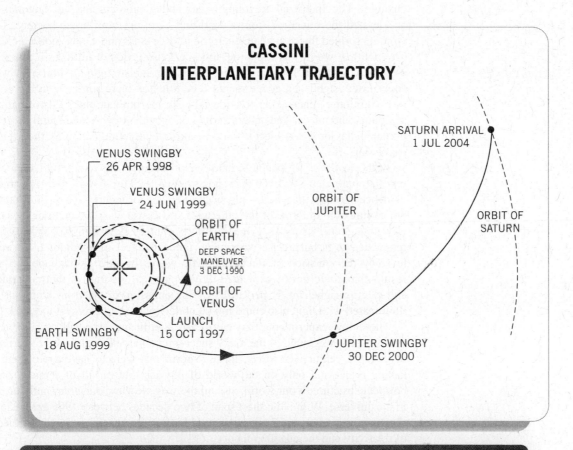

CASSINI
INTERPLANETARY TRAJECTORY

VENUS SWINGBY
26 APR 1998

VENUS SWINGBY
24 JUN 1999

ORBIT OF
EARTH

DEEP SPACE
MANEUVER
3 DEC 1990

ORBIT OF
VENUS

LAUNCH
15 OCT 1997

EARTH SWINGBY
18 AUG 1999

ORBIT OF
JUPITER

SATURN ARRIVAL
1 JUL 2004

ORBIT OF
SATURN

JUPITER SWINGBY
30 DEC 2000

Classical physics, as postulated by Newton, has enabled us to send space probes on trajectories involving many complicated maneuvers, such as the Cassini mission to Saturn, which was launched in 1997 and gained speed for its trip to Saturn by performing four "gravity-assist" flybys of Venus (twice), Earth, and Jupiter. The spacecraft arrived at Saturn in 2004 and is expected to continue to send data through at least 2017. Planning and executing such interplanetary voyages are great triumphs for Newtonian physics, but when objects move at speeds close to the speed of light or when we examine matter on the atomic or subatomic scale, Newtonian mechanics is not adequate to explain our observations, as we discuss in this chapter.

If you were a physicist living at the end of the 19th century, you probably would have been pleased with the progress that physics had made in understanding the laws that govern the processes of nature. Newton's laws of mechanics, including gravitation, had been carefully tested, and their success had provided a framework for understanding the interactions among objects. Electricity and magnetism had been unified by Maxwell's theoretical work, and the electromagnetic waves predicted by Maxwell's equations had been discovered and investigated in the experiments conducted by Hertz. The laws of thermodynamics and kinetic theory had been particularly successful in providing a unified explanation of a wide variety of phenomena involving heat and temperature. These three successful theories—mechanics, electromagnetism, and thermodynamics—form the basis for what we call "classical physics."

Beyond your 19th-century physics laboratory, the world was undergoing rapid changes. The Industrial Revolution demanded laborers for the factories and accelerated the transition from a rural and agrarian to an urban society. These workers formed the core of an emerging middle class and a new economic order. The political world was changing, too—the rising tide of militarism, the forces of nationalism and revolution, and the gathering strength of Marxism would soon upset established governments. The fine arts were similarly in the middle of revolutionary change, as new ideas began to dominate the fields of painting, sculpture, and music. The understanding of even the very fundamental aspects of human behavior was subject to serious and critical modification by the Freudian psychologists.

In the world of physics, too, there were undercurrents that would soon cause revolutionary changes. Even though the overwhelming majority of experimental evidence agreed with classical physics, several experiments gave results that were not explainable in terms of the otherwise successful classical theories. Classical electromagnetic theory suggested that a medium is needed to propagate electromagnetic waves, but precise experiments failed to detect this medium. Experiments to study the emission of electromagnetic waves by hot, glowing objects gave results that could not be explained by the classical theories of thermodynamics and electromagnetism. Experiments on the emission of electrons from surfaces illuminated with light also could not be understood using classical theories.

These few experiments may not seem significant, especially when viewed against the background of the many successful and well-understood experiments of the 19th century. However, these experiments were to have a profound and lasting effect, not only on the world of physics, but on all of science, on the political structure of our world, and on the way we view ourselves and our place in the universe. Within the short span of two decades between 1905 and 1925, the shortcomings of classical physics would lead to the special and general theories of relativity and the quantum theory.

The designation *modern physics* usually refers to the developments that began in about 1900 and led to the relativity and quantum theories, including the applications of those theories to understanding the atom, the atomic nucleus and the particles of which it is composed, collections of atoms in molecules and solids, and, on a cosmic scale, the origin and evolution of the universe. Our discussion of modern physics in this text touches on each of these areas.

We begin our study in this chapter with a brief review of some important principles of classical physics, and we discuss some situations in which classical physics offers either inadequate or incorrect conclusions. These situations are not necessarily those that originally gave rise to the relativity and quantum theories, but they do help us understand why classical physics fails to give us a complete picture of nature.

1.1 REVIEW OF CLASSICAL PHYSICS

Although there are many areas in which modern physics differs radically from classical physics, we frequently find the need to refer to concepts of classical physics. Here is a brief review of some of the concepts of classical physics that we may need.

Mechanics

A particle of mass m moving with velocity v has a *kinetic energy* defined by

$$K = \tfrac{1}{2} mv^2 \tag{1.1}$$

and a *linear momentum* $\vec{\mathbf{p}}$ defined by

$$\vec{\mathbf{p}} = m\vec{\mathbf{v}} \tag{1.2}$$

In terms of the linear momentum, the kinetic energy can be written

$$K = \frac{p^2}{2m} \tag{1.3}$$

When one particle collides with another, we analyze the collision by applying two fundamental conservation laws:

I. **Conservation of Energy.** The total energy of an isolated system (on which no net external force acts) remains constant. In the case of a collision between particles, this means that the total energy of the particles *before* the collision is equal to the total energy of the particles *after* the collision.

II. **Conservation of Linear Momentum.** The total linear momentum of an isolated system remains constant. For the collision, the total linear momentum of the particles *before* the collision is equal to the total linear momentum of the particles *after* the collision. Because linear momentum is a vector, application of this law usually gives us two equations, one for the x components and another for the y components.

These two conservation laws are of the most basic importance to understanding and analyzing a wide variety of problems in classical physics. Problems 1–4 and 11–14 at the end of this chapter review the use of these laws.

The importance of these conservation laws is both so great and so fundamental that, even though in Chapter 2 we learn that the special theory of relativity modifies Eqs. 1.1, 1.2, and 1.3, the laws of conservation of energy and linear momentum remain valid.

Example 1.1

A helium atom ($m = 6.6465 \times 10^{-27}$ kg) moving at a speed of $v_{He} = 1.518 \times 10^6$ m/s collides with an atom of nitrogen ($m = 2.3253 \times 10^{-26}$ kg) at rest. After the collision, the helium atom is found to be moving with a velocity of $v'_{He} = 1.199 \times 10^6$ m/s at an angle of $\theta_{He} = 78.75°$ relative to the direction of the original motion of the helium atom. (a) Find the velocity (magnitude and direction) of the nitrogen atom after the collision. (b) Compare the kinetic energy before the collision with the total kinetic energy of the atoms after the collision.

Solution

(a) The law of conservation of momentum for this collision can be written in vector form as $\vec{p}_{initial} = \vec{p}_{final}$, which is equivalent to

$$p_{x,initial} = p_{x,final} \quad \text{and} \quad p_{y,initial} = p_{y,final}$$

The collision is shown in Figure 1.1. The initial values of the total momentum are, choosing the x axis to be the direction of the initial motion of the helium atom,

$$p_{x,initial} = m_{He}v_{He} \quad \text{and} \quad p_{y,initial} = 0$$

The final total momentum can be written

$$p_{x,final} = m_{He}v'_{He} \cos\theta_{He} + m_N v'_N \cos\theta_N$$
$$p_{y,final} = m_{He}v'_{He} \sin\theta_{He} + m_N v'_N \sin\theta_N$$

The expression for $p_{y,final}$ is written in general form with a + sign even though we expect that θ_{He} and θ_N are on opposite sides of the x axis. If the equation is written in this way, θ_N will come out to be negative. The law of

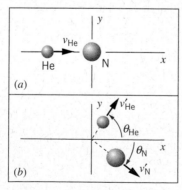

FIGURE 1.1 Example 1.1. (a) Before collision; (b) after collision.

conservation of momentum gives, for the x components, $m_{He}v_{He} = m_{He}v'_{He} \cos\theta_{He} + m_N v'_N \cos\theta_N$, and for the y components, $0 = m_{He}v'_{He} \sin\theta_{He} + m_N v'_N \sin\theta_N$. Solving for the unknown terms, we find

$$v'_N \cos\theta_N = \frac{m_{He}(v_{He} - v'_{He} \cos\theta_{He})}{m_N}$$
$$= \{(6.6465 \times 10^{-27} \text{ kg})[1.518 \times 10^6 \text{ m/s} - (1.199 \times 10^6 \text{ m/s})(\cos 78.75°)]\} \times (2.3253 \times 10^{-26} \text{ kg})^{-1}$$
$$= 3.6704 \times 10^5 \text{ m/s}$$

$$v'_N \sin\theta_N = -\frac{m_{He}v'_{He} \sin\theta_{He}}{m_N}$$
$$= -(6.6465 \times 10^{-27} \text{ kg})(1.199 \times 10^6 \text{ m/s}) \times (\sin 78.75°)(2.3253 \times 10^{-26} \text{ kg})^{-1}$$
$$= -3.3613 \times 10^5 \text{ m/s}$$

We can now solve for v'_N and θ_N:

$$v'_N = \sqrt{(v'_N \sin\theta_N)^2 + (v'_N \cos\theta_N)^2}$$
$$= \sqrt{(-3.3613 \times 10^5 \text{ m/s})^2 + (3.6704 \times 10^5 \text{ m/s})^2}$$
$$= 4.977 \times 10^5 \text{ m/s}$$
$$\theta_N = \tan^{-1} \frac{v'_N \sin\theta_N}{v'_N \cos\theta_N}$$
$$= \tan^{-1}\left(\frac{-3.3613 \times 10^5 \text{ m/s}}{3.6704 \times 10^5 \text{ m/s}}\right) = -42.48°$$

(b) The initial kinetic energy is
$$K_{initial} = \tfrac{1}{2}m_{He}v_{He}^2$$
$$= \tfrac{1}{2}(6.6465 \times 10^{-27} \text{ kg})(1.518 \times 10^6 \text{ m/s})^2$$
$$= 7.658 \times 10^{-15} \text{ J}$$

and the total final kinetic energy is
$$K_{final} = \tfrac{1}{2}m_{He}v_{He}'^2 + \tfrac{1}{2}m_N v_N'^2$$
$$= \tfrac{1}{2}(6.6465 \times 10^{-27} \text{ kg})(1.199 \times 10^6 \text{ m/s})^2 + \tfrac{1}{2}(2.3253 \times 10^{-26} \text{ kg})(4.977 \times 10^5 \text{ m/s})^2$$
$$= 7.658 \times 10^{-15} \text{ J}$$

Note that the initial and final kinetic energies are equal. This is the characteristic of an *elastic* collision, in which no energy is lost to, for example, internal excitation of the particles.

Example 1.2

An atom of uranium ($m = 3.9529 \times 10^{-25}$ kg) at rest decays spontaneously into an atom of helium ($m = 6.6465 \times 10^{-27}$ kg) and an atom of thorium ($m = 3.8864 \times 10^{-25}$ kg). The helium atom is observed to move in the positive x direction with a velocity of 1.423×10^7 m/s (Figure 1.2). (a) Find the velocity (magnitude and direction) of the thorium atom. (b) Find the total kinetic energy of the two atoms after the decay.

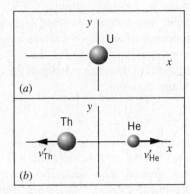

FIGURE 1.2 Example 1.2. (a) Before decay; (b) after decay.

Solution

(a) Here we again use the law of conservation of momentum. The initial momentum before the decay is zero, so the total momentum of the two atoms after the decay must also be zero:

$$p_{x,\text{initial}} = 0 \qquad p_{x,\text{final}} = m_{\text{He}}v'_{\text{He}} + m_{\text{Th}}v'_{\text{Th}}$$

Setting $p_{x,\text{initial}} = p_{x,\text{final}}$ and solving for v'_{Th}, we obtain

$$v'_{\text{Th}} = -\frac{m_{\text{He}}v'_{\text{He}}}{m_{\text{Th}}}$$

$$= -\frac{(6.6465 \times 10^{-27} \text{ kg})(1.423 \times 10^7 \text{ m/s})}{3.8864 \times 10^{-25} \text{ kg}}$$

$$= -2.432 \times 10^5 \text{ m/s}$$

The thorium atom moves in the negative x direction.

(b) The total kinetic energy after the decay is:

$$K = \tfrac{1}{2}m_{\text{He}}v'^2_{\text{He}} + \tfrac{1}{2}m_{\text{Th}}v'^2_{\text{Th}}$$

$$= \tfrac{1}{2}(6.6465 \times 10^{-27} \text{ kg})(1.423 \times 10^7 \text{ m/s})^2$$

$$+ \tfrac{1}{2}(3.8864 \times 10^{-25} \text{ kg})(-2.432 \times 10^5 \text{ m/s})^2$$

$$= 6.844 \times 10^{-13} \text{ J}$$

Clearly kinetic energy is not conserved in this decay, because the initial kinetic energy of the uranium atom was zero. However total energy *is conserved*—if we write the total energy as the sum of kinetic energy and nuclear energy, then the total initial energy (kinetic + nuclear) is equal to the total final energy (kinetic + nuclear). Clearly the gain in kinetic energy occurs as a result of a loss in nuclear energy. This is an example of the type of radioactive decay called alpha decay, which we discuss in more detail in Chapter 12.

Another application of the principle of conservation of energy occurs when a particle moves subject to an external force F. Corresponding to that external force there is often a potential energy U, defined such that (for one-dimensional motion)

$$F = -\frac{dU}{dx} \tag{1.4}$$

The total energy E is the sum of the kinetic and potential energies:

$$E = K + U \tag{1.5}$$

As the particle moves, K and U may change, but E remains constant. (In Chapter 2, we find that the special theory of relativity gives us a new definition of total energy.)

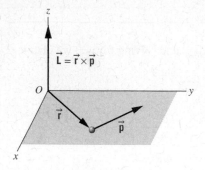

FIGURE 1.3 A particle of mass m, located with respect to the origin O by position vector $\vec{r}$ and moving with linear momentum $\vec{p}$, has angular momentum $\vec{L}$ about O.

When a particle moving with linear momentum $\vec{p}$ is at a displacement $\vec{r}$ from the origin O, its *angular momentum* $\vec{L}$ about the point O is defined (see Figure 1.3) by

$$\vec{L} = \vec{r} \times \vec{p} \qquad (1.6)$$

There is a conservation law for angular momentum, just as with linear momentum. In practice this has many important applications. For example, when a charged particle moves near, and is deflected by, another charged particle, the total angular momentum of the system (the two particles) remains constant if no net external torque acts on the system. If the second particle is so much more massive than the first that its motion is essentially unchanged by the influence of the first particle, the angular momentum of the first particle remains constant (because the second particle acquires no angular momentum). Another application of the conservation of angular momentum occurs when a body such as a comet moves in the gravitational field of the Sun—the elliptical shape of the comet's orbit is necessary to conserve angular momentum. In this case $\vec{r}$ and $\vec{p}$ of the comet must simultaneously change so that $\vec{L}$ remains constant.

Velocity Addition

Another important aspect of classical physics is the rule for combining velocities. For example, suppose a jet plane is moving at a velocity of $v_{PG} = 650$ m/s, as measured by an observer on the ground. The subscripts on the velocity mean "velocity of the plane relative to the ground." The plane fires a missile in the forward direction; the velocity of the missile relative to the plane is $v_{MP} = 250$ m/s. According to the observer on the ground, the velocity of the missile is: $v_{MG} = v_{MP} + v_{PG} = 250$ m/s $+ 650$ m/s $= 900$ m/s.

We can generalize this rule as follows. Let $\vec{v}_{AB}$ represent the velocity of A relative to B, and let $\vec{v}_{BC}$ represent the velocity of B relative to C. Then the velocity of A relative to C is

$$\vec{v}_{AC} = \vec{v}_{AB} + \vec{v}_{BC} \qquad (1.7)$$

This equation is written in vector form to allow for the possibility that the velocities might be in different directions; for example, the missile might be fired not in the direction of the plane's velocity but in some other direction. This seems to be a very "common-sense" way of combining velocities, but we will see later in this chapter (and in more detail in Chapter 2) that this common-sense rule can lead to contradictions with observations when we apply it to speeds close to the speed of light.

A common application of this rule (for speeds small compared with the speed of light) occurs in collisions, when we want to analyze conservation of momentum and energy in a frame of reference that is different from the one in which the collision is observed. For example, let's analyze the collision of Example 1.1 in a frame of reference that is moving with the center of mass. Suppose the initial velocity of the He atom defines the positive x direction. The velocity of the center of mass (relative to the laboratory) is then $v_{CL} = (v_{He}m_{He} + v_N m_N)/(m_{He} + m_N) = 3.374 \times 10^5$ m/s. We would like to find the initial velocity of the He and N relative to the center of mass. If we start with $v_{HeL} = v_{HeC} + v_{CL}$ and $v_{NL} = v_{NC} + v_{CL}$, then

$$v_{HeC} = v_{HeL} - v_{CL} = 1.518 \times 10^6 \text{ m/s} - 3.374 \times 10^5 \text{ m/s} = 1.181 \times 10^6 \text{ m/s}$$

$$v_{NC} = v_{NL} - v_{CL} = 0 - 3.374 \times 10^5 \text{ m/s} = -0.337 \times 10^6 \text{ m/s}$$

In a similar fashion we can calculate the final velocities of the He and N. The resulting collision as viewed from this frame of reference is illustrated in Figure 1.4. There is a special symmetry in this view of the collision that is not apparent from the same collision viewed in the laboratory frame of reference (Figure 1.1); each velocity simply changes direction leaving its magnitude unchanged, and the atoms move in opposite directions. The angles in this view of the collision are different from those of Figure 1.1, because the velocity addition in this case applies only to the x components and leaves the y components unchanged, which means that the angles must change.

Electricity and Magnetism

The electrostatic force (Coulomb force) exerted by a charged particle q_1 on another charge q_2 has magnitude

$$F = \frac{1}{4\pi\varepsilon_0}\frac{|q_1||q_2|}{r^2} \tag{1.8}$$

The direction of F is along the line joining the particles (Figure 1.5). In the SI system of units, the constant $1/4\pi\varepsilon_0$ has the value

$$\frac{1}{4\pi\varepsilon_0} = 8.988 \times 10^9 \ \mathrm{N \cdot m^2/C^2}$$

The corresponding potential energy is

$$U = \frac{1}{4\pi\varepsilon_0}\frac{q_1 q_2}{r} \tag{1.9}$$

In all equations derived from Eq. 1.8 or 1.9 as starting points, *the quantity $1/4\pi\varepsilon_0$ must appear*. In some texts and reference books, you may find electrostatic quantities in which this constant does not appear. In such cases, the centimeter-gram-second (cgs) system has probably been used, in which the constant $1/4\pi\varepsilon_0$ is *defined* to be 1. You should always be very careful in making comparisons of electrostatic quantities from different references and check that the units are identical.

An electrostatic potential difference ΔV can be established by a distribution of charges. The most common example of a potential difference is that between the two terminals of a battery. When a charge q moves through a potential difference ΔV, the change in its electrical potential energy ΔU is

$$\Delta U = q\Delta V \tag{1.10}$$

At the atomic or nuclear level, we usually measure charges in terms of the basic charge of the electron or proton, whose magnitude is $e = 1.602 \times 10^{-19}$ C. If such charges are accelerated through a potential difference ΔV that is a few volts, the resulting loss in potential energy and corresponding gain in kinetic energy will be of the order of 10^{-19} to 10^{-18} J. To avoid working with such small numbers, it is common in the realm of atomic or nuclear physics to measure energies in *electron-volts* (eV), defined to be the energy of a charge equal in magnitude to that of the electron that passes through a potential difference of 1 volt:

$$\Delta U = q\Delta V = (1.602 \times 10^{-19} \ \mathrm{C})(1 \ \mathrm{V}) = 1.602 \times 10^{-19} \ \mathrm{J}$$

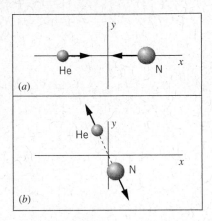

(a)

(b)

FIGURE 1.4 The collision of Figure 1.1 viewed from a frame of reference moving with the center of mass. (*a*) Before collision. (*b*) After collision. In this frame the two particles always move in opposite directions, and for elastic collisions the magnitude of each particle's velocity is unchanged.

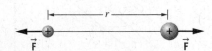

FIGURE 1.5 Two charged particles experience equal and opposite electrostatic forces along the line joining their centers. If the charges have the same sign (both positive or both negative), the force is repulsive; if the signs are different, the force is attractive.

and thus

$$1\,eV = 1.602 \times 10^{-19}\,J$$

Some convenient multiples of the electron-volt are

$$keV = kilo\ electron\text{-}volt = 10^3\,eV$$
$$MeV = mega\ electron\text{-}volt = 10^6\,eV$$
$$GeV = giga\ electron\text{-}volt = 10^9\,eV$$

(In some older works you may find reference to the BeV, for billion electron-volts; this is a source of confusion, for in the United States a billion is 10^9 while in Europe a billion is 10^{12}.)

Often we wish to find the potential energy of two basic charges separated by typical atomic or nuclear dimensions, and we wish to have the result expressed in electron-volts. Here is a convenient way of doing this. First we express the quantity $e^2/4\pi\varepsilon_0$ in a more convenient form:

$$\frac{e^2}{4\pi\varepsilon_0} = (8.988 \times 10^9\,N\cdot m^2/C^2)(1.602 \times 10^{-19}\,C)^2 = 2.307 \times 10^{-28}\,N\cdot m^2$$

$$= (2.307 \times 10^{-28}\,N\cdot m^2)\left(\frac{1}{1.602 \times 10^{-19}\,J/eV}\right)\left(\frac{10^9\,nm}{m}\right)$$

$$= 1.440\,eV\cdot nm$$

With this useful combination of constants it becomes very easy to calculate electrostatic potential energies. For two electrons separated by a typical atomic dimension of 1.00 nm, Eq. 1.9 gives

$$U = \frac{1}{4\pi\varepsilon_0}\frac{e^2}{r} = \frac{e^2}{4\pi\varepsilon_0}\frac{1}{r} = (1.440\,eV\cdot nm)\left(\frac{1}{1.00\,nm}\right) = 1.44\,eV$$

For calculations at the nuclear level, the femtometer is a more convenient unit of distance and MeV is a more appropriate energy unit:

$$\frac{e^2}{4\pi\varepsilon_0} = (1.440\,eV\cdot nm)\left(\frac{1\,m}{10^9\,nm}\right)\left(\frac{10^{15}\,fm}{1\,m}\right)\left(\frac{1\,MeV}{10^6\,eV}\right) = 1.440\,MeV\cdot fm$$

It is remarkable (and convenient to remember) that the quantity $e^2/4\pi\varepsilon_0$ has the same value of 1.440 whether we use typical atomic energies and sizes (eV·nm) or typical nuclear energies and sizes (MeV·fm).

A magnetic field $\vec{B}$ can be produced by an electric current i. For example, the magnitude of the magnetic field at the center of a circular current loop of radius r is (see Figure 1.6a)

$$B = \frac{\mu_0 i}{2r} \tag{1.11}$$

The SI unit for magnetic field is the tesla (T), which is equivalent to a newton per ampere-meter. The constant μ_0 is

$$\mu_0 = 4\pi \times 10^{-7}\,N\cdot s^2/C^2$$

Be sure to remember that i is in the direction of the conventional (*positive*) current, opposite to the actual direction of travel of the negatively charged electrons that typically produce the current in metallic wires. The direction of $\vec{B}$ is chosen according to the right-hand rule: if you hold the wire in the right hand with the

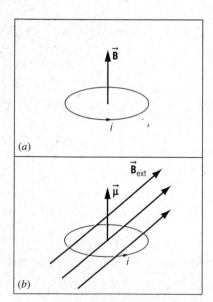

FIGURE 1.6 (*a*) A circular current loop produces a magnetic field $\vec{B}$ at its center. (*b*) A current loop with magnetic moment $\vec{\mu}$ in an external magnetic field $\vec{B}_{ext}$. The field exerts a torque on the loop that will tend to rotate it so that $\vec{\mu}$ lines up with $\vec{B}_{ext}$.

thumb pointing in the direction of the current, the fingers point in the direction of the magnetic field.

It is often convenient to define the *magnetic moment* $\vec{\mu}$ of a current loop:

$$|\vec{\mu}| = iA \tag{1.12}$$

where A is the geometrical area enclosed by the loop. The direction of $\vec{\mu}$ is perpendicular to the plane of the loop, according to the right-hand rule.

When a current loop is placed in a uniform *external* magnetic field $\vec{\mathbf{B}}_{ext}$ (as in Figure 1.6b), there is a torque $\vec{\tau}$ on the loop that tends to line up $\vec{\mu}$ with $\vec{\mathbf{B}}_{ext}$:

$$\vec{\tau} = \vec{\mu} \times \vec{\mathbf{B}}_{ext} \tag{1.13}$$

Another way to describe this interaction is to assign a potential energy to the magnetic moment $\vec{\mu}$ in the external field $\vec{\mathbf{B}}_{ext}$:

$$U = -\vec{\mu} \cdot \vec{\mathbf{B}}_{ext} \tag{1.14}$$

When the field $\vec{\mathbf{B}}_{ext}$ is applied, $\vec{\mu}$ rotates so that its energy tends to a minimum value, which occurs when $\vec{\mu}$ and $\vec{\mathbf{B}}_{ext}$ are parallel.

It is important for us to understand the properties of magnetic moments, because particles such as electrons or protons have magnetic moments. Although we don't imagine these particles to be tiny current loops, their magnetic moments do obey Eqs. 1.13 and 1.14.

A particularly important aspect of electromagnetism is *electromagnetic waves*. In Chapter 3 we discuss some properties of these waves in more detail. Electromagnetic waves travel in free space with speed c (the speed of light), which is related to the electromagnetic constants ε_0 and μ_0:

$$c = (\varepsilon_0 \mu_0)^{-1/2} \tag{1.15}$$

The speed of light has the exact value of $c = 299{,}792{,}458$ m/s.

Electromagnetic waves have a frequency f and wavelength λ, which are related by

$$c = \lambda f \tag{1.16}$$

The wavelengths range from the very short (nuclear gamma rays) to the very long (radio waves). Figure 1.7 shows the electromagnetic spectrum with the conventional names assigned to the different ranges of wavelengths.

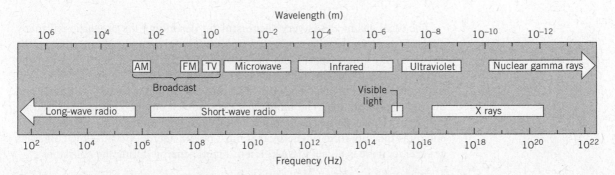

FIGURE 1.7 The electromagnetic spectrum. The boundaries of the regions are not sharply defined.

Kinetic Theory of Matter

An example of the successful application of classical physics to the structure of matter is the understanding of the properties of gases at relatively low pressures and high temperatures (so that the gas is far from the region of pressure and temperature where it might begin to condense into a liquid). Under these conditions, most real gases can be modeled as ideal gases and are well described by the *ideal gas equation of state*

$$PV = NkT \tag{1.17}$$

where P is the pressure, V is the volume occupied by the gas, N is the number of molecules, T is the temperature, and k is the *Boltzmann constant*, which has the value

$$k = 1.381 \times 10^{-23} \text{ J/K}$$

In using this equation and most of the equations in this section, the temperature must be measured in units of kelvins (K). Be careful not to confuse the symbol K for the unit of temperature with the symbol K for kinetic energy.

The ideal gas equation of state can also be expressed as

$$PV = nRT \tag{1.18}$$

where n is the number of moles and R is the *universal gas constant* with a value of

$$R = 8.315 \text{ J/mol} \cdot \text{K}$$

One mole of a gas is the quantity that contains a number of fundamental entities (atoms or molecules) equal to Avogadro's constant N_A, where

$$N_A = 6.022 \times 10^{23} \text{ per mole}$$

That is, one mole of helium contains N_A atoms of He, one mole of nitrogen contains N_A molecules of N_2 (and thus $2N_A$ atoms of N), and one mole of water vapor contains N_A molecules of H_2O (and thus $2N_A$ atoms of H and N_A atoms of O).

Because $N = nN_A$ (number of molecules equals number of moles times number of molecules per mole), the relationship between the Boltzmann constant and the universal gas constant is

$$R = kN_A \tag{1.19}$$

The ideal gas model is very successful for describing the properties of many gases. It assumes that the molecules are of negligibly small volume (that is, the gas is mostly empty space) and move randomly throughout the volume of the container. The molecules make occasional collisions with one another and with the walls of the container. The collisions obey Newton's laws and are elastic and of very short duration. The molecules exert forces on one another only during collisions. Under these assumptions, there is no potential energy so that kinetic energy is the only form of energy that must be considered. Because the collisions are elastic, there is no net loss or gain of kinetic energy during the collisions.

Individual molecules may speed up or slow down due to collisions, but the average kinetic energy of all the molecules in the container does not change. The average kinetic energy of a molecule in fact depends only on the temperature:

$$K_{av} = \tfrac{3}{2}kT \text{ (per molecule)} \tag{1.20}$$

For rough estimates, the quantity kT is often used as a measure of the mean kinetic energy per particle. For example, at room temperature ($20°C = 293$ K), the mean kinetic energy per particle is approximately 4×10^{-21} J (about 1/40 eV), while in the interior of a star where $T \sim 10^7$ K, the mean energy is approximately 10^{-16} J (about 1000 eV).

Sometimes it is also useful to discuss the average kinetic energy of a mole of the gas:

average K per mole = average K per molecule $\times$ number of molecules per mole

Using Eq. 1.19 to relate the Boltzmann constant to the universal gas constant, we find the average molar kinetic energy to be

$$K_{av} = \tfrac{3}{2}RT \text{ (per mole)} \tag{1.21}$$

It should be apparent from the context of the discussion whether K_{av} refers to the average per molecule or the average per mole.

1.2 THE FAILURE OF CLASSICAL CONCEPTS OF SPACE AND TIME

In 1905, Albert Einstein proposed the special theory of relativity, which is in essence a new way of looking at space and time, replacing the "classical" space and time that were the basis of the physical theories of Galileo and Newton. Einstein's proposal was based on a "thought experiment," but in subsequent years experimental data have clearly indicated that the classical concepts of space and time are incorrect. In this section we examine how experimental results support the need for a new approach to space and time.

The Failure of the Classical Concept of Time

In high-energy collisions between two protons, many new particles can be produced, one of which is a *pi meson* (also known as a *pion*). When the pions are produced at rest in the laboratory, they are observed to have an average lifetime (the time between the production of the pion and its decay into other particles) of 26.0 ns (nanoseconds, or 10^{-9} s). On the other hand, pions in motion are observed to have a very different lifetime. In one particular experiment, pions moving at a speed of 2.737×10^8 m/s (91.3% of the speed of light) showed a lifetime of 63.7 ns.

Let us imagine this experiment as viewed by two different observers (Figure 1.8). Observer #1, at rest in the laboratory, sees the pion moving relative to the laboratory at a speed of 91.3% of the speed of light and measures its

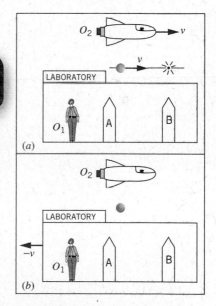

FIGURE 1.8 (*a*) The pion experiment according to O_1. Markers A and B respectively show the locations of the pion's creation and decay. (*b*) The same experiment as viewed by O_2, relative to whom the pion is at rest and the laboratory moves with velocity $-v$.

lifetime to be 63.7 ns. Observer #2 is moving relative to the laboratory at exactly the same velocity as the pion, so according to observer #2 the pion is at rest and has a lifetime of 26.0 ns. The two observers measure different values for the time interval between the same two events—the formation of the pion and its decay.

According to Newton, time is the same for all observers. Newton's laws are based on this assumption. The pion experiment clearly shows that time is *not* the same for all observers, which indicates the need for a new theory that relates time intervals measured by different observers who are in motion with respect to each other.

The Failure of the Classical Concept of Space

The pion experiment also leads to a failure of the classical ideas about space. Suppose observer #1 erects two markers in the laboratory, one where the pion is created and another where it decays. The distance D_1 between the two markers is equal to the speed of the pion multiplied by the time interval from its creation to its decay: $D_1 = (2.737 \times 10^8 \text{ m/s})(63.7 \times 10^{-9} \text{ s}) = 17.4 \text{ m}$. To observer #2, traveling at the same velocity as the pion, the laboratory appears to be rushing by at a speed of $2.737 \times 10^8 \text{ m/s}$ and the time between passing the first and second markers, showing the creation and decay of the pion in the laboratory, is 26.0 ns. According to observer #2, the distance between the markers is $D_2 = (2.737 \times 10^8 \text{ m/s})(26.0 \times 10^{-9} \text{ s}) = 7.11 \text{ m}$. Once again, we have two observers in relative motion measuring different values for the same interval, in this case the distance between the two markers in the laboratory. The physical theories of Galileo and Newton are based on the assumption that space is the same for all observers, and so length measurements should not depend on relative motion. The pion experiment again shows that this cornerstone of classical physics is not consistent with modern experimental data.

The Failure of the Classical Concept of Velocity

Classical physics places no limit on the maximum velocity that a particle can reach. One of the basic equations of kinematics, $v = v_0 + at$, shows that if a particle experiences an acceleration a for a long enough time t, velocities as large as desired can be achieved, perhaps even exceeding the speed of light. For another example, when an aircraft flying at a speed of 200 m/s relative to an observer on the ground launches a missile at a speed of 250 m/s relative to the aircraft, a ground-based observer would measure the missile to travel at a speed of 200 m/s + 250 m/s = 450 m/s, according to the classical velocity addition rule (Eq. 1.7). We can apply that same reasoning to a spaceship moving at a speed of 2.0×10^8 m/s (relative to an observer on a space station), which fires a missile at a speed of 2.5×10^8 m/s relative to the spacecraft. We would expect that the observer on the space station would measure a speed of 4.5×10^8 m/s for the missile. This speed exceeds the speed of light (3.0×10^8 m/s). Allowing speeds greater than the speed of light leads to a number of conceptual and logical difficulties, such as the reversal of the normal order of cause and effect for some observers.

Here again modern experimental results disagree with the classical ideas. Let's go back again to our experiment with the pion, which is moving through the laboratory at a speed of 2.737×10^8 m/s. The pion decays into another particle, called a *muon*, which is emitted in the forward direction (the direction of the pion's velocity) with a speed of 0.813×10^8 m/s relative to the pion. According to Eq. 1.7, an observer in the laboratory should observe the muon to be moving with

1.3 | The Failure of the Classical Theory of Particle Statistics 13

a velocity of 2.737×10^8 m/s $+ 0.813 \times 10^8$ m/s $= 3.550 \times 10^8$ m/s, exceeding the speed of light. The observed velocity of the muon, however, is 2.846×10^8 m/s, below the speed of light. Clearly the classical rule for velocity addition fails in this experiment.

The properties of time and space and the rules for combining velocities are essential concepts of the classical physics of Newton. These concepts are derived from observations at low speeds, which were the only speeds available to Newton and his contemporaries. In Chapter 2, we shall discover how the special theory of relativity provides the correct procedure for comparing measurements of time, distance, and velocity by different observers and thereby removes the failures of classical physics at high speed (while reducing to the classical laws at low speed, where we know the Newtonian framework works very well).

1.3 THE FAILURE OF THE CLASSICAL THEORY OF PARTICLE STATISTICS

Thermodynamics and statistical mechanics were among the great triumphs of 19th-century physics. Describing the behavior of complex systems of many particles was shown to be possible using a small number of aggregate or average properties—for example, temperature, pressure, and heat capacity. Perhaps the crowning achievement in this field was the development of relationships between *macroscopic* properties, such as temperature, and *microscopic* properties, such as the molecular kinetic energy.

Despite these great successes, this statistical approach to understanding the behavior of gases and solids also showed a spectacular failure. Although the classical theory gave the correct heat capacities of gases at high temperatures, it failed miserably for many gases at low temperatures. In this section we summarize the classical theory and explain how it fails at low temperatures. This failure directly shows the inadequacy of classical physics and the need for an approach based on quantum theory, the second of the great theories of modern physics.

The Distribution of Molecular Energies

In addition to the average kinetic energy, it is also important to analyze the distribution of kinetic energies—that is, what fraction of the molecules in the container has kinetic energies between any two values K_1 and K_2. For a gas in thermal equilibrium at absolute temperature T (in kelvins), the distribution of molecular energies is given by the *Maxwell-Boltzmann distribution*:

$$N(E) = \frac{2N}{\sqrt{\pi}} \frac{1}{(kT)^{3/2}} E^{1/2} e^{-E/kT} \tag{1.22}$$

In this equation, N is the total number of molecules (a pure number) while $N(E)$ is the distribution function (with units of energy^{-1}) defined so that $N(E)dE$ is the number of molecules dN in the energy interval dE at E (or, in other words, the number of molecules with energies between E and $E + dE$):

$$dN = N(E)\, dE \tag{1.23}$$

The distribution $N(E)$ is shown in Figure 1.9. The number dN is represented by the area of the narrow strip between E and $E + dE$. If we divide the entire horizontal

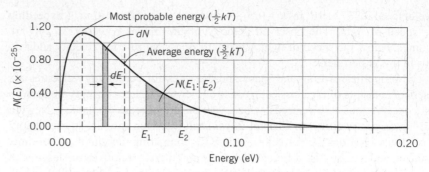

FIGURE 1.9 The Maxwell-Boltzmann energy distribution function, shown for one mole of gas at room temperature (300 K).

axis into an infinite number of such small intervals and add the areas of all the resulting narrow strips, we obtain the total number of molecules in the gas:

$$\int_0^\infty dN = \int_0^\infty N(E)\,dE = \int_0^\infty \frac{2N}{\sqrt{\pi}} \frac{1}{(kT)^{3/2}} E^{1/2} e^{-E/kT}\,dE = N \qquad (1.24)$$

The final step in this calculation involves the definite integral $\int_0^\infty x^{1/2} e^{-x} dx$, which you can find in tables of integrals. Also using calculus techniques (see Problem 8), you can show that the peak of the distribution function (the most probable energy) is $\frac{1}{2}kT$.

The average energy in this distribution of molecules can also be found by dividing the distribution into strips. To find the contribution of each strip to the energy of the gas, we multiply the number of molecules in each strip, $dN = N(E)dE$, by the energy E of the molecules in that strip, and then we add the contributions of all the strips by integrating over all energies. This calculation would give the *total* energy of the gas; to find the average we divide by the total number of molecules N:

$$E_{av} = \frac{1}{N} \int_0^\infty E N(E)\,dE = \int_0^\infty \frac{2}{\sqrt{\pi}} \frac{1}{(kT)^{3/2}} E^{3/2} e^{-E/kT}\,dE \qquad (1.25)$$

Once again, the definite integral can be found in integral tables. The result of carrying out the integration is

$$E_{av} = \tfrac{3}{2}kT \qquad (1.26)$$

Equation 1.26 gives the average energy of a molecule in the gas and agrees precisely with the result given by Eq. 1.20 for the ideal gas in which kinetic energy is the only kind of energy the gas can have.

Occasionally we are interested in finding the number of molecules in our distribution with energies between any two values E_1 and E_2. If the interval between E_1 and E_2 is very small, Eq. 1.23 can be used, with $dE = E_2 - E_1$ and with $N(E)$ evaluated at the midpoint of the interval. This approximation works very well when the interval is small enough that $N(E)$ is either approximately flat or linear over the interval. If the interval is large enough that this approximation is not valid, then it is necessary to integrate to find the number of molecules in the interval:

$$N(E_1 : E_2) = \int_{E_1}^{E_2} N(E)\,dE = \int_{E_1}^{E_2} \frac{2N}{\sqrt{\pi}} \frac{1}{(kT)^{3/2}} E^{1/2} e^{-E/kT}\,dE \qquad (1.27)$$

This number is represented by the shaded area in Figure 1.9. This integral cannot be evaluated directly and must be found numerically.

Example 1.3

(a) In one mole of a gas at a temperature of 650 K ($kT = 8.97 \times 10^{-21}$ J $= 0.0560$ eV), calculate the number of molecules with energies between 0.0105 eV and 0.0135 eV. (b) In this gas, calculate the fraction of the molecules with energies in the range of $\pm 2.5\%$ of the most probable energy ($\frac{1}{2}kT$).

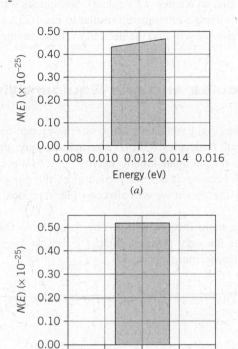

FIGURE 1.10 Example 1.3.

Solution

(a) Figure 1.10a shows the distribution $N(E)$ in the region between $E_1 = 0.0105$ eV and $E_2 = 0.0135$ eV. Because the graph is very close to linear in this region, we can use Eq. 1.23 to find the number of molecules in this range. We take dE to be the width of the range, $dE = E_2 - E_1 = 0.0135$ eV $- 0.0105$ eV $= 0.0030$ eV, and for E we use the energy at the midpoint of the range (0.0120 eV):

$$dN = N(E)\, dE$$

$$= \frac{2N}{\sqrt{\pi}} \frac{1}{(kT)^{3/2}} E^{1/2} e^{-E/kT}\, dE$$

$$= 2(6.022 \times 10^{23})(0.0120 \text{ eV})^{1/2}\, \pi^{-1/2}(0.0560 \text{ eV})^{-3/2}$$

$$\times e^{-(0.0120\,\text{eV})/(0.0560\,\text{eV})}(0.0030 \text{ eV})$$

$$= 1.36 \times 10^{22}$$

(b) Figure 1.10b shows the distribution in this region. To find the fraction of the molecules in this energy range, we want dN/N. The most probable energy is $\frac{1}{2}kT$ or 0.0280 eV, and $\pm 2.5\%$ of this value corresponds to ± 0.0007 eV or a range from 0.0273 eV to 0.0287 eV. The fraction is

$$\frac{dN}{N} = \frac{N(E)\, dE}{N} = \frac{2}{\sqrt{\pi}} \frac{1}{(kT)^{3/2}} E^{1/2} e^{-E/kT}\, dE$$

$$= 2(0.0280 \text{ eV})^{1/2}\, \pi^{-1/2}(0.0560 \text{ eV})^{-3/2}$$

$$\times e^{-(0.0280\,\text{eV})/(0.0560\,\text{eV})}(0.0014 \text{ eV})$$

$$= 0.0121$$

Note from these examples how we use a distribution function. We do *not* use Eq. 1.22 to calculate the *number* of molecules at a particular energy. In this way $N(E)$ differs from many of the functions you have encountered previously in your study of physics and mathematics. We always use the distribution function to calculate how many events occur in a certain *interval* of values rather than at an exact particular value. There are two reasons for this: (1) Asking the question in the form of how many molecules have a certain value of the energy implies that the energy is known exactly (to an infinite number of decimal places), and there is zero probability to find a molecule with that exact value of the energy. (2) Any measurement apparatus accepts a finite range of energies (or speeds) rather than a single exact value, and thus asking about intervals is a better representation of what can be measured in the laboratory.

Note that $N(E)$ has dimensions of energy^{-1}—it gives the number of molecules *per unit energy interval* (for example, number of molecules per eV). To get an actual number that can be compared with measurement, $N(E)$ must be multiplied by an energy interval. In our study of modern physics, we will encounter many different types of distribution functions whose use and interpretation are similar to that of $N(E)$. These functions generally give a number or a probability per some sort of unit interval (for example, probability per unit volume), and to use the distribution function to calculate an outcome we must always multiply by an appropriate interval (for example, an element of volume). Sometimes we will be able to deal with small intervals using a relationship similar to Eq. 1.23, as we did in Example 1.3, but in other cases we will find the need to evaluate an integral, as we did in Eq. 1.27.

Polyatomic Molecules and the Equipartition of Energy

So far we have been considering gases with only one atom per molecule (monatomic gases). For "point" molecules with no internal structure, only one form of energy is important: translational kinetic energy $\frac{1}{2}mv^2$. (We call this "translational" kinetic energy because it describes motion as the gas particles move from one location to another. Soon we will also consider rotational kinetic energy.)

Let's rewrite Eq. 1.26 in a more instructive form by recognizing that, with translational kinetic energy as the only form of energy, $E = K = \frac{1}{2}mv^2$. With $v^2 = v_x^2 + v_y^2 + v_z^2$, we can write the energy as

$$E = \tfrac{1}{2}mv_x^2 + \tfrac{1}{2}mv_y^2 + \tfrac{1}{2}mv_z^2 \tag{1.28}$$

The average energy is then

$$\tfrac{1}{2}m(v_x^2)_{\text{av}} + \tfrac{1}{2}m(v_y^2)_{\text{av}} + \tfrac{1}{2}m(v_z^2)_{\text{av}} = \tfrac{3}{2}kT \tag{1.29}$$

For a gas molecule there is no difference between the x, y, and z directions, so the three terms on the left are equal and each term is equal to $\frac{1}{2}kT$. The three terms on the left represent three *independent* contributions to the energy of the molecule—the motion in the x direction, for example, is not affected by the y or z motions.

We define a *degree of freedom* of the gas as each independent contribution to the energy of a molecule, corresponding to one quadratic term in the expression for the energy. There are three quadratic terms in Eq. 1.28, so in this case there are three degrees of freedom. As you can see from Eq. 1.29, each of the three degrees of freedom of a gas molecule contributes an energy of $\frac{1}{2}kT$ to its average energy. The relationship we have obtained in this special case is an example of the application of a general theorem, called the *equipartition of energy theorem*:

> *When the number of particles in a system is large and Newtonian mechanics is obeyed, each molecular degree of freedom corresponds to an average energy of $\frac{1}{2}kT$.*

The average energy per molecule is then the number of degrees of freedom times $\frac{1}{2}kT$, and the total energy is obtained by multiplying the average energy per molecule by the number of molecules N: $E_{\text{total}} = NE_{\text{av}}$. We will refer to this total energy as the *internal energy* E_{int} to indicate that it represents the random motions of the gas molecules (in contrast, for example, to the energy involved with the motion of the entire container of gas molecules).

$$E_{\text{int}} = N(\tfrac{3}{2}kT) = \tfrac{3}{2}NkT = \tfrac{3}{2}nRT \quad \text{(translation only)} \qquad (1.30)$$

where Eq. 1.19 has been used to express Eq. 1.30 in terms of either the number of molecules or the number of moles.

The situation is different for a *diatomic* gas (two atoms per molecule), illustrated in Figure 1.11. There are still three degrees of freedom associated with the translational motion of the molecule, but now two additional forms of energy are permitted—rotational and vibrational.

First we consider the rotational motion. The molecule shown in Figure 1.11 can rotate about the x' and y' axes (but not about the z' axis, because the rotational inertia about that axis is zero for diatomic molecules in which the atoms are treated as points). Using the general form of $\tfrac{1}{2}I\omega^2$ for the rotational kinetic energy, we can write the energy of the molecule as

$$E = \tfrac{1}{2}mv_x^2 + \tfrac{1}{2}mv_y^2 + \tfrac{1}{2}mv_z^2 + \tfrac{1}{2}I_{x'}\omega_{x'}^2 + \tfrac{1}{2}I_{y'}\omega_{y'}^2 \qquad (1.31)$$

Here we have 5 quadratic terms in the energy, and thus 5 degrees of freedom. According to the equipartition theorem, the average total energy per molecule is $5 \times \tfrac{1}{2}kT = \tfrac{5}{2}kT$, and the total internal energy of n moles of the gas is

$$E_{\text{int}} = \tfrac{5}{2}nRT \quad \text{(translation + rotation)} \qquad (1.32)$$

If the molecule can also vibrate, we can imagine the rigid rod connecting the atoms in Figure 1.11 to be replaced by a spring. The two atoms can then vibrate in opposite directions along the z' axis, with the center of mass of the molecule remaining fixed. The vibrational motion adds two quadratic terms to the energy, corresponding to the vibrational potential energy $(\tfrac{1}{2}kz'^2)$ and the vibrational kinetic energy $(\tfrac{1}{2}mv_{z'}^2)$. Including the vibrational motion, there are now 7 degrees of freedom, so that

$$E_{\text{int}} = \tfrac{7}{2}nRT \quad \text{(translation + rotation + vibration)} \qquad (1.33)$$

Heat Capacities of an Ideal Gas

Now we examine where the classical molecular distribution theory, which gives a very good accounting of molecular behavior under most circumstances, fails to agree with one particular class of experiments. Suppose we have a container of gas with a fixed volume. We transfer energy to the gas, perhaps by placing the container in contact with a system at a higher temperature. All of this transferred energy increases the internal energy of the gas by an amount ΔE_{int}, and there is an accompanying increase in temperature ΔT.

We define the *molar heat capacity* for this constant-volume process as

$$C_V = \frac{\Delta E_{\text{int}}}{n\,\Delta T} \qquad (1.34)$$

(The subscript V reminds us that we are doing this measurement at constant volume.) From Eqs. 1.30, 1.32, and 1.33, we see that the molar heat capacity depends on the type of gas:

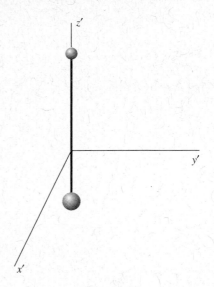

FIGURE 1.11 A diatomic molecule, with the origin at the center of mass. Rotations can occur about the x' and y' axes, and vibrations can occur along the z' axis.

$$C_V = \tfrac{3}{2}R \quad \text{(monatomic or nonrotating, nonvibrating diatomic ideal gas)}$$
$$C_V = \tfrac{5}{2}R \quad \text{(rotating diatomic ideal gas)} \qquad\qquad (1.35)$$
$$C_V = \tfrac{7}{2}R \quad \text{(rotating and vibrating diatomic ideal gas)}$$

When we add energy to the gas, the equipartition theorem tells us that the added energy will on the average be distributed uniformly among all the possible forms of energy (corresponding to the number of degrees of freedom). However, *only the translational kinetic energy contributes to the temperature* (as shown by Eqs. 1.20 and 1.21). Thus, if we add 7 units of energy to a diatomic gas with rotating and vibrating molecules, on the average only 3 units go into translational kinetic energy and so 3/7 of the added energy goes into increasing the temperature. (To measure the temperature rise, the gas molecules must collide with the thermometer, so energy in the rotational and vibrational motions is not recorded by the thermometer.) Put another way, to obtain the same temperature increase ΔT, a mole of diatomic gas requires 7/3 times the energy that is needed for a mole of monatomic gas.

Comparison with Experiment   How well do these heat capacity values agree with experiment? For monatomic gases, the agreement is very good. The equipartition theorem predicts a value of $C_V = 3R/2 = 12.5\,\text{J/mol}\cdot\text{K}$, which should be the same for all monatomic gases and the same at all temperatures (as long as the conditions of the ideal gas model are fulfilled). The heat capacity of He gas is $12.5\,\text{J/mol}\cdot\text{K}$ at $100\,\text{K}, 300\,\text{K}$ (room temperature), and $1000\,\text{K}$, so in this case our calculation is in perfect agreement with experiment. Other inert gases (Ne, Ar, Xe, etc.) have identical values, as do vapors of metals (Cu, Na, Pb, Bi, etc.) and the monatomic (dissociated) state of elements that normally form diatomic molecules (H, N, O, Cl, Br, etc.). So over a wide variety of different elements and a wide range of temperatures, classical statistical mechanics is in excellent agreement with experiment.

The situation is much less satisfactory for diatomic molecules. For a rotating and vibrating diatomic molecule, the classical calculation gives $C_V = 7R/2 = 29.1\,\text{J/mol}\cdot\text{K}$. Table 1.1 shows some values of the heat capacities for different diatomic gases over a range of temperatures.

TABLE 1.1 Heat Capacities of Diatomic Gases

Element	C_V (J/mol · K)		
	100 K	300 K	1000 K
H_2	18.7	20.5	21.9
N_2	20.8	20.8	24.4
O_2	20.8	21.1	26.5
F_2	20.8	23.0	28.8
Cl_2	21.0	25.6	29.1
Br_2	22.6	27.8	29.5
I_2	24.8	28.6	29.7
Sb_2		28.1	29.0
Te_2		28.2	29.0
Bi_2		28.6	29.1

At high temperatures, many of the diatomic gases do indeed approach the expected value of $7R/2$, but at lower temperatures the values are much smaller. For example, fluorine seems to behave as if it has 5 degrees of freedom ($C_V = 20.8$ J/mol · K) at 100 K and 7 degrees of freedom ($C_V = 29.1$ J/mol · K) at 1000 K.

Hydrogen behaves as if it has 5 degrees of freedom at room temperature, but at high enough temperature (3000 K), the heat capacity of H_2 approaches 29.1 J/mol · K, corresponding to 7 degrees of freedom, while at lower temperatures (40 K) the heat capacity is 12.5 J/mol · K, corresponding to 3 degrees of freedom. The temperature dependence of the heat capacity of H_2 is shown in Figure 1.12. There are three plateaus in the graph, corresponding to heat capacities for 3, 5, and 7 degrees of freedom. At the lowest temperatures, the rotational and vibrational motions are "frozen" and do not contribute to the heat capacity. At about 100 K, the molecules have enough energy to allow rotational motion to occur, and by about 300 K the heat capacity is characteristic of 5 degrees of freedom. Starting about 1000 K, the vibrational motion can occur, and by about 3000 K there are enough molecules above the vibrational threshold to allow 7 degrees of freedom.

What's going on here? The classical calculation demands that C_V should be constant, independent of the type of gas or the temperature. The equipartition of energy theorem, which is very successful in predicting many thermodynamic properties, fails miserably in accounting for the heat capacities. This theorem requires that the energy added to a gas must on average be divided equally among all the different forms of energy, and classical physics does not permit a threshold energy for any particular type of motion. How is it possible for 2 degrees of freedom, corresponding to the rotational or vibrational motions, to be "turned on" as the temperature is increased?

The solution to this dilemma can be found in *quantum mechanics*, according to which there is indeed a minimum or threshold energy for the rotational and vibrational motions. We discuss this behavior in Chapters 5 and 9. In Chapter 11 we discuss the failure of the equipartition theorem to account for the heat capacities of solids and the corresponding need to replace the classical Maxwell-Boltzmann energy distribution function with a different distribution that is consistent with quantum mechanics.

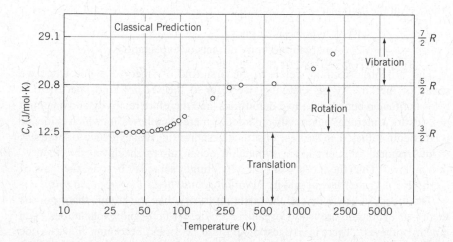

FIGURE 1.12 The heat capacity of molecular hydrogen at different temperatures. The data points disagree with the classical prediction.

1.4 THEORY, EXPERIMENT, LAW

When you first began to study science, perhaps in your elementary or high school years, you may have learned about the "scientific method," which was supposed to be a sort of procedure by which scientific progress was achieved. The basic idea of the "scientific method" was that, on reflecting over some particular aspect of nature, the scientist would invent a *hypothesis* or *theory,* which would then be tested by *experiment* and if successful would be elevated to the status of *law.* This procedure is meant to emphasize the importance of doing experiments as a way of testing hypotheses and rejecting those that do not pass the tests. For example, the ancient Greeks had some rather definite ideas about the motion of objects, such as projectiles, in the Earth's gravity. Yet they tested none of these by experiment, so convinced were they that the power of logical deduction *alone* could be used to discover the hidden and mysterious laws of nature and that once logic had been applied to understanding a problem, no experiments were necessary. If theory and experiment were to disagree, they would argue, then there must be something wrong with the experiment! This dominance of analysis and faith was so pervasive that it was another 2000 years before Galileo, using an inclined plane and a crude timer (equipment surely within the abilities of the early Greeks to construct), discovered the laws of motion, which were later organized and analyzed by Newton.

In the case of modern physics, none of the fundamental concepts is obvious from reason alone. Only by doing often difficult and necessarily precise experiments do we learn about these unexpected and fascinating effects associated with such modern physics topics as relativity and quantum physics. These experiments have been done to unprecedented levels of precision—of the order of one part in 10^6 or better—and it can certainly be concluded that modern physics was tested far better in the 20th century than classical physics was tested in all of the preceding centuries.

Nevertheless, there is a persistent and often perplexing problem associated with modern physics, one that stems directly from your previous acquaintance with the "scientific method." This concerns the use of the word "theory," as in "theory of relativity" or "quantum theory," or even "atomic theory" or "theory of evolution." There are two contrasting and conflicting definitions of the word "theory" in the dictionary:

1. A hypothesis or guess.
2. An organized body of facts or explanations.

The "scientific method" refers to the first kind of "theory," while when we speak of the "theory of relativity" we refer to the second kind. Yet there is often confusion between the two definitions, and therefore relativity and quantum physics are sometimes incorrectly regarded as mere hypotheses, on which evidence is still being gathered, in the hope of someday submitting that evidence to some sort of international (or intergalactic) tribunal, which in turn might elevate the "theory" into a "law." Thus the "theory of relativity" might someday become the "law of relativity," like the "law of gravity." *Nothing could be further from the truth!*

The theory of relativity and the quantum theory, like the atomic theory or the theory of evolution, are truly "organized bodies of facts and explanations" and *not* "hypotheses." There is no question of these "theories" becoming "laws"—the "facts" (experiments, observations) of relativity and quantum physics, like those of atomism or evolution, are accepted by virtually all scientists today. The experimental evidence for all of these processes is so compelling that no one who approaches them in the spirit of free and open inquiry can doubt the observational evidence or their inferences. Whether these collections of evidence are called theories or laws is merely a question of semantics and has nothing to do with their scientific merits. Like all scientific principles, they will continue to develop and change as new discoveries are made; that is the essence of scientific progress.

Chapter Summary

		Section			Section
Classical kinetic energy	$K = \frac{1}{2}mv^2 = \frac{p^2}{2m}$	1.1	Magnetic field of a current loop	$B = \frac{\mu_0 i}{2r}$	1.1
Classical linear momentum	$\vec{p} = m\vec{v}$	1.1	Potential energy of magnetic dipole	$U = -\vec{\mu} \cdot \vec{B}_{ext}$	1.1
Classical angular momentum	$\vec{L} = \vec{r} \times \vec{p}$	1.1	Average kinetic energy in a gas	$K_{av} = \frac{3}{2}kT \text{(per molecule)}$ $= \frac{3}{2}RT \text{(per mole)}$	1.1
Classical conservation laws	In an isolated system, the energy, linear momentum, and angular momentum remain constant.	1.1	Maxwell-Boltzmann distribution	$N(E) = \frac{2N}{\sqrt{\pi}} \frac{1}{(kT)^{3/2}} E^{1/2} e^{-E/kT}$	1.3
Electric force and potential energy of two interacting charges	$F = \frac{1}{4\pi\varepsilon_0} \frac{|q_1||q_2|}{r^2}$ $U = \frac{1}{4\pi\varepsilon_0} \frac{q_1 q_2}{r}$	1.1	Equipartition of energy	Energy per degree of freedom $= \frac{1}{2}kT$	1.3
Relationship between electric potential energy and potential	$\Delta U = q\Delta V$	1.1			

Questions

1. Under what conditions can you apply the law of conservation of energy? Conservation of linear momentum? Conservation of angular momentum?
2. Which of the conserved quantities are scalars and which are vectors? Is there a difference in how we apply conservation laws for scalar and vector quantities?
3. What other conserved quantities (besides energy, linear momentum, and angular momentum) can you name?
4. What is the difference between potential and potential energy? Do they have different dimensions? Different units?
5. In Section 1.1 we defined the electric force between two charges and the magnetic field of a current. Use these quantities to define the electric field of a single charge and the magnetic force on a moving electric charge.
6. Other than from the ranges of wavelengths shown in Figure 1.7, can you think of a way to distinguish radio waves from infrared waves? Visible from infrared? That is,

could you design a radio that could be tuned to infrared waves? Could living beings "see" in the infrared region?
7. Suppose we have a mixture of an equal number N of molecules of two different gases, whose molecular masses are m_1 and m_2, in complete thermal equilibrium at temperature T. How do the distributions of molecular energies of the two gases compare? How do their average kinetic energies per molecule compare?
8. In most gases (as in the case of hydrogen) the rotational motion begins to occur at a temperature well below the temperature at which vibrational motion occurs. What does this tell us about the properties of the gas molecules?
9. Suppose it were possible for a pitcher to throw a baseball faster than the speed of light. Describe how the flight of the ball from the pitcher's hand to the catcher's glove would look to the umpire standing behind the catcher.

10. At low temperatures the molar heat capacity of carbon dioxide (CO_2) is about $5R/2$, and it rises to about $7R/2$ at room temperature. However, unlike the gases discussed in Section 1.3, the heat capacity of CO_2 continues to rise as the temperature increases, reaching $11R/2$ at 1000 K. How can you explain this behavior?

11. If we double the temperature of a gas, is the number of molecules in a narrow interval dE around the most probable energy about the same, double, or half what it was at the original temperature?

Problems

1.1 Review of Classical Physics

1. A hydrogen atom ($m = 1.674 \times 10^{-27}$ kg) is moving with a velocity of 1.1250×10^7 m/s. It collides elastically with a helium atom ($m = 6.646 \times 10^{-27}$ kg) at rest. After the collision, the hydrogen atom is found to be moving with a velocity of -6.724×10^6 m/s (in a direction opposite to its original motion). Find the velocity of the helium atom after the collision in two different ways: (a) by applying conservation of momentum; (b) by applying conservation of energy.

2. A helium atom ($m = 6.6465 \times 10^{-27}$ kg) collides elastically with an oxygen atom ($m = 2.6560 \times 10^{-26}$ kg) at rest. After the collision, the helium atom is observed to be moving with a velocity of 6.636×10^6 m/s in a direction at an angle of $84.7°$ relative to its original direction. The oxygen atom is observed to move at an angle of $-40.4°$. (a) Find the speed of the oxygen atom. (b) Find the speed of the helium atom before the collision.

3. A beam of helium-3 atoms ($m = 3.016$ u) is incident on a target of nitrogen-14 atoms ($m = 14.003$ u) at rest. During the collision, a proton from the helium-3 nucleus passes to the nitrogen nucleus, so that following the collision there are two atoms: an atom of "heavy hydrogen" (deuterium, $m = 2.014$ u) and an atom of oxygen-15 ($m = 15.003$ u). The incident helium atoms are moving at a velocity of 6.346×10^6 m/s. After the collision, the deuterium atoms are observed to be moving forward (in the same direction as the initial helium atoms) with a velocity of 1.531×10^7 m/s. (a) What is the final velocity of the oxygen-15 atoms? (b) Compare the total kinetic energies before and after the collision.

4. An atom of beryllium ($m = 8.00$ u) splits into two atoms of helium ($m = 4.00$ u) with the release of 92.2 keV of energy. If the original beryllium atom is at rest, find the kinetic energies and speeds of the two helium atoms.

5. A 4.15-volt battery is connected across a parallel-plate capacitor. Illuminating the plates with ultraviolet light causes electrons to be emitted from the plates with a speed of 1.76×10^6 m/s. (a) Suppose electrons are emitted near the center of the negative plate and travel perpendicular to that plate toward the opposite plate. Find the speed of the electrons when they reach the positive plate. (b) Suppose instead that electrons are emitted perpendicular to the positive plate. Find their speed when they reach the negative plate.

1.2 The Failure of Classical Concepts of Space and Time

6. Observer A, who is at rest in the laboratory, is studying a particle that is moving through the laboratory at a speed of $0.624c$ and determines its lifetime to be 159 ns. (a) Observer A places markers in the laboratory at the locations where the particle is produced and where it decays. How far apart are those markers in the laboratory? (b) Observer B, who is traveling parallel to the particle at a speed of $0.624c$, observes the particle to be at rest and measures its lifetime to be 124 ns. According to B, how far apart are the two markers in the laboratory?

1.3 The Failure of the Classical Theory of Particle Statistics

7. A sample of argon gas is in a container at $35.0°$ C and 1.22 atm pressure. The radius of an argon atom (assumed spherical) is 0.710×10^{-10} m. Calculate the fraction of the container volume actually occupied by the atoms.

8. By differentiating the expression for the Maxwell-Boltzmann energy distribution, show that the peak of the distribution occurs at an energy of $\frac{1}{2}kT$.

9. A container holds N molecules of nitrogen gas at $T = 280$ K. Find the number of molecules with kinetic energies between 0.0300 eV and 0.0312 eV.

10. A sample of 2.37 moles of an ideal diatomic gas experiences a temperature increase of 65.2 K at constant volume. (a) Find the increase in internal energy if only translational and rotational motions are possible. (b) Find the increase in internal energy if translational, rotational, and vibrational motions are possible. (c) How much of the energy calculated in (a) and (b) is translational kinetic energy?

General Problems

11. An atom of mass $m_1 = m$ moving in the x direction with speed $v_1 = v$ collides elastically with an atom of mass $m_2 = 3m$ at rest. After the collision the first atom moves in the y direction. Find the direction of motion of the second atom and the speeds of both atoms (in terms of v) after the collision.

12. An atom of mass $m_1 = m$ moves in the positive x direction with speed $v_1 = v$. It collides with and sticks to an atom of

mass $m_2 = 2m$ moving in the positive y direction with speed $v_2 = 2v/3$. Find the resultant speed and direction of motion of the combination, and find the kinetic energy lost in this inelastic collision.

13. Suppose the beryllium atom of Problem 4 were not at rest, but instead moved in the positive x direction and had a kinetic energy of 40.0 keV. One of the helium atoms is found to be moving in the positive x direction. Find the direction of motion of the second helium, and find the velocity of each of the two helium atoms. Solve this problem in two different ways: (a) by direct application of conservation of momentum and energy; (b) by applying the results of Problem 4 to a frame of reference moving with the original beryllium atom and then switching to the reference frame in which the beryllium is moving.

14. Suppose the beryllium atom of Problem 4 moves in the positive x direction and has kinetic energy 60.0 keV. One helium atom is found to move at an angle of $30°$ with respect to the x axis. Find the direction of motion of the second helium atom and find the velocity of each helium atom. Work this problem in two ways as you did the previous problem. (*Hint:* Consider one helium to be emitted with velocity components v_x and v_y in the beryllium rest frame. What is the relationship between v_x and v_y? How do v_x and v_y change when we move in the x direction at speed v?)

15. A gas cylinder contains argon atoms ($m = 40.0$ u). The temperature is increased from 293 K ($20°$C) to 373 K ($100°$C). (a) What is the change in the average kinetic energy per atom? (b) The container is resting on a table in the Earth's gravity. Find the change in the vertical position of the container that produces the same change in the average energy per atom found in part (a).

16. Calculate the fraction of the molecules in a gas that are moving with translational kinetic energies between $0.02kT$ and $0.04kT$.

17. For a molecule of O_2 at room temperature (300 K), calculate the average angular velocity for rotations about the x' or y' axes. The distance between the O atoms in the molecule is 0.121 nm.

Chapter 2

THE SPECIAL THEORY OF RELATIVITY

This 12-foot tall statue of Albert Einstein is located at the headquarters of the National Academy of Sciences in Washington DC. The page in his hand shows three equations that he discovered: the fundamental equation of general relativity, which revolutionized our understanding of gravity; the equation for the photoelectric effect, which opened the path to the development of quantum mechanics; and the equation for mass-energy equivalence, which is the cornerstone of his special theory of relativity.

Einstein's special theory of relativity and Planck's quantum theory burst forth on the physics scene almost simultaneously during the first decade of the 20th century. Both theories caused profound changes in the way we view our universe at its most fundamental level.

In this chapter we study the special theory of relativity.* This theory has a completely undeserved reputation as being so exotic that few people can understand it. On the contrary, special relativity is basically a system of kinematics and dynamics, based on a set of postulates that are different from those of classical physics. The resulting formalism is not much more complicated than Newton's laws, but it does lead to several predictions that seem to go against our common sense. Even so, the special theory of relativity has been carefully and thoroughly tested by experiment and found to be correct in all its predictions.

We first review the classical relativity of Galileo and Newton, and then we show why Einstein proposed to replace it. We then discuss the mathematical aspects of special relativity, the predictions of the theory, and finally the experimental tests.

2.1 CLASSICAL RELATIVITY

A "theory of relativity" is in effect a way for observers in different frames of reference to compare the results of their observations. For example, consider an observer in a car parked by a highway near a large rock. To this observer, the rock is at rest. Another observer, who is moving along the highway in a car, sees the rock rush past as the car drives by. To this observer, the rock appears to be moving. A theory of relativity provides the conceptual framework and mathematical tools that enable the two observers to transform a statement such as "rock is at rest" in one frame of reference to the statement "rock is in motion" in another frame of reference. More generally, relativity gives a means for expressing the laws of physics in different frames of reference.

The mathematical basis for comparing the two descriptions is called a *transformation*. Figure 2.1 shows an abstract representation of the situation. Two observers

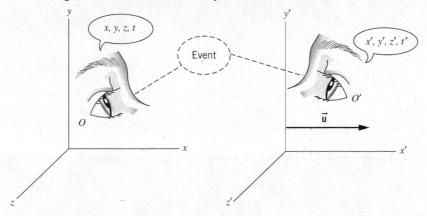

FIGURE 2.1 Two observers O and O' observe the same event. O' moves relative to O with a constant velocity $\vec{\mathbf{u}}$.

---

*The *general* theory of relativity, which is covered briefly in Chapter 15, deals with "curved" coordinate systems, in which gravity is responsible for the curvature. Here we discuss the *special* case of the more familiar "flat" coordinate systems.

O and O' are each at rest in their own frames of reference but move relative to one another with constant velocity $\vec{u}$. (O and O' refer both to the observers and their reference frames or coordinate systems.) They observe the same *event,* which happens at a particular point in space and a particular time, such as a collision between two particles. According to O, the space and time coordinates of the event are x, y, z, t, while according to O' the coordinates of the *same event* are x', y', z', t'. The two observers use calibrated meter sticks and synchronized clocks, so any differences between the coordinates of the two events are due to their different frames of reference and not to the measuring process. We simplify the discussion by assuming that the relative velocity $\vec{u}$ always lies along the common xx' direction, as shown in Figure 2.1, and we let $\vec{u}$ represent the velocity of O' as measured by O (and thus O' would measure velocity $-\vec{u}$ for O).

In this discussion we make a particular choice of the kind of reference frames inhabited by O and O'. We assume that each observer has the capacity to test Newton's laws and finds them to hold in that frame of reference. For example, each observer finds that an object at rest or moving with a constant velocity remains in that state unless acted upon by an external force (Newton's first law, the law of inertia). Such frames of reference are called *inertial frames.* An observer in interstellar space floating in a nonrotating rocket with the engines off would be in an inertial frame of reference. An observer at rest on the surface of the Earth is *not* in an inertial frame, because the Earth is rotating about its axis and orbiting about the Sun; however, the accelerations associated with those motions are so small that we can usually regard our reference frame as approximately inertial. (The noninertial reference frame at the Earth's surface does produce important and often spectacular effects, such as the circulation of air around centers of high or low pressure.) An observer in an accelerating car, a rotating merry-go-round, or a descending roller coaster is *not* in an inertial frame of reference!

We now derive the classical or *Galilean* transformation that relates the coordinates x, y, z, t to x', y', z', t'. We assume as a postulate of classical physics that $t = t'$, that is, time is the same for all observers. We also assume for simplicity that the coordinate systems are chosen so that their origins coincide at $t = 0$. Consider an object in O' at the coordinates x', y', z' (Figure 2.2). According to O, the y and z coordinates are the same as those in O'. Along the x direction, O would observe the object at $x = x' + ut$. We therefore have the *Galilean coordinate transformation*

$$x' = x - ut \qquad y' = y \qquad z' = z \qquad (2.1)$$

To find the velocities of the object as observed by O and O', we take the derivatives of these expressions with respect to t' on the left and with respect to t on the right (which we can do because we have assumed $t' = t$). This gives the *Galilean velocity transformation*

$$v'_x = v_x - u \qquad v'_y = v_y \qquad v'_z = v_z \qquad (2.2)$$

In a similar fashion, we can take the derivatives of Eq. 2.2 with respect to time and obtain relationships between the accelerations

$$a'_x = a_x \qquad a'_y = a_y \qquad a'_z = a_z \qquad (2.3)$$

Equation 2.3 shows again that Newton's laws hold for both observers. As long as u is constant ($du/dt = 0$), the observers measure identical accelerations and agree on the results of applying $\vec{F} = m\vec{a}$.

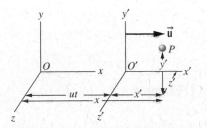

FIGURE 2.2 An object or event at point P is at coordinates x', y', z' with respect to O'. The x coordinate measured by O is $x = x' + ut$. The y and z coordinates in O are the same as those in O'.

Example 2.1

Two cars are traveling at constant speed along a road in the same direction. Car A moves at 60 km/h and car B moves at 40 km/h, each measured relative to an observer on the ground (Figure 2.3a). What is the speed of car A relative to car B?

Solution

Let O be the observer on the ground, who observes car A to move at $v_x = 60$ km/h. Assume O' to be moving with car B at $u = 40$ km/h. Then

$$v'_x = v_x - u = 60 \text{ km/h} - 40 \text{ km/h}$$
$$= 20 \text{ km/h}$$

Figure 2.3b shows the situation as observed by O'.

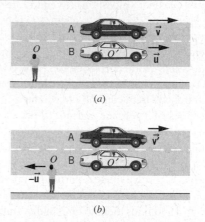

FIGURE 2.3 Example 2.1. (a) As observed by O at rest on the ground. (b) As observed by O' in car B.

Example 2.2

An airplane is flying due east relative to still air at a speed of 320 km/h. There is a 65 km/h wind blowing toward the north, as measured by an observer on the ground. What is the velocity of the plane measured by the ground observer?

Solution

Let O be the observer on the ground, and let O' be an observer who is moving with the wind, for example a balloonist (Figure 2.4). Then $u = 65$ km/h, and (because our equations are set up with $\vec{u}$ in the xx' direction) we must choose the xx' direction to be to the north. In this case we know the velocity with respect to O'; taking the y direction to the east, we have $v'_x = 0$ and $v'_y = 320$ km/h. Using Eq. 2.2 we obtain

$$v_x = v'_x + u = 0 + 65 \text{ km/h} = 65 \text{ km/h}$$
$$v_y = v'_y = 320 \text{ km/h}$$

Relative to the ground, the plane flies in a direction determined by $\phi = \tan^{-1}(65 \text{ km/h})/(320 \text{ km/h}) = 11.5°$, or 11.5° north of east.

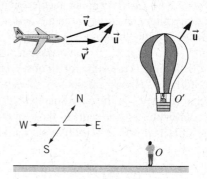

FIGURE 2.4 Example 2.2. As observed by O at rest on the ground, the balloon drifts north with the wind, while the plane flies north of east.

Example 2.3

A swimmer capable of swimming at a speed c in still water is swimming in a stream in which the current is u (which we assume to be less than c). Suppose the swimmer swims upstream a distance L and then returns downstream to the starting point. Find the time necessary to make the round trip, and compare it with the time to swim across the stream a distance L and return.

Solution

Let the frame of reference of O be the ground and the frame of reference of O' be the water, moving at speed u (Figure 2.5a). The swimmer always moves at speed c relative to the water, and thus $v'_x = -c$ for the upstream swim. (Remember that u always defines the *positive* x direction.) According to Eq. 2.2, $v'_x = v_x - u$, so $v_x = v'_x + u = u - c$. (As expected, the velocity relative to the ground has magnitude smaller than c; it is also *negative*, since the swimmer is swimming in the negative x direction, so $|v_x| = c - u$.) Therefore, $t_{up} = L/(c - u)$. For the downstream swim, $v'_x = c$, so $v_x = u + c, t_{down} = L/(c + u)$, and the total time is

$$t = \frac{L}{c + u} + \frac{L}{c - u} = \frac{L(c - u) + L(c + u)}{c^2 - u^2}$$

$$= \frac{2Lc}{c^2 - u^2} = \frac{2L}{c} \frac{1}{1 - u^2/c^2} \qquad (2.4)$$

To swim directly across the stream, the swimmer's efforts must be directed somewhat upstream to counter the effect of the current (Figure 2.5b). That is, in the frame of reference of O we would like to have $v_x = 0$, which requires $v'_x = -u$ according to Eq. 2.2. Since the speed relative to the water is always c, $\sqrt{v'^2_x + v'^2_y} = c$; thus $v'_y = \sqrt{c^2 - v'^2_x} = \sqrt{c^2 - u^2}$, and the round-trip time is

$$t = 2t_{across} = \frac{2L}{\sqrt{c^2 - u^2}} = \frac{2L}{c} \frac{1}{\sqrt{1 - u^2/c^2}} \qquad (2.5)$$

Notice the difference *in form* between this result and the result for the upstream-downstream swim, Eq. 2.4.

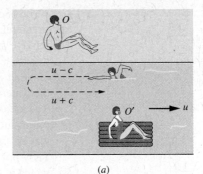

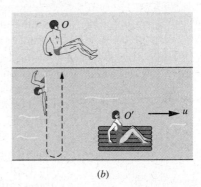

(a)

(b)

FIGURE 2.5 Example 2.3. The motion of a swimmer as seen by observer O at rest on the bank of the stream. Observer O' moves with the stream at speed u.

2.2 THE MICHELSON-MORLEY EXPERIMENT

We have seen how Newton's laws remain valid with respect to a Galilean transformation that relates the description of the motion of an object in one reference frame to that in another reference frame. It is then interesting to ask whether the same transformation rules apply to the motion of a light beam. According to the Galilean transformation, a light beam moving relative to observer O' in the x' direction at speed $c = 299,792,458$ m/s would have a speed of $c + u$ relative to O. Direct high-precision measurements of the speed of light beams have become possible in recent years (as we discuss later in this chapter), but in the 19th century it was necessary to devise a more indirect measurement of the speed of light according to different observers in relative motion.

Suppose the swimmer in Example 2.3 is replaced by a light beam. Observer O' is in a frame of reference in which the speed of light is c, and the frame of reference of observer O' is in motion relative to observer O. What is the speed of light as measured by observer O? If the Galilean transformation is correct, we should expect to see a difference between the speed of the light beam according to O and O' and therefore a time difference between the upstream-downstream and cross-stream times, as in Example 2.3.

Albert A. Michelson (1852–1931, United States). He spent 50 years doing increasingly precise experiments with light, for which he became the first U.S. citizen to win the Nobel Prize in physics (1907).

Physicists in the 19th century postulated just such a situation—a preferred frame of reference in which the speed of light has the precise value of c and other frames in relative motion in which the speed of light would differ, according to the Galilean transformation. The preferred frame, like that of observer O' in Example 2.3, is one that is at rest with respect to the medium in which light propagates at c (like the water of that example). What is the medium of propagation for light waves? It was inconceivable to physicists of the 19th century that a wave disturbance could propagate without a medium (consider mechanical waves such as sound or seismic waves, for example, which propagate due to mechanical forces in the medium). They postulated the existence of an invisible, massless medium, called the *ether*, which filled all space, was undetectable by any mechanical means, and existed solely for the propagation of light waves. It seemed reasonable then to obtain evidence for the ether by measuring the velocity of the Earth moving through the ether. This could be done in the geometry of Figure 2.5 by measuring the difference between the upstream-downstream and cross-stream times for a light wave. The calculation based on Galilean relativity would then give the relative velocity $\vec{u}$ between O (in the Earth's frame of reference) and the ether.

The first detailed and precise search for the preferred frame was performed in 1887 by the American physicist Albert A. Michelson and his associate E. W. Morley. Their apparatus consisted of a specially designed Michelson interferometer, illustrated in Figure 2.6. A monochromatic beam of light is split in two; the two beams travel different paths and are then recombined. Any phase difference between the combining beams causes bright and dark bands or "fringes" to appear, corresponding, respectively, to constructive and destructive interference, as shown in Figure 2.7.

There are two contributions to the phase difference between the beams. The first contribution comes from the path difference $AB - AC$; one of the beams may travel a longer distance. The second contribution, which would still be present even if the path lengths were equal, comes from the time difference between the upstream-downstream and cross-stream paths (as in Example 2.3) and indicates the motion of the Earth through the ether. Michelson and Morley used a clever method to isolate this second contribution—they rotated the entire apparatus by 90°! The rotation doesn't change the first contribution to the phase difference (because the lengths AB and AC don't change), but the second contribution changes sign, because what was an upstream-downstream path before the rotation becomes a cross-stream path after the rotation. As the apparatus is rotated through 90°, the fringes should change from bright to dark and back again as the phase difference changes. Each change from bright to dark represents a phase change of 180° (a half cycle), which corresponds to a time difference of a half period (about 10^{-15} s for visible light). Counting the number of fringe changes thus gives a measure of the time difference between the paths, which in turn gives the relative velocity u. (See Problem 3.)

When Michelson and Morley performed their experiment, there was no observable change in the fringe pattern—they deduced a shift of less than 0.01 fringe, corresponding to a speed of the Earth through the ether of at most 5 km/s. As a last resort, they reasoned that perhaps the orbital motion of the Earth just happened to cancel out the overall motion through the ether. If this were true,

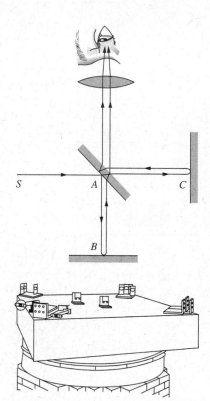

FIGURE 2.6 (Top) Beam diagram of Michelson interferometer. Light from source S is split at A by the half-silvered mirror; one part is reflected by the mirror at B and the other is reflected at C. The beams are then recombined for observation of the interference. (Bottom) Michelson's apparatus. To improve sensitivity, the beams were reflected to travel each leg of the apparatus eight times, rather than just twice. To reduce vibrations from the surroundings, the interferometer was mounted on a 1.5-m square stone slab floating in a pool of mercury.

six months later (when the Earth would be moving in its orbit in the opposite direction) the cancellation should not occur. When they repeated the experiment six months later, they again obtained a null result. In no experiment were Michelson and Morley able to detect the motion of the Earth through the ether.

In summary, we have seen that there is a direct chain of reasoning that leads from Galileo's principle of inertia, through Newton's laws with their implicit assumptions about space and time, ending with the failure of the Michelson-Morley experiment to observe the motion of the Earth relative to the ether. Although several explanations were offered for the unobservability of the ether and the corresponding failure of the upstream-downstream and cross-stream velocities to add in the expected way, the most novel, revolutionary, and ultimately successful explanation is given by Einstein's special theory of relativity, which requires a serious readjustment of our traditional concepts of space and time, and therefore alters some of the very foundations of physics.

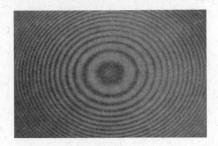

FIGURE 2.7 Interference fringes as observed with the Michelson interferometer of Figure 2.6. When the path length ACA changes by one-half wavelength relative to ABA, all light areas turn dark and all dark areas turn light.

2.3 EINSTEIN'S POSTULATES

The *special theory of relativity* is based on two postulates proposed by Albert Einstein in 1905:

The principle of relativity: *The laws of physics are the same in all inertial reference frames.*

The principle of the constancy of the speed of light: *The speed of light in free space has the same value c in all inertial reference frames.*

The first postulate declares that the laws of physics are absolute, universal, and the same for all inertial observers. Laws that hold for one inertial observer cannot be violated for *any* inertial observer.

The second postulate is more difficult to accept because it seems to go against our "common sense," which is based on the Galilean kinematics we observe in everyday experiences. Consider three observers A, B, and C. Observer B is at rest, while A and C move away from B in opposite directions each at a speed of $c/4$. B fires a light beam in the direction of A. According to the Galilean transformation, if B measures a speed of c for the light beam, then A measures a speed of $c - c/4 = 3c/4$, while C measures a speed of $c + c/4 = 5c/4$. Einstein's second postulate, on the other hand, requires all three observers to measure the same speed of c for the light beam! This postulate immediately explains the failure of the Michelson-Morley experiment—the upstream-downstream and cross-stream speeds are identical (both are equal to c), so there is no phase difference between the two beams.

The two postulates also allow us to dispose of the ether hypothesis. The first postulate does not permit a preferred frame of reference (all inertial frames are equivalent), and the second postulate does not permit only a single frame of reference in which light moves at speed c, because light moves at speed c in *all* frames. The ether, as a preferred reference frame in which light has a unique speed, is therefore unnecessary.

Albert Einstein (1879–1955, Germany-United States). A gentle philosopher and pacifist, he was the intellectual leader of two generations of theoretical physicists and left his imprint on nearly every field of modern physics.

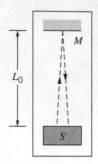

FIGURE 2.8 The clock ticks at intervals Δt_0 determined by the time for a light flash to travel the distance $2L_0$ from the light source S to the mirror M and back to the source where it is detected. (We assume the emission and detection occur at the same location, so the beam travels perpendicular to the mirror).

2.4 CONSEQUENCES OF EINSTEIN'S POSTULATES

Among their many consequences, Einstein's postulates require a new consideration of the fundamental nature of time and space. In this section we discuss how the postulates affect measurements of time and length intervals by observers in different frames of reference.

The Relativity of Time

To demonstrate the relativity of time, we use the timing device illustrated in Figure 2.8. It consists of a flashing light source S that is a distance L_0 from a mirror M. A flash of light from the source is reflected by the mirror, and when the light returns to S the clock ticks and triggers another flash. The time interval between ticks is the distance $2L_0$ (assuming the light travels perpendicular to the mirror) divided by the speed c:

$$\Delta t_0 = 2L_0/c \tag{2.6}$$

This is the time interval that is measured when the clock is at rest with respect to the observer.

We consider two observers: O is at rest on the ground, and O' moves with speed u. Each observer carries a timing device. Figure 2.9 shows a sequence of events that O observes for the clock carried by O'. According to O, the flash is emitted when the clock of O' is at A, reflected when it is at B, and detected at C. In this interval Δt, O observes the clock to move forward a distance of $u\Delta t$ from the point at which the flash was emitted, and O concludes that the light beam travels a distance $2L$, where $L = \sqrt{L_0^2 + (u\Delta t/2)^2}$, as shown in Figure 2.9. Because O observes the light beam to travel at speed c (as required by Einstein's second postulate) the time interval measured by O is

$$\Delta t = \frac{2L}{c} = \frac{2\sqrt{L_0^2 + (u\Delta t/2)^2}}{c} \tag{2.7}$$

Substituting for L_0 from Eq. 2.6 and solving Eq. 2.7 for Δt, we obtain

$$\Delta t = \frac{\Delta t_0}{\sqrt{1 - u^2/c^2}} \tag{2.8}$$

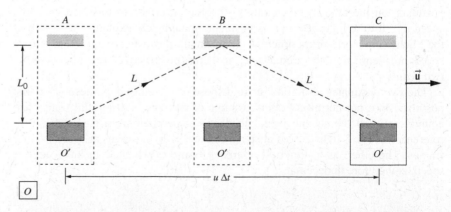

FIGURE 2.9 In the frame of reference of O, the clock carried by O' moves with speed u. The dashed line, of length $2L$, shows the path of the light beam according to O.

According to Eq. 2.8, observer O measures a longer time interval than O' measures. This is a general result of special relativity, which is known as *time dilation*. An observer O' is at rest relative to a device that produces a time interval Δt_0. For this observer, the beginning and end of the time interval occur at the same location, and so the interval Δt_0 is known as the *proper time*. An observer O, relative to whom O' is in motion, measures a longer time interval Δt for the same device. The dilated time interval Δt is always longer than the proper time interval Δt_0, no matter what the magnitude or direction of $\vec{u}$.

This is a real effect that applies not only to clocks based on light beams but also to time itself; all clocks run more slowly according to an observer in relative motion, biological clocks included. Even the growth, aging, and decay of living systems are slowed by the time dilation effect. However, note that under normal circumstances ($u \ll c$), there is no measurable difference between Δt and Δt_0, so we don't notice the effect in our everyday activities. Time dilation has been verified experimentally with decaying elementary particles as well as with precise atomic clocks carried aboard aircraft. Some experimental tests are discussed in the last section of this chapter.

Example 2.4

Muons are elementary particles with a (proper) lifetime of 2.2 μs. They are produced with very high speeds in the upper atmosphere when cosmic rays (high-energy particles from space) collide with air molecules. Take the height L_0 of the atmosphere to be 100 km in the reference frame of the Earth, and find the minimum speed that enables the muons to survive the journey to the surface of the Earth.

Solution
The birth and decay of the muon can be considered as the "ticks" of a clock. In the frame of reference of the Earth (observer O) this clock is moving, and therefore its ticks are slowed by the time dilation effect. If the muon is moving at a speed that is close to c, the time necessary for it to travel from the top of the atmosphere to the surface of the Earth is

$$\Delta t = \frac{L_0}{c} = \frac{100\,\text{km}}{3.00 \times 10^8 \text{ m/s}} = 333 \ \mu\text{s}$$

If the muon is to be observed at the surface of the Earth, it must live for at least 333 μs in the Earth's frame of reference. In the muon's frame of reference, the interval between its birth and decay is a proper time interval of 2.2 μs. The time intervals are related by Eq. 2.8:

$$333 \ \mu\text{s} = \frac{2.2 \ \mu\text{s}}{\sqrt{1 - u^2/c^2}}$$

Solving, we find

$$u = 0.999978c$$

If it were not for the time dilation effect, muons would not survive to reach the Earth's surface. The observation of these muons is a direct verification of the time dilation effect of special relativity.

The Relativity of Length

For this discussion, the moving timing device of O' is turned sideways, so that the light travels parallel to the direction of motion of O'. Figure 2.10 shows the sequence of events that O observes for the moving clock. According to O, the length of the clock (distance between the light source and the mirror) is L; as we shall see, this length is different from the length L_0 measured by O', relative to whom the clock is at rest.

The flash of light is emitted when the clock of O' is at A and reaches the mirror (position B) at time Δt_1 later. In this time interval, the light travels a distance

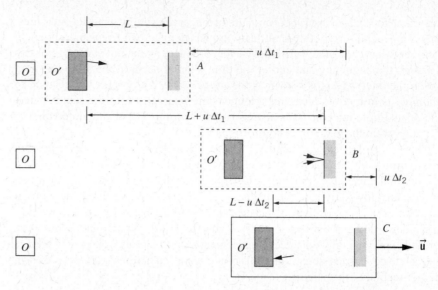

FIGURE 2.10 Here the clock carried by O' emits its light flash in the direction of motion.

$c \Delta t_1$, equal to the length L of the clock plus the additional distance $u \Delta t_1$ that the mirror moves forward in this interval. That is,

$$c \Delta t_1 = L + u \Delta t_1 \qquad (2.9)$$

The flash of light travels from the mirror to the detector in a time Δt_2 and covers a distance of $c \Delta t_2$, equal to the length L of the clock less the distance $u \Delta t_2$ that the clock moves forward in this interval:

$$c \Delta t_2 = L - u \Delta t_2 \qquad (2.10)$$

Solving Eqs. 2.9 and 2.10 for Δt_1 and Δt_2, and adding to find the total time interval, we obtain

$$\Delta t = \Delta t_1 + \Delta t_2 = \frac{L}{c - u} + \frac{L}{c + u} = \frac{2L}{c} \frac{1}{1 - u^2/c^2} \qquad (2.11)$$

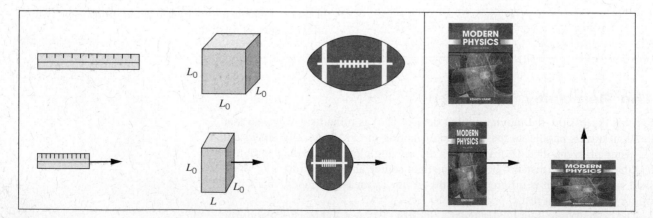

FIGURE 2.11 Some length-contracted objects. Notice that the shortening occurs only in the direction of motion.

From Eq. 2.8,

$$\Delta t = \frac{\Delta t_0}{\sqrt{1 - u^2/c^2}} = \frac{2L_0}{c} \frac{1}{\sqrt{1 - u^2/c^2}} \qquad (2.12)$$

Setting Eqs. 2.11 and 2.12 equal to one another and solving, we obtain

$$L = L_0\sqrt{1 - u^2/c^2} \qquad (2.13)$$

Equation 2.13 summarizes the effect known as *length contraction*. Observer O', who is at rest with respect to the object, measures the *rest length* L_0 (also known as the *proper length,* in analogy with the proper time). All observers relative to whom O' is in motion measure a shorter length, but only along the direction of motion; length measurements transverse to the direction of motion are unaffected (Figure 2.11).

For ordinary speeds ($u \ll c$), the effects of length contraction are too small to be observed. For example, a rocket of length 100 m traveling at the escape speed from Earth (11.2 km/s) would appear to an observer on Earth to contract only by about two atomic diameters!

Length contraction suggests that objects in motion are measured to have a shorter length than they do at rest. The objects do not actually shrink; there is merely a difference in the length measured by different observers. For example, to observers on Earth a high-speed rocket ship would appear to be contracted along its direction of motion (Figure 2.12a), but to an observer on the ship it is the passing Earth that appears to be contracted (Figure 2.12b).

These representations of length-contracted objects are somewhat idealized. The actual appearance of a rapidly moving object is determined by the time at which light leaves the various parts of the object and enters the eye or the camera. The result is that the object appears distorted in shape and slightly rotated.

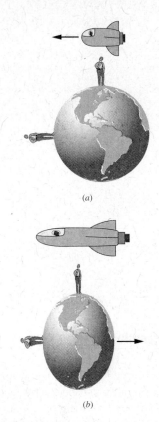

(a)

(b)

FIGURE 2.12 (a) The Earth views the passing contracted rocket. (b) From the rocket's frame of reference, the Earth appears contracted.

Example 2.5

Consider the point of view of an observer who is moving toward the Earth at the same velocity as the muon. In this reference frame, what is the apparent thickness of the Earth's atmosphere?

Solution

In this observer's reference frame, the muon is at rest and the Earth is rushing toward it at a speed of $u = 0.999978c$, as we found in Example 2.4. To an observer on the Earth, the height of the atmosphere is its rest length L_0 of 100 km.

To the observer in the muon's rest frame, the moving Earth has an atmosphere of height given by Eq. 2.13:

$$L = L_0\sqrt{1 - u^2/c^2}$$

$$= (100 \text{ km})\sqrt{1-(0.999978)^2} = 0.66 \text{ km} = 660 \text{ m}$$

This distance is small enough for the muons to reach the Earth's surface within their lifetime.

Note that what appears as a *time dilation* in one frame of reference (the observer on Earth) can be regarded as a *length contraction* in another frame of reference (the observer traveling with the muon). For another example of this effect, let's review again the example of the pion decay discussed in Section 1.2. A pion at rest has a lifetime of 26.0 ns. According to observer O_1 at rest in the laboratory frame of reference, a pion moving through the laboratory at a speed of $0.913c$ has a longer lifetime, which can be calculated to be 63.7 ns (using Eq. 2.8 for the time dilation). According to observer O_2, who is traveling through the laboratory at the same velocity as the pion, the pion appears to be at rest and has its proper lifetime of 26.0 ns. Thus O_1 sees a time dilation effect.

O_1 erects two markers in the laboratory, at the locations where the pion is created and decays. To O_1, the distance between those markers is the pion's speed times its lifetime, which works out to be 17.4 m. Suppose O_1 places a stick of length 17.4 m in the laboratory connecting the two markers. That stick is at rest in the laboratory reference frame and so has its proper length in that frame. In the reference frame of O_2, the stick is moving at a speed of 0.913c and has a shorter length of 7.1 m, which we can find using the length contraction formula (Eq. 2.13). So O_2 measures a distance of 7.1 m between the locations in the laboratory where the pion was created and where it decayed.

Note that O_1 measures the proper length and the dilated time, while O_2 measures the proper time and the contracted length. The proper time and proper length must always be referred to specific observers, who might not be in the same reference frame. The proper time is always measured by an observer according to whom the beginning of the time interval and the end of the time interval occur at the same location. If the time interval is the lifetime of the pion, then O_2 (relative to whom the pion does not move) sees its creation and decay at the same location and thus measures the proper time interval. The proper length, on the other hand, is always measured by an observer according to whom the measuring stick is at rest (O_1 in this case).

Example 2.6

An observer O is standing on a platform of length $D_0 = 65$ m on a space station. A rocket passes at a relative speed of 0.80c moving parallel to the edge of the platform. The observer O notes that the front and back of the rocket simultaneously line up with the ends of the platform at a particular instant (Figure 2.13a). (a) According to O, what is the time necessary for the rocket to pass a particular point on the platform? (b) What is the rest length L_0 of the rocket? (c) According to an observer O' on the rocket, what is the length D of the platform? (d) According to O', how long does it take for observer O to pass the entire length of the rocket? (e) According to O, the ends of the rocket simultaneously line up with the ends of the platform. Are these events simultaneous to O'?

Solution

(a) According to O, the length L of the rocket matches the length D_0 of the platform. The time for the rocket to pass a particular point is measured by O to be

$$\Delta t_0 = \frac{L}{0.80c} = \frac{65 \text{ m}}{2.40 \times 10^8 \text{ m/s}} = 0.27 \ \mu s$$

This is a proper time interval, because O measures the interval between two events that occur at the same point in the frame of reference of O (the front of the rocket passes a point, and then the back of the rocket passes the same point).

(b) O measures the contracted length L of the rocket. We can find its proper length L_0 using Eq. 2.13:

$$L_0 = \frac{L}{\sqrt{1 - u^2/c^2}} = \frac{65 \text{ m}}{\sqrt{1 - (0.80)^2}} = 108 \text{ m}$$

(c) According to O the platform is at rest, so 65 m is its proper length D_0. According to O', the contracted length of

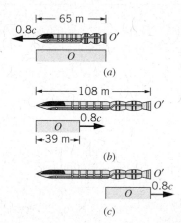

FIGURE 2.13 Example 2.6. (a) From the reference frame of O at rest on the platform, the passing rocket lines up simultaneously with the front and back of the platform. (b, c) From the reference frame O' in the rocket, the passing platform lines up first with the front of the rocket and later with the rear. Note the differing effects of length contraction in the two reference frames.

the platform is therefore

$$D = D_0\sqrt{1 - u^2/c^2} = (65 \text{ m})\sqrt{1 - (0.80)^2} = 39 \text{ m}$$

(*d*) For O to pass the entire length of the rocket, O' concludes that O must move a distance equal to its rest length, or 108 m. The time needed to do this is

$$\Delta t' = \frac{108 \text{ m}}{0.80c} = 0.45 \text{ } \mu s$$

Note that this is *not* a proper time interval for O', who determines this time interval using one clock at the front of the rocket to measure the time at which O passes the front of the rocket, and another clock on the rear of the rocket to measure the time at which O passes the rear of the rocket. The two events therefore occur at different points in O' and so cannot be separated by a proper time in O'. The corresponding time interval measured by O for the same two events, which we calculated in part (*a*), *is* a proper time interval for O, because the two events *do* occur at the same point in O.

The time intervals measured by O and O' should be related by the time dilation formula, as you should verify.

(*e*) According to O', the rocket has a rest length of $L_0 = 108$ m and the platform has a contracted length of $D = 39$ m. There is thus no way that O' could observe the two ends of both to align simultaneously. The sequence of events according to O' is illustrated in Figures 2.13*b* and *c*. The time interval $\Delta t'$ in O' between the two events that are simultaneous in O can be calculated by noting that, according to O', the time interval between the situations shown in Figures 2.13*b* and *c* must be that necessary for the platform to move a distance of 108 m − 39 m = 69 m, which takes a time

$$\Delta t' = \frac{69 \text{ m}}{0.80c} = 0.29 \text{ } \mu s$$

This result illustrates the relativity of simultaneity: two events at different locations that are simultaneous to O (the lining up of the two ends of the rocket with the two ends of the platform) *cannot* be simultaneous to O'.

Relativistic Velocity Addition

The timing device is now modified as shown in Figure 2.14. A source P emits particles that travel at speed v' according to an observer O' at rest with respect to the device. The flashing bulb F is triggered to flash when a particle reaches it. The flash of light makes the return trip to the detector D, and the clock ticks. The time interval Δt_0 between ticks measured by O' is composed of two parts: one for the particle to travel the distance L_0 at speed v' and another for the light to travel the same distance at speed c:

$$\Delta t_0 = L_0/v' + L_0/c \tag{2.14}$$

According to observer O, relative to whom O' moves at speed u, the sequence of events is similar to that shown in Figure 2.10. The emitted particle, which travels at speed v according to O, reaches F in a time interval Δt_1 after traveling the distance $v \Delta t_1$ equal to the (contracted) length L plus the additional distance $u \Delta t_1$ moved by the clock in that interval:

$$v \Delta t_1 = L + u \Delta t_1 \tag{2.15}$$

In the interval Δt_2, the light beam travels a distance $c \Delta t_2$ equal to the length L less the distance $u \Delta t_2$ moved by the clock in that interval:

$$c \Delta t_2 = L - u \Delta t_2 \tag{2.16}$$

We now solve Eqs. 2.15 and 2.16 for Δt_1 and Δt_2, add to find the total interval Δt between ticks according to O, use the time dilation formula, Eq. 2.8, to relate this result to Δt_0 from Eq. 2.14, and finally use the length contraction formula, Eq. 2.13, to relate L to L_0. After doing the algebra, we find the result

$$v = \frac{v' + u}{1 + v'u/c^2} \tag{2.17}$$

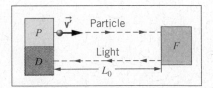

FIGURE 2.14 In this timing device, a particle is emitted by P at a speed v'. When the particle reaches F, it triggers the emission of a flash of light that travels to the detector D.

Equation 2.17 is the *relativistic velocity addition law* for velocity components that are in the direction of u. Later in this chapter we use a different method to derive the corresponding results for motion in other directions.

We can also regard Eq. 2.17 as a velocity transformation, enabling us to convert a velocity v' measured by O' to a velocity v measured by O. The corresponding classical law was given by Eq. 2.2: $v = v' + u$. The difference between the classical and relativistic results is the denominator of Eq. 2.17, which reduces to 1 in cases when the speeds are small compared with c. Example 2.7 shows how this factor prevents the measured speeds from exceeding c.

Equation 2.17 gives an important result when O' observes a light beam. For $v' = c$,

$$v = \frac{c + u}{1 + cu/c^2} = c \tag{2.18}$$

That is, when $v' = c$, then $v = c$, *independent of the value of u*. All observers measure the same speed c for light, exactly as required by Einstein's second postulate.

Example 2.7

A spaceship moving away from the Earth at a speed of $0.80c$ fires a missile parallel to its direction of motion (Figure 2.15). The missile moves at a speed of $0.60c$ relative to the ship. What is the speed of the missile as measured by an observer on the Earth?

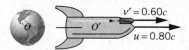

FIGURE 2.15 Example 2.7. A spaceship moves away from Earth at a speed of $0.80c$. An observer O' on the spaceship fires a missile and measures its speed to be $0.60c$ relative to the ship.

Solution

Here O' is on the ship and O is on Earth; O' moves with a speed of $u = 0.80c$ relative to O. The missile moves at speed $v' = 0.60c$ relative to O', and we seek its speed v relative to O. Using Eq. 2.17, we obtain

$$v = \frac{v' + u}{1 + v'u/c^2} = \frac{0.60c + 0.80c}{1 + (0.60c)(0.80c)/c^2}$$

$$= \frac{1.40c}{1.48} = 0.95c$$

According to classical kinematics (the numerator of Eq. 2.17), an observer on the Earth would see the missile moving at $0.60c + 0.80c = 1.40c$, thereby exceeding the maximum relative speed of c permitted by relativity. You can see how Eq. 2.17 brings about this speed limit. Even if v' were $0.9999\ldots c$ and u were $0.9999\ldots c$, the relative speed v measured by O would remain less than c.

The Relativistic Doppler Effect

In the classical Doppler effect for sound waves, an observer moving relative to a source of waves (sound, for example) detects a frequency different from that emitted by the source. The frequency f' heard by the observer O is related to the frequency f emitted by the source S according to

$$f' = f\frac{v \pm v_O}{v \mp v_S} \tag{2.19}$$

where v is the speed of the waves in the medium (such as still air, in the case of sound waves), v_S is the speed of the source *relative to the medium*, and v_O is the speed of the observer *relative to the medium*. The upper signs in the numerator and denominator are chosen whenever S moves toward O or O moves toward S, while the lower signs apply whenever O and S move away from one another.

The classical Doppler shift for motion of the source differs from that for motion of the observer. For example, suppose the source emits sound waves at $f = 1000$ Hz. If the source moves at 30 m/s toward the observer who is at rest in the medium (which we take to be air, in which sound moves at $v = 340$ m/s), then $f' = 1097$ Hz, while if the source is at rest in the medium and the observer moves toward the source at 30 m/s, the frequency is 1088 Hz. Other possibilities in which the relative speed between S and O is 30 m/s, such as each moving toward the other at 15 m/s, give still different frequencies.

Here we have a situation in which it is not the relative speed of the source and observer that determines the Doppler shift—it is the speed of each with respect to the medium. This cannot occur for light waves, since there is no medium (no "ether") and no preferred reference frame by Einstein's first postulate. We therefore require a different approach to the Doppler effect for light waves, an approach that does not distinguish between source motion and observer motion, but involves only the relative motion between the source and the observer.

Consider a source of waves that is at rest in the reference frame of observer O. Observer O' moves relative to the source at speed u. We consider the situation from the frame of reference of O', as shown in Figure 2.16. Suppose O observes the source to emit N waves at frequency f. According to O, it takes an interval $\Delta t_0 = N/f$ for these N waves to be emitted; this is a proper time interval in the frame of reference of O. The corresponding time interval to O' is $\Delta t'$, during which O moves a distance $u\,\Delta t'$. The wavelength according to O' is the total length interval occupied by these waves divided by the number of waves:

$$\lambda' = \frac{c\,\Delta t' + u\,\Delta t'}{N} = \frac{c\,\Delta t' + u\,\Delta t'}{f\,\Delta t_0} \tag{2.20}$$

The frequency according to O' is $f' = c/\lambda'$, so

$$f' = f\frac{\Delta t_0}{\Delta t'}\frac{1}{1 + u/c} \tag{2.21}$$

and using the time dilation formula, Eq. 2.8, to relate $\Delta t'$ and Δt_0, we obtain

$$f' = f\frac{\sqrt{1 - u^2/c^2}}{1 + u/c} = f\sqrt{\frac{1 - u/c}{1 + u/c}} \tag{2.22}$$

This is the formula for the *relativistic Doppler shift*, for the case in which the waves are observed in a direction parallel to $\vec{\mathbf{u}}$. Note that, unlike the classical formula, it does *not* distinguish between source motion and observer motion; the

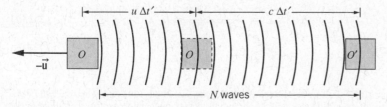

FIGURE 2.16 A source of waves, in the reference frame of O, moves at speed u away from observer O'. In the time $\Delta t'$ (according to O'), O moves a distance $u\,\Delta t'$ and emits N waves.

relativistic Doppler effect depends only on the relative speed u between the source and observer.

Equation 2.22 assumes that the source and observer are separating. If the source and observer are approaching one another, replace u by $-u$ in the formula.

Example 2.8

A distant galaxy is moving away from the Earth at such high speed that the blue hydrogen line at a wavelength of 434 nm is recorded at 600 nm, in the red range of the spectrum. What is the speed of the galaxy relative to the Earth?

Solution
Using Eq. 2.22 with $f = c/\lambda$ and $f' = c/\lambda'$, we obtain

$$\frac{c}{\lambda'} = \frac{c}{\lambda}\sqrt{\frac{1 - u/c}{1 + u/c}}$$

$$\frac{c}{600\,\text{nm}} = \frac{c}{434\,\text{nm}}\sqrt{\frac{1 - u/c}{1 + u/c}}$$

Solving, we find

$$u/c = 0.31$$

Thus the galaxy is moving away from Earth at a speed of $0.31c = 9.4 \times 10^7$ m/s. Evidence obtained in this way indicates that nearly all the galaxies we observe are moving away from us. This suggests that the universe is expanding, and is usually taken to provide evidence in favor of the Big Bang theory of cosmology (see Chapter 15).

2.5 THE LORENTZ TRANSFORMATION

We have seen that the Galilean transformation of coordinates, time, and velocity is not consistent with Einstein's postulates. Although the Galilean transformation agrees with our "common-sense" experience at low speeds, it does not agree with experiment at high speeds. We therefore need a new set of transformation equations that replaces the Galilean set and that is capable of predicting such relativistic effects as time dilation, length contraction, velocity addition, and the Doppler shift.

As before, we seek a transformation that enables observers O and O' in relative motion to compare their measurements of the space and time coordinates of the same event. The transformation equations relate the measurements of O (namely, x, y, z, t) to those of O' (namely, x', y', z', t'). This new transformation must have several properties: It must be linear (depending only on the first power of the space and time coordinates), which follows from the homogeneity of space and time; it must be consistent with Einstein's postulates; and it must reduce to the Galilean transformation when the relative speed between O and O' is small. We again assume that the velocity of O' relative to O is in the positive xx' direction.

This new transformation consistent with special relativity is called the *Lorentz transformation**. Its equations are

$$x' = \frac{x - ut}{\sqrt{1 - u^2/c^2}} \tag{2.23a}$$

$$y' = y \tag{2.23b}$$

---

*H. A. Lorentz (1853–1928) was a Dutch physicist who shared the 1902 Nobel Prize for his work on the influence of magnetic fields on light. In an unsuccessful attempt to explain the failure of the Michelson-Morley experiment, Lorentz developed the transformation equations that are named for him in 1904, a year *before* Einstein published his special theory of relativity. For a derivation of the Lorentz transformation, see R. Resnick and D. Halliday, *Basic Concepts in Relativity* (New York, Macmillan, 1992).

$$z' = z \qquad (2.23c)$$

$$t' = \frac{t - (u/c^2)x}{\sqrt{1 - u^2/c^2}} \qquad (2.23d)$$

It is often useful to write these equations in terms of *intervals* of space and time by replacing each coordinate by the corresponding interval (replace x by Δx, x' by $\Delta x'$, t by Δt, t' by $\Delta t'$).

These equations are written assuming that O' moves *away from* O in the xx' direction. If O' moves *toward* O, replace u with $-u$ in the equations.

The first three equations reduce directly to the Galilean transformation for space coordinates, Eqs. 2.1, when $u \ll c$. The fourth equation, which links the time coordinates, reduces to $t' = t$, which is a fundamental postulate of the Galilean-Newtonian world.

We now use the Lorentz transformation equations to derive some of the predictions of special relativity. The problems at the end of the chapter guide you in some other derivations. The results derived here are identical with those we obtained previously using Einstein's postulates, which shows that the equations of the Lorentz transformation are consistent with the postulates of special relativity.

Length Contraction

A rod of length L_0 is at rest in the reference frame of observer O'. The rod extends along the x' axis from x_1' to x_2'; that is, O' measures the proper length $L_0 = x_2' - x_1'$. Observer O, relative to whom the rod is in motion, measures the ends of the rod to be at coordinates x_1 and x_2. For O to determine the length of the moving rod, O must make a *simultaneous* determination of x_1 and x_2, and then the length is $L = x_2 - x_1$. Suppose the first event is O' setting off a flash bulb at one end of the rod at x_1' and t_1', which O observes at x_1 and t_1, and the second event is O' setting off a flash bulb at the other end at x_2' and t_2', which O observes at x_2 and t_2. The equations of the Lorentz transformation relate these coordinates, specifically,

$$x_1' = \frac{x_1 - ut_1}{\sqrt{1 - u^2/c^2}} \qquad x_2' = \frac{x_2 - ut_2}{\sqrt{1 - u^2/c^2}} \qquad (2.24)$$

Subtracting these equations, we obtain

$$x_2' - x_1' = \frac{x_2 - x_1}{\sqrt{1 - u^2/c^2}} - \frac{u(t_2 - t_1)}{\sqrt{1 - u^2/c^2}} \qquad (2.25)$$

O' must arrange to set off the flash bulbs so that the flashes appear to be simultaneous to O. (They will *not* be simultaneous to O', as we discuss later in this section.) This enables O to make a simultaneous determination of the coordinates of the endpoints of the rod. If O observes the flashes to be simultaneous, then $t_2 = t_1$, and Eq. 2.25 reduces to

$$x_2' - x_1' = \frac{x_2 - x_1}{\sqrt{1 - u^2/c^2}} \qquad (2.26)$$

With $x_2' - x_1' = L_0$ and $x_2 - x_1 = L$, this becomes

$$L = L_0\sqrt{1 - u^2/c^2} \qquad (2.27)$$

which is identical with Eq. 2.13, which we derived earlier using Einstein's postulates.

Velocity Transformation

If O observes a particle to travel with velocity v (components v_x, v_y, v_z), what velocity v' does O' observe for the particle? The relationship between the velocities measured by O and O' is given by the *Lorentz velocity transformation*:

$$v'_x = \frac{v_x - u}{1 - v_x u/c^2} \tag{2.28a}$$

$$v'_y = \frac{v_y \sqrt{1 - u^2/c^2}}{1 - v_x u/c^2} \tag{2.28b}$$

$$v'_z = \frac{v_z \sqrt{1 - u^2/c^2}}{1 - v_x u/c^2} \tag{2.28c}$$

By solving Eq. 2.28a for v_x, you can show that it is identical to Eq. 2.17, a result we derived previously based on Einstein's postulates. Note that, in the limit of low speeds ($u \ll c$), the Lorentz velocity transformation reduces to the Galilean velocity transformation, Eq. 2.2. Note also that $v'_y \neq v_y$, even though $y' = y$. This occurs because of the way the Lorentz transformation handles the time coordinate.

We can derive these transformation equations for velocity from the Lorentz coordinate transformation. By way of example, we derive the velocity transformation for $v'_y = dy'/dt'$. Differentiating the coordinate transformation $y' = y$, we obtain $dy' = dy$. Similarly, differentiating the time coordinate transformation (Eq. 2.23d), we obtain

$$dt' = \frac{dt - (u/c^2)dx}{\sqrt{1 - u^2/c^2}}$$

So

$$v'_y = \frac{dy'}{dt'} = \frac{dy}{[dt - (u/c^2)\,dx]/\sqrt{1 - u^2/c^2}} = \sqrt{1 - u^2/c^2}\,\frac{dy}{dt - (u/c^2)\,dx}$$

$$= \sqrt{1 - u^2/c^2}\,\frac{dy/dt}{1 - (u/c^2)\,dx/dt} = \frac{v_y \sqrt{1 - u^2/c^2}}{1 - uv_x/c^2}$$

Similar methods can be used to obtain the transformation equations for v'_x and v'_z. These derivations are left as exercises (Problem 14).

Simultaneity and Clock Synchronization

Under ordinary circumstances, synchronizing one clock with another is a simple matter. But for scientific work, where timekeeping at a precision below the nanosecond range is routine, clock synchronization can present some significant challenges. At very least, we need to correct for the time that it takes for the signal showing the reading on one clock to be transmitted to the other clock. However, for observers who are in motion with respect to each other, special relativity gives yet another way that clocks may appear to be out of synchronization.

Consider the device shown in Figure 2.17. Two clocks are located at $x = 0$ and $x = L$. A flash lamp is located at $x = L/2$, and the clocks are set running when they

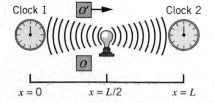

FIGURE 2.17 A flash of light, emitted from a point midway between the two clocks, starts the two clocks simultaneously according to O. Observer O' sees clock 2 start ahead of clock 1.

receive the flash of light from the lamp. The light takes the same interval of time to reach the two clocks, so the clocks start together precisely at a time $L/2c$ after the flash is emitted, and the clocks are exactly synchronized.

Now let us examine the same situation from the point of view of the moving observer O'. In the frame of reference of O, two events occur: the receipt of a light signal by clock 1 at $x_1 = 0, t_1 = L/2c$ and the receipt of a light signal by clock 2 at $x_2 = L, t_2 = L/2c$. Using Eq. 2.23d, we find that O' observes clock 1 to receive its signal at

$$t_1' = \frac{t_1 - (u/c^2)x_1}{\sqrt{1 - u^2/c^2}} = \frac{L/2c}{\sqrt{1 - u^2/c^2}} \tag{2.29}$$

while clock 2 receives its signal at

$$t_2' = \frac{t_2 - (u/c^2)x_2}{\sqrt{1 - u^2/c^2}} = \frac{L/2c - (u/c^2)L}{\sqrt{1 - u^2/c^2}} \tag{2.30}$$

Thus t_2' is smaller than t_1' and clock 2 appears to receive its signal earlier than clock 1, so that the clocks start at times that differ by

$$\Delta t' = t_1' - t_2' = \frac{uL/c^2}{\sqrt{1 - u^2/c^2}} \tag{2.31}$$

according to O'. Keep in mind that this is *not* a time dilation effect—time dilation comes from the *first* term of the Lorentz transformation (Eq. 2.23d) for t', while the lack of synchronization arises from the *second* term. O' observes *both* clocks to run slow, due to time dilation; O' *also* observes clock 2 to be ahead of clock 1.

We therefore reach the following conclusion: two events that are simultaneous in one reference frame are not simultaneous in another reference frame moving with respect to the first, unless the two events occur at the same point in space. (If $L = 0$, Eq. 2.31 shows that the clocks are synchronized in all reference frames.) Clocks that appear to be synchronized in one frame of reference will not necessarily be synchronized in another frame of reference in relative motion.

It is important to note that this clock synchronization effect does not depend on the *location* of observer O' but only on the *velocity* of O'. In Figure 2.17, the location of O' could have been drawn far to the left side of clock 1 or far to the right side of clock 2, and the result would be the same. In those different locations, the propagation time of the light signal showing clock 1 starting will differ from the propagation time of the light signal showing clock 2 starting. However, O' is assumed to be an "intelligent" observer who is aware of the locations where the light signals showing the two clocks starting are received relative to the locations of the clocks. O' corrects for this time difference, which is due only to the propagation time of the light signals, and *even after making that correction the clocks still do not appear to be synchronized!*

Although the location of O' does not appear in Eq. 2.31, the *direction* of the velocity of O' is important—if O' is moving in the opposite direction, the observed starting order of the two clocks is reversed.

Example 2.9

Two rockets are leaving their space station along perpendicular paths, as measured by an observer on the space station. Rocket 1 moves at $0.60c$ and rocket 2 moves at $0.80c$, both measured relative to the space station. What is the velocity of rocket 2 as observed by rocket 1?

Solution

Observer O is the space station, observer O' is rocket 1 (moving at $u = 0.60c$), and each observes rocket 2, moving (according to O) in a direction perpendicular to rocket 1. We take this to be the y direction of the reference frame of O. Thus O observes rocket 2 to have velocity components $v_x = 0$, $v_y = 0.80c$, as shown in Figure 2.18a.

We can find v'_x and v'_y using the Lorentz velocity transformation:

$$v'_x = \frac{v_x - u}{1 - v_x u/c^2} = \frac{0 - 0.60c}{1 - 0(0.60c)/c^2} = -0.60c$$

$$v'_y = \frac{v_y\sqrt{1 - u^2/c^2}}{1 - v_x u/c^2}$$

$$= \frac{0.80c\sqrt{1 - (0.60c)^2/c^2}}{1 - 0(0.60c)/c^2} = 0.64c$$

Thus, according to O', the situation looks like Figure 2.18b.

The speed of rocket 2 according to O' is $\sqrt{(0.60c)^2 + (0.64c)^2} = 0.88c$, less than c. According to

the Galilean transformation, v'_y would be identical with v_y, and thus the speed would be $\sqrt{(0.60c)^2 + (0.80c)^2} = c$. Once again, the Lorentz transformation prevents relative speeds from reaching or exceeding the speed of light.

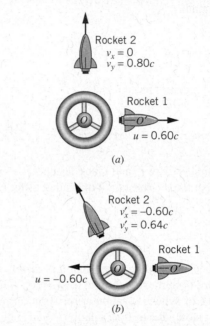

FIGURE 2.18 Example 2.9. (a) As viewed from the reference frame of O. (b) As viewed from the reference frame of O'.

Example 2.10

In Example 2.6, two events that were simultaneous to O (the lining up of the front and back of the rocket ship with the ends of the platform) were not simultaneous to O'. Find the time interval between these events according to O'.

Solution

According to O, the two simultaneous events are separated by a distance of $L = 65$ m. For $u = 0.80c$, Eq. 2.31 gives

$$\Delta t' = \frac{uL/c^2}{\sqrt{1 - u^2/c^2}}$$

$$= \frac{(0.80)(65 \text{ m})/(3.00 \times 10^8 \text{ m/s})}{\sqrt{1 - (0.80)^2}} = 0.29 \text{ } \mu\text{s}$$

which agrees with the result calculated in part (e) of Example 2.6.

2.6 THE TWIN PARADOX

We now turn briefly to what has become known as the twin paradox. Suppose there is a pair of twins on Earth. One, whom we shall call Casper, remains on Earth, while his twin sister Amelia sets off in a rocket ship on a trip to a distant planet. Casper, based on his understanding of special relativity, knows that his sister's clocks will run slow relative to his own and that therefore she should be younger than he when she returns, as our discussion of time dilation would suggest. However, recalling that discussion, we know that for two observers in relative motion, *each* thinks the *other's* clocks are running slow. We could therefore study this problem from the point of view of Amelia, according to whom Casper and the Earth (accompanied by the solar system and galaxy) make a round-trip journey away from her and back again. Under such circumstances, she will think it is her brother's clocks (which are now in motion relative to her own) that are running slow, and will therefore expect her brother to be younger than she when they meet again. While it is possible to disagree over whose clocks are running slow relative to his or her own, which is merely a problem of frames of reference, when Amelia returns to Earth (or when the Earth returns to Amelia), all observers must agree as to which twin has aged less rapidly. This is the paradox—each twin expects the other to be younger.

The resolution of this paradox lies in considering the asymmetric role of the two twins. The laws of special relativity apply only to inertial frames, those moving relative to one another at constant velocity. We may supply Amelia's rockets with sufficient thrust so that they accelerate for a very short length of time, bringing the ship to a speed at which it can coast to the planet, and thus during her outward journey Amelia spends all but a negligible amount of time in a frame of reference moving at constant speed relative to Casper. However, in order to return to Earth, she must decelerate and reverse her motion. Although this also may be done in a very short time interval, Amelia's return journey occurs in a completely different inertial frame than her outward journey. It is Amelia's jump from one inertial frame to another that causes the asymmetry in the ages of the twins. Only Amelia has the necessity of jumping to a new inertial frame to return, and therefore *all observers will agree* that it is Amelia who is "really" in motion, and that it is her clocks that are "really" running slow; therefore she is indeed the younger twin on her return.

Let us make this discussion more quantitative with a numerical example. We assume, as discussed above, that the acceleration and deceleration take negligible time intervals, so that all of Amelia's aging is done during the coasting. For simplicity, we assume the distant planet is at rest relative to the Earth; this does not change the problem, but it avoids the need to introduce yet another frame of reference. Suppose the planet to be 6 light-years distant from Earth, and suppose Amelia travels at a speed of $0.6c$. Then according to Casper it takes his sister 10 years (10 years $\times 0.6c = 6$ light-years) to reach the planet and 10 years to return, and therefore she is gone for a total of 20 years. (However, Casper doesn't know his sister has reached the planet until the light signal carrying news of her arrival reaches Earth. Since light takes 6 years to make the journey, it is 16 years after her departure when Casper sees his sister's arrival at the planet. Four years later she returns to Earth.) From the frame of reference of Amelia aboard the rocket, the distance to the planet is contracted by a factor of $\sqrt{1 - (0.6)^2} = 0.8$, and is therefore 0.8×6 light-years = 4.8 light-years. At a speed of $0.6c$, Amelia will measure 8 years for the trip to the planet, for a total round trip time of 16 years. Thus Casper ages 20 years while Amelia ages only 16 years and is indeed the younger on her return.

We can confirm this analysis by having Casper send a light signal to his sister each year on his birthday. We know that the frequency of the signal as received by Amelia will be Doppler shifted. During the outward journey, she will receive signals at the rate of

$$(1/\text{year})\sqrt{\frac{1 - u/c}{1 + u/c}} = 0.5/\text{year}$$

During the return journey, the Doppler-shifted rate will be

$$(1/\text{year})\sqrt{\frac{1 + u/c}{1 - u/c}} = 2/\text{year}$$

Thus for the first 8 years, during Amelia's trip to the planet, she receives 4 signals, and during the return trip of 8 years, she receives 16 signals, for a total of 20. She receives 20 signals, indicating her brother has celebrated 20 birthdays during her 16-year journey.

Spacetime Diagrams

A particularly helpful way of visualizing the journeys of Casper and Amelia uses a *spacetime* diagram. Figure 2.19 shows an example of a spacetime diagram for motion that involves only one spatial direction.

In your introductory physics course, you probably became familiar with plotting motion on a graph in which distance appeared on the vertical axis and time on the horizontal axis. On such a graph, a straight line represents motion at constant velocity; the slope of the line is equal to the velocity. Note that the axes of the spacetime diagram are switched from the traditional graph of particle motion, with time on the vertical axis and space on the horizontal axis.

On a spacetime diagram, the graph that represents the motion of a particle is called its *worldline*. The *inverse* of the slope of the particle's worldline gives its velocity. Equivalently, the velocity is given by the tangent of the angle that the worldline makes with the *vertical* axis (rather than with the horizontal axis, as would be the case with a conventional plot of distance *vs.* time). Usually, the units of x and t are chosen so that motion at the speed of light is represented by a line with a 45° slope. A vertical line represents a particle that is at the same spatial locations at all times—that is, a particle at rest. Permitted motions with constant velocity are then represented by straight lines between the vertical and the 45° line representing the maximum velocity.

Let's draw the worldlines of Casper and Amelia according to Casper's frame of reference. Casper's worldline is a vertical line, because he is at rest in this frame (Figure 2.20). In Casper's frame of reference, 20 years pass between Amelia's departure and her return, so we can follow Casper's vertical worldline for 20 years.

Amelia is traveling at a speed of $0.6c$, so her worldline makes an angle with the vertical whose tangent is 0.6 (31°). In Casper's frame of reference, the planet visited by Amelia is 6 light-years from Earth. Amelia travels a distance of 6 light-years in a time of 10 years (according to Casper) so that $v = 6$ light-years/10 years $= 0.6c$.

The birthday signals that Casper sends to Amelia at the speed of light are represented by the series of 45° lines in Figure 2.20. Amelia receives 4 birthday signals during her outbound journey (the 4th arrives just as she reaches the planet) and 16 birthday signals during her return journey (the 16th is sent and received just as she returns to Earth).

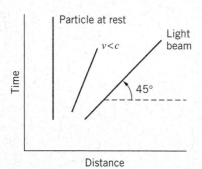

FIGURE 2.19 A spacetime diagram.

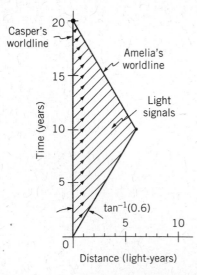

FIGURE 2.20 Casper's spacetime diagram, showing his worldline and Amelia's.

It is left as an exercise (Problems 22 and 24) to consider the situation if it is Amelia who is sending the signals.

2.7 RELATIVISTIC DYNAMICS

We have seen how Einstein's postulates have led to a new "relative" interpretation of such previously absolute concepts as length and time, and that the classical concept of absolute velocity is not valid. It is reasonable then to ask how far this revolution is to go in changing our interpretation of physical concepts. Dynamical quantities, such as momentum and kinetic energy, depend on length, time, and velocity. Do classical laws of momentum and energy conservation remain valid in Einstein's relativity?

Let's test the conservation laws by examining the collision shown in Figure 2.21a. Two particles collide elastically as observed in the reference frame of O'. Particle 1 of mass $m_1 = 2m$ is initially at rest, and particle 2 of mass $m_2 = m$ is moving in the negative x direction with an initial velocity of $v'_{2i} = -0.750c$. Using the classical law of momentum conservation to analyze this collision, O' would calculate the particles to be moving with final velocities $v'_{1f} = -0.500c$ and $v'_{2f} = +0.250c$. According to O', the total initial and final momenta of the particles would be:

$$p'_i = m_1 v'_{1i} + m_2 v'_{2i} = (2m)(0) + (m)(-0.750c) = -0.750mc$$

$$p'_f = m_1 v'_{1f} + m_2 v'_{2f} = (2m)(-0.500c) + (m)(0.250c) = -0.750mc$$

The initial and final momenta are equal according to O', demonstrating that momentum is conserved.

Suppose that the reference frame of O' moves at a velocity of $u = +0.550c$ in the x direction relative to observer O, as in Figure 2.21b. How would observer O analyze this collision? We can find the initial and final velocities of the two particles according to O using the velocity transformation of Eq. 2.17, which gives

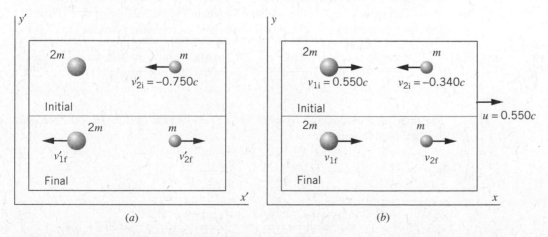

(a) (b)

FIGURE 2.21 (a) A collision between two particles as observed from the reference frame of O'. (b) The same collision observed from the reference frame of O.

the initial velocities shown in the figure and the final velocities $v_{1f} = +0.069c$ and $v_{2f} = +0.703c$. Observer O can now calculate the initial and final values of the total momentum of the two particles:

$$p_i = m_1 v_{1i} + m_2 v_{2i} = (2m)(+0.550c) + (m)(-0.340c) = +0.760mc$$

$$p_f = m_1 v_{1f} + m_2 v_{2f} = (2m)(+0.069c) + (m)(+0.703c) = +0.841mc$$

Momentum is therefore *not conserved* according to observer O.

This collision experiment has shown that that the law of conservation of linear momentum, with momentum defined as $\vec{\mathbf{p}} = m\vec{\mathbf{v}}$, does not satisfy Einstein's first postulate (the law must be the same in all inertial frames). We cannot have a law that is valid for some observers but not for others. Therefore, *if we are to retain the conservation of momentum as a general law consistent with Einstein's first postulate, we must find a new definition of momentum*. This new definition of momentum must have two properties: (1) It must yield a law of conservation of momentum that satisfies the principle of relativity; that is, if momentum is conserved according to an observer in one inertial frame, then it is conserved according to observers in all inertial frames. (2) At low speeds, the new definition must reduce to $\vec{\mathbf{p}} = m\vec{\mathbf{v}}$, which we know works perfectly well in the nonrelativistic case.

These requirements are satisfied by defining the relativistic momentum for a particle of mass m moving with velocity $\vec{\mathbf{v}}$ as

$$\vec{\mathbf{p}} = \frac{m\vec{\mathbf{v}}}{\sqrt{1 - v^2/c^2}} \tag{2.32}$$

In terms of components, we can write Eq. 2.32 as

$$p_x = \frac{mv_x}{\sqrt{1 - v^2/c^2}} \quad \text{and} \quad p_y = \frac{mv_y}{\sqrt{1 - v^2/c^2}} \tag{2.33}$$

The velocity v that appears in the denominator of these expressions is always the velocity of the particle as measured in a particular inertial frame. It is not the velocity of an inertial frame. The velocity in the numerator can be any of the components of the velocity vector.

We can now reanalyze the collision shown in Figure 2.21 using the relativistic definition of momentum. The initial relativistic momentum according to O' is

$$p_i' = \frac{m_1 v_{1i}'}{\sqrt{1 - v_{1i}'^2/c^2}} + \frac{m_2 v_{2i}'}{\sqrt{1 - v_{2i}'^2/c^2}} = \frac{(2m)(0)}{\sqrt{1 - 0^2}} + \frac{(m)(-0.750c)}{\sqrt{1 - (0.750)^2}} = -1.134mc$$

The final velocities according to O' are $v_{1f}' = -0.585c$ and $v_{2f}' = +0.294c$, and the total final momentum is

$$p_f' = \frac{m_1 v_{1f}'}{\sqrt{1 - v_{1f}'^2/c^2}} + \frac{m_2 v_{2f}'}{\sqrt{1 - v_{2f}'^2/c^2}}$$

$$= \frac{(2m)(-0.585c)}{\sqrt{1 - (0.585)^2}} + \frac{(m)(0.294c)}{\sqrt{1 - (0.294)^2}} = -1.134mc$$

Thus $p'_i = p'_f$, and observer O' concludes that momentum is conserved. According to O, the initial relativistic momentum is

$$p_i = \frac{m_1 v_{1i}}{\sqrt{1 - v_{1i}^2/c^2}} + \frac{m_2 v_{2i}}{\sqrt{1 - v_{2i}^2/c^2}} = \frac{(2m)(+0.550c)}{\sqrt{1 - (0.550)^2}} + \frac{(m)(-0.340c)}{\sqrt{1 - (0.340)^2}} = 0.956mc$$

Using the velocity transformation, the final velocities measured by O are $v_{1f} = -0.051c$ and $v_{2f} = +0.727c$, and so O calculates the final momentum to be

$$p_f = \frac{m_1 v_{1f}}{\sqrt{1 - v_{1f}^2/c^2}} + \frac{m_2 v_{2f}}{\sqrt{1 - v_{2f}^2/c^2}} = \frac{(2m)(-0.051c)}{\sqrt{1 - (0.051)^2}} + \frac{(m)(+0.727c)}{\sqrt{1 - (0.727)^2}} = 0.956mc$$

Observer O also concludes that $p_i = p_f$ and that the law of conservation of momentum is valid. Defining momentum according to Eq. 2.32 gives conservation of momentum in *all* reference frames, as required by the principle of relativity.

Example 2.11

What is the momentum of a proton moving at a speed of $v = 0.86c$?

Solution
Using Eq. 2.32, we obtain

$$p = \frac{mv}{\sqrt{1 - v^2/c^2}}$$

$$= \frac{(1.67 \times 10^{-27} \text{ kg})(0.86)(3.00 \times 10^8 \text{ m/s})}{\sqrt{1 - (0.86)^2}}$$

$$= 8.44 \times 10^{-19} \text{ kg} \cdot \text{m/s}$$

The units of kg·m/s are generally not convenient in solving problems of this type. Instead, we manipulate Eq. 2.32 to obtain

$$pc = \frac{mvc}{\sqrt{1 - v^2/c^2}} = \frac{mc^2(v/c)}{\sqrt{1 - v^2/c^2}} = \frac{(938 \text{ MeV})(0.86)}{\sqrt{1 - (0.86)^2}}$$

$$= 1580 \text{ MeV}$$

Here we have used the proton's *rest energy* mc^2, which is defined later in this section. The momentum is obtained from this result by dividing by the symbol c (not its numerical value), which gives

$$p = 1580 \text{ MeV}/c$$

The units of MeV/c for momentum are often used in relativistic calculations because, as we show later, the quantity pc often appears in these calculations. You should be able to convert MeV/c to kg·m/s and show that the two results obtained for p are equivalent.

Relativistic Kinetic Energy

Like the classical definition of momentum, the classical definition of kinetic energy also causes difficulties when we try to compare the interpretations of different observers. According to O', the initial and final kinetic energies in the collision shown in Figure 2.21a are:

$$K'_i = \tfrac{1}{2}m_1 v_{1i}^2 + \tfrac{1}{2}m_2 v_{2i}^2 = (0.5)(2m)(0)^2 + (0.5)(m)(-0.750c)^2 = 0.281mc^2$$

$$K'_f = \tfrac{1}{2}m_1 v_{1f}^2 + \tfrac{1}{2}m_2 v_{2f}^2 = (0.5)(2m)(-0.500c)^2 + (0.5)(m)(0.250c)^2 = 0.281mc^2$$

and so energy is conserved according to O'. The initial and final kinetic energies observed from the reference frame of O (as in Figure 2.21b) are

$$K_i = \tfrac{1}{2}m_1 v_{1i}^2 + \tfrac{1}{2}m_2 v_{2i}^2 = (0.5)(2m)(0.550c)^2 + (0.5)(m)(-0.340c)^2 = 0.360mc^2$$

$$K_f = \tfrac{1}{2}m_1 v_{1f}^2 + \tfrac{1}{2}m_2 v_{2f}^2 = (0.5)(2m)(0.069c)^2 + (0.5)(m)(0.703c)^2 = 0.252mc^2$$

Thus energy is *not conserved* in the reference frame of O if we use the classical formula for kinetic energy. This leads to a serious inconsistency—an elastic collision for one observer would not be elastic for another observer. As in the case of momentum, if we want to preserve the law of conservation of energy for all observers, we must replace the classical formula for kinetic energy with an expression that is valid in the relativistic case (but that reduces to the classical formula for low speeds).

We can derive the relativistic expression for the kinetic energy of a particle using essentially the same procedure used to derive the classical expression, starting with the particle form of the work-energy theorem (see Problem 28). The result of this calculation is

$$K = \frac{mc^2}{\sqrt{1 - v^2/c^2}} - mc^2 \tag{2.34}$$

Using Eq. 2.34, you can show that both O and O' will conclude that kinetic energy is conserved. In fact, all observers will agree on the applicability of the energy conservation law using the relativistic definition for kinetic energy.

Equation 2.34 looks very different from the classical result $K = \tfrac{1}{2}mv^2$, but, as you should show (see Problem 32), Eq. 2.34 reduces to the classical expression in the limit of low speeds ($v \ll c$).

The classical expression for kinetic energy also violates the second relativity postulate by allowing speeds in excess of the speed of light. There is no limit (in either classical or relativistic dynamics) to the energy we can give to a particle. Yet, if we allow the kinetic energy to increase without limit, the classical expression $K = \tfrac{1}{2}mv^2$ implies that the velocity must correspondingly increase without limit, thereby violating the second postulate. You can also see from the first term of Eq. 2.34 that $K \to \infty$ as $v \to c$. Thus we can increase the relativistic kinetic energy of a particle without limit, and its speed will not exceed c.

Relativistic Total Energy and Rest Energy

We can also express Eq. 2.34 as

$$K = E - E_0 \tag{2.35}$$

where the *relativistic total energy* E is defined as

$$E = \frac{mc^2}{\sqrt{1 - v^2/c^2}} \tag{2.36}$$

and the *rest energy* E_0 is defined as

$$E_0 = mc^2 \tag{2.37}$$

The rest energy is in effect the relativistic total energy of a particle measured in a frame of reference in which the particle is at rest.

Sometimes m in Eq. 2.37 is called the *rest mass* m_0 and is distinguished from the "relativistic mass," which is defined as $m_0/\sqrt{1 - v^2/c^2}$. We choose not to use relativistic mass, because it can be a misleading concept. Whenever we refer to mass, we always mean rest mass.

Equation 2.37 suggests that mass can be expressed in units of energy divided by c^2, such as MeV/c^2. For example, a proton has a rest energy of 938 MeV and thus a mass of 938 MeV/c^2. Just like expressing momentum in units of MeV/c, expressing mass in units of MeV/c^2 turns out to be very useful in calculations.

The relativistic total energy is given by Eq. 2.35 as

$$E = K + E_0 \tag{2.38}$$

Collisions of particles at high energies often result in the production of new particles, and thus the final rest energy may not be equal to the initial rest energy (see Example 2.18). Such collisions must be analyzed using conservation of total relativistic energy E; kinetic energy will *not* be conserved when the rest energy changes in a collision. In the special example of the elastic collision considered in this section, the identities of the particles did not change, and so kinetic energy was conserved. In general, collisions do not conserve kinetic energy—it is the relativistic total energy that is conserved in collisions.

Manipulation of Eqs. 2.32 and 2.36 gives a useful relationship among the total energy, momentum, and rest energy:

$$E = \sqrt{(pc)^2 + (mc^2)^2} \tag{2.39}$$

Figure 2.22 shows a useful mnemonic device for remembering this relationship, which has the form of the Pythagorean theorem for the sides of a right triangle.

When a particle travels at a speed close to the speed of light (say, $v > 0.99c$), which often occurs in high-energy particle accelerators, the particle's kinetic energy is much greater than its rest energy; that is, $K \gg E_0$. In this case, Eq. 2.39 can be written, to a very good approximation,

$$E \cong pc \tag{2.40}$$

This is called the *extreme relativistic approximation* and is often useful for simplifying calculations. As v approaches c, the angle in Figure 2.22 between the bottom leg of the triangle (representing mc^2) and the hypotenuse (representing E) approaches 90°. Imagine in this case a very tall triangle, in which the vertical leg (pc) and the hypotenuse (E) are nearly the same length.

For massless particles (such as photons), Eq. 2.39 becomes exactly

$$E = pc \tag{2.41}$$

All massless particles travel at the speed of light; otherwise, by Eqs. 2.34 and 2.36 their kinetic and total energies would be zero.

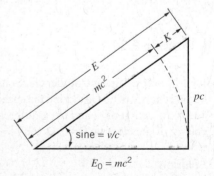

FIGURE 2.22 A useful mnemonic device for recalling the relationships among $E_0, p, K,$ and E. Note that to put all variables in energy units, the quantity pc must be used.

Example 2.12

What are the kinetic and relativistic total energies of a proton ($E_0 = 938 \, \text{MeV}$) moving at a speed of $v = 0.86c$?

Solution
In Example 2.11 we found the momentum of this particle to be $p = 1580 \, \text{MeV}/c$. The total energy can be found from Eq. 2.39:

$$E = \sqrt{(pc)^2 + (mc^2)^2} = \sqrt{(1580 \, \text{MeV})^2 + (938 \, \text{MeV})^2}$$

$$= 1837 \, \text{MeV}$$

The kinetic energy follows from Eq. 2.35:

$$K = E - E_0$$
$$= 1837 \, \text{MeV} - 938 \, \text{MeV}$$
$$= 899 \, \text{MeV}$$

We also could have solved this problem by finding the kinetic energy directly from Eq. 2.34.

Example 2.13

Find the velocity and momentum of an electron ($E_0 = 0.511 \, \text{MeV}$) with a kinetic energy of 10.0 MeV.

Solution
The total energy is $E = K + E_0 = 10.0 \, \text{MeV} + 0.511 \, \text{MeV} = 10.51 \, \text{MeV}$. We then can find the momentum from Eq. 2.39:

$$p = \frac{1}{c}\sqrt{E^2 - (mc^2)^2} = \frac{1}{c}\sqrt{(10.51 \, \text{MeV})^2 - (0.511 \, \text{MeV})^2}$$

$$= 10.5 \, \text{MeV}/c$$

Note that in this problem we could have used the extreme relativistic approximation, $p \cong E/c$, from Eq. 2.40. The error we would make in this case would be only 0.1%.

The velocity can be found by solving Eq. 2.36 for v.

$$\frac{v}{c} = \sqrt{1 - \left(\frac{mc^2}{E}\right)^2} = \sqrt{1 - \left(\frac{0.511 \, \text{MeV}}{10.51 \, \text{MeV}}\right)^2}$$

$$= 0.9988 \tag{2.42}$$

Example 2.14

In the Stanford Linear Collider electrons are accelerated to a kinetic energy of 50 GeV. Find the speed of such an electron as (*a*) a fraction of *c*, and (*b*) a difference from *c*. The rest energy of the electron is $0.511 \, \text{MeV} = 0.511 \times 10^{-3} \, \text{GeV}$.

Solution
(*a*) First we solve Eq. 2.34 for v, obtaining

$$v = c\sqrt{1 - \frac{1}{(1 + K/mc^2)^2}} \tag{2.43}$$

and thus

$$v = c\sqrt{1 - \frac{1}{[1 + (50 \, \text{GeV})/(0.511 \times 10^{-3} \, \text{GeV})]^2}}$$

$$= 0.999\,999\,999\,948c$$

Calculators cannot be trusted to 12 significant digits. Here is a way to avoid this difficulty. We can write Eq. 2.43 as $v = c(1 + x)^{1/2}$, where $x = -1/(1 + K/mc^2)^2$. Because $K \gg mc^2$, we have $x \ll 1$, and we can use the binomial expansion to write $v \cong c(1 + \frac{1}{2}x)$, or

$$v \cong c\left[1 - \frac{1}{2(1 + K/mc^2)^2}\right]$$

which gives

$$v \cong c(1 - 5.2 \times 10^{-11})$$

This leads to the same value of v given above.

(b) From the above result, we have

$$c - v = 5.2 \times 10^{-11} c$$
$$= 0.016 \text{ m/s}$$
$$= 1.6 \text{ cm/s}$$

Example 2.15

At a distance equal to the radius of the Earth's orbit $(1.5 \times 10^{11} \text{ m})$, the Sun's radiation has an intensity of about $1.4 \times 10^3 \text{ W/m}^2$. Find the rate at which the mass of the Sun is decreasing.

Solution

If we assume that the Sun's radiation is distributed uniformly over the surface area $4\pi r^2$ of a sphere of radius 1.5×10^{11} m, then the total radiative power emitted by the Sun is

$$4\pi (1.5 \times 10^{11}\text{m})^2 (1.4 \times 10^3 \text{ W/m}^2)$$
$$= 4.0 \times 10^{26} \text{ W} = 4.0 \times 10^{26} \text{ J/s}$$

By conservation of energy, we know that the energy lost by the Sun through radiation must be accounted for by a corresponding loss in its rest energy. The change in mass Δm corresponding to a change in rest energy ΔE_0 of 4.0×10^{26} J each second is

$$\Delta m = \frac{\Delta E_0}{c^2} = \frac{4.0 \times 10^{26} \text{ J}}{9.0 \times 10^{16} \text{ m}^2/\text{s}^2} = 4.4 \times 10^9 \text{ kg}$$

The Sun loses mass at a rate of about 4 billion kilograms per second! If this rate were to remain constant, the Sun (with a present mass of 2×10^{30} kg) would shine "only" for another 10^{13} years.

2.8 CONSERVATION LAWS IN RELATIVISTIC DECAYS AND COLLISIONS

In all decays and collisions, we must apply the law of conservation of momentum. The only difference between applying this law for collisions at low speed (as we did in Example 1.1) and at high speed is the use of the relativistic expression for momentum (Eq. 2.32) instead of Eq. 1.2. The law of conservation of momentum for relativistic motion can be stated in exactly the same way as for classical motion:

In an isolated system of particles, the total linear momentum remains constant.

In the classical case, kinetic energy is the only form of energy that is present in elastic collisions, so conservation of energy is equivalent to conservation of kinetic energy. In inelastic collisions or decay processes, the kinetic energy does not remain constant. Total energy is conserved in classical inelastic collisions, but we did not account for the other forms of energy that might be important. This missing energy is usually stored in the particles, perhaps as atomic or nuclear energy.

In the relativistic case, the internal stored energy contributes to the rest energy of the particles. Usually rest energy and kinetic energy are the only two forms of energy that we consider in atomic or nuclear processes (later we'll add the energy of radiation to this balance). A loss of kinetic energy in a collision is thus accompanied by a gain in rest energy, but the total relativistic energy (kinetic energy + rest energy) of all the particles involved in the process doesn't change. For example, in a reaction in which new particles are produced, the loss in kinetic energy of the original reacting particles gives the increase in rest energy of the product particles. On the other hand, in a nuclear decay process such as alpha decay, the initial nucleus gives up some rest energy to account for the kinetic energy carried by the decay products.

The law of energy conservation in the relativistic case is:

> *In an isolated system of particles, the relativistic total energy (kinetic energy plus rest energy) remains constant.*

In applying this law to relativistic collisions, we don't have to worry whether the collision is elastic or inelastic, because the inclusion of the rest energy accounts for any loss in kinetic energy.

The following examples illustrate applications of the conservation laws for relativistic momentum and energy.

Example 2.16

A neutral K meson (mass $497.7\,\text{MeV}/c^2$) is moving with a kinetic energy of $77.0\,\text{MeV}$. It decays into a pi meson (mass $139.6\,\text{MeV}/c^2$) and another particle of unknown mass. The pi meson is moving in the direction of the original K meson with a momentum of $381.6\,\text{MeV}/c$. (*a*) Find the momentum and total relativistic energy of the unknown particle. (*b*) Find the mass of the unknown particle.

Solution

(*a*) The total energy and momentum of the K meson are

$$E_K = K_K + m_K c^2 = 77.0\,\text{MeV} + 497.7\,\text{MeV} = 574.7\,\text{MeV}$$

$$p_K = \frac{1}{c}\sqrt{E_K^2 - (m_K c^2)^2}$$

$$= \frac{1}{c}\sqrt{(574.7\,\text{MeV})^2 - (497.7\,\text{MeV})^2}$$

$$= 287.4\,\text{MeV}/c$$

and for the pi meson

$$E_\pi = \sqrt{(cp_\pi)^2 + (m_\pi c^2)^2}$$

$$= \sqrt{(381.6\,\text{MeV})^2 + (139.6\,\text{MeV})^2}$$

$$= 406.3\,\text{MeV}$$

Conservation of relativistic momentum ($p_{\text{initial}} = p_{\text{final}}$) gives $p_K = p_\pi + p_x$ (where x represents the unknown particle), so

$$p_x = p_K - p_\pi = 287.4\,\text{MeV}/c - 381.6\,\text{MeV}/c$$

$$= -94.2\,\text{MeV}/c$$

and conservation of total relativistic energy ($E_{\text{initial}} = E_{\text{final}}$) gives $E_K = E_\pi + E_x$, so

$$E_x = E_K - E_\pi = 574.7\,\text{MeV} - 406.3\,\text{MeV}$$

$$= 168.4\,\text{MeV}$$

(*b*) We can find the mass by solving Eq. 2.39 for mc^2:

$$m_x c^2 = \sqrt{E_x^2 - (cp_x)^2}$$

$$= \sqrt{(168.4\,\text{MeV})^2 - (94.2\,\text{MeV})^2}$$

$$= 139.6\,\text{MeV}$$

Thus the unknown particle has a mass of $139.6\,\text{MeV}/c^2$, and its mass shows that it is another pi meson.

Example 2.17

In the reaction $K^- + p \rightarrow \Lambda^0 + \pi^0$, a charged K meson (mass $493.7\,\text{MeV}/c^2$) collides with a proton ($938.3\,\text{MeV}/c^2$) at rest, producing a lambda particle ($1115.7\,\text{MeV}/c^2$) and a neutral pi meson ($135.0\,\text{MeV}/c^2$), as represented in Figure 2.23. The initial kinetic energy of the K meson is $152.4\,\text{MeV}$. After the interaction, the pi meson has a kinetic energy of $254.8\,\text{MeV}$. (a) Find the kinetic energy of the lambda. (b) Find the directions of motion of the lambda and the pi meson.

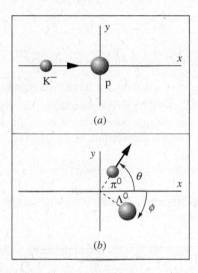

FIGURE 2.23 Example 2.17. (a) A K^- meson collides with a proton at rest. (b) After the collision, a π^0 meson and a Λ^0 are produced.

Solution

(a) The initial and final total energies are

$$E_{\text{initial}} = E_K + E_p = K_K + m_K c^2 + m_p c^2$$
$$E_{\text{final}} = E_\Lambda + E_\pi = K_\Lambda + m_\Lambda c^2 + K_\pi + m_\pi c^2$$

In these two equations, the value of every quantity is known except the kinetic energy of the lambda. Using conservation of total relativistic energy, we set $E_{\text{initial}} = E_{\text{final}}$ and solve for K_Λ:

$$
\begin{aligned}
K_\Lambda &= K_K + m_K c^2 + m_p c^2 - m_\Lambda c^2 - K_\pi - m_\pi c^2 \\
&= 152.4\,\text{MeV} + 493.7\,\text{MeV} + 938.3\,\text{MeV} \\
&\quad - 1115.7\,\text{MeV} - 254.8\,\text{MeV} - 135.0\,\text{MeV} \\
&= 78.9\,\text{MeV}
\end{aligned}
$$

(b) To find the directional information we must apply conservation of momentum. The initial momentum is just that of the K meson. From its total energy, $E_K = K_K + m_K c^2 = 152.4\,\text{MeV} + 493.7\,\text{MeV} = 646.1\,\text{MeV}$, we can find the momentum:

$$
\begin{aligned}
p_{\text{initial}} = p_K &= \frac{1}{c}\sqrt{(E_K)^2 - (m_K c^2)^2} \\
&= \frac{1}{c}\sqrt{(646.1\,\text{MeV})^2 - (493.7\,\text{MeV})^2} \\
&= 416.8\,\text{MeV}/c
\end{aligned}
$$

A similar procedure applied to the two final particles gives $p_\Lambda = 426.9\,\text{MeV}/c$ and $p_\pi = 365.7\,\text{MeV}/c$. The total momentum of the two final particles is $p_{x,\text{final}} = p_\Lambda \cos\theta + p_\pi \cos\phi$ and $p_{y,\text{final}} = p_\Lambda \sin\theta - p_\pi \sin\phi$. Conservation of momentum in the x and y directions gives

$$p_\Lambda \cos\theta + p_\pi \cos\phi = p_{\text{initial}} \quad \text{and} \quad p_\Lambda \sin\theta - p_\pi \sin\phi = 0$$

Here we have two equations with two unknowns (θ and ϕ). We can eliminate θ by writing the first equation as $p_\Lambda \cos\theta = p_{\text{initial}} - p_\pi \cos\phi$, then squaring both equations and adding them. The resulting equation can be solved for ϕ:

$$
\begin{aligned}
\phi &= \cos^{-1}\left(\frac{p_{\text{initial}}^2 + p_\pi^2 - p_\Lambda^2}{2p_\pi p_{\text{initial}}}\right) \\
&= \cos^{-1}\left(\frac{\begin{array}{c}(416.8\,\text{MeV}/c)^2 + (365.7\,\text{MeV}/c)^2 \\ -(426.9\,\text{MeV}/c)^2\end{array}}{2(365.7\,\text{MeV}/c)(416.8\,\text{MeV}/c)}\right) \\
&= 65.7°
\end{aligned}
$$

From the conservation of momentum equation for the y components, we have

$$
\begin{aligned}
\theta &= \sin^{-1}\left(\frac{p_\pi \sin\phi}{p_\Lambda}\right) \\
&= \sin^{-1}\left(\frac{(365.7\,\text{MeV}/c)(\sin 65.7°)}{426.9\,\text{MeV}/c}\right) = 51.3°
\end{aligned}
$$

Example 2.18

The discovery of the antiproton $\bar{p}$ (a particle with the same rest energy as a proton, 938 MeV, but with the opposite electric charge) took place in 1956 through the following reaction:

$$p + p \rightarrow p + p + p + \bar{p}$$

in which accelerated protons were incident on a target of protons at rest in the laboratory. The minimum incident kinetic energy needed to produce the reaction is called the *threshold* kinetic energy, for which the final particles move together as if they were a single unit (Figure 2.24). Find the threshold kinetic energy to produce antiprotons in this reaction.

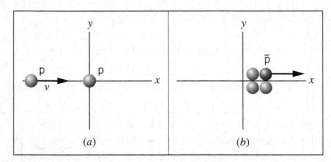

(a) $\qquad$ (b)

FIGURE 2.24 Example 2.18. (a) A proton moving with velocity v collides with another proton at rest. (b) The reaction produces three protons and an antiproton, which move together as a unit.

Solution

This problem can be solved by a straightforward application of energy and momentum conservation. Let E_p and p_p represent the total energy and momentum of the incident proton. Thus the initial total energy of the two protons is $E_p + m_p c^2$. Let E_p' and p_p' represent the total energy and momentum of *each* of the four final particles (which move together and thus have the same energy and momentum). We can then apply conservation of total energy:

$$E_p + m_p c^2 = 4E_p'$$

and conservation of momentum:

$$p_p = 4p_p'$$

We can write the momentum equation as $\sqrt{E_p^2 - (m_p c^2)^2} = 4\sqrt{E_p'^2 - (m_p c^2)^2}$, so now we have two equations in two unknowns (E_p and E_p'). We eliminate E_p', for example by solving the energy conservation equation for E_p' and substituting into the momentum equation. The result is

$$E_p = 7m_p c^2$$

from which we can calculate the kinetic energy of the incident proton:

$$K_p = E_p - m_p c^2 = 6m_p c^2 = 6(938\,\text{MeV}) = 5628\,\text{MeV}$$
$$= 5.628\,\text{GeV}$$

The Bevatron accelerator at the Lawrence Berkeley Laboratory was designed with this experiment in mind, so that it could produce a beam of protons whose energy exceeded 5.6 GeV. The discovery of the antiproton in this reaction was honored with the award of the 1959 Nobel Prize to the experimenters, Emilio Segrè and Owen Chamberlain.

2.9 EXPERIMENTAL TESTS OF SPECIAL RELATIVITY

Because special relativity provided such a radical departure from the notions of space and time in classical physics, it is important to perform detailed experimental tests that can clearly distinguish between the predictions of special relativity and those of classical physics. Many tests of increasing precision have been done since the theory was originally presented, and in every case the predictions of special relativity are upheld. Here we discuss a few of these tests.

Universality of the Speed of Light

The second relativity postulate asserts that the speed of light has the same value c for all observers. This leads to several types of experimental tests, of which we discuss two: (1) Does the speed of light change with the direction of travel? (2) Does the speed of light change with relative motion between source and observer?

The Michelson-Morley experiment provides a test of the first type. This experiment compared the upstream-downstream and cross-stream speeds of light and concluded that they were equal within the experimental error. Equivalently, we may say that the experiment showed that there is no preferred reference frame (no ether) relative to which the speed of light must be measured. If there is an ether, the speed of the Earth through the ether is less than 5 km/s, which is much smaller than the Earth's orbital speed about the Sun, 30 km/s. We can express their result as a difference Δc between the upstream-downstream and cross-stream speeds; the experiment showed that $\Delta c/c < 3 \times 10^{-10}$.

To reconcile the result of the Michelson-Morley experiment with classical physics, Lorentz proposed the "ether drag" hypothesis, according to which the motion of the Earth through the ether caused an electromagnetic drag that contracted the arm of the interferometer in the direction of motion. This contraction was just enough to compensate for the difference in the upstream-downstream and cross-stream times predicted by the Galilean transformation. This hypothesis succeeds only when the two arms of the interferometer are of the same length. To test this hypothesis, a similar experiment was done in 1932 by Kennedy and Thorndike; in their experiment, the lengths of the interferometer arms differed by about 16 cm, the maximum distance over which light sources available at that time could remain coherent. The Kennedy-Thorndike experiment in effect tests the second question, whether the speed of light changes due to relative motion. Their result was $\Delta c/c < 3 \times 10^{-8}$, which excludes the Lorentz contraction hypothesis as an explanation for the Michelson-Morley experiment.

In recent years, these fundamental experiments have been repeated with considerably improved precision using lasers as light sources. Experimenters working at the Joint Institute for Laboratory Astrophysics in Boulder, Colorado, built an apparatus that consisted of two He-Ne lasers on a rotating granite platform. By electronically stabilizing the lasers, they improved the sensitivity of their apparatus by several orders of magnitude. Again expressing the result as a difference between the speeds along the two arms of the apparatus, this experiment corresponds to $\Delta c/c < 8 \times 10^{-15}$, an improvement of about 5 orders of magnitude over the original Michelson-Morley experiment. In a similar repetition of the Kennedy-Thorndike experiment using He-Ne lasers, they obtained $\Delta c/c < 1 \times 10^{-10}$, an improvement over the original experiment by a factor of 300. [See A. Brillet and J. L. Hall, *Physical Review Letters* **42**, 549 (1979); D. Hils and J. L. Hall, *Physical Review Letters* **64**, 1697 (1990).] A considerable improvement in the Kennedy-Thorndike type of experiment has been made possible by comparing the oscillation frequency of a crystal with the frequency of a hydrogen maser (a maser is similar to a laser, but it uses microwaves rather than visible light). The experimenters measured for nearly one year, looking for a change in the relative frequencies as the Earth's velocity changed. No effect was observed, leading to a limit of $\Delta c/c < 2 \times 10^{-12}$. [See P. Wolf et al., *Physical Review Letters* **90**, 060402 (2003).]

Another way of testing the second question is to measure the speed of a light beam emitted by a source in motion. Suppose we observe this beam along the direction of motion of the moving source, which might be moving toward us or away from us. In the rest frame of the source, the emitted light travels at speed c. We can express the speed of light in our reference frame as $c' = c + \Delta c$, where Δc is zero according to special relativity ($c' = c$) or is $\pm u$ according to classical physics ($c' = c \pm u$ in the Galilean transformation, depending on whether the motion is toward or away from the observer).

In one experiment of this type, the decay of pi mesons (pions) into gamma rays (a form of electromagnetic waves traveling at c) was observed. When pions (produced in laboratories with large accelerators) emit these gamma rays, they are traveling at speeds close to the speed of light, relative to the laboratory. Thus if Galilean relativity were valid, we should expect to find gamma rays emitted in the direction of motion of the decaying pions traveling at a speed c' in the laboratory of nearly $2c$, rather than always with c as predicted by special relativity. The observed laboratory speed of these gamma rays in one experiment was $(2.9977 \pm 0.0004) \times 10^8$ m/s when the decaying pions were moving at $u/c = 0.99975$. These results give $\Delta c/c < 2 \times 10^{-4}$, and thus $c' = c$ as expected from special relativity. This experiment shows directly that an object moving at a speed of nearly c relative to the laboratory emits "light" that travels at a speed of c relative to both the object *and* the laboratory, giving direct evidence for Einstein's second postulate. [See T. Alvager et al., *Physics Letters* **12**, 260 (1964).]

Another experiment of this type is to study the X rays emitted by a binary pulsar, a rapidly pulsating source of X rays in orbit about another star, which would eclipse the pulsar as it rotated in its orbit. If the speed of light (in this case, X rays) were to change as the pulsar moved first toward and later away from the Earth in its orbit, the beginning and end of the eclipse would not be equally spaced in time from the midpoint of the eclipse. No such effect is observed, and from these observations it is concluded that $\Delta c/c < 2 \times 10^{-12}$, in agreement with predictions of special relativity. These experiments were done at $u/c = 10^{-3}$. [See K. Brecher, *Physical Review Letters* **39**, 1051 (1977).]

A different type of test of the limit by which the speed of light changes with direction of travel can be done using the clocks carried aboard the network of Earth satellites that make up the Global Positioning System (GPS). By comparing the readings of clocks on the GPS satellites with clocks on the ground at different times of day (as the satellites move relative to the ground stations), it is possible to test whether the change in the direction of travel affects the apparent synchronization of the clocks. No effect was observed, and the experimenters were able to set a limit of $\Delta c/c < 5 \times 10^{-9}$ for the difference between the one-way and round-trip speeds of light. [See P. Wolf and G. Petit, *Physical Review A* **56**, 4405 (1997).]

Time Dilation

We have already discussed the time dilation effect on the decay of muons produced by cosmic rays. Muon decay can also be studied in the laboratory. Muons can be produced following collisions in high-energy accelerators, and the decay of the muons can be followed by observing their decay products (ordinary electrons). These muons can either be trapped and decay at rest, or they can be placed in a beam and decay in flight. When muons are observed at rest, their decay lifetime is $2.198\,\mu s$. (As we discuss in Chapter 12, decays generally follow an exponential law. The lifetime is the time after which a fraction $1/e = 0.368$ of the original muons remain.) This is the *proper lifetime,* measured in a frame of reference in which the muon is at rest. In one particular experiment, muons were trapped in a ring and circulated at a momentum of $p = 3094$ MeV/c. The decays in flight occurred with a lifetime of $64.37\,\mu s$ (measured in the laboratory frame of reference). For muons of this momentum, Eq. 2.8 gives a dilated lifetime of (see Problem 43) $64.38\,\mu s$, which is in excellent agreement with the measured value and confirms the time dilation effect. [See J. Bailey et al., *Nature* **268**, 301 (1977).]

Another similar experiment was done with pions. The proper lifetime, measured for pions at rest, is known to be $26.0\,ns$. In one experiment, pions were observed in flight at $u/c = 0.913$, and their lifetime was measured to be $63.7\,ns$. (Pions decay to muons, so we can follow the exponential radioactive decay of the pions by observing the muons emitted as a result of the decay.) For pions moving at this speed, the expected dilated lifetime is in exact agreement with the measured value, once again confirming the time dilation effect. [See D. S. Ayres et al., *Physical Review D* **3**, 1051 (1971).]

The Doppler Effect

Confirmation of the relativistic Doppler effect first came from experiments done in 1938 by Ives and Stilwell. They sent a beam of hydrogen atoms, generated in a gas discharge, down a tube at a speed u, as shown in Figure 2.25. They could simultaneously observe light emitted by the atoms in a direction parallel to u (atom 1) and opposite to u (atom 2, reflected from the mirror). Using a spectrograph, the experimenters were able to photograph the characteristic spectral lines from these atoms and also, on the same photographic plate, from atoms at rest. If the classical Doppler formula were valid, the wavelengths of the lines from atoms 1 and 2 would be placed at symmetric intervals $\Delta\lambda_1 = \pm\lambda_0(u/c)$ on either side of the line from the atoms at rest (wavelength λ_0), as in Figure 2.25b. The relativistic Doppler formula, on the other hand, gives a small additional asymmetric shift $\Delta\lambda_2 = +\frac{1}{2}\lambda_0(u/c)^2$, as in Figure 2.25c (computed for $u \ll c$, so

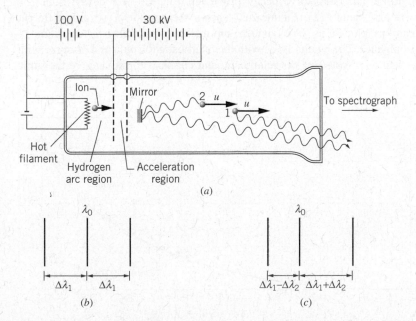

FIGURE 2.25 (a) Apparatus used in the Ives-Stilwell experiment. (b) Line spectrum expected from classical Doppler effect. (c) Line spectrum expected from relativistic Doppler effect.

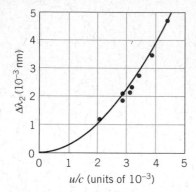

FIGURE 2.26 Results of the Ives-Stilwell experiment. According to classical theory, $\Delta\lambda_2 = 0$, while according to special relativity, $\Delta\lambda_2$ depends on $(u/c)^2$. The solid line, which represents the relativistic formula, gives excellent agreement with the data points.

that higher-order terms in u/c can be neglected). Figure 2.26 shows the results of Ives and Stilwell for one of the hydrogen lines (the blue line of the Balmer series at $\lambda_0 = 486$ nm). The agreement between the observed values and those predicted by the relativistic formula is impressive.

Recent experiments with lasers have verified the relativistic formula at greater accuracy. These experiments are based on the absorption of laser light by an atom; when the radiation is absorbed, the atom changes from its lowest-energy state (the ground state) to one of its excited states. The experiment consists essentially of comparing the laser wavelength needed to excite atoms at rest with that needed for atoms in motion. One experiment used a beam of hydrogen atoms with kinetic energy 800 MeV (corresponding to $u/c = 0.84$) produced in a high-energy proton accelerator. An ultraviolet laser was used to excite the atoms. This experiment verified the relativistic Doppler effect to an accuracy of about 3×10^{-4}. [See D. W. MacArthur et al., *Physical Review Letters* **56**, 282 (1986).] In another experiment, a beam of neon atoms moving with a speed of $u = 0.0036c$ was irradiated with light from a tunable dye laser. This experiment verified the relativistic Doppler shift to a precision of 2×10^{-6}. [See R. W. McGowan et al., *Physical Review Letters* **70**, 251 (1993).] A more recent study used two tunable dye lasers parallel and antiparallel to a beam of lithium atoms moving at $0.064c$. The results of this experiment agreed with the relativistic Doppler formula to within a precision of 2×10^{-7}, improving on the best previous results by an order of magnitude. [See G. Saathoff et al., *Physical Review Letters* **91**, 190403 (2003).]

Relativistic Momentum and Energy

The earliest direct confirmation of the relativistic relationship for energy and momentum came just a few years after Einstein's 1905 paper. Simultaneous measurements were made of the momentum and velocity of high-energy electrons emitted in certain radioactive decay processes (nuclear beta decay, which is discussed in Chapter 12). Figure 2.27 shows the results of several different investigations plotted as p/mv, which should have the value 1 according to classical physics. The results agree with the relativistic formula and disagree with the classical one. Note that the relativistic and classical formulas give the same

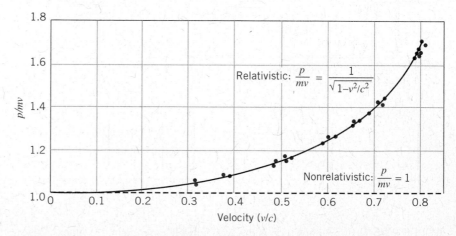

FIGURE 2.27 The ratio p/mv is plotted for electrons of various speeds. The data agree with the relativistic result and not at all with the nonrelativistic result ($p/mv = 1$).

results at low speeds, and in fact the two cannot be distinguished for speeds below $0.1\,c$, which accounts for our failure to observe these effects in experiments with ordinary laboratory objects.

Other more recent experiments, in which the kinetic energies of fast electrons were measured, are shown in Figure 2.28. Once again, the data at high speeds agree with special relativity and disagree with the classical equations. In a more extreme example, experimenters at the Stanford Linear Accelerator Center measured the speed of 20 GeV electrons, whose speed is within 5×10^{-10} of the speed of light (or about 0.15 m/s less than c). The measurement was not capable of this level of precision, but it did determine that the speed of the electrons was within 2×10^{-7} of the speed of light (60 m/s). [See Z. G. T. Guiragossian et al., *Physical Review Letters* **34**, 335 (1975).]

Nearly every time the nuclear or particle physicist enters the laboratory, a direct or indirect test of the momentum and energy relationships of special relativity is made. Principles of special relativity must be incorporated in the design of the high-energy accelerators used by nuclear and particle physicists, so even the construction of these projects gives testimony to the validity of the formulas of special relativity.

For example, consider the capture of a neutron by an atom of hydrogen to form an atom of deuterium or "heavy hydrogen." Energy is released in this process, mostly in the form of electromagnetic radiation (gamma rays). The energy of the gamma rays is measured to be 2.224 MeV. Where does this energy come from?

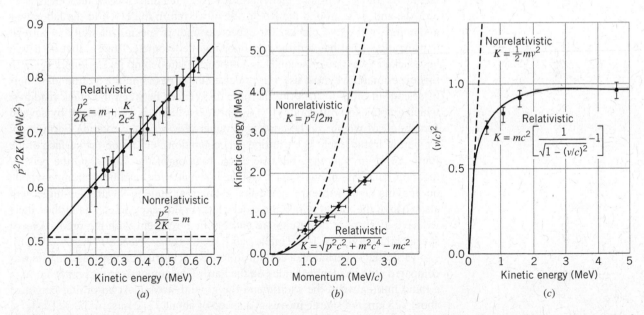

FIGURE 2.28 Confirmation of relativistic kinetic energy relationships. In (a) and (b) the momentum and energy of radioactive decay electrons were measured simultaneously. In these two independent experiments, the data were plotted in different ways, but the results are clearly in good agreement with the relativistic relationships and in poor agreement with the classical, nonrelativistic relationships. In (c) electrons were accelerated to a fixed energy through a large electric field (up to 4.5 million volts, as shown) and the velocities of the electrons were determined by measuring the flight time over 8.4 m. Notice that at small kinetic energies ($K \ll mc^2$), the relativistic and nonrelativistic relationships become identical. [*Sources*: (a) K. N. Geller and R. Kollarits, *Am. J. Phys.* **40**, 1125 (1972); (b) S. Parker, *Am. J. Phys.* **40**, 241 (1972); (c) W. Bertozzi, *Am. J. Phys.* **32**, 551 (1964).].

It comes from the difference in mass when the hydrogen and neutron combine to form deuterium. The difference between the initial and final masses is:

$$\Delta m = m(\text{hydrogen}) + m(\text{neutron}) - m(\text{deuterium})$$
$$= 1.007825 \text{ u} + 1.008665 \text{ u} - 2.014102 \text{ u} = 0.002388 \text{ u}$$

The initial mass of hydrogen plus neutron is greater than the final mass of deuterium by 0.002388 u. The energy equivalent of this change in mass is

$$\Delta E = (\Delta m)c^2 = 2.224 \text{ MeV}$$

which is equal to the energy released as gamma rays.

Similar experiments have been done to test the $E = mc^2$ relationship by measuring the energy released as gamma rays following the capture of neutrons by atoms of silicon and sulfur, and comparing the gamma-ray energies with the difference between the initial and final masses. These experiments are consistent with $E = mc^2$ to a precision of about 4×10^{-7}. [See S. Rainville et al., *Nature* **438**, 1096 (2006).]

Twin Paradox

Although we cannot perform the experiment to test the twin paradox as we have described it, we can do an equivalent experiment. We take two clocks in our laboratory and synchronize them carefully. We then place one of the clocks in an airplane and fly it around the Earth. When we return the clock to the laboratory and compare the two clocks, we expect to find, if special relativity is correct, that the clock that has left the laboratory is the "younger" one—that is, it will have ticked away fewer seconds and appear to run behind its stationary twin. In this experiment, we must use very precise clocks based on the atomic vibrations of cesium in order to measure the time differences between the clock readings, which amount to only about 10^{-7} s. This experiment is complicated by several factors, all of which can be computed rather precisely: the rotating Earth is *not* an inertial frame (there is a centripetal acceleration), clocks on the surface of the Earth are *already* moving because of the rotation of the Earth, and the *general* theory of relativity predicts that a change in the gravitational field strength, which our moving clock will experience as it changes altitude in its airplane flight, will also change the rate at which the clock runs. In this experiment, as in the others we have discussed, the results are entirely in agreement with the predictions of special relativity. [See J. C. Hafele and R. E. Keating, *Science* **177**, 166 (1972).]

In a similar experiment, a cesium atomic clock carried on the space shuttle was compared with an identical clock on the Earth. The comparison was made through a radio link between the shuttle and the ground station. At an orbital height of about 328 km, the shuttle moves at a speed of about 7712 m/s, or $2.5 \times 10^{-5}c$. A clock moving at this speed runs slower than an identical clock at rest by the time dilation factor. For every second the clock is in orbit, it loses 330 ps relative to the clock on Earth; equivalently, it loses about 1.8 μs per orbit. These time intervals can be measured with great precision, and the predicted asymmetric aging was verified to a precision of about 0.1%. [See E. Sappl, *Naturwissenschaften* **77**, 325 (1990).]

Chapter Summary

		Section
Galilean relativity	$x' = x - ut, v'_x = v_x - u$	2.1
Einstein's postulates	(1) The laws of physics are the same in all inertial frames. (2) The speed of light has the same value c in all inertial frames.	2.3
Time dilation	$\Delta t = \dfrac{\Delta t_0}{\sqrt{1 - u^2/c^2}}$ (Δt_0 = proper time)	2.4
Length contraction	$L = L_0\sqrt{1 - u^2/c^2}$ (L_0 = proper length)	2.4
Velocity addition	$v = \dfrac{v' + u}{1 + v'u/c^2}$	2.4
Doppler effect (source and observer separating)	$f' = f\sqrt{\dfrac{1 - u/c}{1 + u/c}}$	2.4
Lorentz transformation	$x' = \dfrac{x - ut}{\sqrt{1 - u^2/c^2}},$ $y' = y, \ z' = z,$ $t' = \dfrac{t - (u/c^2)x}{\sqrt{1 - u^2/c^2}}$	2.5

		Section
Lorentz velocity transformation	$v'_x = \dfrac{v_x - u}{1 - v_x u/c^2},$ $v'_y = \dfrac{v_y\sqrt{1 - u^2/c^2}}{1 - v_x u/c^2},$ $v'_z = \dfrac{v_z\sqrt{1 - u^2/c^2}}{1 - v_x u/c^2}$	2.5
Clock synchronization	$\Delta t' = \dfrac{uL/c^2}{\sqrt{1 - u^2/c^2}}$	2.5
Relativistic momentum	$\vec{p} = \dfrac{m\vec{v}}{\sqrt{1 - v^2/c^2}}$	2.7
Relativistic kinetic energy	$K = \dfrac{mc^2}{\sqrt{1 - v^2/c^2}} - mc^2$	2.7
Rest energy	$E_0 = mc^2$	2.7
Relativistic total energy	$E = K + E_0 = \dfrac{mc^2}{\sqrt{1 - v^2/c^2}}$	2.7
Momentum-energy relationship	$E = \sqrt{(pc)^2 + (mc^2)^2}$	2.7
Extreme relativistic approximation	$E \cong pc$	2.7
Conservation laws	In an isolated system of particles, the total momentum and the relativistic total energy remain constant.	2.8

Questions

1. Explain in your own words what is meant by the term "relativity." Are there different theories of relativity?

2. Suppose the two observers and the rock described in the first paragraph of Section 2.1 were isolated in interstellar space. Discuss the two observers' differing perceptions of the motion of the rock. Is there any experiment they can do to determine whether the rock is moving in any absolute sense?

3. Describe the situation of Figure 2.4 as it would appear from the reference frame of O'.

4. Does the Michelson-Morley experiment show that the ether does not exist or that it is merely unnecessary?

5. Suppose we made a pair of shears in which the cutting blades were many orders of magnitude longer than the handle. Let us in fact make them so long that, when we move the handles at angular velocity ω, a point on the tip of the blade has a tangential velocity $v = \omega r$ that is greater than c. Does this contradict special relativity? Justify your answer.

6. Light travels through water at a speed of about 2.25×10^8 m/s. Is it possible for a particle to travel through water at a speed v greater than 2.25×10^8 m/s?

7. Is it possible to have particles that travel at the speed of light? What does Eq. 2.36 require of such particles?

8. How does relativity combine space and time coordinates into spacetime?

9. Einstein developed the relativity theory after trying unsuccessfully to imagine how a light beam would look to an observer traveling with the beam at speed c. Why is this so difficult to imagine?

10. Explain in your own words the terms *time dilation* and *length contraction*.

11. Does the Moon's disk appear to be a different size to a space traveler approaching it at $v = 0.99c$, compared with the view of a person at rest at the same location?

12. According to the time dilation effect, would the life expectancy of someone who lives at the equator be longer or shorter than someone who lives at the North Pole? By how much?

13. Criticize the following argument. "Here is a way to travel faster than light. Suppose a star is 10 light-years away. A radio signal sent from Earth would need 20 years to make the round trip to the star. If I were to travel to the star in my rocket at $v = 0.8c$, to me the distance to the star is contracted by $\sqrt{1 - (0.8)^2}$ to 6 light-years, and at that speed it would take me 6 light-years/0.8c = 7.5 years to travel there. The round trip takes me only 15 years, and therefore I travel faster than light, which takes 20 years."

14. Is it possible to synchronize clocks that are in motion relative to each other? Try to design a method to do so. Which observers will believe the clocks to be synchronized?

15. Suppose event A causes event B. To one observer, event A comes before event B. Is it possible that in another frame of reference event B could come before event A? Discuss.

16. Is mass a conserved quantity in classical physics? In special relativity?

17. "In special relativity, mass and energy are equivalent." Discuss this statement and give examples.

18. Which is more massive, an object at low temperature or the same object at high temperature? A spring at its natural length or the same spring under compression? A container of gas at low pressure or at high pressure? A charged capacitor or an uncharged one?

19. Could a collision be elastic in one frame of reference and inelastic in another?

20. (a) What properties of nature would be different if there were a relativistic transformation law for electric charge? (b) What experiments could be done to prove that electric charge does *not* change with velocity?

Problems

2.1 Classical Relativity

1. You are piloting a small airplane in which you want to reach a destination that is 750 km due north of your starting location. Once you are airborne, you find that (due to a strong but steady wind) to maintain a northerly course you must point the nose of the plane at an angle that is 22° west of true north. From previous flights on this route in the absence of wind, you know that it takes you 3.14 h to make the journey. With the wind blowing, you find that it takes 4.32 h. A fellow pilot calls you to ask about the wind velocity (magnitude and direction). What is your report?

2. A moving sidewalk 95 m in length carries passengers at a speed of 0.53 m/s. One passenger has a normal walking speed of 1.24 m/s. (a) If the passenger stands on the sidewalk without walking, how long does it take her to travel the length of the sidewalk? (b) If she walks at her normal walking speed on the sidewalk, how long does it take to travel the full length? (c) When she reaches the end of the sidewalk, she suddenly realizes that she left a package at the opposite end. She walks rapidly back along the sidewalk at double her normal walking speed to retrieve the package. How long does it take her to reach the package?

2.2 The Michelson-Morley Experiment

3. A shift of one fringe in the Michelson-Morley experiment corresponds to a change in the round-trip travel time along one arm of the interferometer by one period of vibration of light (about 2×10^{-15} s) when the apparatus is rotated by 90°. Based on the results of Example 2.3, what velocity through the ether would be deduced from a shift of one fringe? (Take the length of the interferometer arm to be 11 m.)

2.4 Consequences of Einstein's Postulates

4. The distance from New York to Los Angeles is about 5000 km and should take about 50 h in a car driving at 100 km/h. (a) How much shorter than 5000 km is the distance according to the car travelers? (b) How much less than 50 h do they age during the trip?

5. How fast must an object move before its length appears to be contracted to one-half its proper length?

6. An astronaut must journey to a distant planet, which is 200 light-years from Earth. What speed will be necessary if the astronaut wishes to age only 10 years during the round trip?

7. The proper lifetime of a certain particle is 100.0 ns. (*a*) How long does it live in the laboratory if it moves at $v = 0.960c$? (*b*) How far does it travel in the laboratory during that time? (*c*) What is the distance traveled in the laboratory according to an observer moving with the particle?

8. High-energy particles are observed in laboratories by photographing the tracks they leave in certain detectors; the length of the track depends on the speed of the particle and its lifetime. A particle moving at $0.995c$ leaves a track 1.25 mm long. What is the proper lifetime of the particle?

9. Carry out the missing steps in the derivation of Eq. 2.17.

10. Two spaceships approach the Earth from opposite directions. According to an observer on the Earth, ship A is moving at a speed of $0.753c$ and ship B at a speed of $0.851c$. What is the velocity of ship A as observed from ship B? Of ship B as observed from ship A?

11. Rocket A leaves a space station with a speed of $0.826c$. Later, rocket B leaves in the same direction with a speed of $0.635c$. What is the velocity of rocket A as observed from rocket B?

12. One of the strongest emission lines observed from distant galaxies comes from hydrogen and has a wavelength of 122 nm (in the ultraviolet region). (*a*) How fast must a galaxy be moving away from us in order for that line to be observed in the visible region at 366 nm? (*b*) What would be the wavelength of the line if that galaxy were moving toward us at the same speed?

13. A physics professor claims in court that the reason he went through the red light ($\lambda = 650$ nm) was that, due to his motion, the red color was Doppler shifted to green ($\lambda = 550$ nm). How fast was he going?

2.5 The Lorentz Transformation

14. Derive the Lorentz velocity transformations for v'_x and v'_z.

15. Observer O fires a light beam in the y direction ($v_y = c$). Use the Lorentz velocity transformation to find v'_x and v'_y and show that O' also measures the value c for the speed of light. Assume that O' moves relative to O with velocity u in the x direction.

16. A light bulb at point x in the frame of reference of O blinks on and off at intervals $\Delta t = t_2 - t_1$. Observer O', moving relative to O at speed u, measures the interval to be $\Delta t' = t'_2 - t'_1$. Use the Lorentz transformation expressions to derive the time dilation expression relating Δt and $\Delta t'$.

17. A neutral K meson at rest decays into two π mesons, which travel in opposite directions along the x axis with speeds of $0.828c$. If instead the K meson were moving in the positive x direction with a velocity of $0.486c$, what would be the velocities of the two π mesons?

18. A rod in the reference frame of observer O makes an angle of $31°$ with the x axis. According to observer O', who is in motion in the x direction with velocity u, the rod makes an angle of $46°$ with the x axis. Find the velocity u.

19. According to observer O, two events occur separated by a time interval $\Delta t = +0.465$ μs and at locations separated by $\Delta x = +53.4$ m. (*a*) According to observer O', who is in motion relative to O at a speed of $0.762c$ in the positive x direction, what is the time interval between the two events? (*b*) What is the spatial separation between the two events, according to O'?

20. According to observer O, a blue flash occurs at $x_b = 10.4$ m when $t_b = 0.124$ μs, and a red flash occurs at $x_r = 23.6$ m when $t_r = 0.138$ μs. According to observer O', who is in motion relative to O at velocity u, the two flashes appear to be simultaneous. Find the velocity u.

2.6 The Twin Paradox

21. Suppose the speed of light were 1000 mi/h. You are traveling on a flight from Los Angeles to Boston, a distance of 3000 mi. The plane's speed is a constant 600 mi/h. You leave Los Angeles at 10:00 A.M., as indicated by your wristwatch and by a clock in the airport. (*a*) According to your watch, what time is it when you land in Boston? (*b*) In the Boston airport is a clock that is synchronized to read exactly the same time as the clock in the Los Angeles airport. What time does that clock read when you land in Boston? (*c*) The following day when the Boston clock that records Los Angeles time reads 10:00 A.M., you leave Boston to return to Los Angeles on the same airplane. When you land in Los Angeles, what are the times read on your watch and on the airport clock?

22. Suppose rocket traveler Amelia has a clock made on Earth. Every year on her birthday she sends a light signal to brother Casper on Earth. (*a*) At what rate does Casper receive the signals during Amelia's outward journey? (*b*) At what rate does he receive the signals during her return journey? (*c*) How many of Amelia's birthday signals does Casper receive during the journey that he measures to last 20 years?

23. Suppose Amelia traveled at a speed of $0.80c$ to a star that (according to Casper on Earth) is 8.0 light-years away. Casper ages 20 years during Amelia's round trip. How much younger than Casper is Amelia when she returns to Earth?

24. Make a drawing similar to Figure 2.20 showing the world-lines of Casper and Amelia from Casper's frame of reference. Divide the world line for Amelia's outward journey into 8 equal segments (for the 8 birthdays that Amelia celebrates). For each birthday, draw a line that represents a light signal that Amelia sends to Casper on her birthday. Do the same for Amelia's return journey. (*a*) According to Casper's time, when does he receive the signal showing Amelia celebrating her 8th birthday after leaving Earth? (*b*) How long does it take for Casper to receive the signals showing Amelia celebrating birthdays 9 through 16?

2.7 Relativistic Dynamics

25. (*a*) Using the relativistically correct final velocities for the collision shown in Figure 2.21a ($v'_{1f} = -0.585c$, $v'_{2f} = +0.294c$), show that relativistic kinetic energy is conserved

according to observer O'. (b) Using the relativistically correct final velocities for the collision shown in Figure 2.21b ($v_{1f} = -0.051c, v_{2f} = +0.727c$), show that relativistic kinetic energy is conserved according to observer O.

26. Find the momentum, kinetic energy, and total energy of a proton moving at a speed of 0.756c.

27. An electron is moving with a kinetic energy of 1.264 MeV. What is its speed?

28. The work-energy theorem relates the change in kinetic energy of a particle to the work done on it by an external force: $\Delta K = W = \int F\, dx$. Writing Newton's second law as $F = dp/dt$, show that $W = \int v\, dp$ and integrate by parts using the relativistic momentum to obtain Eq. 2.34.

29. For what range of velocities of a particle of mass m can we use the classical expression for kinetic energy $\frac{1}{2}mv^2$ to within an accuracy of 1%?

30. For what range of velocities of a particle of mass m can we use the extreme relativistic approximation $E = pc$ to within an accuracy of 1%?

31. Use Eqs. 2.32 and 2.36 to derive Eq. 2.39.

32. Use the binomial expansion $(1 + x)^n = 1 + nx + [n(n-1)/2!]x^2 + \cdots$ to show that Eq. 2.34 for the relativistic kinetic energy reduces to the classical expression $\frac{1}{2}mv^2$ when $v \ll c$. This important result shows that our familiar expressions are correct at low speeds. By evaluating the first term in the expansion beyond $\frac{1}{2}mv^2$, find the speed necessary before the classical expression is off by 0.01%.

33. (a) According to observer O, a certain particle has a momentum of 817 MeV/c and a total relativistic energy of 1125 MeV. What is the rest energy of this particle? (b) An observer O' in a different frame of reference measures the momentum of this particle to be 953 MeV/c. What does O' measure for the total relativistic energy of the particle?

34. An electron is moving at a speed of 0.81c. By how much must its kinetic energy increase to raise its speed to 0.91c?

35. What is the change in mass when 1 g of copper is heated from 0 to 100°C? The specific heat capacity of copper is 0.40 J/g·K.

36. Find the kinetic energy of an electron moving at a speed of (a) $v = 1.00 \times 10^{-4}c$; (b) $v = 1.00 \times 10^{-2}c$; (c) $v = 0.300c$; (d) $v = 0.999c$.

37. An electron and a proton are each accelerated starting from rest through a potential difference of 10.0 million volts. Find the momentum (in MeV/c) and the kinetic energy (in MeV) of each, and compare with the results of using the classical formulas.

38. In a nuclear reactor, each atom of uranium (of atomic mass 235 u) releases about 200 MeV when it fissions. What is the change in mass when 1.00 kg of uranium-235 is fissioned?

2.8 Conservation Laws in Relativistic Decays and Collisions

39. A π meson of rest energy 139.6 MeV moving at a speed of 0.906c collides with and sticks to a proton of rest energy 938.3 MeV that is at rest. (a) Find the total relativistic energy of the resulting composite particle. (b) Find the total linear momentum of the composite particle. (c) Using the results of (a) and (b), find the rest energy of the composite particle.

40. An electron and a positron (an antielectron) make a head-on collision, each moving at $v = 0.99999c$. In the collision the electrons disappear and are replaced by two muons ($mc^2 = 105.7$ MeV), which move off in opposite directions. What is the kinetic energy of each of the muons?

41. It is desired to create a particle of mass 9700 MeV/c² in a head-on collision between a proton and an antiproton (each having a mass of 938.3 MeV/c²) traveling at the same speed. What speed is necessary for this to occur?

42. A particle of rest energy mc^2 is moving with speed v in the positive x direction. The particle decays into two particles, each of rest energy 140 MeV. One particle, with kinetic energy 282 MeV, moves in the positive x direction, and the other particle, with kinetic energy 25 MeV, moves in the negative x direction. Find the rest energy of the original particle and its speed.

2.9 Experimental Tests of Special Relativity

43. In the muon decay experiment discussed in Section 2.9 as a verification of time dilation, the muons move in the lab with a momentum of 3094 MeV/c. Find the dilated lifetime in the laboratory frame. (The proper lifetime is 2.198 μs.)

44. Derive the relativistic expression $p^2/2K = m + K/2c^2$, which is plotted in Figure 2.28a.

General Problems

45. Suppose we want to send an astronaut on a round trip to visit a star that is 200 light-years distant and at rest with respect to Earth. The life support systems on the spacecraft enable the astronaut to survive at most 20 years. (a) At what speed must the astronaut travel to make the round trip in 20 years of spacecraft time? (b) How much time passes on Earth during the round trip?

46. A "cause" occurs at point 1 (x_1, t_1) and its "effect" occurs at point 2 (x_2, t_2). Use the Lorentz transformation to find $t_2' - t_1'$, and show that $t_2' - t_1' > 0$; that is, O' can never see the "effect" coming before its "cause."

47. Observer O sees a red flash of light at the origin at $t = 0$ and a blue flash of light at $x = 3.26$ km at a time $t = 7.63$ μs. What are the distance and the time interval between the flashes according to observer O', who moves relative to O in the direction of increasing x with a speed of 0.625c?

Assume that the origins of the two coordinate systems line up at $t = t' = 0$.

48. Several spacecraft (A, B, C, and D) leave a space station at the same time. Relative to an observer on the station, A travels at $0.60c$ in the x direction, B at $0.50c$ in the y direction, C at $0.50c$ in the negative x direction, and D at $0.50c$ at $45°$ between the y and negative x directions. Find the velocity components, directions, and speeds of B, C, and D as observed from A.

49. Observer O sees a light turn on at $x = 524$ m when $t = 1.52$ μs. Observer O' is in motion at a speed of $0.563c$ in the positive x direction. The two frames of reference are synchronized so that their origins match up ($x = x' = 0$) at $t = t' = 0$. (a) At what time does the light turn on according to O'? (b) At what location does the light turn on in the reference frame of O'?

50. Suppose an observer O measures a particle of mass m moving in the x direction to have speed v, energy E, and momentum p. Observer O', moving at speed u in the x direction, measures v', E', and p' for the same object. (a) Use the Lorentz velocity transformation to find E' and p' in terms of m, u, and v. (b) Reduce $E'^2 - (p'c)^2$ to its simplest form and interpret the result.

51. Repeat Problem 50 for the mass moving in the y direction according to O. The velocity u of O' is still along the x direction.

52. Consider again the situation described in Section 2.6. Amelia's friend Bernice leaves Earth at the same time as Amelia and travels in the same direction at the same speed, but Bernice continues in the original direction when Amelia reaches the planet and turns her ship around. (a) From Bernice's frame of reference, Casper is moving at a velocity of $-0.60c$. Draw Casper's worldline in Bernice's frame of reference. (b) Casper celebrates 20 birthdays during Amelia's journey. In Bernice's frame of reference, how long does it take for Casper to celebrate 20 birthdays? (c) In Bernice's frame of reference, draw a worldline representing Amelia's outbound journey to the planet. (d) Calculate Amelia's velocity during her return journey as observed from Bernice's frame of reference, and draw a worldline showing Amelia's return journey. Amelia's and Casper's worldlines should intersect when Amelia return to Earth.

(e) Divide Casper's worldline into 20 segments, representing his birthdays. He sends a light signal to Amelia on each birthday. Amelia receives a light signal from Casper just as she arrives at the planet. On which birthday did Casper send this signal? (f) Amelia sends Casper a light signal on her 8th birthday. Draw a line on your diagram representing this light signal. When does Casper receive this signal?

53. Electrons are accelerated to high speeds by a two-stage machine. The first stage accelerates the electrons from rest to $v = 0.99c$. The second stage accelerates the electrons from $0.99c$ to $0.999c$. (a) How much energy does the first stage add to the electrons? (b) How much energy does the second stage add in increasing the velocity by only 0.9%?

54. A beam of 1.35×10^{11} electrons/s moving at a speed of $0.732c$ strikes a block of copper that is used as a beam stop. The copper block is a cube measuring 2.54 cm on edge. What is the temperature increase of the block after one hour?

55. An electron moving at a speed of $v_i = 0.960c$ in the positive x direction collides with another electron at rest. After the collision, one electron is observed to move with a speed of $v_{1f} = 0.956c$ at an angle of $\theta_1 = 9.7°$ with the x axis. (a) Use conservation of momentum to find the velocity (magnitude and direction) of the second electron. (b) Based only on the original data given in the problem, use conservation of energy to find the speed of the second electron.

56. A pion has a rest energy of 135 MeV. It decays into two gamma ray photons, bursts of electromagnetic radiation that travel at the speed of light. A pion moving through the laboratory at $v = 0.98c$ decays into two gamma ray photons of equal energies, making equal angles θ with the original direction of motion. Find the angle θ and the energies of the two gamma ray photons.

57. Consider again the decay described in Example 2.16 and determine the energies of the two pi mesons emitted in the decay of the K meson by first making a Lorentz transformation to a reference frame in which the initial K meson is at rest. When a K meson at rest decays into two pi mesons, they move in opposite directions with equal and opposite velocities, so they share the decay energy equally. Find the energies and velocities of the two pi mesons in the K meson's rest frame. Then transform back to the lab frame to find their kinetic energies.

Chapter 3

THE PARTICLELIKE PROPERTIES OF ELECTROMAGNETIC RADIATION

Thermal emission, the radiation emitted by all objects due to their temperatures, laid the groundwork for the development of quantum mechanics around the beginning of the 20th century. Today we use thermography for many applications, including the study of heat loss by buildings, medical diagnostics, night vision and other surveillance, and monitoring potential volcanoes.

We now turn to a discussion of *wave mechanics,* the second theory on which modern physics is based. One consequence of wave mechanics is the breakdown of the classical distinction between particles and waves. In this chapter we consider the three early experiments that provided evidence that light, which we usually regard as a wave phenomenon, has properties that we normally associate with particles. Instead of spreading its energy smoothly over a wave front, the energy is delivered in concentrated bundles like particles; a discrete bundle (*quantum*) of electromagnetic energy is known as a *photon.*

Before we begin to discuss the experimental evidence that supports the existence of the photon and the particlelike properties of light, we first review some of the properties of electromagnetic waves.

3.1 REVIEW OF ELECTROMAGNETIC WAVES

An electromagnetic field is characterized by its electric field $\vec{\mathbf{E}}$ and magnetic field $\vec{\mathbf{B}}$. For example, the electric field at a distance r from a point charge q at the origin is

$$\vec{\mathbf{E}} = \frac{1}{4\pi\varepsilon_0}\frac{q}{r^2}\hat{\mathbf{r}} \tag{3.1}$$

where $\hat{\mathbf{r}}$ is a unit vector in the radial direction. The magnetic field at a distance r from a long, straight, current-carrying wire along the z axis is

$$\vec{\mathbf{B}} = \frac{\mu_0 i}{2\pi r}\hat{\boldsymbol{\phi}} \tag{3.2}$$

where $\hat{\boldsymbol{\phi}}$ is the unit vector in the azimuthal direction (in the xy plane) in cylindrical coordinates.

If the charges are accelerated, or if the current varies with time, an electromagnetic wave is produced, in which $\vec{\mathbf{E}}$ and $\vec{\mathbf{B}}$ vary not only with $\vec{\mathbf{r}}$ but also with t. The mathematical expression that describes such a wave may have many different forms, depending on the properties of the source of the wave and of the medium through which the wave travels. One special form is the *plane wave,* in which the wave fronts are planes. (A point source, on the other hand, produces spherical waves, in which the wave fronts are spheres.) A plane electromagnetic wave traveling in the positive z direction is described by the expressions

$$\vec{\mathbf{E}} = \vec{\mathbf{E}}_0 \sin(kz - \omega t), \qquad \vec{\mathbf{B}} = \vec{\mathbf{B}}_0 \sin(kz - \omega t) \tag{3.3}$$

where the *wave number* k is found from the wavelength λ ($k = 2\pi/\lambda$) and the *angular frequency* ω is found from the frequency f ($\omega = 2\pi f$). Because λ and f are related by $c = \lambda f$, k and ω are also related by $c = \omega/k$.

The polarization of the wave is represented by the vector $\vec{\mathbf{E}}_0$; the plane of polarization is determined by the direction of $\vec{\mathbf{E}}_0$ and the direction of propagation, the z axis in this case. Once we specify the direction of travel and the polarization $\vec{\mathbf{E}}_0$, the direction of $\vec{\mathbf{B}}_0$ is fixed by the requirements that $\vec{\mathbf{B}}$ must be perpendicular to both $\vec{\mathbf{E}}$ and the direction of travel, and that the vector product $\vec{\mathbf{E}} \times \vec{\mathbf{B}}$ point in the direction of travel. For example if $\vec{\mathbf{E}}_0$ is in the x direction ($\vec{\mathbf{E}}_0 = E_0\hat{\mathbf{i}}$, where $\hat{\mathbf{i}}$

is a unit vector in the x direction), then $\vec{\mathbf{B}}_0$ must be in the y direction ($\vec{\mathbf{B}}_0 = B_0\hat{\mathbf{j}}$). Moreover, the magnitude of $\vec{\mathbf{B}}_0$ is determined by

$$B_0 = \frac{E_0}{c} \tag{3.4}$$

where c is the speed of light.

An electromagnetic wave transmits energy from one place to another; the energy flux is specified by the *Poynting vector* $\vec{\mathbf{S}}$:

$$\vec{\mathbf{S}} = \frac{1}{\mu_0}\vec{\mathbf{E}} \times \vec{\mathbf{B}} \tag{3.5}$$

For the plane wave, this reduces to

$$\vec{\mathbf{S}} = \frac{1}{\mu_0}E_0B_0\sin^2(kz - \omega t)\hat{\mathbf{k}} \tag{3.6}$$

where $\hat{\mathbf{k}}$ is a unit vector in the z direction. The Poynting vector has dimensions of power (energy per unit time) per unit area—for example, J/s/m^2 or W/m^2. Figure 3.1 shows the orientation of the vectors $\vec{\mathbf{E}}, \vec{\mathbf{B}}$, and $\vec{\mathbf{S}}$ for this special case.

Let us imagine the following experiment. We place a detector of electromagnetic radiation (a radio receiver or a human eye) at some point on the z axis, and we determine the electromagnetic power that this plane wave delivers to the receiver. The receiver is oriented with its sensitive area A perpendicular to the z axis, so that the maximum signal is received; we can therefore drop the vector representation of $\vec{\mathbf{S}}$ and work only with its magnitude S. The power P entering the receiver is then

$$P = SA = \frac{1}{\mu_0}E_0B_0A\sin^2(kz - \omega t) \tag{3.7}$$

which we can rewrite using Eq. 3.4 as

$$P = \frac{1}{\mu_0 c}E_0^2 A\sin^2(kz - \omega t) \tag{3.8}$$

There are two important features of this expression that you should recognize:

1. The intensity (the average power per unit area) is proportional to E_0^2. This is a general property of waves: *the intensity is proportional to the square of the amplitude*. We will see later that this same property also characterizes the waves that describe the behavior of material particles.

2. The intensity fluctuates with time, with the frequency $2f = 2(\omega/2\pi)$. We don't usually observe this rapid fluctuation—visible light, for example, has a frequency of about 10^{15} oscillations per second, and because our eye doesn't respond that quickly, we observe the time average of many (perhaps 10^{13}) cycles. If T is the observation time (perhaps 10^{-2} s in the case of the eye) then the average power is

$$P_{\text{av}} = \frac{1}{T}\int_0^T P\,dt \tag{3.9}$$

and using Eq. 3.8 we obtain the intensity I:

$$I = \frac{P_{\text{av}}}{A} = \frac{1}{2\mu_0 c}E_0^2 \tag{3.10}$$

because the average value of $\sin^2\theta$ is $^1/_2$.

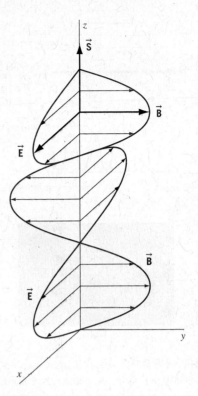

FIGURE 3.1 An electromagnetic wave traveling in the z direction. The electric field $\vec{\mathbf{E}}$ lies in the xz plane and the magnetic field $\vec{\mathbf{B}}$ lies in the yz plane.

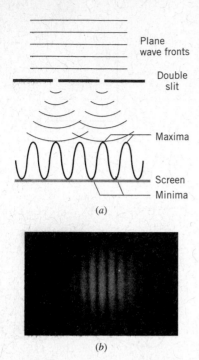

Plane wave fronts

Double slit

Maxima

Screen

Minima

(a)

(b)

FIGURE 3.2   (a) Young's double-slit experiment. A plane wave front passes through both slits; the wave is diffracted at the slits, and inter-ference occurs where the diffracted waves overlap on the screen. (b) The interference fringes observed on the screen.

Interference and Diffraction

The property that makes waves a unique physical phenomenon is the *principle of superposition,* which, for example, allows two waves to meet at a point, to cause a combined disturbance at the point that might be greater or less than the disturbance produced by either wave alone, and finally to emerge from the point of "collision" with all of the properties of each wave totally unchanged by the collision. To appreciate this important distinction between material objects and waves, imagine trying that trick with two automobiles!

This special property of waves leads to the phenomena of *interference* and *diffraction.* The simplest and best-known example of interference is *Young's double-slit experiment,* in which a monochromatic plane wave is incident on a barrier in which two narrow slits have been cut. (This experiment was first done with light waves, but in fact any wave will do as well, not only other electromagnetic waves, such as microwaves, but also mechanical waves, such as water or sound waves. We assume that the experiment is being done with light waves.)

Figure 3.2 illustrates this experimental arrangement. The plane wave is *diffracted* by each of the slits, so that the light passing through each slit covers a much larger area on the screen than the geometric shadow of the slit. This causes the light from the two slits to overlap on the screen, producing the interference. If we move away from the center of the screen just the right distance, we reach a point at which a wave crest passing through one slit arrives at exactly the same time as the previous wave crest that passed through the other slit. When this occurs, the intensity is a maximum, and a bright region appears on the screen. This is *constructive interference,* and it occurs continually at the point on the screen that is exactly one wavelength further from one slit than from the other. That is, if X_1 and X_2 are the distances from the point on the screen to the two slits, then a condition for maximum constructive interference is $|X_1 - X_2| = \lambda$. Constructive interference occurs when any wave crest from one slit arrives simultaneously with another from the other slit, whether it is the next, or the fourth, or the forty-seventh. The general condition for complete constructive interference is that the difference between X_1 and X_2 be an integral number of wavelengths:

$$|X_1 - X_2| = n\lambda \qquad n = 0, 1, 2, \ldots \qquad (3.11)$$

It is also possible for the crest of the wave from one slit to arrive at a point on the screen simultaneously with the trough (valley) of the wave from the other slit. When this happens, the two waves cancel, giving a dark region on the screen. This is known as *destructive interference.* (The existence of destructive interference at intensity minima immediately shows that we must add the electric field vectors $\vec{E}$ of the waves from the two slits, and not their powers P, because P can never be negative.) Destructive interference occurs whenever the distances X_1 and X_2 are such that the phase of one wave differs from the other by one-half cycle, or by one and one-half cycles, two and one-half cycles, and so forth:

$$|X_1 - X_2| = \tfrac{1}{2}\lambda, \tfrac{3}{2}\lambda, \tfrac{5}{2}\lambda, \ldots = (n + \tfrac{1}{2})\lambda \qquad n = 0, 1, 2, \ldots \qquad (3.12)$$

We can find the locations on the screen where the interference maxima occur in the following way. Let d be the separation of the slits, and let D be the distance from the slits to the screen. If y_n is the distance from the center of the screen to the nth maximum, then from the geometry of Figure 3.3 we find (assuming $X_1 > X_2$)

$$X_1^2 = D^2 + \left(\frac{d}{2} + y_n\right)^2 \quad \text{and} \quad X_2^2 = D^2 + \left(\frac{d}{2} - y_n\right)^2 \quad (3.13)$$

Subtracting these equations and solving for y_n, we obtain

$$y_n = \frac{X_1^2 - X_2^2}{2d} = \frac{(X_1 + X_2)(X_1 - X_2)}{2d} \quad (3.14)$$

In experiments with light, D is of order 1 m, and y_n and d are typically at most 1 mm; thus $X_1 \cong D$ and $X_2 \cong D$, so $X_1 + X_2 \cong 2D$, and to a good approximation

$$y_n = (X_1 - X_2)\frac{D}{d} \quad (3.15)$$

Using Eq. 3.11 for the values of $(X_1 - X_2)$ at the maxima, we find

$$y_n = n\frac{\lambda D}{d} \quad (3.16)$$

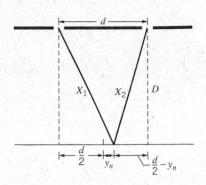

FIGURE 3.3 The geometry of the double-slit experiment.

Crystal Diffraction of X Rays

Another device for observing the interference of light waves is the *diffraction grating,* in which the wave fronts pass through a barrier that has *many* slits (often thousands or tens of thousands) and then recombine. The operation of this device is illustrated in Figure 3.4; interference maxima corresponding to different wavelengths appear at different angles θ, according to

$$d \sin\theta = n\lambda \quad (3.17)$$

where d is the slit spacing and n is the order number of the maximum $(n = 1, 2, 3, \ldots)$.

The advantage of the diffraction grating is its superior resolution—it enables us to get very good separation of wavelengths that are close to one another, and thus it is a very useful device for measuring wavelengths. Notice, however, that in order to get reasonable values of the angle θ—for example, $\sin\theta$ in the range of 0.3 to 0.5—we must have d of the order of a few times the wavelength. For visible light this is not particularly difficult, but for radiations of very short wavelength, mechanical construction of a grating is not possible. For example, for X rays with a wavelength of the order of 0.1 nm, we would need to construct a grating in which the slits were less than 1 nm apart, which is roughly the same as the spacing between the atoms of most materials.

The solution to this problem has been known since the pioneering experiments of Laue and Bragg:* use the atoms themselves as a diffraction grating! A beam of X rays sees the regular spacings of the atoms in a crystal as a sort of three-dimensional diffraction grating.

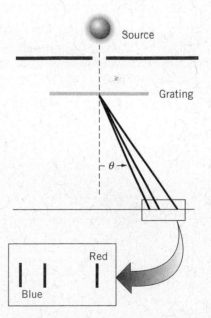

FIGURE 3.4 The use of a diffraction grating to analyze light into its constituent wavelengths.

---

*Max von Laue (1879–1960, Germany) developed the method of X-ray diffraction for the study of crystal structures, for which he received the 1914 Nobel Prize. Lawrence Bragg (1890–1971, England) developed the Bragg law for X-ray diffraction while he was a student at Cambridge University. He shared the 1915 Nobel Prize with his father, William Bragg, for their research on the use of X rays to determine crystal structures.

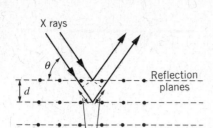

FIGURE 3.5 A beam of X rays reflected from a set of crystal planes of spacing d. The beam reflected from the second plane travels a distance $2d \sin \theta$ greater than the beam reflected from the first plane.

Consider the set of atoms shown in Figure 3.5, which represents a small portion of a two-dimensional slice of the crystal. The X rays are reflected from individual atoms in all directions, but in only one direction will the scattered "wavelets" constructively interfere to produce a reflected beam, and in this case we can regard the reflection as occurring from a plane drawn through the row of atoms. (This situation is identical with the reflection of light from a mirror—only in one direction will there be a beam of reflected light, and in that direction we can regard the reflection as occurring on a plane with the angle of incidence equal to the angle of reflection.)

Suppose the rows of atoms are a distance d apart in the crystal. Then a portion of the beam is reflected from the front plane, and a portion is reflected from the second plane, and so forth. The wave fronts of the beam reflected from the second plane lag behind those reflected from the front plane, because the wave reflected from the second plane must travel an additional distance of $2d \sin \theta$, where θ is the angle of incidence as *measured from the face of the crystal*. (Note that this is different from the usual procedure in optics, in which angles are defined with respect to the *normal* to the surface.) If this path difference is a whole number of wavelengths, the reflected beams interfere constructively and give an intensity maximum; thus the basic expression for the interference maxima in X-ray diffraction from a crystal is

$$2d \sin \theta = n\lambda \qquad n = 1, 2, 3, \ldots \qquad (3.18)$$

This result is known as *Bragg's law* for X-ray diffraction. Notice the factor of 2 that appears in Eq. 3.18 but does *not* appear in the otherwise similar expression of Eq. 3.17 for the ordinary diffraction grating.

Example 3.1

A single crystal of table salt (NaCl) is irradiated with a beam of X rays of wavelength 0.250 nm, and the first Bragg reflection is observed at an angle of 26.3°. What is the atomic spacing of NaCl?

Solution

Solving Bragg's law for the spacing d, we have

$$d = \frac{n\lambda}{2 \sin \theta} = \frac{0.250 \, \text{nm}}{2 \sin 26.3°} = 0.282 \, \text{nm}$$

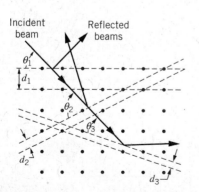

FIGURE 3.6 An incident beam of X rays can be reflected from many different crystal planes.

Our drawing of Figure 3.5 was very arbitrary—we had no basis for choosing which set of atoms to draw the reflecting planes through. Figure 3.6 shows a larger section of the crystal. As you can see, there are many possible reflecting planes, each with a different value of θ and d. (Of course, d_i and θ_i are related and cannot be varied independently.) If we used a beam of X rays of a single wavelength, it might be difficult to find the proper angle and set of planes to observe the interference. However, if we use a beam of X rays of a continuous range of wavelengths, for each d_i and θ_i interference will occur for a certain wavelength λ_i, and so there will be a pattern of interference maxima appearing at different angles of reflection as shown in Figure 3.6. The pattern of interference maxima depends on the spacing and the type of arrangement of the atoms in the crystal.

Figure 3.7 shows sample patterns (called *Laue patterns*) that are obtained from X-ray scattering from two different crystals. The bright dots correspond to interference maxima for wavelengths from the range of incident wavelengths that happen to satisfy Eq. 3.18. The three-dimensional pattern is more complicated

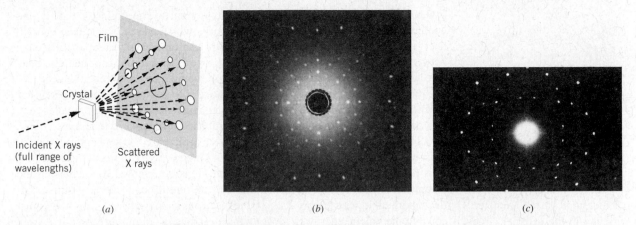

FIGURE 3.7 (*a*) Apparatus for observing X-ray scattering by a crystal. An interference maximum (dot) appears on the film whenever a set of crystal planes happens to satisfy the Bragg condition for a particular wavelength. (*b*) Laue pattern of TiO_2 crystal. (*c*) Laue pattern of a polyethylene crystal. The differences between the two Laue patterns are due to the differences in the geometric structure of the two crystals.

than our two-dimensional drawings, but the individual dots have the same interpretation. Figure 3.8 shows the pattern obtained from a sample that consists of many tiny crystals, rather than one single crystal. (It looks like Figure 3.7*b* or 3.7*c* rotated rapidly about its center.) From such pictures it is also possible to deduce crystal structures and lattice spacing.

All of the examples we have discussed in this section depend on the wave properties of electromagnetic radiation. However, as we now begin to discuss, there are other experiments that cannot be explained if we regard electromagnetic radiation as waves.

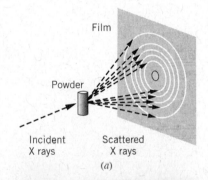

3.2 THE PHOTOELECTRIC EFFECT

We'll now turn to our discussion of the first of three experiments that cannot be explained by the wave theory of light. When a metal surface is illuminated with light, electrons can be emitted from the surface. This phenomenon, known as the *photoelectric effect,* was discovered by Heinrich Hertz in 1887 in the process of his research into electromagnetic radiation. The emitted electrons are called *photoelectrons.*

A sample experimental arrangement for observing the photoelectric effect is illustrated in Figure 3.9. Light falling on a metal surface (the emitter) can release electrons, which travel to the collector. The experiment must be done in an evacuated tube, so that the electrons do not lose energy in collisions with molecules of the air. Among the properties that can be measured are the rate of electron emission and the maximum kinetic energy of the photoelectrons.*

The rate of electron emission can be measured as an electric current *i* by an ammeter in the external circuit. The maximum kinetic energy of the electrons

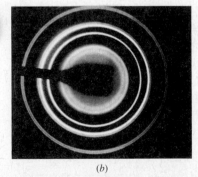

FIGURE 3.8 (*a*) Apparatus for observing X-ray scattering from a powdered or polycrystalline sample. Because the individual crystals have many different orientations, each scattered ray of Figure 3.7 becomes a cone which forms a circle on the film. (*b*) Diffraction pattern (known as *Debye-Scherrer* pattern) of polycrystalline gold.

---

*The electrons can be emitted with many different kinetic energies, depending on how tightly bound they are to the metal. Here we are concerned only with the *maximum* kinetic energy, which depends on the energy needed to remove the least tightly bound electron from the surface of the metal.

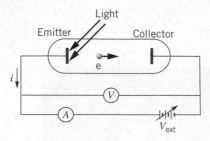

FIGURE 3.9 Apparatus for observing the photoelectric effect. The flow of electrons from the emitter to the collector is measured by the ammeter A as a current i in the external circuit. A variable voltage source V_{ext} establishes a potential difference between the emitter and collector, which is measured by the voltmeter V.

can be measured by applying a negative potential to the collector that is just enough to repel the most energetic electrons, which then do not have enough energy to "climb" the potential energy hill. That is, if the potential difference between the emitter and the collector is ΔV (a negative quantity), then electrons traveling from the emitter to the collector would gain a potential energy of $\Delta U = q \, \Delta V = -e \, \Delta V$ (a positive quantity) and would lose the same amount of kinetic energy. Electrons leaving the emitter with a kinetic energy smaller than this ΔU cannot reach the collector and are pushed back toward the emitter.

As the magnitude of the potential difference is increased, at some point even the most energetic electrons do not have enough kinetic energy to reach the collector. This potential, called the *stopping potential* V_s, is determined by increasing the magnitude of the voltage until the ammeter current drops to zero. At this point the maximum kinetic energy K_{max} of the electrons as they leave the emitter is just equal to the kinetic energy eV_s lost by the electrons in "climbing" the hill:

$$K_{max} = eV_s \tag{3.19}$$

where e is the magnitude of the electric charge of the electron. Typical values of V_s are a few volts.*

In the classical picture, the surface of the metal is illuminated by an electromagnetic wave of intensity I. The surface absorbs energy from the wave until the energy exceeds the binding energy of the electron to the metal, at which point the electron is released. The minimum quantity of energy needed to remove an electron is called the *work function* ϕ of the material. Table 3.1 lists some values of the work function of different materials. You can see that the values are typically a few electron-volts.

The Classical Theory of the Photoelectric Effect

What does the classical wave theory predict about the properties of the emitted photoelectrons?

1. *The maximum kinetic energy of the electrons should be proportional to the intensity of the radiation.* As the brightness of the light source is increased, more energy is delivered to the surface (the electric field is greater) and the electrons should be released with greater kinetic energies. Equivalently, increasing the intensity of the light source increases the electric field $\vec{E}$ of the wave, which also increases the force $\vec{F} = -e\vec{E}$ on the electron and its kinetic energy when it eventually leaves the surface.

2. *The photoelectric effect should occur for light of any frequency or wavelength.* According to the wave theory, as long as the light is intense enough to release electrons, the photoelectric effect should occur no matter what the frequency or wavelength.

3. *The first electrons should be emitted in a time interval of the order of seconds after the radiation begins to strike the surface.* In the wave theory, the energy of the wave is uniformly distributed over the wave front. If the electron absorbs energy directly from the wave, the amount of energy delivered to any

TABLE 3.1 **Some Photoelectric Work Functions**

Material	ϕ (eV)
Na	2.28
Al	4.08
Co	3.90
Cu	4.70
Zn	4.31
Ag	4.73
Pt	6.35
Pb	4.14

*The potential difference ΔV read by the voltmeter is not equal to the stopping potential when the emitter and collector are made of different materials. In that case a correction must be applied to account for the *contact potential difference* between the emitter and collector.

electron is determined by how much radiant energy is incident on the surface area in which the electron is confined. Assuming this area is about the size of an atom, a rough calculation leads to an estimate that the time lag between turning on the light and observing the first photoelectrons should be of the order of seconds (see Example 3.2).

Example 3.2

A laser beam with an intensity of 120 W/m² (roughly that of a small helium-neon laser) is incident on a surface of sodium. It takes a minimum energy of 2.3 eV to release an electron from sodium (the work function ϕ of sodium). Assuming the electron to be confined to an area of radius equal to that of a sodium atom (0.10 nm), how long will it take for the surface to absorb enough energy to release an electron?

Solution

The average power P_{av} delivered by the wave of intensity I to an area A is IA. An atom on the surface displays a "target area" of $A = \pi r^2 = \pi (0.10 \times 10^{-9} \text{ m})^2 = 3.1 \times 10^{-20} \text{ m}^2$. If the entire electromagnetic power is delivered to the electron, energy is absorbed at the rate

$\Delta E/\Delta t = P_{av}$. The time interval Δt necessary to absorb an energy $\Delta E = \phi$ can be expressed as

$$\Delta t = \frac{\Delta E}{P_{av}} = \frac{\phi}{IA}$$
$$= \frac{(2.3 \text{ eV})(1.6 \times 10^{-19} \text{ J/eV})}{(120 \text{ W/m}^2)(3.1 \times 10^{-20} \text{ m}^2)} = 0.10 \text{ s}$$

In reality, electrons in metals are not always bound to individual atoms but instead can be free to roam throughout the metal. However, no matter what reasonable estimate we make for the area over which the energy is absorbed, the characteristic time for photoelectron emission is estimated to have a magnitude of the order of seconds, in a range easily accessible to measurement.

The experimental characteristics of the photoelectric effect were well known by the year 1902. How do the predictions of the classical theory compare with the experimental results?

1. *For a fixed value of the wavelength or frequency of the light source, the maximum kinetic energy of the emitted photoelectrons (determined from the stopping potential) is totally independent of the intensity of the light source.* Figure 3.10 shows a representation of the experimental results. Doubling the intensity of the source leaves the stopping potential unchanged, indicating no change in the maximum kinetic energy of the electrons. This experimental result disagrees with the wave theory, which predicts that the maximum kinetic energy should depend on the intensity of the light.

2. *The photoelectric effect does not occur at all if the frequency of the light source is below a certain value.* This value, which is characteristic of the kind of metal surface used in the experiment, is called the *cutoff frequency* f_c. Above f_c, any light source, no matter how weak, will cause the emission of photoelectrons; below f_c, no light source, no matter how strong, will cause the emission of photoelectrons. This experimental result also disagrees with the predictions of the wave theory.

3. *The first photoelectrons are emitted virtually instantaneously (within 10^{-9} s) after the light source is turned on.* The wave theory predicts a measurable time delay, so this result also disagrees with the wave theory.

These three experimental results all suggest the complete failure of the wave theory to account for the photoelectric effect.

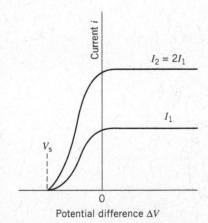

FIGURE 3.10 The photoelectric current i as a function of the potential difference ΔV for two different values of the intensity of the light. When the intensity I is doubled, the current is doubled (twice as many photoelectrons are emitted), but the stopping potential V_s remains the same.

The Quantum Theory of the Photoelectric Effect

A successful theory of the photoelectric effect was developed in 1905 by Albert Einstein. Five years earlier, in 1900, the German physicist Max Planck had developed a theory to explain the wavelength distribution of light emitted by hot, glowing objects (called *thermal radiation*, which is discussed in the next section of this chapter). Based partly on Planck's ideas, Einstein proposed that the energy of electromagnetic radiation is not continuously distributed over the wave front, but instead is concentrated in localized bundles or *quanta* (also known as *photons*). The energy of a photon associated with an electromagnetic wave of frequency f is

$$E = hf \tag{3.20}$$

where h is a proportionality constant known as *Planck's constant*. The photon energy can also be related to the wavelength of the electromagnetic wave by substituting $f = c/\lambda$, which gives

$$E = \frac{hc}{\lambda} \tag{3.21}$$

We often speak about photons as if they were particles, and as concentrated bundles of energy they have particlelike properties. Like the electromagnetic waves, photons travel at the speed of light, and so they must obey the relativistic relationship $p = E/c$. Combining this with Eq. 3.21, we obtain

$$p = \frac{h}{\lambda} \tag{3.22}$$

Photons carry linear momentum as well as energy, and thus they share this characteristic property of particles.

Because a photon travels at the speed of light, it must have zero mass. Otherwise its energy and momentum would be infinite. Similarly, a photon's rest energy $E_0 = mc^2$ must also be zero.

In Einstein's interpretation, a photoelectron is released as a result of an encounter with a *single photon*. The entire energy of the photon is delivered instantaneously to a *single photoelectron*. If the photon energy hf is greater than the work function ϕ of the material, the photoelectron will be released. If the photon energy is smaller than the work function, the photoelectric effect will not occur. This explanation thus accounts for two of the failures of the wave theory: the existence of the cutoff frequency and the lack of any measurable time delay.

If the photon energy hf exceeds the work function, the excess energy appears as the kinetic energy of the electron:

$$K_{\text{max}} = hf - \phi \tag{3.23}$$

The intensity of the light source does not appear in this expression! For a fixed frequency, doubling the intensity of the light means that twice as many photons strike the surface and twice as many photoelectrons are released, but they all have precisely the same maximum kinetic energy.

You can think of Eq. 3.23 as giving a relationship between energy quantities in analogy to making a purchase at a store. The quantity hf represents the payment you hand to the cashier, the quantity ϕ represents the cost of the object, and K_{max} represents the change you receive. In the photoelectric effect, hf is the amount of energy that is available to "purchase" an electron from the surface, the work function ϕ is the "cost" of removing the least tightly bound electron from the surface, and the difference between the available energy and the removal cost is the leftover energy that appears as the kinetic energy of the emitted electron. (The more tightly bound electrons have a greater "cost" and so emerge with smaller kinetic energies.)

A photon that supplies an energy equal to ϕ, exactly the minimum amount needed to remove an electron, corresponds to light of frequency equal to the cutoff frequency f_c. At this frequency, there is no excess energy for kinetic energy, so Eq. 3.23 becomes $hf_c = \phi$, or

$$f_c = \frac{\phi}{h} \qquad (3.24)$$

The corresponding cutoff wavelength $\lambda_c = c/f_c$ is

$$\lambda_c = \frac{hc}{\phi} \qquad (3.25)$$

The cutoff wavelength represents the *largest* wavelength for which the photoelectric effect can be observed for a surface with the work function ϕ.

The photon theory appears to explain all of the observed features of the photoelectric effect. The most detailed test of the theory was done by Robert Millikan in 1915. Millikan measured the maximum kinetic energy (stopping potential) for different frequencies of the light and obtained a plot of Eq. 3.23. A sample of his results is shown in Figure 3.11. From the slope of the line, Millikan obtained a value for Planck's constant of

$$h = 6.57 \times 10^{-34}\,\text{J}\cdot\text{s}$$

In part for his detailed experiments on the photoelectric effect, Millikan was awarded the 1923 Nobel Prize in physics. Einstein was awarded the 1921 Nobel Prize for his photon theory as applied to the photoelectric effect.

As we discuss in the next section, the wavelength distribution of thermal radiation also yields a value for Planck's constant, which is in good agreement with Millikan's value derived from the photoelectric effect. Planck's constant is one of the fundamental constants of nature; just as c is the characteristic constant of relativity, h is the characteristic constant of quantum mechanics. The value of Planck's constant has been measured to great precision in a variety of experiments. The presently accepted value is

$$h = 6.6260696 \times 10^{-34}\,\text{J}\cdot\text{s}$$

This is an experimentally determined value, with a relative uncertainty of about 5×10^{-8} (± 3 units in the last digit).

Robert A. Millikan (1868–1953, United States). Perhaps the best experimentalist of his era, his work included the precise determination of Planck's constant using the photoelectric effect (for which he received the 1923 Nobel Prize) and the measurement of the charge of the electron (using his famous "oil-drop" apparatus).

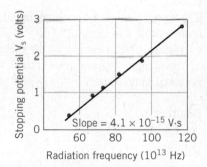

FIGURE 3.11 Millikan's results for the photoelectric effect in sodium. The slope of the line is h/e; the experimental determination of the slope gives a way of determining Planck's constant. The intercept should give the cutoff frequency; however, in Millikan's time the contact potentials of the electrodes were not known precisely and so the vertical scale is displaced by a few tenths of a volt. The slope not affected by this correction.

Example 3.3

(a) What are the energy and momentum of a photon of red light of wavelength 650 nm? (b) What is the wavelength of a photon of energy 2.40 eV?

Solution

(a) Using Eq. 3.21 we obtain

$$E = \frac{hc}{\lambda} = \frac{(6.63 \times 10^{-34} \text{ J} \cdot \text{s})(3.00 \times 10^8 \text{ m/s})}{650 \times 10^{-9} \text{ m}}$$

$$= 3.06 \times 10^{-19} \text{ J}$$

Converting to electron-volts, we have

$$E = \frac{3.06 \times 10^{-19} \text{ J}}{1.60 \times 10^{-19} \text{ J/eV}} = 1.91 \text{ eV}$$

This type of problem can be simplified if we express the combination hc in units of eV · nm:

$$E = \frac{hc}{\lambda} = \frac{1240 \text{ eV} \cdot \text{nm}}{650 \text{ nm}} = 1.91 \text{ eV}$$

The momentum is found in a similar way, using Eq. 3.22

$$p = \frac{h}{\lambda} = \frac{1}{c}\frac{hc}{\lambda} = \frac{1}{c}\left(\frac{1240 \text{ eV} \cdot \text{nm}}{650 \text{ nm}}\right) = 1.91 \text{ eV}/c$$

The momentum could also be found directly from the energy:

$$p = \frac{E}{c} = \frac{1.91 \text{ eV}}{c} = 1.91 \text{ eV}/c$$

(It may be helpful to review the discussion in Example 2.11 about these units of momentum.)

(b) Solving Eq. 3.21 for λ, we find

$$\lambda = \frac{hc}{E} = \frac{1240 \text{ eV} \cdot \text{nm}}{2.40 \text{ eV}} = 517 \text{ nm}$$

Example 3.4

The work function for tungsten metal is 4.52 eV. (a) What is the cutoff wavelength λ_c for tungsten? (b) What is the maximum kinetic energy of the electrons when radiation of wavelength 198 nm is used? (c) What is the stopping potential in this case?

Solution

(a) Equation 3.25 gives

$$\lambda_c = \frac{hc}{\phi} = \frac{1240 \text{ eV} \cdot \text{nm}}{4.52 \text{ eV}} = 274 \text{ nm}$$

in the ultraviolet region.

(b) At the shorter wavelength,

$$K_{max} = hf - \phi = \frac{hc}{\lambda} - \phi$$

$$= \frac{1240 \text{ eV} \cdot \text{nm}}{198 \text{ nm}} - 4.52 \text{ eV}$$

$$= 1.74 \text{ eV}$$

(c) The stopping potential is the voltage corresponding to K_{max}:

$$V_s = \frac{K_{max}}{e} = \frac{1.74 \text{ eV}}{e} = 1.74 \text{ V}$$

3.3 THERMAL RADIATION

The second type of experiment we discuss that cannot be explained by the classical wave theory is *thermal radiation*, which is the electromagnetic radiation emitted by all objects because of their temperature. At room temperature the thermal radiation is mostly in the infrared region of the spectrum, where our eyes are not sensitive. As we heat objects to higher temperatures, they may emit visible light.

A typical experimental arrangement is shown in Figure 3.12. An object is maintained at a temperature T_1. The radiation emitted by the object is detected by an apparatus that is sensitive to the wavelength of the radiation. For example, a dispersive medium such as a prism can be used so that different wavelengths appear at different angles θ. By moving the radiation detector to different angles θ we can measure the intensity* of the radiation at a specific wavelength. The detector is not a geometrical point (hardly an efficient detector!) but instead subtends a small range of angles $\Delta\theta$, so what we really measure is the amount of radiation in some range $\Delta\theta$ at θ, or, equivalently, in some range $\Delta\lambda$ at λ.

Many experiments were done in the late 19th century to study the wavelength spectrum of thermal radiation. These experiments, as we shall see, gave results that totally disagreed with the predictions of the classical theories of thermodynamics and electromagnetism; instead, the successful analysis of the experiments provided the first evidence of the quantization of energy, which would eventually be seen as the basis for the new quantum theory.

Let's first review the experimental results. The goal of these experiments was to measure the intensity of the radiation emitted by the object as a function of wavelength. Figure 3.13 shows a typical set of experimental results when the object is at a temperature $T_1 = 1000$ K. If we now change the temperature of the object to a different value T_2, we obtain a different curve, as shown in Figure 3.13 for $T_2 = 1250$ K. If we repeat the measurement for many different temperatures, we obtain systematic results for the radiation intensity that reveal two important characteristics:

1. The total intensity radiated over all wavelengths (that is, the area under each curve) increases as the temperature is increased. This is not a surprising result: we commonly observe that a glowing object glows brighter and thus radiates more energy as we increase its temperature. From careful measurement, we find that the total intensity increases as the fourth power of the absolute or kelvin temperature:

$$I = \sigma T^4 \tag{3.26}$$

where we have introduced the proportionality constant σ. Equation 3.26 is called *Stefan's law* and the constant σ is called the *Stefan-Boltzmann constant*. Its value can be determined from experimental results such as those illustrated in Figure 3.13:

$$\sigma = 5.67037 \times 10^{-8} \text{ W/m}^2 \cdot \text{K}^4$$

2. The wavelength λ_{max} at which the emitted intensity reaches its maximum value decreases as the temperature is increased, in inverse proportion to the temperature: $\lambda_{max} \propto 1/T$. From results such as those of Figure 3.13, we can determine the proportionality constant, so that

$$\lambda_{max} T = 2.8978 \times 10^{-3} \text{ m} \cdot \text{K} \tag{3.27}$$

This result is known as *Wien's displacement law*; the term "displacement" refers to the way the peak is moved or displaced as the temperature is

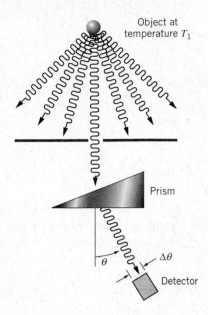

FIGURE 3.12 Measurement of the spectrum of thermal radiation. A device such as a prism is used to separate the wavelengths emitted by the object.

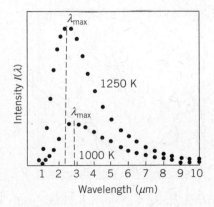

FIGURE 3.13 A possible result of the measurement of the radiation intensity over many different wavelengths. Each different temperature of the emitting body gives a different peak λ_{max}.

---

*As always, intensity means energy per unit time per unit area (or power per unit area), as in Eq. 3.10. Previously, "unit area" referred to the wave front, such as would be measured if we recorded the waves with an antenna of a certain area. Here, "unit area" indicates the electromagnetic radiation emitted from each unit area of the surface of the object whose thermal emissions are being observed.

varied. Wien's law is qualitatively consistent with our common observation that heated objects first begin to glow with a red color, and at higher temperatures the color becomes more yellow. As the temperature is increased, the wavelength at which most of the radiation is emitted moves from the longer-wavelength (red) part of the visible region toward medium wavelengths. The term "white hot" refers to an object that is hot enough to produce the mixture of all wavelengths in the visible region to make white light.

Example 3.5

(a) At what wavelength does a room-temperature ($T = 20°C$) object emit the maximum thermal radiation? (b) To what temperature must we heat it until its peak thermal radiation is in the red region of the spectrum ($\lambda = 650$ nm)? (c) How many times as much thermal radiation does it emit at the higher temperature?

Solution

(a) Using the absolute temperature, $T_1 = 273 + 20 = 293$ K, Wien's displacement law gives

$$\lambda_{max} = \frac{2.8978 \times 10^{-3} \text{ m} \cdot \text{K}}{T_1}$$
$$= \frac{2.8978 \times 10^{-3} \text{ m} \cdot \text{K}}{293 \text{ K}} = 9.89 \, \mu\text{m}$$

This is in the infrared region of the electromagnetic spectrum.

(b) For $\lambda_{max} = 650$ nm, we again use Wien's displacement law to find the new temperature T_2 :

$$T_2 = \frac{2.8978 \times 10^{-3} \text{ m} \cdot \text{K}}{\lambda_{max}}$$
$$= \frac{2.8978 \times 10^{-3} \text{ m} \cdot \text{K}}{650 \times 10^{-9} \text{ m}}$$
$$= 4460 \text{ K}$$

(c) The total intensity of radiation is proportional to T^4, so the ratio of the total thermal emissions will be

$$\frac{I_2}{I_1} = \frac{\sigma T_2^4}{\sigma T_1^4} = \frac{(4460 \text{ K})^4}{(293 \text{ K})^4}$$
$$= 5.37 \times 10^4$$

Be sure to notice the use of absolute (kelvin) temperatures in this example.

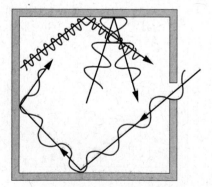

FIGURE 3.14 A cavity filled with electromagnetic radiation in thermal equilibrium with its walls at temperature T. Some radiation escapes through the hole, which represents an ideal blackbody.

The theoretical analysis of the emission of thermal radiation from an arbitrary object is extremely complicated. It depends on details of the surface properties of the object, and it also depends on how much radiation the object reflects from its surroundings. To simplify our analysis, we consider a special type of object called a *blackbody*, which absorbs all radiation incident on it and reflects none of the incident radiation.

To simplify further, we consider a special type of blackbody: a hole in a hollow metal box whose walls are in thermal equilibrium at temperature T. The box is filled with electromagnetic radiation that is emitted and reflected by the walls. A small hole in one wall of the box allows some of the radiation to escape (Figure 3.14). *It is the hole, and not the box itself, that is the blackbody.* Radiation from outside that is incident on the hole gets lost inside the box and has a negligible chance of reemerging from the hole; thus no reflections occur from the blackbody (the hole). The radiation that emerges from the hole is just a sample of the radiation inside the box, so understanding the nature of the radiation inside the box allows us to understand the radiation that leaves through the hole.

Let's consider the radiation inside the box. It has an energy density (energy per unit volume) per unit wavelength interval $u(\lambda)$. That is, if we could look into the interior of the box and measure the energy density of the electromagnetic radiation with wavelengths between λ and $\lambda + d\lambda$ in a small volume element, the result would be $u(\lambda)d\lambda$. For the radiation in this wavelength interval, what is the corresponding intensity (power per unit area) emerging from the hole? At any particular instant, half of the radiation in the box will be moving away from the hole. The other half of the radiation is moving toward the hole at velocity of magnitude c but directed over a range of angles. Averaging over this range of angles to evaluate the energy flowing perpendicular to the surface of the hole introduces another factor of $1/2$, so the contribution of the radiation in this small wavelength interval to the intensity passing through the hole is

$$I(\lambda) = \frac{c}{4}u(\lambda) \tag{3.28}$$

The quantity $I(\lambda)d\lambda$ is the radiant intensity in the small interval $d\lambda$ at the wavelength λ. This is the quantity whose measurement gives the results displayed in Figure 3.13. Each data point represents a measurement of the intensity in a small wavelength interval. The goal of the theoretical analysis is to find a mathematical function $I(\lambda)$ that gives a smooth fit through the data points of Figure 3.13.

If we wish to find the total intensity emitted in the region between wavelengths λ_1 and λ_2, we divide the region into narrow intervals $d\lambda$ and add the intensities in each interval, which is equivalent to the integral between those limits:

$$I(\lambda_1 : \lambda_2) = \int_{\lambda_1}^{\lambda_2} I(\lambda) \, d\lambda \tag{3.29}$$

This is similar to Eq. 1.27 for determining the number of molecules with energies between two limits. The total emitted intensity can be found by integrating over all wavelengths:

$$I = \int_0^\infty I(\lambda) \, d\lambda \tag{3.30}$$

This total intensity should work out to be proportional to the 4th power of the temperature, as required by Stefan's law (Eq. 3.26).

Classical Theory of Thermal Radiation

Before discussing the quantum theory of thermal radiation, let's see what the classical theories of electromagnetism and thermodynamics can tell us about the dependence of I on λ. The complete derivation is not given here, only a brief outline of the theory.* The derivation involves first computing the amount of radiation (number of waves) at each wavelength and then finding the contribution of each wave to the total energy in the box.

---

*For a more complete derivation, see R. Eisberg and R. Resnick, *Quantum Theory of Atoms, Molecules, Solids, Nuclei, and Particles*, 2nd edition (Wiley, 1985), pp. 9–13.

1. *The box is filled with electromagnetic standing waves.* If the walls of the box are metal, radiation is reflected back and forth with a node of the electric field at each wall (the electric field must vanish inside a conductor). This is the same condition that applies to other standing waves, like those on a stretched string or a column of air in an organ pipe.

2. *The number of standing waves with wavelengths between λ and $\lambda + d\lambda$ is*

$$N(\lambda)\, d\lambda = \frac{8\pi V}{\lambda^4}\, d\lambda \qquad (3.31)$$

where V is the volume of the box. For one-dimensional standing waves, as on a stretched string of length L, the allowed wavelength are $\lambda = 2L/n$, ($n = 1, 2, 3, \ldots$). The number of possible standing waves with wavelengths between λ_1 and λ_2 is $n_2 - n_1 = 2L(1/\lambda_2 - 1/\lambda_1)$. In the small interval from λ to $\lambda + d\lambda$, the number of standing waves is $N(\lambda)d\lambda = |dn/d\lambda|d\lambda = (2L/\lambda^2)d\lambda$. Equation 3.31 can be obtained by extending this approach to three dimensions.

3. *Each individual wave contributes an average energy of kT to the radiation in the box.* This result follows from an analysis similar to that of Section 1.3 for the statistical mechanics of gas molecules. In this case we are interested in the statistics of the oscillating atoms in the walls of the cavity, which are responsible for setting up the standing electromagnetic waves in the cavity. For a one-dimensional oscillator, the energies are distributed according to the Maxwell-Boltzmann distribution:*

$$N(E) = \frac{N}{kT}e^{-E/kT} \qquad (3.32)$$

Recall from Section 1.3 that $N(E)$ is defined so that the number of oscillators with energies between E and $E + dE$ is $dN = N(E)dE$, and thus the total number of oscillators at all energies is $\int dN = \int_0^\infty N(E)dE$, which (as you should show) works out to N. The average energy per oscillator is then found in the same way as the average energy of a gas molecule (Eq. 1.25):

$$E_{av} = \frac{1}{N}\int_0^\infty E\, N(E)\, dE = \frac{1}{kT}\int_0^\infty E\, e^{-E/kT}\, dE \qquad (3.33)$$

which does indeed work out to $E_{av} = kT$.

Putting all these ingredients together, we can find the energy density of radiation in the wavelength interval $d\lambda$ inside the cavity: energy density = (number of standing waves per unit volume) × (average energy per standing wave) or

$$u(\lambda)\, d\lambda = \frac{N(\lambda)\, d\lambda}{V}kT = \frac{8\pi}{\lambda^4}kT\, d\lambda \qquad (3.34)$$

The corresponding intensity per unit wavelength interval $d\lambda$ is

$$I(\lambda) = \frac{c}{4}u(\lambda) = \frac{c}{4}\frac{8\pi}{\lambda^4}kT = \frac{2\pi c}{\lambda^4}kT \qquad (3.35)$$

This result is known as the *Rayleigh-Jeans formula*; based firmly on the classical theories of electromagnetism and thermodynamics, it represents our best attempt

---

*The exponential part of this expression is that same as that of Eq. 1.22 for gas molecules, but the rest of the equation is different, because the statistical behavior of one-dimensional oscillators is different from that of gas molecules moving in three dimensions. We'll consider these calculations in greater detail in Chapter 10.

to apply classical physics to understanding the problem of blackbody radiation. In Figure 3.15 the intensity calculated from the Rayleigh-Jeans formula is compared with typical experimental results. The intensity calculated with Eq. 3.35 approaches the data at long wavelengths, but at short wavelengths, the classical theory (which predicts $u \to \infty$ as $\lambda \to 0$) fails miserably. The failure of the Rayleigh-Jeans formula at short wavelengths is known as the *ultraviolet catastrophe* and represents a serious problem for classical physics, because the theories of thermodynamics and electromagnetism on which the Rayleigh-Jeans formula is based have been carefully tested in many other circumstances and found to give extremely good agreement with experiment. It is apparent in the case of blackbody radiation that the classical theories do not work, and that a new kind of physical theory is needed.

Quantum Theory of Thermal Radiation

The new physics that gave the correct interpretation of thermal radiation was proposed by the German physicist Max Planck in 1900. The ultraviolet catastrophe occurs because the Rayleigh-Jeans formula predicts too much intensity at short wavelengths (or equivalently at high frequencies). What is needed is a way to make $u \to 0$ as $\lambda \to 0$, or as $f \to \infty$. Again considering the electromagnetic standing waves to result from the oscillations of atoms in the walls of the cavity, Planck tried to find a way to reduce the number of high-frequency standing waves by reducing the number of high-frequency oscillators. He did this by a bold assumption that formed the cornerstone of a new physical theory, *quantum physics*. Associated with this theory is a new version of mechanics, known as *wave mechanics* or *quantum mechanics*. We discuss the methods of wave mechanics in Chapter 5; for now we show how Planck's theory provided the correct interpretation of the emission spectrum of thermal radiation.

Planck suggested that an oscillating atom can absorb or emit energy only in discrete bundles. This bold suggestion was necessary to keep the average energy of a low-frequency (long-wavelength) oscillator equal to kT (in agreement with the Rayleigh-Jeans law at long wavelength), but it also made the average energy of a high-frequency (low-wavelength) oscillator approach zero. Let's see how Planck managed this remarkable feat.

In Planck's theory, each oscillator can emit or absorb energy only in quantities that are integer multiples of a certain basic quantity of energy ε,

$$E_n = n\varepsilon \qquad n = 1, 2, 3, \ldots \tag{3.36}$$

where n is the number of quanta. Furthermore, the energy of each of the quanta is determined by the frequency

$$\varepsilon = hf \tag{3.37}$$

where h is the constant of proportionality, now known as Planck's constant. From the mathematical standpoint, the difference between Planck's calculation and the classical calculation using Maxwell-Boltzmann statistics is that the energy of an oscillator at a certain wavelength or frequency is no longer a continuous variable—it is a discrete variable that takes only the values given by Eq. 3.36. The integrals in the classical calculation are then replaced by sums, and the number of oscillators with energy E_n is then

$$N_n = N(1 - e^{-\varepsilon/kT})e^{-n\varepsilon/kT} \tag{3.38}$$

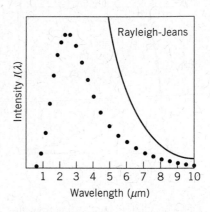

FIGURE 3.15 The failure of the classical Rayleigh-Jeans formula to fit the observed intensity. At long wavelengths the theory approaches the data, but at short wavelengths the classical formula fails miserably.

Max Planck (1858–1947, Germany). His work on the spectral distribution of radiation, which led to the quantum theory, was honored with the 1918 Nobel Prize. In his later years, he wrote extensively on religious and philosophical topics.

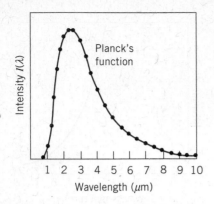

FIGURE 3.16 Planck's function fits the observed data perfectly.

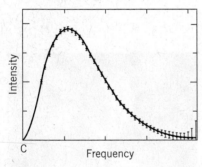

FIGURE 3.17 Data from the COBE satellite, launched in 1989 to determine the temperature of the cosmic microwave background radiation from the early universe. The data points exactly fit the Planck function corresponding to a temperature of 2.725 K. To appreciate the remarkable precision of this experiment, note that the sizes of the error bars have been increased by a factor of 400 to make them visible! (Source: NASA Office of Space Science)

(Compare this result with Eq. 3.32 for the continuous case.) Here N_n represents the number of oscillators with energy E_n, while N is the total number. You should be able to show that $\sum_{n=0}^{\infty} N_n = N$, again giving the total number of oscillators when summed over all possible energies. Planck's calculation then gives the average energy:

$$E_{av} = \frac{1}{N} \sum_{n=0}^{\infty} N_n E_n = (1 - e^{-\varepsilon/kT}) \sum_{n=0}^{\infty} (n\varepsilon) e^{-n\varepsilon/kT} \qquad (3.39)$$

which gives (see Problem 14)

$$E_{av} = \frac{\varepsilon}{e^{\varepsilon/kT} - 1} = \frac{hf}{e^{hf/kT} - 1} = \frac{hc/\lambda}{e^{hc/\lambda kT} - 1} \qquad (3.40)$$

Note from this equation that $E_{av} \cong kT$ at small f (large λ) but that $E_{av} \to 0$ at large f (small λ). Thus the small-wavelength oscillators carry a vanishingly small energy, and the ultraviolet catastrophe is solved!

Based on Planck's result, the intensity of the radiation then becomes (using Eqs. 3.28 and 3.31):

$$I(\lambda) = \frac{c}{4} \left(\frac{8\pi}{\lambda^4} \right) \left[\frac{hc/\lambda}{e^{hc/\lambda kT} - 1} \right] = \frac{2\pi hc^2}{\lambda^5} \frac{1}{e^{hc/\lambda kT} - 1} \qquad (3.41)$$

(An alternative approach to deriving this result is given in Section 10.6.) The perfect agreement between experiment and Planck's formula is illustrated in Figure 3.16.

In Problems 15 and 16 at the end of this chapter you will demonstrate that Planck's formula can be used to deduce Stefan's law and Wien's displacement law. In fact, deducing Stefan's law from Planck's formula results in a relationship between the Stefan-Boltzmann constant and Planck's constant:

$$\sigma = \frac{2\pi^5 k^4}{15 c^2 h^3} \qquad (3.42)$$

By determining the value of the Stefan-Boltzmann constant from the intensity data available in 1900, Planck was able to determine a value of the constant h:

$$h = 6.56 \times 10^{-34} \text{ J} \cdot \text{s}$$

which agrees very well with the value of h that Millikan deduced 15 years later based on the analysis of data from the photoelectric effect. The good agreement of these two values is remarkable, because they are derived from very different kinds of experiments—one involves the *emission* and the other the *absorption* of electromagnetic radiation. This suggests that the quantization property is not an accident arising from the analysis of one particular experiment, but is instead a property of the electromagnetic field itself. Along with many other scientists of his era, Planck was slow to accept this interpretation. However, later experimental evidence (including the Compton effect) proved to be so compelling that it left no doubt about Einstein's photon theory and the particlelike structure of the electromagnetic field.

Planck's formula still finds important applications today in the measurement of temperature. By measuring the intensity of radiation emitted by an object at a particular wavelength (or, as in actual experiments, in a small interval of wavelengths), Eq. 3.41 can be used to deduce the temperature of the object. Note that only one measurement, at any wavelength, is all that is required to obtain the temperature. A *radiometer* is a device for measuring the intensity of thermal radiation at selected wavelengths, enabling a determination of temperature. Radiometers in orbiting satellites are used to measure the temperature of the land and sea areas of the Earth and of the upper surface of clouds. Other orbiting radiometers have been aimed toward "empty space" to measure the temperature of the radiation from the early history of the universe (Figure 3.17).

▌ Example 3.6

You are using a radiometer to observe the thermal radiation from an object that is heated to maintain its temperature at 1278 K. The radiometer records radiation in a wavelength interval of 12.6 nm. By changing the wavelength at which you are measuring, you set the radiometer to record the most intense radiation emission from the object. What is the intensity of the emitted radiation in this interval?

Solution

The wavelength setting for the most intense radiation is determined from Wien's displacement law:

$$\lambda_{max} = \frac{2.8978 \times 10^{-3} \text{ m} \cdot \text{K}}{T} = \frac{2.8978 \times 10^{-3} \text{ m} \cdot \text{K}}{1278 \text{ K}}$$

$$= 2.267 \times 10^{-6} \text{ m} = 2267 \text{ nm}$$

The given temperature corresponds to $kT = (8.6174 \times 10^{-5} \text{ eV/K})(1278 \text{ K}) = 0.1101 \text{ eV}$. The radiation intensity in this small wavelength interval is

$$I(\lambda)d\lambda = \frac{2\pi hc^2}{\lambda^5} \frac{1}{e^{hc/\lambda kT} - 1} d\lambda$$

$$= 2\pi(6.626 \times 10^{-34} \text{ J} \cdot \text{s})(2.998 \times 10^8 \text{ m/s})^2$$
$$\times (12.6 \times 10^{-9} \text{ m})(2.267 \times 10^{-6} \text{ m})^{-5}$$
$$\times (e^{(1240 \text{ eV·nm})/(2267 \text{ nm})(0.1101 \text{ eV})} - 1)^{-1}$$

$$= 552 \text{ W/m}^2$$

3.4 THE COMPTON EFFECT

Another way for radiation to interact with matter is by means of the Compton effect, in which radiation scatters from loosely bound, nearly free electrons. Part of the energy of the radiation is given to the electron; the remainder of the energy is reradiated as electromagnetic radiation. According to the wave picture, the scattered radiation is less energetic than the incident radiation (the difference going into the kinetic energy of the electron) but has the same wavelength. As we will see, the photon concept leads to a very different prediction for the scattered radiation.

The scattering process is analyzed simply as an interaction (a "collision" in the classical sense of particles) between a single photon and an electron, which

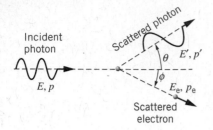

Incident
photon

Scattered photon

θ E', p'

ϕ

E, p

E_e, p_e

Scattered
electron

FIGURE 3.18   The geometry of Compton scattering.

Arthur H. Compton (1892–1962, United States). His work on X-ray scattering verified Einstein's photon theory and earned him the 1927 Nobel Prize. He was a pioneer in research with X rays and cosmic rays. During World War II he directed a portion of the U.S. atomic bomb research.

we assume to be at rest. Figure 3.18 shows the process. Initially, the photon has energy E and linear momentum p given by

$$E = hf = \frac{hc}{\lambda} \quad \text{and} \quad p = \frac{E}{c} \tag{3.43}$$

The electron, initially at rest, has rest energy $m_e c^2$. After the scattering, the photon has energy $E' = hc/\lambda'$ and momentum $p' = E'/c$, and it moves in a direction at an angle θ with respect to the direction of the incident photon. The electron has total final energy E_e and momentum p_e and moves in a direction at an angle ϕ with respect to the initial photon. (To allow for the possibility of high-energy incident photons giving energetic scattered electrons, we use relativistic kinematics for the electron.) The conservation laws for total relativistic energy and momentum are then applied:

$$E_{\text{initial}} = E_{\text{final}}: \qquad E + m_e c^2 = E' + E_e \tag{3.44a}$$

$$p_{x,\text{initial}} = p_{x,\text{final}}: \qquad p = p_e \cos\phi + p' \cos\theta \tag{3.44b}$$

$$p_{y,\text{initial}} = p_{y,\text{final}}: \qquad 0 = p_e \sin\phi - p' \sin\theta \tag{3.44c}$$

We have three equations with four unknowns (θ, ϕ, E_e, E'; p_e and p' are not independent unknowns) that cannot be solved uniquely, but we can eliminate any two of the four unknowns by solving the equations simultaneously. If we choose to measure the energy and direction of the scattered photon, we eliminate E_e and ϕ. The angle ϕ is eliminated by first rewriting the momentum equations:

$$p_e \cos\phi = p - p' \cos\theta \quad \text{and} \quad p_e \sin\phi = p' \sin\theta \tag{3.45}$$

Squaring these equations and adding the results, we obtain

$$p_e^2 = p^2 - 2pp' \cos\theta + p'^2 \tag{3.46}$$

The relativistic relationship between energy and momentum is, according to Eq. 2.39, $E_e^2 = c^2 p_e^2 + m_e^2 c^4$. Substituting in this equation for E_e from Eq. 3.44a and for p_e^2 from Eq. 3.46, we obtain

$$(E + m_e c^2 - E')^2 = c^2(p^2 - 2pp' \cos\theta + p'^2) + m_e^2 c^4 \tag{3.47}$$

and after a bit of algebra, we find

$$\frac{1}{E'} - \frac{1}{E} = \frac{1}{m_e c^2}(1 - \cos\theta) \tag{3.48}$$

In terms of wavelength, this equation can also be written as

$$\lambda' - \lambda = \frac{h}{m_e c}(1 - \cos\theta) \tag{3.49}$$

where λ is the wavelength of the incident photon and λ' is the wavelength of the scattered photon. The quantity $h/m_e c$ is known as the *Compton wavelength of the electron* and has a value of 0.002426 nm; however, keep in mind that it is *not* a true wavelength but rather is a *change* of wavelength.

Equations 3.48 and 3.49 give the change in energy or wavelength of the photon, as a function of the *scattering angle* θ. Because the quantity on the right-hand side is never negative, E' is always less than E, so that the scattered photon has less energy than the original incident photon; the difference $E - E'$ is just the kinetic energy given to the electron, $E_e - m_e c^2$. Similarly, λ' is greater than λ, meaning the scattered photon always has a longer wavelength than the incident photon; the change in wavelength ranges from 0 at $\theta = 0°$ to twice the Compton wavelength at $\theta = 180°$. Of course the descriptions in terms of energy and wavelength are equivalent, and the choice of which to use is merely a matter of convenience.

Using $E_e = K_e + m_e c^2$, where K_e is the kinetic energy of the electron, conservation of energy (Eq. 3.44a) can also be written as $E + m_e c^2 = E' + K_e + m_e c^2$. Solving for K_e, we obtain

$$K_e = E - E' \tag{3.50}$$

That is, the kinetic energy acquired by the electron is equal to the difference between the initial and final photon energies.

We can also find the direction of the electron's motion by dividing the two momentum relationships in Equation 3.45:

$$\tan \phi = \frac{p_e \sin \phi}{p_e \cos \phi} = \frac{p' \sin \theta}{p - p' \cos \theta} = \frac{E' \sin \theta}{E - E' \cos \theta} \tag{3.51}$$

where the last result comes from using $p = E/c$ and $p' = E'/c$.

▌Example 3.7

X rays of wavelength 0.2400 nm are Compton-scattered, and the scattered beam is observed at an angle of 60.0° relative to the incident beam. Find: (a) the wavelength of the scattered X rays, (b) the energy of the scattered X-ray photons, (c) the kinetic energy of the scattered electrons, and (d) the direction of travel of the scattered electrons.

Solution

(a) λ' can be found immediately from Eq. 3.49:

$$\lambda' = \lambda + \frac{h}{m_e c}(1 - \cos \theta)$$
$$= 0.2400 + (0.00243 \text{ nm})(1 - \cos 60°)$$
$$= 0.2412 \text{ nm} \quad .$$

(b) The energy E' can be found directly from λ' :

$$E' = \frac{hc}{\lambda'} = \frac{1240 \text{ eV} \cdot \text{nm}}{0.2412 \text{ nm}} = 5141 \text{ eV}$$

(c) The initial photon energy E is $hc/\lambda = 5167$ eV, so

$$K_e = E - E' = 5167 \text{ eV} - 5141 \text{ eV} = 26 \text{ eV}$$

(d) From Eq. 3.51,

$$\phi = \tan^{-1} \frac{E' \sin \theta}{E - E' \cos \theta}$$
$$= \tan^{-1} \frac{(5141 \text{ eV})(\sin 60°)}{(5167 \text{ eV}) - (5141 \text{ eV})(\cos 60°)}$$
$$= 59.7°$$

The first experimental demonstration of this type of scattering was done by Arthur Compton in 1923. A diagram of his experimental arrangement is shown in Figure 3.19. A beam of X rays of a single wavelength λ is incident on a scattering target, for which Compton used carbon. (Although no scattering target contains actual "free" electrons, the outer or valence electrons in many materials are very weakly attached to the atom and behave like nearly free electrons. The binding energies of these electrons in the atom are so small compared with the energies of the incident X-ray photons that they can be regarded as nearly "free" electrons.) A movable detector measured the energy of the scattered X rays at various angles θ.

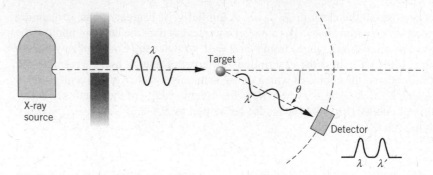

FIGURE 3.19 Schematic diagram of Compton-scattering apparatus. The wavelength λ' of the scattered X rays is measured by the detector, which can be moved to different positions θ. The wavelength difference $\lambda' - \lambda$ varies with θ.

Compton's original results are illustrated in Figure 3.20. At each angle, two peaks appear, corresponding to scattered X-ray photons with two different energies or wavelengths. The wavelength of one peak does not change as the angle is varied; this peak corresponds to scattering that involves "inner" electrons of the atom, which are more tightly bound to the atom so that the photon can scatter with no loss of energy. The wavelength of the other peak, however, varies strongly with angle; as can be seen from Figure 3.21, this variation is exactly as the Compton formula predicts.

Similar results can be obtained for the scattering of gamma rays, which are higher-energy (shorter wavelength) photons emitted in various radioactive decays. Compton also measured the variation in wavelength of scattered gamma rays, as illustrated in Figure 3.22. The change in wavelength in the

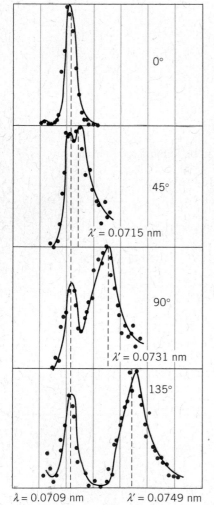

FIGURE 3.20 Compton's original results for X-ray scattering.

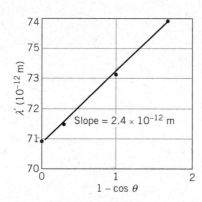

FIGURE 3.21 The scattered X-ray wavelengths λ', from Figure 3.20, for different scattering angles. The expected slope is 2.43×10^{-12} m, in agreement with the measured slope of Compton's data points.

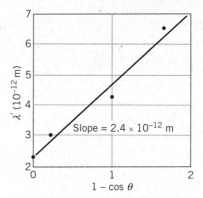

FIGURE 3.22 Compton's results for gamma-ray scattering. The wavelengths are much smaller than for X-rays, but the slope is the same as in Figure 3.21, which the Compton formula, Eq. 3.49, predicts.

gamma-ray measurements is identical with the change in wavelength in the X-ray measurements, as Eq. 3.49 predicts—the change in wavelength does not depend on the incident wavelength.

3.5 OTHER PHOTON PROCESSES

Although thermal radiation, the photoelectric effect, and Compton scattering provided the earliest experimental evidence in support of the quantization (particlelike behavior) of electromagnetic radiation, there are numerous other experiments that can also be interpreted correctly only if we assume the existence of photons as discrete quanta of electromagnetic radiation. In this section we discuss some of these processes, which cannot be understood if we consider only the wave nature of electromagnetic radiation. As you study the descriptions of these processes, note how photons interact with atoms or electrons by delivering energy in discrete bundles, in contrast to the wave interpretation in which the energy can be regarded as arriving continuously.

Interactions of Photons with Atoms

The emission of electromagnetic radiation from atoms takes place in discrete amounts characterized by one or more photons. When an atom emits a photon of energy E, the atom loses an equivalent amount of energy. Consider an atom at rest that has an initial energy E_i. The atom emits a photon of energy E. After the emission, the atom is left with a final energy E_f, which we will take as the energy associated with the internal structure of the atom. Because of conservation of momentum, the final atom must have a momentum that is equal and opposite to the momentum of the emitted photon, so the atom must also have a "recoil" kinetic energy K. (Normally this kinetic energy is very small.) Conservation of energy then gives

$$E_i = E_f + K + E \quad \text{or} \quad E = (E_i - E_f) - K \tag{3.52}$$

The energy of the emitted photon is equal to the net energy lost by the atom, minus a negligibly small contribution to the recoil kinetic energy of the atom.

In the reverse process, an atom can *absorb* a photon of energy E. If the atom is initially at rest, it must again acquire a small recoil kinetic energy in order to conserve momentum. Now conservation of energy gives

$$E_i + E = E_f + K \quad \text{or} \quad E_f - E_i = E - K \tag{3.53}$$

The energy available to add to the atom's internal supply of energy is the photon energy, less a recoil kinetic energy that is usually negligible.

Photon emission and absorption experiments are among the most important techniques for acquiring information about the internal structure of atoms, as we discuss in Chapter 6.

Bremsstrahlung and X-ray Production

When an electric charge, such as an electron, is accelerated or decelerated, it radiates electromagnetic energy; according to the quantum interpretation, we would say that it emits photons. Suppose we have a beam of electrons, which has been accelerated through a potential difference ΔV, so that the electrons experience a loss in potential energy of $-e \, \Delta V$ and thus acquire a kinetic energy of $K = e \, \Delta V$ (Figure 3.23). When the electrons strike a target they are slowed down and eventually come to rest, because they make collisions with the atoms of the target material. In such a collision, momentum is transferred to the atom, the electron slows down, and photons are emitted. The recoil kinetic energy of the atom is small (because the atom is so massive) and can safely be neglected. If the electron has a kinetic energy K before the encounter and if it leaves after the collision with a smaller kinetic energy K', then the photon energy $hf = hc/\lambda$ is

$$hf = \frac{hc}{\lambda} = K - K' \tag{3.54}$$

The amount of energy lost, and therefore the energy and wavelength of the emitted photon, are not uniquely determined, because K is the only known energy in Eq. 3.54. An electron usually will make many collisions, and therefore emit many different photons, before it is brought to rest; the photons then will range all the way from very small energies (large wavelengths) corresponding to small energy losses, up to a maximum photon energy hf_{max} equal to K, corresponding to an electron that loses all of its kinetic energy K in a single encounter (that is, when $K' = 0$). The smallest emitted wavelength λ_{min} is therefore determined by the maximum possible energy loss,

$$\lambda_{min} = \frac{hc}{K} = \frac{hc}{e \, \Delta V} \tag{3.55}$$

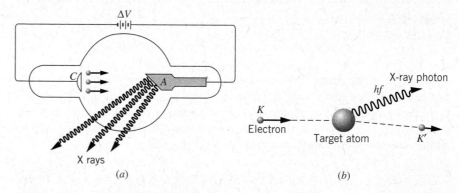

(a)   (b)

FIGURE 3.23   (a) Apparatus for producing bremsstrahlung. Electrons from a cathode C are accelerated to the anode A through the potential difference ΔV. When an electron encounters a target atom of the anode, it can lose energy, with the accompanying emission of an X-ray photon. (b) A schematic representation of the bremsstrahlung process.

For typical accelerating voltages in the range of 10,000 V, $\lambda_{\min}$ is in the range of a few tenths of nm, which corresponds to the X-ray region of the spectrum. This *continuous* distribution of X rays (which is very different from the *discrete* X-ray energies that are emitted in atomic transitions; more about these in Chapter 8) is called *bremsstrahlung*, which is German for braking, or decelerating, radiation. Some sample bremsstrahlung spectra are illustrated in Figure 3.24.

Symbolically we can write the bremsstrahlung process as

$$\text{electron} \rightarrow \text{electron} + \text{photon}$$

This is just the reverse process of the photoelectric effect, which is

$$\text{electron} + \text{photon} \rightarrow \text{electron}$$

However, neither process occurs for free electrons. In both cases there must be a heavy atom in the neighborhood to take care of the recoil momentum.

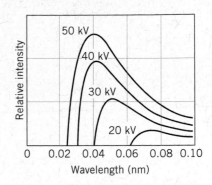

FIGURE 3.24 Some typical bremsstrahlung spectra. Each spectrum is labeled with the value of the accelerating voltage ΔV.

Pair Production and Annihilation

Another process that can occur when photons encounter atoms is *pair production*, in which the photon loses all its energy and in the process two particles are created: an electron and a positron. (A positron is a particle that is identical in mass to the electron but has a positive electric charge; more about *antiparticles* in Chapter 14.) Here we have an example of the creation of rest energy. The electron did not exist before the encounter of the photon with the atom (it was *not* an electron that was part of the atom). The photon energy hf is converted into the relativistic total energies E_+ and E_- of the positron and electron:

$$hf = E_+ + E_- = (m_e c^2 + K_+) + (m_e c^2 + K_-) \tag{3.56}$$

Because K_+ and K_- are always positive, the photon must have an energy of at least $2m_e c^2 = 1.02\,\text{MeV}$ in order for this process to occur; such high-energy photons are in the region of *nuclear gamma rays*. Symbolically,

$$\text{photon} \rightarrow \text{electron} + \text{positron}$$

This process, like bremsstrahlung, will not occur unless there is an atom nearby to supply the necessary recoil momentum. The reverse process,

$$\text{electron} + \text{positron} \rightarrow \text{photon}$$

also occurs; this process is known as *electron-positron annihilation* and can occur for free electrons and positrons as long as at least two photons are created. In this process the electron and positron disappear and are replaced by two photons. Conservation of energy requires that

$$(m_e c^2 + K_+) + (m_e c^2 + K_-) = E_1 + E_2 \tag{3.57}$$

where E_1 and E_2 are the photon energies. Usually the kinetic energies K_+ and K_- are negligibly small, so we can assume the positron and electron to be essentially at rest. Momentum conservation then requires the two photons to have equal and opposite momenta and thus equal energies. The two annihilation photons have equal energies of 0.511 MeV $(= m_e c^2)$ and move in exactly opposite directions.

3.6 WHAT IS A PHOTON?

We can describe photons by giving a few of their basic properties:

- like an electromagnetic wave, photons move with the speed of light;
- they have zero mass and rest energy;
- they carry energy and momentum, which are related to the frequency and wavelength of the electromagnetic wave by $E = hf$ and $p = h/\lambda$;
- they can be created or destroyed when radiation is emitted or absorbed;
- they can have particlelike collisions with other particles such as electrons.

In this chapter we have described some experiments that favor the photon interpretation of electromagnetic radiation, according to which the energy of the radiation is concentrated in small bundles. Other experiments, such as interference and diffraction, favor the wave interpretation, according to which the energy of the radiation is spread over its entire wavefront. For example, the explanation of the double-slit interference experiment requires that the wavefront be divided so that some of its intensity can pass through each slit. A particle must choose to go through one slit or the other; only a wave can go through both.

If we regard the wave and particle pictures as valid but exclusive alternatives, we must assume that the light emitted by a source must travel *either* as waves *or* as particles. How does the source know what kind of light (particles or waves) to emit? Suppose we place a double-slit apparatus on one side of the source and a photoelectric cell on the other side. Light emitted toward the double slit behaves like a wave and light emitted toward the photocell behaves like particles. How did the source know in which direction to aim the waves and in which direction to aim the particles?

Perhaps nature has a sort of "secret code" in which the kind of experiment we are doing is signaled back to the source so that it knows whether to emit particles or waves. Let us repeat our dual experiment with light from a distant galaxy, light that has been traveling toward us for a time roughly equal to the age of the universe (13×10^9 years). Surely the kind of experiment we are doing could not be signaled back to the limits of the known universe in the time it takes us to remove the double-slit apparatus from the laboratory table and replace it with the photoelectric apparatus. Yet we find that the starlight can produce both the double-slit interference and also the photoelectric effect.

Figure 3.25 shows a recent experiment that was designed to test whether this dual nature is an intrinsic property of light or of our apparatus. A light beam from a laser goes through a beam splitter, which separates the beam into two components (A and B). The mirrors reflect the two component beams so that they can recombine to form an interference pattern. In path A there is a switch that can deflect the beam into a detector. If the switch is off, beam A is not deflected and will combine with beam B to produce the interference pattern. If the switch is on,

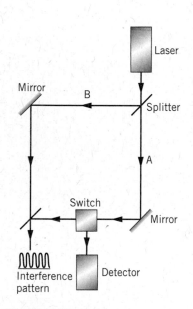

FIGURE 3.25 Apparatus for delayed choice experiment. Photons from the laser strike the beam splitter and can then travel paths A or B. The switch in path A can deflect the beam into a detector. If the switch is off, the beam on path A recombines with the beam on path B to form an interference pattern. [*Source:* A. Shimony, "The Reality of the Quantum World," *Scientific American* **258**, 46 (January 1988)].

beam A is deflected and observed in the detector, indicating that the light traveled a definite path, as would be characteristic of a particle. To put this another way, if the switch is off, the light beam is observed as a wave; if it is on, the light beam is observed as particles.

If light behaves like *particles,* the beam splitter sends it along *either* path A or path B; either path can be randomly chosen for the particle, but each particle can travel only one path. If light behaves like a *wave,* on the other hand, the beam splitter sends it along *both* paths, dividing its intensity between the two. Perhaps the beam splitter can somehow sense whether the switch is open or closed, so that it knows whether we are doing a particle-type or a wave-type experiment. If this were true, then the beam splitter would "know" whether to send all of the intensity down one path (so that we would observe a particle) or to split the intensity between the paths (so that we would observe a wave). However, in this experiment the experimenters used a very fast optical switch whose response time was shorter than the time it takes for light to travel through the apparatus to the switch. That is, the state of the switch could be changed *after* the light had already passed through the beam splitter, and so it was impossible for the beam splitter to "know" how the switch was set and thus whether a particle-type or a wave-type experiment was being done. This kind of experiment is called a "delayed choice" experiment, because the experimenter makes the choice of what kind of experiment to do after the light is already traveling on its way to the observation apparatus.

In this experiment, the investigators discovered that whenever they had the switch off, they observed the interference pattern characteristic of waves. When they had the switch on, they observed particles in the detector and no interference pattern. That is, whenever they did a wave-type experiment they observed waves, and whenever they did a particle-type experiment they observed particles. The wave and particle natures are both present simultaneously in the light, and this dual nature is clearly associated with the light and is not characteristic of the apparatus.

Many other experiments of this type have been done, and they all produce similar results. We are therefore trapped into an uncomfortable conclusion: Light is not *either* particles *or* waves; it is somehow *both* particles *and* waves, and only shows one or the other aspect, depending on the kind of experiment we are doing. A particle-type experiment shows the particle nature, while a wave-type experiment shows the wave nature. Our failure to classify light as *either* particle *or* wave is not so much a failure to understand the nature of light as it is a failure of our limited vocabulary (based on experiences with ordinary particles and waves) to describe a phenomenon that is more elegant and mysterious than either simple particles or waves.

Wave-Particle Duality

The dilemma of the dual particle+wave nature of light, which is called *wave-particle duality*, cannot be resolved with a simple explanation; physicists and philosophers have struggled with this problem ever since the quantum theory was introduced. The best we can do is to say that neither the wave nor the particle picture is wholly correct all of the time, that both are needed for a complete description of physical phenomena, and that in fact the two are *complementary* to one another.

Suppose we use a photographic film to observe the double-slit interference pattern. The film responds to individual photons. When a single photon is absorbed by the film, a single grain of the photographic emulsion is darkened; a complete picture requires a large number of grains to be darkened.

Let us imagine for the moment that we could see individual grains of the film as they absorbed photons and darkened, and let us do the double-slit experiment with a light source that is so weak that there is a relatively long time interval between photons. We would see first one grain darken, then another, and so forth, until after a large number of photons we would see the interference pattern begin to emerge. Some areas of the film (the interference maxima) show evidence for the arrival of a large number of photons, while in other areas (the interference minima) few photons arrive.

Alternatively, the wave picture of the double-slit experiment suggests that we could find the net electric field of the wave that strikes the screen by superimposing the electric fields of the portions of the incident wave fronts that pass through the two slits; the intensity or power in that combined wave could then be found by a procedure similar to Eqs. 3.7 through 3.10, and we would expect that the resultant intensity should show maxima and minima just like the observed double-slit interference pattern.

In summary, the correct explanation of the origin and appearance of the interference pattern comes from the wave picture, and the correct interpretation of the evolution of the pattern on the film comes from the photon picture; the two explanations, which according to our limited vocabulary and common-sense experience cannot simultaneously be correct, must somehow be taken together to give a complete description of the properties of electromagnetic radiation.

Keep in mind that "photon" and "wave" represent descriptions of the behavior of electromagnetic radiation when it encounters material objects. It is not correct to think of light as being "composed" of photons, just as we don't think of light as being "composed" of waves. The explanation in terms of photons applies to some interactions of radiation with matter, while the explanation in terms of waves applies to other interactions. For example, when we say that an atom "emits" a photon, we don't mean that there is a supply of photons stored within the atom; instead, we mean that the atom has given up a quantity of its internal energy to create an equivalent amount of energy in the form of electromagnetic radiation.

In the case of the double-slit experiment, we might reason as follows: the interaction between a "source" of radiation and the electromagnetic field is quantized, so that we can think of the emission of radiation by the atoms of the source in terms of individual photons. The interaction at the opposite end of the experiment, the photographic film, is also quantized, and we have the similarly useful view of atoms absorbing radiation as individual photons. In between, the electromagnetic radiation propagates smoothly and continuously as a wave and can show wave-type behavior (interference or diffraction) when it encounters the double slit.

Where the wave has large intensity, the film reveals the presence of many photons; where the wave has small intensity, few photons are observed. Recalling that the intensity of the wave is proportional to the square of its amplitude, we then have

$$\text{probability to observe photons} \propto |\text{electric field amplitude}|^2$$

It is this expression that provides the ultimate connection between the wave behavior and the particle behavior, and we will see in the next two chapters that a similar expression connects the wave and the particle aspects of those objects, such as electrons, which have been previously considered to behave as classical particles.

Chapter Summary

		Section			Section
Double-slit maxima	$y_n = n\dfrac{\lambda D}{d}$ $n = 0, 1, 2, 3, \ldots$	3.1	Rayleigh-Jeans formula	$I(\lambda) = \dfrac{2\pi c}{\lambda^4} kT$	3.3
Bragg's law for X-ray diffraction	$2d \sin\theta = n\lambda$ $n = 1, 2, 3, \cdots$	3.1	Planck's blackbody distribution	$I(\lambda) = \dfrac{2\pi hc^2}{\lambda^5} \dfrac{1}{e^{hc/\lambda kT} - 1}$	3.3
Energy of photon	$E = hf = hc/\lambda$	3.2	Compton scattering	$\dfrac{1}{E'} - \dfrac{1}{E} = \dfrac{1}{m_e c^2}(1 - \cos\theta),$	3.4
Maximum kinetic energy of photoelectrons	$K_{max} = eV_s = hf - \phi$	3.2		$\lambda' - \lambda = \dfrac{h}{m_e c}(1 - \cos\theta)$	
Cutoff wavelength	$\lambda_c = hc/\phi$	3.2	Bremsstrahlung	$\lambda_{min} = hc/K = hc/e\Delta V$	3.5
Stefan's law	$I = \sigma T^4$	3.3	Pair production	$hf = E_+ + E_- =$ $(m_e c^2 + K_+) + (m_e c^2 + K_-)$	3.5
Wien's displacement law	$\lambda_{max} T = 2.8978 \times 10^{-3}$ m·K	3.3	Electron-positron annihilation	$(m_e c^2 + K_+) + (m_e c^2 + K_-)$ $= E_1 + E_2$	3.5

Questions

1. The diameter of an atomic nucleus is about 10×10^{-15} m. Suppose you wanted to study the diffraction of photons by nuclei. What energy of photons would you choose? Why?

2. How is the wave nature of light unable to account for the observed properties of the photoelectric effect?

3. In the photoelectric effect, why do some electrons have kinetic energies smaller than K_{max}?

4. Why doesn't the photoelectric effect work for free electrons?

5. What does the work function tell us about the properties of a metal? Of the metals listed in Table 3.1, which has the least tightly bound electrons? Which has the most tightly bound?

6. Electric current is charge flowing per unit time. If we increase the kinetic energy of the photoelectrons (by increasing the energy of the incident photons), shouldn't the current increase, because the charge flows more rapidly? Why doesn't it?

7. What might be the effects on a photoelectric effect experiment if we were to double the frequency of the incident light? If we were to double the wavelength? If we were to double the intensity?

8. In the photoelectric effect, how can a photon moving in one direction eject an electron moving in a different direction? What happens to conservation of momentum?

9. In Figure 3.10, why does the photoelectric current rise slowly to its saturation value instead of rapidly, when the potential difference is greater than V_s? What does this figure indicate about the experimental difficulties that might arise from trying to determine V_s in this way?

10. Suppose that the frequency of a certain light source is just above the cutoff frequency of the emitter, so that the photoelectric effect occurs. To an observer in relative motion, the frequency might be Doppler shifted to a lower value that is below the cutoff frequency. Would this moving observer conclude that the photoelectric effect does not occur? Explain.

11. Why do cavities that form in a wood fire seem to glow brighter than the burning wood itself? Is the temperature in such cavities hotter than the surface temperature of the exposed burning wood?

12. What are the fields of classical physics on which the classical theory of blackbody radiation is based? Why don't

we believe that the "ultraviolet catastrophe" suggests that something is wrong with one of those classical theories?

13. In what region of the electromagnetic spectrum do room-temperature objects radiate? What problems would we have if our eyes were sensitive in that region?

14. How does the total intensity of thermal radiation vary when the temperature of an object is doubled?

15. Compton-scattered photons of wavelength λ' are observed at $90°$. In terms of λ', what is the scattered wavelength observed at $180°$?

16. The Compton-scattering formula suggests that objects viewed from different angles should show scattered light of different wavelengths. Why don't we observe a change in color of objects as we change the viewing angle?

17. You have a monoenergetic source of X rays of energy 84 keV, but for an experiment you need 70 keV X rays. How would you convert the X-ray energy from 84 to 70 keV?

18. TV sets with picture tubes can be significant emitters of X rays. What is the origin of these X rays? Estimate their wavelengths.

19. The X-ray peaks of Figure 3.20 are not sharp but are spread over a range of wavelengths. What reasons might account for that spreading?

20. A beam of photons passes through a block of matter. What are the three ways discussed in this chapter that the photons can lose energy in interacting with the material?

21. Of the photon processes discussed in this chapter (photoelectric effect, thermal radiation, Compton scattering, bremsstrahlung, pair production, electron-positron annihilation), which conserve momentum? Energy? Mass? Number of photons? Number of electrons? Number of electrons minus number of positrons?

Problems

3.1 Review of Electromagnetic Waves

1. A double-slit experiment is performed with sodium light ($\lambda = 589.0$ nm). The slits are separated by 1.05 mm, and the screen is 2.357 m from the slits. Find the separation between adjacent maxima on the screen.

2. In Example 3.1, what angle of incidence will produce the second-order Bragg peak?

3. Monochromatic X rays are incident on a crystal in the geometry of Figure 3.5. The first-order Bragg peak is observed when the angle of incidence is $34.0°$. The crystal spacing is known to be 0.347 nm. (a) What is the wavelength of the X rays? (b) Now consider a set of crystal planes that makes an angle of $45°$ with the surface of the crystal (as in Figure 3.6). For X rays of the same wavelength, find the angle of incidence measured from the surface of the crystal that produces the first-order Bragg peak. At what angle from the surface does the emerging beam appear in this case?

4. A certain device for analyzing electromagnetic radiation is based on the Bragg scattering of the radiation from a crystal. For radiation of wavelength 0.149 nm, the first-order Bragg peak appears centered at an angle of $15.15°$. The aperture of the analyzer passes radiation in the angular range of $0.015°$. What is the corresponding range of wavelengths passing through the analyzer?

3.2 The Photoelectric Effect

5. Find the momentum of (a) a 10.0-MeV gamma ray; (b) a 25-keV X ray; (c) a 1.0-μm infrared photon; (d) a 150-MHz radio-wave photon. Express the momentum in kg·m/s and eV/c.

6. Radio waves have a frequency of the order of 1 to 100 MHz. What is the range of energies of these photons? Our bodies are continuously bombarded by these photons. Why are they not dangerous to us?

7. (a) What is the wavelength of an X-ray photon of energy 10.0 keV? (b) What is the wavelength of a gamma-ray photon of energy 1.00 MeV? (c) What is the range of energies of photons of visible light with wavelengths 350 to 700 nm?

8. What is the cutoff wavelength for the photoelectric effect using an aluminum surface?

9. A metal surface has a photoelectric cutoff wavelength of 325.6 nm. It is illuminated with light of wavelength 259.8 nm. What is the stopping potential?

10. When light of wavelength λ illuminates a copper surface, the stopping potential is V. In terms of V, what will be the stopping potential if the same wavelength is used to illuminate a sodium surface?

11. The cutoff wavelength for the photoelectric effect in a certain metal is 254 nm. (a) What is the work function for that metal? (b) Will the photoelectric effect be observed for $\lambda > 254$ nm or for $\lambda < 254$ nm?

12. A surface of zinc is illuminated and photoelectrons are observed. (a) What is the largest wavelength that will cause photoelectrons to be emitted? (b) What is the stopping potential when light of wavelength 220.0 nm is used?

3.3 Blackbody Radiation

13. (a) Show that in the classical result for the energy distribution of the cavity wall oscillators (Eq. 3.32), the total number of oscillators at all energies is N. (b) Show that $E_{av} = kT$ for the classical oscillators.

14. (a) Writing the discrete Maxwell-Boltzmann distribution for Planck's cavity wall oscillators as $N_n = Ae^{-E_n/kT}$ (where A is a constant to be determined), show that the

condition $\sum_{n=0}^{\infty} N_n = N$ gives $A = N(1 - e^{-\varepsilon/kT})$ as in Eq. 3.38. [*Hint:* Use $\sum_{n=0}^{\infty} e^{nx} = (1 - e^x)^{-1}$]. (*b*) By taking the derivative with respect to x of the equation given in the hint, show that $\sum_{n=0}^{\infty} ne^{nx} = e^x/(1 - e^x)^2$. (*c*) Use this result to derive Eq. 3.40 from Eq. 3.39. (*d*) Show that $E_{av} \cong kT$ at large λ and $E_{av} \to 0$ for small λ.

15. By differentiating Eq. 3.41 show that $I(\lambda)$ has its maximum as expected according to Wien's displacement law, Eq. 3.27.

16. Integrate Eq. 3.41 to obtain Eq. 3.26. Use the definite integral $\int_0^{\infty} x^3 dx/(e^x - 1) = \pi^4/15$ to obtain Eq. 3.42 relating the Stefan-Boltzmann constant to Planck's constant.

17. Use the numerical value of the Stefan-Boltzmann constant to find the numerical value of Planck's constant from Eq. 3.42.

18. The surface of the Sun has a temperature of about 6000 K. At what wavelength does the Sun emit its peak intensity? How does this compare with the peak sensitivity of the human eye?

19. The universe is filled with thermal radiation, which has a blackbody spectrum at an effective temperature of 2.7 K (see Chapter 15). What is the peak wavelength of this radiation? What is the energy (in eV) of quanta at the peak wavelength? In what region of the electromagnetic spectrum is this peak wavelength?

20. (*a*) Assuming the human body (skin temperature 34°C) to behave like an ideal thermal radiator, find the wavelength where the intensity from the body is a maximum. In what region of the electromagnetic spectrum is radiation with this wavelength? (*b*) Making whatever (reasonable) assumptions you may need, estimate the power radiated by a typical person isolated from the surroundings. (*c*) Estimate the radiation power *absorbed* by a person in a room in which the temperature is 20°C.

21. A cavity is maintained at a temperature of 1650 K. At what rate does energy escape from the interior of the cavity through a hole in its wall of diameter 1.00 mm?

22. An analyzer for thermal radiation is set to accept wavelengths in an interval of 1.55 nm. What is the intensity of the radiation in that interval at a wavelength of 875 nm emitted from a glowing object whose temperature is 1675 K?

23. (*a*) Assuming the Sun to radiate like an ideal thermal source at a temperature of 6000 K, what is the intensity of the solar radiation emitted in the range 550.0 nm to 552.0 nm? (*b*) What fraction of the total solar radiation does this represent?

3.4 The Compton Effect

24. Show how Eq. 3.48 follows from Eq. 3.47.

25. Incident photons of energy 10.39 keV are Compton scattered, and the scattered beam is observed at 45.00° relative to the incident beam. (*a*) What is the energy of the scattered photons at that angle? (*b*) What is the kinetic energy of the scattered electrons?

26. X-ray photons of wavelength 0.02480 nm are incident on a target and the Compton-scattered photons are observed at 90.0°. (*a*) What is the wavelength of the scattered photons? (*b*) What is the momentum of the incident photons? Of the scattered photons? (*c*) What is the kinetic energy of the scattered electrons? (*d*) What is the momentum (magnitude and direction) of the scattered electrons?

27. High-energy gamma rays can reach a radiation detector by Compton scattering from the surroundings, as shown in Figure 3.26. This effect is known as *back-scattering*. Show that, when $E \gg m_e c^2$, the back-scattered photon has an energy of approximately 0.25 MeV, independent of the energy of the original photon, when the scattering angle is nearly 180°.

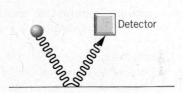

FIGURE 3.26 Problem 27.

28. Gamma rays of energy 0.662 MeV are Compton scattered. (*a*) What is the energy of the scattered photon observed at a scattering angle of 60.0°? (*b*) What is the kinetic energy of the scattered electrons?

3.5 Other Photon Processes

29. Suppose an atom of iron at rest emits an X-ray photon of energy 6.4 keV. Calculate the "recoil" momentum and kinetic energy of the atom. (*Hint:* Do you expect to need classical or relativistic kinetic energy for the atom? Is the kinetic energy likely to be much smaller than the atom's rest energy?)

30. What is the minimum X-ray wavelength produced in bremsstrahlung by electrons that have been accelerated through 2.50×10^4 V?

31. An atom absorbs a photon of wavelength 375 nm and immediately emits another photon of wavelength 580 nm. What is the net energy absorbed by the atom in this process?

General Problems

32. A certain green light bulb emits at a single wavelength of 550 nm. It consumes 55 W of electrical power and is 75% efficient in converting electrical energy into light. (*a*) How many photons does the bulb emit in one hour? (*b*) Assuming the emitted photons to be distributed uniformly in space, how many photons per second strike a 10 cm by 10 cm paper held facing the bulb at a distance of 1.0 m?

33. When sodium metal is illuminated with light of wavelength 4.20×10^2 nm, the stopping potential is found to be 0.65 V; when the wavelength is changed to 3.10×10^2 nm, the

stopping potential is 1.69 V. Using *only these data* and the values of the speed of light and the electronic charge, find the work function of sodium and a value of Planck's constant.

34. A photon of wavelength 192 nm strikes an aluminum surface along a line perpendicular to the surface and releases a photoelectron traveling in the opposite direction. Assume the recoil momentum is taken up by a single aluminum atom on the surface. Calculate the recoil kinetic energy of the atom. Would this recoil energy significantly affect the kinetic energy of the photoelectron?

35. A certain cavity has a temperature of 1150 K. (*a*) At what wavelength will the intensity of the radiation inside the cavity have its maximum value? (*b*) As a fraction of the maximum intensity, what is the intensity at twice the wavelength found in part (*a*)?

36. In Compton scattering, calculate the maximum kinetic energy given to the scattered electron for a given photon energy.

37. The COBE satellite was launched in 1989 to study the cosmic background radiation and measure its temperature. By measuring at many different wavelengths, researchers were able to show that the background radiation exactly followed the spectral distribution expected for a blackbody. At a wavelength of 0.133 cm, the radiant intensity is 1.440×10^{-7} W/m^2 in a wavelength interval of 0.00833 cm. What is the temperature of the radiation that would be deduced from these data?

38. The WMAP satellite launched in 2001 studied the cosmic microwave background radiation and was able to chart small fluctuations in the temperature of different regions of the background radiation. These fluctuations in temperature correspond to regions of large and small density in the early universe. The satellite was able to measure differences in temperature of 2×10^{-5} K at a temperature of 2.7250 K. At the peak wavelength, what is the difference in the radiation intensity per unit wavelength interval between the "hot" and "cold" regions of the background radiation?

39. You have been hired as an engineer on a NASA project to design a microwave spectrometer for an orbital mission to measure the cosmic background radiation, which has a blackbody spectrum with an effective temperature of 2.725 K. (*a*) The spectrometer is to scan the sky between wavelengths of 0.50 mm and 5.0 mm, and at each wavelength it accepts radiation in a wavelength range of 3.0×10^{-4} mm. What maximum and minimum radiation intensity do you expect to find in this region? (*b*) The photon detector in the spectrometer is in the form of a disk of diameter 0.86 cm. How many photons per second will the spectrometer record at its maximum and minimum intensities?

40. A photon of wavelength 7.52 pm scatters from a free electron at rest. After the interaction, the electron is observed to be moving in the direction of the original photon. Find the momentum of the electron.

41. A hydrogen atom is moving at a speed of 125.0 m/s. It absorbs a photon of wavelength 97 nm that is moving in the opposite direction. By how much does the speed of the atom change as a result of absorbing the photon?

42. Before a positron and an electron annihilate, they form a sort of "atom" in which each orbits about their common center of mass with identical speeds. As a result of this motion, the photons emitted in the annihilation show a small Doppler shift. In one experiment, the Doppler shift in energy of the photons was observed to be 2.41 keV. (*a*) What would be the speed of the electron or positron before the annihilation to produce this Doppler shift? (*b*) The positrons form these atom-like structures with the nearly "free" electrons in a solid. Assuming the positron and electron must have about the same speed to form this structure, find the kinetic energy of the electron. This technique, called "Doppler broadening," is an important method for learning about the energies of electrons in materials.

43. Prove that it is *not* possible to conserve both momentum and total relativistic energy in the following situation: A free electron moving at velocity $\vec{v}$ emits a photon and then moves at a slower velocity $\vec{v}'$.

44. A photon of energy E interacts with an electron at rest and undergoes pair production, producing a positive electron (positron) and an electron (in addition to the original electron):

$$\text{photon} + e^- \rightarrow e^+ + e^- + e^-$$

The two electrons and the positron move off with identical momenta in the direction of the initial photon. Find the kinetic energy of the three final particles and find the energy E of the photon. (*Hint:* Conserve momentum and total relativistic energy.)

Chapter 4

THE WAVELIKE PROPERTIES OF PARTICLES

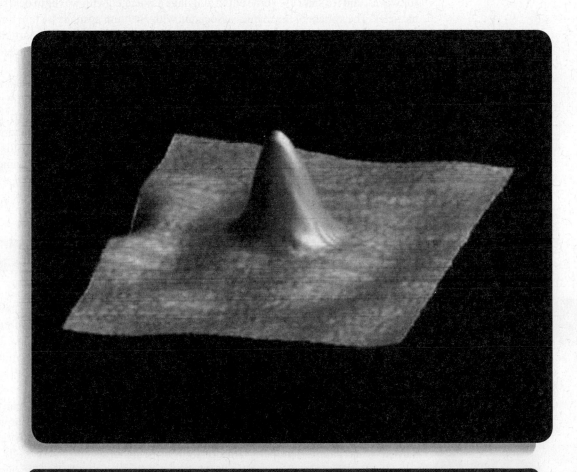

Just as we produce images from light waves that scatter from objects, we can also form images from "particle waves". The electron microscope produces images from electron waves that enable us to visualize objects on a scale that is much smaller than the wavelength of light. The ability to observe individual human cells and even sub-cellular objects such as chromosomes has revolutionized our understanding of biological processes. It is even possible to form images of a single atom, such as this cobalt atom on a gold surface. The ripples on the surface show electrons from gold atoms reacting to the presence of the intruder.

In classical physics, the laws describing the behavior of waves and particles are fundamentally different. Projectiles obey particle-type laws, such as Newtonian mechanics. Waves undergo interference and diffraction, which cannot be explained by the Newtonian mechanics associated with particles. The energy carried by a particle is confined to a small region of space; a wave, on the other hand, distributes its energy throughout space in its wavefronts. In describing the behavior of a particle we often want to specify its location, but this is not so easy to do for a wave. How would you describe the exact location of a sound wave or a water wave?

In contrast to this clear distinction found in classical physics, quantum physics requires that particles sometimes obey the rules that we have previously established for waves, and we shall use some of the language associated with waves to describe particles. The system of mechanics associated with quantum systems is sometimes called "wave mechanics" because it deals with the wavelike behavior of particles. In this chapter we discuss the experimental evidence in support of this wavelike behavior for particles such as electrons.

As you study this chapter, notice the frequent references to such terms as the *probability* of the outcome of a measurement, the *average* of many repetitions of a measurement, and the *statistical* behavior of a system. These terms are fundamental to quantum mechanics, and you cannot begin to understand quantum behavior until you feel comfortable with discarding such classical notions as fixed trajectories and certainty of outcome, while substituting the quantum mechanical notions of probability and statistically distributed outcomes.

4.1 DE BROGLIE'S HYPOTHESIS

Progress in physics often can be characterized by long periods of experimental and theoretical drudgery punctuated occasionally by flashes of insight that cause profound changes in the way we view the universe. Frequently the more profound the insight and the bolder the initial step, the simpler it seems in historical perspective, and the more likely we are to sit back and wonder, "Why didn't I think of that?" Einstein's special theory of relativity is one example of such insight; the hypothesis of the Frenchman Louis de Broglie is another.*

In the previous chapter we discussed the double-slit experiment (which can be understood only if light behaves as a wave) and the photoelectric and Compton effects (which can be understood only if light behaves as a particle). Is this dual particle-wave nature a property only of light or of material objects as well? In a bold and daring hypothesis in his 1924 doctoral dissertation, de Broglie chose the latter alternative. Examining Eq. 3.20, $E = hf$, and Eq. 3.22, $p = h/\lambda$, we find some difficulty in applying the first equation in the case of particles, for we cannot be sure whether E should be the kinetic energy, total energy, or total relativistic energy (all, of course, are identical for light). No such difficulties arise from the second relationship. De Broglie suggested, lacking any experimental evidence in

Louis de Broglie (1892–1987, France). A member of an aristocratic family, his work contributed substantially to the early development of the quantum theory.

---

*De Broglie's name should be pronounced "deh-BROY" or "deh-BROY-eh," but it is often said as "deh-BROH-lee."

support of his hypothesis, that associated with any material particle moving with momentum p there is a wave of wavelength λ, related to p according to

$$\lambda = \frac{h}{p} \qquad (4.1)$$

where h is Planck's constant. The wavelength λ of a particle computed according to Eq. 4.1 is called its *de Broglie wavelength*.

Example 4.1

Compute the de Broglie wavelength of the following: (*a*) A 1000-kg automobile traveling at 100 m/s (about 200 mi/h). (*b*) A 10-g bullet traveling at 500 m/s. (*c*) A smoke particle of mass 10^{-9} g moving at 1 cm/s. (*d*) An electron with a kinetic energy of 1 eV. (*e*) An electron with a kinetic energy of 100 MeV.

Solution

(*a*) Using the classical relation between velocity and momentum,

$$\lambda = \frac{h}{p} = \frac{h}{mv} = \frac{6.6 \times 10^{-34}\, \text{J} \cdot \text{s}}{(10^3\, \text{kg})(100\, \text{m/s})} = 6.6 \times 10^{-39}\, \text{m}$$

(*b*) As in part (*a*),

$$\lambda = \frac{h}{mv} = \frac{6.6 \times 10^{-34}\, \text{J} \cdot \text{s}}{(10^{-2}\, \text{kg})(500\, \text{m/s})} = 1.3 \times 10^{-34}\, \text{m}$$

(*c*)

$$\lambda = \frac{h}{mv} = \frac{6.6 \times 10^{-34}\, \text{J} \cdot \text{s}}{(10^{-12}\, \text{kg})(10^{-2}\, \text{m/s})} = 6.6 \times 10^{-20}\, \text{m}$$

(*d*) The rest energy (mc^2) of an electron is 5.1×10^5 eV. Because the kinetic energy (1 eV) is much less than the rest energy, we can use nonrelativistic kinematics.

$$p = \sqrt{2mK}$$

$$= \sqrt{2(9.1 \times 10^{-31}\, \text{kg})(1\ \text{eV})(1.6 \times 10^{-19}\, \text{J/eV})}$$

$$= 5.4 \times 10^{-25}\, \text{kg} \cdot \text{m/s}$$

Then,

$$\lambda = \frac{h}{p} = \frac{6.6 \times 10^{-34}\, \text{J} \cdot \text{s}}{5.4 \times 10^{-25}\, \text{kg} \cdot \text{m/s}}$$

$$= 1.2 \times 10^{-9}\, \text{m} = 1.2\, \text{nm}$$

We can also find this solution in the following way, using $p = \sqrt{2mK}$ and $hc = 1240$ eV $\cdot$ nm.

$$cp = c\sqrt{2mK} = \sqrt{2(mc^2)K}$$

$$= \sqrt{2(5.1 \times 10^5\ \text{eV})(1\ \text{eV})} = 1.0 \times 10^3\ \text{eV}$$

$$\lambda = \frac{h}{p} = \frac{hc}{pc} = \frac{1240\ \text{eV} \cdot \text{nm}}{1.0 \times 10^3\ \text{eV}} = 1.2\, \text{nm}$$

This method may seem artificial at first, but with practice it becomes quite useful, especially because energies are usually given in electron-volts in atomic and nuclear physics.

(*e*) In this case, the kinetic energy is much greater than the rest energy, and so we are in the extreme relativistic realm, where $K \cong E \cong pc$, as in Eq. 2.40. The wavelength is

$$\lambda = \frac{hc}{pc} = \frac{1240\ \text{MeV} \cdot \text{fm}}{100\ \text{MeV}} = 12\, \text{fm}$$

Note that the wavelengths computed in parts (*a*), (*b*), and (*c*) are far too small to be observed in the laboratory. Only in the last two cases, in which the wavelength is of the same order as atomic or nuclear sizes, do we have any chance of observing the wavelength. *Because of the smallness of h, only for particles of atomic or nuclear size will the wave behavior be observable.*

Two questions immediately follow. First, just what sort of wave is it that has this de Broglie wavelength? That is, what does the amplitude of the de Broglie wave measure? We'll discuss the answer to this question later in this chapter. For now, we assume that, associated with the particle as it moves, there is a de Broglie wave of wavelength λ, which shows itself *when a wave-type experiment (such as diffraction) is performed on it.* The outcome of the wave-type experiment depends on this wavelength. The de Broglie wavelength, which characterizes the wave-type behavior of particles, is central to the quantum theory.

The second question then occurs: Why was this wavelength not directly observed before de Broglie's time? As parts (*a*), (*b*), and (*c*) of Example 4.1 showed, for ordinary objects the de Broglie wavelength is very small. Suppose we tried to demonstrate the wave nature of these objects through a double-slit type of experiment. Recall from Eq. 3.16 that the spacing between adjacent fringes in a double-slit experiment is $\Delta y = \lambda D / d$. Putting in reasonable values for the slit separation d and slit-to-screen distance D, you will find that there is no achievable experimental configuration that can produce an observable separation of the fringes (see Problem 9). *There is no experiment that can be done to reveal the wave nature of macroscopic (laboratory-sized) objects.* Experimental verification of de Broglie's hypothesis comes only from experiments with objects on the atomic scale, which are discussed in the next section.

4.2 EXPERIMENTAL EVIDENCE FOR DE BROGLIE WAVES

The indications of wave behavior come mostly from interference and diffraction experiments. Double-slit interference, which was reviewed in Section 3.1, is perhaps the most familiar type of interference experiment, but the experimental difficulties of constructing double slits to do interference experiments with beams of atomic or subatomic particles were not solved until long after the time of de Broglie's hypothesis. We discuss these experiments later in this section. First we'll discuss diffraction experiments with electrons.

Particle Diffraction Experiments

Diffraction of light waves is discussed in most introductory physics texts and is illustrated in Figure 4.1 for light diffracted by a single slit. For light of wavelength λ incident on a slit of width a, the diffraction minima are located at angles given by

$$a \sin \theta = n\lambda \qquad n = 1, 2, 3, \ldots \qquad (4.2)$$

on either side of the central maximum. Note that most of the light intensity falls in the central maximum.

The experiments that first verified de Broglie's hypothesis involve *electron diffraction,* not through an artificially constructed single slit (as for the diffraction pattern in Figure 4.1) but instead through the atoms of a crystal. The outcomes of these experiments resemble those of the similar X-ray diffraction experiments illustrated in Section 3.1.

In an electron diffraction experiment, a beam of electrons is accelerated from rest through a potential difference ΔV, acquiring a nonrelativistic kinetic energy $K = e \Delta V$ and a momentum $p = \sqrt{2mK}$. Wave mechanics would describe the beam of electrons as a wave of wavelength $\lambda = h/p$. The beam strikes a

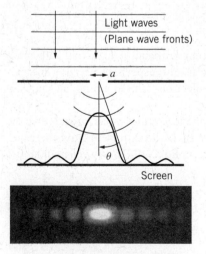

FIGURE 4.1   Light waves (represented as plane wave fronts) are incident on a narrow slit of width a. Diffraction causes the waves to spread after passing through the slit, and the intensity varies along the screen. The photograph shows the resulting intensity pattern.

crystal, and the scattered beam is photographed (Figure 4.2). The similarity between electron diffraction patterns (Figure 4.2) and X-ray diffraction patterns (Figure 3.7) strongly suggests that the electrons are behaving as waves.

The "rings" produced in X-ray diffraction of polycrystalline materials (Figure 3.8b) are also produced in electron diffraction, as shown in Figure 4.3, again providing strong evidence for the similarity in the wave behavior of electrons and X rays. Experiments of the type illustrated in Figure 4.3 were first done in 1927 by G. P. Thomson, who shared the 1937 Nobel Prize for this work. (Thomson's father, J. J. Thomson, received the 1906 Nobel Prize for his discovery of the electron and measurement of its charge-to-mass ratio. Thus it can be said that Thomson, the father, discovered the particle nature of the electron, while Thomson, the son, discovered its wave nature.)

An electron diffraction experiment gave the first experimental confirmation of the wave nature of electrons (and the quantitative confirmation of the de Broglie relationship $\lambda = h/p$) soon after de Broglie's original hypothesis. In 1926, at the Bell Telephone Laboratories, Clinton Davisson and Lester Germer were investigating the reflection of electron beams from the surface of nickel crystals. A schematic view of their apparatus is shown in Figure 4.4. A beam of electrons from a heated filament is accelerated through a potential difference ΔV. After passing through a small aperture, the beam strikes a single crystal of nickel. Electrons are scattered in all directions by the atoms of the crystal, some of them striking a detector, which can be moved to any angle ϕ relative to the incident beam and which measures the intensity of the electron beam scattered at that angle.

Figure 4.5 shows the results of one of the experiments of Davisson and Germer. When the accelerating voltage is set at 54 V, there is an intense reflection of the beam at the angle $\phi = 50°$. Let's see how these results give confirmation of the de Broglie wavelength.

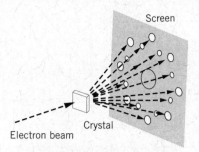

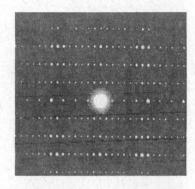

FIGURE 4.2 (Top) Electron diffraction apparatus. (Bottom) Electron diffraction pattern. Each bright dot is a region of constructive interference, as in the X-ray diffraction patterns of Figure 3.7. The target is a crystal of $Ti_2Nb_{10}O_{29}$.

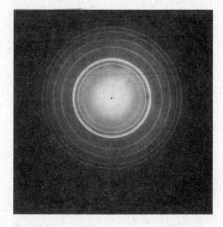

FIGURE 4.3 Electron diffraction of polycrystalline beryllium. Note the similarity between this pattern and the pattern for X-ray diffraction of a polycrystalline material (Figure 3.8b).

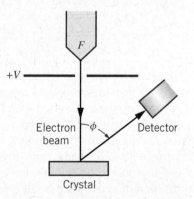

FIGURE 4.4 Apparatus used by Davisson and Germer to study electron diffraction. Electrons leave the filament F and are accelerated by the voltage V. The beam strikes a crystal and the scattered beam is detected at an angle ϕ relative to the incident beam. The detector can be moved in the range 0 to 90°.

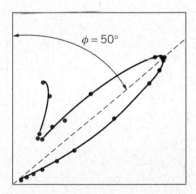

FIGURE 4.5 Results of Davisson and Germer. Each point on the plot represents the relative intensity when the detector in Figure 4.4 is located at the corresponding angle ϕ measured from the vertical axis. Constructive interference causes the intensity of the reflected beam to reach a maximum at $\phi = 50°$ for $V = 54$ V.

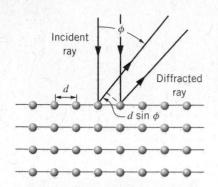

FIGURE 4.6 The crystal surface acts like a diffraction grating with spacing d.

FIGURE 4.7 Diffraction of neutrons by a sodium chloride crystal.

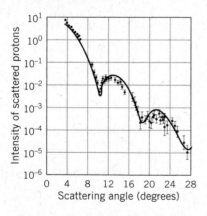

FIGURE 4.8 Diffraction of 1-GeV protons by oxygen nuclei. The pattern of maxima and minima is similar to that of single-slit diffraction of light waves. [*Source:* H. Palevsky et al., *Physical Review Letters* **18**, 1200 (1967).]

Each of the atoms of the crystal can act as a scatterer, so the scattered *electron waves* can interfere, and we have a crystal diffraction grating for the electrons. Figure 4.6 shows a simplified representation of the nickel crystal used in the Davisson-Germer experiment. Because the electrons were of low energy, they did not penetrate very far into the crystal, and it is sufficient to consider the diffraction to take place in the plane of atoms on the surface. The situation is entirely similar to using a reflection-type diffraction grating for light; the spacing d between the rows of atoms on the crystal is analogous to the spacing between the slits in the optical grating. The maxima for a diffraction grating occur at angles ϕ such that the path difference between adjacent rays $d \sin \phi$ is equal to a whole number of wavelengths:

$$d \sin \phi = n\lambda \qquad n = 1, 2, 3, \ldots \qquad (4.3)$$

where n is the order number of the maximum.

From independent data, it is known that the spacing between the rows of atoms in a nickel crystal is $d = 0.215\,\text{nm}$. The peak at $\phi = 50°$ must be a first-order peak ($n = 1$), because no peaks were observed at smaller angles. If this is indeed an interference maximum, the corresponding wavelength is, from Eq. 4.3,

$$\lambda = d \sin \phi = (0.215\,\text{nm})(\sin 50°) = 0.165\,\text{nm}$$

We can compare this value with that expected on the basis of the de Broglie theory. An electron accelerated through a potential difference of 54 V has a kinetic energy of 54 eV and therefore a momentum of

$$p = \sqrt{2mK} = \frac{1}{c}\sqrt{2mc^2K} = \frac{1}{c}\sqrt{2(511,000\,\text{eV})(54\,\text{eV})} = \frac{1}{c}(7430\,\text{eV})$$

The de Broglie wavelength is $\lambda = h/p = hc/pc$. Using $hc = 1240\,\text{eV} \cdot \text{nm}$,

$$\lambda = \frac{hc}{pc} = \frac{1240\,\text{eV} \cdot \text{nm}}{7430\,\text{eV}} = 0.167\,\text{nm}$$

This is in excellent agreement with the value found from the diffraction maximum, and provides strong evidence in favor of the de Broglie theory. For this experimental work, Davisson shared the 1937 Nobel Prize with G. P. Thomson.

The wave nature of particles is not exclusive to electrons; *any* particle with momentum p has de Broglie wavelength h/p. Neutrons are produced in nuclear reactors with kinetic energies corresponding to wavelengths of roughly 0.1 nm; these also should be suitable for diffraction by crystals. Figure 4.7 shows that diffraction of neutrons by a salt crystal produces the same characteristic patterns as the diffraction of electrons or X rays. Clifford Shull shared the 1994 Nobel Prize for the development of the neutron diffraction technique.

To study the nuclei of atoms, much smaller wavelengths are needed, of the order of 10^{-15} m. Figure 4.8 shows the diffraction pattern produced by the scattering of 1-GeV kinetic energy protons by oxygen nuclei. Maxima and minima of the diffracted intensity appear in a pattern similar to the single-slit diffraction shown in Figure 4.1. (The intensity at the minima does not fall to zero because nuclei do not have a sharp boundary. The determination of nuclear sizes from such diffraction patterns is discussed in Chapter 12.)

Example 4.2

Protons of kinetic energy 1.00 GeV were diffracted by oxygen nuclei, which have a radius of 3.0 fm, to produce the data shown in Figure 4.8. Calculate the expected angles where the first three diffraction minima should appear.

Solution

The total relativistic energy of the protons is $E = K + mc^2 = 1.00\,\text{GeV} + 0.94\,\text{GeV} = 1.94\,\text{GeV}$ is, so their momentum is

$$p = \frac{1}{c}\sqrt{E^2 - (mc^2)^2}$$
$$= \frac{1}{c}\sqrt{(1.94\,\text{GeV})^2 - (0.94\,\text{GeV})^2} = 1.70\,\text{GeV}/c$$

The corresponding de Broglie wavelength is

$$\lambda = \frac{h}{p} = \frac{hc}{pc} = \frac{1240\,\text{MeV}\cdot\text{fm}}{1700\,\text{MeV}} = 0.73\,\text{fm}$$

We can represent the oxygen nuclei as circular disks, for which the diffraction formula is a bit different from Eq. 4.2:

$a \sin\theta = 1.22n\lambda$, where a is the diameter of the diffracting object. Based on this formula, the first diffraction minimum ($n = 1$) should appear at the angle

$$\sin\theta = \frac{1.22n\lambda}{a} = \frac{(1.22)(1)(0.73\,\text{fm})}{6.0\,\text{fm}} = 0.148$$

or $\theta = 8.5°$. Because the sine of the diffraction angle is proportional to the index n, the $n = 2$ minimum should appear at the angle where $\sin\theta = 2 \times 0.148 = 0.296$ ($\theta = 17.2°$), and the $n = 3$ minimum where $\sin\theta = 3 \times 0.148 = 0.444$ ($\theta = 26.4°$).

From the data in Figure 4.8, we see the first diffraction minimum at an angle of about $10°$, the second at about $18°$, and the third at about $27°$, all in very good agreement with the expected values. The data don't exactly follow the formula for diffraction by a disk, because nuclei don't behave quite like disks. In particular, they have diffuse rather than sharp edges, which prevents the intensity at the diffraction minima from falling to zero and also alters slightly the locations of the minima.

Double-Slit Experiments with Particles

The definitive evidence for the wave nature of *light* was deduced from the double-slit experiment performed by Thomas Young in 1801 (discussed in Section 3.1). In principle, it should be possible to do double-slit experiments with *particles* and thereby directly observe their wavelike behavior. However, the technological difficulties of producing double slits for particles are formidable, and such experiments did not become possible until long after the time of de Broglie. The first double-slit experiment with electrons was done in 1961. A diagram of the apparatus is shown in Figure 4.9. The electrons from a hot filament were accelerated through 50 kV (corresponding to $\lambda = 5.4$ pm) and then passed through a double slit of separation 2.0 μm and width 0.5 μm. A photograph of the resulting intensity pattern is shown in Figure 4.10. The similarity with the double-slit pattern for light (Figure 3.2) is striking.

A similar experiment can be done for neutrons. A beam of neutrons from a nuclear reactor can be slowed to a room-temperature "thermal" energy distribution (average $K \approx kT \approx 0.025$ eV), and a specific wavelength can be selected by a scattering process similar to Bragg diffraction (see Eq. 3.18 and Problem 32 at the end of the present chapter). In one experiment, neutrons of kinetic energy 0.00024 eV and de Broglie wavelength 1.85 nm passed through a gap of diameter 148 μm in a material that absorbs virtually all of the neutrons incident on it (Figure 4.11). In the center of the gap was a boron wire (also highly absorptive for neutrons) of diameter 104 μm. The neutrons could pass on either side of the wire through slits of width 22 μm. The intensity of neutrons that pass through this double slit was observed by sliding

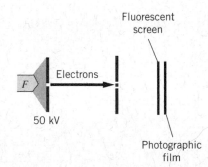

FIGURE 4.9   Double-slit apparatus for electrons. Electrons from the filament F are accelerated through 50 kV and pass through the double slit. They produce a visible pattern when they strike a fluorescent screen (like a TV screen), and the resulting pattern is photographed. A photograph is shown in Figure 4.10. [See C. Jonsson, *American Journal of Physics* **42**, 4 (1974).]

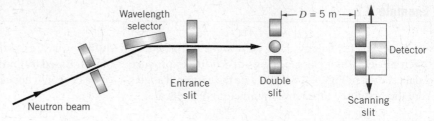

FIGURE 4.10   Double-slit interference pattern for electrons.

FIGURE 4.11 Double-slit apparatus for neutrons. Thermal neutrons from a reactor are incident on a crystal; scattering through a particular angle selects the energy of the neutrons. After passing through the double slit, the neutrons are counted by the scanning slit assembly, which moves laterally.

another slit across the beam and measuring the intensity of neutrons passing through this "scanning slit." Figure 4.12 shows the resulting pattern of intensity maxima and minima, which leaves no doubt that interference is occurring and that the neutrons have a corresponding wave nature. The wavelength can be deduced from the slit separation using Eq. 3.16 to obtain the spacing between adjacent maxima, $\Delta y = y_{n+1} - y_n$. Estimating the spacing Δy from Figure 4.12 to be about 75 μm, we obtain

$$\lambda = \frac{d\Delta y}{D} = \frac{(126\,\mu\mathrm{m})(75\,\mu\mathrm{m})}{5\,\mathrm{m}} = 1.89\,\mathrm{nm}$$

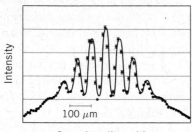

FIGURE 4.12   Intensity pattern observed for double-slit interference with neutrons. The spacing between the maxima is about 75 μm. [*Source:* R. Gahler and A. Zeilinger, *American Journal of Physics* **59**, 316 (1991).]

This result agrees very well with the de Broglie wavelength of 1.85 nm selected for the neutron beam.

It is also possible to do a similar experiment with atoms. In this case, a source of helium atoms formed a beam (of velocity corresponding to a kinetic energy of 0.020 eV) that passed through a double slit of separation 8 μm and width 1 μm. Again a scanning slit was used to measure the intensity of the beam passing through the double slit. Figure 4.13 shows the resulting intensity pattern. Although the results are not as dramatic as those for electrons and neutrons, there is clear evidence of interference maxima and minima, and the separation of the maxima gives a wavelength that is consistent with the de Broglie wavelength (see Problem 8).

Diffraction can be observed with even larger objects. Figure 4.14 shows the pattern produced by fullerene molecules (C_{60}) in passing through a diffraction grating with a spacing of $d = 100$ nm. The diffraction pattern was observed at a distance of 1.2 m from the grating. Estimating the separation of the maxima in Figure 4.14 as 50 μm, we get the angular separation of the maxima to be $\theta \approx \tan\theta = (50\,\mu\mathrm{m})/(1.2\,\mathrm{m}) = 4.2 \times 10^{-5}$ rad, and thus $\lambda = d\sin\theta = 4.2$ pm. For C_{60} molecules with a speed of 117 m/s used in this experiment, the expected de Broglie wavelength is 4.7 pm, in good agreement with our estimate from the diffraction pattern.

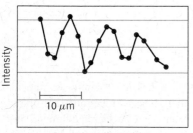

FIGURE 4.13   Intensity pattern observed for double-slit interference with helium atoms. [*Source:* O. Carnal and J. Mlynek, *Physical Review Letters* **66**, 2689 (1991).]

In this chapter we have discussed several interference and diffraction experiments using different particles—electrons, protons, neutrons, atoms, and molecules. These experiments are not restricted to any particular type of particle or to any particular type of observation. They are examples of a *general* phenomenon, the wave nature of particles, that was unobserved before 1920 because the necessary experiments had not yet been done. Today this wave nature is used as a basic tool by scientists. For example, neutron diffraction

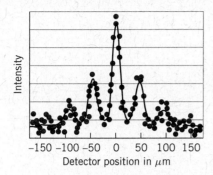

FIGURE 4.14 Diffraction grating pattern produced by C_{60} molecules. [*Source:* O. Nairz, M. Arndt, and A. Zeilinger, *American Journal of Physics* **71**, 319 (2003).]

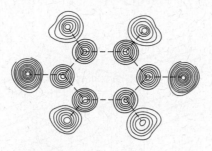

FIGURE 4.15   The atomic structure of solid benzene as deduced from neutron diffraction. The circles indicate contours of constant density. The black circles show the locations of the six carbon atoms that form the familiar benzene ring. The blue circles show the locations of the hydrogen atoms.

FIGURE 4.16   Electron microscope image of bacteria on the surface of a human tongue. The magnification here is about a factor of 5000.

gives detailed information on the structure of solid crystals and of complex molecules (Figure 4.15). The electron microscope uses electron waves to illuminate and form an image of objects; because the wavelength can be made thousands of times smaller than that of visible light, it is possible to resolve and observe small details that are not observable with visible light (Figure 4.16).

Through Which Slit Does the Particle Pass?

When we do a double-slit experiment with particles such as electrons, it is tempting to try to determine through which slit the particle passes. For example, we could surround each slit with an electromagnetic loop that causes a meter to deflect whenever a charged particle or perhaps a particle with a magnetic moment passes through the loop (Figure 4.17). If we fired the particles through the slits at a slow enough rate, we could track each particle as it passed through one slit or the other and then appeared on the screen.

   If we performed this imaginary experiment, the result would no longer be an interference pattern on the screen. Instead, we would observe a pattern similar to that shown in Figure 4.17, with "hits" in front of each slit, but no interference fringes. No matter what sort of device we use to determine through which slit the particle passes, the interference pattern will be destroyed. The classical *particle* must pass through one slit or the other; only a *wave* can reveal interference, which depends on parts of the wavefront passing through *both* slits and then recombining.

   When we ask through which slit the particle passed, we are investigating only the *particle* aspects of its behavior, and we cannot observe its wave nature (the interference pattern). Conversely, when we study the wave nature, we cannot simultaneously observe the particle nature. The electron will behave as a particle *or* a wave, but we cannot observe *both* aspects of its behavior simultaneously. This curious aspect of quantum mechanics was also discussed for photons in Section 3.6, where we discovered that experiments can reveal either the particle nature of the photon or its wave nature, but not both aspects simultaneously.

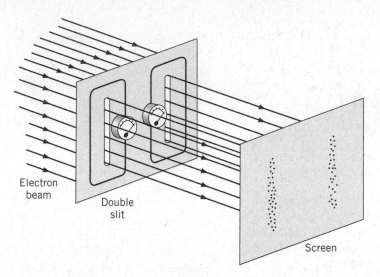

FIGURE 4.17 Apparatus to record passage of electrons through slits. Each slit is surrounded by a loop with a meter that signals the passage of an electron through the slit. No interference fringes are seen on the screen.

This is the basis for the *principle of complementarity,* which asserts that the complete description of a photon or a particle such as an electron cannot be made in terms of only particle properties or only wave properties, but that both aspects of its behavior must be considered. Moreover, the particle and wave natures cannot be observed simultaneously, and the type of behavior that we observe depends on the kind of experiment we are doing: a particle-type experiment shows only particle like behavior, and a wave-type experiment shows only wavelike behavior.

4.3 UNCERTAINTY RELATIONSHIPS FOR CLASSICAL WAVES

In quantum mechanics, we want to use de Broglie waves to describe particles. In particular, the amplitude of the wave will tell us something about the location of the particle. Clearly a pure sinusoidal wave, as in Figure 4.18a, is not much use in locating a particle—the wave extends from $-\infty$ to $+\infty$, so the particle might be found anywhere in that region. On the other hand, a narrow wave pulse like Figure 4.18b does a pretty good job of locating the particle in a small region of space, but this wave does not have an easily identifiable wavelength. In the first case, we know the wavelength exactly but have no knowledge of the location of the particle, while in the second case we have a good idea of the location of the particle but a poor knowledge of its wavelength. Because wavelength is associated with momentum by the de Broglie relationship (Eq. 4.1), a poor knowledge of the wavelength is associated with a poor knowledge of the particle's momentum. For a classical particle, we would like to know both its location and its momentum as precisely as possible. For a quantum particle, we are going to have to make some compromises—the better we know its momentum (or wavelength), the less we know about its location. We can improve our knowledge of its location only at the expense of our knowledge of its momentum.

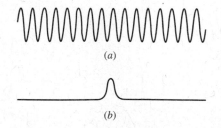

(a)

(b)

FIGURE 4.18 (*a*) A pure sine wave, which extends from $-\infty$ to $+\infty$. (*b*) A narrow wave pulse.

This competition between knowledge of location and knowledge of wavelength is not restricted to de Broglie waves—classical waves show the same effect. All real waves can be represented as *wave packets*—disturbances that are localized to a finite region of space. We will discuss more about constructing wave packets in Section 4.5. In this section we will examine this competition between specifying the location and the wavelength of classical waves more closely.

Figure 4.19*a* shows a very small wave packet. The disturbance is well localized to a small region of space of length Δx. (Imagine listening to a very short burst of sound, of such brief duration that it is hard for you to recognize the pitch or frequency of the wave.) Let's try to measure the wavelength of this wave packet. Placing a measuring stick along the wave, we have some difficulty defining exactly where the wave starts and where it ends. Our measurement of the wavelength is therefore subject to a small *uncertainty* $\Delta\lambda$. Let's represent this uncertainty as a fraction ε of the wavelength λ, so that $\Delta\lambda \sim \varepsilon\lambda$. The fraction ε is certainly less than 1, but it is probably greater than 0.01, so we estimate that $\varepsilon \sim 0.1$ to within an order of magnitude. (In our discussion of uncertainty, we use the $\sim$ symbol to indicate a rough order-of-magnitude estimate.) That is, the uncertainty in our measurement of the wavelength might be roughly 10% of the wavelength.

The size of this wave disturbance is roughly one wavelength, so $\Delta x \approx \lambda$. For this discussion we want to examine the product of the size of the wave packet and the uncertainty in the wavelength, Δx times $\Delta\lambda$ with $\Delta x \approx \lambda$ and $\Delta\lambda \sim \varepsilon\lambda$:

$$\Delta x \Delta\lambda \sim \varepsilon\lambda^2 \tag{4.4}$$

This expression shows the inverse relationship between the size of the wave packet and the uncertainty in the wavelength: for a given wavelength, the smaller the size of the wave packet, the greater the uncertainty in our knowledge of the wavelength. That is, as Δx gets smaller, $\Delta\lambda$ must become larger.

Making a larger wave packet doesn't help us at all. Figure 4.19*b* shows a larger wave packet with the same wavelength. Suppose this larger wave packet contains

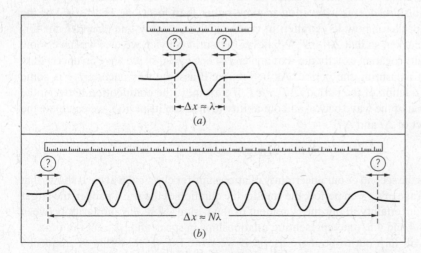

FIGURE 4.19 (*a*) Measuring the wavelength of a wave represented by a small wave packet of length roughly one wavelength. (*b*) Measuring the wavelength of a wave represented by a large wave packet consisting of N waves.

N cycles of the wave, so that $\Delta x \approx N\lambda$. Again using our measuring stick, we try to measure the size of N wavelengths, and dividing this distance by N we can then determine the wavelength. We still have the same uncertainty of $\varepsilon\lambda$ in locating the start and end of this wave packet, but when we divide by N to find the wavelength, the uncertainty in one wavelength becomes $\Delta\lambda \sim \varepsilon\lambda/N$. For this larger wave packet, the product of Δx and $\Delta\lambda$ is $\Delta x \Delta\lambda \sim (N\lambda)(\varepsilon\lambda/N) = \varepsilon\lambda^2$, exactly the same as in the case of the smaller wave packet. Equation 4.4 is a fundamental property of classical waves, independent of the type of wave or the method used to measure its wavelength. This is the first of the *uncertainty relationships* for classical waves.

Example 4.3

In a measurement of the wavelength of water waves, 10 wave cycles are counted in a distance of 196 cm. Estimate the minimum uncertainty in the wavelength that might be obtained from this experiment.

Solution

With 10 wave crests in a distance of 196 cm, the wavelength is about (196 cm)/10 = 19.6 cm. We can take $\varepsilon \sim 0.1$ as a good order-of-magnitude estimate of the typical precision

that might be obtained. From Eq. 4.4, we can find the uncertainty in wavelength:

$$\Delta\lambda \sim \frac{\varepsilon\lambda^2}{\Delta x} = \frac{(0.1)(19.6\,\text{cm})^2}{196\,\text{cm}} = 0.2\,\text{cm}$$

With an uncertainty of 0.2 cm, the "true" wavelength might range from 19.5 cm to 19.7 cm, so we might express this result as 19.6 ± 0.1 cm.

The Frequency-Time Uncertainty Relationship

We can take a different approach to uncertainty for classical waves by imagining a measurement of the period rather than the wavelength of the wave that comprises our wave packet. Suppose we have a timing device that we use to measure the duration of the wave packet, as in Figure 4.20. Here we are plotting the wave disturbance as a function of time rather than location. The "size" of the wave packet is now its duration in time, which is roughly one period T for this wave packet, so that $\Delta t \approx T$. Whatever measuring device we use, we have some difficulty locating exactly the start and end of one cycle, so we have an uncertainty ΔT in measuring the period. As before, we'll assume this uncertainty is some small fraction of the period: $\Delta T \sim \varepsilon T$. To examine the competition between the duration of the wave packet and our ability to measure its period, we calculate the product of Δt and ΔT:

$$\Delta t \Delta T \sim \varepsilon T^2 \tag{4.5}$$

This is the second of our uncertainty relationships for classical waves. It shows that for a wave of a given period, the smaller the duration of the wave packet, the larger is the uncertainty in our measurement of the period. Note the similarity between Eqs. 4.4 and 4.5, one representing relationships in space and the other in time.

It will turn out to be more useful if we write Eq. 4.5 in terms of frequency instead of period. Given that period T and frequency f are related by $f = 1/T$, how is Δf related to ΔT? The correct relationship is certainly *not* $\Delta f = 1/\Delta T$, which would imply that a very small uncertainty in the period would lead to a very large

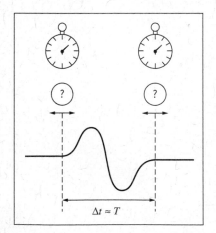

FIGURE 4.20 Measuring the period of a wave represented by a small wave packet of duration roughly one period.

$\Delta t \approx T$

uncertainty in the frequency. Instead, they should be directly related—the better we know the period, the better we know the frequency. Here is how we obtain the relationship: Beginning with $f = 1/T$, we take differentials on both sides:

$$df = -\frac{1}{T^2} dT$$

Next we convert the infinitesimal differentials to finite intervals, and because we are interested only in the magnitude of the uncertainties we can ignore the minus sign:

$$\Delta f = \frac{1}{T^2} \Delta T \qquad (4.6)$$

Combining Eqs. 4.5 and 4.6, we obtain

$$\Delta f \Delta t \sim \varepsilon \qquad (4.7)$$

Equation 4.7 shows that the longer the duration of the wave packet, the more precisely we can measure its frequency.

Example 4.4

An electronics salesman offers to sell you a frequency-measuring device. When hooked up to a sinusoidal signal, it automatically displays the frequency of the signal, and to account for frequency variations, the frequency is remeasured once each second and the display is updated. The salesman claims the device to be accurate to 0.01 Hz. Is this claim valid?

Solution

Based on Eq. 4.7, and again estimating ε to be about 0.1, we know that a measurement of frequency in a time $\Delta t = 1 s$

must have an associated uncertainty of about

$$\Delta f \sim \frac{\varepsilon}{\Delta t} = \frac{0.1}{1 s}$$
$$= 0.1 \, Hz$$

It appears that the salesman may be exaggerating the precision of this device.

4.4 HEISENBERG UNCERTAINTY RELATIONSHIPS

The uncertainty relationships discussed in the previous section apply to *all* waves, and we should therefore apply them to de Broglie waves. We can use the basic de Broglie relationship $p = h/\lambda$ to relate the uncertainty in the momentum Δp to the uncertainty in wavelength $\Delta \lambda$, using the same procedure that we used to obtain Eq. 4.6. Starting with $p = h/\lambda$, we take differentials on both sides and obtain $dp = (-h/\lambda^2)d\lambda$. Now we change the differentials into differences, ignoring the minus sign:

$$\Delta p = \frac{h}{\lambda^2} \Delta \lambda \qquad (4.8)$$

Werner Heisenberg (1901–1976, Germany). Best known for the uncertainty principle, he also developed a complete formulation of the quantum theory based on matrices.

An uncertainty in the momentum of the particle is directly related to the uncertainty in the wavelength associated with the particle's de Broglie wave packet.

Combining Eq. 4.8 with Eq. 4.4, we obtain

$$\Delta x \Delta p \sim \varepsilon h \qquad (4.9)$$

Just like Eq. 4.4, this equation suggests an inverse relationship between Δx and Δp. The smaller the size of the wave packet of the particle, the larger is the uncertainty in its momentum (and thus in its velocity).

Quantum mechanics provides a formal procedure for calculating Δx and Δp for wave packets corresponding to different physical situations and for different schemes for confining a particle. One outcome of these calculations gives the wave packet with the smallest possible value of the product $\Delta x \Delta p$, which turns out to be $h/4\pi$, as we will discuss in the next chapter. Thus $\varepsilon = 1/4\pi$ in this case. All other wave packets will have larger values for $\Delta x \Delta p$.

The combination $h/2\pi$ occurs frequently in quantum mechanics and is given the special symbol $\hbar$ ("h-bar")

$$\hbar = \frac{h}{2\pi} = 1.05 \times 10^{-34}\,\text{J}\cdot\text{s} = 6.58 \times 10^{-16}\,\text{eV}\cdot\text{s}$$

In terms of $\hbar$, we can write the uncertainty relationship as

$$\Delta x \Delta p_x \geqslant \tfrac{1}{2}\hbar \qquad (4.10)$$

The x subscript has been added to the momentum to remind us that Eq. 4.10 applies to motion in a given direction and relates the uncertainties in position and momentum in that direction only. Similar and independent relationships can be applied in the other directions as necessary; thus $\Delta y \Delta p_y \geqslant \hbar/2$ or $\Delta z \Delta p_z \geqslant \hbar/2$.

Equation 4.10 is the first of the *Heisenberg uncertainty relationships*. It sets the limit of the best we can possibly do in an experiment to measure *simultaneously* the location and the momentum of a particle. Another way of interpreting this equation is to say that the more we try to confine a particle, the less we know about its momentum.

Because the limit of $\hbar/2$ represents the minimum value of the product $\Delta x \Delta p_x$, in most cases we will do worse than this limit. It is therefore quite acceptable to take

$$\Delta x \Delta p_x \sim \hbar \qquad (4.11)$$

as a rough estimate of the relationship between the uncertainties in location and momentum.

As an example, let's consider a beam of electrons incident on a single slit, as in Figure 4.21. We know this experiment as single-slit diffraction, which produces the characteristic diffraction pattern illustrated in Figure 4.1. We'll assume that the particles are initially moving in the y direction and that we know their momentum in that direction as precisely as possible. If the electrons initially have no component of their momentum in the x direction, we know p_x exactly (it is exactly zero), so that $\Delta p_x = 0$; thus we know nothing about the x coordinates of the electrons ($\Delta x = \infty$). This situation represents a very wide beam of electrons, only a small fraction of which pass through the slit.

At the instant that some of the electrons pass through the slit, we know quite a bit more about their x location. In order to pass through the slit, the uncertainty

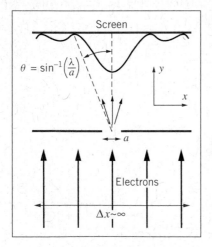

FIGURE 4.21 Single-slit diffraction of electrons. A wide beam of electrons is incident on a narrow slit. The electrons that pass through the slit acquire a component of momentum in the x direction.

in their x location is no larger than a, the width of the slit; thus $\Delta x = a$. This improvement in our knowledge of the electron's location comes at the expense of our knowledge of its momentum, however. According to Eq. 4.11, the uncertainty in the x component of its momentum is now $\Delta p_x \sim \hbar/a$. Measurements beyond the slit no longer show the particle moving precisely in the y direction (for which $p_x = 0$); the momentum now has a small x component as well, with values distributed about zero but now with a range of roughly $\pm\hbar/a$. In passing through the slit, a particle acquires on the average an x component of momentum of roughly $\hbar/a$, according to the uncertainty principle.

Let us now find the angle θ that specifies where a particle with this value of p_x lands on the screen. For small angles, $\sin\theta \approx \tan\theta$ and so

$$\sin\theta \approx \tan\theta = \frac{p_x}{p_y} = \frac{\hbar/a}{p_y} = \frac{\lambda}{2\pi a}$$

using $\lambda = h/p_y$ for the de Broglie wavelength of the electrons. The first minimum of the diffraction pattern of a single slit is located at $\sin\theta = \lambda/a$, which is larger than the spread of angles into which most of the particles are diffracted. The calculation shows that the distribution of transverse momentum given by the uncertainty principle is roughly equivalent to the spreading of the beam into the central diffraction peak, and it illustrates again the close connection between wave behavior and uncertainty in particle location.

The diffraction (spreading) of a beam following passage through a slit is just the effect of the uncertainty principle on our attempt to specify the location of the particle. As we make the slit narrower, p_x increases and the beam spreads even more. In trying to obtain more precise knowledge of the location of the particle by making the slit narrower, we have lost knowledge of the direction of its travel. This trade-off between observations of position and momentum is the essence of the Heisenberg uncertainty principle.

We can also apply the second of our classical uncertainty relationships (Eq. 4.7) to de Broglie waves. If we assume the energy-frequency relationship for light, $E = hf$, can be applied to particles, then we immediately obtain $\Delta E = h\Delta f$. Combining this with Eq. 4.7, we obtain

$$\Delta E \Delta t \sim \varepsilon h \qquad (4.12)$$

Once again, the minimum uncertainty wave packet gives $\varepsilon = 1/4\pi$, and so

$$\Delta E \Delta t \geqslant \tfrac{1}{2}\hbar \qquad (4.13)$$

This is the second of the Heisenberg uncertainty relationships. It tells us that the more precisely we try to determine the time coordinate of a particle, the less precisely we know its energy. For example, if a particle has a very short lifetime between its creation and decay ($\Delta t \to 0$), a measurement of its rest energy (and thus its mass) will be very imprecise ($\Delta E \to \infty$). Conversely, the rest energy of a stable particle (one with an infinite lifetime, so that $\Delta t = \infty$) can in principle be measured with unlimited precision ($\Delta E = 0$).

As in the case of the first Heisenberg relationship, we can take

$$\Delta E \Delta t \sim \hbar \qquad (4.14)$$

as a reasonable estimate for most wave packets.

The Heisenberg uncertainty relationships are the mathematical representations of the *Heisenberg uncertainty principle,* which states:

It is not possible to make a simultaneous determination of the position and the momentum of a particle with unlimited precision,

and

It is not possible to make a simultaneous determination of the energy and the time coordinate of a particle with unlimited precision.

These relationships give an estimate of the minimum uncertainty that can result from any experiment; measurement of the position and momentum of a particle will give a spread of values of widths Δx and Δp_x. We may, for other reasons, do much worse than Eqs. 4.10 and 4.13, but *we can do no better.*

These relationships have a profound impact on our view of nature. It is quite acceptable to say that there is an uncertainty in locating the position of a water wave. It is quite another matter to make the same statement about a de Broglie wave, because *there is an implied corresponding uncertainty in the position of the particle.* Equations 4.10 and 4.13 say that *nature imposes a limit on the accuracy with which we can do experiments.* To emphasize this point, the Heisenberg relationships are sometimes called "indeterminacy" rather than "uncertainty" principles, because the idea of uncertainty may suggest an experimental limit that can be reduced by using better equipment or technique. In actuality, these coordinates are *indeterminate* to the limits provided by Eqs. 4.10 and 4.13—no matter how hard we try, it is simply not possible to measure more precisely.

Example 4.5

An electron moves in the x direction with a speed of 3.6×10^6 m/s. We can measure its speed to a precision of 1%. With what precision can we simultaneously measure its x coordinate?

Solution

The electron's momentum is

$$p_x = mv_x = (9.11 \times 10^{-31}\,\text{kg})(3.6 \times 10^6\,\text{m/s})$$
$$= 3.3 \times 10^{-24}\,\text{kg} \cdot \text{m/s}$$

The uncertainty Δp_x is 1% of this value, or $3.3 \times 10^{-26}\,\text{kg} \cdot \text{m/s}$. The uncertainty in position is then

$$\Delta x \sim \frac{\hbar}{\Delta p_x} = \frac{1.05 \times 10^{-34}\,\text{J} \cdot \text{s}}{3.3 \times 10^{-26}\,\text{kg} \cdot \text{m/s}}$$
$$= 3.2\,\text{nm}$$

which is roughly 10 atomic diameters.

Example 4.6

Repeat the calculations of the previous example in the case of a pitched baseball ($m = 0.145$ kg) moving at a speed of 95 mi/h (42.5 m/s). Again assume that its speed can be measured to a precision of 1%.

Solution

The baseball's momentum is

$$p_x = mv_x = (0.145\,\text{kg})(42.5\,\text{m/s}) = 6.16\,\text{kg} \cdot \text{m/s}$$

The uncertainty in momentum is $6.16 \times 10^{-2}\,\text{kg} \cdot \text{m/s}$, and the corresponding uncertainty in position is

$$\Delta x \sim \frac{\hbar}{\Delta p_x} = \frac{1.05 \times 10^{-34}\,\text{J} \cdot \text{s}}{6.16 \times 10^{-2}\,\text{kg} \cdot \text{m/s}} = 1.7 \times 10^{-33}\,\text{m}$$

This uncertainty is 19 orders of magnitude smaller than the size of an atomic nucleus. The uncertainty principle cannot be blamed for the batter missing the pitch! Once again we see that, because of the small magnitude of Planck's constant, quantum effects are not observable for ordinary objects.

A Statistical Interpretation of Uncertainty

A diffraction pattern, such as that shown in Figure 4.21, is the result of the passage of many particles or photons through the slit. So far, we have been discussing the behavior of only one particle. Let's imagine that we do an experiment in which a large number of particles passes (one at a time) through the slit, and we measure the transverse (x component) momentum of each particle after it passes through the slit. We can do this experiment simply by placing a detector at different locations on the screen where we observe the diffraction pattern. Because the detector actually accepts particles over a finite region on the screen, it measures in a range of deflection angles or equivalently in a range of transverse momentum. The result of the experiment might look something like Figure 4.22. The vertical scale shows the number of particles with momentum in each interval corresponding to different locations of the detector on the screen. The values are symmetrically arranged about zero, which indicates that the mean or average value of p_x is zero. The *width* of the distribution is characterized by Δp_x.

Figure 4.22 resembles a statistical distribution, and in fact the precise definition of Δp_x is similar to that of the standard deviation σ_A of a quantity A that has a mean or average value A_{av}:

$$\sigma_A = \sqrt{(A^2)_{\mathrm{av}} - (A_{\mathrm{av}})^2}$$

If there are N individual measurements of A, then $A_{\mathrm{av}} = N^{-1}\Sigma A_i$ and $(A^2)_{\mathrm{av}} = N^{-1}\Sigma A_i^2$.

By analogy, we can make a rigorous definition of the uncertainty in momentum as

$$\Delta p_x = \sqrt{(p_x^2)_{\mathrm{av}} - (p_{x,\mathrm{av}})^2} \tag{4.15}$$

The average value of the transverse momentum for the situation shown in Figure 4.22 is zero, so

$$\Delta p_x = \sqrt{(p_x^2)_{\mathrm{av}}} \tag{4.16}$$

which gives in effect a root-mean-square value of p_x. This can be taken to be a rough measure of the magnitude of p_x. Thus it is often said that Δp_x gives a measure of the magnitude of the momentum of the particle. As you can see from Figure 4.22, this is indeed true.*

---

*The relationship between the value of Δp_x calculated from Eq. 4.16 and the width of the distribution shown in Figure 4.22 depends on the exact shape of the distribution. You should consider the value from Eq. 4.16 as a rough order-of-magnitude estimate of the width of the distribution.

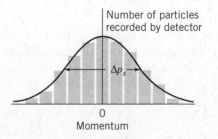

FIGURE 4.22   Results that might be obtained from measuring the number of electrons in a given time interval at different locations on the screen of Figure 4.21. The distribution is centered around $p_x = 0$ and has a width that is characterized by Δp_x.

Example 4.7

In nuclear beta decay, electrons are observed to be ejected from the atomic nucleus. Suppose we assume that electrons are somehow trapped within the nucleus, and that occasionally one escapes and is observed in the laboratory. Take the diameter of a typical nucleus to be 1.0×10^{-14} m, and use the uncertainty principle to estimate the range of kinetic energies that such an electron must have.

Solution

If the electron were trapped in a region of width $\Delta x \approx 10^{-14}$ m, the corresponding uncertainty in its momentum would be

$$\Delta p_x \sim \frac{\hbar}{\Delta x} = \frac{1}{c} \frac{\hbar c}{\Delta x} = \frac{1}{c} \frac{197 \, \text{MeV} \cdot \text{fm}}{10 \, \text{fm}} = 19.7 \, \text{MeV}/c$$

Note the use of $\hbar c = 197 \, \text{MeV} \cdot \text{fm}$ in this calculation. This momentum is clearly in the relativistic regime for electrons, so we must use the relativistic formula to find the kinetic

energy for a particle of momentum $19.7 \, \text{MeV}/c$:

$$K = \sqrt{p^2 c^2 + (mc^2)^2} - mc^2$$
$$= \sqrt{(19.7 \, \text{MeV})^2 + (0.5 \, \text{MeV})^2} - 0.5 \, \text{MeV} = 19 \, \text{MeV}$$

where we have used Eq. 4.16 to relate Δp_x to p_x^2. This result gives the spread of kinetic energies corresponding to a spread in momentum of $19.7 \, \text{MeV}/c$.

Electrons emitted from the nucleus in nuclear beta decay typically have kinetic energies of about 1 MeV, much smaller than the typical spread in energy required by the uncertainty principle for electrons confined inside the nucleus. This suggests that beta-decay electrons of such low energies cannot be confined in a region of the size of the nucleus, and that another explanation must be found for the electrons observed in nuclear beta decay. (As we discuss in Chapter 12, these electrons cannot preexist in the nucleus, which would violate the uncertainty principle, but are "manufactured" by the nucleus at the instant of the decay.)

Example 4.8

(a) A charged pi meson has a rest energy of 140 MeV and a lifetime of 26 ns. Find the energy uncertainty of the pi meson, expressed in MeV and also as a fraction of its rest energy. (b) Repeat for the uncharged pi meson, with a rest energy of 135 MeV and a lifetime of 8.3×10^{-17} s. (c) Repeat for the rho meson, with a rest energy of 765 MeV and a lifetime of 4.4×10^{-24} s.

Solution

(a) If the pi meson lives for 26 ns, we have only that much time in which to measure its rest energy, and Eq. 4.8 tells us that *any* energy measurement done in a time Δt is uncertain by an amount of at least

$$\Delta E = \frac{\hbar}{\Delta t} = \frac{6.58 \times 10^{-16} \, \text{eV} \cdot \text{s}}{26 \times 10^{-9} \, \text{s}}$$
$$= 2.5 \times 10^{-8} \, \text{eV}$$
$$= 2.5 \times 10^{-14} \, \text{MeV}$$

$$\frac{\Delta E}{E} = \frac{2.5 \times 10^{-14} \, \text{MeV}}{140 \, \text{MeV}}$$
$$= 1.8 \times 10^{-16}$$

(b) In a similar way,

$$\Delta E = \frac{\hbar}{\Delta t} = \frac{6.58 \times 10^{-16} \, \text{eV} \cdot \text{s}}{8.3 \times 10^{-17} \, \text{s}} = 7.9 \, \text{eV}$$
$$= 7.9 \times 10^{-6} \, \text{MeV}$$

$$\frac{\Delta E}{E} = \frac{7.9 \times 10^{-6} \, \text{MeV}}{135 \, \text{MeV}} = 5.9 \times 10^{-8}$$

(c) For the rho meson,

$$\Delta E = \frac{\hbar}{\Delta t} = \frac{6.58 \times 10^{-16} \, \text{eV} \cdot \text{s}}{4.4 \times 10^{-24} \, \text{s}} = 1.5 \times 10^8 \, \text{eV}$$
$$= 150 \, \text{MeV}$$

$$\frac{\Delta E}{E} = \frac{150 \, \text{MeV}}{765 \, \text{MeV}} = 0.20$$

In the first case, the uncertainty principle does not give a large enough effect to be measured—particle masses cannot be measured to a precision of 10^{-16} (about 10^{-6} is the best precision that we can obtain). In the second example, the uncertainty principle contributes at about the level of 10^{-7}, which approaches the limit of our measuring instruments and therefore might be observable in the laboratory. In the third example, we see that the uncertainty principle can contribute substantially to the precision of our knowledge of the rest energy of the rho meson; measurements of its rest energy will show a statistical distribution centered about 765 MeV with a spread of 150 MeV, and no matter how precise an instrument we use to measure the rest energy, we can never reduce that spread.

The lifetime of a very short-lived particle such as the rho meson cannot be measured directly. In practice we reverse the procedure of the calculation of this example—we measure the rest energy, which gives a distribution similar to Figure 4.22, and from the "width" ΔE of the distribution we deduce the lifetime using Eq. 4.8. This procedure is discussed in Chapter 14.

Example 4.9

Estimate the minimum velocity that would be measured for a billiard ball ($m \approx 100\,\text{g}$) confined to a billiard table of dimension 1 m.

Solution

For $\Delta x \approx 1\,\text{m}$, we have

$$\Delta p_x \sim \frac{\hbar}{\Delta x} = \frac{1.05 \times 10^{-34}\,\text{J}\cdot\text{s}}{1\,\text{m}} = 1 \times 10^{-34}\,\text{kg}\cdot\text{m/s}$$

so

$$\Delta v_x = \frac{\Delta p_x}{m} = \frac{1 \times 10^{-34}\,\text{kg}\cdot\text{m/s}}{0.1\,\text{kg}} = 1 \times 10^{-33}\,\text{m/s}$$

Thus quantum effects might result in motion of the billiard ball with a speed distribution having a spread of about 1×10^{-33} m/s. At this speed, the ball would move a distance of 1% of the diameter of an atomic nucleus in a time equal to the age of the universe! Once again, we see that quantum effects are not observable with macroscopic objects.

4.5 WAVE PACKETS

In Section 4.3, we described measurements of the wavelength or frequency of a wave packet, which we consider to be a finite group of oscillations of a wave. That is, the wave amplitude is large over a finite region of space or time and is very small outside that region.

Before we begin our discussion, it is necessary to keep in mind that we are discussing *traveling waves*, which we imagine as moving in one direction with a uniform speed. (We'll discuss the speed of the wave packet later.) As the wave packet moves, individual locations in space will oscillate with the frequency or wavelength that characterizes the wave packet. When we show a static picture of a wave packet, it doesn't matter that some points within the packet appear to have positive displacement, some have negative displacement, and some may even have zero displacement. As the wave travels, those locations are in the process of oscillating, and our drawings may "freeze" that oscillation. What is important is the locations in space where the overall wave packet has a large oscillation amplitude and where it has a very small amplitude.*

In this section we will examine how to build a wave packet by adding waves together. A pure sinusoidal wave is of no use in representing a particle—the wave

---

*By analogy, think of a radio wave traveling from the station to your receiver. At a particular instant of time, some points in space may have instantaneous electromagnetic field values of zero, but that doesn't affect your reception of the signal. What is important is the overall amplitude of the traveling wave.

extends from $-\infty$ to $+\infty$, so the particle could be found anywhere. We would like the particle to be represented by a wave packet that describes how the particle is *localized* to a particular region of space, such as an atom or a nucleus.

The key to the process of building a wave packet involves adding together waves of different wavelength. We represent our waves as $A \cos kx$, where k is the wave number ($k = 2\pi/\lambda$) and A is the amplitude. For example, let's add together two waves:

$$y(x) = A_1 \cos k_1 x + A_2 \cos k_2 x = A_1 \cos(2\pi x/\lambda_1) + A_2 \cos(2\pi x/\lambda_2) \quad (4.17)$$

This sum is illustrated in Figure 4.23a for the case $A_1 = A_2$ and $\lambda_1 = 9, \lambda_2 = 11$. This combined wave shows the phenomenon known as *beats* in the case of sound waves. So far we don't have a result that looks anything like the wave packet we are after, but you can see that by adding together two different waves we have reduced the amplitude of the wave packet at some locations. This pattern repeats endlessly from $-\infty$ to $+\infty$, so the particle is still not localized.

Let's try a more detailed sum. Figure 4.23b shows the result of adding 5 waves with wavelengths 9, 9.5, 10, 10.5, 11. Here we have been a bit more successful in restricting the amplitude of the wave packet in some regions. By adding even more waves with a larger range of wavelengths, we can obtain still narrower regions of large amplitude: Figure 4.23c shows the result of adding 9 waves with wavelengths $8, 8.5, 9, \ldots, 12$, and Figure 4.23d shows the result of adding 13 waves of wavelengths $7, 7.5, 8, \ldots, 13$. Unfortunately, all of these patterns (including the regions of large amplitude) repeat endlessly from $-\infty$ to $+\infty$, so even though we have obtained increasingly large regions where the wave packet has small amplitude, we haven't yet created a wave packet that might represent a particle localized to a particular region. If these wave packets did represent particles, then the particle would not be confined to any finite region.

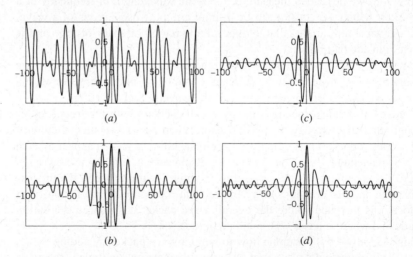

(a)

(c)

(b)

(d)

FIGURE 4.23 (a) Adding two waves of wavelengths 9 and 11 gives beats. (b) Adding 5 waves with wavelengths ranging from 9 to 11. (c) Adding 9 waves with wavelengths ranging from 8 to 12. (d) Adding 13 waves with wavelengths from 7 to 13. All of the patterns repeat from $-\infty$ to $+\infty$.

The regions of large amplitude in Figures 4.23*b,c,d* do show how adding more waves of a greater range of wavelengths helps to restrict the size of the wave packet. The region of large amplitude in Figure 4.23*b* ranges from about $x = -40$ to $+40$, while in Figure 4.23*c* it is from about $x = -20$ to $+20$ and in Figure 4.23*d* from about $x = -15$ to $+15$. This shows again the inverse relationship between Δx and $\Delta \lambda$ expected for wave packets given by Eq. 4.4: as the range of wavelengths increases from 2 to 4 to 6, the size of the "allowed" regions decreases from about 80 to 40 to 30. Once again we find that to restrict the size of the wave packet we must sacrifice the precise knowledge of the wavelength.

Note that for all four of these wave patterns, the disturbance seems to have a wavelength of about 10, equal to the central wavelength of the range of values of the functions we constructed. We can therefore regard these functions as a cosine wave with a wavelength of 10 that is shaped or *modulated* by the other cosine waves included in the function. For example, for the case of $A_1 = A_2 = A$, Eq. 4.17 can be rewritten after a bit of trigonometric manipulation as

$$y(x) = 2A \cos\left(\frac{\pi x}{\lambda_1} - \frac{\pi x}{\lambda_2}\right) \cos\left(\frac{\pi x}{\lambda_1} + \frac{\pi x}{\lambda_2}\right) \qquad (4.18)$$

If λ_1 and λ_2 are close together (that is, if $\Delta \lambda = \lambda_2 - \lambda_1 \ll \lambda_1, \lambda_2$), this can be approximated as

$$y(x) = 2A \cos\left(\frac{\Delta \lambda \pi x}{\lambda_{\text{av}}^2}\right) \cos\left(\frac{2\pi x}{\lambda_{\text{av}}}\right) \qquad (4.19)$$

where $\lambda_{\text{av}} = (\lambda_1 + \lambda_2)/2 \approx \lambda_1$ or λ_2. The second cosine term represents a wave with a wavelength of 10, and the first cosine term provides the shaping envelope that produces the beats.

Any finite combination of waves with discrete wavelengths will produce patterns that repeat between $-\infty$ to $+\infty$, so this method of adding waves will not work in constructing a finite wave packet. To construct a wave packet with a finite width, we must replace the first cosine term in Eq. 4.19 with a function that is large in the region where we want to confine the particle but that falls to zero as $x \to \pm\infty$. For example, the simplest function that has this property is $1/x$, so we might imagine a wave packet whose mathematical form is

$$y(x) = \frac{2A}{x} \sin\left(\frac{\Delta \lambda \pi x}{\lambda_0^2}\right) \cos\left(\frac{2\pi x}{\lambda_0}\right) \qquad (4.20)$$

Here λ_0 represents the central wavelength, replacing λ_{av}. (In going from Eq. 4.19 to Eq. 4.20, the cosine modulating term has been changed to a sine; otherwise the function would blow up at $x = 0$.) This function is plotted in Figure 4.24*a*. It looks more like the kind of function we are seeking—it has large amplitude only in a small region of space, and the amplitude drops rapidly to zero outside that region. Another function that has this property is the *Gaussian* modulating function:

$$y(x) = A e^{-2(\Delta \lambda \pi x / \lambda_0^2)^2} \cos\left(\frac{2\pi x}{\lambda_0}\right) \qquad (4.21)$$

which is shown in Figure 4.24*b*.

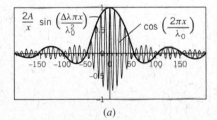

(*a*)

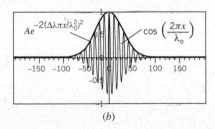

(*b*)

FIGURE 4.24   (*a*) A wave packet in which the modulation envelope decreases in amplitude like $1/x$. (*b*) A wave packet with a Gaussian modulating function. Both curves are drawn for $\lambda_0 = 10$ and $\Delta \lambda = 0.58$, which corresponds approximately to Figure 4.23*b*.

Both of these functions show the characteristic inverse relationship between an arbitrarily defined size of the wave packet Δx and the wavelength range parameter $\Delta\lambda$ that is used in constructing the wave packet. For example, consider the wave packet shown in Figure 4.24a. Let's arbitrarily define the width of the wave packet as the distance over which the amplitude of the central region falls by $1/2$. That occurs roughly where the argument of the sine has the value $\pm\pi/2$, which gives $\Delta x \Delta\lambda \sim \lambda_0^2$, consistent with our classical uncertainty estimate.

These wave packets can also be constructed by adding together waves of differing amplitude and wavelength, but the wavelengths form a continuous rather than a discrete set. It is a bit easier to illustrate this if we work with wave number $k = 2\pi/\lambda$ rather than wavelength. So far we have been adding waves in the form of $A \cos kx$, so that

$$y(x) = \sum A_i \cos k_i x \tag{4.22}$$

where $k_i = 2\pi/\lambda_i$. The waves plotted in Figure 4.23 represent applications of the general formula of Eq. 4.22 carried out over different numbers of discrete waves. If we have a continuous set of wave numbers, the sum in Eq. 4.22 becomes an integral:

$$y(x) = \int A(k) \cos kx \, dk \tag{4.23}$$

where the integral is carried out over whatever range of wave numbers is permitted (possibly infinite).

For example, suppose we have a range of wave numbers from $k_0 - \Delta k/2$ to $k_0 + \Delta k/2$ that is a continuous distribution of wave numbers of width Δk centered at k_0. If all of the waves have the same amplitude A_0, then from Eq. 4.23 the form of the wave packet can be shown to be (see Problem 24 at the end of the chapter)

$$y(x) = \frac{2A_0}{x} \sin\left(\frac{\Delta k}{2}x\right) \cos k_0 x \tag{4.24}$$

This is identical with Eq. 4.20 with $k_0 = 2\pi/\lambda_0$ and $\Delta k = 2\pi\Delta\lambda/\lambda_0^2$. This relationship between Δk and $\Delta\lambda$ follows from a procedure similar to what was used to obtain Eq. 4.6. With $k = 2\pi/\lambda$, taking differentials gives $dk = -(2\pi/\lambda^2)d\lambda$. Replacing the differentials with differences and ignoring the minus sign gives the relationship between Δk and $\Delta\lambda$.

A better approximation of the shape of the wave packet can be found by letting $A(k)$ vary according to a Gaussian distribution $A(k) = A_0 e^{-(k-k_0)^2/2(\Delta k)^2}$. This gives a range of wave numbers that has its largest contribution at the central wave number k_0 and falls off to zero for larger or smaller wave numbers with a characteristic width of Δk. Applying Eq. 4.23 to this case, with k ranging from $-\infty$ to $+\infty$ gives (see Problem 25)

$$y(x) = A_0 \Delta k \sqrt{2\pi} e^{-(\Delta kx)^2/2} \cos k_0 x \tag{4.25}$$

which shows how the form of Eq. 4.21 originates.

By specifying the distribution of wavelengths, we can construct a wave packet of any desired shape. A wave packet that restricts the particle to a region in space of width Δx will have a distribution of wavelengths characterized by a width $\Delta\lambda$. The smaller we try to make Δx, the larger will be the spread $\Delta\lambda$ of the wavelength distribution. The mathematics of this process gives a result that is consistent with the uncertainty relationship for classical waves (Eq. 4.4).

4.6 THE MOTION OF A WAVE PACKET

Let's consider again the "beats" wave packet represented by Eq. 4.17 and illustrated in Figure 4.23a. We now want to turn our "static" waves into traveling waves. It will again be more convenient for this discussion to work with the wave number k instead of the wavelength. To turn a static wave $y(x) = A \cos kx$ into a traveling wave moving in the positive x direction, we replace kx with $kx - \omega t$, so that the traveling wave is written as $y(x, t) = A \cos(kx - \omega t)$. (For motion in the negative x direction, we would replace kx with $kx + \omega t$.) Here ω is the *circular frequency* of the wave: $\omega = 2\pi f$. The combined traveling wave then would be represented as

$$y(x, t) = A_1 \cos(k_1 x - \omega_1 t) + A_2 \cos(k_2 x - \omega_2 t) \qquad (4.26)$$

For any individual wave, the wave speed is related to its frequency and wavelength according to $v = \lambda f$. In terms of the wave number and circular frequency, we can write this as $v = (2\pi/k)(\omega/2\pi)$, so $v = \omega/k$. This quantity is sometimes called the *phase speed* and represents the speed of one particular phase or component of the wave packet. In general, each individual component may have a different phase speed. As a result, the shape of the wave packet may change with time.

For Figure 4.23a, we chose $A_1 = A_2$ and $\lambda_1 = 9, \lambda_2 = 11$. Let's choose $v_1 = 6$ units/s and $v_2 = 4$ units/s. Figure 4.25 shows the waveform at a time of $t = 1$ s. In that time, wave 1 will have moved a distance of 6 units in the positive x direction and wave 2 will have moved a distance of 4 units in the positive x direction. However, the combined wave moves a much greater distance in that time: the center of the beat that was formerly at $x = 0$ has moved to $x = 15$ units. How is it possible that the combined waveform moves faster than either of its component waves?

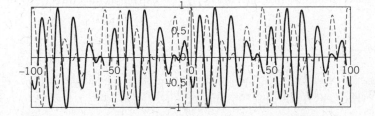

FIGURE 4.25 The solid line shows the waveform of Figure 4.23a at $t = 1$ s, and the dashed line shows the same waveform at $t = 0$. Note that the peak that was originally at $x = 0$ has moved to $x = 15$ at $t = 1$ s.

To produce the peak at $x = 0$ and $t = 0$, the two component waves were exactly in phase—their two maxima lined up exactly to produce the combined maximum of the wave. At $x = 15$ units and $t = 1$ s, two individual maxima line up once again to produce a combined maximum. They are *not* the same two maxima that lined up to produce the maximum at $t = 0$, but it happens that two other maxima are in phase at $x = 15$ units and $t = 1$ s to produce the combined maximum. If we were to watch an animation of the wave, we would see the maximum originally at $x = 0$ move gradually to $x = 15$ between $t = 0$ and $t = 1$ s.

We can understand how this occurs by writing Eq. 4.26 in a form similar to Eq. 4.18 using trigonometric identities. The result is (again assuming $A_1 = A_2 = A$, as we did in Eq. 4.18):

$$y(x, t) = 2A \cos\left(\frac{\Delta k}{2}x - \frac{\Delta \omega}{2}t\right) \cos\left(\frac{k_1 + k_2}{2}x - \frac{\omega_1 + \omega_2}{2}t\right) \qquad (4.27)$$

As in Eq. 4.18, the second term in Eq. 4.27 represents the rapid variation of the wave within the envelope given by the first term. It is the first term that dictates the overall shape of the waveform, so it is this term that determines the speed of travel of the waveform. For a wave that is written as $\cos(kx - \omega t)$, the speed is ω/k. For this wave envelope, the speed is $(\Delta \omega/2)/(\Delta k/2) = \Delta \omega/\Delta k$. This speed is called the *group speed* of the wave packet. As we have seen, the group speed of the wave packet can be very different from the phase speed of the component waves. For more complicated situations than the two-component "beat" waveform, the group speed can be generalized by turning the differences into differentials:

$$v_{\text{group}} = \frac{d\omega}{dk} \qquad (4.28)$$

The group speed depends on the relationship between frequency and wavelength for the component waves. If the phase speed of all component waves is the same and is independent of frequency or wavelength (as, for example, light waves in empty space), then the group speed is identical to the phase speed and the wave packet keeps its original shape as it travels. In general, the propagation of a component wave depends on the properties of the medium, and different component waves will travel with different speeds. Light waves in glass or sound waves in most solids travel with a speed that varies with frequency or wavelength, and so their wave packets change shape as they travel. De Broglie waves in general have different phase speeds, so that their wave packets expand as they travel.

Example 4.10

Certain ocean waves travel with a phase velocity $v_{\text{phase}} = \sqrt{g\lambda/2\pi}$, where g is the acceleration due to gravity. What is the group velocity of a "wave packet" of these waves?

Solution
With $k = 2\pi/\lambda$, we can write the phase velocity as a function of k as

$$v_{\text{phase}} = \sqrt{g/k}$$

But with $v_{\text{phase}} = \omega/k$, we have $\omega/k = \sqrt{g/k}$, so $\omega = \sqrt{gk}$ and Eq. 4.28 gives

$$v_{\text{group}} = \frac{d\omega}{dk} = \frac{d}{dk}\sqrt{gk} = \frac{1}{2}\sqrt{\frac{g}{k}} = \frac{1}{2}\sqrt{\frac{g\lambda}{2\pi}}$$

Note that the group speed of the wave packet increases as the wavelength increases.

The Group Speed of deBroglie Waves

Suppose we have a localized particle, represented by a group of de Broglie waves. For each component wave, the energy of the particle is related to the frequency of the de Broglie wave by $E = hf = \hbar\omega$, and so $dE = \hbar\,d\omega$. Similarly, the momentum of the particle is related to the wavelength of the de Broglie wave by $p = h/\lambda = \hbar k$, so $dp = \hbar\,dk$. The group speed of the de Broglie wave then can be expressed as

$$v_{\text{group}} = \frac{d\omega}{dk} = \frac{dE/\hbar}{dp/\hbar} = \frac{dE}{dp} \tag{4.29}$$

For a classical particle having only kinetic energy $E = K = p^2/2m$, we can find dE/dp as

$$\frac{dE}{dp} = \frac{d}{dp}\left(\frac{p^2}{2m}\right) = \frac{p}{m} = v \tag{4.30}$$

which is the velocity of the particle.

Combining Eqs. 4.29 and 4.30 we obtain an important result:

$$v_{\text{group}} = v_{\text{particle}} \tag{4.31}$$

The speed of a particle is equal to the group speed of the corresponding wave packet. The wave packet and the particle move together—wherever the particle goes, its de Broglie wave packet moves along with it like a shadow. If we do a wave-type experiment on the particle, the de Broglie wave packet is always there to reveal the wave behavior of the particle. A particle can never escape its wave nature!

The Spreading of a Moving Wave Packet

Suppose we have a wave packet that represents a confined particle at $t = 0$. For example, the particle might have passed through a single-slit apparatus. Its initial uncertainty in position is Δx_0 and its initial uncertainty in momentum is Δp_{x0}. The wave packet moves in the x direction with velocity v_x, but that velocity is not precisely known—the uncertainty in its momentum gives a corresponding uncertainty in velocity: $\Delta v_{x0} = \Delta p_{x0}/m$. Because there is an uncertainty in the velocity of the wave packet, we can't be sure where it will be located at time t. That is, its location at time t is $x = v_x t$, with velocity $v_x = v_{x0} \pm \Delta v_{x0}$. Thus there are two contributions to the uncertainty in its location at time t: the initial uncertainty Δx_0 and an additional amount equal to $\Delta v_{x0} t$ that represents the spreading of the wave packet. We'll assume that these two contributions add quadratically, like experimental uncertainties, so that the total uncertainty in the location of the particle is

$$\Delta x = \sqrt{(\Delta x_0)^2 + (\Delta v_{x0} t)^2} = \sqrt{(\Delta x_0)^2 + (\Delta p_{x0} t/m)^2} \tag{4.32}$$

Taking $\Delta p_{x0} = \hbar/\Delta x_0$ according to the uncertainty principle, we have

$$\Delta x = \sqrt{(\Delta x_0)^2 + (\hbar t/m\Delta x_0)^2} \tag{4.33}$$

If we try to make the wave packet very small at $t = 0$ (Δx_0 is small), then the second term under the square root makes the wave packet expand rapidly, because Δx_0 appears in the denominator of that term. *The more successful we are at confining*

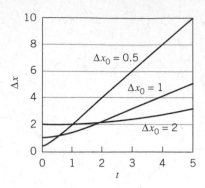

FIGURE 4.26 The smaller the initial wave packet, the more quickly it grows.

a wave packet, the more quickly it spreads. This reminds us of the single-slit experiment discussed in Section 4.4: the narrower we make the slit, the more the waves diverge after passing through the slit. Figure 4.26 shows how the size of the wave packet expands with time for two different initial sizes, and you can see that the smaller initial wave packet grows more rapidly than the larger initial packet.

4.7 PROBABILITY AND RANDOMNESS

Any single measurement of the position or momentum of a particle can be made with as much precision as our experimental skill permits. How then does the wavelike behavior of a particle become observable? How does the uncertainty in position or momentum affect our experiment?

Suppose we prepare an atom by attaching an electron to a nucleus. (For this example we regard the nucleus as being fixed in space.) Some time after preparing our atom, we measure the position of the electron. We then repeat the procedure, preparing the atom *in an identical way,* and find that a remeasurement of the position of the electron yields a value different from that found in our first measurement. In fact, each time we repeat the measurement, we may obtain a different outcome. If we repeat the measurement a large number of times, we find ourselves led to a conclusion that runs counter to a basic notion of classical physics—*systems that are prepared in identical ways do not show identical subsequent behavior.* What hope do we then have of constructing a mathematical theory that has any usefulness at all in predicting the outcome of a measurement, if that outcome is completely random?

The solution to this dilemma lies in the consideration of the *probability* of obtaining any given result from an experiment whose possible results are subject to the laws of statistics. We cannot predict the outcome of a single flip of a coin or roll of the dice, because any *single result* is as likely as any other *single result.* We can, however, predict the *distribution* of a large number of individual measurements. For example, on a single flip of a coin, we cannot predict whether the outcome will be "heads" or "tails"; the two are equally likely. If we make a large number of trials, we expect that approximately 50% will turn up "heads" and 50% will yield "tails"; even though we cannot predict the result of any single toss of the coin, we can predict reasonably well the result of a large number of tosses.

Our study of systems governed by the laws of quantum physics leads us to a similar situation. We cannot predict the outcome of any *single* measurement of the position of the electron in the atom we prepared, but if we do a large number of measurements, we ought to find a statistical distribution of results. We cannot develop a mathematical theory that predicts the result of a single measurement, but we do have a mathematical theory that predicts the statistical behavior of a system (or of a large number of identical systems). The quantum theory provides this mathematical procedure, which enables us to calculate the average or probable outcome of measurements and the distribution of individual outcomes about the average. This is not such a disadvantage as it may seem, for in the realm of quantum physics, we seldom do measurements with, for example, a single atom. If we were studying the emission of light by a radiant system or the properties of a solid or the scattering of nuclear particles, we would be dealing with a large number of atoms, and so our concept of statistical averages is very useful.

In fact, such concepts are not as far removed from our daily lives as we might think. For example, what is meant when the TV weather forecaster "predicts" a 50% chance of rain tomorrow? Will it rain 50% of the time, or over 50% of the city? The proper interpretation of the forecast is that the existing set of atmospheric conditions will, in a large number of similar cases, result in rain in about half the cases. A surgeon who asserts that a patient has a 50% chance of surviving an operation means exactly the same thing—experience with a large number of similar cases suggests recovery in about half.

Quantum mechanics uses similar language. For example, if we say that the electron in a hydrogen atom has a 50% probability of circulating in a clockwise direction, we mean that in observing a large collection of similarly prepared atoms we find 50% to be circulating clockwise. Of course a *single measurement* shows *either* clockwise or counterclockwise circulation. (Similarly, it *either* rains or it doesn't; the patient *either* lives or dies.)

Of course, one could argue that the flip of a coin or the roll of the dice is not a random process, but that the apparently random nature of the outcome simply reflects our lack of knowledge of the state of the system. For example, if we knew exactly how the dice were thrown (magnitude and direction of initial velocity, initial orientation, rotational speed) and precisely what the laws are that govern their bouncing on the table, we should be able to predict exactly how they would land. (Similarly, if we knew a great deal more about atmospheric physics or physiology, we could predict with certainty whether or not it will rain tomorrow or an individual patient will survive.) When we instead analyze the outcomes in terms of probabilities, we are really admitting our inability to do the analysis exactly. There is a school of thought that asserts that the same situation exists in quantum physics. According to this interpretation, we could predict *exactly* the behavior of the electron in our atom if only we knew the nature of a set of so-called "hidden variables" that determine its motion. However, experimental evidence disagrees with this theory, and so we must conclude that the random behavior of a system governed by the laws of quantum physics is a fundamental aspect of nature and not a result of our limited knowledge of the properties of the system.

The Probability Amplitude

What does the amplitude of the de Broglie wave represent? In any wave phenomenon, a physical quantity such as displacement or pressure varies with location and time. What is the physical property that varies as the de Broglie wave propagates?

A localized particle is represented by a wave packet. If a particle is confined to a region of space of dimension Δx, its wave packet has large amplitude only in a region of space of dimension Δx and has small amplitude elsewhere. That is, the amplitude is large where the particle is likely to be found and small where the particle is less likely to be found. *The probability of finding the particle at any point depends on the amplitude of its de Broglie wave at that point.* In analogy with classical physics, in which the intensity of any wave is proportional to the square of its amplitude, we have

probability to observe particles $\propto$ | de Broglie wave amplitude |^2

Compare this with the similar relationship for photons discussed in Section 3.6:

probability to observe photons $\propto$ | electric field amplitude |^2

Just as the electric field amplitude of an electromagnetic wave indicates regions of high and low probability for observing photons, the de Broglie wave performs

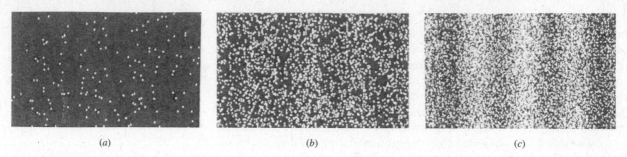

(a) (b) (c)

FIGURE 4.27 The buildup of an electron interference pattern as increasing numbers of electrons are detected: (a) 100 electrons; (b) 3000 electrons; (c) 70,000 electrons. (Reprinted with permission from Akira Tonomura, Hitachi, Ltd, T. Matsuda and T. Kawasaki, Advanced Research Laboratory. From American Journal of Physics 57, 117. (Copyright 1989) American Association of Physics Teachers.)

the same function for particles. Figure 4.27 illustrates this effect, as individual electrons in a double-slit type of experiment eventually produce the characteristic interference fringes. The path of each electron is guided by its de Broglie wave toward the allowed regions of high probability. This statistical effect is not apparent for a small number of electrons, but it becomes quite apparent when a large number of electrons has been detected.

In the next chapter we discuss the mathematical framework for computing the wave amplitudes for a particle in various situations, and we also develop a more rigorous mathematical definition of the probability.

Chapter Summary

		Section			Section
De Broglie wavelength	$\lambda = h/p$	4.1	Statistical momentum uncertainty	$\Delta p_x = \sqrt{(p_x^2)_{av} - (p_{x,av})^2}$	4.4
Single slit diffraction	$a \sin \theta = n\lambda \quad n = 1, 2, 3, \ldots$	4.2			
Classical position-wavelength uncertainty	$\Delta x \Delta \lambda \sim \varepsilon \lambda^2$	4.3	Wave packet (discrete k)	$y(x) = \sum A_i \cos k_i x$	4.5
Classical frequency-time uncertainty	$\Delta f \Delta t \sim \varepsilon$	4.3	Wave packet (continuous k)	$y(x) = \int A(k) \cos kx \, dk$	4.5
Heisenberg position-momentum uncertainty	$\Delta x \Delta p_x \sim \hbar$	4.4	Group speed of wave packet	$v_{group} = \dfrac{d\omega}{dk}$	4.6
Heisenberg energy-time uncertainty	$\Delta E \Delta t \sim \hbar$	4.4			

Questions

1. When an electron moves with a certain de Broglie wavelength, does any aspect of the electron's motion vary with that wavelength?

2. Imagine a different world in which the laws of quantum physics still apply, but which has $h = 1$ J·s. What might be some of the difficulties of life in such a world? (See *Mr.*

Tompkins in Paperback by George Gamow for an imaginary account of such a world.)

3. Suppose we try to measure an unknown frequency f by listening for beats between f and a known (and controllable) frequency f'. (We assume f' is known to arbitrarily small uncertainty.) The beat frequency is $|f' - f|$. If we hear no

beats, then we conclude that $f = f'$. (a) How long must we listen to hear "no" beats? (b) If we hear no beats in one second, how accurately have we determined f? (c) If we hear no beats in 10 s, how accurately? In 100 s? (d) How is this experiment related to Eq. 4.7?

4. What difficulties does the uncertainty principle cause in trying to pick up an electron with a pair of forceps?

5. Does the uncertainty principle apply to nature itself or only to the results of experiments? That is, is it the position and momentum that are *really* uncertain, or merely our knowledge of them? What is the difference between these two interpretations?

6. The uncertainty principle states in effect that the more we try to confine an object, the faster we are likely to find it moving. Is this why you can't seem to keep money in your pocket for long? Make a numerical estimate.

7. Consider a collection of gas molecules trapped in a container. As we move the walls of the container closer together (compressing the gas) the molecules move faster (the temperature increases). Does the gas behave this way because of the uncertainty principle? Justify your answer with some numerical estimates.

8. Many nuclei are unstable and undergo radioactive decay to other nuclei. The lifetimes for these decays are typically of the order of days to years. Do you expect that the uncertainty principle will cause a measurable effect in the precision to which we can measure the masses of atoms of these nuclei?

9. Just as the classical limit of relativity can be achieved by letting $c \to \infty$, the classical limit of quantum behavior is achieved by letting $h \to 0$. Consider the following in the $h \to 0$ limit and explain how they behave classically: the size of the energy quantum of an electromagnetic wave, the de Broglie wavelength of an electron, the Heisenberg uncertainty relationships.

10. Assume the electron beam in a television tube is accelerated through a potential difference of 25 kV and then passes through a deflecting capacitor of interior width 1 cm. Are diffraction effects important in this case? Justify your answer with a calculation.

11. The structure of crystals can be revealed by X-ray diffraction (Figures 3.7 and 3.8), electron diffraction (Figure 4.2), and neutron diffraction (Figure 4.7). In what ways do these experiments reveal similar structure? In what ways are they different?

12. Often it happens in physics that great discoveries are made inadvertently. What would have happened if Davisson and Germer had their accelerating voltage set below 32 V?

13. Suppose we cover one slit in the two-slit electron experiment with a very thin sheet of fluorescent material that emits a photon of light whenever an electron passes through. We then fire electrons one at a time at the double slit; whether or not we see a flash of light tells us which slit the electron went through. What effect does this have on the interference pattern? Why?

14. In another attempt to determine through which slit the electron passes, we suspend the double slit itself from a very fine spring balance and measure the "recoil" momentum of the slit as a result of the passage of the electron. Electrons that strike the screen near the center must cause recoils in opposite directions depending on which slit they pass through. Sketch such an apparatus and describe its effect on the interference pattern. (*Hint:* Consider the uncertainty principle $\Delta x \Delta p_x \sim \hbar$ as applied to the motion of the slits suspended from the spring. How precisely do we know the position of the slit?)

15. It is possible for v_{phase} to be greater than c? Can v_{group} be greater than c?

16. In a nondispersive medium, $v_{group} = v_{phase}$; this is another way of saying that all waves travel with the same phase velocity, no matter what their wavelengths. Is this true for (a) de Broglie waves? (b) Light waves in glass? (c) Light waves in vacuum? (d) Sound waves in air? What difficulties would be encountered in attempting communication (by speech or by radio signals for example) in a strongly dispersive medium?

Problems

4.1 De Broglie's Hypothesis

1. Find the de Broglie wavelength of (a) a 5-MeV proton; (b) a 50-GeV electron; (c) an electron moving at $v = 1.00 \times 10^6$ m/s.

2. The neutrons produced in a reactor are known as *thermal neutrons,* because their kinetic energies have been reduced (by collisions) until $K = \frac{3}{2}kT$, where T is room temperature (293 K). (a) What is the kinetic energy of such neutrons? (b) What is their de Broglie wavelength? Because this wavelength is of the same order as the lattice spacing of the atoms of a solid, neutron diffraction (like X-ray and electron diffraction) is a useful means of studying solid lattices.

3. By doing a nuclear diffraction experiment, you measure the de Broglie wavelength of a proton to be 9.16 fm. (a) What is the speed of the proton? (b) Through what potential difference must it be accelerated to achieve that speed?

4. A proton is accelerated from rest through a potential difference of -2.36×10^5 V. What is its de Broglie wavelength?

4.2 Experimental Evidence for de Broglie Waves

5. Find the potential difference through which electrons must be accelerated (as in an electron microscope, for example) if we wish to resolve: (*a*) a virus of diameter 12 nm; (*b*) an atom of diameter 0.12 nm; (*c*) a proton of diameter 1.2 fm.

6. In an electron microscope we wish to study particles of diameter about 0.10 μm (about 1000 times the size of a single atom). (*a*) What should be the de Broglie wavelength of the electrons? (*b*) Through what potential difference should the electrons be accelerated to have that de Broglie wavelength?

7. In order to study the atomic nucleus, we would like to observe the diffraction of particles whose de Broglie wavelength is about the same size as the nuclear diameter, about 14 fm for a heavy nucleus such as lead. What kinetic energy should we use if the diffracted particles are (*a*) electrons? (*b*) Neutrons? (*c*) Alpha particles ($m = 4$ u)?

8. In the double-slit interference pattern for helium atoms (Figure 4.13), the kinetic energy of the beam of atoms was 0.020 eV. (*a*) What is the de Broglie wavelength of a helium atom with this kinetic energy? (*b*) Estimate the de Broglie wavelength of the atoms from the fringe spacing in Figure 4.13, and compare your estimate with the value obtained in part (*a*). The distance from the double slit to the scanning slit is 64 cm.

9. Suppose we wish to do a double-slit experiment with a beam of the smoke particles of Example 4.1c. Assume we can construct a double slit whose separation is about the same size as the particles. Estimate the separation between the fringes if the double slit and the screen were on opposite coasts of the United States.

10. In the Davisson-Germer experiment using a Ni crystal, a second-order beam is observed at an angle of 55°. For what accelerating voltage does this occur?

11. A certain crystal is cut so that the rows of atoms on its surface are separated by a distance of 0.352 nm. A beam of electrons is accelerated through a potential difference of 175 V and is incident normally on the surface. If all possible diffraction orders could be observed, at what angles (relative to the incident beam) would the diffracted beams be found?

4.3 Uncertainty Relationships for Classical Waves

12. Suppose a traveling wave has a speed v (where $v = \lambda f$). Instead of measuring waves over a distance Δx, we stay in one place and count the number of wave crests that pass in a time Δt. Show that Eq. 4.7 is equivalent to Eq. 4.4 for this case.

13. Sound waves travel through air at a speed of 330 m/s. A whistle blast at a frequency of about 1.0 kHz lasts for 2.0 s. (*a*) Over what distance in space does the "wave train" representing the sound extend? (*b*) What is the wavelength of the sound? (*c*) Estimate the precision with which an observer could measure the wavelength. (*d*) Estimate the precision with which an observer could measure the frequency.

14. A stone tossed into a body of water creates a disturbance at the point of impact that lasts for 4.0 s. The wave speed is 25 cm/s. (*a*) Over what distance on the surface of the water does the group of waves extend? (*b*) An observer counts 12 wave crests in the group. Estimate the precision with which the wavelength can be determined.

15. A radar transmitter emits a pulse of electromagnetic radiation with wavelength 0.225 m. The pulses have a duration of 1.17 μs. The receiver is set to accept a range of frequencies about the central frequency. To what range of frequencies should the receiver be set?

16. Estimate the signal processing time that would be necessary if you want to design a device to measure frequencies to a precision of no worse than 10,000 Hz.

4.4 Heisenberg Uncertainty Relationships

17. The speed of an electron is measured to within an uncertainty of 2.0×10^4 m/s. What is the size of the smallest region of space in which the electron can be confined?

18. An electron is confined to a region of space of the size of an atom (0.1 nm). (*a*) What is the uncertainty in the momentum of the electron? (*b*) What is the kinetic energy of an electron with a momentum equal to Δp? (*c*) Does this give a reasonable value for the kinetic energy of an electron in an atom?

19. The Σ^* particle has a rest energy of 1385 MeV and a lifetime of 2.0×10^{-23} s. What would be a typical range of outcomes of measurements of the Σ^* rest energy?

20. A pi meson (pion) and a proton can briefly join together to form a Δ particle. A measurement of the energy of the πp system (Figure 4.28) shows a peak at 1236 MeV, corresponding to the rest energy of the Δ particle, with an experimental spread of 120 MeV. What is the lifetime of the Δ?

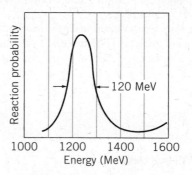

FIGURE 4.28 Problem 20.

21. A nucleus emits a gamma ray of energy 1.0 MeV from a state that has a lifetime of 1.2 ns. What is the uncertainty in the energy of the gamma ray? The best gamma-ray detectors can measure gamma-ray energies to a precision of no better than a few eV. Will this uncertainty be directly measurable?

22. In special conditions (see Section 12.9), it is possible to measure the energy of a gamma-ray photon to 1 part in 10^{15}. For a photon energy of 50 keV, estimate the maximum lifetime that could be determined by a direct measurement of the spread of the photon energy.

23. Alpha particles are emitted in nuclear decay processes with typical energies of 5 MeV. In analogy with Example 4.7, deduce whether the alpha particle can exist inside the nucleus.

4.5 Wave Packets

24. Use a distribution of wave numbers of constant amplitude in a range Δk about k_0:

$$A(k) = A_0 \qquad k_0 - \frac{\Delta k}{2} \leqslant k \leqslant k_0 + \frac{\Delta k}{2}$$

$$= 0 \qquad \text{otherwise}$$

and obtain Eq. 4.24 from Eq. 4.23.

25. Use the distribution of wave numbers $A(k) = A_0 e^{-(k-k_0)^2/2(\Delta k)^2}$ for $k = -\infty$ to $+\infty$ to derive Eq. 4.25.

26. Do the trigonometric manipulation necessary to obtain Eq. 4.18.

4.6 The Motion of a Wave Packet

27. Show that the data used in Figure 4.25 are consistent with Eq. 4.27; that is, use $\lambda_1 = 9$ and $\lambda_2 = 11$, $v_1 = 6$ and $v_2 = 4$ to show that $v_{\text{group}} = 15$.

28. (a) Show that the group velocity and phase velocity are related by:

$$v_{\text{group}} = v_{\text{phase}} - \lambda \frac{dv_{\text{phase}}}{d\lambda}$$

(b) When white light travels through glass, the phase velocity of each wavelength depends on the wavelength. (This is the origin of dispersion and the breaking up of white light into its component colors—different wavelengths travel at different speeds and have different indices of refraction.) How does v_{phase} depend on λ? Is $dv_{\text{phase}}/d\lambda$ positive or negative? Therefore, is $v_{\text{group}} > v_{\text{phase}}$ or $< v_{\text{phase}}$?

29. Certain surface waves in a fluid travel with phase velocity $\sqrt{b/\lambda}$, where b is a constant. Find the group velocity of a packet of surface waves, in terms of the phase velocity.

30. By a calculation similar to that of Eq. 4.30, show that $dE/dp = v$ remains valid when E represents the relativistic kinetic energy of the particle.

General Problems

31. A free electron bounces elastically back and forth in one dimension between two walls that are $L = 0.50$ nm apart. (a) Assuming that the electron is represented by a de Broglie standing wave with a node at each wall, show that the permitted de Broglie wavelengths are $\lambda_n = 2L/n$ ($n = 1, 2, 3, \ldots$). (b) Find the values of the kinetic energy of the electron for $n = 1, 2$, and 3.

32. A beam of thermal neutrons (see Problem 2) emerges from a nuclear reactor and is incident on a crystal as shown in Figure 4.29. The beam is Bragg scattered, as in Figure 3.5, from a crystal whose scattering planes are separated by 0.247 nm. From the continuous energy spectrum of the beam we wish to select neutrons of energy 0.0105 eV. Find the Bragg-scattering angle that results in a scattered beam of this energy. Will other energies also be present in the scattered beam at that angle?

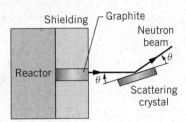

FIGURE 4.29 Problem 32.

33. (a) Find the de Broglie wavelength of a nitrogen molecule in air at room temperature (293 K). (b) The density of air at room temperature and atmospheric pressure is 1.292 kg/m^3. Find the average distance between air molecules at this temperature and compare with the de Broglie wavelength. What do you conclude about the importance of quantum effects in air at room temperature? (c) Estimate the temperature at which quantum effects might become important.

34. In designing an experiment, you want a beam of photons and a beam of electrons with the same wavelength of 0.281 nm, equal to the separation of the Na and Cl ions in a crystal of NaCl. Find the energy of the photons and the kinetic energy of the electrons.

35. A nucleus of helium with mass 5 u breaks up from rest into a nucleus of ordinary helium (mass = 4 u) plus a neutron (mass = 1 u). The rest energy liberated in the break-up is 0.89 MeV, which is shared (not equally) by the products. (a) Using energy and momentum conservation, find the kinetic energy of the neutron. (b) The lifetime of the original nucleus is 1.0×10^{-21} s. What range of values of the neutron kinetic energy might we measure in the laboratory as a result of the uncertainty relationship?

36. In a metal, the conduction electrons are not attached to any one atom, but are relatively free to move throughout the entire metal. Consider a cube of copper measuring 1.0 cm on each edge. (a) What is the uncertainty in any one component of the momentum of an electron confined to the metal? (b) Estimate the average kinetic energy of an electron in the metal. (Assume $\Delta p = [(\Delta p_x)^2 + (\Delta p_y)^2 + (\Delta p_z)^2]^{1/2}$.) (c) Assuming the heat capacity of copper to be 24.5 J/mole·K, would the contribution of this motion to the internal energy of the copper be important at room temperature? What do you conclude from this? (See also Problem 38.)

37. A proton or a neutron can sometimes "violate" conservation of energy by emitting and then reabsorbing a pi meson, which has a mass of $135\,\text{MeV}/c^2$. This is possible as long as the pi meson is reabsorbed within a short enough time Δt consistent with the uncertainty principle. (a) Consider $p \rightarrow p + \pi$. By what amount ΔE is energy conservation violated? (Ignore any kinetic energies.) (b) For how long a time Δt can the pi meson exist? (c) Assuming the pi meson to travel at very nearly the speed of light, how far from the proton can it go? (This procedure, as we discuss in Chapter 12, gives us an estimate of the *range* of the nuclear force, because protons and neutrons are held together in the nucleus by exchanging pi mesons.)

38. In a crystal, the atoms are a distance L apart; that is, each atom must be localized to within a distance of at most L. (a) What is the minimum uncertainty in the momentum of the atoms of a solid that are $0.20\,\text{nm}$ apart? (b) What is the average kinetic energy of such an atom of mass 65 u? (c) What would a collection of such atoms contribute to the internal energy of a typical solid, such as copper? Is this contribution important at room temperature? (See also Problem 36.)

39. An apparatus is used to prepare an atomic beam by heating a collection of atoms to a temperature T and allowing the beam to emerge through a hole of diameter d in one side of the oven. The beam then travels through a straight path of length L. Show that the uncertainty principle causes the diameter of the beam at the end of the path to be larger than d by an amount of order $L\hbar/d\sqrt{3mkT}$, where m is the mass of an atom. Make a numerical estimate for typical values of $T = 1500\,\text{K}, m = 7\,\text{u}$ (lithium atoms), $d = 3\,\text{mm}, L = 2\,\text{m}$.

Chapter 5

THE SCHRÖDINGER EQUATION

Quantum mechanics provides a mathematical framework in which the description of a process often includes different and possibly contradictory outcomes. A favorite illustration of that situation is the case of Schrödinger's cat. The cat is confined in a chamber with a radioactive atom, the decay of which will trigger the release of poison from a vial. Because we don't know exactly when that decay will occur, until an observation of the condition of the cat is made the quantum-mechanical description of the cat must include both "cat alive" and "cat dead" components.

The future behavior of a particle in a classical (nonrelativistic, nonquantum) situation may be predicted with absolute certainty using Newton's laws. If a particle interacts with its environment through a known force $\vec{F}$ (which might be associated with a potential energy U), we can do the mathematics necessary to solve Newton's second law, $\vec{F} = d\vec{p}/dt$ (a second-order, linear differential equation), and find the particle's location $\vec{r}(t)$ and velocity $\vec{v}(t)$ at all future times t. The mathematics may be difficult, and in fact it may not be possible to solve the equations in closed form (in which case an approximate solution can be obtained with the help of a computer). Aside from any such mathematical difficulties, the physics of the problem consists of writing down the original equation $\vec{F} = d\vec{p}/dt$ and interpreting its solutions $\vec{r}(t)$ and $\vec{v}(t)$. For example, a satellite or planet moving under the influence of a $1/r^2$ gravitational force can be shown, after the equations have been solved, to follow exactly an elliptical path.

In the case of nonrelativistic quantum physics, the basic equation to be solved is a second-order differential equation known as the *Schrödinger equation*. Like Newton's laws, the Schrödinger equation is written for a particle interacting with its environment, although we describe the interaction in terms of the potential energy rather than the force. Unlike Newton's laws, the Schrödinger equation does not give the trajectory of the particle; instead, its solution gives the *wave function* of the particle, which carries information about the particle's wavelike behavior. In this chapter we introduce the Schrödinger equation, obtain some of its solutions for certain potential energies, and learn how to interpret those solutions.

5.1 BEHAVIOR OF A WAVE AT A BOUNDARY

In studying wave motion, we often must analyze what occurs when a wave moves from one region or medium to a different region or medium in which the properties of the wave may change. For example, when a light wave moves from air into glass, its wavelength and the amplitude of its electric field both decrease. At every such boundary, a portion of the incident wave intensity is transmitted into the second medium and a portion is reflected back into the first medium.

Let's consider the case of a light wave incident on a glass plate, as in Figure 5.1a. At boundary A, the light wave moves from air (region 1) into glass (region 2), while at B the light wave moves from glass into air (region 3). The wavelength in air in region 3 is the same as the original wavelength of the incident wave in region 1, but the amplitude in region 3 is less than the amplitude in region 1, because some of the intensity is reflected at A and at B.

Other types of waves show similar behavior. For example, Figure 5.1b shows a surface water wave that moves into a region of shallower depth. In that region, its wavelength is smaller (but its amplitude is larger) compared with the original incident wave. When the wave enters region 3, in which the depth is the same as in region 1, the wavelength returns to its original value, but the amplitude of the wave is smaller in region 3 than in region 1 because some of the intensity was reflected at the two boundaries.

The same type of behavior occurs for de Broglie waves that characterize particles. Consider, for example, the apparatus in Figure 5.1c. Electrons are incident from the left and move inside a narrow metal tube that is at ground potential ($V = 0$). Another narrow tube in region 2 is connected to the negative terminal of a battery, which maintains it at a uniform potential of $-V_0$. Region 3 is connected to region 1 at ground potential. The gaps between the tubes can in

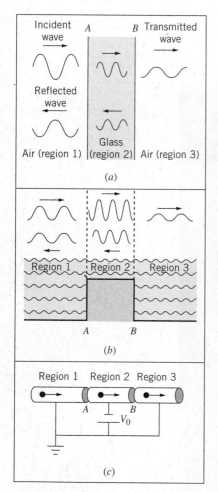

FIGURE 5.1 (a) A light wave in air is incident on a slab of glass, showing transmitted and reflected waves at the two boundaries (A and B). (b) A surface wave in water incident on a region of smaller depth similarly has transmitted and reflected waves. (c) The de Broglie waves of electrons moving from a region of constant zero potential to a region of constant negative potential V_0 also have transmitted and reflected components.

principle be made so small that we can regard the changes in potential at A and B as occurring suddenly. In region 1, the electrons have kinetic energy K, momentum $p = \sqrt{2mK}$, and de Broglie wavelength $\lambda = h/p$. In region 2, the potential energy for the electrons is $U = qV = (-e)(-V_0) = +eV_0$. We assume that the original kinetic energy of the electrons in region 1 is greater than eV_0, so that the electrons move into region 2 with a smaller kinetic energy (equal to $K - eV_0$), a smaller momentum, and thus a greater wavelength. When the electrons move from region 2 into region 3, they gain back the lost kinetic energy and move with their original kinetic energy K and thus with their original wavelength. As in the case of the light wave or the water wave, the amplitude of the de Broglie wave in region 3 is smaller than in region 1, meaning that the current of electrons in region 3 is smaller than the incident current, because some of the electrons are reflected at the boundaries at A and B.

We can thus identify a total of 5 waves moving in the three regions: (1) a wave moving to the right in region 1 (the incident wave); (2) a wave moving to the left in region 1 (representing the net combination of waves reflected from boundary A plus waves reflected from boundary B and then transmitted through boundary A back into region 1); (3) a wave moving to the right in region 2 (representing waves transmitted through boundary A plus waves reflected at B and then reflected again at A); (4) a wave moving to the left in region 2 (waves reflected at B); and (5) a wave moving to the right in region 3 (the transmitted waves at boundary B). Because we are assuming that waves are incident from region 1, it is not possible to have a wave moving to the left in region 3.

Penetration of the Reflected Wave

Another property of classical waves that carries over into quantum waves is penetration of a totally reflected wave into a forbidden region. When a light wave is completely reflected from a boundary, an exponentially decreasing wave called the *evanescent wave* penetrates into the second medium. Because 100% of the light wave intensity is reflected, the evanescent wave carries no energy and so cannot be directly observed in the second medium. But if we make the second medium very thin (perhaps equal to a few wavelengths of light) the light wave can emerge on the opposite side of the second medium. We'll discuss this phenomenon in more detail at the end of this chapter.

The same effect occurs with de Broglie waves. Suppose we increase the battery voltage in Figure 5.1c so that the potential energy in region 2 (equal to eV_0) is greater than the initial kinetic energy in region 1. The electrons do not have enough energy to enter region 2 (they would have negative kinetic there) and so all electrons are reflected back into region 1.

Like light waves, de Broglie waves can also penetrate into the forbidden region with exponentially decreasing amplitudes. However, because de Broglie waves are associated with the motion of electrons, that means that electrons must also penetrate a short distance into the forbidden region. The electrons cannot be directly observed in that region, because they have negative kinetic energy there. Nor can we do any experiment that would reveal their "real" existence in the forbidden region, such as measuring the speed of their passage through that region or detecting the magnetic field that their motion might produce.

One explanation for the penetration of the electrons into the forbidden region relies on the uncertainty principle—because we can't know exactly the energy of the incident electrons, we can't say with certainty that they don't have enough kinetic energy to penetrate into the forbidden region. For short enough time Δt, the

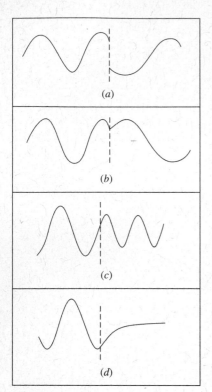

FIGURE 5.2   (*a*) A discontinuous wave. (*b*) A continuous wave with a discontinuous slope. (*c*) Two sine waves join smoothly. (*d*) A sine wave and an exponential join smoothly.

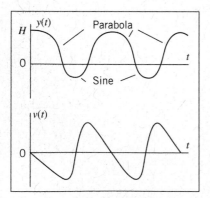

FIGURE 5.3 The position and velocity of a ball dropped from a height H above a springlike rubber sheet at $y = 0$.

energy uncertainty $\Delta E \sim \hbar/\Delta t$ might allow the electron to travel a short distance into the forbidden region, but this extra energy does not "belong to" the electron in any permanent sense. Later in this chapter we'll discuss a more mathematical approach to this explanation of penetration into the forbidden region.

Continuity at the Boundaries

When a wave such as a light wave or a water wave crosses a boundary as in Figure 5.1, the mathematical function that describes the wave must have two properties at each boundary:

1. The wave function must be continuous.
2. The slope of the wave function must be continuous, except when the boundary height is infinite.

Figure 5.2*a* shows a discontinuous wave function; the wave displacement changes suddenly at a single location. This type of behavior is not allowed. Figure 5.2*b* shows a continuous wave function (there are no gaps) with a discontinuous slope. This type of behavior is also not allowed, unless the boundary is of infinite height. Figures 5.2*c, d* show how two sine curves and an exponential and a sine can be joined so that both the function and the slope are continuous.

Across any non-infinite boundary, the wave must be smooth—no gaps in the function and no sharp changes in slope. When we solve for the mathematical form of a wave function, there are usually undetermined parameters, such as the amplitude and phase of the wave. In order to make the wave smooth at the boundary, we obtain the values of those coefficients by applying the two *boundary conditions* to make the function and its slope continuous. For example, at boundary A in Figure 5.1, we first evaluate the total wave function in region 1 at A and set it equal to the wave function in region 2 at A. This guarantees that the total wave function is continuous at A. We then take the derivative of the wave function in region 1, evaluate it at A, and set that equal to the derivative of the wave function in region 2 evaluated at A. This step makes the slope in region 1 match the slope in region 2 at boundary A. These two steps give us two equations relating the parameters of the waves and allow us to find relationships between the amplitudes and phases of the waves in regions 1 and 2. The process must be repeated at every boundary, such as at B in Figure 5.1 to match the waves in regions 2 and 3.

We can understand the exception to the continuity of the slope for infinite boundaries with an example from classical physics. Imagine a ball dropped from a height $y = H$ above a stretched rubber sheet at $y = 0$. The ball falls freely under gravity until it strikes the sheet, which we assume behaves like an elastic spring. The sheet stretches as the ball is brought to rest, after which the restoring force propels the ball upward. The motion of the ball might be represented by Figure 5.3. Above the sheet ($y > 0$) the motion is represented by parabolas, and while the ball is in contact with the sheet ($y < 0$) the motion is described by sine curves. Note how the curves join smoothly at $y = 0$, and note how both $y(t)$ and its derivative $v(t)$ are continuous.

On the other hand, imagine a ball hitting a steel surface, which we assume to be perfectly rigid. The ball rebounds elastically, and at the instant it is in contact with the surface its velocity reverses direction. The motion of the ball is represented in Figure 5.4. At the points of contact with the surface, there is a sudden change in the velocity, corresponding to an infinite acceleration and thus to an infinite force. The function $y(t)$ is continuous, but its slope is not—the function has no gaps, but it does have sharp "points" where the slope changes suddenly.

The assumption of the perfectly rigid surface is an idealization that we make to help us understand the situation and also to help simplify the mathematics. In reality the steel surface will flex slightly and ultimately behave somewhat like a much stiffer version of the rubber sheet. In quantum mechanics we will also sometimes use an assumption of a perfectly rigid or impenetrable boundary to help us understand and simplify the analysis of a more complicated physical situation.

In this section we have established several properties of classical waves that also apply to quantum waves:

1. When a wave crosses a boundary between two regions, part of the wave intensity is reflected and part is transmitted.
2. When a wave encounters a boundary to a region from which it is forbidden, the wave will penetrate perhaps by a few wavelengths before reflecting.
3. At a finite boundary, the wave and its slope are continuous. At an infinite boundary, the wave is continuous but its slope is discontinuous.

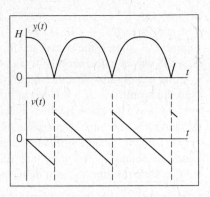

FIGURE 5.4 The position and velocity of a ball dropped from a height H above a rigid surface.

Example 5.1

In the geometry of Figure 5.1, the wave in region 1 is given by $y_1(x) = C_1 \sin(2\pi x/\lambda_1 - \phi_1)$, where $C_1 = 11.5$, $\lambda_1 = 4.97$ cm, and $\phi_1 = -65.3°$. In region 2, the wavelength is $\lambda_2 = 10.5$ cm. The boundary A is located at $x = 0$, and the boundary B is located at $x = L$, where $L = 20.0$ cm. Find the wave functions in regions 2 and 3.

Solution

The general form of the wave in region 2 can be represented in a form similar to that of the wave in region 1: $y_2(x) = C_2 \sin(2\pi x/\lambda_2 - \phi_2)$. To find the complete wave function in region 2, we must find the amplitude C_2 and the phase ϕ_2 by applying the boundary conditions on the function and its slope at boundary A ($x = 0$). Setting $y_1(x = 0) = y_2(x = 0)$ gives

$$-C_1 \sin\phi_1 = -C_2 \sin\phi_2$$

The slopes can be found from the derivative of the general form $dy/dx = (2\pi/\lambda)C\cos(2\pi x/\lambda - \phi)$ evaluated at $x = 0$:

$$\frac{2\pi}{\lambda_1}C_1 \cos\phi_1 = \frac{2\pi}{\lambda_2}C_2 \cos\phi_2$$

Dividing the first equation by the second eliminates C_2 and allows us to solve for ϕ_2:

$$\phi_2 = \tan^{-1}\left(\frac{\lambda_1}{\lambda_2}\tan\phi_1\right)$$

$$= \tan^{-1}\left(\frac{4.97\,\text{cm}}{10.5\,\text{cm}}\tan(-65.3°)\right)$$

$$= -45.8°$$

We can solve for C_2 using the result from applying the first boundary condition:

$$C_2 = C_1\frac{\sin\phi_1}{\sin\phi_2} = 11.5\frac{\sin(-65.3°)}{\sin(-45.8°)} = 14.6$$

To find the wave function in region 3, which we assume to have the same form $y_3(x) = C_3 \sin(2\pi x/\lambda_1 - \phi_3)$, we must apply the boundary conditions on y_2 and y_3 at $x = L$. Applying the two boundary conditions in the same way we did at $x = 0$, we obtain

$$C_2 \sin\left(\frac{2\pi L}{\lambda_2} - \phi_2\right) = C_3 \sin\left(\frac{2\pi L}{\lambda_1} - \phi_3\right)$$

$$\frac{2\pi}{\lambda_2}C_2 \cos\left(\frac{2\pi L}{\lambda_2} - \phi_2\right) = \frac{2\pi}{\lambda_1}C_3 \cos\left(\frac{2\pi L}{\lambda_1} - \phi_3\right)$$

Proceeding as we did before, we divide these two equations to find $\phi_3 = 60.9°$, and then from either equation obtain $C_3 = 7.36$. Our two solutions are then $y_2(x) = 14.6 \sin(2\pi x/10.5 + 45.8°)$ and $y_3(x) = 7.36 \sin(2\pi x/4.97 + 14.6°)$, with x measured in cm. Figure 5.5 shows the wave in all three regions. Note how the waves join smoothly at the boundaries.

How is it possible that the amplitude of y_2 can be greater than the amplitude of y_1? Keep in mind that y_1 represents the total wave in region 1, which includes the incident wave and the reflected wave. Depending on the phase difference between them, when the incident and reflected waves are added to obtain y_1, the amplitude of the resultant can be smaller than the amplitude of either wave.

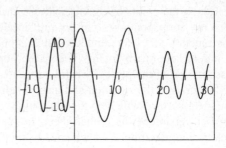

FIGURE 5.5 Example 5.1.

5.2 CONFINING A PARTICLE

A free particle (that is, a particle on which no forces act anywhere) is by definition not confined, so it can be located anywhere. It has, as we discussed in Chapter 4, a definite wavelength, momentum, and energy (for which we can choose any value).

A confined particle, on the other hand, is represented by a wave packet that makes it likely to be found only in a region of space of size Δx. We construct such a wave packet by adding together different sine or cosine waves to obtain the desired mathematical shape.

In quantum mechanics, we often want to analyze the behavior of confined particles, for example an electron that is attached to a specific atom or molecule. We'll consider the properties of atomic electrons beginning in Chapter 6, but for now let's look at a simpler problem: an electron moving in one dimension and confined by a series of electric fields. Figure 5.6 shows how the apparatus of Figure 5.1c might be modified for this purpose. The center section is grounded (so that $V = 0$) and the two side sections are connected to batteries so that they are at potentials of $-V_0$ relative to the center section. As before, we assume that the gaps between the center section and the side sections can be made as narrow as possible, so we can regard the potential energy as changing instantaneously at the boundaries A and B. This arrangement is often called a *potential energy well*.

The potential energy of an electron in this situation is then 0 in the center section and $U_0 = qV = (-e)(-V_0) = +eV_0$ in the two side sections as shown in Figure 5.6. To confine the electron, we want to consider cases in which it moves in the center section with a kinetic energy K that is less than U_0. For example, the electron might have a kinetic energy of 5 eV in the center section, and the side sections might have potential energies of 10 eV. The electron thus does not have enough energy to "climb" the potential energy hill between the center section and the side sections, and (at least from the classical point of view) the electron is confined to the center section.

We'll discuss the full solution to this problem later in this chapter, but for now let's simplify even further and consider the case of an infinitely high potential energy barrier at A and B. This is a good approximation to the situation in which the kinetic energy of the electron in the center section is much smaller than the potential energy supplied by the batteries. In this case the penetration

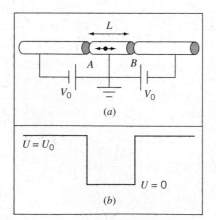

FIGURE 5.6 (a) Apparatus for confining an electron to the center region of length L. (b) The potential energy of an electron in this apparatus.

into the forbidden region, which we discussed in Section 5.1, cannot occur. The probability to find the electron in either of the side regions is therefore precisely zero everywhere in those regions, and thus the wave amplitude is zero everywhere in those regions, including at the boundaries (locations A and B). For the wave function to be continuous, the wave function in the center section must have values of zero at A and B.

Of all the possible waves that might be used to describe the particle in this center section, the continuity condition restricts us to waves that have zero amplitude at the boundaries. Some of those waves are illustrated in Figure 5.7. Note that the wave function is continuous, but its slope is not (there are sharp points in the function at locations A and B). This is an example of the exception to the second boundary condition—the slope may be discontinuous at an infinite barrier.

In contrast to the free particle for which the wavelength could have any value, *only certain values of the wavelength are allowed*. The de Broglie relationship then tells us that only certain values of the momentum are allowed, and consequently only certain values of the energy are allowed. The energy is not a continuous variable, free to take on any arbitrary value; instead, the energy is a discrete variable that is restricted to a certain set of values. This is known as *quantization of energy*.

You can see directly from Figure 5.7 that the allowed wavelengths are $2L, L, 2L/3, \ldots$, where L is the length of the center section. We can write these wavelengths as

$$\lambda_n = \frac{2L}{n} \qquad n = 1, 2, 3, \ldots \tag{5.1}$$

This set of wavelengths is identical to the wavelengths of the classical problem of standing waves on a string stretched between two points. From the de Broglie relationship $\lambda = h/p$ we obtain

$$p_n = n\frac{h}{2L} \tag{5.2}$$

The energy of the particle in the center section is only kinetic energy $p^2/2m$, and so

$$E_n = n^2 \frac{h^2}{8mL^2} \tag{5.3}$$

These are the allowed or quantized values of the energy of the electron.

A wave packet describing the electron in this region must be a combination of waves with the allowed values of the wavelengths. However, it is not necessary to construct a wave packet from a combination of waves to describe this confined particle. Even a single one of these waves represents the confined particle, because the wave function must be zero in the forbidden regions. So the waveforms shown in Figure 5.7 can represent wave packets of this confined electron, each wave packet consisting of only a single wave.

The appearance of energy quantization accompanies every attempt to confine a particle to a finite region of space. Quantization of energy is one of the principal features of the quantum theory, and studying the quantized energy levels of systems (such as by observing the energies of emitted photons) is an important technique of experimental physics that gives us information about the properties of atoms and nuclei.

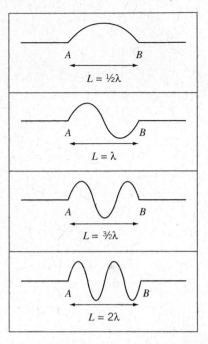

FIGURE 5.7 Some possible waves that might be used to describe an electron confined by an infinite potential energy barrier to a region of length L.

Applying the Uncertainty Principle to a Confined Particle

In Chapter 4 we constructed wave packets and showed how the uncertainty principle related the size of the wave packet to the range of wavelengths that was used in its construction. Let's now see how the Heisenberg uncertainty relationships apply in the case of a confined particle.

In the arrangement of Figure 5.6 (with infinitely high barriers on each side), the particle is known to be somewhere in the center section of the apparatus, and thus $\Delta x \sim L$ is a reasonable estimate of the uncertainty in its location. To find the uncertainty in its momentum, we use the rigorous definition of uncertainty given in Eq. 4.15: $\Delta p_x = \sqrt{(p_x^2)_{av} - (p_{x,av})^2}$. The particle moving in the center section can be considered to be moving to the left or to the right with equal probability (just as the classical standing-wave problem can be analyzed as the superposition of identical waves moving to the left and to the right). Thus $p_{x,av} = 0$. If the particle is moving with a momentum given by Eq. 5.2, $p_x^2 = (nh/L)^2$ and so $\Delta p_x = nh/L$. Combining the uncertainties in position and momentum, we have

$$\Delta x \Delta p_x \sim L \frac{nh}{L} = nh \tag{5.4}$$

The product of the uncertainties is certainly greater than $\hbar/2$, and so the result of confining the particle is entirely consistent with the Heisenberg uncertainty relationship. Note that even the smallest possible value of the product of the uncertainties (which is obtained for $n = 1$) is still much larger than the minimum value given by the uncertainty principle.

Later in this chapter, we will use a more rigorous way to evaluate the uncertainty in position using a formula similar to Eq. 4.15 to find the uncertainty in position, and we will find that the result does not differ very much from the estimate of Eq. 5.4.

5.3 THE SCHRÖDINGER EQUATION

The differential equation whose solution gives us the wave behavior of particles is called the *Schrödinger equation*. It was developed in 1926 by Austrian physicist Erwin Schrödinger. The equation cannot be derived from any previous laws or postulates; like Newton's equations of motion or Maxwell's equations of electromagnetism, it is a new and independent result whose correctness can be determined only by comparing its predictions with experimental results. For nonrelativistic motion, the Schrödinger equation gives results that correctly account for observations at the atomic and subatomic level.

We can justify the form of the Schrödinger equation by examining the solution expected for the free particle, which should give a wave whose shape at any particular time, specified by the *wave function* $\psi(x)$, is that of a simple de Broglie wave, such as $\psi(x) = A \sin kx$, where A is the amplitude of the wave and $k = 2\pi/\lambda$. If we are looking for a differential equation, then we need to take some derivatives:

$$\frac{d\psi}{dx} = kA \cos kx, \qquad \frac{d^2\psi}{dx^2} = -k^2 A \sin kx = -k^2 \psi(x)$$

Erwin Schrödinger (1887–1961, Austria). Although he disagreed with the probabilistic interpretation that was later given to his work, he developed the mathematical theory of wave mechanics that for the first time permitted the wave behavior of physical systems to be calculated.

Note that the second derivative gives the original function again. With the kinetic energy $K = p^2/2m = (h/\lambda)^2/2m = \hbar^2 k^2/2m$, we can then write

$$\frac{d^2\psi}{dx^2} = -k^2\psi(x) = -\frac{2m}{\hbar^2}K\psi(x) = -\frac{2m}{\hbar^2}(E - U)\psi(x)$$

where $E = K + U$ is the nonrelativistic total energy of the particle. For a free particle, $U = 0$ so $E = K$; however, we are using the free particle solution to try to extend to the more general case in which there is a potential energy $U(x)$. The equation then becomes

$$-\frac{\hbar^2}{2m}\frac{d^2\psi}{dx^2} + U(x)\psi(x) = E\psi(x) \tag{5.5}$$

Equation 5.5 is the *time-independent Schrödinger equation* for one-dimensional motion.

The solution to Eq. 5.5 gives the shape of the wave at time $t = 0$. The mathematical function that describes a one-dimensional *traveling* wave must involve both x and t. This wave is represented by the function $\Psi(x,t)$:

$$\Psi(x,t) = \psi(x)e^{-i\omega t} \tag{5.6}$$

The time dependence is given by the complex exponential function $e^{-i\omega t}$ with $\omega = E/\hbar$. (You can find a few useful formulas involving complex numbers in Appendix B.) We'll discuss the time-dependent part later in this chapter. For now, we'll concentrate on the time-independent function $\psi(x)$.

We assume that we know the potential energy $U(x)$, and we wish to obtain the wave function $\psi(x)$ and the energy E *for that potential energy.* This is a general example of a type of problem known as an *eigenvalue* problem; we find that it is possible to obtain solutions to the equation only for particular values of E, which are known as the *energy eigenvalues.*

The general procedure for solving the Schrödinger equation is as follows:

1. Begin by writing Eq. 5.5 with the appropriate $U(x)$. Note that if the potential energy changes discontinuously [$U(x)$ may be represented by a discontinuous function; $\psi(x)$ may *not*], we may need to write different equations for different regions of space. Examples of this sort are given in Section 5.4.
2. Using general mathematical techniques suited to the form of the equation, find a mathematical function $\psi(x)$ that is a solution to the differential equation. Because there is no one specific technique for solving differential equations, we will study several examples to learn how to find solutions.
3. In general, several solutions may be found. By applying boundary conditions some of these may be eliminated and some arbitrary constants may be determined. It is generally the application of the boundary conditions that selects out the allowed energies.
4. If you are seeking solutions for a potential energy that changes discontinuously, you must apply the continuity conditions on $\psi(x)$ (and usually on $d\psi/dx$) at the boundary between different regions.

Because the Schrödinger equation is linear, any constant multiplying a solution is also a solution. The method to determine the amplitude of the wave function is discussed in the next section.

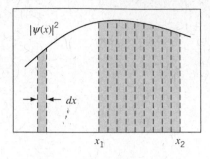

FIGURE 5.8 The probability to find the particle in a small region of width dx is equal to the area of the strip under the $|\psi(x)|^2$ curve. The total probability to find the particle between x_1 and x_2 is the sum of the areas of the strips, equal to the integral between those limits.

Probabilities and Normalization

The remaining steps in the procedure for applying the Schrödinger recipe depend on the physical interpretation of the solution to the differential equation. Our original goal in solving the Schrödinger equation was to obtain the wave properties of the particle. What does the amplitude of $\psi(x)$ represent, and what is the physical variable that is waving? It is certainly not a displacement, as in the case of a water wave or a wave on a stretched piano wire, nor is it a pressure wave, as in the case of sound. *It is a very different kind of wave, whose squared absolute amplitude gives the probability for finding the particle in a given region of space.*

If we define $P(x)$ as *the probability density* (probability per unit length, in one dimension), then according to the Schrödinger recipe

$$P(x)\,dx = |\psi(x)|^2\,dx \tag{5.7}$$

as indicated in Figure 5.8. In Eq. 5.7, $|\psi(x)|^2\,dx$ gives the probability to find the particle in the interval dx at x (that is, between x and $x + dx$).* Because the wave function $\psi(x)$ might be a complex function, it is necessary to square its absolute magnitude to make sure that the probability is a positive real number.

The squared magnitude of the general time-dependent wave function (Eq. 5.6) is:

$$|\Psi(x,t)|^2 = |\psi(x)|^2 |e^{-i\omega t}|^2 = |\psi(x)|^2 \tag{5.8}$$

where the last step can be taken because the magnitude of the time-dependent factor is 1. For this reason, the probability density associated with a solution to the Schrödinger equation (for any allowed value of E) is independent of time. These special quantum states are called *stationary states*.

This interpretation of $|\psi(x)|^2$ helps us to understand the continuity condition of $\psi(x)$. We must not allow the probability to change discontinuously, but, like any well-behaved wave, the probability to locate the particle varies smoothly and continuously.

This interpretation of $\psi(x)$ now permits us to complete the Schrödinger recipe and to illustrate how to use the wave function to calculate quantities that we can measure in the laboratory. Steps 1 through 4 were given previously; the recipe continues:

5. For a wave function describing a single particle, the probability summed over all locations must give 100%—that is, the particle must be located *somewhere* between $x = -\infty$ and $x = +\infty$. The probability to find the particle in a small interval was given in Eq. 5.7. The total probability to find the particle in all such intervals must be exactly 1:

$$\int_{-\infty}^{+\infty} |\psi(x)|^2\,dx = 1 \tag{5.9}$$

The Schrödinger equation is linear, which means that if $\psi(x)$ is a solution then any constant times $\psi(x)$ is also a solution. For the probability to be a meaningful concept, this constant must be chosen so that Eq. 5.9 is satisfied.

---

*It is not correct to speak of "the probability to find the particle at the point x." A single point is a mathematical abstraction with no physical dimension. The probability of finding a particle *at a point* is zero, but there can be a nonzero probability of finding the particle *in an interval*.

A wave function with its multiplicative constant chosen in this way is said to be *normalized*, and Eq. 5.9 is known as the *normalization condition*.

6. Because the solution to the Schrödinger equation represents a probability, any solution that becomes infinite must be discarded—it makes no sense to have an infinite probability to find a particle in any interval. In practice, we "discard" a solution by setting its multiplicative constant equal to zero. For example, if the mathematical solution to the differential equation yields $\psi(x) = Ae^{kx} + Be^{-kx}$ for the *entire* region $x > 0$, then we must require $A = 0$ for the solution to be physically meaningful; otherwise $|\psi(x)|^2$ would become infinite as x goes to infinity. On the other hand, if this solution is to be valid in the *entire* region $x < 0$, then we must set $B = 0$. However, if the solution is to be valid only in a small portion of the range of x—say, $0 < x < L$—then we cannot set either $A = 0$ or $B = 0$.

7. Suppose the interval between two points x_1 and x_2 is divided into a series of infinitesimal intervals of width dx (Figure 5.8). To find the total probability for the particle to be located between x_1 and x_2, which we represent as $P(x_1 : x_2)$, we calculate the sum of all the probabilities $P(x)\, dx$ in each interval dx. This sum can be expressed as an integral:

$$P(x_1 : x_2) = \int_{x_1}^{x_2} P(x)\, dx = \int_{x_1}^{x_2} |\psi(x)|^2\, dx \qquad (5.10)$$

If the wave function has been properly normalized, Eq. 5.10 will always yield a probability that lies between 0 and 1.

8. Because we can no longer speak with certainty about the position of the particle, we can no longer guarantee the outcome of a single measurement of any physical quantity that depends on its position. Instead, we can find the *average* outcome of a large number of measurements. For example, suppose we wish to find the average location of a particle by measuring its coordinate x. From a large number of measurements, we find the value x_1 a certain number of times n_1, x_2 a number of times n_2, etc., and in the usual way we can find the average value

$$x_{av} = \frac{n_1 x_1 + n_2 x_2 + \cdots}{n_1 + n_2 + \cdots} = \frac{\sum n_i x_i}{\sum n_i} \qquad (5.11)$$

The number of times n_i that we measure each x_i is proportional to the probability $P(x_i)dx$ to find the particle in the interval dx at x_i. Making this substitution and changing the sums to integrals, we have

$$x_{av} = \frac{\displaystyle\int_{-\infty}^{+\infty} P(x)x\, dx}{\displaystyle\int_{-\infty}^{+\infty} P(x)\, dx} = \int_{-\infty}^{+\infty} |\psi(x)|^2 x\, dx \qquad (5.12)$$

where the last step can be made if the wave function is normalized, in which case the denominator of Eq. 5.12 is equal to one.

By analogy, the average value of any function of x can be found:

$$[f(x)]_{av} = \int_{-\infty}^{+\infty} P(x)f(x)\, dx = \int_{-\infty}^{+\infty} |\psi(x)|^2 f(x)\, dx \qquad (5.13)$$

Average values calculated according to Eq. 5.12 or 5.13 are known as *expectation values*.

5.4 APPLICATIONS OF THE SCHRÖDINGER EQUATION

Solutions for Constant Potential Energy

First let's examine the solutions to the Schrödinger equation for the special case of a constant potential energy, equal to U_0. Then Eq. 5.5 becomes

$$-\frac{\hbar^2}{2m}\frac{d^2\psi}{dx^2} + U_0\psi(x) = E\psi(x) \tag{5.14}$$

or (assuming for now that $E > U_0$)

$$\frac{d^2\psi}{dx^2} = -k^2\psi(x) \quad \text{with} \quad k = \sqrt{\frac{2m(E - U_0)}{\hbar^2}} \tag{5.15}$$

The parameter k in this equation is equal to the wave number $2\pi/\lambda$.

The solution to Eq. 5.15 is a function of x that, when differentiated twice, gives back the original function multiplied by the negative constant $-k^2$. The function that has this property is $\sin kx$ or $\cos kx$. The most general solution to the equation is

$$\psi(x) = A\sin kx + B\cos kx \tag{5.16}$$

The constants A and B must be determined by applying the continuity and normalization requirements. We can demonstrate that Eq. 5.16 satisfies Eq. 5.15 by taking two derivatives:

$$\frac{d\psi}{dx} = kA\cos kx - kB\sin kx$$

$$\frac{d^2\psi}{dx^2} = -k^2A\sin kx - k^2B\cos kx = -k^2(A\sin kx + B\cos kx) = -k^2\psi(x)$$

so the original equation is indeed satisfied.

To analyze the penetration of a particle into a forbidden region, we must consider the case in which the energy E of the particle is smaller than the potential energy U_0. For this case we can rewrite Eq. 5.14 as

$$\frac{d^2\psi}{dx^2} = k'^2\psi(x) \quad \text{with} \quad k' = \sqrt{\frac{2m(U_0 - E)}{\hbar^2}} \tag{5.17}$$

In this case the general solution in the forbidden regions is

$$\psi(x) = Ae^{k'x} + Be^{-k'x} \tag{5.18}$$

Once again, we can demonstrate that Eq. 5.18 is a solution of Eq. 5.17 by taking two derivatives:

$$\frac{d\psi}{dx} = k'Ae^{k'x} - k'Be^{-k'x}$$

$$\frac{d^2\psi}{dx^2} = k'^2Ae^{k'x} + k'^2Be^{-k'x} = k'^2(Ae^{k'x} + Be^{-k'x}) = k'^2\psi(x)$$

We will use Eqs. 5.16 and 5.18 as our solutions to the Schrödinger equation for constant potential energy in the allowed ($E > U_0$) and forbidden ($E < U_0$) regions.

The Free Particle

For a free particle, the force is zero and so the potential energy is constant. We may choose any value for that constant, so for convenience we'll choose $U_0 = 0$. The solution is given by Eq. 5.16, $\psi(x) = A \sin kx + B \cos kx$. The energy of the particle is

$$E = \frac{\hbar^2 k^2}{2m} \tag{5.19}$$

Our solution has placed no restrictions on k, so the energy is permitted to have any value (in the language of quantum physics, we say that the energy is *not* quantized). We note that Eq. 5.19 is the kinetic energy of a particle with momentum $p = \hbar k$ or, equivalently, $p = h/\lambda$. This is as we would have expected, because the free particle can be represented by a de Broglie wave with any wavelength.

Solving for A and B presents some difficulties because the normalization integral, Eq. 5.9, cannot be evaluated from $-\infty$ to $+\infty$ for this wave function. We therefore cannot determine probabilities for the free particle from the wave function of Eq. 5.16.

It is also instructive to write the wave function in terms of complex exponentials, using $\sin kx = (e^{ikx} - e^{-ikx})/2i$ and $\cos kx = (e^{ikx} + e^{-ikx})/2$:

$$\psi(x) = A\left(\frac{e^{ikx} - e^{-ikx}}{2i}\right) + B\left(\frac{e^{ikx} + e^{-ikx}}{2}\right) = A'e^{ikx} + B'e^{-ikx} \tag{5.20}$$

where $A' = A/2i + B/2$ and $B' = -A/2i + B/2$. To interpret this solution in terms of waves we form the complete time-dependent wave function using Eq. 5.6:

$$\Psi(x,t) = (A'e^{ikx} + B'e^{-ikx})e^{-i\omega t} = A'e^{i(kx-\omega t)} + B'e^{-i(kx+\omega t)} \tag{5.21}$$

The dependence of the first term on $kx - \omega t$ identifies this term as representing a wave moving to the right (in the positive x direction) with amplitude A', and the second term involving $kx + \omega t$ represents a wave moving to the left (in the negative x direction) with amplitude B'.

If we want the wave to represent a beam of particles moving in the $+x$ direction, then we must set $B' = 0$. The probability density associated with this wave is then, according to Eq. 5.7,

$$P(x) = |\psi(x)|^2 = |A'|^2 e^{ikx} e^{-ikx} = |A'|^2 \tag{5.22}$$

The probability density is constant, meaning the particles are equally likely to be found anywhere along the x axis. This is consistent with our discussion of the free-particle de Broglie wave in Chapter 4—a wave of precisely defined wavelength extends from $x = -\infty$ to $x = +\infty$ and thus gives a completely unlocalized particle.

Infinite Potential Energy Well

Now we'll consider the formal solution to the problem we discussed in Section 5.2: a particle is trapped in the region between $x = 0$ and $x = L$ by infinitely high potential energy barriers. Imagine an apparatus like that of Figure 5.6, in which the particle moves freely in this region and makes elastic collisions with the perfectly rigid barriers that confine it. This problem is sometimes called "a particle in a box." For now we'll assume that the particle moves in only one dimension; later we'll expand to two and three dimensions.

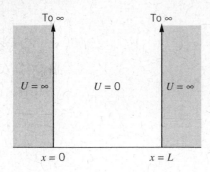

FIGURE 5.9 The potential energy of a particle that moves freely ($U = 0$) in the region $0 \leq x \leq L$ but is completely excluded ($U = \infty$) from the regions $x < 0$ and $x > L$.

The potential energy may be expressed as:

$$U(x) = 0 \qquad 0 \leq x \leq L$$
$$= \infty \qquad x < 0, x > L \qquad (5.23)$$

The potential energy is shown in Figure 5.9. We are free to choose any constant value for U in the region $0 \leq x \leq L$; we choose it to be zero for convenience.

Because the potential energy is different in the regions inside and outside the well, we must find separate solutions in each region. We can analyze the outside region in either of two ways. If we examine Eq. 5.5 for the region outside the well, we find that the only way to keep the equation from becoming meaningless when $U \to \infty$ is to require $\psi = 0$, so that $U\psi$ will not become infinite. Alternatively, we can go back to the original statement of the problem. If the walls at the boundaries of the well are perfectly rigid, the particle must always be in the well, and the probability for finding it outside must be zero. To make the probability zero everywhere outside the well, we must make $\psi = 0$ everywhere outside. Thus we have

$$\psi(x) = 0 \qquad x < 0, x > L \qquad (5.24)$$

The Schrödinger equation for $0 \leq x \leq L$, when $U(x) = 0$, is identical with Eq. 5.14 with $U_0 = 0$ and has the same solution:

$$\psi(x) = A \sin kx + B \cos kx \qquad 0 \leq x \leq L \qquad (5.25)$$

with

$$k = \sqrt{\frac{2mE}{\hbar^2}} \qquad (5.26)$$

Our solution is not yet complete, for we have not evaluated A or B, nor have we found the allowed values of the energy E. To do this, we must apply the requirement that $\psi(x)$ is continuous across any boundary. In this case, we require that our solutions for $x < 0$ and $x > 0$ match up at $x = 0$; similarly, the solutions for $x > L$ and $x < L$ must match at $x = L$.

Let us begin at $x = 0$. At $x < 0$, we have found that $\psi = 0$, and so we must set $\psi(x)$ of Eq. 5.25 to zero at $x = 0$.

$$\psi(0) = A \sin 0 + B \cos 0 = 0 \qquad (5.27)$$

which gives $B = 0$. Because $\psi = 0$ for $x > L$, the second boundary condition is $\psi(L) = 0$, so

$$\psi(L) = A \sin kL + B \cos kL = 0 \qquad (5.28)$$

We have already found $B = 0$, so we must now have $A \sin kL = 0$. Either $A = 0$, in which case $\psi = 0$ *everywhere*, $\psi^2 = 0$ everywhere, and there is no particle (a meaningless solution) or else $\sin kL = 0$, which is true only when $kL = \pi, 2\pi, 3\pi, \ldots$, or

$$kL = n\pi \qquad n = 1, 2, 3, \ldots \qquad (5.29)$$

With $k = 2\pi/\lambda$, we have $\lambda = 2L/n$; this is identical with the result obtained in introductory mechanics for the wavelengths of the standing waves in a string of length L fixed at both ends, which we already obtained in Section 5.2 (Eq. 5.1). *Thus the solution to the Schrödinger equation for a particle trapped in a linear region of length L is a series of standing de Broglie waves!* Not all wavelengths are permitted; only certain values, determined from Eq. 5.29, may occur.

From Eq. 5.26 we find that, because only certain values of k are permitted by Eq. 5.29, only certain values of E may occur—*the energy is quantized!* Solving Eq. 5.29 for k and substituting into Eq. 5.26, we obtain

$$E_n = \frac{\hbar^2 k^2}{2m} = \frac{\hbar^2 \pi^2 n^2}{2mL^2} = \frac{h^2 n^2}{8mL^2} \qquad n = 1, 2, 3, \ldots \qquad (5.30)$$

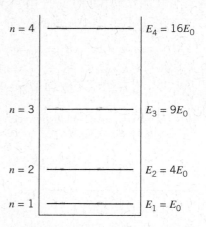

FIGURE 5.10 The first four energy levels in a one-dimensional infinite potential energy well.

For convenience, let $E_0 = \hbar^2\pi^2/2mL^2 = h^2/8mL^2$; this unit of energy is determined by the mass of the particle and the width of the well. Then $E_n = n^2 E_0$, and the only allowed energies for the particle are $E_0, 4E_0, 9E_0, 16E_0$, etc. All intermediate values, such as $3E_0$ or $6.2E_0$, are forbidden. Figure 5.10 shows the allowed energy levels. The lowest energy state, for which $n = 1$, is known as the *ground state,* and the states with higher energies ($n > 1$) are known as *excited states.*

Because the energy is purely kinetic in this case, our result means that only certain speeds are permitted for the particle. This is very different from the case of the classical trapped particle, in which the particle can be given any initial velocity and will move forever, back and forth, at the same speed. In the quantum case, this is not possible; only certain initial speeds can result in sustained states of motion; these special conditions represent the "stationary states." Average values calculated according to Eq. 5.13 likewise do not change with time.

From one energy state, the particle can make jumps or transitions to another energy state by absorbing or releasing an amount of energy equal to the energy difference between the two states. By absorbing energy the particle will move to a higher energy state, and by releasing energy it moves to a lower energy state. A similar effect occurs for electrons in atoms, in which the absorbed or released energy is usually in the form of a photon of visible light or other electromagnetic radiation. For example, from the state with $n = 3$ ($E_3 = 9E_0$), the particle might absorb an energy of $\Delta E = 7E_0$ and jump upward to the $n = 4$ state ($E_4 = 16E_0$) or might release energy of $\Delta E = 5E_0$ and jump downward to the $n = 2$ state ($E_2 = 4E_0$).

Example 5.2

An electron is trapped in a one-dimensional region of length 1.00×10^{-10} m (a typical atomic diameter). (*a*) Find the energies of the ground state and first two excited states. (*b*) How much energy must be supplied to excite the electron from the ground state to the second excited state? (*c*) From the second excited state, the electron drops down to the first excited state. How much energy is released in this process?

Solution

(*a*) The basic quantity of energy needed for this calculation is

$$E_0 = \frac{h^2}{8mL^2} = \frac{(hc)^2}{8mc^2 L^2}$$

$$= \frac{(1240 \, \text{eV} \cdot \text{nm})^2}{8(511,000 \, \text{eV})(0.100 \, \text{nm})^2} = 37.6 \, \text{eV}$$

With $E_n = n^2 E_0$, we can find the energy of the states:

$$n = 1 : \quad E_1 = E_0 = 37.6\,\text{eV}$$
$$n = 2 : \quad E_2 = 4E_0 = 150.4\,\text{eV}$$
$$n = 3 : \quad E_3 = 9E_0 = 338.4\,\text{eV}$$

(b) The energy difference between the ground state and the second excited state is

$$\Delta E = E_3 - E_1 = 338.4\,\text{eV} - 37.6\,\text{eV} = 300.8\,\text{eV}$$

This is the energy that must be absorbed for the electron to make this jump.

(c) The energy difference between the second and first excited states is

$$\Delta E = E_3 - E_2 = 338.4\,\text{eV} - 150.4\,\text{eV} = 188.0\,\text{eV}$$

This is the energy that is released when the electron makes this jump.

To complete the solution for $\psi(x)$, we must determine the constant A by using the normalization condition given in Eq. 5.9, $\int_{-\infty}^{+\infty} |\psi(x)|^2 dx = 1$. The integrand is zero in the regions $-\infty < x \leq 0$ and $L \leq x < +\infty$, so all that remains is

$$\int_0^L A^2 \sin^2 \frac{n\pi x}{L}\, dx = 1 \tag{5.31}$$

from which we find $A = \sqrt{2/L}$. The complete wave function for $0 \leq x \leq L$ is then

$$\psi_n(x) = \sqrt{\frac{2}{L}} \sin \frac{n\pi x}{L} \qquad n = 1, 2, 3, \ldots \tag{5.32}$$

In Figure 5.11, the wave functions and probability densities ψ^2 are illustrated for the lowest several states.

In the ground state, the particle has the greatest probability to be found near the middle of the well ($x = L/2$), and the probability falls off to zero between the center and the sides of the well. This is very different from the behavior of a classical particle—a classical particle moving at constant speed would be found with equal probability at every location inside the well. The quantum particle also has constant speed but yet is still found with differing probability at various locations in the well. It is the wave nature of the quantum particle that is responsible for this very nonclassical behavior.

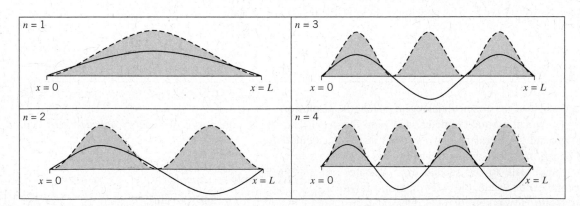

FIGURE 5.11 The wave functions (solid lines) and probability densities (shaded regions) of the first four states in the one-dimensional infinite potential energy well.

Another example of nonclassical behavior occurs for the first excited state. The probability density has a maximum at $x = L/4$ and another maximum at $x = 3L/4$. Between the two maxima, there is zero probability to find the particle in the center of the well at $x = L/2$. How can the particle travel from $x = L/4$ to $x = 3L/4$ without ever being at $x = L/2$? Of course, no classical particle could behave in such a way, but it is a common behavior for waves. For example, the first overtone of a vibrating string has a node at its midpoint and antinodes (vibrational maxima) at the $1/4$ and $3/4$ locations.

The calculation of probabilities and average values is illustrated by the following examples.

Example 5.3

Consider again an electron trapped in a one-dimensional region of length 1.00×10^{-10} m $= 0.100$ nm. (a) In the ground state, what is the probability of finding the electron in the region from $x = 0.0090$ nm to 0.0110 nm? (b) In the first excited state, what is the probability of finding the electron between $x = 0$ and $x = 0.025$ nm?

Solution
(a) When the interval is small, it is often simpler to use Eq. 5.7 to find the probability, instead of using the integration method. The width of the small interval is $dx = 0.0110$ nm $- 0.0090$ nm $= 0.0020$ nm. Evaluating the wave function at the midpoint of the interval ($x = 0.0100$ nm), we can use the $n = 1$ wave function with Eq. 5.7 to find

$$P(x)\, dx = |\psi_1(x)|^2\, dx = \frac{2}{L} \sin^2 \frac{\pi x}{L}\, dx$$

$$= \frac{2}{0.100\,\text{nm}} \sin^2 \frac{\pi(0.0100\,\text{nm})}{0.100\,\text{nm}} (0.002\,\text{nm})$$

$$= 0.0038 = 0.38\%$$

(b) For this wide interval, we must use the integration method to find the probability:

$$P(x_1 : x_2) = \int_{x_1}^{x_2} |\psi_2(x)|^2\, dx$$

$$= \frac{2}{L} \int_{x_1}^{x_2} \sin^2 \frac{2\pi x}{L}\, dx$$

$$= \left(\frac{x}{L} - \frac{1}{4\pi} \sin \frac{4\pi x}{L} \right)\Bigg|_{x_1}^{x_2}$$

Evaluating this expression using the limits $x_1 = 0$ and $x_2 = 0.025$ nm gives a probability of 0.25 or 25%. This result is of course what we would expect by inspection of the graph of ψ^2 for $n = 2$ in Figure 5.11. The interval from $x = 0$ to $x = L/4$ contains 25% of the total area under the ψ^2 curve.

Example 5.4

Show that the average value of x is $L/2$, independent of the quantum state.

Solution
We use Eq. 5.12; because $\psi = 0$ except for $0 \le x \le L$, the limits of integration are 0 and L:

$$x_{av} = \int_0^L |\psi(x)|^2 x\, dx = \frac{2}{L} \int_0^L \sin^2 \frac{n\pi x}{L} x\, dx$$

This can be integrated by parts or found in integral tables; the result is

$$x_{av} = \frac{L}{2}$$

Note that, as required, this result is independent of n. Thus a measurement of the average position of the particle yields no information about its quantum state.

Let's now look at how the uncertainty principle applies to the motion of this trapped particle. By solving Problems 34 and 35, you will find that the uncertainties in position and momentum for a particle in an infinite potential well are $\Delta x = L\sqrt{1/12 - 1/2\pi^2 n^2}$ and $\Delta p = hn/2L$. The product of the uncertainties is

$$\Delta x \Delta p = \frac{hn}{2}\sqrt{\frac{1}{12} - \frac{1}{2\pi^2 n^2}} = \frac{h}{2}\sqrt{\frac{n^2}{12} - \frac{1}{2\pi^2}}$$

Clearly the product of the uncertainties grows as n grows. The minimum value occurs for $n = 1$, in which case $\Delta x \Delta p = 0.090h = 0.57\hbar$. The ground state represents a fairly "compact" wave packet, but it is somewhat less compact than the minimum possible limit of $0.50\hbar$ (Eq. 4.10). You can see from Figure 5.11 how the wave becomes less compact (spreads out more) as n increases. Even for $n = 2$, the product of the uncertainties grows quickly to $1.67\hbar$.

Finite Potential Energy Well

Because the infinite potential energy well is an idealization of a technique for confining a particle, we should examine the solution when the barriers at the sides of the well are finite rather than infinite. The potential energy well can be described by

$$U(x) = 0 \qquad 0 \leq x \leq L$$
$$= U_0 \quad x < 0, x > L \qquad (5.33)$$

and is sketched in Figure 5.12. We look for solutions in which the particle is confined to this well, and thus the energies that we deduce for the particle must be less than U_0.

The solution in the center region (between $x = 0$ and $x = L$) is exactly the same as it was for the infinite well (Eq. 5.25):

$$\psi(x) = A \sin kx + B \cos kx \qquad (0 \leq x \leq L) \qquad (5.34)$$

although the values that we deduced previously for the coefficients A and B are not valid in this calculation. The region $x < 0$ is an example of a situation in which the energy E of the particle is less than the potential energy U_0, and so we must use the solution in the form of Eq. 5.18, $\psi(x) = Ce^{k'x} + De^{-k'x}$ with k' given in Eq. 5.17. This region includes $x = -\infty$, for which the term with the coefficient D becomes infinite. Because we cannot allow the probability to become infinite, we must discard this term by setting $D = 0$. The solution for $x < 0$ is then

$$\psi(x) = Ce^{k'x} \qquad (x < 0) \qquad (5.35)$$

In the region $x > L$, the energy E is once again smaller than U_0, and so the solution is also in the form of Eq. 5.18, $\psi(x) = Fe^{k'x} + Ge^{-k'x}$. Here the region now includes $x = +\infty$, for which the term with the coefficient F would become infinite. We must prevent that possibility by setting $F = 0$, so the solution in this region is

$$\psi(x) = Ge^{-k'x} \qquad (x > L) \qquad (5.36)$$

We now have 4 coefficients to determine (A, B, C, G) along with the energy E. For this determination, we have 4 equations from the boundary conditions (the continuity of both ψ and $d\psi/dx$ at both $x = 0$ and $x = L$) and one equation

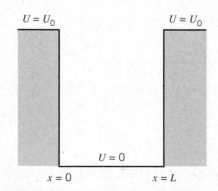

FIGURE 5.12 The potential energy of a particle that is confined to the region $0 \leq x \leq L$ by finite barriers U_0 at $x = 0$ and $x = L$.

from the normalization condition. As you might imagine, solving 5 equations in 5 unknowns presents a straightforward but very tedious algebraic challenge. Moreover, the resulting solution for the energy values cannot be obtained in terms of a direct equation such as Eq. 5.30, but instead must be found numerically by solving a transcendental equation. The result is a series of increasing energy values, but the number of energy values is finite rather than infinite, because the energy cannot be allowed to exceed the value of U_0.

As we did for the infinite potential energy well in Example 5.2, let's consider a well of width $L = 0.100$ nm. We'll choose the depth of the well to be $U_0 = 400$ eV. Applying the boundary conditions at $x = 0$ and $x = L$, we can eliminate all of the coefficients and find an equation that involves only k and k' (both of which depend on the energy E). Solving that equation numerically, we find four possible values of the energy: $E_1 = 26$ eV, $E_2 = 104$ eV, $E_3 = 227$ eV, $E_4 = 375$ eV. Here the subscript just numbers the energy values, starting at the ground state; there is no simple functional dependence of the energies on the quantum number n as there was for the infinite well. The allowed energy levels are shown in Figure 5.13.

The probability densities (square of the wave functions) for these four states are shown in Figure 5.14. In some ways they are similar to the probability densities in the infinite well—note that each state has n maxima in its probability density, just like the infinite well (see Figure 5.11). Unlike the infinite well, these probability densities show the property of penetration into the classically forbidden region. Look carefully at the continuity of the wave function and its slope at $x = 0$ and $x = L$; see how smoothly the sine and cosine function inside the well joins the exponentials in the forbidden regions.

The energy levels of the finite well are smaller than those of the infinite well of the same width (38 eV, 150 eV, 338 eV, 602 eV), and the differences increase as we go to higher states. This is consistent with the uncertainty principle—because of the penetration into the forbidden region, Δx is larger for the finite well and thus Δp_x must be smaller. As a result, the kinetic energies are smaller for the finite well. From Figure 5.14 we see that the penetration distance increases as we go up in energy, so the difference between Δx for the finite well and the infinite well increases and the energy discrepancy also increases.

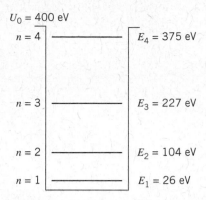

FIGURE 5.13 The energy levels in a potential energy well of depth 400 eV. There are only four energy states in this well.

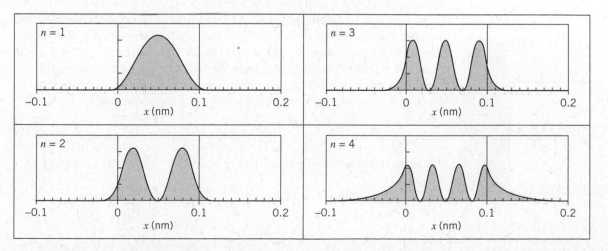

FIGURE 5.14 The probability densities of the four states in the one-dimensional potential energy well of width 0.100 nm and depth 400 eV.

For an energy close to the top of the well such as E_4, a smaller uncertainty ΔE is necessary to reach the top of the well, giving a larger $\Delta t \sim \hbar/\Delta E$ and thus a larger penetration distance. At the bottom of the well, the state E_1 requires much more energy to reach the top of the well and thus needs a much larger ΔE; the smaller resulting Δt gives a smaller penetration distance into the forbidden region.

Two-Dimensional Infinite Potential Energy Well

When we extend the previous calculation to two and three dimensions, the principal features of the solution remain the same, but an important new feature is introduced. In this section we show how this occurs; this new feature, known as *degeneracy,* will turn out to be very important in our study of atomic physics.

To begin with, we need a Schrödinger equation that is valid in more than one dimension; our previous version, Eq. 5.5, included only one spatial dimension. If the potential energy is a function of x and y, we expect that ψ also depends on both x and y, and the derivatives with respect to x must be replaced by derivatives with respect to x and y. In two dimensions, we then have*

$$-\frac{\hbar^2}{2m}\left(\frac{\partial^2 \psi(x,y)}{\partial x^2} + \frac{\partial^2 \psi(x,y)}{\partial y^2}\right) + U(x,y)\psi(x,y) = E\psi(x,y) \tag{5.37}$$

The two-dimensional potential energy well is:

$$U(x,y) = 0 \qquad 0 \leq x \leq L; 0 \leq y \leq L$$
$$= \infty \qquad \text{otherwise} \tag{5.38}$$

The particle is confined by infinitely high barriers to the square region with the vertices $(x,y) = (0,0), (L,0), (L,L), (0,L)$, as shown in Figure 5.15. A classical analog might be a small disk sliding without friction on a tabletop and colliding elastically with walls at $x = 0, x = L, y = 0$, and $y = L$. (For simplicity, we have made the allowed region square; we could have made it rectangular by setting $U = 0$ when $0 \leq x \leq a$ and $0 \leq y \leq b$.)

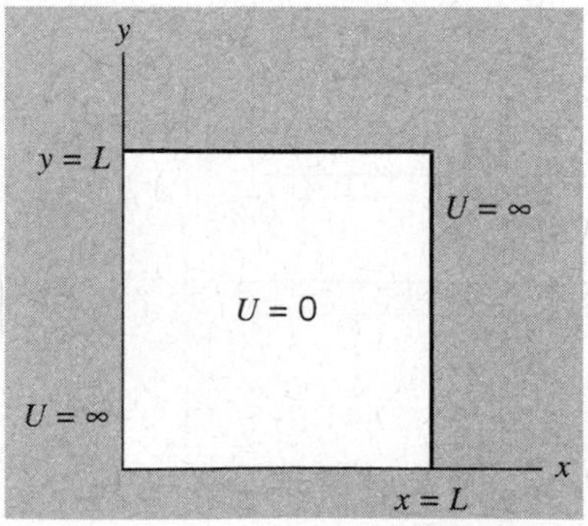

Solving *partial* differential equations requires a technique more involved than we need to consider, so we will not give the details of the solution. We suspect that, as in the previous case, $\psi(x,y) = 0$ outside the allowed region, in order to make the probability zero there. Inside the well, we consider solutions that are *separable*; that is, our function of x and y can be expressed as the product of one function that depends only on x and another that depends only on y:

$$\psi(x,y) = f(x)g(y) \tag{5.39}$$

where the functions f and g are similar to Eq. 5.16:

$$f(x) = A \sin k_x x + B \cos k_x x, \qquad g(y) = C \sin k_y y + D \cos k_y y \tag{5.40}$$

FIGURE 5.15 A particle moves freely in the two-dimensional region $0 < x < L, 0 < y < L$, but encounters infinite barriers beyond that region.

*The first two terms on the left side of this equation require *partial* derivatives; for well-behaved functions, these involve taking the derivative with respect to one variable while keeping the other constant. Thus if $f(x,y) = x^2 + xy + y^2$, then $\partial f/\partial x = 2x + y$ and $\partial f/\partial y = x + 2y$.

The wave number k of the one-dimensional problem has become the separate wave numbers k_x for $f(x)$ and k_y for $g(y)$. We show later how these are related. (See also Problem 18 at the end of this chapter.)

The continuity condition on $\psi(x, y)$ requires that the solutions inside and outside match at the boundary. Because $\psi = 0$ everywhere outside, the continuity condition then requires that $\psi = 0$ everywhere on the boundary. That is,

$$\psi(0, y) = 0 \quad \text{and} \quad \psi(L, y) = 0 \quad \text{for all } y$$
$$\psi(x, 0) = 0 \quad \text{and} \quad \psi(x, L) = 0 \quad \text{for all } x$$

In analogy with the one-dimensional problem, the condition at $x = 0$ gives $f(0) = 0$, which requires $B = 0$ in Eq. 5.40. Similarly, the condition at $y = 0$ gives $g(0) = 0$, which requires $D = 0$. The condition $f(L) = 0$ requires that $\sin k_x L = 0$, and thus that $k_x L$ be an integer multiple of π; the condition $g(L) = 0$ similarly requires that $k_y L$ be an integer multiple of π. These two integers do not necessarily need to be the same, so we call them n_x and n_y. Making all these substitutions into Eq. 5.39, we obtain

$$\psi(x, y) = A' \sin \frac{n_x \pi x}{L} \sin \frac{n_y \pi y}{L} \tag{5.41}$$

where we have combined A and C into A'. The coefficient A' is once again found by the normalization condition, which in two dimensions becomes

$$\iint \psi^2 dx \, dy = 1 \tag{5.42}$$

For our case this gives

$$\int_0^L dy \int_0^L A'^2 \sin^2 \frac{n_x \pi x}{L} \sin^2 \frac{n_y \pi y}{L} dx = 1 \tag{5.43}$$

from which follows

$$A' = \frac{2}{L} \tag{5.44}$$

(The solutions to this problem, which are standing de Broglie waves on a two-dimensional surface, are similar to the solutions of the classical problem of the vibrations of a stretched membrane such as a drumhead.)

Finally, we can substitute our solution for $\psi(x, y)$ back into Eq. 5.41 to find the energy:

$$E = \frac{\hbar^2 \pi^2}{2mL^2}(n_x^2 + n_y^2) = \frac{h^2}{8mL^2}(n_x^2 + n_y^2) \tag{5.45}$$

Compare this result with Eq. 5.30. Once again we let $E_0 = \hbar^2 \pi^2 / 2mL^2 = h^2 / 8mL^2$ so that $E = E_0(n_x^2 + n_y^2)$. In Figure 5.16 the energies of the excited states are shown. You can see how different the energies are from those of the one-dimensional case shown in Figure 5.10.

Figure 5.17 shows the probability density ψ^2 for several different combinations of the *quantum numbers* n_x and n_y. The probability has maxima and minima, just like the probability in the one-dimensional problem. For example, if we gave

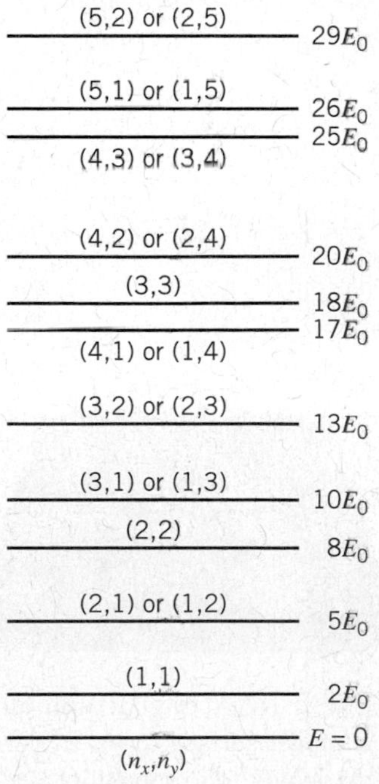

Level	Energy
(5,2) or (2,5)	$29E_0$
(5,1) or (1,5)	$26E_0$
(4,3) or (3,4)	$25E_0$
(4,2) or (2,4)	$20E_0$
(3,3)	$18E_0$
(4,1) or (1,4)	$17E_0$
(3,2) or (2,3)	$13E_0$
(3,1) or (1,3)	$10E_0$
(2,2)	$8E_0$
(2,1) or (1,2)	$5E_0$
(1,1)	$2E_0$
(n_x, n_y)	$E = 0$

FIGURE 5.16 The lower permitted energy levels of a particle confined to an infinite two-dimensional potential energy well.

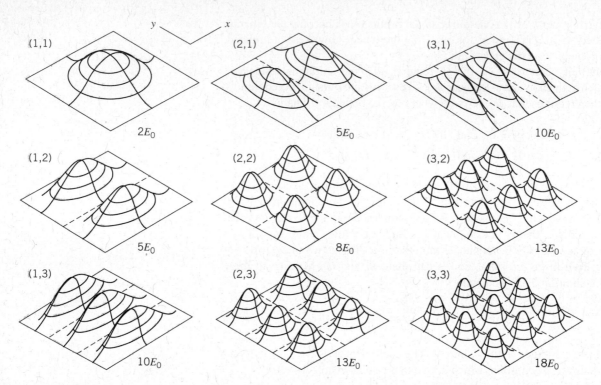

FIGURE 5.17 The probability density for some of the lower energy levels of a particle confined to the infinite two-dimensional potential energy well. The individual plots are labeled with the quantum numbers (n_x, n_y) and with the value of the energy E.

FIGURE 5.18 A ring of iron atoms on a copper surface forms a "corral" within which the probability density of trapped electrons is clearly visible. This image was obtained with a scanning tunneling electron microscope. (Image originally created by IBM Corporation.)

the particle an energy of $8E_0$ and then made a large number of measurements of its position, we would expect to find it most often near the four points $(x, y) = (L/4, L/4), (L/4, 3L/4), (3L/4, L/4)$ and $(3L/4, 3L/4)$; we expect *never* to find it near $x = L/2$ or $y = L/2$. The *shape* of the probability density tells us something about the quantum numbers and therefore about the energy. Thus if we measured the probability density and found six maxima, as shown in Figure 5.17, we would deduce that the particle had an energy of $13E_0$ with $n_x = 2$ and $n_y = 3$, or else $n_x = 3, n_y = 2$.

Recently it has become possible to photograph the probability densities of electrons confined in a two-dimensional region. The tip of an electron microscope was used to place 48 individual iron atoms on a metal surface in a ring or "corral" of radius 7.13 nm that formed the walls of a potential well, as shown in Figure 5.18. Inside the ring, the waves of probability density for electrons trapped in the potential well are clearly visible. The potential well is circular, rather than square, but otherwise the analysis follows the procedures described in this section; when the Schrödinger equation is solved in cylindrical polar coordinates with the potential energy for a circular well, the calculated probability density gives a close match with the observed one. These beautiful results are a dramatic confirmation of the wave functions obtained for the two-dimensional potential energy well.

Degeneracy   Occasionally it happens that two different sets of quantum numbers n_x and n_y have exactly the same energy. This situation is known as *degeneracy,* and the energy levels are said to be *degenerate*. For example, the energy level at $E = 13E_0$ is degenerate, because both $n_x = 2, n_y = 3$ and $n_x = 3, n_y = 2$ have

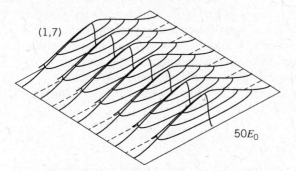

(1,7)

$50E_0$

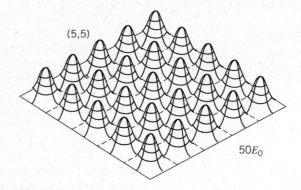

(5,5)

$50E_0$

FIGURE 5.19 Two very different probability densities with exactly the same energy.

$E = 13E_0$. This degeneracy arises from interchanging n_x and n_y (which is the same as interchanging the x and y axes), so the probability distributions in the two cases are not very different. However, consider the state with $E = 50E_0$, for which there are three sets of quantum numbers: $n_x = 7, n_y = 1; n_x = 1, n_y = 7;$ and $n_x = 5, n_y = 5$. The first two sets of quantum numbers result from the interchange of n_x and n_y and so have similar probability distributions, but the third represents a *very* different state of motion, as shown in Figure 5.19. The level at $E = 13E_0$ is said to be *two-fold* degenerate, while the level at $E = 50E_0$ is *three-fold* degenerate; we could also say that one level has a degeneracy of 2, while the other has a degeneracy of 3.

Degeneracy occurs in general whenever a system is labeled by two or more quantum numbers; as we have seen in the above calculation, different combinations of quantum numbers often can give the same value of the energy. The number of different quantum numbers required by a given physical problem turns out to be exactly equal to the number of dimensions in which the problem is being solved—one-dimensional problems need only one quantum number, two-dimensional problems need two, and so forth. When we get to three dimensions, as in Problem 19 at the end of this chapter and especially in the hydrogen atom in Chapter 7, we find that the effects of degeneracy become more significant; in the case of atomic physics, the degeneracy is a major contributor to the structure and properties of atoms.

5.5 THE SIMPLE HARMONIC OSCILLATOR

Another situation that can be analyzed using the Schrödinger equation is the one-dimensional simple harmonic oscillator. The classical oscillator is an object of mass m attached to a spring of force constant k. The spring exerts a restoring force $F = -kx$ on the object, where x is the displacement from its equilibrium position. Using Newton's laws, we can analyze the oscillator and show that it has a (circular or angular) frequency $\omega_0 = \sqrt{k/m}$ and a period $T = 2\pi\sqrt{m/k}$. The maximum distance of the oscillating object from its equilibrium position is x_0, the amplitude of the oscillation. The oscillator has its maximum kinetic energy at $x = 0$; its kinetic energy vanishes at the *turning points* $x = \pm x_0$. At the turning points the oscillator comes to rest for an instant and then reverses its direction of motion. The motion is, of course, confined to the region $-x_0 \leq x \leq +x_0$.

Why analyze the motion of such a system using quantum mechanics? Although we never find in nature an example of a one-dimensional quantum oscillator, there are systems that behave approximately as one—a vibrating diatomic molecule, for example. In fact, any system in a smoothly varying potential energy well near its minimum behaves approximately like a simple harmonic oscillator.

A force $F = -kx$ has the associated potential energy $U = \frac{1}{2}kx^2$, and so we have the Schrödinger equation:

$$-\frac{\hbar^2}{2m}\frac{d^2\psi}{dx^2} + \frac{1}{2}kx^2\psi = E\psi \qquad (5.46)$$

(Because we are working in one dimension, U and ψ are functions only of x.) There are no boundaries between different regions of potential energy here, so the wave function must fall to zero for both $x \to +\infty$ and $x \to -\infty$. The simplest function that satisfies these conditions, which turns out to be the correct ground state wave function, is $\psi(x) = Ae^{-ax^2}$. The constant a and the energy E can be found by substituting this function into Eq. 5.46. We begin by evaluating $d^2\psi/dx^2$.

$$\frac{d\psi}{dx} = -2ax(Ae^{-ax^2})$$

$$\frac{d^2\psi}{dx^2} = -2a(Ae^{-ax^2}) - 2ax(-2ax)Ae^{-ax^2} = (-2a + 4a^2x^2)Ae^{-ax^2}$$

Substituting into Eq. 5.46 and canceling the common factor Ae^{-ax^2} yields

$$\frac{\hbar^2 a}{m} - \frac{2a^2\hbar^2}{m}x^2 + \frac{1}{2}kx^2 = E \qquad (5.47)$$

Equation 5.47 is *not* an equation to be solved for x, because we are looking for a solution that is valid for *any* x, not just for one specific value. In order for this to hold for *any* x, the coefficients of x^2 must cancel and the remaining constants must be equal. (That is, consider the equation $bx^2 = c$. It will be true for *any* and *all* x only if both $b = 0$ and $c = 0$.) Thus

$$-\frac{2a^2\hbar^2}{m} + \frac{1}{2}k = 0 \qquad \text{and} \qquad \frac{\hbar^2 a}{m} = E \qquad (5.48)$$

which yield

$$a = \frac{\sqrt{km}}{2\hbar} \qquad \text{and} \qquad E = \frac{1}{2}\hbar\sqrt{k/m} \qquad (5.49)$$

We can also write the energy in terms of the classical frequency $\omega_0 = \sqrt{k/m}$ as

$$E = \frac{1}{2}\hbar\omega_0 \qquad (5.50)$$

The coefficient A is found from the normalization condition (see Problem 20 at the end of the chapter). The result, which is valid *only* for this ground-state wave function, is $A = (m\omega_0/\hbar\pi)^{1/4}$. The complete wave function of the ground state is then

$$\psi(x) = \left(\frac{m\omega_0}{\hbar\pi}\right)^{1/4} e^{-(\sqrt{km}/2\hbar)x^2} \qquad (5.51)$$

The probability density for this wave function is illustrated in Figure 5.20. Note that, as in the case of the finite potential energy well, the probability density can penetrate into the forbidden region beyond the classical turning points at $x = \pm x_0$ (in this region the potential energy is greater than E).

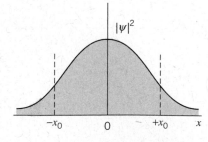

FIGURE 5.20 The probability density for the ground state of the simple harmonic oscillator. The classical turning points are at $x = \pm x_0$.

The solution we have found corresponds only to the *ground state* of the oscillator. The general solution is of the form $\psi_n(x) = Af_n(x)e^{-ax^2}$, where $f_n(x)$ is a polynomial in which the highest power of x is x^n. The corresponding energies are

$$E_n = \left(n + \frac{1}{2}\right)\hbar\omega_0 \qquad n = 0, 1, 2, \ldots \qquad (5.52)$$

These levels are shown in Figure 5.21. Note that they are *uniformly spaced*, in contrast to the one-dimensional infinite potential energy well. Probability densities are shown in Figure 5.22. All of the solutions have the property of penetration of probability density into the forbidden region beyond the classical turning points. The probability density oscillates, somewhat like a sine wave, between the turning points, and decreases like e^{-ax^2} to zero beyond the turning points. Note the great similarity between the probability densities for the quantum oscillator and those of the finite potential energy well (Figure 5.14).

A sequence of vibrational excited states similar to Figure 5.21 is commonly found in diatomic molecules such as HCl (see Chapter 9). The spacing between the states is typically $0.1-1$ eV; the states are observed when photons (in the infrared region of the spectrum) are emitted or absorbed as the molecule jumps from one state to another. A similar sequence is observed in nuclei, where the spacing is $0.1-1$ MeV and the radiations are in the gamma-ray region of the spectrum.

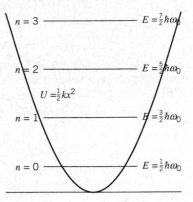

FIGURE 5.21 Energy levels of the simple harmonic oscillator. Note that the levels have equal spacings and that the distance between the classical turning points increases with energy.

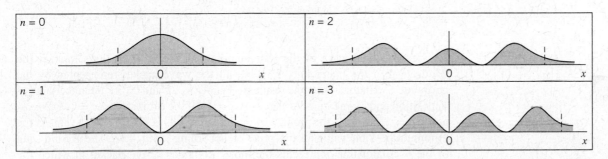

FIGURE 5.22 Probability densities for the simple harmonic oscillator. Note how the distance between the classical turning points (marked by the short vertical lines) increases with energy. Compare with the probability densities for the finite potential energy well (Figure 5.14).

Example 5.5

An electron is bound to a region of space by a springlike force with an effective spring constant of $k = 95.7$ eV/nm^2. (*a*) What is its ground-state energy? (*b*) How much energy must be absorbed for the electron to jump from the ground state to the second excited state?

Solution

(*a*) The ground-state energy is

$$E = \frac{1}{2}\hbar\omega_0 = \frac{1}{2}\hbar\sqrt{\frac{k}{m}} = \frac{1}{2}\hbar c\sqrt{\frac{k}{mc^2}}$$

$$= \frac{1}{2}(197\,\text{eV}\cdot\text{nm})\sqrt{\frac{95.7\,\text{eV/nm}^2}{0.511 \times 10^6\,\text{eV}}}$$

$$= 1.35\,\text{eV}$$

(*b*) The difference between adjacent energy levels is $\hbar\omega_0 = 2.70$ eV for all energy levels, so the energy that must be absorbed to go from the ground state to the second excited state is $\Delta E = 2 \times 2.70\,\text{eV} = 5.40\,\text{eV}$.

Example 5.6

For the electron of Example 5.5 in its ground state, what is the probability to find it in a narrow interval of width 0.004 nm located halfway between the equilibrium position and the classical turning point?

Solution

First we need to find the location of the turning point. At the classical turning points $x = \pm x_0$, the kinetic energy is zero and so the total energy is all potential. Thus $E = \frac{1}{2}kx_0^2$, and so

$$x_0 = \sqrt{\frac{2E}{k}} = \sqrt{\frac{2(1.35\,\text{eV})}{95.7\,\text{eV/nm}^2}} = 0.168\,\text{nm}$$

Evaluating the parameters of the wave function Ae^{-ax^2} (the normalization constant A and the exponential coefficient a), we have

$$A = \left(\frac{m\omega_0}{\hbar\pi}\right)^{1/4} = \left(\frac{mc^2\hbar\omega_0}{\hbar^2 c^2\pi}\right)^{1/4}$$

$$= \left(\frac{(0.511\times 10^6\,\text{eV})(2.70\,\text{eV})}{(197\,\text{eV}\cdot\text{nm})^2\pi}\right)^{1/4} = 1.83\,\text{nm}^{-1/2}$$

$$a = \frac{\sqrt{km}}{2\hbar} = \frac{\sqrt{kmc^2}}{2\hbar c}$$

$$= \frac{\sqrt{(95.7\,\text{eV/nm}^2)(0.511\times 10^6\,\text{eV})}}{2(197\,\text{eV}\cdot\text{nm})}$$

$$= 17.74\,\text{nm}^{-2}$$

The probability in the interval $dx = 0.004$ nm at $x = x_0/2 = 0.084$ nm is then

$$P(x)\,dx = |\psi(x)|^2\,dx = A^2 e^{-2ax^2}\,dx$$

$$= (1.83\,\text{nm}^{-1/2})^2 e^{-2(17.74\,\text{nm}^{-2})(0.084\,\text{nm})^2}(0.004\,\text{nm})$$

$$= 0.0104 = 1.04\%$$

As we did in the case of the infinite potential energy well, let's look at the application of the uncertainty principle to the wave packet represented by the harmonic oscillator. Using the results of Problems 22 and 23 for the uncertainties in position and momentum for the ground state of the oscillator, $\Delta x = \sqrt{\hbar/2m\,\omega_0}$ and $\Delta p = \sqrt{\hbar\omega_0 m/2}$, the product of the uncertainties is $\Delta x \Delta p = \hbar/2$. This is the minimum possible value for this product, according to Eq. 4.10. The ground state of the oscillator thus represents the most "compact" wave packet in which the product of the uncertainties has its smallest value. You can see from Figure 5.22 that the excited states of the oscillator are much less compact (more spread out) than the ground state.

5.6 STEPS AND BARRIERS

In this general type of problem, we analyze what happens when a particle moving (again in one dimension) in a region of constant potential energy suddenly moves into a region of different, but also constant, potential energy. We will not discuss in detail the solutions to these problems, but the methods of solution of each are so similar that we can outline the steps to take in the solution. In this discussion, we let E be the (fixed) total energy of the particle and U_0 will be the value of the constant potential energy. In these calculations, the particle is *not* confined, so the energy is not quantized—we are free to choose any value for the particle energy.

Potential Energy Step, $E > U_0$

Consider the potential energy step shown in Figure 5.23:

$$U(x) = 0 \qquad x < 0$$
$$= U_0 \qquad x \geq 0 \qquad\qquad (5.53)$$

If the total energy E of the particle is greater than U_0, then we can write the solutions to the Schrödinger equation in the two regions based on the general form of Eq. 5.16:

$$\psi_0(x) = A \sin k_0 x + B \cos k_0 x \qquad k_0 = \sqrt{\frac{2mE}{\hbar^2}} \qquad x < 0 \quad (5.54a)$$

$$\psi_1(x) = C \sin k_1 x + D \cos k_1 x \qquad k_1 = \sqrt{\frac{2m}{\hbar^2}(E - U_0)} \quad x > 0 \quad (5.54b)$$

Relationships among the four coefficients, A, B, C, and D, may be found by applying the condition that $\psi(x)$ and $\psi'(x) = d\psi/dx$ must be continuous at the boundary; thus $\psi_0(0) = \psi_1(0)$ and $\psi_0'(0) = \psi_1'(0)$. A typical solution might look like Figure 5.24. Note the smooth transition between the solutions at $x = 0$, which results from applying the continuity conditions.

The coefficients A, B, C, and D are in general complex, so to visualize the complete wave we need both the real and imaginary parts of ψ. We can use the equation $e^{i\theta} = \cos\theta + i\sin\theta$ to transform these solutions from sines and cosines to complex exponentials:

$$\psi_0(x) = A' e^{ik_0 x} + B' e^{-ik_0 x} \qquad x < 0 \qquad (5.55a)$$
$$\psi_1(x) = C' e^{ik_1 x} + D' e^{-ik_1 x} \qquad x > 0 \qquad (5.55b)$$

The coefficients A', B', C', D' can be found from the coefficients A, B, C, D. The time dependent wave functions are obtained by multiplying each term by $e^{-i\omega t}$, which gives

$$\Psi_0(x,t) = A' e^{i(k_0 x - \omega t)} + B' e^{-i(k_0 x + \omega t)} \qquad (5.56a)$$
$$\Psi_1(x,t) = C' e^{i(k_1 x - \omega t)} + D' e^{-i(k_1 x + \omega t)} \qquad (5.56b)$$

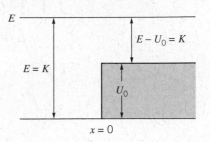

FIGURE 5.23 A step of height U_0. Particles are incident from the left with energy E. The kinetic energy is equal to E in the region $x < 0$ and is reduced to $E - U_0$ in the region $x > 0$.

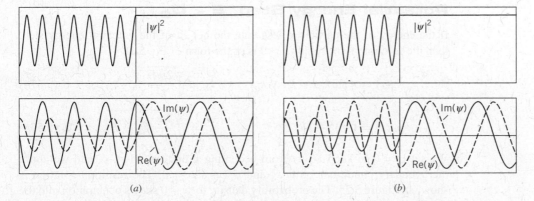

FIGURE 5.24 Wave function for electrons incident from the left on a potential energy step for $E > U_0$. The probability density and the real and imaginary parts of the wavefunction are shown for (a) $t = 0$ and (b) $t = 1/4$ period. The vertical line marks the location of the step.

We can then make the following identification of the component waves, recalling that $(kx - \omega t)$ is the phase of a wave moving in the positive x direction, while $(kx + \omega t)$ is the phase of a wave moving in the negative x direction, and assuming that *the squared magnitude of each coefficient gives the intensity of the corresponding component wave*. In the region $x < 0$, Eq. 5.56a describes the superposition of a wave $e^{i(k_0 x - \omega t)}$ of intensity $|A'|^2$ moving in the positive x direction (from $-\infty$ to 0) and a wave $e^{-i(k_0 x + \omega t)}$ of intensity $|B'|^2$ moving in the negative x direction. Suppose we had intended our solution to describe particles that are incident from the left on this step. Then $|A'|^2$ gives the intensity of the incident wave (more exactly, the de Broglie wave describing the incident beam of particles) and $|B'|^2$ gives the intensity of the reflected wave. The ratio $|B'|^2/|A'|^2$ tells us the reflected fraction of the incident wave intensity.

In the region $x > 0$, Eq. 5.56b describes the transmitted wave $e^{i(k_1 x - \omega t)}$ of intensity $|C'|^2$ moving to the right and a wave $e^{-i(k_1 x + \omega t)}$ of intensity $|D'|^2$ moving to the left. If particles are incident from $-\infty$, it is not possible to have particles in the region $x > 0$ moving to the left, so in this particular experimental situation we are justified in setting D' to zero.

Figure 5.24a shows that the probability density has the same value everywhere in the region $x > 0$. You can see this immediately from Eq. 5.56b with $D' = 0$; taking the squared magnitude of the remaining term gives a constant result, independent of x and t. This is consistent with what we expect for the de Broglie wave of free particles; the particles can be found anywhere in the region $x > 0$ with equal probability.

In the region $x < 0$, the incident and reflected waves combine to produce a standing wave, for which the probability density has fixed maxima and minima. The probability density in this region does not vary with time, as suggested by the plots for the two different times ($t = 0$ and $t = \frac{1}{4}$ period) shown in Figure 5.24.

To illustrate the propagation of the de Broglie wave, it is instructive also to plot the real and imaginary parts of the wave function, which are shown in Figure 5.24. Here you can see the change in wavelength (corresponding to the change in kinetic energy or momentum) in crossing the step. You can also see something of the time dependence—the wave propagates in both regions, but it does so in a way that the real and imaginary parts combine to give a probability density that remains unchanged in time.

Potential Energy Step, $E < U_0$

If the energy of the particle is less than the height of the potential energy step, then the solution in the region $x > 0$ is of the form of Eq. 5.18:

$$\psi_0(x) = A \sin k_0 x + B \cos k_0 x \qquad k_0 = \sqrt{\frac{2mE}{\hbar^2}} \qquad x < 0 \quad (5.57a)$$

$$\psi_1(x) = C e^{k_1 x} + D e^{-k_1 x} \qquad k_1 = \sqrt{\frac{2m}{\hbar^2}(U_0 - E)} \quad x > 0 \quad (5.57b)$$

We set $C = 0$ to keep $\psi_1(x)$ from becoming infinite as $x \to \infty$, and we apply the boundary conditions on $\psi(x)$ and $\psi'(x)$ at $x = 0$. The resulting solution is shown in Figure 5.25. The probability density for $x > 0$ shows penetration into the classically forbidden region. All particles are reflected from the barrier; that is, if we write $\Psi_0(x,t)$ in the form of Equation 5.56a, then we must have $|A'| = |B'|$, indicating that the waves moving to the right (the incident wave) and to the left (the reflected wave) in the region $x < 0$ have equal amplitudes.

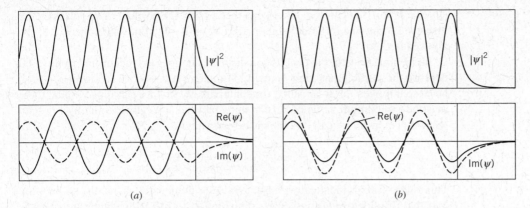

FIGURE 5.25 Wave function for electrons incident from the left on a potential energy step for $E < U_0$. The probability density and the real and imaginary parts of the wavefunction are shown for (a) $t = 0$ and (b) $t = 1/4$ period. The vertical line marks the location of the step.

Figure 5.25 shows the probability density at two different times, illustrating that the probability density does not change with time. In the region $x < 0$ we again have standing waves with fixed maxima and minima. Viewing the real and imaginary parts at two different times ($t = 0$ and $t = 1/4$ period) shows that the wave is propagating, even though the probability density does not change with time.

Penetration into the forbidden region is associated with the wave nature of the particle and also with the uncertainty in the particle's energy or location. The probability density in the $x > 0$ region is $|\psi_1|^2$, which according to Eq. 5.57b is proportional to $e^{-2k_1 x}$. If we define a representative penetration distance Δx to be the distance over which the probability drops by $1/e$, then $e^{-2k_1 \Delta x} = e^{-1}$ and so

$$\Delta x = \frac{1}{2k_1} = \frac{1}{2} \frac{\hbar}{\sqrt{2m(U_0 - E)}} \qquad (5.58)$$

To be able to enter the region with $x > 0$, the *particle* must gain an energy of at least $U_0 - E$ in order to get over the potential energy step; it must in addition gain some kinetic energy if it is to move in the region $x > 0$. Of course, it is a violation of conservation of energy for the particle to spontaneously gain *any* amount of energy, but according to the uncertainty relationship $\Delta E \Delta t \sim \hbar$ conservation of energy does not apply at times smaller than Δt except to within an amount $\Delta E \sim \hbar/\Delta t$. That is, if the particle "borrows" an amount of energy ΔE and "returns" the borrowed energy within a time $\Delta t \sim \hbar/\Delta E$, we observers will still believe energy is conserved. Suppose the particle borrows an energy sufficient to give it a kinetic energy of K in the forbidden region. How far into the forbidden region does the particle penetrate?

The "borrowed" energy is $(U_0 - E) + K$; the energy $(U_0 - E)$ gets the particle to the top of the step, and the extra kinetic energy K gives it its motion. The energy must be returned within a time

$$\Delta t = \frac{\hbar}{U_0 - E + K} \qquad (5.59)$$

The particle moves with speed $v = \sqrt{2K/m}$, and so the distance it can travel is

$$\Delta x = \frac{1}{2} v \Delta t = \frac{1}{2} \sqrt{\frac{2K}{m}} \frac{\hbar}{U_0 - E + K} \qquad (5.60)$$

(The factor of 1/2 is present because in the time Δt the particle must penetrate the distance Δx into the forbidden region and return through that same distance to the allowed region.)

In the limit $K \to 0$, the penetration distance Δx goes to 0 according to Eq. 5.60 because the particle has zero velocity; similarly, $\Delta x \to 0$ in the limit $K \to \infty$, because it moves for a vanishing time interval Δt. In between those limits, there must be a maximum value of Δx for some particular K. Differentiating Eq. 5.60 with respect to K, we can find the maximum value

$$\Delta x_{\max} = \frac{1}{2} \frac{\hbar}{\sqrt{2m(U_0 - E)}} \tag{5.61}$$

This value of Δx is identical with Eq. 5.58! This demonstrates that the penetration into the forbidden region given by the solution to the Schrödinger equation is entirely consistent with the uncertainty relationship. (The agreement between Eqs. 5.58 and 5.61 is really somewhat accidental, because the factor $1/e$ used to obtain Eq. 5.58 was chosen arbitrarily. What we have really demonstrated is that the estimates of uncertainty given by the Heisenberg relationships are consistent with the wave properties of the particle obtained from the Schrödinger equation. This should not be surprising, because the uncertainty principle can be derived as a consequence of the Schrödinger equation.)

Potential Energy Barrier

Consider now the potential energy barrier shown in Figure 5.26:

$$
\begin{aligned}
U(x) &= 0 & x &< 0 \\
&= U_0 & 0 &\leq x \leq L \\
&= 0 & x &> L
\end{aligned}
\tag{5.62}
$$

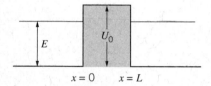

FIGURE 5.26 A barrier of height U_0 and width L.

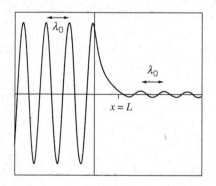

FIGURE 5.27 The real part of the wave function of a particle of energy $E < U_0$ encountering a barrier (the particle is incident from the left in the figure). The wavelength λ_0 is the same on both sides of the barrier, but the amplitude beyond the barrier is much less than the original amplitude.

Particles with energy E less than U_0 are incident from the left. Our experience then leads us to expect solutions of the form shown in Figure 5.27—sinusoidal oscillation in the region $x < 0$ (an incident wave and a reflected wave), exponentials in the region $0 \leq x \leq L$, and sinusoidal oscillations in the region $x > L$ (the transmitted wave). Note that the intensity of the transmitted wave ($x > L$) is much smaller than the intensity of the incident + reflected waves ($x < 0$), which means that most of the particles are reflected and few are transmitted through the barrier. Also note that the wavelengths are the same on either side of the barrier (because the kinetic energies are the same).

The intensity of the transmitted wave, which can be found by application of the continuity conditions, depends on the energy of the particle and on the height and thickness of the barrier. Classically, the particles should never appear at $x > L$, because they do not have sufficient energy to overcome the barrier. This situation is an example of *barrier penetration,* sometimes called *quantum mechanical tunneling.* Particles can not be *observed* while they are in the classically forbidden region $0 \leq x \leq L$, but can "tunnel" *through* that region and be observed at $x > L$.

Every particle incident on the barrier of Figure 5.26 is either reflected or transmitted; the number of incident particles is equal to the number reflected back to $x < 0$ plus the number transmitted to $x > L$. None are "trapped" or ever seen in the forbidden region $0 < x < L$. How can the incident *particle* get from $x < 0$ to $x > L$? As a classical particle, *it can't!* However, the wave representing the particle *can* penetrate through the barrier, which allows the particle to be observed in the classically *allowed* region $x > L$.

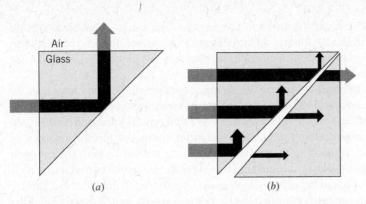

(a)   (b)

FIGURE 5.28 (a) Total internal reflection of light waves at a glass-air boundary. (b) Frustrated total internal reflection. The thicker the air gap, the smaller the probability to penetrate. Note that the light beam does not appear *in* the gap.

This phenomenon of penetration of a forbidden region is a well-known property of classical waves. Quantum physics provides a new aspect to this phenomenon by associating a particle with the wave, and thus allowing a particle to pass through a classically forbidden region. An example of the penetration effect for classical waves occurs for total internal reflection* of light waves. Figure 5.28a shows a light beam in glass incident on a boundary with air. The beam is totally reflected in the glass. However, if a second piece of glass is brought close to the first, as in Figure 5.28b, the beam can appear in the second piece of glass. This effect is called *frustrated total internal reflection*. The intensity of the beam in the second piece, represented by the widths of the arrows in Figure 5.28b, decreases rapidly as the thickness of the gap increases.

Just like our unobservable quantum wave, which penetrates a few wavelengths into the forbidden region, an unobservable light wave of exponentially decreasing amplitude, the *evanescent wave*, penetrates into the air even when the light wave undergoes total reflection in the glass. The evanescent wave carries no energy away from the interface, so it cannot be directly observed in the air, but it can be observed in another medium such as a second piece of glass placed close to the first. Evanescent waves have applications in microscopy, where they enable the production of images of individual molecules.

Although the potential energy barrier of Figure 5.26 is rather artificial, there are many practical examples of quantum tunneling:

1. **Alpha Decay.** An atomic nucleus consists of protons and neutrons in a constant state of motion; occasionally these particles form themselves into an aggregate of two protons and two neutrons, called an alpha particle. In one form of radioactive decay, the nucleus can emit an alpha particle, which can be detected in the laboratory. However, in order to escape from the nucleus the alpha particle must penetrate a barrier of the form shown in Figure 5.29. The probability for the alpha particle to penetrate the barrier, and be detected in the laboratory, can be computed based on the energy of the alpha particle

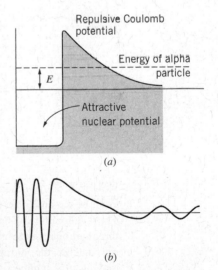

(a)

(b)

FIGURE 5.29 (a) A nuclear potential energy barrier is penetrated by an alpha particle of energy E. (b) A representation of the real part of the wave function of the alpha particle. The probability to penetrate the barrier depends strongly on the energy of the alpha particle.

---

*Total internal reflection occurs when a light beam is incident on a boundary between two substances, such as glass and air, from the side with the higher index of refraction. If the angle of incidence inside the glass exceeds a certain critical value, the light beam is totally reflected back into the glass.

and the height and thickness of the barrier. The decay probability can be measured in the laboratory, and it is found to be in excellent agreement with the value obtained from a quantum-mechanical calculation based on barrier penetration.

2. **Ammonia Inversion.** Figure 5.30 is a representation of the ammonia molecule NH_3. If we were to try to move the nitrogen atom along the axis of the molecule, toward the plane of the hydrogen atoms, we find repulsion caused by the three hydrogen atoms, which produces a potential energy of the form shown in Figure 5.31. According to classical mechanics, unless we give the nitrogen atom sufficient energy, it should not be able to surmount the barrier and appear on the other side of the plane of hydrogens. According to quantum mechanics, the nitrogen can tunnel through the barrier and appear on the other side of the molecule. In fact, the nitrogen atom actually tunnels back and forth with a frequency in excess of 10^{10} oscillations per second.

3. **The Tunnel Diode.** A tunnel diode is an electronic device that uses the phenomenon of tunneling. Schematically, the potential energy of an electron in a tunnel diode can be represented by Figure 5.32. The current that flows through the device is produced by electrons tunneling through the barrier. The rate of tunneling, and therefore the current, can be regulated merely by changing the height of the barrier, which can be done with an applied voltage.

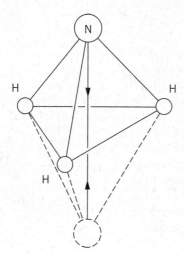

FIGURE 5.30 A schematic diagram of the ammonia molecule. The Coulomb repulsion of the three hydrogens establishes a barrier against the nitrogen atom moving to a symmetric position (shown in dashed lines) on the opposite side of the plane of hydrogens.

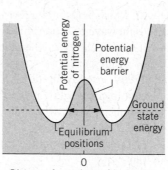

FIGURE 5.31 The potential energy seen by the nitrogen atom in an ammonia molecule. The nitrogen can penetrate the barrier and move from one equilibrium position to another.

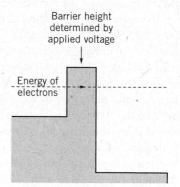

FIGURE 5.32 The potential energy barrier seen by an electron in a tunnel diode. The conductivity of the device is determined by the electron's probability to penetrate the barrier, which depends on the height of the barrier.

This can be done rapidly, so that switching frequencies in excess of 10^9 Hz can be obtained. Ordinary semiconductor diodes depend on the diffusion of electrons across a junction, and therefore operate on much longer time scales (that is, at lower frequencies).

4. **The Scanning Tunneling Microscope.** Images of individual atoms on the surface of materials (such as Figure 5.18) can be made with the scanning tunneling microscope. Electrons are trapped in a surface by a potential energy barrier (the work function of the material). When a needlelike probe is placed within about 1 nm of the surface (Figure 5.33), electrons can tunnel through the barrier between the surface and the probe and produce a current that can be recorded in an external circuit. The current is very sensitive to the width of the barrier (the distance from the probe to the surface). In practice, a feedback mechanism keeps the current constant by moving the tip up and down. The motion of the tip gives a map of the surface that reveals details smaller than 0.01 nm, about 1/100 the diameter of an atom! For the development of the scanning tunneling microscope, Gerd Binnig and Heinrich Rohrer were awarded the 1986 Nobel Prize in physics.

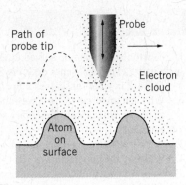

FIGURE 5.33 In a scanning tunneling microscope, a needlelike probe is scanned over a surface. The probe is moved vertically so that the distance between the probe and the surface remains constant as the probe scans laterally.

Chapter Summary

		Section
Time-independent Schrödinger equation	$-\dfrac{\hbar^2}{2m}\dfrac{d^2\psi}{dx^2} + U(x)\psi(x) = E\psi(x)$	5.3
Time-dependent Schrödinger equation	$\Psi(x,t) = \psi(x)e^{-i\omega t}$	5.3
Probability density	$P(x) = |\psi(x)|^2$	5.3
Normalization condition	$\int_{-\infty}^{+\infty} |\psi(x)|^2 \, dx = 1$	5.3
Probability in interval x_1 to x_2	$P(x_1 : x_2) = \int_{x_1}^{x_2} |\psi(x)|^2 \, dx$	5.3
Average or expectation value of $f(x)$	$[f(x)]_{av} = \int_{-\infty}^{+\infty} |\psi(x)|^2 f(x) \, dx$	5.3
Constant potential energy, $E > U_0$	$\psi(x) = A\sin kx + B\cos kx,$ $k = \sqrt{2m(E - U_0)/\hbar^2}$	5.4
Constant potential energy, $E < U_0$	$\psi(x) = Ae^{k'x} + Be^{-k'x},$ $k' = \sqrt{2m(U_0 - E)/\hbar^2}$	5.4

		Section
Infinite potential energy well	$\psi_n(x) = \sqrt{\dfrac{2}{L}}\sin\dfrac{n\pi x}{L},$ $E_n = \dfrac{h^2 n^2}{8mL^2}\ (n = 1, 2, 3, \ldots)$	5.4
Two-dimensional infinite well	$\psi(x,y) = \dfrac{2}{L}\sin\dfrac{n_x\pi x}{L}\sin\dfrac{n_y\pi y}{L}$ $E = \dfrac{h^2}{8mL^2}(n_x^2 + n_y^2)$	5.4
Simple harmonic oscillator ground state	$\psi(x) = (m\omega_0/\hbar\pi)^{1/4}\,e^{-(\sqrt{km}/2\hbar)x^2}$	5.5
Simple harmonic oscillator energies	$E_n = (n + \tfrac{1}{2})\hbar\omega_0\ (n = 0, 1, 2, \ldots)$	5.5
Potential energy step, $E > U_0$	$\psi_0(x < 0) = A\sin k_0 x + B\cos k_0 x$ $\psi_1(x > 0) = C\sin k_1 x + D\cos k_1 x$	5.6
Potential energy step, $E < U_0$	$\psi_0(x < 0) = A\sin k_0 x + B\cos k_0 x$ $\psi_1(x > 0) = Ce^{k_1 x} + De^{-k_1 x}$	5.6

Questions

1. Newton's laws can be solved to give the future behavior of a particle. In what sense does the Schrödinger equation also do this? In what sense does it not?

2. Why is it important for a wave function to be normalized? Is an unnormalized wave function a solution to the Schrödinger equation?

3. What is the physical meaning of $\int_{-\infty}^{+\infty} |\psi|^2 dx = 1$?

4. What are the dimensions of $\psi(x)$? Of $\psi(x, y)$?

5. None of the following are permitted as solutions of the Schrödinger equation. Give the reasons in each case.

   (a) $\psi(x) = A \cos kx \qquad x < 0$
   $\psi(x) = B \sin kx \qquad x > 0$
   (b) $\psi(x) = Ax^{-1}e^{-kx} \qquad -L \leq x \leq L$
   (c) $\psi(x) = A \sin^{-1} kx$
   (d) $\psi(x) = A \tan kx \qquad x > 0$

6. What happens to the probability density in the infinite well when $n \to \infty$? Is this consistent with classical physics?

7. How would the solution to the infinite potential energy well be different if the well extended from $x = x_0$ to $x = x_0 + L$, where x_0 is a nonzero value of x? Would any of the measurable properties be different?

8. How would the solution to the one-dimensional infinite potential energy well be different if the potential energy were not zero for $0 \leq x \leq L$ but instead had a constant value U_0? What would be the energies of the excited states? What would be the wavelengths of the standing de Broglie waves? Sketch the behavior of the lowest two wave functions.

9. Assuming a pendulum to behave like a quantum oscillator, what are the energy differences between the quantum states of a pendulum of length 1 m? Are such differences observable?

10. For the potential energy barrier (Figure 5.26), is the wavelength for $x > L$ the same as the wavelength for $x < 0$? Is the amplitude the same?

11. Suppose particles were incident on the potential energy step from the *positive x* direction. Which of the four coefficients of Eq. 5.56 would be set to zero? Why?

12. The energies of the excited states of the systems we have discussed in this chapter have been exact—there is no energy uncertainty. What does this suggest about the lifetime of particles in those excited states? Left on its own, will a particle ever make transitions from one state to another?

13. Explain how the behavior of a particle in a one-dimensional infinite well can be considered in terms of standing de Broglie waves.

14. How would you design an experiment to observe barrier penetration with sound waves? What range of thicknesses would you choose for the barrier?

15. If U_0 were negative in Figure 5.26, how would the wave functions appear for $E > 0$?

16. Does Eq. 5.2 imply that we know the momentum of the particle exactly? If so, what does the uncertainty principle indicate about our knowledge of its location? How can you reconcile this with our knowledge that the particle *must* be in the well?

17. Do sharp boundaries and discontinuous jumps of potential energy occur in nature? If not, how would our analysis of potential energy steps and barriers be different?

Problems

5.1 Behavior of a Wave at a Boundary

1. A ball falls from rest at a height H above a lake. Let $y = 0$ at the surface of the lake. As it falls, it experiences a gravitational force $-mg$. When it enters the water, it experiences a buoyant force B so the net force in the water is $B - mg$. (a) Write expressions for $v(t)$ and $y(t)$ while the ball is falling in air. (b) In the water, let $v_2(t) = at + b$ and $y_2(t) = \frac{1}{2}at^2 + bt + c$ where $a = (B - mg)/m$. Use the continuity conditions at the surface of the water to find the constants b and c.

2. A wave has the form $y = A \cos(2\pi x/\lambda + \pi/3)$ when $x < 0$. For $x > 0$, the wavelength is $\lambda/2$. By applying continuity conditions at $x = 0$, find the amplitude (in terms of A) and phase of the wave in the region $x > 0$. Sketch the wave, showing both $x < 0$ and $x > 0$.

5.2 Confining a Particle

3. The lowest energy of a particle in an infinite one-dimensional well is 4.4 eV. If the width of the well is doubled, what is its lowest energy?

4. An electron is trapped in an infinite well of width 0.120 nm. What are the three longest wavelengths permitted for the electron's de Broglie waves?

5. An electron is trapped in a one-dimensional region of width 0.050 nm. Find the three smallest possible values allowed for the energy of the electron.

6. What is the minimum energy of a neutron ($mc^2 = 940$ MeV) confined to a region of space of nuclear dimensions (1.0×10^{-14} m)?

5.3 The Schrödinger Equation

7. In the region $0 \leq x \leq a$, a particle is described by the wave function $\psi_1(x) = -b(x^2 - a^2)$. In the region $a \leq x \leq w$, its wave function is $\psi_2(x) = (x - d)^2 - c$. For $x \geq w, \psi_3(x) = 0$. (a) By applying the continuity conditions at $x = a$, find c and d in terms of a and b. (b) Find w in terms of a and b.

8. A particle is described by the wave function $\psi(x) = b(a^2 - x^2)$ for $-a \leq x \leq +a$ and $\psi(x) = 0$ for $x \leq -a$ and $x \geq +a$, where a and b are positive real constants. (a) Using the normalization condition, find b in terms of a. (b) What is the probability to find the particle at $x = +a/2$ in a small interval of width $0.010a$? (c) What is the probability for the particle to be found between $x = +a/2$ and $x = +a$?

9. In a certain region of space, a particle is described by the wave function $\psi(x) = Cxe^{-bx}$ where C and b are real constants. By substituting into the Schrödinger equation, find the potential energy in this region and also find the energy of the particle. (Hint: Your solution must give an energy that is a constant everywhere in this region, independent of x.)

10. A particle is represented by the following wave function:

$$\begin{aligned} \psi(x) &= 0 & x &< -L/2 \\ &= C(2x/L + 1) & -L/2 &< x < 0 \\ &= C(-2x/L + 1) & 0 &< x < +L/2 \\ &= 0 & x &> +L/2 \end{aligned}$$

(a) Use the normalization condition to find C. (b) Evaluate the probability to find the particle in an interval of width $0.010L$ at $x = L/4$ (that is, between $x = 0.245L$ and $x = 0.255L$. (No integral is necessary for this calculation.) (c) Evaluate the probability to find the particle between $x = 0$ and $x = +L/4$. (d) Find the average value of x and the rms value of x: $x_{\text{rms}} = \sqrt{(x^2)_{\text{av}}}$.

5.4 Applications of the Schrödinger Equation

11. A particle in an infinite well is in the ground state with an energy of $1.26 \, \text{eV}$. How much energy must be added to the particle to reach the second excited state ($n = 3$)? The third excited state ($n = 4$)?

12. An electron is trapped in an infinitely deep one-dimensional well of width $0.251 \, \text{nm}$. Initially the electron occupies the $n = 4$ state. (a) Suppose the electron jumps to the ground state with the accompanying emission of a photon. What is the energy of the photon? (b) Find the energies of other photons that might be emitted if the electron takes other paths between the $n = 4$ state and the ground state.

13. Show that Eq. 5.31 gives the value $A = \sqrt{2/L}$.

14. A particle is trapped in an infinite one-dimensional well of width L. If the particle is in its ground state, evaluate the probability to find the particle (a) between $x = 0$ and

$x = L/3$; (b) between $x = L/3$ and $x = 2L/3$; (c) between $x = 2L/3$ and $x = L$.

15. A particle is confined between rigid walls separated by a distance $L = 0.189 \, \text{nm}$. The particle is in the second excited state ($n = 3$). Evaluate the probability to find the particle in an interval of width $1.00 \, \text{pm}$ located at: (a) $x = 0.188 \, \text{nm}$; (b) $x = 0.031 \, \text{nm}$; (c) $x = 0.079 \, \text{nm}$. (Hint: No integrations are required for this problem; use Eq. 5.7 directly.) What would be the corresponding results for a classical particle?

16. What is the next level (above $E = 50E_0$) of the two-dimensional particle in a box in which the degeneracy is greater than 2?

17. A particle is confined to a two-dimensional box of length L and width $2L$. The energy values are $E = (\hbar^2\pi^2/2mL^2)(n_x^2 + n_y^2/4)$. Find the two lowest degenerate levels.

18. Show by direct substitution that Eq. 5.39 gives a solution to the two-dimensional Schrödinger equation, Eq. 5.37. Find the relationship between k_x, k_y, and E.

19. A particle is confined to a three-dimensional region of space of dimensions L by L by L. The energy levels are $(\hbar^2\pi^2/2mL^2)(n_x^2 + n_y^2 + n_z^2)$, where n_x, n_y, and n_z are integers ≥ 1. Sketch an energy level diagram, showing the energies, quantum numbers, and degeneracies for the lowest 10 energy levels.

5.5 The Simple Harmonic Oscillator

20. Using the normalization condition, show that the constant A has the value $(m\omega_0/\hbar\pi)^{1/4}$ for the one-dimensional simple harmonic oscillator in its ground state.

21. (a) At the classical turning points $\pm x_0$ of the simple harmonic oscillator, $K = 0$ and so $E = U$. From this relationship, show that $x_0 = (\hbar\omega_0/k)^{1/2}$ for an oscillator in its ground state. (b) Find the turning points in the first and second excited states.

22. Use the ground-state wave function of the simple harmonic oscillator to find x_{av}, $(x^2)_{\text{av}}$, and Δx. Use the normalization constant $A = (m\omega_0/\hbar\pi)^{1/4}$.

23. (a) Using a symmetry argument rather than a calculation, determine the value of p_{av} for a simple harmonic oscillator. (b) Conservation of energy for the harmonic oscillator can be used to relate p^2 to x^2. Use this relation, along with the value of $(x^2)_{\text{av}}$ from Problem 22, to find $(p^2)_{\text{av}}$ for the oscillator in its ground state. (c) Using the results of parts a and b, show that $\Delta p = \sqrt{\hbar\omega_0 m/2}$.

24. The ground state energy of an oscillating electron is $1.24 \, \text{eV}$. How much energy must be added to the electron to move it to the second excited state? The fourth excited state?

25. Compare the probabilities for an oscillating particle in its ground state to be found in a small interval of width dx at the center of the well and at the classical turning points.

5.6 Steps and Barriers

26. Find the value of K at which Eq. 5.60 has its maximum value, and show that Eq. 5.61 is the maximum value of Δx.

27. For a particle with energy $E < U_0$ incident on the potential energy step, use ψ_0 and ψ_1 from Eqs. 5.57, and evaluate the constants B and D in terms of A by applying the boundary conditions at $x = 0$.

28. Using the wave functions of Eq. 5.55 for the potential energy step, apply the boundary conditions of ψ and $d\psi/dx$ to find B' and C' in terms of A', for the potential step when particles are incident from the negative x direction. Evaluate the ratios $|B'|^2/|A'|^2$ and $|C'|^2/|A'|^2$ and interpret.

29. (a) Write down the wave functions for the three regions of the potential energy barrier (Figure 5.26) for $E < U_0$. You will need six coefficients in all. Use complex exponential notation. (b) Use the boundary conditions at $x = 0$ and at $x = L$ to find four relationships among the six coefficients. (Do not try to solve these relationships.) (c) Suppose particles are incident on the barrier from the left. Which coefficient should be set to zero? Why?

30. Repeat Problem 29 for the potential energy barrier when $E > U_0$, and sketch a representative probability density that shows several cycles of the wave function. In your sketch, make sure the amplitude and wavelength in each region accurately describe the situation.

General Problems

31. An electron is trapped in a one-dimensional well of width 0.132 nm. The electron is in the $n = 10$ state. (a) What is the energy of the electron? (b) What is the uncertainty in its momentum? (Hint: Use Eq. 4.10.) (c) What is the uncertainty in its position? How do these results change as $n \to \infty$? Is this consistent with classical behavior?

32. Sketch the form of a possible solution to the Schrödinger equation for each of the potential energies shown in Figure 5.34. The potential energies go to infinity at the boundaries. In each case show several cycles of the wave function. In your sketches, pay attention to the continuity conditions (where applicable) and to changes in the wavelength and amplitude.

33. Show that the average value of x^2 in the one-dimensional infinite potential energy well is $L^2(1/3 - 1/2n^2\pi^2)$.

34. Use the result of Problem 33 to show that, for the infinite one-dimensional well, defining $\Delta x = \sqrt{(x^2)_{av} - (x_{av})^2}$ gives $\Delta x = L\sqrt{1/12 - 1/2\pi^2 n^2}$.

35. (a) In the infinite one-dimensional well, what is p_{av}? (Use a symmetry argument.) (b) What is $(p^2)_{av}$? [Hint: What is $(p^2/2m)_{av}$?] (c) Defining $\Delta p = \sqrt{(p^2)_{av} - (p_{av})^2}$, show that $\Delta p = hn/2L$.

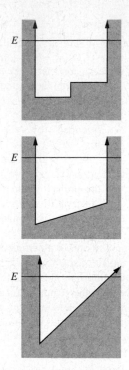

FIGURE 5.34 Problem 32.

36. The first excited state of the harmonic oscillator has a wave function of the form $\psi(x) = Axe^{-ax^2}$. Follow the method outlined in Section 5.5 to find a and the energy E. Find the constant A from the normalization condition.

37. Using the normalization constant A from Problem 20 and the value of a from Eq. 5.49, evaluate the probability to find an oscillator in the ground state beyond the classical turning points $\pm x_0$. This problem cannot be solved in closed, analytic form. Develop an approximate, numerical method using a graph, calculator, or computer. Consider a particle bound to an atomic-sized region ($x_0 = 0.1$ nm) with an effective force constant of 1.0 eV/nm^2.

38. A two-dimensional harmonic oscillator has energy $E = \hbar\omega_0(n_x + n_y + 1)$, where n_x and n_y are integers beginning with zero. (a) Justify this result based on the energy of the one-dimensional oscillator. (b) Sketch an energy-level diagram similar to Figure 5.21, showing the lowest 4 energy levels. For each level, show the value of E (in units of $\hbar\omega_0$), the quantum numbers n_x and n_y, and the degeneracy. (c) Show that the degeneracy of each level is equal to $n_x + n_y + 1$.

Chapter 6

THE RUTHERFORD-BOHR MODEL OF THE ATOM

This model of the atom, based on the work of Rutherford and Bohr, shows electrons circulating about the nucleus like planets circulating about the Sun. It can be a useful model for some purposes, but it does not represent even approximately the structure of real atoms. In Chapters 7 and 8 we will learn more about the behavior and properties of electrons in atoms.

Our goal in this chapter is to understand some of the details of atomic structure that can be learned from experimental studies of atoms. In particular, we discuss two types of experiments that are important in the development of our theory of atomic structure: the scattering of charged particles by atoms, which tells us about the distribution of electric charge in atoms, and the emission or absorption of radiation by atoms, which tells us about their excited states.

We use the information obtained from these experiments to develop an *atomic model*, which helps us understand and explain the properties of atoms. A model is usually an oversimplified picture of a more complex system, which provides some insight into its operation but may not be sufficiently detailed to explain *all* of its properties.

In this chapter, we discuss the experiments that led to the *Rutherford-Bohr model* (also known simply as the Bohr model), which is based on the familiar "planetary" structure in which the electrons orbit about the nucleus like planets about the Sun. Even though this model is not strictly valid from the standpoint of wave mechanics, it does help us understand many atomic properties, especially the excited states of the simplest atom, hydrogen. In Chapter 7, we show how wave mechanics changes our picture of the hydrogen atom, and in Chapter 8 we consider the structure of more complicated atoms.

6.1 BASIC PROPERTIES OF ATOMS

Before we begin to construct a model of the atom, it is helpful to summarize some of the basic properties of atoms.

1. *Atoms are very small,* about 0.1 nm (0.1×10^{-9} m) in radius. Thus any effort to "see" an atom using visible light ($\lambda = 500$ nm) is hopeless owing to diffraction effects. We can make a crude estimate of the *maximum* size of an atom in the following way. Consider a cube of elemental matter—for example, iron. Iron has a density of about 8 g/cm^3 and a molar mass of 56 g. One mole of iron (56 g) contains Avogadro's number of atoms, about 6×10^{23}. Thus 6×10^{23} atoms occupy about 7 cm^3 and so 1 atom occupies about 10^{-23} cm^3. If we assume the atoms of a solid are packed together in the most efficient possible way, like hard spheres in contact, then the diameter of one atom is about $\sqrt[3]{10^{-23} \text{cm}^3} = 2 \times 10^{-8}$ cm = 0.2 nm.

2. *Atoms are stable*—they do not spontaneously break apart into smaller pieces or collapse; therefore the internal forces that hold the atom together must be in equilibrium. This immediately tells us that the forces that pull the parts of an atom together must be opposed in some way; otherwise atoms would collapse.

3. *Atoms contain negatively charged electrons, but are electrically neutral.* If we disturb an atom or collection of atoms with sufficient force, electrons are emitted. We learn this fact from studying the Compton effect and the photoelectric effect. We also learned in Chapter 4 that even though electrons are emitted from the nuclei of atoms in certain radioactive decay processes, they don't "exist" in those nuclei but are manufactured there by some process. Electrons were excluded from the nucleus based on the uncertainty principle, which forbids emitted electrons of the energies observed in the laboratory from existing in the nucleus (see Example 4.7). The uncertainty principle places no such restriction on the existence of electrons in a volume as large as an atom (see Problem 1).

    We can also easily observe that bulk matter is electrically neutral, and we assume that this is likewise a property of the atoms. Experiments with beams of individual atoms support this assumption. From these experimental facts we deduce that an atom with Z negatively charged electrons must also contain a net positive charge of Ze.

4. *Atoms emit and absorb electromagnetic radiation.* This radiation may take many forms—visible light ($\lambda \sim 500$ nm), X rays ($\lambda \sim 1$ nm), ultraviolet rays ($\lambda \sim 10$ nm), infrared rays ($\lambda \sim 0.1$ μm), and so forth. In fact it is from observation of these emitted and absorbed radiations, which can be measured with great precision, that we learn most of what we know about atoms. In a typical emission measurement, an electric current is passed through a glass tube containing a small sample of the gas phase of the element under study, and radiation is emitted when an excited atom returns to its ground state. The absorption wavelengths can be measured by passing a beam of white light through a sample of the gas and noting which colors are removed from the white light by absorption in the gas. One particularly curious feature of the atomic radiations is that atoms don't always emit and absorb radiations at the same wavelengths—some wavelengths present in the *emission* experiment do not also appear in the *absorption* experiment. Any successful theory of atomic structure must be able to account for these emission and absorption wavelengths.

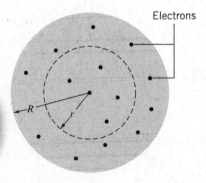

FIGURE 6.1 The Thomson model of the atom. Z electrons are imbedded in a uniform sphere of positive charge Ze and radius R. An imaginary spherical surface of radius r contains a fraction r^3/R^3 of the positive charge.

6.2 SCATTERING EXPERIMENTS AND THE THOMSON MODEL

An early model of the structure of the atom was proposed (in 1904) by J. J. Thomson, who was known for his previous identification of the electron and measurement of its charge-to-mass ratio e/m. The Thomson model incorporates many of the known properties of atoms: size, mass, number of electrons, and electrical neutrality. In this model, an atom contains Z electrons that are embedded in a uniform sphere of positive charge (Figure 6.1). The total positive charge of the sphere is Ze, the mass of the sphere is essentially the mass of the atom (the electrons don't contribute significantly to the total mass), and the radius R of the sphere is the radius of the atom. (This model is sometimes known as the "plum-pudding" model, because the electrons are distributed throughout the atom like raisins in a plum pudding.) As we will see, the Thomson model gives predictions that disagree with experiment, and so it is not the correct way of understanding the structure of atoms.

One way of studying atoms is by probing the distribution of electric charge in their interior, which we can do by bombarding the atom with charged particles and observing the angle by which particles are deflected from their original direction. This type of experiment is called a *scattering experiment*. Ideally we would like to do this experiment with a single atom, such as is represented in Figure 6.2. The scattering angle θ depends on the *impact parameter b*, which measures the distance from the center of the atom that a projectile would pass if it were not deflected. Each different value of the impact parameter results in a different value of the scattering angle.

The particle is deflected from its original trajectory by the electrical forces exerted on the particle by the atom. For a positively charged particle, these forces are: (1) a repulsive force due to the positive charge of the atom, and (2) an attractive force due to the negatively charged electrons. We assume that the mass of the deflected particle is much greater than the mass of an electron but also much less than the mass of the atom. In the encounter between the projectile and an electron,

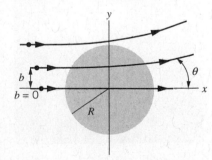

FIGURE 6.2 A positively charged particle is deflected by an angle θ as it passes through a positively charged sphere, representing a Thomson model atom. The scattering angle depends on the value of the impact parameter b, which varies from 0 to R.

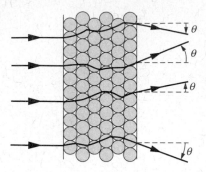

FIGURE 6.3 Scattering by a thin foil. Some individual scatterings tend to increase θ, while others tend to decrease θ.

the forces exerted on each by the other are equal and opposite (by Newton's third law), and so the principal victim of the encounter is the much less massive electron; the effect on the projectile is negligible. (Imagine rolling a bowling ball through a field of Ping-Pong balls!) We thus need consider only the positively charged atom as a cause of the deflection of the particle. By the same argument, we neglect any possible motion of the more massive atom caused by the passage of the projectile. The basic experiment, then, is the scattering of a positively charged projectile by the stationary positively charged massive part of the atom.

In practice we cannot do the experiment with one atom. Instead, we bombard a thin foil, as in Figure 6.3. The scattering angle θ that we observe in the laboratory is the result of scattering by many atoms, with impact parameters that we do not know and cannot control. Let's assume that for a single atom the average scattering angle is θ_{av}, which represents an average over all possible impact parameters from zero up to the atomic radius R. For a typical foil thickness of $1\ \mu m\ (10^{-6}\ m)$, the projectile is scattered by about 10^4 atoms.

The total scattering angle θ is determined by statistical considerations, because some of the individual scatterings move the projectile toward larger scattering angles and some toward smaller angles, as represented in Figure 6.3. This is an example of a "random walk" problem—for N scatterings, the most likely observed net scattering angle θ is related to the average individual scattering angle by

$$\theta \simeq \sqrt{N}\theta_{av} \tag{6.1}$$

According to the Thomson model, the average scattering angle for a single atom is on the order of $0.01°$, and for a foil that is 10^4 atoms thick the net scattering angle should be about $1°$. This is consistent with experimental observations.

The most critical test of the Thomson model, which it fails completely, occurs when we examine the probability for scattering at large angles. If each individual scattering deflects the projectile through an angle of around $0.01°$, then to observe projectiles scattered through a total angle greater than $90°$, we must have about 10^4 successive scatterings, *all* of which push the projectile toward larger angles. Because the probabilities of individual scatterings toward either larger or smaller angles are equal, the probability of having 10^4 successive scatterings toward larger angles, like the probability of finding 10^4 successive heads in tossing a coin, is about $(1/2)^{10,000} = 10^{-3000}$.

An experiment to observe this scattering was performed by Hans Geiger and Ernest Marsden in the laboratory of Ernest Rutherford at Manchester University in 1910. For projectiles they used alpha particles, which are nuclei of helium (of charge $+2e$) emitted in radioactive decay. Their results showed that the probability of an alpha particle scattering at angles greater than $90°$ was about 10^{-4}. This remarkable discrepancy between the expected value based on the Thomson model (10^{-3000}) and the observed value (10^{-4}) was described by Rutherford in this way:

It was quite the most incredible event that ever happened to me in my life. It was as incredible as if you fired a 15-inch shell at a piece of tissue paper and it came back and hit you.

The analysis of the results of such scattering experiments led Rutherford to propose that the mass and positive charge of the atom are not distributed uniformly over the volume of the atom, but instead are concentrated in an extremely small region, about 10^{-14} m in diameter, at the center of the atom. In Section 6.3 we will see how this proposal is consistent with the large-angle scattering results.

Ernest Rutherford (1871–1937, England). Founder of nuclear physics, he is known for his pioneering work on alpha-particle scattering and radioactive decays. His inspiring leadership influenced a generation of British nuclear and atomic scientists.

Scattering in the Thomson Model (Optional)

Let's assume that a projectile of positive charge ze is incident on an atom of radius R that we represent according to the Thomson model as a uniform sphere of positive charge Ze. The force on the projectile when it is a distance r from the center of the atom can be computed using Gauss's law (see Problem 2):

$$F = \frac{zZe^2}{4\pi\varepsilon_0 R^3} r \tag{6.2}$$

Before discussing the scattering, we should note that this equation can also describe (if we put $z = 1$) the force on an electron embedded in the Thomson atom at a distance r from its center. This force can be written $F = kr$ with $k = Ze^2/4\pi\varepsilon_0 R^3$. This linear restoring force permits the electrons to oscillate about their equilibrium positions just like a mass on a spring subject to the linear restoring force $F = kx$. We therefore expect the electrons in the Thomson atom to oscillate about their equilibrium positions with a frequency $f = (2\pi)^{-1}\sqrt{k/m}$, where k is the force constant. Because an oscillating electric charge radiates electromagnetic waves whose frequency is identical to the oscillation frequency, we might expect, based on the Thomson model, that the radiation emitted by atoms would show this characteristic frequency. This turns out *not* to be true (see Problem 3); the calculated frequencies do not correspond to the frequencies observed for radiation emitted by atoms.

The exact calculation of the scattering angle for different values of the impact parameter in the Thomson model of the atom is fairly complicated, but for our purposes we want only an estimate of the average value of the angle. As we will find out later, it's not very important if our estimate is off by a small factor.

Initially the projectile moves in the x direction in the geometry of Figure 6.2, but the atom exerts a force in the y direction that produces a small component of momentum p_y in that direction. Using Newton's second law we can find the momentum from the *impulse* received by the projectile due to the electrostatic force:

$$p_y = \int F_y \, dt \tag{6.3}$$

Rather than carry out this complicated integral for a force that is changing in magnitude and direction as the projectile travels, we'll estimate the average scattering angle by choosing an average value for the impact parameter, namely $b = R/2$ (representing the middle trajectory of Figure 6.2), and we'll assume the force acts in the y direction for a time Δt determined by the projectile's flight along a line of length roughly equal to R. This underestimates the amount of time during which the force acts but overestimates the effect of the force (which doesn't act purely in the y direction along the entire trajectory), so to some extent these two effects should cancel one another.

Making these approximations, we obtain

$$p_y \cong F\Delta t \cong \frac{zZe^2(R/2)}{4\pi\varepsilon_0 R^3}\frac{R}{v} = \frac{zZe^2}{8\pi\varepsilon_0 Rv} \tag{6.4}$$

The angle θ is small, so we can make the approximation $\tan\theta \cong \theta$, and we can assume that p_x changes very little from its initial value mv, and so the average scattering angle is

$$\theta_{av} \cong \tan\theta_{av} = \frac{p_y}{p_x} = \frac{p_y}{mv} = \frac{zZe^2}{8\pi\varepsilon_0 Rv}\frac{1}{mv} = \frac{zZe^2}{16\pi\varepsilon_0 RK} \tag{6.5}$$

using the nonrelativistic kinetic energy $K = \frac{1}{2}mv^2$. This gives an estimate of the scattering angle when the impact parameter b is equal to half the radius R. Smaller values of b will give smaller deflection angles, and larger values of b will give larger angles, so this is a reasonable estimate for the average scattering angle for a Thomson model atom.

Example 6.1

Using the Thomson model, estimate the average scattering angle when alpha particles ($z = 2$) with kinetic energy 3 MeV are scattered from gold ($Z = 79$). The atomic radius of gold is 0.179 nm.

Solution

Using $e^2/4\pi\varepsilon_0 = 1.44$ eV·nm, we have

$$\theta_{av} \cong \frac{zZe^2}{16\pi\varepsilon_0 RK} = \frac{1}{4}\frac{e^2}{4\pi\varepsilon_0}\frac{zZ}{RK}$$

$$= \frac{0.25(1.44\,\text{eV} \cdot \text{nm})(2)(79)}{(0.179\,\text{nm})(3 \times 10^6\,\text{eV})}$$

$$= 1 \times 10^{-4}\,\text{rad} = 0.01°$$

Even though this result represents a rough estimate of the average scattering angle in the Thomson model of the atom, its accuracy does not affect our conclusions about the failure of the model. Even if our estimate were too small by as much as a factor of 10 (which is highly unlikely), we would be comparing an expected probability of 10^{-300} (instead of 10^{-3000}) with the observed 10^{-4}, still a spectacular disagreement. Any reasonable estimate shows the complete failure of the Thomson model to account for these scattering experiments.

6.3 THE RUTHERFORD NUCLEAR ATOM

In analyzing the scattering of alpha particles, Rutherford concluded that the most likely way an alpha particle ($m = 4$ u) can be deflected through large angles is by a *single* collision with a more massive object. Rutherford therefore proposed that the charge and mass of the atom were concentrated at its center, in a region called the *nucleus*. Figure 6.4 illustrates the scattering geometry in this case. The projectile, of charge ze, experiences a repulsive force due to the positively charged nucleus:

$$F = \frac{1}{4\pi\varepsilon_0}\frac{|q_1||q_2|}{r^2} = \frac{(ze)(Ze)}{4\pi\varepsilon_0 r^2} \qquad (6.6)$$

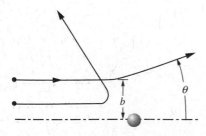

FIGURE 6.4   Scattering by a nuclear atom. The path of the scattered particle is a hyperbola. Smaller impact parameters give larger scattering angles.

(Compare this with Eq. 6.2, which describes a projectile that is inside the sphere of charge Ze and so feels only a portion of the positive charge. We assume now that the projectile is always outside the nucleus, so it feels the full nuclear charge Ze.) The atomic electrons, with their small mass, do not appreciably affect the path of the projectile and we neglect their effect on the scattering. We also assume that the nucleus is so much more massive than the projectile that it does not move during the scattering process; because no recoil motion is given to the nucleus, the initial and final kinetic energies K of the projectile are equal.

As Figure 6.4 shows, for each impact parameter b, there is a certain scattering angle θ, and we need the relationship between b and θ. The projectile can be shown* to follow a hyperbolic path; in polar coordinates r and ϕ, the equation of the hyperbola is

$$\frac{1}{r} = \frac{1}{b}\sin\phi + \frac{zZe^2}{8\pi\varepsilon_0 b^2 K}(\cos\phi - 1) \tag{6.7}$$

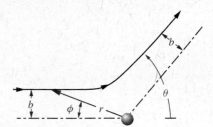

FIGURE 6.5 The hyperbolic trajectory of a scattered particle.

As shown in Figure 6.5, the initial position of the particle is $\phi = 0, r \to \infty$, and the final position is $\phi = \pi - \theta, r \to \infty$. Using the coordinates at the final position, Eq. 6.7 reduces to

$$b = \frac{zZe^2}{8\pi\varepsilon_0 K}\cot\tfrac{1}{2}\theta = \frac{zZ}{2K}\frac{e^2}{4\pi\varepsilon_0}\cot\tfrac{1}{2}\theta \tag{6.8}$$

(This result is written in this form so that $e^2/4\pi\varepsilon_0 = 1.44\,\text{eV}\cdot\text{nm}$ or $\text{MeV}\cdot\text{fm}$ can be easily inserted.) A projectile that approaches the nucleus with impact parameter b will be scattered at an angle θ; projectiles approaching with smaller values of b will be scattered through larger angles, as shown in Figure 6.4.

We divide our study of the scattering of charged projectiles by nuclei (which is commonly called *Rutherford scattering*) into three parts: (1) calculation of the fraction of projectiles scattered at angles greater than some value of θ, (2) the Rutherford scattering formula and its experimental verification, and (3) the closest approach of a projectile to the nucleus.

1. The Fraction of Projectiles Scattered at Angles Greater than θ. From Figure 6.4 we see immediately that every projectile with impact parameters less than a given value of b will be scattered at angles greater than its corresponding θ. What is the chance of a projectile having an impact parameter less than a given value of b? Suppose the foil were one atom thick—a single layer of atoms packed tightly together, as in Figure 6.6. Each atom is represented by a circular disc, of area πR^2. If the foil contains N atoms, its total area is $N\pi R^2$. For scattering at angles greater than θ, the impact parameter must fall between zero and b—that is, the projectile must approach the atom within a circular disc of area πb^2. If the projectiles are spread uniformly over the area of the disc, then the fraction of projectiles that fall within that area is just $\pi b^2/\pi R^2$.

A real scattering foil may be thousands or tens of thousands of atoms thick. Let t be the thickness of the foil and A its area, and let ρ and M be the density and molar mass of the material of which the foil is made. The volume of the foil is then At, its mass is ρAt, the number of moles is $\rho At/M$, and the number of atoms or nuclei per unit volume is

$$n = N_A\frac{\rho At}{M}\frac{1}{At} = \frac{N_A\rho}{M} \tag{6.9}$$

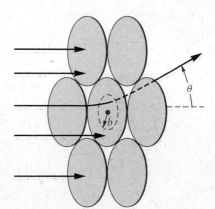

FIGURE 6.6 Scattering geometry for many atoms. For impact parameter b, the scattering angle is θ. If the particle enters the atom within the disc of area πb^2, its scattering angle will be larger than θ.

where N_A is Avogadro's number (the number of atoms per mole). As seen by an incident projectile, the number of nuclei per unit area is $nt = N_A\rho t/M$; that is, on the average, each nucleus contributes an area $(N_A\rho t/M)^{-1}$ to the field of view

---

*See, for example, R. M. Eisberg and R. Resnick, *Quantum Physics of Atoms, Molecules, Solids, Nuclei, and Particles*, 2nd ed. (New York, Wiley, 1985).

of the projectile. For scattering at angles greater than θ, it must once again be true that the projectile must fall within an area πb^2 of the center of an atom; the fraction scattered at angles greater than θ is just the fraction that approaches an atom within the area πb^2:

$$f_{<b} = f_{>\theta} = nt\pi b^2 \tag{6.10}$$

assuming that the incident particles are spread uniformly over the area of the foil.

Example 6.2

A gold foil ($\rho = 19.3\,\mathrm{g/cm^3}$, $M = 197\,\mathrm{g/mole}$) has a thickness of 2.0×10^{-4} cm. It is used to scatter alpha particles of kinetic energy 8.0 MeV. (a) What fraction of the alpha particles is scattered at angles greater than 90°? (b) What fraction of the alpha particles is scattered at angles between 90° and 45°?

Solution

(a) For this case the number of nuclei per unit volume can be computed as

$$n = \frac{N_A \rho}{M} = \frac{(6.02 \times 10^{23}\,\text{atoms/mole})(19.3\,\text{g/cm}^3)}{(197\,\text{g/mole})(1\,\text{m}/10^2\,\text{cm})^3}$$

$$= 5.9 \times 10^{28}\,\text{m}^{-3}$$

For scattering at 90°, the impact parameter b can be found from Eq. 6.8:

$$b = \frac{zZ}{2K} \frac{e^2}{4\pi\varepsilon_0} \cot \frac{1}{2}\theta$$

$$= \frac{(2)(79)}{2(8.0\,\text{MeV})}(1.44\,\text{MeV·fm})\cot 45°$$

$$= 14\,\text{fm} = 1.4 \times 10^{-14}\,\text{m}$$

and using Eq. 6.10 we then have

$$f_{>90°} = nt\pi b^2$$

$$= (5.9 \times 10^{28}\,\text{m}^{-3})(2.0 \times 10^{-6}\,\text{m})\pi(1.4 \times 10^{-14}\,\text{m})^2$$

$$= 7.5 \times 10^{-5}$$

(b) Repeating the calculation for $\theta = 45°$, we find

$$b = \frac{zZ}{2K} \frac{e^2}{4\pi\varepsilon_0} \cot \frac{1}{2}\theta$$

$$= \frac{(2)(79)}{2(8.0\,\text{MeV})}(1.44\,\text{MeV} \cdot \text{fm})\cot 22.5°$$

$$= 34\,\text{fm} = 3.4 \times 10^{-14}\,\text{m}$$

$$f_{>45°} = nt\pi b^2$$

$$= (5.9 \times 10^{28}\,\text{m}^{-3})(2.0 \times 10^{-6}\,\text{m})\pi(3.4 \times 10^{-14}\,\text{m})^2$$

$$= 4.4 \times 10^{-4}$$

If a total fraction of 4.4×10^{-4} is scattered at angles greater than 45°, and of that, 7.5×10^{-5} is scattered at angles greater than 90°, the fraction scattered *between* 45° and 90° must be

$$4.4 \times 10^{-4} - 7.5 \times 10^{-5} = 3.6 \times 10^{-4}$$

2. The Rutherford Scattering Formula and Its Experimental Verification. In order to find the probability that a projectile will be scattered into a small angular range at θ (between θ and $\theta + d\theta$), we require that the impact parameter lie within a small range of values db at b (see Figure 6.7). The fraction, df, is then

$$df = nt(2\pi b\, db) \tag{6.11}$$

from Eq. 6.10. Differentiating Eq. 6.8 we find db in terms of $d\theta$:

$$db = \frac{zZ}{2K} \frac{e^2}{4\pi\varepsilon_0}(-\csc^2 \tfrac{1}{2}\theta)(\tfrac{1}{2}\,d\theta) \tag{6.12}$$

and so

$$|df| = \pi nt \left(\frac{zZ}{2K}\right)^2 \left(\frac{e^2}{4\pi\varepsilon_0}\right)^2 \csc^2 \tfrac{1}{2}\theta \cot \tfrac{1}{2}\theta \, d\theta \qquad (6.13)$$

(This minus sign in Eq. 6.12 is not important—it just tells us that θ increases as b decreases.) Suppose we place a detector for the scattered projectiles at the angle θ a distance r from the nucleus. The probability for a projectile to be scattered into the detector depends on df, which gives the probability for scattered particles to pass through the ring of radius $r \sin \theta$ and width $r \, d\theta$. The area of the ring is $dA = (2\pi r \sin \theta)r \, d\theta$. In order to calculate the rate at which projectiles are scattered *into the detector* we must know the probability *per unit area* for scattering into the ring. This is $|df|/dA$, which we call $N(\theta)$, and, after some manipulation, we find:

$$N(\theta) = \frac{nt}{4r^2} \left(\frac{zZ}{2K}\right)^2 \left(\frac{e^2}{4\pi\varepsilon_0}\right)^2 \frac{1}{\sin^4 \tfrac{1}{2}\theta} \qquad (6.14)$$

This is the *Rutherford scattering formula*.

In Rutherford's laboratory, Hans Geiger and Ernest Marsden tested the predictions of this formula in a remarkable series of experiments involving the scattering of alpha particles ($z = 2$) from a variety of thin metal foils. In those days before electronic recording and processing equipment was available, Geiger and Marsden observed and recorded the alpha particles by counting the scintillations (flashes of light) produced when the alpha particles struck a zinc sulfide screen. A schematic view of their apparatus is shown in Figure 6.8. In all, four predictions of the Rutherford scattering formula were tested:

(a) $N(\theta) \propto t$. With a source of 8-MeV alpha particles from radioactive decay, Geiger and Marsden used scattering foils of varying thicknesses t while keeping the scattering angle θ fixed at about 25°. Their results are summarized in Figure 6.9, and the linear dependence of $N(\theta)$ on t is apparent. This is also evidence that, even at this moderate scattering angle, *single* scattering is much more important than *multiple* scattering. (In a random statistical theory of multiple scattering, the probability for scattering at a large angle would be proportional to the square root of the number of single scatterings, and we would expect $N(\theta) \propto t^{1/2}$. Figure 6.9 shows clearly that this is not true.)

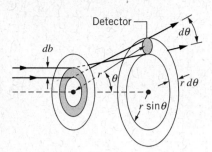

FIGURE 6.7 Particles entering the ring between b and $b + db$ are distributed uniformly along a ring of angular width $d\theta$. A detector is at a distance r from the scattering foil.

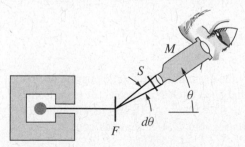

FIGURE 6.8 Schematic diagram of alpha-particle scattering experiment. A radioactive source of alpha particles is in a shield with a small hole. Alpha particles strike the foil F and are scattered into the angular range $d\theta$. Each time a scattered particle strikes the screen S a flash of light is emitted and observed with the movable microscope M.

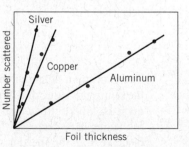

FIGURE 6.9 The dependence of scattering rate on foil thickness for three different scattering foils.

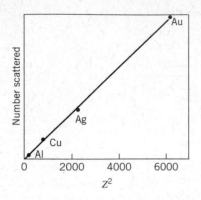

FIGURE 6.10   The dependence of scattering rate on the nuclear charge Z for foils of different materials. The data are plotted against Z^2.

This result emphasizes a significant difference between scattering by a Thomson model atom and a Rutherford nuclear atom: In the Thomson model, the projectile is scattered by *every* atom along its path as it passes through the foil (see Figure 6.3), while in the Rutherford nuclear model the nucleus is so tiny that the chance of even a single significant encounter is small and the chance of encountering more than one nucleus is negligible.

(b) $N(\theta) \propto Z^2$. In this experiment, Geiger and Marsden used a variety of different scattering materials, of approximately (*but not exactly*) the same thickness. This proportionality is therefore much more difficult to test than the previous one, since it involves the comparison of *different thicknesses* of *different materials*. However, as shown in Figure 6.10, the results are consistent with the proportionality of $N(\theta)$ to Z^2.

(c) $N(\theta) \propto K^{-2}$. In order to test this prediction of the Rutherford scattering formula, Geiger and Marsden kept the thickness of the scattering foil constant and varied the speed of the alpha particles. They accomplished this by slowing down the alpha particles emitted from the radioactive source by passing them through thin sheets of mica. From independent measurements they knew the effect of different thicknesses of mica on the velocity of the alpha particles. The results of the experiment are shown in Figure 6.11; once again we see excellent agreement with the expected relationship.

(d) $N(\theta) \propto \sin^{-4} \frac{1}{2}\theta$. This dependence of N on θ is perhaps the most important and distinctive feature of the Rutherford scattering formula. It also produces the largest variation in N over the range accessible by experiment. In the tests discussed so far, N varied by perhaps an order of magnitude; in this case N varies by about *five* orders of magnitude from the smaller to the larger angles. Geiger and Marsden used a gold foil and varied θ from 5 to 150°, to obtain the relationship between N and θ plotted in Figure 6.12. The agreement with the Rutherford formula is again very good.

Thus all predictions of the Rutherford scattering formula were confirmed by experiment, and the "nuclear atom" was verified.

3.   The Closest Approach of a Projectile to the Nucleus. A positively charged projectile slows down as it approaches a nucleus, exchanging part of its initial kinetic energy for the electrostatic potential energy due to the nuclear repulsion. The closer the projectile gets to the nucleus, the more potential energy it gains, because

$$U = \frac{1}{4\pi\varepsilon_0} \frac{q_1 q_2}{r} = \frac{1}{4\pi\varepsilon_0} \frac{zZe^2}{r} \tag{6.15}$$

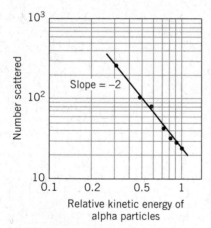

FIGURE 6.11   The dependence of scattering rate on the kinetic energy of the incident alpha particles for scattering by a single foil. The slope of -2 on the log-log scale shows that $N \propto K^{-2}$, as expected from the Rutherford formula.

The maximum potential energy, and thus the minimum kinetic energy, occurs at the minimum value of r. We assume that $U = 0$ when the projectile is far from the nucleus, where it has total energy $E = K = \frac{1}{2}mv^2$. As the projectile approaches the nucleus, K decreases and U increases, but $U + K$ remains constant. At the distance $r_{\min}$, the speed is $v_{\min}$ and:

$$E = \frac{1}{2}mv_{\min}^2 + \frac{1}{4\pi\varepsilon_0} \frac{zZe^2}{r_{\min}} = \frac{1}{2}mv^2 \tag{6.16}$$

(See Figure 6.13.)

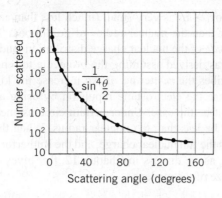

FIGURE 6.12 The dependence of scattering rate on the scattering angle θ, using a gold foil. The $\sin^{-4}(\theta/2)$ dependence is exactly as predicted by the Rutherford formula.

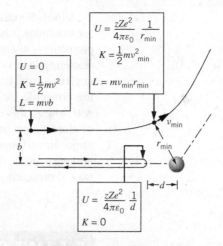

FIGURE 6.13 Closest approach of the projectile to the nucleus.

Angular momentum is also conserved. Far from the nucleus, the angular momentum L is mvb, and at r_{min}, the angular momentum is $mv_{min}r_{min}$, so

$$mvb = mv_{min}r_{min} \tag{6.17}$$

which gives $v_{min} = bv/r_{min}$. Substituting this result into Eq. 6.16, we find

$$\frac{1}{2}mv^2 = \frac{1}{2}m\left(\frac{b^2v^2}{r_{min}^2}\right) + \frac{1}{4\pi\varepsilon_0}\frac{zZe^2}{r_{min}} \tag{6.18}$$

This expression can be solved for the value of r_{min}.

Notice that the kinetic energy of the projectile is not zero at r_{min}, *unless $b = 0$*. (See Figure 6.13.) In this case, the projectile would lose all of its kinetic energy, and thus get closest to the nucleus. At this point its distance from the nucleus is *d, the distance of closest approach*. We find this distance by solving Eq. 6.18 for r_{min} when $b = 0$, and obtain

$$d = \frac{1}{4\pi\varepsilon_0}\frac{zZe^2}{K} \tag{6.19}$$

Example 6.3

Find the distance of closest approach of an 8.0-MeV alpha particle incident on a gold foil.

Solution

$$d = \frac{zZe^2}{4\pi\varepsilon_0}\frac{1}{K} = (2)(79)(1.44\ \text{MeV·fm})\frac{1}{8.0\ \text{MeV}} = 28\ \text{fm}$$

Although a distance of 28 fm is very small (much less than an atomic radius, for example) it is larger than the nuclear radius of gold (about 7 fm). Thus the projectile is always *outside* of the nuclear charge distribution, and the Rutherford scattering law, which was derived assuming the projectile to remain outside the nucleus, correctly describes the scattering. If we increase the kinetic energy of the projectile, or decrease the electrostatic repulsion by using a target nucleus with low Z, this may not be the case. Under certain circumstances, the distance of closest approach can be less than the nuclear radius. When this happens, the projectile no longer feels the full nuclear charge, and the Rutherford scattering law no longer holds. In fact, as we discuss in Chapter 12, this gives us a convenient way of measuring the size of the nucleus.

6.4 LINE SPECTRA

The radiation from atoms can be classified into continuous spectra and discrete or line spectra. In a continuous spectrum, all wavelengths from some minimum, perhaps 0, to some maximum, perhaps approaching ∞, are emitted. The radiation from a hot, glowing object is an example of this category. White light is a mixture of all of the different colors of visible light; an object that glows white hot is emitting light at all wavelengths of the visible spectrum. If, on the other hand, we force an electric discharge in a tube containing a small amount of the gas or vapor of a certain element, such as mercury, sodium, or neon, light is emitted at a few discrete wavelengths and not at any others. Examples of such emission "line" spectra are shown in Figure 6.14. The strong 436 nm (blue) and 546 nm (green) lines in the mercury emission spectrum give mercury-vapor street lights their blue-green tint; the strong yellow line at 590 nm in the sodium spectrum (which is actually a *doublet*—two very closely spaced lines) gives sodium-vapor street lights a softer, yellowish color. The intense red lines of neon are responsible for the red color of "neon signs."

Another possible experiment is to pass a beam of white light, containing all wavelengths, through a sample of a gas. When we do so, we find that certain

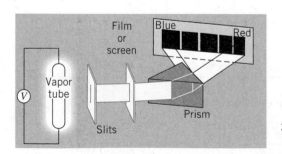
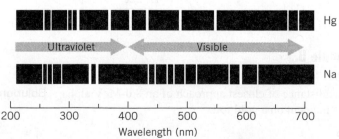

FIGURE 6.14 Apparatus for observing emission spectra. Light is emitted when an electric discharge is created in a tube containing a vapor of an element. The light passes through a dispersive medium, such as a prism or a diffraction grating, which displays the individual component wavelengths at different positions. Sample line spectra are shown for mercury and sodium in the visible and near ultraviolet.

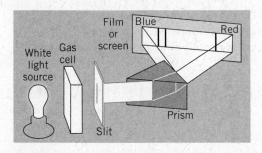

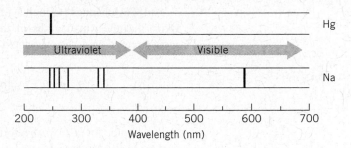

FIGURE 6.15 Apparatus for observing absorption spectra. A light source produces a continuous range of wavelengths, some of which are absorbed by a gaseous element. The light is dispersed, as in Figure 6.14. The result is a continuous "rainbow" spectrum, with dark lines at wavelengths where the light was absorbed by the gas.

wavelengths have been absorbed from the light, and again a line spectrum results. In this case there are dark lines, superimposed on the bright continuous spectrum, at the wavelengths where the absorption occurred. These wavelengths correspond to many (*but not all*) of the wavelengths seen in the emission spectrum. Examples of absorption spectra are shown in Figure 6.15.

In general, the interpretation of line spectra is very difficult in complex atoms, and so we will deal for now with the line spectra of the simplest atom, hydrogen. Regularities appear in both the emission and absorption spectra, as shown in Figure 6.16. Notice that, as with the mercury and sodium spectra, some lines present in the emission spectrum are missing from the absorption spectrum.

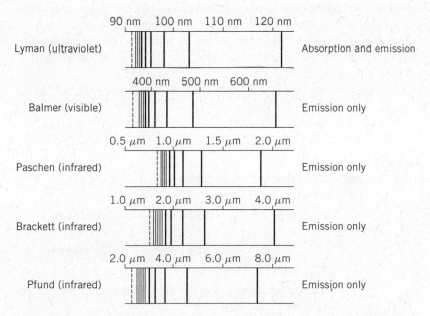

FIGURE 6.16 Emission and absorption spectral series of hydrogen. Note the regularities in the spacing of the spectral lines. The lines get closer together as the limit of each series (dashed line) is approached. Only the Lyman series appears in the absorption spectrum; all series are present in the emission spectrum.

In 1885 Johannes Balmer, a Swiss schoolteacher, noticed (mostly by trial and error) that the wavelengths of the group of emission lines of hydrogen in the visible region could be calculated very accurately from the formula

$$\lambda = (364.5 \, \text{nm}) \frac{n^2}{n^2 - 4} \qquad (n = 3, 4, 5, \ldots) \tag{6.20}$$

For example, for $n = 3$, the formula gives $\lambda = 656.1$ nm, which corresponds exactly to the longest wavelength of the series of hydrogen lines in the visible region (see Figure 6.16). This formula is now known as the *Balmer formula* and the series of lines that it fits is called the *Balmer series*. The wavelength 364.5 nm, corresponding to $n \to \infty$, is called the *series limit* (which is shown as the dashed line at the left end of the Balmer series in Figure 6.16).

It was soon discovered that all of the groupings of lines in the hydrogen spectrum could be fit with a similar formula of the form

$$\lambda = \lambda_{\text{limit}} \frac{n^2}{n^2 - n_0^2} \qquad (n = n_0 + 1, n_0 + 2, n_0 + 3 \ldots) \tag{6.21}$$

where λ_{limit} is the wavelength of the appropriate series limit. For the Balmer series, $n_0 = 2$. The other series are today known as Lyman ($n_0 = 1$), Paschen ($n_0 = 3$), Brackett ($n_0 = 4$), and Pfund ($n_0 = 5$). These series of hydrogen spectral lines are shown in Figure 6.16.

Another interesting property of the hydrogen wavelengths is summarized in the *Ritz combination principle*. If we convert the hydrogen emission wavelengths to frequencies, we find the curious property that certain pairs of frequencies added together give other frequencies that appear in the spectrum.

Any successful model of the hydrogen atom must be able to explain the occurrence of these interesting arithmetic regularities in the emission spectra.

Example 6.4

The series limit of the Paschen series ($n_0 = 3$) is 820.1 nm. What are the three longest wavelengths of the Paschen series?

Solution

From Eq. 6.21,

$$\lambda = (820.1 \, \text{nm}) \frac{n^2}{n^2 - 3^2} \qquad (n = 4, 5, 6, \ldots)$$

The three longest wavelengths are:

$$n = 4: \quad \lambda = (820.1 \, \text{nm}) \frac{4^2}{4^2 - 3^2} = 1875 \, \text{nm}$$

$$n = 5: \quad \lambda = (820.1 \, \text{nm}) \frac{5^2}{5^2 - 3^2} = 1281 \, \text{nm}$$

$$n = 6: \quad \lambda = (820.1 \, \text{nm}) \frac{6^2}{6^2 - 3^2} = 1094 \, \text{nm}$$

These transitions are in the infrared region of the electromagnetic spectrum.

Example 6.5

Show that the longest wavelength of the Balmer series and the longest *two* wavelengths of the Lyman series satisfy the Ritz combination principle. For the Lyman series, $\lambda_{limit} = 91.13$ nm.

Solution

Using Eq. 6.20 with $n = 3$, we find the longest wavelength of the Balmer series to be 656.1 nm. Converting this to a frequency, we obtain

$$f = \frac{c}{\lambda} = \frac{2.998 \times 10^8 \text{ m/s}}{(656.1 \text{ nm})(10^{-9} \text{ m/nm})} = 4.57 \times 10^{14} \text{ Hz}$$

Using Eq. 6.21 for $n = 2$ and 3 with $n_0 = 1$, we find the longest two wavelengths of the Lyman series and their corresponding frequencies to be

$$n = 2: \quad \lambda = (91.13 \text{ nm})\frac{2^2}{2^2 - 1^2} = 121.5 \text{ nm}$$

$$f = \frac{c}{\lambda} = \frac{2.998 \times 10^8 \text{ m/s}}{(121.5 \text{ nm})(10^{-9} \text{ m/nm})}$$

$$= 24.67 \times 10^{14} \text{ Hz}$$

$$n = 3: \quad \lambda = (91.13 \text{ nm})\frac{3^2}{3^2 - 1^2} = 102.5 \text{ nm}$$

$$f = \frac{c}{\lambda} = \frac{2.998 \times 10^8 \text{ m/s}}{(102.5 \text{ nm})(10^{-9} \text{ m/nm})}$$

$$= 29.24 \times 10^{14} \text{ Hz}$$

Adding the smallest frequency of the Lyman series to the smallest frequency of the Balmer series gives the next smallest Lyman frequency:

$$24.67 \times 10^{14} \text{ Hz} + 4.57 \times 10^{14} \text{ Hz} = 29.24 \times 10^{14} \text{ Hz}$$

demonstrating the Ritz combination principle.

6.5 THE BOHR MODEL

Following Rutherford's proposal that the mass and positive charge are concentrated in a very small region at the center of the atom, the Danish physicist Niels Bohr in 1913 (while working in Rutherford's laboratory) suggested that the atom resembled a miniature planetary system, with the electrons circulating about the nucleus like planets circulating about the Sun. The atom thus doesn't collapse under the influence of the electrostatic Coulomb force of the nucleus on the electrons for the same reason that the solar system doesn't collapse under the influence of the gravitational force of the Sun on the planets. In both cases, the attractive force provides the centripetal acceleration necessary to maintain the orbital motion.

As we discuss later, the Bohr model does not give a correct view of the actual structure and properties of atoms, but it represents an important first step in achieving an understanding of atoms. The correct view requires methods of quantum mechanics, which we discuss in Chapter 7.

We consider for simplicity the hydrogen atom, with a single electron circulating about a nucleus that has a single positive charge, as in Figure 6.17. The radius of the circular orbit is r, and the electron (of mass m) moves with constant tangential speed v. The attractive Coulomb force provides the centripetal acceleration v^2/r, so

$$F = \frac{1}{4\pi\varepsilon_0}\frac{|q_1||q_2|}{r^2} = \frac{1}{4\pi\varepsilon_0}\frac{e^2}{r^2} = \frac{mv^2}{r} \tag{6.22}$$

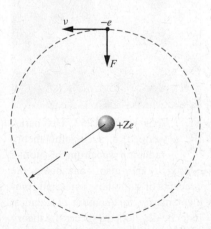

FIGURE 6.17 The Bohr model of the atom ($Z = 1$ for hydrogen).

Manipulating this equation, we can find the kinetic energy of the electron (we are assuming the more massive nucleus to remain at rest—more about this later):

$$K = \frac{1}{2}mv^2 = \frac{1}{8\pi\varepsilon_0}\frac{e^2}{r} \tag{6.23}$$

The potential energy of the electron-nucleus system is the Coulomb potential energy:

$$U = \frac{1}{4\pi\varepsilon_0}\frac{q_1 q_2}{r} = -\frac{1}{4\pi\varepsilon_0}\frac{e^2}{r} \tag{6.24}$$

The total energy $E = K + U$ is obtained by adding Eqs. 6.23 and 6.24:

$$E = K + U = \frac{1}{8\pi\varepsilon_0}\frac{e^2}{r} + \left(-\frac{1}{4\pi\varepsilon_0}\frac{e^2}{r}\right) = -\frac{1}{8\pi\varepsilon_0}\frac{e^2}{r} \tag{6.25}$$

We have ignored one serious difficulty with this model thus far. Classical physics requires that an accelerated electric charge, such as our orbiting electron, must continuously radiate electromagnetic energy. As it radiates this energy, its total energy would decrease, the electron would spiral in toward the nucleus, and the atom would collapse. To overcome this difficulty, Bohr made a bold and daring hypothesis—he proposed that there are certain special states of motion, called *stationary states,* in which the electron may exist without radiating electromagnetic energy. In these states, according to Bohr, the angular momentum L of the electron takes values that are integer multiples of $\hbar$. In stationary states, the angular momentum of the electron may have magnitude $\hbar, 2\hbar, 3\hbar, \ldots$, but never such values as $2.5\hbar$ or $3.1\hbar$. This is called the *quantization of angular momentum.*

In a circular orbit, the position vector $\vec{r}$ that locates the electron relative to the nucleus is always perpendicular to its linear momentum $\vec{p}$. The angular momentum, which is defined as $\vec{L} = \vec{r} \times \vec{p}$, has magnitude $L = rp = mvr$ when $\vec{r}$ is perpendicular to $\vec{p}$. Thus Bohr's postulate is

$$mvr = n\hbar \tag{6.26}$$

where n is an integer ($n = 1, 2, 3, \ldots$). We can use this expression with Eq. 6.23 for the kinetic energy

$$\frac{1}{2}mv^2 = \frac{1}{2}m\left(\frac{n\hbar}{mr}\right)^2 = \frac{1}{8\pi\varepsilon_0}\frac{e^2}{r} \tag{6.27}$$

to find a series of allowed values of the radius r:

$$r_n = \frac{4\pi\varepsilon_0 \hbar^2}{me^2}n^2 = a_0 n^2 \qquad (n = 1, 2, 3, \ldots) \tag{6.28}$$

where the *Bohr radius* a_0 is defined as

$$a_0 = \frac{4\pi\varepsilon_0 \hbar^2}{me^2} = 0.0529\,\text{nm} \tag{6.29}$$

Niels Bohr (1885–1962, Denmark). He developed a successful theory of the radiation spectrum of atomic hydrogen and also contributed the concepts of stationary states and complementarity to quantum mechanics. Later he developed a successful theory of nuclear fission. The institute of theoretical physics he founded in Copenhagen attracts scholars from around the world.

This important result is very different from what we expect from classical physics. A satellite may be placed into Earth orbit at any desired radius by boosting it to the appropriate altitude and then supplying the proper tangential speed. This is not true for an electron's orbit—only certain radii are allowed by the Bohr model. The radius of the electron's orbit may be $a_0, 4a_0, 9a_0, 16a_0$, and so forth, but *never* $3a_0$ or $5.3a_0$.

Substituting Eq. 6.28 for r into Eq. 6.25 gives the energy:

$$E_n = -\frac{me^4}{32\pi^2\varepsilon_0^2\hbar^2}\frac{1}{n^2} = \frac{-13.60\,\text{eV}}{n^2} \qquad (n = 1, 2, 3, \ldots) \qquad (6.30)$$

The *energy levels* calculated from Eq. 6.30 are shown in Figure 6.18. The electron's energy is *quantized*—only certain energy values are possible. In its lowest level, with $n = 1$, the electron has energy $E_1 = -13.60\,\text{eV}$ and orbits with a radius of $r_1 = 0.0529\,\text{nm}$. This state is the *ground state*. The higher states ($n = 2$ with $E_2 = -3.40\,\text{eV}$, $n = 3$ with $E_3 = -1.51\,\text{eV}$, etc.) are the *excited states*.

The *excitation energy* of an excited state n is the energy above the ground state, $E_n - E_1$. Thus the first excited state ($n = 2$) has excitation energy

$$\Delta E = E_2 - E_1 = -3.40\,\text{eV} - (-13.60\,\text{eV}) = 10.20\,\text{eV}$$

the second excited state has excitation energy

$$\Delta E = E_3 - E_1 = -1.51\,\text{eV} - (-13.60\,\text{eV}) = 12.09\,\text{eV}$$

and so forth. The excitation energy can also be regarded as the amount of energy that the atom must absorb for the electron to make an upward jump. For example, if the atom absorbs an energy of 10.20 eV when the electron is in the ground state ($n = 1$), the electron will jump upward to the first excited state ($n = 2$).

The magnitude of an electron's energy $|E_n|$ is sometimes called its *binding energy*; for example, the binding energy of an electron in the $n = 2$ state is 3.40 eV. If the atom absorbs an amount of energy equal to the binding energy of the electron, the electron will be removed from the atom and become a free electron. The atom, minus its electron, is called an *ion*. The amount of energy needed to remove an electron from an atom is also called the *ionization energy*. Usually the ionization energy of an atom indicates the energy to remove an electron from the ground state. If the atom absorbs more energy than the minimum necessary to remove the electron, the excess energy appears as the kinetic energy of the now free electron.

The binding energy can also be regarded as the energy that is released when the atom is assembled from an electron and a nucleus that are initially separated by a large distance. If we bring an electron from a large distance away (where $E = 0$) and place it in orbit in the state n where its energy has the negative value E_n, energy amounting to $|E_n|$ is released, usually in the form of one or more photons.

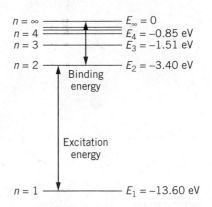

FIGURE 6.18 The energy levels of atomic hydrogen, showing the excitation energy of the electron from $n = 1$ to $n = 2$ and the binding energy of the $n = 2$ electron.

The Hydrogen Wavelengths in the Bohr Model

We previously discussed the emission and absorption spectra of atomic hydrogen, and our discussion of the Bohr model is not complete without an understanding of the origin of these spectra. Bohr postulated that, even though the electron doesn't radiate when it remains in any particular stationary state, it can emit radiation when it moves to a lower energy level. In the lower level, the electron has less energy than in the original level, and the energy difference appears as a quantum of radiation whose energy hf is equal to the energy difference between the levels.

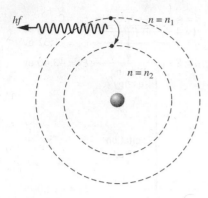

FIGURE 6.19   An electron jumps from the state n_1 to the state n_2 as a photon is emitted.

That is, if the electron jumps from $n = n_1$ to $n = n_2$, as in Figure 6.19, a photon appears with energy

$$hf = E_{n_1} - E_{n_2} \tag{6.31}$$

or, using Eq. 6.30 for the energies,

$$f = \frac{me^4}{64\pi^3 \varepsilon_0^2 \hbar^3} \left(\frac{1}{n_2^2} - \frac{1}{n_1^2} \right) \tag{6.32}$$

The wavelength of the emitted radiation is

$$\lambda = \frac{c}{f} = \frac{64\pi^3 \varepsilon_0^2 \hbar^3 c}{me^4} \left(\frac{n_1^2 n_2^2}{n_1^2 - n_2^2} \right) = \frac{1}{R_\infty} \left(\frac{n_1^2 n_2^2}{n_1^2 - n_2^2} \right) \tag{6.33}$$

where R_∞ is called the *Rydberg constant*

$$R_\infty = \frac{me^4}{64\pi^3 \varepsilon_0^2 \hbar^3 c} \tag{6.34}$$

The presently accepted numerical value is

$$R_\infty = 1.097373 \times 10^7 \text{ m}^{-1}$$

Example 6.6

Find the wavelengths of the transitions from $n_1 = 3$ to $n_2 = 2$ and from $n_1 = 4$ to $n_2 = 2$ in atomic hydrogen.

Solution

For $n_1 = 3$ and $n_2 = 2$, Eq. 6.33 gives

$$\lambda = \frac{1}{R_\infty} \left(\frac{n_1^2 n_2^2}{n_1^2 - n_2^2} \right)$$

$$= \frac{1}{1.097 \times 10^7 \text{ m}^{-1}} \left(\frac{3^2 2^2}{3^2 - 2^2} \right) = 656.1 \text{ nm}$$

and for $n_1 = 4$ and $n_2 = 2$,

$$\lambda = \frac{1}{R_\infty} \left(\frac{n_1^2 n_2^2}{n_1^2 - n_2^2} \right)$$

$$= \frac{1}{1.097 \times 10^7 \text{ m}^{-1}} \left(\frac{4^2 2^2}{4^2 - 2^2} \right)$$

$$= 486.0 \text{ nm}$$

These wavelengths are remarkably close to the values of the two longest wavelengths of the Balmer series (Figure 6.16). In fact, Eq. 6.33 gives

$$\lambda = (364.5 \text{ nm}) \left(\frac{n_1^2}{n_1^2 - 4} \right)$$

for the wavelength of a transition from any state n_1 to $n_2 = 2$. This is identical with Eq. 6.21 for the Balmer series. Thus we see that the radiations identified as the Balmer series correspond to transitions from higher levels to the $n = 2$ level. Similar identifications can be made for other series of radiations, as shown in Figure 6.20. This association between the transitions expected according to the Bohr model and the observed wavelengths (as in Figure 6.16) represents a huge triumph for the model.

The Bohr formulas also explain the Ritz combination principle, according to which certain frequencies in the emission spectrum can be summed to give other

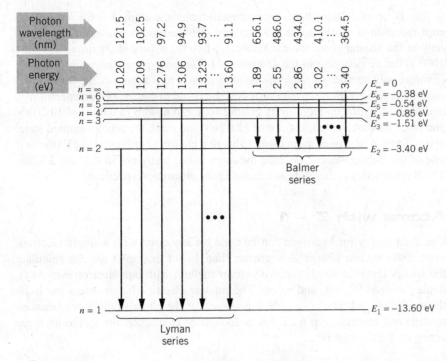

FIGURE 6.20 The transitions of the Lyman and Balmer series in hydrogen. The series limit is shown at the right of each group.

frequencies. Let us consider a transition from a state n_3 to a state n_2, that is followed by a transition from n_2 to n_1. Equation 6.32 can be used for this case to give

$$f_{n_3 \rightarrow n_2} = cR_\infty \left(\frac{1}{n_3^2} - \frac{1}{n_2^2} \right)$$

$$f_{n_2 \rightarrow n_1} = cR_\infty \left(\frac{1}{n_2^2} - \frac{1}{n_1^2} \right)$$

Thus

$$f_{n_3 \rightarrow n_2} + f_{n_2 \rightarrow n_1} = cR_\infty \left(\frac{1}{n_3^2} - \frac{1}{n_2^2} \right) + cR_\infty \left(\frac{1}{n_2^2} - \frac{1}{n_1^2} \right) = cR_\infty \left(\frac{1}{n_3^2} - \frac{1}{n_1^2} \right)$$

which is equal to the frequency of the single photon emitted in a direct transition from n_3 to n_1, so

$$f_{n_3 \rightarrow n_2} + f_{n_2 \rightarrow n_1} = f_{n_3 \rightarrow n_1} \tag{6.35}$$

The Bohr model is thus entirely consistent with the Ritz combination principle. The frequency of an emitted photon is related to its energy by $E = hf$, so the summing of frequencies is equivalent to the summing of energies. We may thus restate the Ritz combination principle in terms of energy: The energy of a photon emitted in a transition that skips or crosses over one or more states is equal to the step-by-step sum of the energies of the transitions connecting all of the individual states. (See Problem 25.)

The Bohr model also helps us understand why the atom doesn't absorb and emit radiation at all the same wavelengths. Isolated atoms are normally found only in the ground state; the excited states live for a very short time (less than 10^{-9} s) before decaying to the ground state. *The absorption spectrum therefore contains only transitions from the ground state.* From Figure 6.20, we see that only the radiations of the Lyman series can be found in the absorption spectrum of hydrogen. A hydrogen atom in its ground state can absorb radiation of 10.20 eV and reach the first excited state, or of 12.09 eV and reach the second excited state, and so forth. A hydrogen atom cannot absorb a photon of energy 1.89 eV (the first line of the Balmer series), because the atom is originally not in the $n = 2$ level. The Balmer series is therefore *not* found in the absorption spectrum.

Atoms with Z > 1

The Bohr theory for hydrogen can be used for any atom with a single electron, even if the nuclear charge Z is greater than 1. For example, we can calculate the energy levels of singly ionized helium (helium with one electron removed), doubly ionized lithium, and so on. The nuclear electric charge enters the Bohr theory in only one place—in the expression for the electrostatic force between nucleus and electron, Eq. 6.22. For a nucleus of charge Ze, the Coulomb force acting on the electron is

$$F = \frac{1}{4\pi\varepsilon_0}\frac{|q_1||q_2|}{r^2} = \frac{1}{4\pi\varepsilon_0}\frac{Ze^2}{r^2} \tag{6.36}$$

That is, where we had e^2 previously, we now have Ze^2. Making the same substitution in the final results, we can find the allowed radii:

$$r_n = \frac{4\pi\varepsilon_0\hbar^2}{Ze^2 m}n^2 = \frac{a_0 n^2}{Z} \tag{6.37}$$

and the energies become

$$E_n = -\frac{m(Ze^2)^2}{32\pi^2\varepsilon_0^2\hbar^2}\frac{1}{n^2} = -(13.60\,\text{eV})\frac{Z^2}{n^2} \tag{6.38}$$

The orbits in the higher-Z atoms are closer to the nucleus and have larger (negative) energies; that is, the electron is more tightly bound to the nucleus.

Example 6.7

Calculate the two longest wavelengths of the Balmer series of triply ionized beryllium ($Z = 4$).

Solution

The radiations of the Balmer series end with the $n = 2$ level, and so the two longest wavelengths are the radiations corresponding to $n = 3 \rightarrow n = 2$ and $n = 4 \rightarrow n = 2$. The energies of the radiations and their corresponding wavelengths are

$$E_3 - E_2 = -(13.60\,\text{eV})(4^2)\left(\frac{1}{3^2} - \frac{1}{2^2}\right) = 30.2\,\text{eV}$$

$$\lambda = \frac{hc}{E} = \frac{1240\,\text{eV·nm}}{30.2\,\text{eV}} = 41.0\,\text{nm}$$

$$E_4 - E_2 = -(13.60 \, \text{eV})(4^2)\left(\frac{1}{4^2} - \frac{1}{2^2}\right) = 40.8 \, \text{eV}$$

$$\lambda = \frac{hc}{E} = \frac{1240 \, \text{eV·nm}}{40.8 \, \text{eV}} = 30.4 \, \text{nm}$$

These radiations are in the ultraviolet region.

Note that we cannot use Eq. 6.33 to find the wavelengths, because that equation applies only to hydrogen ($Z = 1$).

6.6 THE FRANCK-HERTZ EXPERIMENT

Let us imagine the following experiment, performed with the apparatus shown schematically in Figure 6.21. A filament heats the cathode, which then emits electrons. These electrons are accelerated toward the grid by the potential difference V, which we control. Electrons pass through the grid and reach the plate if V exceeds V_0, a small retarding voltage between the grid and the plate. The current of electrons reaching the plate is measured using the ammeter A.

Now suppose the tube is filled with atomic hydrogen gas at a low pressure. As the voltage is increased from zero, more and more electrons reach the plate, and the current rises accordingly. The electrons inside the tube may make collisions with atoms of hydrogen, *but lose no energy in these collisions*—the collisions are perfectly elastic. The only way the electron can give up energy in a collision is if the electron has enough energy to cause the hydrogen atom to make a transition to an excited state. Thus, when the energy of the electrons reaches and barely exceeds 10.2 eV (or when the voltage reaches 10.2 V), the electrons can make *inelastic* collisions, leaving 10.2 eV of energy with the atom (now in the $n = 2$ level), and the original electron moves off with very little energy. If it should pass through the grid, the electron might not have sufficient energy to overcome the small retarding potential and reach the plate. Thus when $V = 10.2$ V, a drop in the current is observed. As V is increased further, we begin to see the effects of multiple collisions. That is, when $V = 20.4$ V, an electron can make an inelastic collision, leaving the atom in the $n = 2$ state. The electron loses 10.2 eV of energy in this process, and so it moves off after the collision with a remaining 10.2 eV of energy, which is sufficient to excite a second hydrogen atom in an inelastic collision. Thus, if a drop in the current is observed at V, similar drops are observed at $2V, 3V, \ldots$.

This experiment should thus give rather direct evidence for the existence of atomic excited states. Unfortunately, it is not easy to do this experiment with hydrogen, because hydrogen occurs naturally in the molecular form H_2, rather than in atomic form. The molecules can absorb energy in a variety of ways, which would confuse the interpretation of the experiment. A similar experiment was done in 1914 by James Franck and Gustav Hertz, using a tube filled with mercury vapor. Their results are shown in Figure 6.22, which gives clear evidence for an excited state at 4.9 eV; whenever the voltage is a multiple of 4.9 V, a drop in the current appears. Coincidentally, the *emission* spectrum of mercury shows an intense ultraviolet line of wavelength 254 nm, which corresponds to an energy of 4.9 eV; this results from a transition between the same 4.9-eV excited state and the ground state. The Franck-Hertz experiment showed that an electron must have a certain minimum energy to make an inelastic collision with an atom; we now interpret that minimum energy as the energy of an excited state of the atom. Franck and Hertz were awarded the 1925 Nobel Prize in physics for this work.

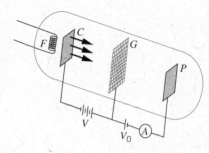

FIGURE 6.21   Franck-Hertz apparatus. Electrons leave the cathode C, are accelerated by the voltage V toward the grid G, and reach the plate P where they are recorded on the ammeter A.

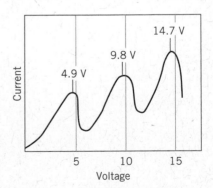

FIGURE 6.22 Result of Franck-Hertz experiment using mercury vapor. The current drops at voltages of 4.9 V, 9.8 V ($= 2 \times 4.9$ V), 14.7 V ($= 3 \times 4.9$ V).

*6.7 THE CORRESPONDENCE PRINCIPLE

We have seen how Bohr's model permits calculations of transition wavelengths in atomic hydrogen that are in excellent agreement with the wavelengths observed in the emission and absorption spectra. However, in order to obtain this agreement, Bohr had to introduce postulates that were radical departures from classical physics. In particular, according to classical physics an accelerated charged particle radiates electromagnetic energy, but an electron in Bohr's atomic model, accelerated as it moves in a circular orbit, does not radiate (unless it jumps to another orbit). Here we have a very different case than we did in our study of special relativity. You will recall, for example, that relativity gives us one expression for the kinetic energy, $K = E - E_0$, and classical physics gives us another, $K = \frac{1}{2}mv^2$; however, we showed that $E - E_0$ reduces to $\frac{1}{2}mv^2$ when $v \ll c$. Thus these two expressions are really not very different—one is merely a special case of the other. The dilemma associated with the accelerated electron is not simply a matter of atomic physics (as an example of quantum physics) being a special case of classical physics. Either the accelerated charge radiates, or it doesn't! Bohr's solution to this serious dilemma was to propose the *correspondence principle,* which states that

> *Quantum theory must agree with classical theory in the limit in which classical theory is known to agree with experiment,*

or equivalently,

> *Quantum theory must agree with classical theory in the limit of large quantum numbers.*

Let us see how we can apply this principle to the Bohr atom. According to classical physics, an electric charge moving in a circle radiates at a frequency equal to its frequency of rotation. For an atomic orbit, the period of revolution is the distance traveled in one orbit, $2\pi r$, divided by the orbital speed $v = \sqrt{2K/m}$, where K is the kinetic energy:

$$T = \frac{2\pi r}{\sqrt{2K/m}} = \frac{\sqrt{16\pi^3\varepsilon_0 mr^3}}{e} \tag{6.39}$$

where we use Eq. 6.23 for the kinetic energy. The frequency f is the inverse of the period:

$$f = \frac{1}{T} = \frac{e}{\sqrt{16\pi^3\varepsilon_0 mr^3}} \tag{6.40}$$

Using Eq. 6.28 for the radii of the allowed orbits, we find

$$f_n = \frac{me^4}{32\pi^3\varepsilon_0^2\hbar^3}\frac{1}{n^3} \tag{6.41}$$

A "classical" electron moving in an orbit of radius r_n would radiate at this frequency f_n.

---

*This is an optional section that may be skipped without loss of continuity.

If we made the radius of the Bohr atom so large that it went from a quantum-sized object (10^{-10} m) to a laboratory-sized object (10^{-3} m), the atom should behave classically. The radius increases with increasing n like n^2, so this classical behavior should occur for n in the range 10^3–10^4. Let us then calculate the frequency of the radiation emitted by such an atom when the electron drops from the orbit n to the orbit $n - 1$. According to Eq. 6.32, the frequency is

$$f = \frac{me^4}{64\pi^3 \varepsilon_0^2 \hbar^3} \left(\frac{1}{(n-1)^2} - \frac{1}{n^2} \right) = \frac{me^4}{64\pi^3 \varepsilon_0^2 \hbar^3} \frac{2n-1}{n^2(n-1)^2} \qquad (6.42)$$

If n is very large, then we can approximate $n - 1$ by n and $2n - 1$ by $2n$, which gives

$$f \cong \frac{me^4}{64\pi^3 \varepsilon_0^2 \hbar^3} \frac{2n}{n^4} = \frac{me^4}{32\pi^3 \varepsilon_0^2 \hbar^3} \frac{1}{n^3} \qquad (6.43)$$

This is identical with Eq. 6.41 for the "classical" frequency. The "classical" electron spirals slowly in toward the nucleus, radiating at the frequency given by Eq. 6.41, while the "quantum" electron jumps from the orbit n to the orbit $n - 1$ and then to the orbit $n - 2$, and so forth, radiating at the frequency given by the identical Eq. 6.43. (When the circular orbits are very large, this jumping from one circular orbit to the next smaller one looks very much like a spiral, as in Figure 6.23.)

In the region of large n, where classical and quantum physics overlap, the classical and quantum expressions for the radiation frequencies are identical. This is an example of an application of Bohr's correspondence principle. The applications of the correspondence principle go far beyond the Bohr atom, and this principle is important in understanding how we get from the domain in which the laws of classical physics are valid to the domain in which the laws of quantum physics are valid.

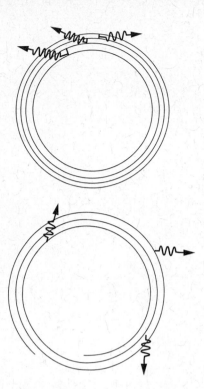

FIGURE 6.23 (Top) A large quantum atom. Photons are emitted in discrete transitions as the electron jumps to lower states. (Bottom) A classical atom. Photons are emitted continuously by the accelerated electron.

6.8 DEFICIENCIES OF THE BOHR MODEL

The Bohr model gives us a picture of how electrons move about the nucleus, and many of our attempts to explain the behavior of atoms refer to this picture, even though it is not strictly correct. Our presentation ignored two effects that must be included to improve the accuracy of the model. Other deficiencies in the model cannot be so easily fixed, because they are inconsistent with the correct quantum-mechanical picture, which is presented in the next chapter.

1. Motion of the Proton. Our model was based on an electron orbiting around a fixed proton, but actually the electron and proton both orbit about their center of mass (just as the Earth and Sun orbit around their center of mass). The kinetic energy should thus include a term describing the motion of the proton. We can account for this effect if the mass that appears in the equation for the energy levels (Eq. 6.30) is not the electron mass but instead is the *reduced mass* of the proton-electron system, calculated from the electron mass m_e and proton mass m_p according to

$$m = \frac{m_e m_p}{m_e + m_p} \qquad (6.44)$$

The reduced mass is just slightly smaller than the electron mass and has the effect of decreasing the energy and frequency or increasing the wavelength by about 0.05%. Equivalently, in Eq. 6.33 for the wavelengths we can replace the Rydberg constant R_∞ (so called because it would be correct if the proton mass were infinite) with the value $R = R_\infty(1 + m_e/m_p)$.

2. Wavelengths in Air. Another small but easily fixable error occurs when we convert the frequencies (Eq. 6.32) calculated directly from the Bohr energy levels to wavelengths (Eq. 6.33). The wavelength measurements are normally done in air, so we should calculate the wavelength as $\lambda = v_{air}/f$, where v_{air} is the speed of light in air. This has the effect of decreasing the calculated wavelengths by about 0.03% (to some extent offsetting the error we made by ignoring the motion of the proton).

3. Angular Momentum. A serious failure of the Bohr model is that it gives incorrect predictions for the angular momentum of the electron. In Bohr's theory, the orbital angular momentum is quantized in integer multiples of $\hbar$, which is correct. However, for the ground state of hydrogen ($n = 1$), the Bohr theory gives $L = \hbar$, while experiment clearly shows $L = 0$.

4. Uncertainty. Another deficiency of the model is that it violates the uncertainty relationship. (In Bohr's defense, remember that the model was developed a decade before the introduction of wave mechanics, with its accompanying ideas of uncertainty.) Suppose the electron orbits in the xy plane. In this case we know exactly its z coordinate (in the xy plane, $z = 0$ and so $\Delta z = 0$) and the z component of its momentum (also precisely zero, so $\Delta p_z = 0$). Such an atom would therefore violate the uncertainty relationship $\Delta z \Delta p_z \geq \hbar$. In fact, as we discuss in the next chapter, quantum mechanics introduces a degree of "fuzziness" to the behavior of electrons in atoms that is not consistent with any orbit in a single plane.

In spite of its successes, the Bohr model is at best an incomplete model. It is useful only for atoms that contain one electron (hydrogen, singly ionized helium, doubly ionized lithium, and so forth), but not for atoms with two or more electrons, because we have considered only the force between electron and nucleus, and not the force between the electrons themselves. Furthermore, if we look very carefully at the emission spectrum, we find that many lines are in fact not single lines, but very closely spaced combinations of two or more lines; the Bohr model is unable to account for these *doublets* of spectral lines. The model is also limited in its usefulness as a basis from which to calculate other properties of the atom; although we can accurately calculate the energies of the spectral lines, we cannot calculate their intensities. For example, how often will an electron in the $n = 3$ state jump directly to the $n = 1$ state, emitting the corresponding photon, and how often will it jump first to the $n = 2$ state and then to the $n = 1$ state, emitting two photons? A complete theory should provide a way to calculate this property.

We do not wish, however, to discard the model completely. The Bohr model provides a useful starting point in our study of atoms, and Bohr introduced several ideas (stationary states, quantization of angular momentum, correspondence principle) that carry over into the correct quantum-mechanical calculation. There are many atomic properties, especially those associated with magnetism, that can be simply modeled on the basis of Bohr orbits. Most remarkably, when we treat the hydrogen atom correctly in the next chapter using quantum mechanics, we find that the energy levels calculated by solving the Schrödinger equation are in fact identical with those of the Bohr model.

Chapter Summary

		Section
Scattering impact parameter	$b = \dfrac{zZ}{2K}\dfrac{e^2}{4\pi\varepsilon_0}\cot\dfrac{1}{2}\theta$	6.3
Fraction scattered at angles $> \theta$	$f_{>\theta} = nt\pi b^2$	6.3
Rutherford scattering formula	$N(\theta) = \dfrac{nt}{4r^2}\left(\dfrac{zZ}{2K}\right)^2\left(\dfrac{e^2}{4\pi\varepsilon_0}\right)^2\dfrac{1}{\sin^4\frac{1}{2}\theta}$	6.3
Distance of closest approach	$d = \dfrac{1}{4\pi\varepsilon_0}\dfrac{zZe^2}{K}$	6.3
Balmer formula	$\lambda = (364.5\text{ nm})\dfrac{n^2}{n^2 - 4}$ $(n = 3, 4, 5, \ldots)$	6.4
Radii of Bohr orbits in hydrogen	$r_n = \dfrac{4\pi\varepsilon_0\hbar^2}{me^2}n^2 = a_0 n^2$ $(n = 1, 2, 3, \ldots)$	6.5
Energies of Bohr orbits in hydrogen	$E_n = -\dfrac{me^4}{32\pi^2\varepsilon_0^2\hbar^2}\dfrac{1}{n^2}$ $= \dfrac{-13.60\text{ eV}}{n^2}$ $(n = 1, 2, 3, \ldots)$	6.5

		Section
Excitation energy of level n	$E_n - E_1$	6.5
Binding (or ionization) energy of level n	$|E_n|$	6.5
Hydrogen wavelengths in Bohr model	$\lambda = \dfrac{64\pi^3\varepsilon_0^2\hbar^3 c}{me^4}\left(\dfrac{n_1^2 n_2^2}{n_1^2 - n_2^2}\right)$ $= \dfrac{1}{R_\infty}\left(\dfrac{n_1^2 n_2^2}{n_1^2 - n_2^2}\right)$	6.5
Single-electron atoms with $Z > 1$	$r_n = \dfrac{a_0 n^2}{Z}$, $E_n = -(13.60\text{ eV})\dfrac{Z^2}{n^2}$	6.5
Reduced mass of proton-electron system	$m = \dfrac{m_e m_p}{m_e + m_p}$	6.8

Questions

1. Does the Thomson model fail at large scattering angles or at small scattering angles? Why?
2. What principles of physics would be violated if we scattered a beam of alpha particles with a single impact parameter from a single target atom at rest?
3. Could we use the Rutherford scattering formula to analyze the scattering of (a) protons incident on iron? (b) Alpha particles incident on lithium ($Z = 3$)? (c) Silver nuclei incident on gold? (d) Hydrogen *atoms* incident on gold? (e) Electrons incident on gold?
4. What determines the angular range $d\theta$ in the alpha-particle scattering experiment (Figure 6.8)?
5. Why didn't Bohr use the concept of de Broglie waves in his theory?
6. In which Bohr orbit does the electron have the largest velocity? Are we justified in treating the electron nonrelativistically in that case?
7. How does an electron in hydrogen get from $r = 4a_0$ to $r = a_0$ without being anywhere in between?
8. How is the quantization of the energy in the hydrogen atom similar to the quantization of the systems discussed in Chapter 5? How is it different? Do the quantizations originate from similar causes?
9. In a Bohr atom, an electron jumps from state n_1, with angular momentum $n_1\hbar$, to state n_2, with angular momentum $n_2\hbar$. How can an isolated system change its angular momentum? (In classical physics, a change in angular momentum requires an external torque.) Can the photon carry away the difference in angular momentum? Estimate the maximum angular momentum, relative to the center of the atom, that the photon can have. Does this suggest another failure of the Bohr model?
10. The product $E_n r_n$ for the hydrogen atom is (1) independent of Planck's constant and (2) independent of the quantum number n. Does this observation have any significance? Is this a classical or a quantum effect?
11. (a) How does a Bohr atom violate the position-momentum uncertainty relationship? (b) How does a Bohr atom violate the energy-time uncertainty relationship? (What is ΔE? What does this imply about Δt? What do you conclude about transitions between levels?)

12. List the assumptions made in deriving the Bohr theory. Which of these are a result of neglecting small quantities? Which of these violate basic principles of relativity or quantum physics?

13. List the assumptions made in deriving the Rutherford scattering formula. Which of these are a result of neglecting small quantities? Which of these violate basic principles of relativity or quantum physics?

14. In both the Rutherford theory and the Bohr theory, we used the classical expression for the kinetic energy. Estimate the velocity of an electron in the Bohr atom and of an alpha particle in a typical scattering experiment, and decide whether the use of the classical formula is justified.

15. In both the Rutherford theory and the Bohr theory, we neglected any wave properties of the particles. Estimate the de Broglie wavelength of an electron in a Bohr atom and compare it with the size of the atom. Estimate the de Broglie wavelength of an alpha particle and compare it with the

size of the nucleus. Is the wave behavior expected to be important in either case?

16. What is the distinction between binding energy and ionization energy? Between binding energy and excitation energy? If you were given the value of the binding energy of a level in hydrogen, could you find its excitation energy without knowing which level it is?

17. Why are the decreases in current in the Franck-Hertz experiment not sharp?

18. As indicated by the Franck-Hertz experiment, the first excited state of mercury is at an energy of 4.9 eV. Do you expect mercury to show absorption lines in the visible spectrum?

19. Is the correspondence principle a necessary part of quantum physics or is it merely an accidental agreement of two formulas? Where do we draw the line between the world of quantum physics and the world of classical, nonquantum physics?

Problems

6.1 Basic Properties of Atoms

1. Electrons in atoms are known to have kinetic energies in the range of a few eV. Show that the uncertainty principle allows electrons of this energy to be confined in a region the size of an atom (0.1 nm).

6.2 Scattering Experiments and the Thomson Model

2. Consider an electron in Figure 6.1 embedded in a sphere of positive charge Ze at a distance r from its center. (a) Using Gauss's law, show that the electric field on the electron due to the positive charge is

$$E = \frac{1}{4\pi\varepsilon_0} \frac{Ze}{R^3} r$$

(b) For this electric field, show that the force on the electron is given by Eq. 6.2.

3. (a) Compute the oscillation frequency of the electron and the expected absorption or emission wavelength in a Thomson-model hydrogen atom. Use $R = 0.053$ nm. Compare with the observed wavelength of the strongest emission and absorption line in hydrogen, 122 nm. (b) Repeat for sodium ($Z = 11$). Use $R = 0.18$ nm. Compare with the observed wavelength, 590 nm.

4. Consider the Thomson model for an atom with 2 electrons. Let the electrons be located along a diameter on opposite sides of the center of the sphere, each a distance x from the center. (a) Show that the configuration is stable if $x = R/2$. (b) Try to construct similar stable configurations for atoms with 3, 4, 5, and 6 electrons.

6.3 The Rutherford Nuclear Atom

5. Alpha particles of kinetic energy 5.00 MeV are scattered at 90° by a gold foil. (a) What is the impact parameter? (b) What is the minimum distance between alpha particles and gold nucleus? (c) Find the kinetic and potential energies at that minimum distance.

6. How much kinetic energy must an alpha particle have before its distance of closest approach to a gold nucleus is equal to the nuclear radius (7.0×10^{-15} m)?

7. What is the distance of closest approach when alpha particles of kinetic energy 6.0 MeV are scattered by a thin copper foil?

8. Protons of energy 5.0 MeV are incident on a silver foil of thickness 4.0×10^{-6} m. What fraction of the incident protons is scattered at angles: (a) Greater than 90°? (b) Greater than 10°? (c) Between 5° and 10°? (d) Less than 5°?

9. Protons are incident on a copper foil 12 μm thick. (a) What should the proton kinetic energy be in order that the distance of closest approach equal the nuclear radius (5.0 fm)? (b) If the proton energy were 7.5 MeV, what is the impact parameter for scattering at 120°? (c) What is the minimum distance between proton and nucleus for this case? (d) What fraction of the protons is scattered beyond 120°?

10. Alpha particles of kinetic energy K are scattered either from a gold foil or a silver foil of identical thickness. What is the ratio of the number of particles scattered at angles greater than 90° by the gold foil to the same number for the silver foil?

11. The maximum kinetic energy given to the target nucleus will occur in a head-on collision with $b = 0$. (Why?) Estimate

the maximum kinetic energy given to the target nucleus when 8.0 MeV alpha particles are incident on a gold foil. Are we justified in neglecting this energy?

12. The maximum kinetic energy that an alpha particle can transmit to an *electron* occurs during a head-on collision. Compute the kinetic energy lost by an alpha particle of kinetic energy 8.0 MeV in a head-on collision with an electron at rest. Are we justified in neglecting this energy in the Rutherford theory?

13. Alpha particles of energy 9.6 MeV are incident on a silver foil of thickness 7.0 μm. For a certain value of the impact parameter, the alpha particles lose exactly half their incident kinetic energy when they reach their minimum separation from the nucleus. Find the minimum separation, the impact parameter, and the scattering angle.

14. Alpha particles of kinetic energy 6.0 MeV are incident at a rate of 3.0×10^7 per second on a gold foil of thickness 3.0×10^{-6} m. A circular detector of diameter 1.0 cm is placed 12 cm from the foil at an angle of $30°$ with the direction of the incident alpha particles. At what rate does the detector measure scattered alpha particles?

6.4 Line Spectra

15. The shortest wavelength of the hydrogen Lyman series is 91.13 nm. Find the three longest wavelengths in this series.

16. One of the lines in the Brackett series (series limit = 1458 nm) has a wavelength of 1944 nm. Find the next higher and next lower wavelengths in this series.

17. The longest wavelength in the Pfund series is 7459 nm. Find the series limit.

6.5 The Bohr Model

18. In the $n = 3$ state of hydrogen, find the electron's velocity, kinetic energy, and potential energy.

19. Use the Bohr theory to find the series wavelength limits of the Lyman and Paschen series of hydrogen.

20. (*a*) Show that the speed of an electron in the nth Bohr orbit of hydrogen is $\alpha c/n$, where α is the fine structure constant, equal to $e^2/4\pi\varepsilon_0\hbar c$. (*b*) What would be the speed in a hydrogenlike atom with a nuclear charge of Ze?

21. An electron is in the $n = 5$ state of hydrogen. To what states can the electron make transitions, and what are the energies of the emitted radiations?

22. Continue Figure 6.20, showing the transitions of the Paschen series and computing their energies and wavelengths.

23. A collection of hydrogen atoms in the ground state is illuminated with ultraviolet light of wavelength 59.0 nm. Find the kinetic energy of the emitted electrons.

24. Find the ionization energy of: (*a*) the $n = 3$ level of hydrogen; (*b*) the $n = 2$ level of He$^+$ (singly ionized helium); (*c*) the $n = 4$ level of Li^{++} (doubly ionized lithium).

25. Use the Bohr formula to find the energy differences $E(n_1 \rightarrow n_2) = E_{n_1} - E_{n_2}$ and show that (*a*) $E(4 \rightarrow 2) = E(4 \rightarrow 3) + E(3 \rightarrow 2)$; (*b*) $E(4 \rightarrow 1) = E(4 \rightarrow 2) + E(2 \rightarrow 1)$. (*c*) Interpret these results based on the Ritz combination principle.

26. Find the shortest and the longest wavelengths of the Lyman series of singly ionized helium.

27. Draw an energy-level diagram showing the lowest four levels of singly ionized helium. Show all possible transitions from the levels and label each transition with its wavelength.

28. A long time ago, in a galaxy far, far away, electric charge had not yet been invented, and atoms were held together by gravitational forces. Compute the Bohr radius and the $n = 2$ to $n = 1$ transition energy in a gravitationally bound hydrogen atom.

29. An alternative development of the Bohr theory begins by assuming that the stationary states are those for which the circumference of the orbit is an integral number of de Broglie wavelengths. (*a*) Show that this condition leads to standing de Broglie waves around the orbit. (*b*) Show that this condition gives the angular momentum condition, Eq. 6.26, used in the Bohr theory.

6.6 The Franck-Hertz Experiment

30. A hypothetical atom has only two excited states, at 4.0 and 7.0 eV, and has a ground-state ionization energy of 9.0 eV. If we used a vapor of such atoms for the Franck-Hertz experiment, for what voltages would we expect to see decreases in the current? List all voltages up to 20 V.

31. The first excited state of sodium decays to the ground state by emitting a photon of wavelength 590 nm. If sodium vapor is used for the Franck-Hertz experiment, at what voltage will the first current drop be recorded?

6.7 The Correspondence Principle

32. Suppose all of the excited levels of hydrogen had lifetimes of 10^{-8} s. As we go to higher and higher excited states, they get closer and closer together, and soon they are so close in energy that the energy uncertainty of each state becomes as large as the energy spacing between states, and we can no longer resolve individual states. Find the value of n for which this occurs. What is the radius of such an atom?

33. Compare the frequency of revolution of an electron with the frequency of the photons emitted in transitions from n to $n - 1$ for (*a*) $n = 10$; (*b*) $n = 100$; (*c*) $n = 1000$; (*d*) $n = 10,000$.

6.8 Deficiencies of the Bohr Model

34. What is the difference in wavelength between the first line of the Balmer series in ordinary hydrogen ($M = 1.007825$ u) and in "heavy" hydrogen ($M = 2.014102$ u)?

General Problems

35. A hydrogen atom is in the $n = 6$ state. (a) Counting all possible paths, how many different photon energies can be emitted if the atom ends up in the ground state? (b) Suppose only $\Delta n = 1$ transitions were allowed. How many different photon energies would be emitted? (c) How many different photon energies would occur in a Thomson-model hydrogen atom?

36. An electron is in the $n = 8$ level of ionized helium. (a) Find the three longest wavelengths that are emitted when the electron makes a transition from the $n = 8$ level to a lower level. (b) Find the shortest wavelength that can be emitted. (c) Find the three longest wavelengths at which the electron in the $n = 8$ level will *absorb* a photon and move to a higher state, if we could somehow keep it in that level long enough to absorb. (d) Find the shortest wavelength that can be absorbed.

37. The lifetimes of the levels in a hydrogen atom are of the order of 10^{-8} s. Find the energy uncertainty of the first excited state and compare it with the energy of the state.

38. The following wavelengths are found among the many radiations emitted by singly ionized helium: 24.30 nm, 25.63 nm, 102.5 nm, 320.4 nm. If we group the transitions in helium as we did in hydrogen by identifying the final state n_0 and initial state n, to which series does each transition belong?

39. Adjacent wavelengths 72.90 nm and 54.00 nm are found in one series of transitions among the radiations emitted by doubly-ionized lithium. Find the value of n_0 for this series and find the next wavelength in the series.

40. When an atom emits a photon in a transition from a state of energy E_1 to a state of energy E_2, the photon energy is not precisely equal to $E_1 - E_2$. Conservation of momentum requires that the atom must recoil, and so some energy must go into recoil kinetic energy K_R. Show that $K_R \cong (E_1 - E_2)^2/2Mc^2$ where M is the mass of the atom. Evaluate this recoil energy for the $n = 2$ to $n = 1$ transition of hydrogen.

41. In a *muonic atom*, the electron is replaced by a negatively charged particle called the *muon*. The muon mass is 207 times the electron mass. (a) Ignoring the correction for finite nuclear mass, what is the shortest wavelength of the Lyman series in a muonic hydrogen atom? In what region of the electromagnetic spectrum does this belong? (b) How large is the correction for the finite nuclear mass in this case? (See the discussion at the beginning of Section 6.8.)

42. Consider an atom in which the single electron is replaced by a negatively charged muon ($m_\mu = 207m_e$). What is the radius of the first Bohr orbit of a muonic lead atom ($Z = 82$)? Compare with the nuclear radius of about 7 fm.

Chapter 7

THE HYDROGEN ATOM IN WAVE MECHANICS

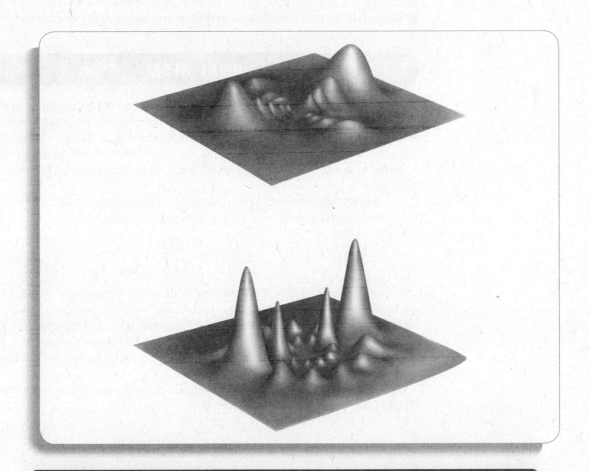

These computer-generated distributions represent the probability to locate the electron in the $n = 8$ state of hydrogen for angular momentum quantum number $l = 2$ (top) and $l = 6$ (bottom). The nucleus is at the center, and the height at any point gives the probability to find the electron in a small volume element at that location in the xz plane. This way of describing the motion of an electron in hydrogen is very different from the circular orbits of the Bohr model.

In this chapter we study the solutions of the Schrödinger equation for the hydrogen atom. We will see that these solutions, which lead to the same energy levels calculated in the Bohr model, differ from the Bohr model by allowing for the uncertainty in localizing the electron.

Other deficiencies of the Bohr model are not so easily eliminated by solving the Schrödinger equation. First, the so-called "fine structure" of the spectral lines (the splitting of the lines into close-lying doublets) cannot be explained by our solutions; the proper explanation of this effect requires the introduction of a new property of the electron, the *intrinsic spin*. Second, the mathematical difficulties of solving the Schrödinger equation for atoms containing two or more electrons are formidable, so we restrict our discussion in this chapter to one-electron atoms, in order to see how wave mechanics enables us to understand some basic atomic properties. In the next chapter we discuss the structure of many-electron atoms.

7.1 A ONE-DIMENSIONAL ATOM

Quantum mechanics gives us a view of the structure of the hydrogen atom that is very different from the Bohr model. In the Bohr model, the electron moves about the proton in a circular orbit. Quantum mechanics, on the other hand, does not allow a fixed radius or a fixed orbital plane but instead describes the electron in terms of a probability density, which leads to an uncertainty in locating the electron.

To analyze the hydrogen atom according to quantum mechanics, we must solve the Schrödinger equation for the Coulomb potential energy of the proton and the electron:

$$U(r) = -\frac{e^2}{4\pi\varepsilon_0 r} \tag{7.1}$$

Eventually we will discuss the solutions to this three-dimensional problem for the hydrogen atom using spherical polar coordinates, but for now let's look at the simpler one-dimensional problem, in which a proton is fixed at the origin ($x = 0$) and an electron moves along the positive x axis. (This doesn't represent a real atom, but it does show how some properties of electron wave functions in atoms emerge from solving the Schrödinger equation.)

In one dimension, the Schrödinger equation for an electron with potential energy $U(x) = -e^2/4\pi\varepsilon_0 x$ would then be

$$-\frac{\hbar^2}{2m}\frac{d^2\psi}{dx^2} - \frac{e^2}{4\pi\varepsilon_0 x}\psi(x) = E\psi(x) \tag{7.2}$$

For a bound state, the wave function must fall to zero as $x \to \infty$. Moreover, in order for the second term on the left side to remain finite at $x = 0$, the wave function must be zero at $x = 0$. The simplest function that satisfies both of these requirements is $\psi(x) = Axe^{-bx}$, where A is the normalization constant. By substituting this trial wave function into Eq. 7.2, we find a solution when $b = me^2/4\pi\varepsilon_0\hbar^2 = 1/a_0$ (where a_0 is the Bohr radius defined in Eq. 6.29). The energy corresponding to this wave function is $E = -\hbar^2 b^2/2m = -me^4/32\pi^2\varepsilon_0^2\hbar^2$, which happens by chance to be identical to the energy of the ground state in the Bohr model (Eq. 6.30 for $n = 1$).

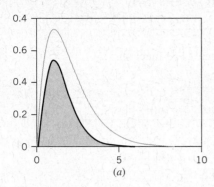
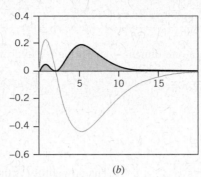
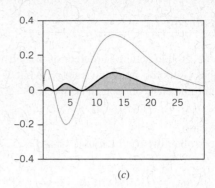

FIGURE 7.1 Wave functions and probability densities (shaded areas) for an electron bound in a one-dimensional Coulomb potential energy. The horizontal axis represents the distance between the proton and electron in units of a_0. (*a*) Ground state. (*b*) First excited state. (*c*) Second excited state.

Figure 7.1*a* shows this wave function and its corresponding probability density $|\psi(x)|^2$. There is clearly an uncertainty in specifying the location of the electron. The most probable region to find the electron is near $x = a_0$, but there is a nonzero probability for the electron to be anywhere in the range $0 < x < \infty$. This is very different from the Bohr model, in which the distance between the proton and electron is fixed at the value a_0.

Also shown in the figure are wave functions and probability densities corresponding to the first and second excited states. The wave functions have the oscillatory or wavelike property that we expect for quantum wave functions. As we go to higher excited states, there are more peaks in the probability density and the region of maximum probability moves to larger distances. These same features emerge from the solution to the three-dimensional problem. From this simple one-dimensional calculation (which *does not* in any way physically represent the real three-dimensional hydrogen atom) you can already see how quantum mechanics will resolve some of the difficulties associated with the Bohr model.

Example 7.1

Find the normalization constant of the ground state wave function for a particle trapped in the one-dimensional Coulomb potential energy.

Solution

The normalization integral (with $b = 1/a_0$) is

$$\int_0^\infty |\psi(x)|^2 \, dx = A^2 \int_0^\infty x^2 e^{-2x/a_0} \, dx = 1$$

The integration is in a standard form that is found in integral tables and that we will have occasion to use frequently in analyzing the hydrogen wave functions:

$$\int_0^\infty x^n e^{-cx} \, dx = \frac{n!}{c^{n+1}} \tag{7.3}$$

Using this standard form with $n = 2$ and $c = 2/a_0$, the normalization integral becomes

$$A^2 \frac{2!}{(2/a_0)^3} = 1 \quad \text{or} \quad A = 2a_0^{-3/2}$$

Example 7.2

In the ground state of an electron bound in a one-dimensional Coulomb potential energy, what is the probability to find the electron located between $x = 0$ and $x = a_0$?

Solution

The probability can be found using Eq. 5.10:

$$P(0 : a_0) = \int_0^{a_0} |\psi(x)|^2 \, dx = \frac{4}{a_0^3} \int_0^{a_0} x^2 e^{-2x/a_0} \, dx$$

with the normalization constant from Example 7.1. The integral is a standard form that we will later find useful for analyzing the hydrogen wave functions:

$$\int x^n e^{-cx} \, dx = -\frac{e^{-cx}}{c}$$
$$\times \left(x^n + \frac{nx^{n-1}}{c} + \frac{n(n-1)x^{n-2}}{c^2} + \cdots + \frac{n!}{c^n} \right) \quad (7.4)$$

The probability is then

$$P(0 : a_0) = \frac{4}{a_0^3} \left[-\frac{e^{-2x/a_0}}{2/a_0} \left(x^2 + \frac{2x}{2/a_0} + \frac{2}{(2/a_0)^2} \right) \right]_0^{a_0}$$

$$= 0.323$$

7.2 ANGULAR MOMENTUM IN THE HYDROGEN ATOM

Angular momentum played a significant role in Bohr's analysis of the structure of the hydrogen atom. Bohr was able to obtain the correct energy levels by assuming that in the orbit with quantum number n, the angular momentum of the electron is equal to $n\hbar$. Bohr's idea about the "quantization of angular momentum" turned out to have some correct features, but his analysis is not consistent with the actual quantum mechanical nature of angular momentum.

Angular Momentum of Classical Orbits

Before considering the angular momentum of an orbiting electron, it is helpful to review how angular momentum affects classical orbits, such as those of planets or comets about the Sun. Classically, the angular momentum of a particle is represented by the vector $\vec{L} = \vec{r} \times \vec{p}$, where $\vec{r}$ is the position vector that locates the particle and $\vec{p}$ is its linear momentum. The direction of $\vec{L}$ is perpendicular to the plane of the orbit. Along with the energy, the angular momentum remains constant as the planet orbits.

The total energy of the orbital motion determines the average distance of the planet from the Sun. For a given total energy, many different orbits are possible, from the nearly circular orbit of the Earth to the highly elongated elliptical orbits of the comets. These orbits differ in their angular momentum L, which is largest for the circular orbit and smallest for the elongated ellipse. Figure 7.2 shows a variety of planetary orbits having the same total energy but different angular momentum. The complete specification of the orbit requires that we give not only the magnitude of the angular momentum vector but also its direction; this direction identifies the plane of the orbit. To completely describe the angular momentum vector requires three numbers; for example, we might give the three components of $\vec{L}$ (L_x, L_y, L_z). Equivalently, we might give the magnitude L of the vector and two angular coordinates that give its direction (similar to latitude and longitude on a sphere).

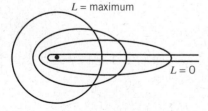

FIGURE 7.2 Planetary orbits of the same energy but different angular momentum L. As L decreases, the elliptical orbits become longer and thinner.

Angular Momentum in Quantum Mechanics

Quantum mechanics gives us a very different view of angular momentum. The angular momentum properties of a three-dimensional wave function are described by two quantum numbers. The first is the *angular momentum quantum number l*. This quantum number determines the length of the angular momentum vector:

$$|\vec{L}| = \sqrt{l(l+1)}\hbar \qquad (l = 0, 1, 2, \ldots) \qquad (7.5)$$

Note that this is very different from the Bohr condition $|\vec{L}| = n\hbar$. In particular, it is possible for the quantum vector to have a length of zero, but in the Bohr model the minimum length is $\hbar$.

The second number that we use to describe angular momentum in quantum mechanics is the *magnetic quantum number m_l*. This quantum number tells us about one component of the angular momentum vector, which we usually choose to be the z component. The relationship between the z component of $\vec{L}$ and the magnetic quantum number is

$$L_z = m_l\hbar \qquad (m_l = 0, \pm 1, \pm 2, \ldots, \pm l) \qquad (7.6)$$

Note that for each value of l there are $2l + 1$ possible values of m_l.

Unlike the classical angular momentum vector, for which we provide an exact specification by giving three numbers, the quantum angular momentum is described by *only two* numbers. Clearly two numbers cannot completely identify a vector in three-dimensional space, so something is missing from our description of the quantum angular momentum. As we discuss later, this missing part of the description of the quantum angular momentum vector is directly related to the application of the uncertainty principle to angular momentum.

Example 7.3

Compute the length of the angular momentum vectors that represent the orbital motion of an electron in a quantum state with $l = 1$ and in another state with $l = 2$.

Solution

Equation 7.5 gives the relationship between the length of the vector and the angular momentum quantum number l.

For $l = 1$,

$$|\vec{L}| = \sqrt{1(1+1)}\hbar = \sqrt{2}\hbar$$

and for $l = 2$,

$$|\vec{L}| = \sqrt{2(2+1)}\hbar = \sqrt{6}\hbar$$

Example 7.4

What are the possible z components of the vector $\vec{L}$ that represents the orbital angular momentum of a state with $l = 2$?

Solution

The possible m_l values for $l = 2$ are +2, +1, 0, −1, −2, and so the $\vec{L}$ vector can have any of five possible z components: $L_z = 2\hbar, \hbar, 0, -\hbar,$ or $-2\hbar$. The length of the vector $\vec{L}$, as we found previously, is $\sqrt{6}\hbar$.

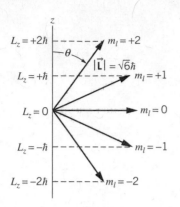

FIGURE 7.3 The orientations in space and z components of a vector with $l = 2$. There are five different possible orientations.

The components of the vector $\vec{L}$ for $l = 2$ are illustrated in Figure 7.3. Each orientation in space of the vector $\vec{L}$ corresponds to a different m_l value. The polar angle θ that the vector $\vec{L}$ makes with the z axis can be found by referring to the figure. With $L_z = |\vec{L}| \cos\theta$, we have

$$\cos\theta = \frac{L_z}{|\vec{L}|} = \frac{m_l}{\sqrt{l(l+1)}} \tag{7.7}$$

using Eq. 7.6 for L_z and Eq. 7.5 for $|\vec{L}|$.

This behavior represents a curious aspect of quantum mechanics called *spatial quantization*—only certain orientations of angular momentum vectors are allowed. The number of these orientations is equal to $2l + 1$ (the number of different possible m_l values) and the magnitudes of their successive z components always differ by $\hbar$. For example, an angular momentum state with $l = 1$ can have m_l values of $+1$, 0, or -1 (corresponding to z components $L_z = +\hbar, 0, -\hbar$) and thus $\cos\theta = +1/\sqrt{2}, 0,$ or $-1/\sqrt{2}$. The $\vec{L}$ vector in this case can have one of only three possible orientations relative to the z axis, corresponding to angles of $45°, 90°,$ or $135°$. This is in contrast to a classical angular momentum vector, which can have any possible orientation in space; that is, the angle between a classical angular momentum vector and the z axis can take any value between 0 and $180°$.

The Angular Momentum Uncertainty Relationship

In quantum mechanics, the maximum amount of permitted information about the angular momentum vector is its length (given by Eq. 7.5) and its z component (given by Eq. 7.6). Because the complete description of a vector requires *three* numbers, we are always missing some information about the angular momentum of a quantum state. If we specify $|\vec{L}|$ and L_z exactly, then we have no information about the other components of $\vec{L}$ (L_x and L_y). Any possible outcome of a measurement of L_x or L_y can therefore occur (as long as $|\vec{L}|^2 = L_x^2 + L_y^2 + L_z^2$). In graphic terms, we can imagine that the tip of the $\vec{L}$ vector rotates or *precesses* about the z axis so that L_z remains fixed but L_x and L_y are undetermined, as in Figure 7.4. This rotation cannot be directly measured; all we can observe is the "smeared out" distribution of values of L_x and L_y.

There is thus an uncertainty or indeterminacy in specifying $\vec{L}$ that is summarized by another form of the uncertainty principle:

$$\Delta L_z \Delta \phi \geq \hbar \tag{7.8}$$

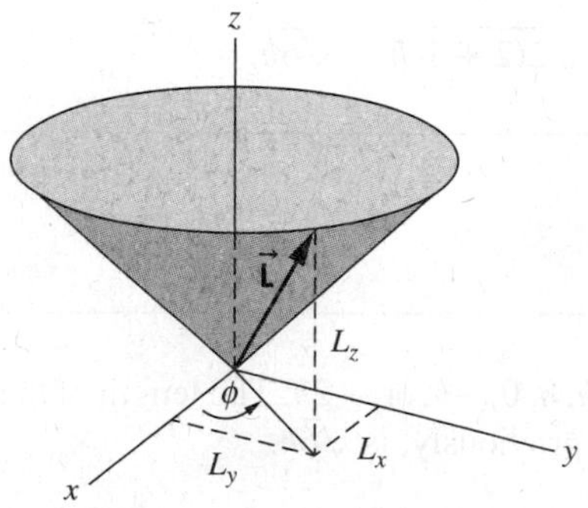

where ϕ is the azimuthal angle shown in Figure 7.4. If we know L_z exactly ($\Delta L_z = 0$), then we have no knowledge at all of the angle ϕ—all values are equally probable. This is equivalent to saying that we know nothing at all about L_x and L_y; whenever one component of $\vec{L}$ is determined, the other components are completely undetermined.

On the other hand, if we try to construct an angular momentum state in which a different component—for example, L_x—is completely specified (so that ϕ would be known), the state becomes a mixture or superposition of different L_z values. In effect, we can reduce the uncertainty in ϕ only at the expense of increasing the uncertainty in L_z. This is exactly the same type of behavior that was described by

FIGURE 7.4 The vector $\vec{L}$ precesses rapidly about the z axis, so that L_z stays constant, but L_x and L_y are indeterminate.

the other forms of the uncertainty principle; for example, reducing the uncertainty in x is always accompanied by an increase in the uncertainty in p_x.

From this discussion you can see why the length of the angular momentum is defined according to Eq. 7.5 and why, for example, we could not have simply defined the length as $|\vec{L}| = l\hbar$. If this were possible, then when m_l had its maximum value ($m_l = +l$), we would have $L_z = m_l\hbar = l\hbar$; the length of the vector would then be equal to its z component, and so it must lie along the z axis with $L_x = L_y = 0$. However, this simultaneous exact knowledge of all three components of $\vec{L}$ violates the angular momentum form of the uncertainty principle, and therefore this situation is not permitted to occur. It is therefore necessary for the length of $\vec{L}$ to be greater than $l\hbar$.

7.3 THE HYDROGEN ATOM WAVE FUNCTIONS

To find the complete spatial description of the electron in a hydrogen atom, we must obtain three-dimensional wave functions. The Schrödinger equation in three-dimensional Cartesian coordinates has the following form:

$$-\frac{\hbar^2}{2m}\left(\frac{\partial^2\psi}{\partial x^2} + \frac{\partial^2\psi}{\partial y^2} + \frac{\partial^2\psi}{\partial z^2}\right) + U(x,y,z)\psi(x,y,z) = E\psi(x,y,z) \qquad (7.9)$$

where ψ is a function of x, y, and z. The usual procedure for solving a partial differential equation of this type is to separate the variables by replacing a function of three variables with the product of three functions of one variable—for example, $\psi(x,y,z) = X(x)Y(y)Z(z)$. However, the Coulomb potential energy (Eq. 7.1) written in Cartesian coordinates, $U(x,y,z) = -e^2/4\pi\varepsilon_0\sqrt{x^2+y^2+z^2}$, does not lead to a separable solution.

For this calculation, it is more convenient to work in spherical polar coordinates (r, θ, ϕ) instead of Cartesian coordinates (x, y, z). The variables of spherical polar coordinates are illustrated in Figure 7.5. This simplification in the solution is at the expense of an increased complexity of the Schrödinger equation, which becomes:

$$-\frac{\hbar^2}{2m}\left[\frac{\partial^2\psi}{\partial r^2} + \frac{2}{r}\frac{\partial\psi}{\partial r} + \frac{1}{r^2\sin\theta}\frac{\partial}{\partial\theta}\left(\sin\theta\frac{\partial\psi}{\partial\theta}\right) + \frac{1}{r^2\sin^2\theta}\frac{\partial^2\psi}{\partial\phi^2}\right] \qquad (7.10)$$

$$+ U(r)\psi(r,\theta,\phi) = E\psi(r,\theta,\phi)$$

where now ψ is a function of the spherical polar coordinates r, θ, and ϕ. When the potential energy depends only on r (and not on θ or ϕ), as is the case for the Coulomb potential energy, we can find solutions that are separable and can be factored as

$$\psi(r,\theta,\phi) = R(r)\Theta(\theta)\Phi(\phi) \qquad (7.11)$$

where the *radial function $R(r)$*, the *polar function $\Theta(\theta)$*, and the *azimuthal function $\Phi(\phi)$* are each functions of a single variable. This procedure gives three differential equations, each of a single variable (r, θ, ϕ).

The quantum state of a particle that moves in a potential energy that depends only on r can be described by angular momentum quantum numbers l and m_l.

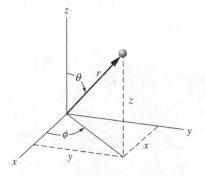

FIGURE 7.5 Spherical polar coordinates for the hydrogen atom. The proton is at the origin and the electron is at a radius r, in a direction determined by the polar angle θ and the azimuthal angle ϕ.

The polar and azimuthal solutions are given by combinations of standard trigono-metric functions. The remaining radial function is then obtained from solving the radial equation:

$$-\frac{\hbar^2}{2m}\left(\frac{d^2R}{dr^2} + \frac{2}{r}\frac{dR}{dr}\right) + \left(-\frac{e^2}{4\pi\varepsilon_0 r} + \frac{l(l+1)\hbar^2}{2mr^2}\right)R(r) = ER(r) \qquad (7.12)$$

The mass that appears in this equation is the *reduced mass* of the proton-electron system defined in Eq. 6.44.

Quantum Numbers and Wave Functions

When we solve a three-dimensional equation such as the Schrödinger equation, three parameters emerge in a natural way as indices or labels for the solutions, just as the single index n emerged from our solution of the one-dimensional infinite well in Section 5.4. These indices are the three *quantum numbers* that label the solutions. The three quantum numbers that emerge from the solutions and their allowed values are:

n	principal quantum number	$1, 2, 3, \ldots$
l	angular momentum quantum number	$0, 1, 2, \ldots, n-1$
m_l	magnetic quantum number	$0, \pm1, \pm2, \ldots, \pm l$

The principal quantum number n is identical to the quantum number n that we obtained in the Bohr model. It determines the quantized energy levels:

$$E_n = -\frac{me^4}{32\pi^2\varepsilon_0^2\hbar^2}\frac{1}{n^2} \qquad (7.13)$$

which is identical to Eq. 6.30. Note that the energy depends only on n and not on the other quantum numbers l or m_l. The permitted values of the angular momentum quantum number l are limited by n (l ranges from 0 to $n-1$) and those of the magnetic quantum number m_l are limited by l.

Complete with quantum numbers, the separated solutions of Eq. 7.10 can be written

$$\psi_{n,l,m_l}(r,\theta,\phi) = R_{n,l}(r)\Theta_{l,m_l}(\theta)\Phi_{m_l}(\phi) \qquad (7.14)$$

The indices (n, l, m_l) are the three quantum numbers that are necessary to describe the solutions. Wave functions corresponding to some values of the quantum numbers are shown in Table 7.1. The wave functions are written in terms of the Bohr radius a_0 defined in Eq. 6.29.

For the ground state ($n = 1$), only $l = 0$ and $m_l = 0$ are allowed. The complete set of quantum numbers for the ground state is then $(n, l, m_l) = (1, 0, 0)$, and the wave function for this state is given in the first line of Table 7.1. The first excited state ($n = 2$) can have $l = 0$ or $l = 1$. For $l = 0$, only $m_l = 0$ is allowed. This state has quantum numbers $(2, 0, 0)$, and its wave function is given in the second line of Table 7.1. For $l = 1$, we can have $m_l = 0$ or ± 1. There are thus three possible sets of quantum numbers: $(2, 1, 0)$ and $(2, 1, \pm 1)$. The wave functions for these states are given in the third and fourth lines of Table 7.1. The second excited state ($n = 3$) can have $l = 0$ ($m_l = 0$), $l = 1$ ($m_l = 0, \pm 1$), or $l = 2$ ($m_l = 0, \pm 1, \pm 2$).

For the $n = 2$ level, there are four different possible sets of quantum numbers and correspondingly four different wave functions. All of these wave functions

TABLE 7.1 Some Hydrogen Atom Wave Functions

n	l	m_l	$R(r)$	$\Theta(\theta)$	$\Phi(\phi)$
1	0	0	$\dfrac{2}{a_0^{3/2}}e^{-r/a_0}$	$\dfrac{1}{\sqrt{2}}$	$\dfrac{1}{\sqrt{2\pi}}$
2	0	0	$\dfrac{1}{(2a_0)^{3/2}}\left(2-\dfrac{r}{a_0}\right)e^{-r/2a_0}$	$\dfrac{1}{\sqrt{2}}$	$\dfrac{1}{\sqrt{2\pi}}$
2	1	0	$\dfrac{1}{\sqrt{3}(2a_0)^{3/2}}\dfrac{r}{a_0}e^{-r/2a_0}$	$\sqrt{\dfrac{3}{2}}\cos\theta$	$\dfrac{1}{\sqrt{2\pi}}$
2	1	± 1	$\dfrac{1}{\sqrt{3}(2a_0)^{3/2}}\dfrac{r}{a_0}e^{-r/2a_0}$	$\mp\dfrac{\sqrt{3}}{2}\sin\theta$	$\dfrac{1}{\sqrt{2\pi}}e^{\pm i\phi}$
3	0	0	$\dfrac{2}{(3a_0)^{3/2}}\left(1-\dfrac{2r}{3a_0}+\dfrac{2r^2}{27a_0^2}\right)e^{-r/3a_0}$	$\dfrac{1}{\sqrt{2}}$	$\dfrac{1}{\sqrt{2\pi}}$
3	1	0	$\dfrac{8}{9\sqrt{2}(3a_0)^{3/2}}\left(\dfrac{r}{a_0}-\dfrac{r^2}{6a_0^2}\right)e^{-r/3a_0}$	$\sqrt{\dfrac{3}{2}}\cos\theta$	$\dfrac{1}{\sqrt{2\pi}}$
3	1	± 1	$\dfrac{8}{9\sqrt{2}(3a_0)^{3/2}}\left(\dfrac{r}{a_0}-\dfrac{r^2}{6a_0^2}\right)e^{-r/3a_0}$	$\mp\dfrac{\sqrt{3}}{2}\sin\theta$	$\dfrac{1}{\sqrt{2\pi}}e^{\pm i\phi}$
3	2	0	$\dfrac{4}{27\sqrt{10}(3a_0)^{3/2}}\dfrac{r^2}{a_0^2}e^{-r/3a_0}$	$\sqrt{\dfrac{5}{8}}(3\cos^2\theta-1)$	$\dfrac{1}{\sqrt{2\pi}}$
3	2	± 1	$\dfrac{4}{27\sqrt{10}(3a_0)^{3/2}}\dfrac{r^2}{a_0^2}e^{-r/3a_0}$	$\mp\sqrt{\dfrac{15}{4}}\sin\theta\cos\theta$	$\dfrac{1}{\sqrt{2\pi}}e^{\pm i\phi}$
3	2	± 2	$\dfrac{4}{27\sqrt{10}(3a_0)^{3/2}}\dfrac{r^2}{a_0^2}e^{-r/3a_0}$	$\dfrac{\sqrt{15}}{4}\sin^2\theta$	$\dfrac{1}{\sqrt{2\pi}}e^{\pm 2i\phi}$

correspond to the same energy, so the $n = 2$ level is *degenerate*. (Degeneracy was introduced in Section 5.4.) The $n = 3$ level is degenerate with nine possible sets of quantum numbers. In general, the level with principal quantum number n has a degeneracy equal to n^2. Figure 7.6 illustrates the labeling of the first three levels.

If different combinations of quantum numbers have exactly the same energy, what is the purpose of listing them separately? First, as we discuss in the last section of this chapter, the levels are not precisely degenerate, but are separated

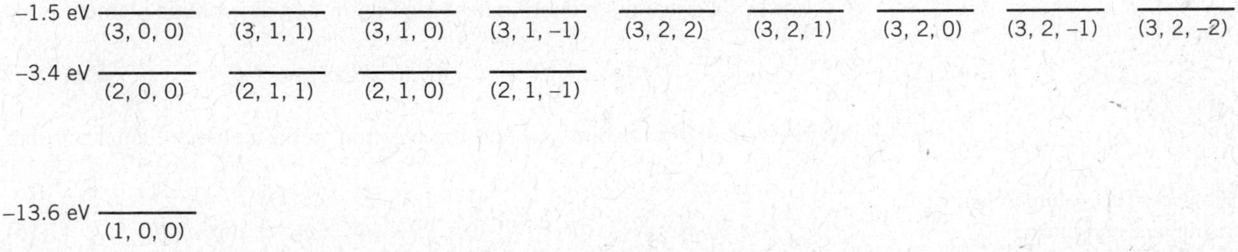

−1.5 eV								
(3, 0, 0)	(3, 1, 1)	(3, 1, 0)	(3, 1, −1)	(3, 2, 2)	(3, 2, 1)	(3, 2, 0)	(3, 2, −1)	(3, 2, −2)

−3.4 eV			
(2, 0, 0)	(2, 1, 1)	(2, 1, 0)	(2, 1, −1)

−13.6 eV
(1, 0, 0)

FIGURE 7.6 The lower energy levels of hydrogen, labeled with the quantum numbers (n, l, m_l). The first excited state is four-fold degenerate and the second excited state is nine-fold degenerate.

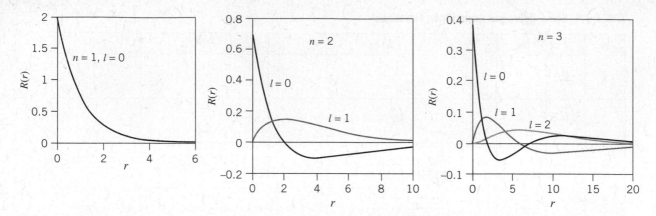

FIGURE 7.7 The radial wave functions of the $n = 1$, $n = 2$, and $n = 3$ states of hydrogen. The radius coordinate is measured in units of a_0.

by a very small energy (about 10^{-5} eV). Second, in the study of the transitions between the levels, we find that the intensities of the individual transitions depend on the quantum numbers of the particular level from which the transition originates. Third, and perhaps most important, *each of these sets of quantum numbers corresponds to a very different wave function, and therefore represents a very different state of motion of the electron.* These states have different spatial probability distributions for locating the electron, and thus can affect many atomic properties—for example, the way two atoms can form molecular bonds.

The radial wave functions for the states listed in Table 7.1 are plotted in Figure 7.7. You can readily see the differences in the motion of the electron for the different states. For example, in the $n = 2$ level, the $l = 0$ and $l = 1$ wave functions have the same energy but their behavior is very different: the $l = 1$ wave function falls to zero at $r = 0$, but the $l = 0$ wave function remains nonzero at $r = 0$. The $l = 0$ electron thus has a much greater probability of being found close to (or even inside) the nucleus, which turns out to play a large role in determining the rates for certain radioactive decay processes.

Probability Densities

As we learned in Chapter 5, the probability of finding the electron in any spatial interval is determined by the square of the wave function. For the hydrogen atom, $|\psi(r, \theta, \phi)|^2$ gives the *volume probability density* (probability per unit volume) at the location (r, θ, ϕ). To compute the actual probability of finding the electron, we multiply the probability per unit volume by the volume element dV located at (r, θ, ϕ). In spherical polar coordinates (see Figure 7.8) the volume element is

$$dV = r^2 \sin\theta \, dr \, d\theta \, d\phi \tag{7.15}$$

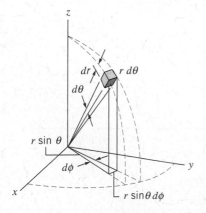

FIGURE 7.8 The volume element in spherical polar coordinates.

and therefore the probability to find the electron in the volume element at that location is

$$|\psi_{n,l,m_l}(r, \theta, \phi)|^2 \, dV = |R_{n,l}(r)|^2 |\Theta_{l,m_l}(\theta)|^2 |\Phi_{m_l}(\phi)|^2 r^2 \sin\theta \, dr \, d\theta \, d\phi \tag{7.16}$$

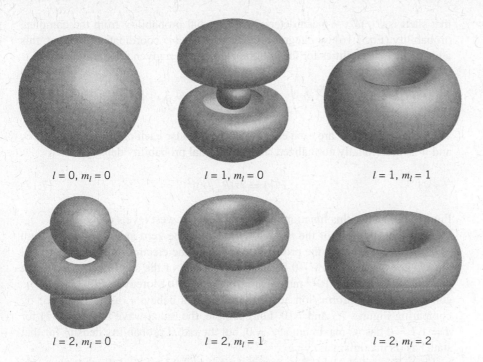

$l = 0, m_l = 0$ $l = 1, m_l = 0$ $l = 1, m_l = 1$

$l = 2, m_l = 0$ $l = 2, m_l = 1$ $l = 2, m_l = 2$

FIGURE 7.9 Representations of $|\psi|^2$ for different sets of quantum numbers. The z axis is the vertical direction. The diagrams represent surfaces on which the probability has the same value.

Some representations of the probability density $|\psi(r, \theta, \phi)|^2$ are shown in Figure 7.9. We can regard these illustrations as representing the "smeared out" distribution of electronic charge in the atom, which results from the uncertainty in the electron's location. They also represent the statistical outcomes of a large number of measurements of the location of the electron in the atom. These spatial distributions have important consequences for the structure of atoms with many electrons, which is discussed in Chapter 8, and also for the joining of atoms into molecules, which is discussed in Chapter 9.

In the next two sections, we will separately examine how the probability density depends on the radial coordinate and on the angular coordinates.

7.4 RADIAL PROBABILITY DENSITIES

Instead of asking about the complete probability density to locate the electron, we might want to know the probability to find the electron at a particular distance from the nucleus, no matter what the values of θ and ϕ might be. That is, imagine a thin spherical shell of radius r and thickness dr. What is the probability to find the electron in the shell between spheres of radius r and $r + dr$? We define the *radial probability density P(r)* so that the probability to find the electron within that shell is $P(r)dr$. We can determine the radial probability from the complete probability (Eq. 7.16) by integrating over the θ and ϕ coordinates. In effect, this adds up the probabilities for the volume elements at a given r for all θ and ϕ.

$$P(r)\,dr = |R_{n,l}(r)|^2 r^2\,dr \int_0^\pi |\Theta_{l,m_l}(\theta)|^2 \sin\theta\,d\theta \int_0^{2\pi} |\Phi_{m_l}(\phi)|^2\,d\phi \qquad (7.17)$$

The θ and ϕ integrals are each equal to unity, because each of the functions R, Θ, and Φ is individually normalized. Thus the radial probability density is

$$P(r) = r^2 |R_{n,l}(r)|^2 \qquad (7.18)$$

Figure 7.10 shows this function for several of the lowest levels of hydrogen.

Note that, because of the r^2 factor, $P(r)$ must be zero at $r = 0$ even though $R(r)$ might not. That is, the probability to locate the electron in a spherical shell always goes to zero as $r \to 0$ because the volume of the shell goes to zero, but the probability density $|\psi|^2$ may be nonzero at $r = 0$. Moreover, $P(r)$ and $|R(r)|^2$ convey different information about the electron's behavior, as you can see by comparing Figures 7.7 and 7.10. For example, the radial wave function $R(r)$ for $n = 1, l = 0$ has its maximum at $r = 0$, but the radial probability density for that state has its maximum at $r = a_0$.

Using the radial probability densities, it is possible to find the average value of the radial coordinate, that is, the average distance between the proton and the electron (see Problems 30 and 31). These values are indicated by markers in Figure 7.10. Notice that the average radial coordinate is about $1.5a_0$ for the $n = 1$ wave function and is much greater, about $5a_0$, for both of the $n = 2$ wave functions. The average radius is greater still, about $12a_0$, for the $n = 3$ states. It appears from these graphs that the average radius depends mostly on n and not very much on l. The principal quantum number n thus determines not only

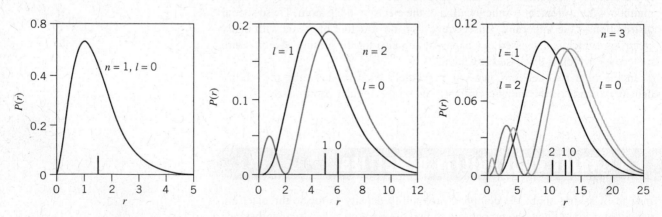

FIGURE 7.10 The radial probability density $P(r)$ for the $n = 1, n = 2$, and $n = 3$ states of hydrogen. The radius coordinate is measured in units of a_0. The markers on the horizontal axis show the values of the average radius r_{av} labeled with the value of l.

the energy level of the electron, it also determines to a great extent the average distance of the electron from the nucleus. As in the Bohr model, this average radius varies roughly as n^2, so that an $n = 2$ electron is on the average about 4 times farther from the nucleus than an $n = 1$ electron, an $n = 3$ electron is about 9 times farther from the nucleus than an $n = 1$ electron, and so forth.

Another measure of the location of the electron is its most probable radius, determined from the location at which $P(r)$ has its maximum value. For each $n, P(r)$ for the state with $l = n - 1$ has only a single maximum, which occurs at the location of the Bohr orbit, $r = n^2 a_0$. The following example illustrates this for the $n = 2$ state.

Example 7.5

Prove that the most likely distance from the origin of an electron in the $n = 2, l = 1$ state is $4a_0$.

Solution
In the $n = 2, l = 1$ level, the radial probability density is

$$P(r) = r^2 |R_{2,1}(r)|^2 = r^2 \frac{1}{24a_0^3} \frac{r^2}{a_0^2} e^{-r/a_0}$$

We wish to find where this function has its maximum; in the usual fashion, we take the first derivative of $P(r)$ and set it equal to zero:

$$\frac{dP(r)}{dr} = \frac{1}{24a_0^5} \frac{d}{dr} (r^4 e^{-r/a_0})$$

$$= \frac{1}{24a_0^5} \left[4r^3 e^{-r/a_0} + r^4 \left(-\frac{1}{a_0} \right) e^{-r/a_0} \right] = 0$$

or

$$\frac{1}{24a_0^5} e^{-r/a_0} \left(4r^3 - \frac{r^4}{a_0} \right) = 0$$

The only solution that yields a maximum is $r = 4a_0$.

Example 7.6

For the $n = 2$ states ($l = 0$ and $l = 1$), compare the probabilities of the electron being found inside the Bohr radius.

Solution
For the $n = 2$, $l = 0$ level, $P(r)dr = r^2 |R_{2,0}(r)|^2 dr = r^2 \frac{1}{8a_0^3} \left(2 - \frac{r}{a_0} \right)^2 e^{-r/a_0} dr$. The total probability of finding the electron between $r = 0$ and $r = a_0$ is

$$P(0 : a_0) = \int_0^{a_0} P(r) \, dr$$

$$= \frac{1}{8a_0^3} \int_0^{a_0} \left(4r^2 - \frac{4r^3}{a_0} + \frac{r^4}{a_0^2} \right) e^{-r/a_0} \, dr$$

Evaluating the integrals using Eq. 7.4, we obtain

$$P(0 : a_0) = 0.034$$

For the $n = 2$, $l = 1$ level $P(r)dr = r^2 |R_{2,1}(r)|^2 dr = r^2 \frac{1}{24a_0^3} \frac{r^2}{a_0^2} e^{-r/a_0} dr$. The total probability between $r = 0$ and $r = a_0$ is

$$P(0 : a_0) = \int_0^{a_0} P(r) \, dr$$

$$= \frac{1}{24a_0^5} \int_0^{a_0} \frac{r^4}{a_0^2} e^{-r/a_0} \, dr = 0.0037$$

The results of Example 7.6 are consistent with the radial probability densities shown in Figure 7.10—in the $l = 0$ state, the probability of finding the electron inside a_0 is about 10 times larger than in the $l = 1$ state, as suggested by the small peak in the radial probability density for $n = 2, l = 0$ at small r. There is clearly more area under the $P(r)$ curve between $r = 0$ and $r = a_0$ for $n = 2, l = 0$ than there is for $n = 2, l = 1$.

Curiously, Figure 7.10 also shows that for $n = 2$ the $l = 0$ radial probability density is also greater than the $l = 1$ probability density at large r. Thus the $l = 0$ electron spends more time close to the nucleus than the $l = 1$ electron and it also spends more time farther away. This is a general result that holds for any value of n: the smaller the l value, the larger is the probability to find the electron both close to the nucleus and far from the nucleus. The classical planetary orbits of Figure 7.2 show the same type of behavior—the orbit with $L = 0$ spends more time close to the central body and also more time far away from the central body, compared with the orbits that have larger L.

7.5 ANGULAR PROBABILITY DENSITIES

In this section, we consider the angular part of the probability density, which is obtained from the squared magnitudes of the angular parts of the wave function:

$$P(\theta, \phi) = |\Theta_{l,m_l}(\theta)\Phi_{m_l}(\phi)|^2 \tag{7.19}$$

Figure 7.11 shows the angular probablity densities for the $l = 0$ and $l = 1$ wave functions listed in Table 7.1.

Note that all of the probability densities are *cylindrically symmetric*—there is no dependence on the azimuthal angle ϕ. The $l = 0$ wave function is also *spherically symmetric*—that is, the probability density is independent of direction.

The $l = 1$ probability densities have two distinct shapes. For $m_l = 0$, the electron is found primarily in two regions of maximum probability along the positive and negative z axis, while for $m_l = \pm 1$, the electron is found primarily near the xy plane. For $m_l = 0$, the electron's angular momentum vector lies in the xy plane (Figure 7.3). Classically, the angular momentum vector is perpendicular to the orbital plane, so it should not be surprising that the electron is most likely to be found in a location away from the xy plane—that is, along the z axis. For

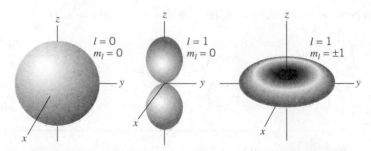

FIGURE 7.11 The angular dependence of the $l = 0$ and $l = 1$ probability densities.

$m_l = \pm 1$, the angular momentum vector has its maximum projection along the z axis; the electron, again orbiting perpendicular to $\vec{L}$, spends most of its time near the xy plane. These probability densities for locating the electron are consistent with the information given by the orientation of the angular momentum vector, and the cylindrical symmetry of the probability densities is consistent with the uncertainty in the knowledge of the orientation of $\vec{L}$ represented in Figure 7.4.

Example 7.7

For the $n = 2, l = 1$ wave functions, find the direction in space at which the maximum probability occurs when $m_l = 0$ and when $m_l = \pm 1$.

Solution

For $l = 1, m_l = 0$ we have $P(\theta, \phi) = |\Theta_{2,0}(\theta)\Phi_0(\phi)|^2 = \frac{3}{4\pi}\cos^2\theta$. To find the location of the maximum, we set $dP/d\theta$ equal to zero:

$$\frac{dP}{d\theta} = \frac{3}{4\pi}(-2\cos\theta\sin\theta) = 0$$

There are two solutions to this equation: one for $\cos\theta = 0$, for which $\theta = \pi/2$, and another for $\sin\theta = 0$, which gives $\theta = 0$ or π. By taking the second derivative, we find that $\theta = \pi/2$ leads to a minimum while $\theta = 0$ or π gives the maximum. There are thus two regions of maximum probability, one along the positive z axis ($\theta = 0$) and another along the negative z axis ($\theta = \pi$), as in Figure 7.11.

For $l = 1, m_l = \pm 1$ the angular probability density is $P(\theta, \phi) = |\Theta_{2,\pm 1}(\theta)\Phi_{\pm 1}(\phi)|^2 = \frac{3}{8\pi}\sin^2\theta$. We can then find the location of the maximum:

$$\frac{dP}{d\theta} = \frac{3}{4\pi}(\sin\theta\cos\theta) = 0$$

Once again there are two solutions: $\theta = 0, \pi$ or $\theta = \pi/2$. However, in this case the maximum occurs for $\theta = \pi/2$ and the probability maximum occurs in the xy plane, as in Figure 7.11.

7.6 INTRINSIC SPIN

One way of observing spatial quantization is to place the atom in an externally applied magnetic field. From the interaction between the magnetic field and the *magnetic dipole moment* of the atom (which is related to the electron's orbital angular momentum), it is possible both to observe the separate components of $\vec{L}$ and also to determine l by counting the number of z components (which, as we have seen, is equal to $2l + 1$). However, when this experiment is done, a surprising result emerges that indicates an unexpected property of the electron, known as *intrinsic spin*.

Orbital Magnetic Dipole Moments

Figure 7.12 shows a classical magnetic dipole moment, which might be produced by a current loop or the orbital motion of a charged object. The classical magnetic dipole moment $\vec{\mu}$ is defined as a vector whose magnitude is equal to the product of the circulating current and the area enclosed by the orbital loop. The direction of $\vec{\mu}$ is perpendicular to the plane of the orbit, determined by the right-hand rule—with the fingers in the direction of the conventional (positive) current, the thumb indicates the direction of $\vec{\mu}$, as shown in Figure 7.12 for a circulating negative charge like an electron.

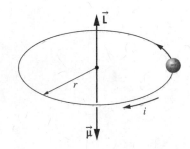

FIGURE 7.12 A circulating negative charge is represented as a current loop. Because the charge is negative, $\vec{L}$ and $\vec{\mu}$ point in opposite directions.

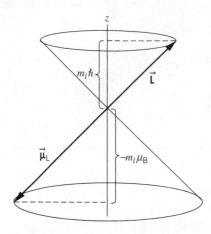

FIGURE 7.13 According to quantum mechanics, the vectors can be considered to precess around the z axis, and so we can specify only the z components of $\vec{L}$ and $\vec{\mu}$.

As we have seen, quantum mechanics forbids exact knowledge of the direction of $\vec{L}$ and therefore of $\vec{\mu}$. Figure 7.13 suggests the relationship between $\vec{L}$ and $\vec{\mu}$ that is consistent with quantum mechanics. Only the z components of these vectors can be specified. Because the electron has a negative charge, $\vec{L}$ and $\vec{\mu}$ have z components of opposite signs.

We can use the Bohr model with a circular orbit to obtain the relationship between $\vec{L}$ and $\vec{\mu}$, which turns out to be identical with the correct quantum mechanical result. We regard the circulating electron as a circular loop of current $i = dq/dt = q/T$, where q is the charge of the electron $(-e)$ and T is the time for one circuit around the loop. If the electron moves with speed $v = p/m$ around a loop of radius r, then $T = 2\pi r/v = 2\pi rm/p$. The magnitude of the magnetic moment is

$$\mu = iA = \frac{q}{2\pi rm/p}\pi r^2 = \frac{q}{2m}rp = \frac{q}{2m}|\vec{L}| \qquad (7.20)$$

with $|\vec{L}| = rp$. Writing Eq. 7.20 in terms of vectors and putting $-e$ for the electronic charge, we obtain

$$\vec{\mu}_L = -\frac{e}{2m}\vec{L} \qquad (7.21)$$

The negative sign, which is present because the electron has a negative charge, indicates that the vectors $\vec{L}$ and $\vec{\mu}_L$ point in opposite directions. The subscript L on $\vec{\mu}_L$ reminds us that this magnetic moment arises from the *orbital* angular momentum $\vec{L}$ of the electron.

The z component of the magnetic moment is

$$\mu_{L,z} = -\frac{e}{2m}L_z = -\frac{e}{2m}m_l\hbar = -\frac{e\hbar}{2m}m_l = -m_l\mu_B \qquad (7.22)$$

The quantity $e\hbar/2m$ is defined to be the *Bohr magneton*

$$\mu_B = \frac{e\hbar}{2m} \qquad (7.23)$$

The value of μ_B is

$$\mu_B = 9.274 \times 10^{-24} \text{ J/T}$$

The Bohr magneton is a convenient unit for expressing atomic magnetic moments, which typically have values of the order of μ_B.

A Dipole in an External Field

Before we consider further the behavior of $\vec{\mu}_L$, we discuss the similar behavior of an *electric dipole,* which consists of two equal and opposite charges q separated by a distance r. The electric dipole moment $\vec{p}$ has magnitude qr and points from the negative charge to the positive charge. As shown in Figure 7.14a, in a uniform *electric* field, the vertical forces $\vec{F}_+$ on the positive charge and $\vec{F}_-$ on the negative charge are of equal magnitude. The dipole experiences a torque that tends to rotate it into alignment with $\vec{E}$, but the *net force* on the dipole is zero. Suppose now that the field is not uniform—for example, the field strength decreases from the bottom of the figure to the top, as in Figure 7.14b. Now the downward force $\vec{F}_-$ acting on the negative charge is greater than the upward force $\vec{F}_+$ on the positive

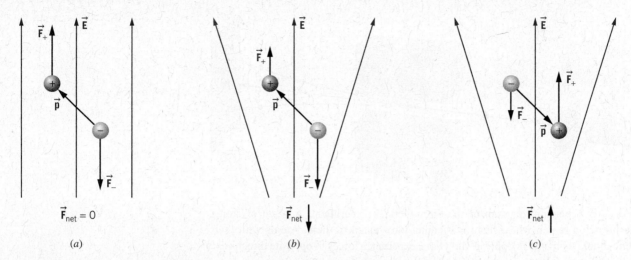

FIGURE 7.14 (a) An electric dipole in a uniform electric field $\vec{E}$ experiences no net force. (b) In a nonuniform electric field (decreasing from the bottom of the figure to the top), the force $\vec{F}_-$ is greater than the force $\vec{F}_+$. There is a net downward force on the dipole. (c) If the dipole moment is reversed, the net force is in the opposite direction.

charge. There is still a net torque that tends to rotate the dipole, but there is also a net force that tends to move the dipole, in this case downward. On the other hand, if we reverse the locations of the two charges (Figure 7.14c), which is equivalent to reversing the electric dipole moment $\vec{p}$, the upward force $\vec{F}_+$ on the positive charge is now greater than the downward force $\vec{F}_-$ on the negative charge, so the net force on the dipole is upward.

We can state this result in another way that will be more applicable to our discussion of *magnetic* dipole moments. Let the field direction define the z axis. Then dipoles with $p_z > 0$ (as in Figure 7.14b) experience a net negative force and move in the negative z direction, while dipoles with $p_z < 0$ (as in Figure 7.14c) experience a net positive force and move in the positive z direction.

A magnetic dipole moment $\vec{\mu}$ behaves in an identical way. (In fact, if we imagine fictitious N and S poles, the behavior of a magnetic moment would be described by illustrations similar to Figure 7.14.) A nonuniform *magnetic* field acting on the *magnetic* moments gives an unbalanced force that causes a displacement. Figure 7.15 illustrates the behavior of magnetic dipole moments having different orientations in a nonuniform magnetic field. The two different orientations give net forces in opposite directions: if μ_z is positive the force on the dipole is negative, and if μ_z is negative the force on the dipole is positive.

The Stern-Gerlach Experiment

Imagine the following experiment, illustrated schematically in Figure 7.16. A beam of hydrogen atoms is prepared in the $n = 2, l = 1$ state. The beam consists of equal numbers of atoms in the $m_l = -1, 0$, and $+1$ states. (We assume we can do the experiment so quickly that the $n = 2$ state doesn't decay to the $n = 1$ state. In practice this may not be possible.) The beam passes through a region in which there is a nonuniform magnetic field. The atoms with $m_l = +1$ ($\mu_{L,z} = -\mu_B$)

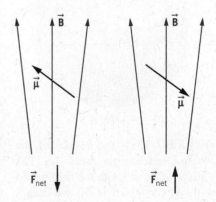

FIGURE 7.15 Two magnetic dipoles in a nonuniform magnetic field. Oppositely directed dipoles experience net forces in opposite directions.

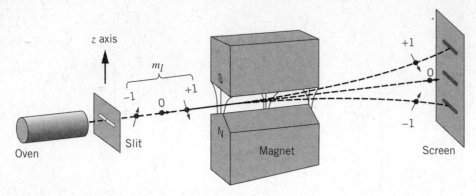

FIGURE 7.16 Schematic diagram of the Stern-Gerlach experiment. A beam of atoms passes through a region where there is a nonuniform magnetic field. Atoms with their magnetic dipole moments in opposite directions experience forces in opposite directions.

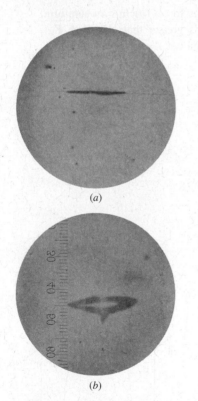

FIGURE 7.17   The results of the Stern-Gerlach experiment. (*a*) The image of the slit with the field turned off. (*b*) With the field on, two images of the slit appear. The small divisions in the scale at the left represent 0.05 mm. [*Source:* W. Gerlach and O. Stern, *Zeitschrift für Physik* **9**, 349 (1922)]

experience a net upward force and are deflected upward, while the atoms with $m_l = -1$ ($\mu_{L,z} = +\mu_B$) are deflected downward. The atoms with $m_l = 0$ are undeflected.

After passing through the field, the beam strikes a screen where it makes a visible image. When the field is off, we expect to see one image of the slit in the center of the screen, because there is no deflection at all. When the field is on, we expect three images of the slit on the screen—one in the center (corresponding to $m_l = 0$), one above the center ($m_l = +1$), and one below the center ($m_l = -1$). If the atom were in the ground state ($l = 0$), we expect to see one image in the screen whether the field was off or on (recall that a $m_l = 0$ atom is not deflected). If we had prepared the beam in a state with $l = 2$, we would see five images with the field on. *The number of images that appears is just the number of different m_l values,* which is equal to $2l + 1$. With the possible values for l of 0, 1, 2, 3, ..., it follows that $2l + 1$ has the values 1, 3, 5, 7, ...; that is, we should always see an *odd number* of images on the screen. However, if we were actually to perform the experiment with hydrogen in the $l = 1$ state, we would find not three but *six* images on the screen! Even more confusing, if we did the experiment with hydrogen in the $l = 0$ state, we would find not one but *two* images on the screen, one representing an upward deflection and one a downward deflection! In the $l = 0$ state, the vector $\mathbf{L}$ has length zero, and so we expect that there is *no magnetic moment* for the magnetic field to deflect. We observe this not to be true—even when $l = 0$, the atom still has a magnetic moment, in contradiction to Eq. 7.21.

The first experiment of this type was done by O. Stern and W. Gerlach in 1921. They used a beam of silver atoms; although the electronic structure of silver is more complicated than that of hydrogen (as we discuss in Chapter 8), the same basic principle applies—the silver atom must have $l = 0, 1, 2, 3, ...$, and so an *odd number* of images is expected to appear on the screen. In fact, they observed the beam to split into *two* components, producing two images of the slits on the screen (see Figure 7.17).

The observation of separated images was the first conclusive evidence of *spatial quantization*; classical magnetic moments would have all possible orientations and would make a continuous smeared-out pattern on the screen, but the observation of a number of discrete images on the screen means that the atomic magnetic moments can take only certain discrete orientations in space. These correspond to the discrete orientations of the magnetic moment (or, equivalently, of the angular momentum).

However, the *number* of discrete images on the screen does not agree with our expectations that it be an odd number. We expect $2l + 1$ images, so for two images we should have $l = \frac{1}{2}$, which is not permitted by the Schrödinger equation. We can resolve this dilemma if there is another contribution to the angular momentum of the atom, the *intrinsic angular momentum* of the electron. An electron in an atom has two kinds of angular momentum, somewhat like the Earth as it both orbits the Sun and rotates on its axis. The electron has an *orbital angular momentum* $\vec{L}$, which characterizes the motion of the electron about the nucleus, and an *intrinsic angular momentum* $\vec{S}$, which behaves as if the electron were spinning about its axis. For this reason, $\vec{S}$ is usually called the intrinsic *spin*. (However, it is not correct to use the classical analogy to think of the electron as a tiny ball of charge spinning about an axis, because the electron is a point particle with no physical size.) The idea of electron spin was proposed by S. A. Goudsmit and G. E. Uhlenbeck in 1925, and P. A. M. Dirac showed in 1928 that *relativistic* quantum theory for the electron gives the electron spin directly as an additional quantum number.

In order to explain the result of the Stern-Gerlach experiment, we must assign to the electron an intrinsic spin quantum number s of $\frac{1}{2}$. The intrinsic spin behaves much like the orbital angular momentum; there is the quantum number s (which we can regard as a label arising from the mathematics), the angular momentum vector $\vec{S}$, a z component S_z, an associated magnetic moment $\vec{\mu}_S$, and a spin magnetic quantum number m_s. Figure 7.18 illustrates the vector properties of $\vec{S}$, and Table 7.2 compares the properties of orbital and spin angular momentum for electrons in atoms.

The inclusion of spin gives a direct explanation for the Stern-Gerlach experiment. The outermost electron in a silver atom occupies a state with $l = 0$. (The other electrons do not contribute to the magnetic properties of the atom.) The magnetic behavior is therefore due entirely to the spin magnetic moment, which has only two possible orientations in the magnetic field (corresponding to $m_s = \pm \frac{1}{2}$) and thus gives the two beams observed emerging from the magnet.

Every fundamental particle has a characteristic intrinsic spin and a corresponding spin magnetic moment. For example, the proton and neutron also have a spin quantum number of $\frac{1}{2}$. The photon has a spin quantum number of 1, while the pi meson (pion) has $s = 0$.

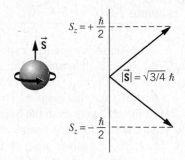

FIGURE 7.18 The spin angular momentum of an electron and the spatial orientation of the spin angular momentum vector.

TABLE 7.2 Orbital and Spin Angular Momentum of Electrons in Atoms

	Orbital	Spin
Quantum number	$l = 0, 1, 2, \ldots$	$s = \frac{1}{2}$
Length of vector	$|\vec{L}| = \sqrt{l(l+1)}\hbar$	$|\vec{S}| = \sqrt{s(s+1)}\hbar = \sqrt{3/4}\,\hbar$
z component	$L_z = m_l \hbar$	$S_z = m_s \hbar$
Magnetic quantum number	$m_l = 0, \pm 1, \pm 2, \ldots, \pm l$	$m_s = \pm \frac{1}{2}$
Magnetic moment	$\vec{\mu}_L = -(e/2m)\vec{L}$	$\vec{\mu}_S = -(e/m)\vec{S}$

Example 7.8

In a Stern-Gerlach type of experiment, the magnetic field varies with distance in the z direction according to $dB_z/dz = 1.4\,\text{T/mm}$. The silver atoms travel a distance $x = 3.5\,\text{cm}$ through the magnet. The most probable speed of the atoms emerging from the oven is $v = 750\,\text{m/s}$. Find the separation of the two beams as they leave the magnet. The mass of a silver atom is 1.8×10^{-25} kg, and its magnetic moment is about 1 Bohr magneton.

Solution

The potential energy of the magnetic moment in the magnetic field is

$$U = -\vec{\mu} \cdot \vec{B} = -\mu_z B_z$$

because the field along the central axis of the magnet has only a z component. The force on the atom can be found from the potential energy according to

$$F_z = -\frac{dU}{dz} = \mu_z \frac{dB_z}{dz}$$

The acceleration of a silver atom of mass m as it passes through the magnet is

$$a = \frac{F_z}{m} = \frac{\mu_z(dB_z/dz)}{m}$$

The vertical deflection Δz of either beam can be found from $\Delta z = \frac{1}{2}at^2$, where t, the time to traverse the magnet, equals x/v. Each beam is deflected by this amount, so the net separation d is $2\Delta z$, or

$$
\begin{aligned}
d &= \frac{\mu_z(dB_z/dz)x^2}{mv^2} \\
&= \frac{(9.27 \times 10^{-24}\,\text{J/T})(1.4 \times 10^3\,\text{T/m})(3.5 \times 10^{-2}\,\text{m})^2}{(1.8 \times 10^{-25}\,\text{kg})(750\,\text{m/s})^2} \\
&= 1.6 \times 10^{-4}\,\text{m} = 0.16\,\text{mm}
\end{aligned}
$$

This is consistent with the separation that can be read from the scale in Figure 7.17.

7.7 ENERGY LEVELS AND SPECTROSCOPIC NOTATION

We previously described all of the possible electronic states in hydrogen by three quantum numbers (n, l, m_l), but as we have seen, a fourth property of the electron, the intrinsic angular momentum or *spin,* requires the introduction of a fourth quantum number. We don't need to specify the spin s, because it is always $1/2$ (we regard it as a fundamental property of the electron, like its electric charge or its mass), but we must specify the value of the quantum number m_s $(+1/2$ or $-1/2)$, which tells us about the z component of the spin. Thus the complete description of the state of an electron in an atom requires the four quantum numbers (n, l, m_l, m_s).

For example, the ground state of hydrogen was previously labeled as $(n, l, m_l) = (1, 0, 0)$. With the addition of m_s, this would become either $(1, 0, 0, +1/2)$ or $(1, 0, 0, -1/2)$. The degeneracy of the ground state is now 2. The first excited state would have eight possible labels: $(2, 0, 0, +1/2)$, $(2, 0, 0, -1/2)$, $(2, 1, +1, +1/2)$, $(2, 1, +1, -1/2)$, $(2, 1, 0, +1/2)$, $(2, 1, 0, -1/2)$, $(2, 1, -1, +1/2)$, and $(2, 1, -1, -1/2)$. There are now two possible labels for each previous single label (each n, l, m_l becomes $n, l, m_l, -1/2$ and $n, l, m_l, -1/2$, so the degeneracy of each level is $2n^2$ instead of n^2.

It is important to know the direction (z component) of the angular momentum vectors when an atom is in a magnetic field, but for most other applications the values of m_l and m_s are of no significance, and it is cumbersome to write them each time we wish to refer to a certain level of an atom. We therefore use a different notation, known as *spectroscopic notation,* to label the levels. In this system we use letters to stand for the different l values: for $l = 0$, we use the letter s (do not confuse this with the quantum number s), for $l = 1$, we use the letter p, and so on. The complete notation is as follows:

Value of l	0	1	2	3	4	5	6
Designation	s	p	d	f	g	h	i

(The first four letters stand for sharp, principal, diffuse, and fundamental, which were terms used to describe atomic spectra before atomic theory was developed.) In spectroscopic notation, the ground state of hydrogen is labeled $1s$, where the value $n = 1$ is specified before the s. Figure 7.19 illustrates the labeling of the hydrogen atom levels in this notation.

Also shown in Figure 7.19 are arrows representing some different photons that can be emitted when the atom makes a transition from one state to a lower state. Some of the missing arrows (such as $4d$ to $3s$) would represent transitions that are not allowed to occur. By solving the Schrödinger equation and using the solutions to compute *transition probabilities*, we find that the transitions most likely to occur are those that change l by one unit. This restriction is called a *selection rule*, and for atomic transitions the selection rule is

$$\Delta l = \pm 1 \tag{7.24}$$

For example, the $3s$ level cannot emit a photon in a transition to the $2s$ level ($\Delta l = 0$), but rather must go to the $2p$ level ($\Delta l = 1$). There is no selection rule for n, so the $3p$ level can go to $2s$ or $1s$ (but not to $2p$).

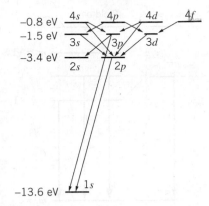

FIGURE 7.19 A partial energy level diagram of hydrogen, showing the spectroscopic notation of the levels and some of the transitions that satisfy the $\Delta l = \pm 1$ selection rule.

*7.8 THE ZEEMAN EFFECT

Consider for the moment a hypothetical (and less interesting) world in which the electron has no spin, and therefore no spin magnetic moment. Suppose we prepared a hydrogen atom in a $2p$ ($l = 1$) level and placed it in an external uniform magnetic field $\vec{\mathbf{B}}$ (supplied by a laboratory electromagnet, for example). The magnetic moment $\vec{\mu}_L$ associated with the orbital angular momentum then interacts with the field, and the energy associated with this interaction is

$$U = -\vec{\mu}_L \cdot \vec{\mathbf{B}} \tag{7.25}$$

That is, magnetic moments aligned in the direction of the field have less energy than those aligned oppositely to the field. Using Eq. 7.22 for the z component of the magnetic moment (assuming that the field is in the z direction), we have

$$U = -\mu_{L,z}B = m_l \mu_B B \tag{7.26}$$

in terms of the Bohr magneton μ_B defined in Eq. 7.23. In the absence of a magnetic field, the $2p$ level has a certain energy E_0 ($-3.4\,\text{eV}$). When the field is turned on, the energy becomes $E_0 + U = E_0 + m_l\mu_B B$; that is, there are now three different possible energies for the level, depending on the value of m_l. Figure 7.20 illustrates this situation.

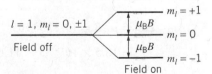

FIGURE 7.20 The splitting of an $l = 1$ level in an external magnetic field. (The effects of the electron's spin angular momentum are ignored.) The energy in a magnetic field is different for different values of m_l.

---

*This is an optional section that may be skipped without loss of continuity.

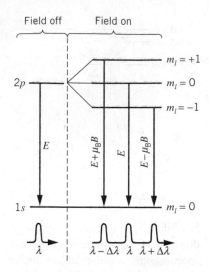

Field off   Field on

$m_l = +1$

$2p$

$m_l = 0$

$m_l = -1$

E $E + \mu_B B$ E $E - \mu_B B$

$1s$ $m_l = 0$

λ $\lambda - \Delta\lambda$ λ $\lambda + \Delta\lambda$

FIGURE 7.21   The normal Zeeman effect. When the field is turned on, the original wavelength λ becomes three separate wavelengths.

Now suppose the atom emits a photon in a transition from the $2p$ state to the $1s$ ground state. In the absence of the magnetic field, a single photon is emitted with an energy of 10.2 eV and a corresponding wavelength of 122 nm. When the magnetic field is present, three photons can be emitted, with energies of 10.2 eV $+ \mu_B B$, 10.2 eV, and 10.2 eV $- \mu_B B$. To determine how a small change in energy ΔE affects the wavelength, we differentiate the expression $E = hc/\lambda$ and obtain

$$dE = -\frac{hc}{\lambda^2} d\lambda \qquad (7.27)$$

Replacing the differentials with small differences, taking absolute magnitudes, and solving for $\Delta\lambda$ gives

$$\Delta\lambda = \frac{\lambda^2}{hc} \Delta E \qquad (7.28)$$

where ΔE is the energy splitting between the levels when the field is on ($\Delta E = \mu_B B$). Figure 7.21 illustrates the three transitions, and shows an example of the result of a measurement of the emitted wavelengths.

In analyzing transitions between different m_l states, often we need to use a second *selection rule:* the only transitions that occur are those that change m_l by 0, +1, or −1:

$$\Delta m_l = 0, \pm 1 \qquad (7.29)$$

Changes in m_l of two or more are not permitted.

Example 7.9

Compute the change in wavelength of the $2p \rightarrow 1s$ photon when a hydrogen atom is placed in a magnetic field of 2.00 T.

Solution

The energy of the photon from $n = 2$ to $n = 1$ is $E = -13.6\,\text{eV}(\frac{1}{2^2} - \frac{1}{1^2}) = 10.2\,\text{eV}$, and its wavelength is $\lambda = hc/E = (1240\,\text{eV} \cdot \text{nm})/(10.2\,\text{eV}) = 122\,\text{nm}$. The energy change ΔE of the levels is

$$\Delta E = \mu_B B = (9.27 \times 10^{-24}\,\text{J/T})(2.00\,\text{T})$$

$$= 18.5 \times 10^{-24}\,\text{J} = 11.6 \times 10^{-5}\,\text{eV}$$

and so, from Eq. 7.28,

$$\Delta\lambda = \frac{\lambda^2}{hc} \Delta E$$

$$= \frac{(122\,\text{nm})^2}{1240\,\text{eV} \cdot \text{nm}} 11.6 \times 10^{-5}\,\text{eV}$$

$$= 0.00139\,\text{nm}$$

Even for a fairly large magnetic field of 2 T, the change in wavelength is very small, but it is easily measurable using an optical spectrometer.

The experiment we have just considered is an example of the *Zeeman effect*—the splitting of a spectral line with a single wavelength into lines with several different wavelengths when the emitting atoms are in an externally applied magnetic field. In the *normal Zeeman effect* a single spectral line splits into three components; this occurs only in atoms without spin. (All electrons of course have spin, unlike the hypothetical spinless electrons we considered; however, in certain atoms with several electrons, the spins can pair off and cancel, so that the atom behaves like a spinless one.) When spin *is* present, we must consider not only the effect of the orbital magnetic moment but also the spin magnetic moment. The resulting pattern of level splittings is more complicated, and spectral lines may split into more than three components. This case is known as the *anomalous Zeeman effect,* an example of which is shown in Figure 7.22.

FIGURE 7.22   The anomalous Zeeman effect in sodium. (Top) The so-called sodium D-lines, a close-lying doublet of wavelengths 589.0 and 589.6 nm in the absence of a magnetic field. (Bottom) Splitting of the lines into six and four components in a magnetic field. This image was photographed by Peter Zeeman in 1897.

*7.9 FINE STRUCTURE

A careful inspection of the emission lines of atomic hydrogen shows that many of them are in fact not single lines but very closely spaced combinations of two lines. In this section we examine the origin of that effect, known *as fine structure.*

In this calculation it is more convenient for us to examine the hydrogen atom from the electron's frame of reference, in which the proton *appears* to travel around the electron, just as the Sun *appears* to travel around the Earth. For convenience, we treat this problem in the context of the Bohr model to obtain an estimate of the effect.

Figure 7.23a shows the atom viewed from the ordinary frame of reference of the proton. We assume the electron to orbit counterclockwise so that the orbital angular momentum $\vec{\mathbf{L}}$ is in the z direction, and we also assume that the spin $\vec{\mathbf{S}}$ (which could point either up or down) is also in the z direction. The same situation is shown in Figure 7.23b from the viewpoint of the electron, with the proton now appearing to move in a circular orbit around the electron.

In the electron reference frame the motion of the proton in a circular orbit of radius r can be considered to be a current loop, which causes a magnetic field $\vec{\mathbf{B}}$ at the electron, as shown in Figure 7.23c. This magnetic field interacts with the spin magnetic moment of the electron, $\vec{\mu}_S = -(e/m)\vec{\mathbf{S}}$. The interaction energy of the magnetic moment $\vec{\mu}_S$ in a magnetic field is

$$U = -\vec{\mu}_S \cdot \vec{\mathbf{B}} \tag{7.30}$$

We choose the z direction to be the direction of $\vec{\mathbf{B}}$; with $\vec{\mu}_S = -(e/m)\vec{\mathbf{S}}$, we have

$$U = \frac{e}{m}\vec{\mathbf{S}} \cdot \vec{\mathbf{B}} = \frac{e}{m}S_z B \tag{7.31}$$

With $S_z = \pm\frac{1}{2}\hbar$, the energy is

$$U = \pm\frac{e\hbar}{2m}B = \pm\mu_B B \tag{7.32}$$

---

*This is an optional section that may be skipped without loss of continuity.

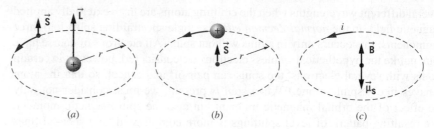

FIGURE 7.23 An electron circulates about the proton in a hydrogen atom. (b) From the point of view of the electron, the proton circulates about the electron. (c) The apparently circulating proton is represented by the current i and causes a magnetic field $\vec{B}$ at the location of the electron.

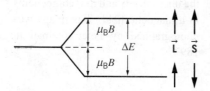

FIGURE 7.24 The fine-structure splitting in hydrogen. The state with $\vec{L}$ and $\vec{S}$ parallel is slightly higher in energy than the state with $\vec{L}$ and $\vec{S}$ antiparallel.

The situation shown in Figure 7.23 has $S_z = +\frac{1}{2}\hbar$, and thus $U = +\mu_B B$. When $\vec{S}$ has the opposite orientation, $U = -\mu_B B$. The effect is to split each level into two, a higher state with $\vec{L}$ and $\vec{S}$ parallel and a lower state with $\vec{L}$ and $\vec{S}$ antiparallel, as shown in Figure 7.24. The energy difference between the states is $\Delta E = 2\mu_B B$.

At this point, the result looks rather similar to that of our previous discussion of the Zeeman effect, but it is important to note one significant difference: the magnetic field B in this case is *not* a field in the laboratory that can be turned on or off; it is, instead, a field that is always present, produced by the relative motion between the proton and the electron.

We can use the Bohr model to make a rough estimate of the magnitude of this energy splitting. A circular loop of radius r carrying current i establishes at its center a magnetic field $B = \mu_0 i/2r$. The current i is the charge carried around the loop ($+e$ in this case) divided by the time T for one orbit. The time for one orbit is the distance traveled ($2\pi r$) divided by the speed v.

$$B = \frac{\mu_0 i}{2r} = \frac{\mu_0}{2r}\frac{e}{T} = \frac{\mu_0}{2r}\frac{ev}{2\pi r} \tag{7.33}$$

The energy difference between the states is then

$$\Delta E = 2\mu_B B = \frac{\mu_0 ev}{2\pi r^2}\mu_B = \frac{\mu_0 e^2 \hbar^2 n}{4\pi m^2 r^3} \tag{7.34}$$

where the last result is obtained by substituting $v = n\hbar/mr$ from Bohr's angular momentum condition (Eq. 6.26) and $\mu_B = e\hbar/2m$ from Eq. 7.23. Finally, substituting from Eq. 6.28 for the radius of the orbits in the Bohr atom, we obtain

$$\Delta E = \frac{\mu_0 e^2 \hbar^2 n}{4\pi m^2}\left(\frac{me^2}{4\pi\varepsilon_0 \hbar^2}\frac{1}{n^2}\right)^3 = \frac{\mu_0 me^8}{256\pi^4\varepsilon_0^3 \hbar^4}\frac{1}{n^5} \tag{7.35}$$

We can rewrite this in a somewhat simpler form by recalling that $c^2 = 1/\varepsilon_0\mu_0$ and using the dimensionless constant α, known as the *fine structure constant*,

$$\alpha = \frac{e^2}{4\pi\varepsilon_0 \hbar c} \tag{7.36}$$

which gives

$$\Delta E = mc^2 \alpha^4 \frac{1}{n^5} \tag{7.37}$$

The value of the fine structure constant is approximately 1/137. For hydrogen in the $n = 2$ level, we expect the energy difference between the state with $\vec{L}$ and $\vec{S}$ parallel and the state with $\vec{L}$ and $\vec{S}$ antiparallel to be

$$\Delta E = (0.511 \text{ MeV}) \left(\frac{1}{137}\right)^4 \frac{1}{2^5} = 4.53 \times 10^{-5} \text{ eV}$$

We can compare this estimate with the experimental value, based on the observed splitting of the first line of the Lyman series, which gives 4.54×10^{-5} eV. We see that in spite of the assumptions we have made, our use of the Bohr model, and our failure to use the hydrogen wave functions to do this calculation, the agreement with the experimental value is remarkably good. (In fact, the agreement is so good as to be embarrassing, for we neglected to consider the important *relativistic* effect of the motion of the electron, which contributes to the fine structure about equally to the *spin-orbit* interaction discussed in this section. We really should regard this calculation as an *order-of-magnitude* estimate, which happens by chance to give a numerical result close to the observed value.)

Chapter Summary

		Section			Section
Orbital angular momentum	$\|\vec{L}\| = \sqrt{l(l+1)}\hbar$ $(l = 0, 1, 2, \ldots)$	7.2	Angular probability density	$P(\theta, \phi) = \|\Theta_{l,m_l}(\theta)\Phi_{m_l}(\phi)\|^2$	7.5
Orbital magnetic quantum number	$L_z = m_l \hbar$ $(m_l = 0, \pm1, \pm2, \ldots, \pm l)$	7.2	Orbital magnetic dipole moment	$\vec{\mu}_L = -(e/2m)\vec{L}$	7.6
Spatial quantization	$\cos\theta = \dfrac{L_z}{\|\vec{L}\|} = \dfrac{m_l}{\sqrt{l(l+1)}}$	7.2	Spin magnetic dipole moment	$\vec{\mu}_S = -(e/m)\vec{S}$	7.6
Angular momentum uncertainty relationship	$\Delta L_z \Delta\phi \geqslant \hbar$	7.2	Spin angular momentum	$\|\vec{S}\| = \sqrt{s(s+1)}\hbar = \sqrt{3/4}\hbar$ (for $s = 1/2$)	7.6
Hydrogen quantum numbers	$n = 1, 2, 3, \ldots$ $l = 0, 1, 2, \ldots, n-1$ $m_l = 0, \pm1, \pm2, \ldots, \pm l$	7.3	Spin magnetic quantum number	$S_z = m_s \hbar \, (m_s = \pm 1/2)$	7.6
			Spectroscopic notation	$s\,(l=0), p(l=1), d(l=2),$ $f(l=3), \ldots$	7.7
Hydrogen energy levels	$E_n = -\dfrac{me^4}{32\pi^2\varepsilon_0^2\hbar^2}\dfrac{1}{n^2}$	7.3	Selection rules for photon emission	$\Delta l = \pm1 \quad \Delta m_l = 0, \pm1$	7.7, 7.8
Hydrogen wave functions	$\psi_{n,l,m_l}(r,\theta,\phi) =$ $R_{n,l}(r)\Theta_{l,m_l}(\theta)\Phi_{m_l}(\phi)$	7.3	Normal Zeeman effect	$\Delta\lambda = \dfrac{\lambda^2}{hc}\Delta E = \dfrac{\lambda^2}{hc}\mu_B B$	7.8
Radial probability density	$P(r) = r^2\|R_{n,l}(r)\|^2$	7.4	Fine-structure estimate	$\Delta E = mc^2\alpha^4/n^5 \quad (\alpha \approx 1/137)$	7.9

Questions

1. How does the quantum-mechanical interpretation of the hydrogen atom differ from the Bohr model?

2. How does a quantized angular momentum vector differ from a classical angular momentum vector?

3. What are the meanings of the quantum numbers n, l, m_l according to (a) the quantum-mechanical calculation; (b) the vector model; (c) the Bohr (orbital) model?

4. List the dynamical quantities that are constant for a specific choice of n and l. List the dynamical quantities that are *not* constant. Compare these lists with the Bohr model.

5. How does the orbital angular momentum differ between the Bohr model and the quantum-mechanical calculation?

6. What does it mean that $\vec{L}$ precesses about the z axis? Can we observe the precession?

7. In the Bohr model, we calculated the total energy from the potential energy and kinetic energy for each orbit. In the quantum-mechanical calculation, is the potential energy constant for any set of quantum numbers? Is the kinetic energy? Is the total energy?

8. What is meant by the term *spatial quantization*? Is space really quantized?

9. A deficiency of the Bohr model is the problem of angular momentum conservation in transitions between levels. Discuss this problem in relation to the quantum-mechanical angular momentum properties of the atom, especially the selection rule Eq. 7.24. The photon can be considered to carry angular momentum $\hbar$.

10. The $2s$ electron has a greater probability to be close to the nucleus than the $2p$ electron and also a greater probability to be farther away (see Figure 7.10). How is this possible?

11. The probability density $\psi^*\psi$ does not depend on ϕ for the wave functions listed in Table 7.1. What is the significance of this?

12. How would the wave functions of Table 7.1 change if the nuclear charge were Ze instead of e? (Recall how we made the same change in the Bohr model in Section 6.5.) What effect would this have on the radial probability densities $P(r)$?

13. Can a hydrogen atom in its ground state absorb a photon (of the proper energy) and end up in the $3d$ state?

14. Is it *correct* to think of the electron as a tiny ball of charge spinning on its axis? Is it *useful*? Is this situation similar to using the Bohr model to represent the electron's orbital motion?

15. The photon has a spin quantum number of 1, but its spin magnetic moment is zero. Explain.

16. What are the similarities and differences between Zeeman splitting and fine-structure splitting?

17. How would the calculated fine structure be different in an atom with a single electron and a nuclear charge of Ze?

18. Does the fine structure, as we have calculated it, have any effect on the $n = 1$ level?

19. How would (a) the Zeeman effect and (b) the fine structure be different in a muonic hydrogen atom? (See Problems 41 and 42 in Chapter 6.) The muon has the same spin as the electron, but is 207 times as massive.

20. Even though our calculation of the fine structure was based on a very simplified model, it does yield a result similar to the more correct calculation: the fine-structure splitting decreases as we go to higher excited states. Give at least two qualitative reasons for this.

Problems

7.1 A One-Dimensional Atom

1. By substituting the wave function $\psi(x) = Axe^{-bx}$ into Eq. 7.2, show that a solution can be obtained only for $b = 1/a_0$, and find the ground-state energy.

2. Show that the probability density for the ground-state solution of the one-dimensional Coulomb potential energy has its maximum at $x = a_0$.

3. An electron in its ground state is trapped in the one-dimensional Coulomb potential energy. What is the probability to find it in the region between $x = 0.99a_0$ and $x = 1.01a_0$?

7.2 Angular Momentum in the Hydrogen Atom

4. An electron is in an angular momentum state with $l = 3$. (a) What is the length of the electron's angular momentum vector? (b) How many different possible z components can the angular momentum vector have? List the possible z components. (c) What are the values of the angle that the $\vec{L}$ vector makes with the z axis?

5. What angles does the $\vec{L}$ vector make with the z axis when $l = 2$?

7.3 The Hydrogen Atom Wave Functions

6. List the 16 possible sets of quantum numbers n, l, m_l of the $n = 4$ level of hydrogen (as in Figure 7.6).

7. (a) What are the possible values of l for $n = 6$? (b) What are the possible values of m_l for $l = 6$? (c) What is the smallest possible value of n for which l can be 4? (d) What is the smallest possible l that can have a z component of $4\hbar$?

8. Show that the (1, 0, 0) and (2, 0, 0) wave functions listed in Table 7.1 are properly normalized.

9. Show by direct substitution that the $n = 2, l = 0, m_l = 0$ and $n = 2, l = 1, m_l = 0$ wave functions of Table 7.1 are both solutions of Eq. 7.10 corresponding to the energy of the first excited state of hydrogen.

10. Show by direct substitution that the wave function corresponding to $n = 1, l = 0, m_l = 0$ is a solution of Eq. 7.10 corresponding to the ground-state energy of hydrogen.

11. Consider a thin spherical shell located between $r = 0.49a_0$ and $0.51a_0$. For the $n = 2, l = 1$ state of hydrogen, find the probability for the electron to be found in a small volume element that subtends a polar angle of $0.11°$ and an azimuthal angle of $0.25°$ if the center of the volume element is located at: (a) $\theta = 0, \phi = 0$; (b) $\theta = 90°, \phi = 0$; (c) $\theta = 90°, \phi = 90°$; (d) $\theta = 45°, \phi = 0$. Do the calculation for all possible m_l values.

7.4 Radial Probability Densities

12. Show that the radial probability density of the $1s$ level has its maximum value at $r = a_0$.

13. Find the values of the radius where the $n = 2, l = 0$ radial probability density has its maximum values.

14. What is the probability of finding a $n = 2, l = 1$ electron between a_0 and $2a_0$?

15. For a hydrogen atom in the ground state, what is the probability to find the electron between $1.00a_0$ and $1.01a_0$? (Hint: It is not necessary to evaluate any integrals to solve this problem.)

7.5 Angular Probability Densities

16. Find the directions in space where the angular probability density for the $l = 2, m_l = \pm1$ electron in hydrogen has its maxima and minima.

17. Find the directions in space where the angular probability density for the $l = 2, m_l = 0$ electron in hydrogen has its maxima and minima.

7.6 Intrinsic Spin

18. (a) Including the electron spin, what is the degeneracy of the $n = 5$ energy level of hydrogen? (b) By adding up the number of states for each value of l permitted for $n = 5$, show that the same degeneracy as part (a) is obtained.

19. For each l value, the number of possible states is $2(2l + 1)$. Show explicitly that the total number of states for each principal quantum number is $\sum_{l=0}^{n-1} 2(2l + 1) = 2n^2$. This gives the degeneracy of each energy level.

20. Explain why each of the following sets of quantum numbers (n, l, m_l, m_s) is not permitted for hydrogen. (a) $(2, 2, -1, +1/2)$ (b) $(3, 1, +2, -1/2)$ (c) $(4, 1, +1, -3/2)$ (d) $(2, -1, +1, +1/2)$

7.7 Energy Levels and Spectroscopic Notation

21. List the excited states (in spectroscopic notation) to which the $4p$ state can make downward transitions.

22. (a) A hydrogen atom is in an excited $5g$ state, from which it makes a series of transitions by emitting photons, ending in the $1s$ state. Show, on a diagram similar to Figure 7.19, the sequence of transitions that can occur. (b) Repeat part (a) if the atom begins in the $5d$ state.

23. (a) List in spectroscopic notation all levels with $n = 7$. (b) An electron is initially in the state with $n = 7, l = 2$. List in spectroscopic notation all lower states to which transitions are allowed.

7.8 The Zeeman Effect

24. Consider the normal Zeeman effect applied to the $3d$ to $2p$ transition. (a) Sketch an energy-level diagram that shows the splitting of the $3d$ and $2p$ levels in an external magnetic field. Indicate all possible transitions from each m_l state of the $3d$ level to each m_l state of the $2p$ level. (b) Which transitions satisfy the $\Delta m_l = \pm1$ or 0 selection rule? (c) Show that there are only three different transition energies emitted.

25. A collection of hydrogen atoms is placed in a magnetic field of 3.50 T. Ignoring the effects of electron spin, find the wavelengths of the three normal Zeeman components (a) of the $3d$ to $2p$ transition; (b) of the $3s$ to $2p$ transition.

7.9 Fine Structure

26. Calculate the wavelengths of the components of the first line of the Lyman series, taking the fine structure of the $2p$ level into account.

27. Calculate the energies and wavelengths of the $3d$ to $2p$ transition, taking into account the fine structure of both levels. How many component wavelengths might there be in the transition?

General Problems

28. Show that the wave function $\psi(x) = A(x + cx^2)e^{-bx}$ gives a solution to the Schrödinger equation for the one-dimensional Coulomb potential energy. Evaluate the constants A, b, c, and find the energy corresponding to this solution.

29. Find the probabilities for the $n = 2, l = 0$ and $n = 2, l = 1$ electron states in hydrogen to be further than $r = 5a_0$ from the nucleus. Which has the greater probability to be far from the nucleus?

30. The mean or average value of the radius r can be found according to $r_{av} = \int_0^\infty rP(r)dr$. Show that the mean value of r for the $1s$ state of hydrogen is $\frac{3}{2}a_0$. Why is this greater than the Bohr radius?

31. Find the value of r_{av} (see Problem 30) for the $2s$ and $2p$ levels.

32. The mean or average value of the potential energy of the electron in a hydrogen atom can be found from

$U_{av} = \int_0^\infty U(r)P(r)dr$. Find U_{av} in the $1s$ state and compare with the potential energy computed with the Bohr model when $n = 1$.

33. Suppose the source of atoms in a Stern-Gerlach experiment were an oven of temperature 1000 K. Assume the magnetic field gradient to be 10 T/m, and take the length of the magnetic field region and the field-free region between magnet and screen to be 1 m each. Make any other assumptions you may need and estimate the separation of the images observed on the screen.

34. For the $1s, 2s$, and $2p$ states of hydrogen, show that $(r^{-1})_{av} = 1/n^2a_0$. This turns out to be a general result for any state of hydrogen. Based on this result, explain why the Bohr model gives such a good estimate for the fine-structure splitting as well as for other magnetic effects due to the circulating electron.

Chapter 8

MANY-ELECTRON ATOMS

This computer-generated drawing shows the structure of an atom of neon, with the electron probability distributions surrounding the central nucleus. The bright inner sphere represents the *1s* electrons, the dark outer sphere is the *2s* electrons, and the lobes are the *2p* electrons. This is a more realistic picture of an atom than the "planetary" view developed in Chapter 6.

Physicists often attack complex problems by trying to separate the more important parts from the less important. For example, in analyzing the motion of the Earth in the Solar System, we can start by ignoring all bodies other than the Sun. With this simplification, we find that the Earth moves about the Sun in an elliptical orbit. Now we can account for the effect of the Moon, which introduces a slight "wobble" about the ellipse. Finally, we can introduce the much weaker effect of the gravitational pull of the other planets.

It is tempting to try to use a similar approach to understand the motion of electrons in atoms with more than one electron. Unfortunately, we can't analyze the motion of an electron in an atom with more than one electron by separating out the more and less influential forces. For example, in a neutral atom with atomic number Z, each electron experiences an electrostatic force due to the nucleus with a charge of $+Ze$, but it also experiences an electrostatic force due to all the other electrons with a total charge of $-(Z-1)e$. The effect of the nucleus is comparable to the effect of the other electrons, which can't be analyzed as a small correction.

We are thus required to consider simultaneously the effect of the nucleus and each of the other electrons. The problem of the mutual interactions of three or more objects is an example of what physicists call the *many-body problem*. Exact, closed-form solutions to the Schrödinger equation cannot be found for such problems. The solutions must be obtained numerically using a computer. In this chapter, we consider an approximate set of energy levels for many-electron atoms, and we try to understand some of the properties of atoms (chemical, electrical, magnetic, optical, etc.) based on those energy levels.

8.1 THE PAULI EXCLUSION PRINCIPLE

Wolfgang Pauli (1900–1958, Switzerland). His exclusion principle gave the basis for understanding atomic structure. He also contributed to the development of quantum theory, to the theory of nuclear beta decay, and to the understanding of symmetry in physical laws.

Let's begin by considering how the Z electrons in an atom might occupy the atomic energy levels. As a first guess, we might expect that all Z electrons will eventually cascade down to the lowest energy level, the $1s$ state. If this were correct, we would expect the properties of the atom to vary rather smoothly compared with its neighbors having $Z \pm 1$ electrons. Indeed, certain of the properties of atoms, such as the energies of the emitted X rays, show this smooth variation. However, other properties do not vary in this way and thus are not consistent with this model of all electrons in the same level. For example, neon (with $Z = 10$) is an *inert gas*; it is practically unreactive and does not form chemical compounds under most conditions. Its neighbors, fluorine ($Z = 9$) and sodium ($Z = 11$), are among the most reactive of the elements and under most conditions will combine with other substances, sometimes violently. As another example, nickel ($Z = 28$) is strongly magnetic (ferromagnetic) and, for a metal, does not have a particularly large electrical conductivity. Copper ($Z = 29$) is an excellent electrical conductor but is not magnetic. Such wide variations in properties between neighboring elements suggest that it is not correct to assume that all electrons occupy the same energy level.

The rule that prevents all of the electrons in an atom from falling into the $1s$ level was proposed by Wolfgang Pauli in 1925, based on a study of the transitions that are present, and those that are expected but *not* present, in the emission spectra of atoms. Simply stated, the *Pauli exclusion principle* is as follows:

No two electrons in a single atom can have the same set of quantum numbers (n, l, m_l, m_s).

The Pauli principle is the most important rule governing the structure of atoms, and no study of the properties of atoms can be attempted without a thorough understanding of this principle.

To illustrate how the Pauli principle works, consider the structure of helium ($Z = 2$). The first electron in helium, in the $1s$ ground state, has quantum numbers $n = 1, l = 0, m_l = 0, m_s = +\frac{1}{2}$ or $-\frac{1}{2}$. The second electron can have the same $n, l,$ and m_l, but it cannot have the same m_s, because the exclusion principle would be violated. Thus if the first $1s$ electron has $m_s = +\frac{1}{2}$, the second $1s$ electron must have $m_s = -\frac{1}{2}$. Now consider an atom of lithium ($Z = 3$). Just as with helium, the first two electrons will have quantum numbers $(n, l, m_l, m_s) = (1, 0, 0, +\frac{1}{2})$ and $(1, 0, 0, -\frac{1}{2})$. According to the exclusion principle, the third electron cannot have the same set of quantum numbers as the first two, so it *cannot go into the n = 1 level,* because there are only two different sets of quantum numbers available in the $n = 1$ level, and both of those sets have already been used. The third electron must therefore go into one of the $n = 2$ levels, and experiments indicate that the $2s$ level is the next available. Without the Pauli principle, lithium would have three electrons in the $1s$ level; with the Pauli principle, we expect that lithium has two electrons in the $1s$ level and one electron in the $2s$ level. These two different possible structures for lithium would give very different physical properties, and the physical properties of lithium indicate that the structure with one electron in the $2s$ level is the correct one.

We can continue this process with beryllium ($Z = 4$). The fourth electron can join the third electron in the $2s$ level, but that now completes the capacity of the $2s$ level—one of the electrons might have quantum numbers $(n, l, m_l, m_s) = (2, 0, 0, +\frac{1}{2})$ and the other might have $(2, 0, 0, -\frac{1}{2})$. There are no other sets of quantum numbers that an additional electron could have in the $2s$ level without duplicating one of the sets that has already been assigned and thus violating the Pauli principle. When we reach boron, with $Z = 5$, the fifth electron must go into a different level—one of the $2p$ levels. We might therefore expect that the properties of boron, with a $2p$ electron, would be different from the properties of lithium or beryllium, which have only $2s$ electrons.

It is this process of first using up all of the possible quantum numbers for one level, and then placing electrons in the next level, that accounts for the variations in the chemical and physical properties of the elements.

Example 8.1

A certain atom has six electrons in the $3d$ level. (a) What is the maximum possible total m_l for the six electrons, and what is the total m_s in that configuration? (b) What is the maximum possible total m_s for the six electrons, and what would be the largest possible total m_l in that configuration?

Solution

(a) For a d state $l = 2$, so the possible m_l values are $+2, +1, 0, -1,$ and -2. At most two electrons can be assigned m_l of $+2$ according to the Pauli principle (one with $m_s = +\frac{1}{2}$ and one with $m_s = -\frac{1}{2}$). Similarly, two electrons can be assigned m_l of $+1$ (again, with $m_s = +\frac{1}{2}$ and $m_s = -\frac{1}{2}$), and the remaining two electrons can be assigned to m_l of 0. That gives a total m_l of $+6$, with a total m_s of 0.

(b) To maximize m_s, we can assign at most five electrons to $m_s = +\frac{1}{2}$ (with corresponding m_l values of $+2, +1, 0, -1,$ and -2). The sixth electron cannot also have $m_s = +\frac{1}{2}$, because its m_l value would be the same as one already assigned, which would violate the Pauli principle by having two electrons with the same m_l and m_s labels. The sixth electron must therefore have $m_s = -\frac{1}{2}$, giving a total m_s of $+2$. The first five electrons give a total m_l of 0, so the largest total m_l would be obtained by assigning the sixth electron to m_l of $+2$, giving a total m_l of $+2$.

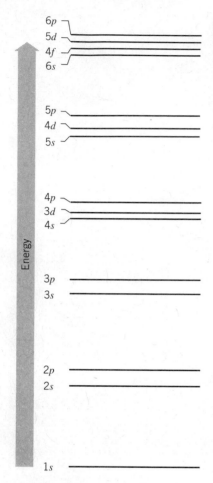

FIGURE 8.1 Atomic subshells, in order of increasing energy. The energy groupings are not to scale, but represent the relative energies of the subshells.

8.2 ELECTRONIC STATES IN MANY-ELECTRON ATOMS

Figure 8.1 illustrates the result of an approximate calculation of the order of the filling of energy levels in many-electron atoms as the atomic number Z increases. The $1s$ level is always the lowest energy level to be filled, and the $2s$ and $2p$ levels are fairly close in energy. The $2s$ level always lies a bit lower in energy than the $2p$ level, and so the $2s$ level is filled before the $2p$. (The fine-structure splitting is very small on the scale of this diagram.) We can understand why the $2s$ level lies lower in energy if we recall Example 7.6 and Figure 7.10. An electron in the $2s$ level has a greater probability to be found at small radii compared with an electron in the $2p$ level. (Penetrating close to the nucleus, the $2s$ electron also is attracted by the full nuclear charge $+Ze$, while the $2p$ electron spends most of its time beyond the orbits of the $1s$ electrons where it is attracted by an effective charge that is less than the full charge of the nucleus. We'll discuss this effect, which is called *electron screening*, in Section 8.3.) These two effects—closer penetration to the nucleus and screening—are responsible for the tighter binding of the $2s$ electrons compared with the $2p$ electrons.

A more extreme example of the *tighter binding* of the *penetrating orbits* occurs for the $n = 3$ levels. The $3s$ electron penetrates the inner orbits (it has a large probability density at small r; see Figure 7.10), and the $3p$ electron penetrates almost as much. The $3d$ electron has negligible penetration of the inner orbits. As a result, the $3s$ and $3p$ levels are more tightly bound and therefore lower in energy than the $3d$ level. A similar effect occurs for the $n = 4$ levels—the tighter binding of the $4s$ and $4p$ electrons pulls their energy levels down so low that they almost coincide with the $3d$ level, as shown in Figure 8.1. The $3d$ and $4s$ levels are very close in energy—for some atoms the $3d$ level is lower and for some atoms the $4s$ is lower. This small energy difference is an important factor that contributes to the large electrical conductivity of copper, as we discuss later in this chapter.

The tighter binding of the penetrating s and p orbits also pulls the $5s$ and $5p$ levels down close to the $4d$ level, and similarly causes the $6s$ and $6p$ levels to appear at roughly the same energy as the $5d$ and $4f$ levels.

As we learned in the case of the hydrogen atom, orbits with the same value of n all lie at about the same average distance from the nucleus. (The electrons in the penetrating orbits spend some of their time closer to the nucleus than the nonpenetrating orbits, but also some of their time further from the nucleus; the average distance from the nucleus of the penetrating orbits is then about the same as the average distance from the nucleus of the nonpenetrating orbits with the same value of n. See Problem 31 in Chapter 7 for a verification of this property for the hydrogen atom.) The set of orbits with a certain value of n, with about the same average distance from the nucleus, is known as an atomic *shell*. The atomic shells are designated by letter, as follows:

n	1	2	3	4	5
Shell	K	L	M	N	O

The levels with a certain value of n and l (for instance, $2s$ or $3d$) are known as *subshells*. According to the Pauli principle, the maximum number of electrons that can be placed in each subshell is $2(2l + 1)$. The $(2l + 1)$ factor comes from the number of different m_l values for each l, because m_l can take the values $0, \pm1, \pm2, \pm3, \ldots, \pm l$. The extra factor of 2 comes from the two different m_s

values; for each m_l, we can have $m_s = +1/2$ or $m_s = -1/2$. According to this scheme, the $1s$ subshell has a capacity of $2(2 \times 0 + 1) = 2$ electrons; the $3d$ subshell has a capacity of $2(2 \times 2 + 1) = 10$ electrons. (Note that this capacity doesn't depend on n; any d subshell has a capacity of 10 electrons.) Table 8.1 shows the ordering and capacity of the subshells.

It is important to keep in mind exactly what is represented by Figure 8.1 and Table 8.1. They give the order of filling of the energy levels, and so they represent only the "outer" or valence electrons. For example, the first 18 electrons fill the levels up through $3p$, and the energy levels (subshells) available to the 19th electron in potassium ($Z = 19$) or calcium ($Z = 20$) are well described by Figure 8.1. However, the energy levels appropriate to the 19th electron in a heavy element such as lead ($Z = 82$) would be very different. In this case it is more correct to describe the atom in terms of shells—all of the $n = 3$ states (the M shell) are grouped together, as are all of the $n = 4$ states (the N shell), and so forth. When we discuss the *inner* structure of the atom, as in the case of X rays, the ordering of Figure 8.1 is not appropriate, and it is more appropriate to group the levels by shells, as we do in Section 8.5.

The Periodic Table

Figure 8.2 shows the periodic table, which is an orderly array of the chemical elements, listed in order of increasing atomic number Z and arranged in such a way that the vertical columns, called *groups,* contain elements with rather similar physical and chemical properties. In this section we discuss the way in which the filling of electronic subshells helps us understand the arrangement of the periodic table. In later sections we examine some of the physical and chemical properties of the elements.

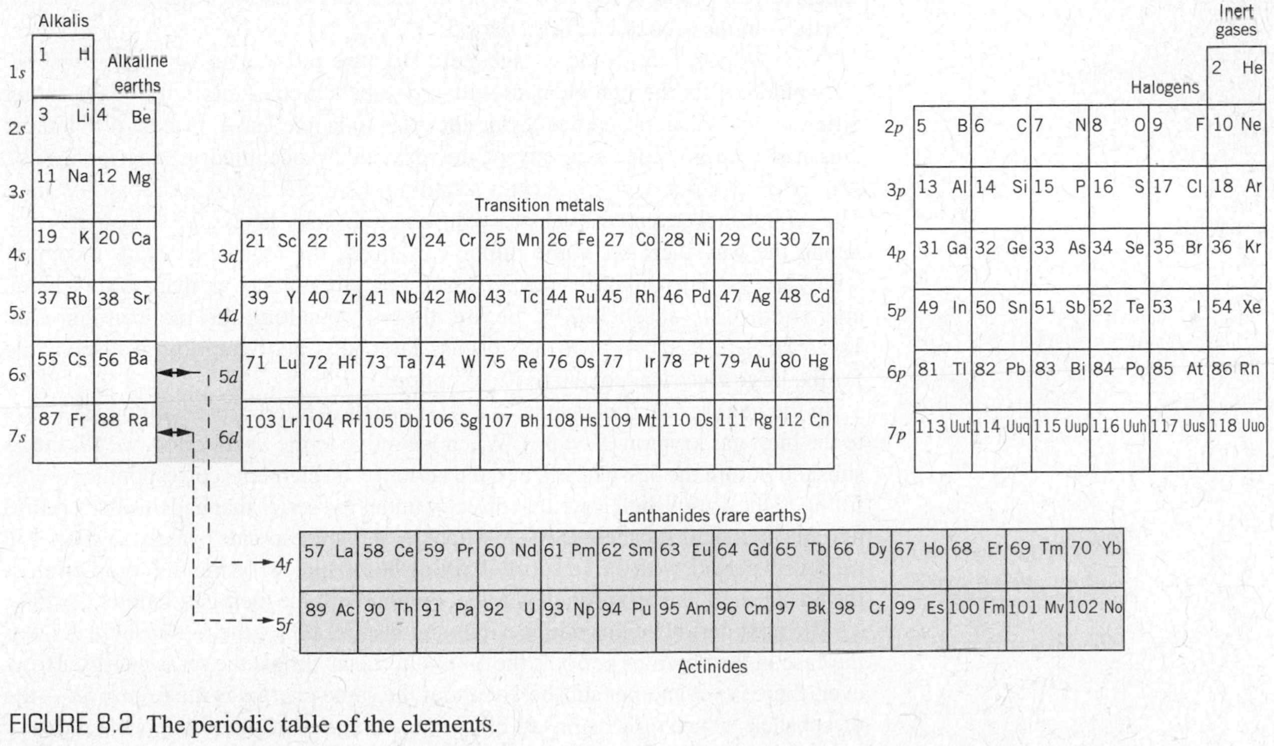

TABLE 8.1 Filling of Atomic Subshells

n	l	Subshell	Capacity $2(2l+1)$
1	0	$1s$	2
2	0	$2s$	2
2	1	$2p$	6
3	0	$3s$	2
3	1	$3p$	6
4	0	$4s$	2
3	2	$3d$	10
4	1	$4p$	6
5	0	$5s$	2
4	2	$4d$	10
5	1	$5p$	6
6	0	$6s$	2
4	3	$4f$	14
5	2	$5d$	10
6	1	$6p$	6
7	0	$7s$	2
5	3	$5f$	14
6	2	$6d$	10

FIGURE 8.2 The periodic table of the elements.

In attempting to understand the ordering of subshells and the periodic table, we must follow two rules for filling the electronic subshells:

1. The capacity of each subshell is $2(2l+1)$. (This is of course just another way of stating the Pauli exclusion principle.)
2. The electrons occupy the lowest energy states available.

To indicate the electron configuration of each element, we use a notation in which the identity of the subshell and the number of electrons in it are listed. The identity of the subshell is indicated in the usual way, and the number of electrons in that subshell is indicated by a superscript. Thus hydrogen has the configuration $1s^1$, for one electron in the $1s$ shell, and helium has the configuration $1s^2$. Helium has both a filled subshell (the $1s$) and a closed major shell (the K shell) and thus is an extraordinarily stable and inert element. With lithium ($Z = 3$), we begin to fill the $2s$ subshell; lithium has the configuration $1s^2 2s^1$. With beryllium ($Z = 4, 1s^2 2s^2$) the $2s$ subshell is full, and the next element must begin filling the $2p$ subshell (boron, $Z = 5, 1s^2 2s^2 2p^1$). The $2p$ subshell has a capacity of six electrons, and with neon ($Z = 10, 1s^2 2s^2 2p^6$) both the $2p$ subshell and the L shell ($n = 2$) are complete.

The next row (or *period*) begins with sodium ($Z = 11, 1s^2 2s^2 2p^6 3s^1$), and the $3s$ and $3p$ subshells are filled in much the same way as the $2s$ and $2p$ subshells, ending with the inert gas argon ($Z = 18, 1s^2 2s^2 2p^6 3s^2 3p^6$). The elements of the third row (period) are chemically similar to the corresponding elements of the second row (period), and so are written directly under them. The next electron might be expected to go into the $3d$ level. However, the highly penetrating orbit of the $4s$ electron causes the $4s$ level to appear at a slightly lower energy than the $3d$ level, so the $4s$ subshell normally fills first. The configurations of potassium ($Z = 19$) and calcium ($Z = 20$) are therefore respectively $1s^2 2s^2 2p^6 3s^2 3p^6 4s^1$ and $1s^2 2s^2 2p^6 3s^2 3p^6 4s^2$. These elements have properties similar to, and therefore appear directly under, the corresponding elements with one and two s-subshell electrons in the second and third periods.

We now begin to fill the $3d$ subshell. Because there is no $1d$ or $2d$ subshell, we would expect the first element with a d-subshell configuration to have rather different chemical properties from the elements we have placed previously; thus it should not appear in any of our previously occupied groups (columns), and so we begin a new group with scandium ($Z = 21, 1s^2 2s^2 2p^6 3s^2 3p^6 4s^2 3d^1$). The $3d$ subshell eventually closes with zinc ($Z = 30, 1s^2 2s^2 2p^6 3s^2 3p^6 4s^2 3d^{10}$). Along the way there are some minor variations; the most important is copper, with $Z = 29$. For this case the $3d$ level lies slightly lower than the $4s$ level, and so the $3d$ subshell fills before the $4s$, resulting in the configuration $1s^2 2s^2 2p^6 3s^2 3p^6 3d^{10} 4s^1$. As we discuss later, this configuration is responsible for the large electrical conductivity of copper.

In the next series of elements, the $4p$ subshell is filled, from gallium ($Z = 31$) to the inert gas krypton ($Z = 36$). When we move to the next period, we fill the $5s$ subshell before the $4d$ subshell, and the series of 10 elements corresponding to the filling of the $4d$ subshell is written directly under the series that had unfilled configurations in the $3d$ subshell. (Silver, with $Z = 47$, corresponds exactly to copper in the fourth period, with the $4d$ subshell filling before the $5s$.) After the completion of the $4d$ subshell, the $5p$ subshell is filled, ending with the inert gas xenon ($Z = 54$).

The next period begins with cesium and barium filling the $6s$ subshell. As was the case in the previous periods, the $5d$ and $6s$ lie at almost the same energy. However, there is yet another subshell at about the same energy as the $6s$ and $5d$—the $4f$ subshell, which now begins to fill, from lanthanum to ytterbium. This series of

TABLE 8.2 Electronic Configurations of Some Elements

H	$1s^1$	Mn	$[Ar]4s^23d^5$	La	$[Xe]6s^25d^1$
He	$1s^2$	Cu	$[Ar]4s^13d^{10}$	Ce	$[Xe]6s^25d^14f^1$
Li	$1s^22s^1$	Zn	$[Ar]4s^23d^{10}$	Pr	$[Xe]6s^24f^3$
Be	$1s^22s^2$	Ga	$[Ar]4s^23d^{10}4p^1$	Gd	$[Xe]6s^25d^14f^7$
B	$1s^22s^22p^1$	Kr	$[Ar]4s^23d^{10}4p^6$	Dy	$[Xe]6s^24f^{10}$
Ne	$1s^22s^22p^6$	Rb	$[Kr]5s^1$	Yb	$[Xe]6s^24f^{14}$
Na	$[Ne]3s^1$	Y	$[Kr]5s^24d^1$	Lu	$[Xe]6s^25d^14f^{14}$
Al	$[Ne]3s^23p^1$	Mo	$[Kr]5s^14d^5$	Re	$[Xe]6s^25d^54f^{14}$
Ar	$[Ne]3s^23p^6$	Ag	$[Kr]5s^14d^{10}$	Au	$[Xe]6s^15d^{10}4f^{14}$
K	$[Ar]4s^1$	In	$[Kr]5s^24d^{10}5p^1$	Hg	$[Xe]6s^25d^{10}4f^{14}$
Sc	$[Ar]4s^23d^1$	Xe	$[Kr]5s^24d^{10}5p^6$	Tl	$[Xe]6s^25d^{10}4f^{14}6p^1$
Cr	$[Ar]4s^13d^5$	Cs	$[Xe]6s^1$	Rn	$[Xe]6s^25d^{10}4f^{14}6p^6$

A symbol in brackets [ ] means that the atom has the configuration of the previous inert gas plus the additional electrons listed.

elements, called the *lanthanides* or *rare earths*, is usually written separately in the periodic table, because there have been no other f-subshell elements under which to write them. The $4f$ subshell has a capacity of 14 electrons, and so there are 14 elements in the lanthanide series. Once the $4f$ subshell is complete, we return to filling the $5d$ subshell, writing those elements in the groups under the corresponding $3d$ and $4d$ elements, and then complete the period with the filling of the $6p$ subshell, ending with the inert gas radon ($Z = 86$). The seventh period is filled much like the sixth, with a series known as the actinides, written under the lanthanides, corresponding to the filling of the $5f$ subshell.

What is most remarkable about this scheme is that the arrangement of the periodic table was known well before the introduction of atomic theory. The elements were organized into groups and periods based on their physical and chemical properties by Dmitri Mendeleev in 1859; understanding that organization in terms of atomic levels is a great triumph for the atomic theory. This way of organizing the elements gives us great insight into their physical and chemical properties, as we discuss in the next sections.

Table 8.2 lists the electronic configurations of some of the elements.

Example 8.2

Copper has the electronic configuration $[Ar]4s^13d^{10}$ in its ground state. By adding a small amount of energy (about 1 eV) to a copper atom, it is possible to move one of the $3d$ electrons to the $4s$ level and change the configuration to $[Ar]4s^23d^9$. By adding still more energy (about 5 eV), one of the $3d$ electrons can be moved to the $4p$ level so that the configuration becomes $[Ar]4s^13d^94p^1$. For each of

these configurations, determine the maximum value of the total m_s of the electrons.

Solution

The electrons in the filled shells (Ar core) have a total m_s of zero. In fact, any filled subshell has equal numbers of electrons in $m_s = +1/2$ and $m_s = -1/2$ states, which also gives a total of zero. In the $4s^13d^{10}$ configuration, only the single

$4s$ electron contributes to m_s, and its maximum value is $+1/2$. In the $4s^2 3d^9$ configuration, the two $4s$ electrons give a total m_s of zero. In the $3d$ subshell, there are 5 different m_l values, so we can have at most 5 electrons with $m_s = +1/2$. The remaining 4 electrons must have $m_s = -1/2$, so we have a total m_s of $5 \times (+1/2) + 4 \times (-1/2) = +1/2$. In the $4s^1 3d^9 4p^1$ configuration, each of the three subshells contributes a maximum m_s of $+1/2$, so the maximum total m_s is $+3/2$.

8.3 OUTER ELECTRONS: SCREENING AND OPTICAL TRANSITIONS

The electronic configurations of the alkali elements (those in the first column of the periodic table) all show a single s electron outside an inert gas core. These elements are very reactive, meaning they can easily give up the s electron to another element to form a chemical bond. For example, lithium ($1s^2 2s^1$) readily gives up its $2s$ electron to form the positive ion Li^+.

It may at first seem somewhat surprising that Li gives up its electron so easily. The ionization energy of Li is 5.39 eV. This is *smaller* than the ionization energy of hydrogen (13.6 eV), even though from Eq. 6.38 we might expect that the energies of electrons in atoms should increase in proportion to Z^2.

We can understand this effect from the diagram of Figure 8.3. The lithium atom can be roughly characterized by an inner atomic shell consisting of two $1s$ electrons and a single electron in the $2s$ subshell. As was the case in the one-electron atoms we considered in Chapters 6 and 7, the principal quantum number n determines the average distance of an electron from the nucleus. Although there is no simple formula that allows us to calculate the average orbital radius in atoms with more than one electron, it is certainly reasonable to expect that the $2s$ electron is most likely to be found much farther from the nucleus than the $1s$ electrons.

The net electric force on the $2s$ electron can be estimated using Gauss's law. Imagine a spherical surface centered at the nucleus having a radius equal to the average orbital radius of the $2s$ electron. The electric field at that distance is determined, according to Gauss's law, by the net charge contained within the sphere. The electrons in the $n = 1$ orbit have nearly a 100% probability of being found within the sphere. Thus the net charge inside the sphere must include the nucleus ($+3e$) and the two $n = 1$ electrons ($-2e$) for a total net charge of $+e$. To a good approximation, for some applications a lithium atom looks very much like a one-electron atom with the electron in the $n = 2$ orbit about a nucleus with an effective charge of $+e$. (Recall from electrostatics that if the charge distribution is spherically symmetric, we can replace an extended charge distribution with a point charge at the center of the sphere.) Equation 6.38 gives the energy of such an electron in the $n = 2$ orbit in an atom with an effective nuclear charge of $Z_{eff}e = +e$ as

$$E_n = (-13.6\,\text{eV}) \frac{Z_{eff}^2}{n^2} = -3.40\,\text{eV} \qquad (8.1)$$

This simple model predicts that the ionization energy of a neutral lithium atom is 3.40 eV. The measured value is 5.39 eV. The agreement is not extremely good, but the estimated value is off by much less than a factor of $Z^2 = 9$, so the calculation is probably on the right track.

The difference between the measured and estimated values can be accounted for by an effect that we have already discussed: the penetration of the s electrons through the inner shells to be occasionally found close to the nucleus. The $2s$

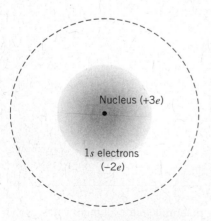

FIGURE 8.3　Electron structure in lithium, as might be seen from the average location of an outer ($2s$) electron. The dashed line represents a spherical Gaussian surface at that location.

electron sometimes finds itself much closer to the nucleus than its average orbital radius, and may occasionally be *inside* the $n = 1$ shell. In this case Gauss's law tells us that the electron feels the full $+3e$ charge of the nucleus, which results in an increase in the binding energy.

Let's instead consider an excited state of lithium, in which the $2s$ electron moves to the $2p$ state. The $2p$ electron penetrates the inner shell hardly at all. The energy of the $2p$ electron in lithium is -3.54 eV, in almost exact agreement with the prediction of our simple model. The small discrepancy might indicate a small degree of penetration of the $2p$ electron inside some of the $1s$ probability distribution, which gives a small increase to the binding energy. If we instead move the outer electron to the $3d$ state, the measured energy is -1.51 eV, in exact agreement with the prediction of Eq. 8.1 for $n = 3$. The $3d$ electron has almost no penetration inside the $1s$ shell, and so that electron is very well described by $Z_{eff} = 1$.

This effect is called *electron screening*. To an outer electron, the charge of the nucleus can be screened or shielded by the electrons in the inner shells. This is one case in which the formulas we derived for the energies of a one-electron atom can be used to determine approximately the energy of an electron in an atom with more than one electron. For the outer electron in lithium, the 3 positive charges in the nucleus are screened by the negative charges of the two inner electrons, giving a net charge of one unit. The less penetrating is the orbit of the outer electron, the more accurate is the prediction of Eq. 8.1. In lithium, for example, the $3d$ orbit has almost no penetration of the inner shells and so the formula gives a very accurate representation of the binding of that electron. The $2p$ orbit in lithium has relatively little penetration, so again the approximate formula gives a good prediction. It is less accurate for the $2s$ electron, which does occasionally penetrate through the inner $1s$ orbits.

Electron screening can also be used in a qualitative way to help understand the ionization energies of atoms. Consider helium, for example. In ionized helium, the single electron has an energy of -54.4 eV in its ground state. If we add a second electron to make neutral helium (with both electrons in the $1s$ state), the ionization energy is 24.6 eV. The screening of one electron by a portion of the probability distribution of the other is responsible for reducing the ionization energy from 54.4 eV when no second electron is present to 24.6 eV when the second electron is present.

Example 8.3

The ground state of helium has the configuration $1s^2$. Use the electron screening model to predict the energies of the following excited states of helium: (a) $1s^1 2s^1$ (measured value -4.0 eV); (b) $1s^1 2p^1$ (-3.4 eV); (c) $1s^1 3d^1$ (-1.5 eV).

Solution

(a) For the outer electron in helium, the nuclear charge of $+2e$ is screened by the single $1s$ electron, so the effective charge seen by the outer electron is $+e$. From Eq. 8.1, we have

$$E_n = (-13.6 \, \text{eV}) \frac{Z_{eff}^2}{n^2} = (-13.6 \, \text{eV}) \frac{1^2}{2^2} = -3.4 \, \text{eV}$$

The measured value is -4.0 eV, suggesting that the $2s$ electron has a small penetration through the $1s$ distribution

and thus experiences a somewhat tighter binding than this simple model predicts.

(b) Because Eq. 8.1 depends on n but not l, the calculation for the $2p$ excited state gives the same result as the calculation for the $2s$ excited state (-3.4 eV). Now the agreement is almost exact, because the $2p$ has less penetration than the $2s$.

(c) For the $3d$ excited state, Eq. 8.1 gives

$$E_n = (-13.6 \, \text{eV}) \frac{Z_{eff}^2}{n^2} = (-13.6 \, \text{eV}) \frac{1^2}{3^2} = -1.5 \, \text{eV}$$

The agreement is again very good, suggesting little penetration of the $3d$ electron inside the $1s$ probability distribution.

Optical Transitions

When we excite one of the *outer* electrons to a higher energy level or remove it completely from the atom, the resulting vacancy can be filled by electrons dropping into the empty state. The energy lost by these electrons usually appears as emitted photons, which are in the visible range of the spectrum and are thus known as *optical* transitions. The binding energies of the outer electrons in a typical atom are of the order of several electron-volts, and so it takes relatively little energy to move an outer electron and produce an optical transition. In fact, it is the absorption and reemission of light by these outer electrons that are responsible for the colors of material objects (although in solids the electron energy levels are usually very different from those in isolated atoms). In contrast with X-ray spectra, which vary slowly and smoothly from one element to the next, optical spectra can show large variations between neighboring elements, especially those that correspond to filled subshells.

Beyond hydrogen, the simplest energy-level diagrams to understand are those of the alkali metals, which have a single s electron outside an inert core. Many of the excited states then correspond to the excitation of this single electron, and the resulting spectra are very similar to the spectrum of hydrogen, because the nuclear charge of $+Ze$ is screened by the other $(Z-1)$ electrons. Figure 8.4 shows the energy levels of Li and Na along with some of the emitted transitions, which follow the same $\Delta l = \pm 1$ selection rule as the transitions in hydrogen (see Figure 7.19).

The ground-state configuration of lithium is $1s^2 2s^1$ and the ground-state configuration of sodium is $1s^2 2s^2 2p^6 3s^1$. The excited states in both cases can be obtained by moving the outer electron to a higher state. For example, the first excited state of Li is $1s^2 2p^1$, with the $2s$ electron moving to the $2p$ level. (The energy necessary to accomplish this can be provided by various means, such as by absorption of a photon or by passage of an electric current through the material as in a gas discharge tube.) The excited electron in the $2p$ state rapidly drops back

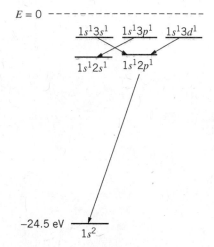

FIGURE 8.5 A small portion of the energy level diagram for helium. Note the $\Delta l = \pm 1$ transitions.

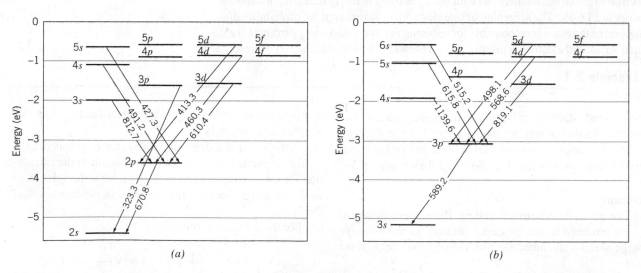

FIGURE 8.4 (*a*) Energy-level diagram of lithium, showing some of the transitions (labeled with wavelength in nm) in the optical region. (*b*) Energy-level diagram for sodium. Because of the fine-structure splitting, the $3p$ level in sodium is actually a very closely spaced pair of levels, so all transitions involving that level show two closely spaced wavelengths. The wavelength shown here is the average of the two. The fine-structure splitting is negligible for the other levels in sodium and for all of the levels in lithium.

to the $2s$ state, with the emission of a photon of wavelength 670.8 nm. The inert core doesn't participate in this excitation or emission, so to a good approximation we can ignore all but the outer electron in studying the levels and transitions in the alkali elements.

The ground-state configuration of helium is $1s^2$. We can produce an excited state by moving one of these electrons up to a higher level, and so some possible excited-state configurations might be $1s^12s^1$, $1s^12p^1$, $1s^13s^1$, and so forth. Photons are emitted when the excited electron drops back to the $1s$ level. The $\Delta l = \pm 1$ selection rule for transitions once again limits those that can occur. Figure 8.5 shows a portion of the energy level diagram for helium.

The phenomenon of *fluorescence* is responsible for the appearance of objects under so-called "black light," which is a source of ultraviolet radiation. Photons in the ultraviolet region, invisible to the human eye, have higher energies than those in the visible region, and hence if an ultraviolet photon is absorbed by an atom, the outer electron (which is responsible for the optical transitions) can be excited to high levels. These electrons make transitions back to their ground state, accompanied by the emission of photons in the visible region. Objects seen in ultraviolet light often show colors in the blue or violet end of the spectrum that are not present when the objects are viewed in sunlight. We can understand this effect by considering the composition of sunlight and the optical excited states of a hypothetical atom shown in Figure 8.6. The intensity of sunlight is concentrated in the center of the visible spectrum, in the yellow region; very little intensity is present in the red or blue ends of the visible spectrum. The "yellow" photons have enough energy to excite the hypothetical atom to levels 1 and 2 shown in Figure 8.6, but not enough to reach level 3 or 4. However, the higher-energy ultraviolet photons have sufficient energy to reach the higher levels, so the light emitted by the atom has a stronger blue component when that atom is excited by ultraviolet light than when excited by sunlight.

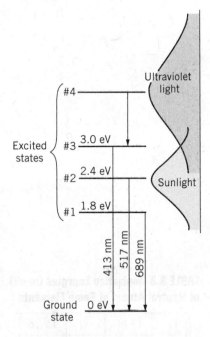

FIGURE 8.6 Excited states of a hypothetical atom. Only excited states 1 and 2 can be easily reached by exposure to sunlight; exposure to ultraviolet light populates state 4, which in turn populates state 3. Under ultraviolet light, a stronger blue or violet (413 nm) color is revealed than under sunlight.

Example 8.4

Calculate the energy difference between the $3d$ and $2p$ states in lithium, and compare with the corresponding energy difference in hydrogen.

Solution

From Figure 8.4, the wavelength of the photon emitted in the $3d$ to $2p$ transition is 610.4 nm. The energy difference is then

$$\Delta E = \frac{hc}{\lambda} = \frac{1240 \text{ eV} \cdot \text{nm}}{610.4 \text{ nm}} = 2.03 \text{ eV}$$

The energy difference between corresponding levels in hydrogen (Figure 6.20) is $E_3 - E_2 = -1.51 \text{ eV} - (-3.40 \text{ eV}) = 1.89 \text{ eV}$. Due to electron screening, we expect the outer electron in lithium to behave similarly to the electron in hydrogen, so the energy differences are in rough agreement.

8.4 PROPERTIES OF THE ELEMENTS

In this section we briefly study the way our knowledge of atomic structure helps us to understand the physical and chemical properties of the elements. Our discussion is based on the following two principles:

1. Filled subshells are normally very stable configurations. An atom with one electron beyond a filled shell will readily give up that electron to another atom

to form a chemical bond. Similarly, an atom lacking one electron from a filled shell will readily accept an additional electron from another atom in forming a chemical bond.

2. Filled subshells do not normally contribute to the chemical or physical properties of an atom. Only the electrons in the unfilled subshells need be considered. (X-ray energies, discussed in the next section, are an exception to this rule.) Sometimes only a single outer electron is the primary factor influencing the physical properties of an element.

We consider a number of different physical properties of the elements, and try to understand those properties based on atomic theory.

1. **Atomic Radii.** The radius of an atom is not a precisely defined quantity, because the electron probability density determines the "size" of an atom. The radii are also difficult to define experimentally, and in fact different kinds of experiments may give different values for the radii. One way of defining the radius is by means of the spacing between the atoms in a crystal containing that element. Figure 8.7 shows how such typical atomic radii vary with Z.

2. **Ionization Energy.** Table 8.3 gives the ionization energies of some of the elements, and Figure 8.8 shows the variation of ionization energy with atomic number Z.

3. **Electrical Resistivity.** In bulk materials, an electric current flows when a potential difference (voltage) is applied across the material. The current i and voltage V are related according to the expression $V = iR$, where R is the electrical resistance of the material. If the material is uniform with length L and cross-sectional area A, then the resistance is

$$R = \rho \frac{L}{A} \qquad (8.2)$$

The *resistivity* ρ is characteristic of the kind of material and is measured in units of $\Omega \cdot m$ (ohm $\cdot$ meter). A good electrical conductor has a small

TABLE 8.3 Ionization Energies (in eV) of Neutral Atoms of Some Elements

H	13.60	Ar	15.76
He	24.59	K	4.34
Li	5.39	Cu	7.72
Be	9.32	Kr	14.00
Ne	21.56	Rb	4.18
Na	5.14	Au	9.22

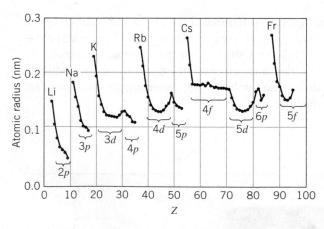

FIGURE 8.7 Atomic radii, determined from atomic separations in ionic crystals. These radii are different from the mean radii of the electron cloud for free atoms.

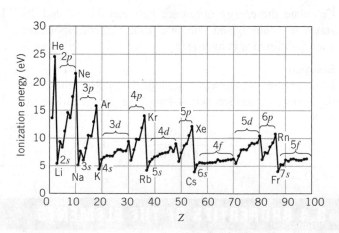

FIGURE 8.8 Ionization energies of neutral atoms of the elements.

resistivity ($\rho = 1.7 \times 10^{-8}$ $\Omega \cdot$m for copper); a poor conductor has a large resistivity ($\rho = 2 \times 10^{15}$ $\Omega \cdot$m for sulfur). From the atomic point of view, current depends on the movement of relatively loosely bound electrons, which can be removed from their atoms by the applied potential difference, and also on the ability of the electrons to travel from one atom to another. Thus elements with s electrons, which are the least tightly bound and which also travel farthest from the nucleus, are expected to have small resistivities.

Figure 8.9 shows the variation of electrical resistivity with atomic number.

4. **Magnetic Susceptibility.** When a material is placed in a magnetic field of intensity B, the material becomes "magnetized" and acquires a magnetization M, which for many materials is proportional to B:

$$\mu_0 M = \chi B \tag{8.3}$$

where χ is a dimensionless constant called the *magnetic susceptibility*. (Materials for which $\chi > 0$ are known as *paramagnetic*, and those for which $\chi < 0$ are called *diamagnetic*; materials that remain permanently magnetized even when B is removed are known as *ferromagnetic,* and χ is undefined for such materials.)

From the atomic point of view, the magnetism of atoms depends on the $\vec{L}$ and $\vec{S}$ of the electrons in unfilled subshells, because the atomic magnetic moments $\vec{\mu}_L$ and $\vec{\mu}_S$ are proportional to $\vec{L}$ and $\vec{S}$ (recall Table 7.2). This effect is responsible for paramagnetic susceptibilities and occurs in all atoms in which $\vec{L}$ or $\vec{S}$ is nonzero. Diamagnetism is caused by the following effect: when a varying magnetic field occurs in an area bounded by an electric circuit, an *induced current* flows in the circuit; the induced current sets up a magnetic field which tends to *oppose* the changes in the applied field (Lenz's law). In the atomic physics case, the electric circuit is the circulating electron, and the induced current consists of a slight speeding up or slowing down of the electron in its orbit when a magnetic field is applied. This produces a

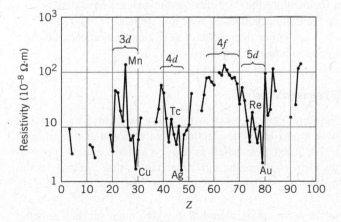

FIGURE 8.9 Electrical resistivities of the elements.

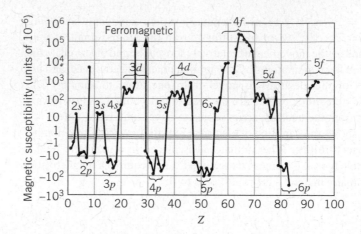

FIGURE 8.10 Magnetic susceptibilities of the elements.

contribution to the magnetization of the material that is opposite to the applied field $\vec{B}$, and so the diamagnetic contribution to χ is negative.

Figure 8.10 shows the magnetic susceptibilities of the elements.

Just by examining Figures 8.7 to 8.10, you can see the remarkable regularities in the properties of the elements. Notice especially how similar the properties of the different sequences of elements are—for example, the electrical resistivity of the d-subshell elements or the magnetic susceptibility of the p-subshell elements. We now look at how the atomic structure is responsible for these properties.

Inert Gases

The inert gases occupy the last column of the periodic table. Because they have only filled subshells, the inert gases do not generally combine with other elements to form compounds; these elements are very reluctant to give up or to accept an electron. At room temperature they are monatomic gases. Their atoms don't easily join together, so the boiling points are very low (typically $-200°C$). Their ionization energies are much larger than those of neighboring elements, because of the extra energy needed to break open a filled subshell.

p-Subshell Elements

The elements of the column (group) next to the inert gases are the halogens (F, Cl, Br, I, At). These atoms lack one electron from a closed shell and have the configuration np^5. A filled p subshell is a very stable configuration, so these elements readily form compounds with other atoms that can provide an extra electron to complete the p subshell. The halogens are therefore extremely reactive.

As we move across the series of six elements in which the p subshell is being filled, the atomic radius *decreases*. This "shrinking" occurs because the nuclear charge is increasing and pulling all of the orbits closer to the nucleus. Notice from Figure 8.7 that the halogens have the smallest radii within each p subshell series. (The ionic crystal radii of the inert gases are not known.)

As we increase the nuclear charge, the p electrons also become more tightly bound; Figure 8.8 shows how the ionization energy increases systematically as the p subshell is filled.

From Figure 8.10 we see that each p subshell series is diamagnetic, with a characteristic negative magnetic susceptibility.

s-Subshell Elements

The elements of the first two columns (groups) are known as the alkalis (configuration ns^1) and alkaline earths (ns^2). The single s electron makes the alkalis quite reactive. The alkaline earths are similarly reactive, in spite of the filled s subshell. This occurs because the s electron wave functions can extend rather far from the nucleus, where the electrons are screened (by $Z - 2$ other electrons) from the nuclear charge and therefore not tightly bound. (Notice from Figure 8.7 that the ns^1 and ns^2 configurations give the largest atomic radii, and from Figure 8.8 that they have the smallest ionization energies.) For the same reasons, the ns^1 and ns^2 elements are relatively good electrical conductors. From Figure 8.10 we see that these elements are paramagnetic; for $l = 0$, there is no diamagnetic contribution to the magnetism.

Transition Metals

The three rows of elements in which the d subshell is filling (Sc to Zn, Y to Cd, Lu to Hg) are known as the *transition metals*. Many of their chemical properties are determined by the outer electrons—those whose wave functions extend furthest from the nucleus. For the transition metals, these are always s electrons, which have a larger mean radius than the d electrons. (Remember that the mean radius depends mostly on n; the s electrons of the transition metals have a larger n than the d electrons. For example, in the first row of transition metals, the $3d$ subshell is filling but the $4s$ subshell is already filled.) As the atomic number increases across the transition metal series, we add one d electron and one unit of nuclear charge; the net effect on the s electron is very small, because the additional d electron screens the s electron from the additional nuclear charge. Properties of the transition metals, which are in large part determined by the outermost electrons, can therefore be very similar, as the small variation in radius and ionization energy shows.

The electrical resistivity of the transition metals shows two interesting features: a sharp rise at the center of the sequence, and a sharp drop near the end (Figure 8.9). The sharp drop near the end of the sequence indicates the small resistivity (large conductivity) of copper, silver, and gold. If we filled the d subshell in the expected sequence, copper would have the configuration $4s^2 3d^9$; however, the filled d subshell is more stable than a filled s subshell, and so one of the s electrons transfers to the d subshell, resulting in the configuration $4s^1 3d^{10}$. This relatively free, single s electron makes copper an excellent conductor. Silver ($5s^1 4d^{10}$) and gold ($6s^1 5d^{10}$) behave similarly.

At the center of the sequence of transition metals there is a sharp rise in the resistivity; apparently a half-filled shell is also a stable configuration, and so Mn ($3d^5$), Tc ($4d^5$), and Re ($5d^5$) have larger resistivities than their neighbors. A similar rise in resistivity is seen at the center of the $4f$ sequence.

The transition metals have similar paramagnetic susceptibilities, due to the large orbital angular momentum of the d electrons and also to the large *number* of d-subshell electrons that can couple their *spin* magnetic moments. These two effects are large enough to overcome the diamagnetism of the orbital motion. It is the d electrons that are also responsible for the ferromagnetism of iron, nickel, and cobalt. As soon as the d subshell is filled, however, the orbital and spin magnetic moments no longer contribute to the magnetic properties (all of the m_l and m_s values, positive as well as negative, are taken); for this reason, copper and zinc are diamagnetic, not paramagnetic like their transition metal neighbors.

Lanthanides (Rare Earths)

The lanthanide (or rare earth) elements are contained in the series of 14 elements from La to Yb; this series is usually drawn at the bottom of the periodic chart of the elements. The rare earths are rather similar to the transition metals in that an "inner" subshell (the $4f$) is being filled after an "outer" subshell (the $6s$) is already filled. For the same reasons discussed above, the chemical properties of the rare earths should be rather similar, because they are determined mainly by the $6s$ electrons; the radii and ionization energies show that this is true.

Because of the larger orbital angular momentum of f-subshell electrons ($l = 3$) and also because of the larger *number* of f-subshell electrons (up to 14) that can align their spin magnetic moments, the paramagnetic susceptibilities of the rare earths are even larger than those of the transition metals. Even the ferromagnetism of the rare earths is substantially stronger than that of the iron group. Generally, we think of iron as the most magnetic of the elements. The internal magnetic field within a magnetized piece of iron is about 28 T. Magnetized holmium metal, a rare earth, has an internal magnetic field of 800 T, roughly 30 times that of iron! Most of the other rare earths have similar magnetic properties. (The rare earth metals do not reveal their ferromagnetic properties at room temperature, but must be cooled to lower temperatures. Holmium must be cooled to 20 K to reveal its ferromagnetic properties.)

Actinides

The actinide series of elements, which corresponds to the filling of the $5f$ subshell, is usually shown in the periodic table directly under the lanthanide series. These elements should have chemical and physical properties similar to those of the rare earths. Unfortunately, most of the actinide elements (those beyond uranium) are radioactive and do not occur in nature. They are artificially produced elements and are available only in microscopic quantities. We are thus unable to determine many of their bulk properties.

8.5 INNER ELECTRONS: ABSORPTION EDGES AND X RAYS

Let's imagine doing the Franck-Hertz experiment (see Section 6.6), in which we accelerate a beam of electrons that then passes through a chamber filled with mercury vapor. However, instead of using accelerating voltages in the range of 10 V, we'll use voltages in the range of 10^5 V. Figure 8.11 shows the current passing through the tube as a function of the accelerating voltage. A sudden drop in the current occurs at 83.1 kV. Low accelerating voltages correspond to interactions that push the outer electrons in the mercury atoms to higher excited states (or ionize the atom). The drop in the current at 83.1 kV occurs when the mercury atom absorbs energy from the electron beam that ionizes the atom by knocking loose one of the tightly bound *inner* electrons. The binding energy of the inner electron in this case is 83.1 keV.

A similar experiment can be done by passing a beam of X rays through a thin film of mercury and measuring the absorption of the photon intensity. If we are able to vary the wavelength of the X rays, the absorption as a function of wavelength might look like Figure 8.12. Photons are absorbed from the beam by the photoelectric effect, in which electrons are knocked loose from mercury atoms. As the photon wavelength is increased (or as the photon energy is decreased), we reach a point at which the photons do not have enough energy to produce at least one component of photoelectrons, and thus there is a sudden decrease in the photon absorption. The wavelength at which this occurs, 0.0149 nm, corresponds to an energy of 83.1 keV, in agreement with the value deduced from electron scattering (Figure 8.11).

The sudden drop in the electron current or in the photoelectron emission is called the *absorption edge*. It corresponds to the release of an inner electron from the atom. In the case of mercury, the most tightly bound ($1s$) electrons have a binding energy of 83.1 keV. In the electron scattering experiment, when the energy of the electrons in the beam exceeds 83.1 keV, the collision of an electron with a mercury atom can transfer an energy of 83.1 keV to the atom and result in the ejection of one of the $1s$ electrons. Similarly, when the photon energy exceeds 83.1 keV (or when its wavelength is below 0.0149 nm), the photons can eject a photoelectron from the $1s$ level, but when the photon energy is below 83.1 keV that is not possible.

As discussed in Section 8.3, the $n = 1$ level is also known as the K shell. So far we have been discussing the K absorption edge in mercury, which corresponds to the release of an electron from the K shell. It is also possible to release a less tightly bound electron from the L shell ($n = 2$), in which case we would speak of the L absorption edge. In mercury, the L absorption edge is about 14 keV. (Because of the fine-structure splitting, there are actually three different states in the L shell with slightly different energies.)

Figure 8.13 shows the K absorption edges of the elements. There is a very noticeable difference between the data shown in Figure 8.13 and those shown in

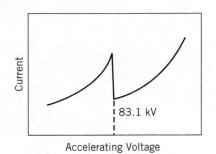

FIGURE 8.11 Electron current passing through mercury vapor as a function of accelerating voltage.

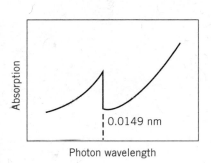

FIGURE 8.12 Absorption of photons by a thin film of mercury as a function of the photon wavelength.

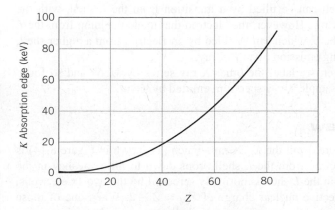

FIGURE 8.13 K absorption edges of the elements.

Figures 8.7–8.10: the K absorption edges show no evidence for any shell effects. Instead, there is a smooth dependence on the atomic number over the entire range of elements. As the nuclear charge increases, the $1s$ electrons are pulled into smaller and more tightly bound orbits, but this is a gradual process that is largely unaffected by the stacking of electrons into higher energy shells. There are no sudden changes in the $1s$ properties as a higher shell is filled and a still higher shell begins filling.

X-Ray Transitions

X rays, as we discussed in Chapter 3, are electromagnetic radiations with wavelengths from approximately 0.01 to 10 nm (energies from 100 eV to 100 keV). In Chapter 3 we discussed the *continuous* X-ray spectrum emitted by accelerated electrons. In this section we are concerned with the *discrete* X-ray line spectra emitted by atoms.

X rays are emitted in transitions between the more tightly bound inner electron energy levels of an atom. Under normal conditions all of the inner shells of an atom are filled, so X-ray transitions do not occur between these levels. However, when we remove one of the inner electrons, such as by ejecting a K electron following electron scattering or a photoelectric process, an electron from a higher subshell will rapidly make a transition to fill that vacancy, emitting an X-ray photon in the process. The energy of the photon is equal to the energy difference of the initial and final atomic levels of the electron that makes the transition.

When we remove a $1s$ electron, we are creating a vacancy in the K shell. The X rays that are emitted in the process of filling this vacancy are known as *K-shell X rays*, or simply *K X rays*. (These X rays are emitted in transitions that come *from* the $L, M, N, \dots$ shells, but they are known by the vacancy that they fill, not by the shell from which they originate.) The K X ray that originates with the $n = 2$ shell (L shell) is known as the K_α X ray, and the K X rays originating from the M shell are known as K_β X rays. Figure 8.14 illustrates these transitions.

If the bombarding electrons or photons knock loose an electron from the L shell, electrons from higher levels will drop down to fill this vacancy. The photons emitted in these transitions are known as L X rays. The lowest-energy X ray of the L series is known as L_α, and the other L X rays are labeled in order of increasing energy as shown in Figure 8.14.

It is possible to have an L X ray emitted directly following the K_α X ray. A vacancy in the K shell can be filled by a transition from the L shell, with the emission of the K_α X ray. However, the electron that made the jump from the L shell left a vacancy there, which can be filled by an electron from a higher shell, with the accompanying emission of an L X ray.

In a similar manner, we label the other X-ray series by M, N, and so forth. Figure 8.15 shows a sample X-ray spectrum emitted by silver.

Moseley's Law

Let us consider in more detail the K_α X ray, which (as shown in Figure 8.14) is emitted when an electron from the L shell drops down to fill a vacancy in the K shell. An electron in the L shell is normally screened by the two $1s$ electrons, and so it sees an effective nuclear charge of $Z_{\text{eff}} = Z - 2$. When one of those $1s$ electrons is removed in the creation of a K-shell vacancy, only the remaining single $1s$ electron shields the L shell, and so $Z_{\text{eff}} = Z - 1$. (In this calculation, we neglect the small screening effect of the outer electrons; their probability densities

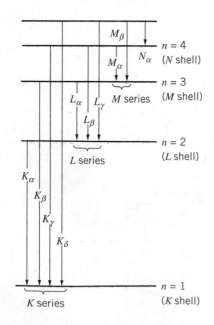

FIGURE 8.14 X-ray series.

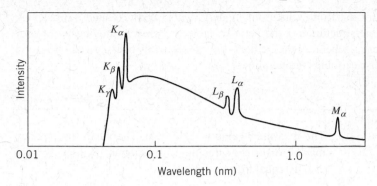

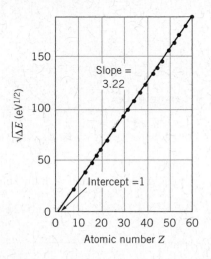

FIGURE 8.15 Characteristic X-ray spectrum of silver, such as might be produced by 30 keV electrons striking a silver target. The continuous distribution is a bremsstrahlung spectrum.

FIGURE 8.16 Moseley plot of square root of K_α X-ray energy as a function of atomic number.

are not zero within the L-shell orbits, but they are sufficiently small that their effect on Z_{eff} can be neglected.) To a very good approximation, the K_α X ray can thus be analyzed as a transition from the $n = 2$ level to the $n = 1$ level in a one-electron atom with $Z_{\text{eff}} = Z - 1$. Using Eq. 6.38 for the Bohr atom, we can find the energy of the K_α transition in an atom of atomic number Z:

$$\Delta E = E_2 - E_1 = (-13.6\,\text{eV})(Z-1)^2 \left(\frac{1}{2^2} - \frac{1}{1^2} \right) = (10.2\,\text{eV})(Z-1)^2 \quad (8.4)$$

Just as was the case for the K absorption edge, the energies of the K_α X rays vary smoothly with atomic number and show no effects of atomic shells. If we plot $\sqrt{\Delta E}$ as a function of Z, we expect to obtain a straight line with slope $(10.2\,\text{eV})^{1/2} = 3.19\,\text{eV}^{1/2}$. Figure 8.16 is an example of such a plot. The measured slope is $3.22\,\text{eV}^{1/2}$, in excellent agreement with what is expected from Eq. 8.4. The straight line intersects the x axis at a value very close to 1, as we expect from Eq. 8.4.

This method gives us a powerful and direct way to determine the atomic number Z of an atom, as was first demonstrated in 1913 by the British physicist H. G. J. Moseley, who measured the K_α (and other) X-ray energies of the elements and thus determined their atomic numbers. The dependence of the X-ray energies on Z given by Eq. 8.4 is known as *Moseley's law*. Moseley was the first to demonstrate the type of linear relationship shown in Figure 8.16; such graphs are now known as *Moseley plots*. His discovery provided the first direct means of measuring the atomic numbers of the elements. Previously, the elements had been ordered in the periodic table according to increasing mass. Moseley found certain elements listed out of order, in which the element of higher Z happened to have the smaller mass (for example, cobalt and nickel or iodine and tellurium). He also found gaps corresponding to yet undiscovered elements; for example, the naturally radioactive element technetium ($Z = 43$) does not exist in nature and was not known at the time of Moseley's work, but Moseley showed the existence of such a gap at $Z = 43$.

The straight-line plot of Figure 8.16 is independent of our assumption regarding the exact value of the screening correction. That is, we could have written $Z_{\text{eff}} = Z - k$, where k is some unknown number, probably close to 1. The only change in our plot would be in the intercept. We would still have a straight line with the same slope.

Moseley's work was of great importance in the development of atomic physics. Working in the same year as Rutherford and Bohr, Moseley not only provided

Henry G. J. Moseley (1887–1915, England). His work on X-ray spectra provided the first link between the chemical periodic table and atomic physics, but his brilliant career was cut short when he died on a World War I battlefield.

confirmation of the Rutherford-Bohr model, he also demonstrated a direct link between atomic structure and the periodic table, which was previously a rather arbitrary ordering scheme of the elements but subsequently became a classification based on their electronic configurations.

Example 8.5

Compute the energy of the K_α X ray of sodium ($Z = 11$).

Solution
The energy can be found with the help of Eq. 8.4,

$$\Delta E = (10.2\,\text{eV})(Z - 1)^2 = (10.2\,\text{eV})(10)^2 = 1.02\,\text{keV}$$

The measured value is 1.04 keV. The small discrepancy may be due to the screening correction in Z_{eff}, which is not exactly equal to 1.

Example 8.6

Some measured X-ray energies in silver ($Z = 47$) are $\Delta E(K_\alpha) = 21.990\,\text{keV}$ and $\Delta E(K_\beta) = 25.145\,\text{keV}$. The binding energy of the K electron in silver is $E(K) = 25.514\,\text{keV}$. From these data, find: (a) the energy of the L_α X ray, and (b) the binding energy of the L electron.

Solution
(a) From Figure 8.14, we see that the energies are related by:

$$\Delta E(L_\alpha) + \Delta E(K_\alpha) = \Delta E(K_\beta)$$

or

$$\Delta E(L_\alpha) = \Delta E(K_\beta) - \Delta E(K_\alpha)$$
$$= 25.145\,\text{keV} - 21.990\,\text{keV} = 3.155\,\text{keV}$$

(b) Again from Figure 8.14, we see that

$$\Delta E(K_\alpha) = E(L) - E(K)$$

or

$$E(L) = E(K) + \Delta E(K_\alpha)$$
$$= -25.514\,\text{keV} + 21.990\,\text{keV} = -3.524\,\text{keV}$$

The binding energy of the L electron is therefore 3.524 keV.

*8.6 ADDITION OF ANGULAR MOMENTA

The properties of an alkali atom such as sodium are determined primarily by the single outer electron; if that electron has quantum numbers (n, l, m_l, m_s) then the entire atom behaves as if it had those same quantum numbers. In atoms with several electrons outside of filled subshells, this is not the case. For example, the electronic configuration of carbon ($Z = 6$) is $1s^2 2s^2 2p^2$. To find the angular momentum of carbon, we must combine the angular momenta of the two $2p$ electrons to find the total orbital angular momentum quantum number L and total magnetic quantum number M_L that characterize the entire atom.

---

*This is an optional section that may be skipped without loss of continuity.

Suppose we have an atom with two electrons outside of filled subshells. These electrons have quantum numbers $(n_1, l_1, m_{l1}, m_{s1})$ and $(n_2, l_2, m_{l2}, m_{s2})$. The total orbital angular momentum of the atom is determined by the vector sum of the orbital angular momenta of the two electrons:

$$\vec{\mathbf{L}} = \vec{\mathbf{L}}_1 + \vec{\mathbf{L}}_2 \qquad (8.5)$$

Each vector is related to its corresponding angular momentum quantum number by

$$|\vec{\mathbf{L}}| = \sqrt{L(L+1)}\hbar \qquad |\vec{\mathbf{L}}_1| = \sqrt{l_1(l_1+1)}\hbar \qquad |\vec{\mathbf{L}}_2| = \sqrt{l_2(l_2+1)}\hbar \qquad (8.6)$$

These vectors do not add like ordinary vectors, but have special addition rules associated with quantized angular momentum. These rules enable us to find L and its associated magnetic quantum number M_L.

1. The maximum value of the total orbital angular momentum quantum number is

$$L_{max} = l_1 + l_2 \qquad (8.7)$$

2. The minimum value of the total orbital angular momentum quantum number is

$$L_{min} = |l_1 - l_2| \qquad (8.8)$$

3. The permitted values of L range from L_{min} to L_{max} in integer steps:

$$L = L_{min}, L_{min} + 1, L_{min} + 2, \ldots, L_{max} \qquad (8.9)$$

4. The z component of the total angular momentum vector is found from the sum of the z components of the individual vectors:

$$L_z = L_{1z} + L_{2z} \qquad (8.10)$$

or, in terms of the magnetic quantum numbers,

$$M_L = m_{l1} + m_{l2} \qquad (8.11)$$

The permitted values of the total magnetic quantum number M_L range from $-L$ to $+L$ in integer steps:

$$M_L = -L, -L + 1, \ldots, -1, 0, +1, \ldots, L - 1, L \qquad (8.12)$$

An identical set of rules holds for coupling the spin angular momentum vectors to give the total spin angular momentum $\vec{\mathbf{S}}$. For two electrons, each of which has $s = \frac{1}{2}$, the total spin quantum number S can be 0 or 1.

All filled subshells have $L = 0$ and $S = 0$, so we don't need to consider filled subshells in analyzing the angular momentum of an atom. For this reason, filled subshells ordinarily do not contribute to the magnetic properties of atoms.

For coupling more than two electrons, the procedure is first to couple the angular momenta of two electrons to give the maximum and minimum values of L. Then couple each allowed L to the angular momentum of the third electron to find the largest maximum and smallest minimum. This continues for all of the electrons in the unfilled subshell.

Example 8.7

Find the total orbital and spin quantum numbers for carbon.

Solution

Carbon has two $2p$ electrons outside filled subshells. Each of these electrons has $l = 1$. According to the rules for adding angular momenta, we have

$$L_{max} = 1 + 1 = 2, \qquad L_{min} = |1 - 1| = 0$$

Thus $L = 0, 1$, or 2. For the spin angular momentum, we have

$$S_{max} = \tfrac{1}{2} + \tfrac{1}{2} = 1, \qquad S_{min} = |\tfrac{1}{2} - \tfrac{1}{2}| = 0$$

and so $S = 0$ or 1. Some combinations of L and S might be forbidden by the Pauli principle. For example, to obtain $L = 2$, the two electrons must both have $m_l = +1$. The two electrons must therefore have different values of m_s, so $S = 1$ is not allowed when $L = 2$.

Example 8.8

Find the total orbital and spin quantum numbers for nitrogen.

Solution

Nitrogen has three $2p$ electrons, each with $l = 1$, outside filled subshells. If we add the first two, we get $L_{max} = 2$ and $L_{min} = 0$, as in Example 8.7, so that $L = 0, 1$, or 2. We now couple the third $l = 1$ electron to each of these values to find the largest maximum and smallest minimum, which give

$$L_{max} = 2 + 1 = 3, \qquad L_{min} = |1 - 1| = 0$$

and so $L = 0, 1, 2$, or 3. For the spin vectors, we again couple the first two to give $S_{max} = 1$ and $S_{min} = 0$. Adding the third $s = 1/2$ electron, we have

$$S_{max} = 1 + \tfrac{1}{2} = \tfrac{3}{2}, \qquad S_{min} = |0 - \tfrac{1}{2}| = \tfrac{1}{2}$$

The resulting values of S are $1/2$ and $3/2$ (from the minimum to the maximum in integer steps). Once again, the Pauli principle may forbid certain combinations of L and S. The state with $L = 3$ cannot exist at all, because all three electrons must have $m_l = +1$, and assigning m_s quantum numbers will then result in two electrons with the same m_l and m_s, which is forbidden by the Pauli principle.

The two $2p$ electrons of carbon can combine to give $L = 0, 1$, or 2 and $S = 0$ or 1. The ground state of carbon will be identified by only one particular choice of L and S. How do we know which of these combinations will be the ground state? The rules for finding the ground state quantum numbers are known as *Hund's rules*:

1.   First find the maximum value of the total spin magnetic quantum number M_S consistent with the Pauli principle. Then

$$S = M_{S,max} \tag{8.13}$$

2.   Next, for that M_S, find the maximum value of M_L consistent with the Pauli principle. Then

$$L = M_{L,max} \tag{8.14}$$

In the case of carbon, the maximum value of M_S is $+1$, obtained when the two valence electrons both have $m_s = +1/2$. Thus $S = 1$. With only two electrons in the $2p$ shell, the Pauli principle places no restrictions on S; in fact, three electrons in the $2p$ shell can be assigned $m_s = +1/2$. Our next task is to find the maximum value of M_L. The maximum value of m_l for the first p electron is $+1$. The second p electron cannot also have $m_l = +1$, because that would give both electrons the same set of quantum numbers, in violation of the Pauli principle. The maximum value of m_l for the second electron is 0, so $M_{L,\max} = +1$ and $L = 1$. The ground state of carbon is therefore characterized by $S = 1$ and $L = 1$.

Example 8.9

Use Hund's rules to find the ground-state quantum numbers of nitrogen.

Solution

The electronic configuration of nitrogen is $1s^2 2s^2 2p^3$. We begin by maximizing the total M_S for the three $2p$ electrons. Three electrons in the p subshell are permitted by the Pauli principle to have $m_s = +1/2$, so the maximum

value of M_S is $3/2$, and therefore S is $3/2$. Each of the three electrons has quantum numbers $(2, 1, m_l, +1/2)$. To maximize M_L we assign the first electron the maximum value of m_l—namely, $+1$. The maximum value of m_l left for the second electron is 0, and the third electron must therefore have $m_l = -1$. The total M_L is $1 + 0 + (-1) = 0$, so $L = 0$. Thus $L = 0, S = 3/2$ are the ground-state quantum numbers for nitrogen.

Example 8.10

Find the ground-state L and S of oxygen ($Z = 8$).

Solution

The electronic configuration of oxygen is $1s^2 2s^2 2p^4$. Because only three electrons in the p subshell can have $m_s = +1/2$, the fourth must have $m_s = -1/2$, so

$M_{S,\max} = 1/2 + 1/2 + 1/2 + (-1/2) = +1$, and it follows that $S = 1$. To find L, we note that, as for nitrogen, the three electrons with $m_s = +1/2$ have $m_l = +1, 0$, and -1, and we maximize M_L by giving the fourth electron $m_l = +1$. Thus $M_{L,\max} = +1$, and $L = 1$.

Let us look now at the energy levels of helium. The ground-state configuration of helium is $1s^2$. Both electrons are s electrons, with $l = 0$, and so the only possible value of L is zero. Because both electrons have $m_l = 0$, the Pauli principle requires that the spin of the two electrons be opposite, so that one has $m_s = +1/2$ and the other has $m_s = -1/2$. The *only* possible total M_S is therefore zero, so the ground state of helium has $L = 0$ and $S = 0$. The first excited state has configuration $1s^1 2s^1$. Both electrons still have $l = 0$, so we must again have $L = 0$. However, the total spin S can now be 0 or 1, because the Pauli principle does not restrict m_s in this case—the two electrons already have different principal quantum numbers n, and so there is nothing to prevent them from having the same m_s. There are, therefore, *two* "first excited states" of helium, one with $L = 0$ and $S = 0$, and another with $L = 0$ and $S = 1$. (Both of these states have configuration $1s^1 2s^1$.) A state with $S = 0$ is called a *singlet* state (because there is only a single possible M_S value), and a state with $S = 1$ is called a *triplet* state (because there are three possible M_S values: $+1, 0, -1$).

The classification of states into singlet and triplet is important when we consider the *selection rules* for transitions between states; these selection rules tell us which transitions are allowed (and therefore likely) to occur and which are not. The selection rules, which involve both L and S, are

$$\Delta L = 0, \pm 1 \qquad\qquad (8.15)$$

$$\Delta S = 0 \qquad\qquad (8.16)$$

(There are no selection rules for n.) Of course, the selection rule $\Delta l = \pm 1$ *for the single electron that makes the transition* still applies. For the two $1s^1 2s^1$ states

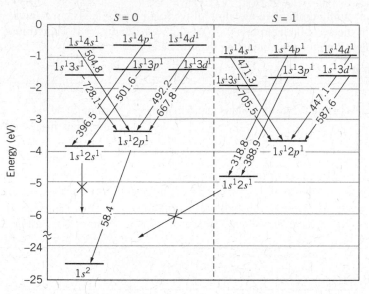

FIGURE 8.17 Energy-level diagram for helium. The states are grouped into singlets ($S = 0$) and triplets ($S = 1$). Some of the transitions in the optical and ultraviolet regions are shown. Transitions marked with an X would violate the $\Delta l = \pm 1$ selection rule.

of helium, the Δl rule does not permit either state to make transitions to the $1s^2$ ground state ($2s$ to $1s$ would be $\Delta l = 0$), and in addition, the ΔS rule forbids the triplet ($S = 1$) states from decaying to the $S = 0$ ground state. These transitions can thus occur only by violating these selection rules. Because that is a very unlikely event, the transitions occur with very low probability. Energy levels that have a low probability of decay must "live" for a long time before they decay; such states are known as *metastable* states.

Figure 8.17 shows the energy levels and transitions in helium. The singlet and triplet levels are grouped separately, because transitions between singlet and triplet levels would violate the $\Delta S = 0$ selection rule.

Figure 8.18 shows the energy-level diagram of carbon. Notice the increasing complexity of the diagram, compared with the alkali metals and even with helium. This follows from the coupling of two electrons, both of whose l values may be different from zero. We have already discussed how the $2p^2$ configuration can give $L = 0, 1$, or 2 and $S = 0$ or 1. Only one of these ($L = 1, S = 1$) is the ground state of carbon; the others are excited states. More excited states can be obtained by promoting one of the $2p$ electrons to a higher level, giving configurations of $2p^1 3s^1$ ($L = 1, S = 0$ or 1), $2p^1 3p^1$ ($L = 0, 1$, or 2; $S = 0$ or 1), $2p^1 3d^1$ ($L = 1, 2$, or 3; $S = 0$ or 1), and so forth. Imagine the difficulty of analyzing the energy level diagram of the rare earths or actinides, which have f subshells ($l = 3$) with as many as 14 electrons!

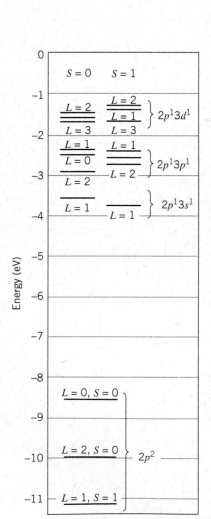

FIGURE 8.18 Energy-level diagram for carbon. Each group of levels is labeled with the electron configuration. Each individual level is labeled with the total L and S.

8.7 LASERS

There are three means by which radiation can interact with the energy levels of atoms (depicted in Figure 8.19). The first two we have already discussed. In the first kind of interaction, an atom in an excited state makes a transition to a lower state, with the emission of a photon. (In all the examples we consider here, the photon energy is equal to the energy difference of the two atomic states.) This is *spontaneous emission*, which we represent as

$$\text{atom}^* \rightarrow \text{atom} + \text{photon}$$

where the asterisk indicates an excited state.

The second interaction, *induced absorption*, is responsible for absorption spectra and resonance absorption. An atom in the ground state absorbs a photon (of the proper energy) and makes a transition to an excited state. Symbolically:

$$\text{atom} + \text{photon} \rightarrow \text{atom}^*$$

The third interaction, which is responsible for the operation of the laser, is *induced* (or *stimulated*) *emission*. In this process, an atom is initially in an excited state. A passing photon of just the right energy (again, equal to the energy difference of the two levels) induces the atom to emit a photon and make a transition to the lower, or ground, state. (Of course, it would eventually have made that transition left on its own, but it makes it *sooner* after being prodded by the passing photon.) Symbolically,

$$\text{atom}^* + \text{photon} \rightarrow \text{atom} + 2 \text{ photons}$$

The significant detail is that the two photons that emerge are traveling in *exactly the same direction* with *exactly the same energy*, and the associated electromagnetic waves are *perfectly in phase (coherent)*.

Suppose we have a collection of atoms, all in the same excited state, as shown in Figure 8.20. A photon passes the first atom, causing induced emission and resulting in two photons. Each of these two photons causes an induced emission process, resulting in four photons. This process continues, doubling the number of photons at each step, until we build up an intense beam of photons, all coherent and moving in the same direction. In its simplest interpretation, this is the basis of operation of the laser. (The word *laser* is an acronym for Light Amplification by Stimulated Emission of Radiation.)

This simple model for a laser will not work, for several reasons. First, it is difficult to keep a collection of atoms in their excited states until they are stimulated to emit the photon (we don't want any *spontaneous* emission). A

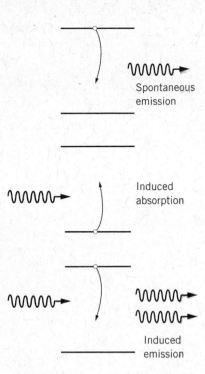

FIGURE 8.19 Interactions of radiation with atomic energy levels.

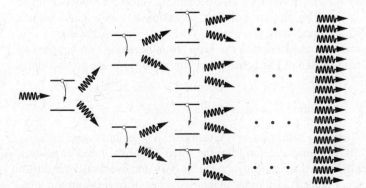

FIGURE 8.20 Buildup of intense beam in a laser. Each emitted photon interacts with an excited atom and produces two photons.

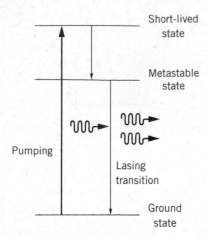

FIGURE 8.21 A three-level atom.

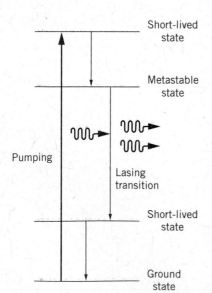

FIGURE 8.22 A four-level atom.

second reason is that atoms that happen to be in their ground state undergo absorption and thus *remove* photons from the beam as it builds up.

To solve these problems, we must achieve a *population inversion*—in a collection of atoms, there must be more atoms in the upper state than in the lower state. This is called an "inversion" because under normal conditions at thermal equilibrium, the lower state always has the greater population. The "inversion" is thus an unnatural situation that must be achieved by artificial means, because it is essential for the operation of the laser.

The first laser, which was constructed by T. H. Maiman in 1960, was based on a three-level atom (Figure 8.21). The laser medium is a solid ruby rod, in which the chromium atoms are responsible for the action of the laser. The atoms, originally in the ground state, are "pumped" into the excited state by an external source of energy (a burst of light from a flash lamp that surrounds the ruby rod). The excited state decays very rapidly (by spontaneous emission) to a lower excited state, which is a metastable state—the atom remains in that level for a relatively long time, perhaps 10^{-3} s, compared with 10^{-8} s for the short-lived states. The transition from the metastable state to the ground state is the "lasing" transition, resulting from stimulated emission by a passing photon.

If the pumping action is successful, there are more atoms in the metastable state than in the ground state, and we have achieved a population inversion. However, as the lasing transition occurs, the population of the ground state is increased, thereby upsetting the population inversion. This excess of population in the ground state allows absorption of the lasing transition, thereby removing photons that might contribute to the lasing action.

The four-level laser illustrated in Figure 8.22 relieves this remaining difficulty. The ground state is pumped to an excited state that decays rapidly to the metastable state, as with the three-level laser. The lasing transition proceeds from the metastable state to yet another excited state, which in turn decays rapidly to the ground state. *The atom in its ground state thus cannot absorb at the energy of the lasing transition*, and we have a workable laser. Because the lower short-lived state decays rapidly, its population is always smaller than that of the metastable state, which maintains the population inversion.

A common example of the four-level laser is the familiar helium-neon laser, which operates with a mixture of helium and neon gas (about 90% helium). The important energy levels of He and Ne are shown in Figure 8.23. An electrical current in the gas "pumps" the helium from its ground state to the excited state at an energy of about 20.6 eV. This is a metastable state of helium—the atom remains in that state for a relatively long time because a $2s$ electron is not permitted to return to the $1s$ level by photon emission. Occasionally, an excited helium atom collides with a ground-state neon atom. When this occurs, the 20.6 eV of excitation energy may be transferred to the neon atom, because neon happens to have an excited state at 20.6 eV, and the helium atom returns to its ground state. Symbolically,

$$\text{helium}^* + \text{neon} \rightarrow \text{helium} + \text{neon}^*$$

where the excited state is indicated by the asterisk. The excited state of neon corresponds to removing one electron from the filled $2p$ subshell and promoting it to the $5s$ subshell. From there it decays to the $3p$ level and eventually returns to the $2p$ ground state. Figure 8.23 illustrates this sequence of events and the level schemes. (The level shown with a dashed line, the neon $3s$ level, is not important

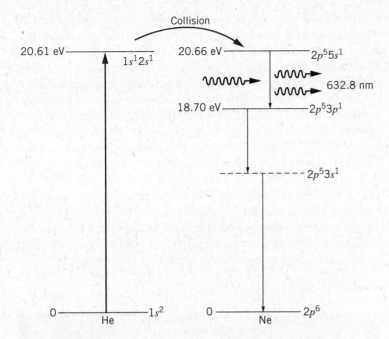

FIGURE 8.23 Sequence of transitions in a He-Ne laser.

for the basic operation of the laser, but it is necessary as an intermediate step in the return to the neon ground state, because the $\Delta l = 0$ transition $3p \to 2p$ is not allowed, but the sequence $3p \to 3s \to 2p$ is permitted.)

At any given time, there are more neon atoms in the $5s$ state than in the $3p$ state, because the good energy matchup of the $5s$ state with the helium excited state gives a high probability of the $5s$ state in neon being excited. The $3p$ state, on the other hand, decays rapidly. This provides the population inversion that is needed for the laser.

In the helium-neon laser, the gases are enclosed in a narrow tube (Figure 8.24). Occasionally a neon atom in the $5s$ state spontaneously emits a photon (at a wavelength of 632.8 nm) parallel to the axis of the tube. This photon causes stimulated emission by other atoms, and a beam of coherent (in-phase) radiation eventually builds up traveling along the tube axis. Mirrors are carefully aligned at the ends of the tube to help in the formation of the coherent wave, as it bounces

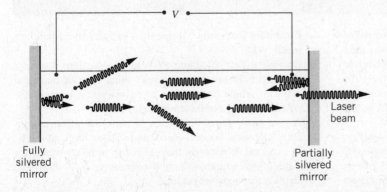

FIGURE 8.24 Schematic diagram of a He-Ne laser.

back and forth between the two ends of the tube, causing additional stimulated emission. One of the mirrors is only partially silvered, allowing a portion of the beam to escape through one end.

The laser is not a particularly efficient device; the small helium-neon lasers you have probably seen used for laboratory or demonstration experiments have a light output of perhaps a few milliwatts; the electric power required to operate such a device may be of the order of 10 to 100 W, and thus the efficiency (power out ÷ power in) of such a device is only about 10^{-4} to 10^{-5}. It is the *coherence* and *directionality* of the laser beam and its *energy density* that make the laser such a useful device—its power can be concentrated in a beam only a few millimeters in diameter, and thus even a small laser can deliver 100 to 1000 W/m². Larger lasers in the megawatt (10^6 W) range are presently readily available, and research laboratories are using lasers in the 100 terawatt (10^{14} W) range for special applications. These powerful lasers do not operate continuously, but are instead *pulsed*, producing short (perhaps 10^{-9} s) pulses at rates of order 100 Hz. (Such a pulse is, in fact, an excellent example of a wave packet.)

Chapter Summary

		Section				Section
Pauli exclusion principle	*No two electrons in a single atom can have the same set of quantum numbers* (n, l, m_l, m_s).	8.1	Energy of screened electron	$E_n = (-13.6\,\text{eV})\dfrac{Z_{\text{eff}}^2}{n^2}$		8.3
Filling order of atomic subshells	$1s, 2s, 2p, 3s, 3d, 4s, 3d,$ $4p, 5s, 4d, 5p, 6s, 4f, 5d,$ $6p, 7s, 5f, 6d$	8.2	Moseley's law for K_α X rays	$\Delta E = (10.2\,\text{eV})(Z - 1)^2$		8.5
Capacity of subshell nl	$2(2l + 1)$	8.2	Adding angular momenta l_1, m_{l1} and l_2, m_{l2}	$L_{\text{max}} = l_1 + l_2,$ $L_{\text{min}} = |l_1 - l_2|,$ $M_L = m_{l1} + m_{l2}$		8.6
			Hund's rules for ground state	First $S = M_{S,\text{max}}$, then $L = M_{L,\text{max}}$		8.6

Questions

1. Continue Figure 8.1 upward, showing the next two major groups. What will be the atomic number of the next inert gas below Rn? What will be the structure of the eighth row (period) of the periodic table? Where do you expect the first g subshell to begin filling? What properties would you expect the g-subshell elements to have? What will be the atomic number of the second inert gas below radon?

2. Why do the $4s$ and $3d$ subshells appear so close in energy, when they belong to different principal quantum numbers n?

3. Would you expect element 107 to be a good conductor or a poor conductor? How about element 111? Do you expect element 112 to be paramagnetic or diamagnetic?

4. Zirconium frequently is present as an impurity in hafnium metal. Why?

5. Do you expect ytterbium (Yb) to become ferromagnetic at sufficiently low temperatures? What type of magnetic behavior would be expected at ordinary temperatures for polonium (Po)? For francium (Fr)?

6. As we move across the series of transition metal or rare earth elements, we add electrons to the d or f subshells. In chemical compounds, these elements often show valence states of +2, which correspond to removing two s electrons. Explain this apparent paradox.

7. Why do the rare earth (lanthanide) elements have such similar chemical properties? What property might you use to distinguish lanthanide atoms from one another?

8. Explain why the Bohr theory gives a poor accounting of optical transitions but does well in predicting the energies of X-ray transitions.

9. What can you conclude about the electronic configuration of an atom that has both $L = 0$ and $S = 0$ in the ground state?

10. Suppose we do a Stern-Gerlach experiment using an atom that has angular momentum quantum numbers L and S in its ground state. Into how many components will the beam split? Do you expect them to be equally spaced?

11. What is the degeneracy of a state of total orbital angular momentum L that has $S = 0$? What is the degeneracy of a state of total spin angular momentum S that has $L = 0$? What is the *total* degeneracy of a state in which both L and S are nonzero?

12. What L and S values must an atom have in order to show the *normal* Zeeman effect? Does this apply only to the ground state or to excited states also? Can an atom show the normal Zeeman effect in some transitions and the anomalous Zeeman effect in other transitions? Could the same atom even show no Zeeman effect at all in some transitions?

13. Based on the rules for coupling electron l and s values to give the total L and S, explain why filled subshells don't contribute to the magnetic properties of an atom.

14. If an atom in its ground state has $S = 0$, can you infer whether it has an even or an odd number of electrons? What if $L = 0$?

15. The L atomic shell actually contains three distinct levels: a $2s$ level and two $2p$ levels (a fine-structure doublet). If we look carefully at the K_α X ray under high resolution, we see two, not three, different components. Explain this discrepancy.

16. The K_α energies computed using Eq. 8.4 are about 0.1% low for $Z = 20$, 1% low for $Z = 40$, and 10% low for $Z = 80$. Why does the simple theory fail for large Z? Could it be because the screening effect has not been handled correctly and that Z_{eff} is not $Z - 1$? If not, can you suggest an alternative reason?

17. The first excited state in sodium is a fine-structure doublet; the wavelengths emitted in the decay of these states are 589.59 nm and 589.00 nm, a difference of 0.59 nm. The excited $4s^1$ state in sodium (see Figure 8.4) decays to the $3p$ doublet with the emission of radiation at the wavelengths 1138.15 nm and 1140.38 nm, a difference of 2.23 nm. Explain how the $3p$ fine structure can give a wavelength difference of 0.59 nm in one case and 2.23 nm in the other case.

18. Suppose we had a three-level atom, like that of Figure 8.21, in which the metastable state were the higher excited state; the lasing transition would then be the upper transition. Does this atom solve the problem of absorption of the lasing transition? Would such an atom make a good laser?

19. How does a laser beam differ from a point source of light? Contrast the change in beam intensity with distance from the source for a laser and a point source.

20. Explain what is meant by a population inversion and why it is necessary for the operation of a laser.

21. How could you demonstrate that laser light is coherent? What would be the result of the same experiment using an ordinary monochromatic source? A white light source?

Problems

8.1 The Pauli Exclusion Principle

1. (a) List the six possible sets of quantum numbers (n, l, m_l, m_s) of a $2p$ electron. (b) Suppose we have an atom such as carbon, which has two $2p$ electrons. Ignoring the Pauli principle, how many different possible combinations of quantum numbers of the two electrons are there? (c) How many of the possible combinations of part (b) are eliminated by applying the Pauli principle? (d) Suppose carbon is in an excited state with configuration $2p^1 3p^1$. Does the Pauli principle restrict the choice of quantum numbers for the electrons? How many different sets of quantum numbers are possible for the two electrons?

2. Nitrogen $(Z = 7)$ has three electrons in the $2p$ level (in addition to two electrons each in the $1s$ and $2s$ levels). (a) Consistent with the Pauli principle, what is the maximum possible value of the total m_s of all seven electrons? (b) List the quantum numbers of the three $2p$ electrons that result in the largest total m_s. (c) If the electrons in the $2p$ level occupy states that maximize m_s, what would be the maximum possible value for the total m_l? (d) What would be the maximum possible total m_l if the three $2p$ electrons were in states that did not maximize m_s?

3. (a) How many different sets of quantum numbers (n, l, m_l, m_s) are possible for an electron in the $4f$ level? (b) Suppose a certain atom has three electrons in the $4f$ level. What is the maximum possible value of the total m_s of the three electrons? (c) What is the maximum possible total m_l of three $4f$ electrons? (d) Suppose an atom has ten electrons in the $4f$ level. What is the maximum possible value of the total m_s of the ten $4f$ electrons? (e) What is the maximum possible total m_l of ten $4f$ electrons?

8.2 Electronic States in Many-Electron Atoms

4. (a) Suppose a beryllium atom $(Z = 4)$ absorbs energy (such as from a beam of photons) that pushes one of the electrons to an excited state. If the photon energy is set at the

minimum necessary for this to occur, from which subshell does the electron make the transition and to which subshell does it jump? (b) Suppose the same experiment is done with neon ($Z = 10$). At the minimum energy for absorption, from which subshell does the electron make the transition and to which subshell does it jump? (c) Would you expect the minimum absorption energy for beryllium to be larger or smaller than the minimum energy for neon? Explain.

5. (a) List all elements with a p^3 configuration. (b) List all elements with a d^7 configuration.

6. Give the electronic configuration of (a) P; (b) V; (c) Sb; (d) Pb.

7. (a) What is the electronic configuration of Fe? (b) In its ground state, what is the maximum possible total m_s of its electrons? (c) When the electrons have their maximum possible total m_s, what is the maximum total m_l? (d) Suppose one of the d electrons is excited to the next highest level. What is the maximum possible total m_s, and when m_s has its maximum total what is the maximum total m_l?

8.3 Outer Electrons: Screening and Optical Transitions

8. The ground state of singly ionized lithium ($Z = 3$) is $1s^2$. Use the electron screening model to predict the energies of the $1s^1 2p^1$ and $1s^1 3d^1$ excited states in singly ionized lithium. Compare your predictions with the measured energies (respectively $-13.4\,\text{eV}$ and $-6.0\,\text{eV}$).

9. The ground state of neutral beryllium ($Z = 4$) is $1s^2 2s^2$. Use the electron screening model to predict the energies of the following excited states: $1s^2 2s^1 3p^1$ (measured $-2.02\,\text{eV}$) and $1s^2 2s^1 4d^1$ ($-0.90\,\text{eV}$).

10. Using the wavelengths given in Figure 8.4, compute the energy difference between the $3d$ and $4d$ states in lithium; do the same for sodium. Compare those values with the corresponding $n = 4$ to $n = 3$ energy difference in hydrogen. Why is the agreement so good, considering the different values of Z?

11. (a) Using the information for lithium given in Figure 8.4, compute the energy difference of the $3p$ and $3d$ states. (b) Compute the energy of the $3s, 4s$, and $5s$ states above the ground state. (c) The ionization energy of lithium in its ground state is $5.39\,\text{eV}$. What is the ionization energy of the $2p$ state? Of the $3s$ state?

8.5 Inner Electrons: Absorption Edges and X Rays

12. A certain element emits a K_α X ray of wavelength $0.1940\,\text{nm}$. Identify the element.

13. Compute the K_α X ray energies of calcium ($Z = 20$), zirconium ($Z = 40$), and mercury ($Z = 80$). Compare with the measured values of $3.69\,\text{keV}$, $15.8\,\text{keV}$, and $70.8\,\text{keV}$. (See Question 16.)

8.6 Addition of Angular Momenta

14. Chromium has the electron configuration $4s^1 3d^5$ beyond the inert argon core. What are the ground-state L and S values?

15. Use Hund's rules to find the ground-state L and S of (a) Ce, configuration $[\text{Xe}]6s^2 4f^1 5d^1$; (b) Gd, configuration $[\text{Xe}]6s^2 4f^7 5d^1$; (c) Pt, configuration $[\text{Xe}]6s^1 4f^{14} 5d^9$.

16. Using Hund's rules, find the ground-state L and S of (a) fluorine ($Z = 9$); (b) magnesium ($Z = 12$); (c) titanium ($Z = 22$); (d) iron ($Z = 26$).

17. A certain excited state of an atom has the configuration $4d^1 5d^1$. What are the possible L and S values?

18. Use the degeneracies of the states with all possible total L and S to find how many different levels the $2p^1 3p^1$ excited state of carbon includes. (See Figure 8.18.) Compare this result with the result of counting the individual m_l and m_s values from Problem 1(d). (See also Question 11.)

8.7 Lasers

19. A small helium-neon laser produces a light beam with an average power of $3.5\,\text{mW}$ and a diameter of $2.4\,\text{mm}$. (a) How many photons per second are emitted by the laser? (b) What is the amplitude of the electric field of the light wave? Compare this result with the electric field at a distance of 1 m from an incandescent light bulb that emits $100\,\text{W}$ of visible light.

General Problems

20. (a) How many different possible ways are there to assign the sets of quantum numbers to the four $2p$ electrons in oxygen ($Z = 8$)? (b) List all possible values of the total m_s for the four electrons. (c) List all possible values of the total m_l of the four electrons. (d) If the total m_s has its largest possible value, what are the possible values of the total m_l? (e) If the total m_l has its largest possible value, what are the possible values of the total m_s?

21. (a) The ionization energy of sodium is $5.14\,\text{eV}$. What is the effective charge seen by the outer electron? (b) If the $3s$ electron of a sodium atom is moved to the $4f$ state, the measured binding energy is $0.85\,\text{eV}$. What is the effective charge seen by an electron in this state?

22. Draw a Moseley plot, similar to Figure 8.16, for the K_β X rays using the following energies in keV:

Ne	0.858	Mn	6.51	Zr	17.7
P	2.14	Zn	9.57	Rh	22.8
Ca	4.02	Br	13.3	Sn	28.4

Determine the slope and compare with the expected value. (Equation 8.4 applies only to K_α X rays; you will need to derive a similar equation for the K_β X rays.) Determine the z-axis intercept and give its interpretation.

23. Draw a Moseley plot, similar to Figure 8.16, for the L_α X rays using the following energies in keV:

Mn	0.721	Rh	2.89
Zn	1.11	Sn	3.71
Br	1.60	Cs	4.65
Zr	2.06	Nd	5.72

Give interpretations of the slope and intercept.

24. Because of the fine-structure splitting of the $3p$ state, the $3p \rightarrow 3s$ transition in sodium actually consists of two closely spaced lines of wavelengths 589.00 nm and 589.59 nm. Assuming a magnetic moment of one Bohr magneton, find the effective magnetic field that produces the fine-structure splitting of the $3p$ state of sodium.

25. (a) What is the longest wavelength of the absorption spectrum of lithium? (b) What is the longest wavelength of the absorption spectrum of helium? In what region of the spectrum does this occur? (c) What are the *shortest* wavelengths in the absorption spectra of helium and lithium? In what region of the electromagnetic spectrum are these?

26. Using the wavelengths given in Figure 8.17, compute the energy difference between the $1s^1 4p^1$ and $1s^1 3p^1$ singlet ($S = 0$) states in helium. Compare this energy difference with the value expected using the Bohr model, assuming that the p electron is screened by the s electron. Repeat the calculation for the $3d$ and $4d$ triplet ($S = 1$) states.

Chapter 9

MOLECULAR STRUCTURE

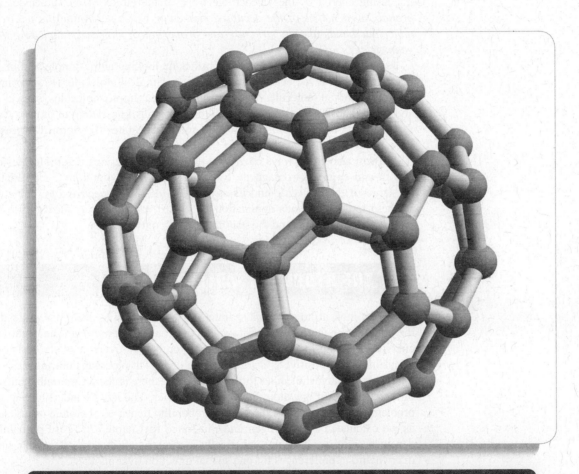

Molecules range from the simple with only two atoms to very complex organic molecules such as DNA. The photo shows a computer model of C_{60}, a spherical arrangement of 60 carbon atoms in pentagons and hexagons, known as a "buckyball."

In this chapter we consider the combination of atoms into molecules, the excited states of molecules, and the ways that molecules can absorb and emit radiation. From a variety of experiments we learn that the spacing of atoms in molecules is of the order of 0.1 nm, and that the binding energy of an atom in a molecule is of the order of electron-volts. This spacing and binding energy are characteristic of electronic orbits, which suggests that the forces that bind molecules together originate with the electrons. The negatively charged electrons provide the binding that overcomes the Coulomb repulsion of the positively charged nuclei of the atoms in the molecule.

When atoms are brought together to form molecules, the atomic states of the electrons change into molecular states. These states are filled in the order of increasing energy by the valence electrons of the atoms of the molecule. *The probability densities of the occupied molecular states determine the nature of molecular bonds and the structure and properties of molecules, including their geometrical shapes.*

Just as we began to study atomic physics by looking at the simplest atom, we begin our study of molecular physics with the simplest molecule, H_2^+, the singly ionized hydrogen molecule. We next turn to other simple molecules, such as H_2 and NaCl, and finally we look at how our previous knowledge of atomic wave functions can help us to understand the molecular states that form the basis of organic chemistry.

We will also study ways other than electronic excitations that molecules can absorb and emit electromagnetic radiation. These radiations give a distinctive signature of the molecule and its structure. *Molecular spectroscopy*, the study of these radiations, finds application in such diverse areas as identification of atmospheric pollutants and the search for life in outer space.

9.1 THE HYDROGEN MOLECULE

Let's first look at how the wave functions of the atomic electrons can lead to the binding together of atoms into stable molecules. Even though the negatively charged electrons provide the attractive force that overcomes the Coulomb repulsion of the positively charged atomic nuclei, it is perhaps not immediately obvious how stable molecules form at all because there is also a Coulomb repulsion of the *electrons* of one atom for those of another. The key to understanding this problem is the existence of the spatial probability densities of atomic orbits, such as we calculated for hydrogen and illustrated in Chapter 7. These probability densities are frequently not spherically symmetric, and very often may show overwhelming preferences for one spatial direction over another.

A complete understanding of the effect of the electrons on molecular binding is in general made difficult by what also complicates atomic structure—there are too many electrons present for us to be able to write down and solve the equations that govern the structure of the atom or molecule. We therefore use the same tactic to study molecular structure that we used for atomic structure: we begin with a molecule that has only one electron. Such a molecule is H_2^+, *the hydrogen molecule ion*, which results when we remove an electron from a molecule of ordinary hydrogen, H_2.

Before we discuss the wave mechanical properties of H_2^+, let's try to guess what holds this molecule together. We first realize that it is *not* correct to think of H_2^+ as an atom of hydrogen (proton plus electron) joined to a second proton.

FIGURE 9.1 The electron wave functions for two hydrogen atoms separated by a large distance.

The atom of hydrogen in such a combination is electrically neutral, so there is no electrostatic Coulomb force to hold the two pieces together. In this kind of molecule, at least, it is apparently not correct to identify the electron as belonging exclusively to one or the other of the components. *The electron must somehow be shared between the two parts.* The electron must spend a significant part of its time in the region between the two protons. In the language of quantum mechanics, the electron's probability density must have a large value in that region.

As we learned in Chapter 7, an electron in the ground state of hydrogen has an energy of $-13.6\,\text{eV}$, a wave function $\psi = (\pi a_0^3)^{-1/2} e^{-r/a_0}$, where a_0 is the Bohr radius, and a probability density proportional to ψ^2. Figure 9.1 shows the wave function for an electron that could be bound to either of two protons separated by a large distance. As we bring the two protons closer together, the wave functions begin to overlap, and we must combine them according to the rules of quantum mechanics—first add the wave functions, then square the result to find the combined probability density. (Note that this gives a very different result from first squaring, then adding.)

We can combine these two wave functions in two different ways, depending on whether they have the same signs or opposite signs. The absolute sign of a wave function is arbitrary. When we calculate the normalization constant of a wave function, we actually compute its square. We could choose either the positive or the negative root; for convenience we usually choose the positive one. When we calculate probability densities ψ^2 for a single wave function, the choice of sign becomes irrelevant. However, when we combine different wave functions, their *relative* signs determine whether the two functions add or subtract, which can result in very different probability densities.

Consider the two different combinations of wave functions shown in Figure 9.2. In one case (Figure 9.2a), the two wave functions have the same sign, and in the other case (Figure 9.2b) they have opposite signs. This has a substantial effect on the probability distributions, which are shown in Figure 9.3. The probability density obtained from squaring $\psi_1 + \psi_2$ (Figure 9.3a) has relatively large values in the region between the two protons. This suggests a concentration of negative charge between the protons, which can supply the Coulomb attraction to pull the two protons together and form a stable molecule. The square of $\psi_1 - \psi_2$ (Figure 9.3b), however, gives a vanishing probability density midway between the protons and thus a small density of negative charge in the region between the protons. There is not enough negative charge to overcome the Coulomb repulsion of the protons, and as a result this combination of wave functions does not lead to the formation of a stable molecule.

Binding Energy of H$_2^+$

There are two contributions to the energy of the H$_2^+$ molecule: the Coulomb repulsion of the two positively charged protons for each other, and the attraction of the combination of the two protons for the negatively charged electron.

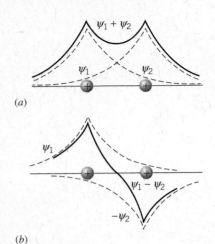

(a)

(b)

FIGURE 9.2   The overlap of two hydrogenic wave functions. The wave functions are indicated by the dashed lines, and their sum by the solid line. In (a), the two wave functions have the same sign, while in (b) they have opposite signs.

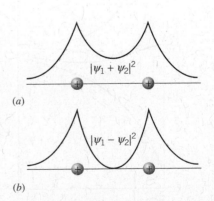

(a)

(b)

FIGURE 9.3 The probability densities corresponding to the two combined wave functions of Figure 9.2.

The Coulomb repulsion energy of the protons is positive, and the energy of the attraction of the electron by the protons is negative. For a stable molecule to form, the total energy must be negative, so the critical question is whether the electrons can provide enough negative energy of attraction to overcome the positive repulsion energy of the protons.

To find the conditions necessary for a stable H_2^+ ion to form, let's look at how the various contributions to the energy of the ion depend on the separation distance R between the two protons. The Coulomb potential energy that characterizes the repulsion of the bare protons is $U_p = e^2/4\pi\varepsilon_0 R$; this function is plotted in Figure 9.4. To find the electron energy as a function of R, we first consider the case when the two protons are very far apart. In this case the electron is in the ground state orbit about one of the protons, for which $E = -13.6\,\text{eV}$. As we bring the protons together, the electron becomes more tightly bound (because it is attracted by both protons) and its energy becomes more negative. As $R \to 0$, the system approaches a single atom with $Z = 2$. For the wave function $\psi_1 + \psi_2$ (Figure 9.2a), the combined wave function has a maximum at $R = 0$ and resembles the ground-state wave function of an atom with $Z = 2$. Recalling the result from Chapter 6 for the electron energy in a hydrogen-like atom,

$$E_n = (-13.6\,\text{eV})\frac{Z^2}{n^2} \tag{9.1}$$

where n is the principal quantum number, we find the energy of a ground-state electron for $Z = 2$ to be $-54.4\,\text{eV}$. The energy corresponding to the sum of the two wave functions, which we label E_+, therefore has the value $-13.6\,\text{eV}$ at large R and approaches the value $-54.4\,\text{eV}$ at small R. The result of an exact calculation of E_+ is shown in Figure 9.4.

For the combination corresponding to the difference between the two wave functions, the energy is once again $-13.6\,\text{eV}$ for large R. As $R \to 0$, the combined wave function approaches 0 (Figure 9.2b). The lowest energy level with a wave function that vanishes at $R = 0$ is the $2p$ state, for which the energy in a $Z = 2$ hydrogenlike atom is $-13.6\,\text{eV}$. The energy E_- corresponding to the wave function $\psi_1 - \psi_2$ therefore has the value $-13.6\,\text{eV}$ for both large R and small R. Its exact form is shown in Figure 9.4.

The total energy of the hydrogen molecule ion is the sum of the proton energy U_p and the electron energy E_+ or E_-. These two sums are also plotted in Figure 9.4. You can see that the combination $U_p + E_-$ has no minimum and therefore no stable bound state. The wave function $\psi_1 - \psi_2$ does not lead to a stable configuration for the hydrogen molecule ion, just as we originally suspected.

The sum $U_p + E_+$ gives the stable configuration of the ion, for which the equilibrium condition occurs at the point where $U_p + E_+$ has its minimum value. The minimum occurs at a separation $R_{eq} = 0.106\,\text{nm}$ and an energy of $-16.3\,\text{eV}$. The binding energy B of H_2^+ is the energy necessary to take apart the ion into H and H^+ and corresponds to the depth of the potential energy minimum of $U_p + E_+$ in Figure 9.4:

$$B = E(H + H^+) - E(H_2^+) = -13.6\,\text{eV} - (-16.3\,\text{eV}) = 2.7\,\text{eV} \tag{9.2}$$

Note that we have defined molecular binding energy as the energy difference between the separate components (H and H^+) and the combined system (H_2^+).

It is interesting to note that the stability is achieved at $R_{eq} = 2a_0$. In Chapter 7 we learned that the radial probability density for the $1s$ state of hydrogen has its

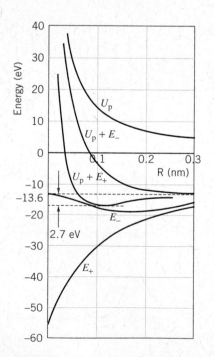

FIGURE 9.4 Dependence of energy on separation distance R for H_2^+.

maximum value at $r = a_0$. Thus the stable configuration of the H_2^+ ion is such that the maximum in the radial probability density for a single H atom would fall exactly in the middle of the molecule! This is once again consistent with our expectations for the structure of H_2^+—the electron must spend most of its time between the two protons.

In summary, from our study of this simple molecule we have learned that an important feature of molecular bonding concerns the *sharing* of a single electron by two atoms of the molecule. This sharing is responsible for the stability of the molecule. With this in mind we can now add a second electron and consider the H_2 molecule.

The H₂ Molecule

Suppose we have two hydrogen atoms separated by a very large distance. Associated with each atom there is a $1s$ electronic state, at an energy of -13.6 eV, because the atoms are so far apart that there is no interaction between the electrons. As we bring the atoms closer together to form a H_2 molecule, the electron wave functions begin to overlap, so that the electrons are "shared" between the two atoms. As we have seen in the previous discussion, this can occur in such a way that the two electron wave functions *add* in the region between the two protons, giving a stable molecule, or *subtract*, leading to no stable molecule. The separate, individual electronic states of the atoms now become *molecular* states.

Notice that, as shown in Figure 9.5, the number of states does not change as the separation R is reduced. When the atoms are separated by a large distance, there are two states, each at -13.6 eV, so the total energy at $R = \infty$ is -27.2 eV. When the separation is reduced, there are still two states, but now at different energies. One state corresponds to the sum of the two wave functions and leads to a stable H_2 molecule; the other state corresponds to the difference of the two wave functions and does not give a stable molecule. The molecular state that leads to a stable molecule is known as a *bonding* state, and the one that does not lead to a stable molecule is an *antibonding* state.

As we found previously for H_2^+, in order to form a molecule, the electron probability distribution must be large in the region between the two protons. In the case of H_2, this is true for both electrons, and it is certainly our expectation, based on the Pauli principle, that for the two electrons *both* to occupy the *molecular* state leading to the large probability in the central region, their spins must be oppositely directed; that is, one must have $m_s = +1/2$ and the other $m_s = -1/2$. As long as the two electrons have opposite spins, they can *both* occupy the bonding state, leading to a stable molecule.

The energy of the bonding state for H_2 is shown in Figure 9.6; as you can see, there is a minimum with $E = -31.7$ eV at $R = 0.074$ nm. The *molecular binding energy* of H_2 is the difference between the energy of the separated neutral H atoms and the energy of the combined system:

$$B = E(H + H) - E(H_2) = 2(-13.6 \text{ eV}) - (-31.7 \text{ eV}) = 4.5 \text{ eV} \quad (9.3)$$

Comparing Figures 9.4 and 9.6, you can immediately see the effect of adding an additional electron to H_2^+: the binding energy is greater (the molecule is more tightly bound), and the protons are drawn closer together. Both of these effects are due to the presence of the increased electron density in the region between the two protons.

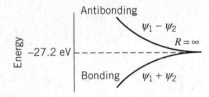

FIGURE 9.5 Energy of different combinations of wave functions in H_2.

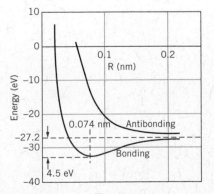

FIGURE 9.6 Bonding and antibonding in H_2.

We can also understand why He does not form the molecule He_2—as two He atoms are brought together, the bonding and antibonding states are formed in much the same way as with H_2. The He_2 molecule would have four electrons; at most two can be in the bonding state, so the other two must be in the antibonding state. The net effect is that no stable molecule forms. (However, He_2^+ is stable, with two bonding electrons and only *one* antibonding electron. The binding energy of He_2^+ is 3.1 eV and the separation is 0.108 nm, remarkably close to the corresponding values of H_2^+.)

9.2 COVALENT BONDING IN MOLECULES

The sharing of electrons in a molecule such as H_2 is the origin of the *covalent* bond; this type of bonding occurs commonly in molecules containing two identical atoms, in which case it is called *homopolar* or *homonuclear* bonding.

The essential features of covalent bonding are:

1. As two atoms are brought together, the electrons interact and the separate atomic states and energy levels are transformed into molecular states.
2. In one of the molecular states, the electron wave functions overlap in such a way as to give a *lower* energy than the separated atoms had; this is the bonding state that leads to the formation of stable molecules.
3. The other molecular state (the antibonding state) has an increased energy relative to the separated atoms and does not lead to the formation of stable molecules.
4. The restrictions of the Pauli principle apply to molecular states just as they do to atomic states; each molecular state has a maximum occupancy of two electrons, corresponding to the two different orientations of electron spin.

Other hydrogenlike atoms with a single s electron can also form stable molecules through covalent bonding. For example, two Li atoms ($Z = 3$, configuration $1s^2 2s^1$) can form a molecule of Li_2. The four $1s$ electrons (two from each atom) fill the $1s$ bonding and antibonding states, and the remaining two $2s$ electrons can both occupy the $2s$ bonding state. The binding energy of Li_2 is 1.10 eV, which is considerably smaller than the binding energy of H_2 (4.52 eV), and the equilibrium separation distance of the atoms in the molecule is 0.267 nm, much larger than that of H_2 (0.074 nm).

Other homonuclear molecules formed from s-state bonds are listed in Table 9.1. It is customary to characterize the molecular bond strength in terms of the *dissociation energy* rather than the binding energy; the two terms are usually equivalent and indicate the energy needed to break the molecule into neutral atoms. The dissociation energy is weakly temperature dependent. Some tabulations list the values at room temperature (as in Table 9.1), while others list values at 0 K. The room-temperature values are higher than the 0 K values by about $1.5kT = 0.04$ eV.

As Z increases, meaning that the s electrons are associated with increasing principal quantum numbers n, the dissociation energy decreases and the equilibrium separation increases. This is consistent with the behavior of the s electron in atoms as n increases—as Figure 8.7 shows, the radius of the orbit of the s electron increases with increasing n for the alkali elements.

TABLE 9.1 Properties of *s*-Bonded Molecules*

Molecule	Dissociation Energy (eV)	Equilibrium Separation (nm)
H_2	4.52	0.074
Li_2	1.10	0.267
Na_2	0.80	0.308
K_2	0.59	0.392
Rb_2	0.47	0.422
Cs_2	0.43	0.450
LiH	2.43	0.160
LiNa	0.91	0.281
NaH	2.09	0.189
KNa	0.66	0.347
NaRb	0.61	0.359

*Values taken from the *Handbook of Chemistry and Physics* and the *American Institute of Physics Handbook*.

We can also form molecular bonds with two different alkali elements. Some of these are listed in Table 9.1. The dissociation energies and equilibrium separations are consistent with those of the corresponding homonuclear molecules. For example, the dissociation energy and equilibrium separation of LiH are midway between those of H_2 and Li_2.

Atoms with valence electrons in *p* states can also form diatomic molecules through covalent bonds—oxygen and nitrogen, for example. There are three atomic *p* states, so there will be six molecular *p* states, and the classification of levels can become quite tedious, but we can understand the structure of molecules composed of atoms with *p* electrons based on the geometry of atomic *p* states.

In Chapter 7 we solved the Schrödinger equation for the H atom and showed the spatial probability distributions for the various possible electronic wave functions. Of course, these solutions for hydrogen will *not* be correct for other atoms, but the essential features of the geometry of the atomic states remains correct. We identified three different *p* states, corresponding to $m_l = -1$, 0, and +1. The probability distributions corresponding to these m_l values were shown in Figure 7.11.

We can imagine these distributions to have a sort of "figure-eight" shape with two distinct lobes of large probability. In the $m_l = 0$ case, the figure eight has its long axis along the *z* axis, and the two lobes of maximum probability occur in the +*z* and −*z* directions. In the $m_l = \pm 1$ cases, the probability distribution can be regarded as occurring from a figure eight probability distribution in the *xy* plane that is rotating about the *z* axis, counterclockwise for $m_l = +1$ and clockwise for $m_l = -1$. Because of the uncertainty principle, we can't observe the two probability lobes in the *xy* plane; all we can observe is the smeared-out "donut-shaped" distribution shown in Figure 7.11.

For our purposes here, it is not as convenient to use the m_l notation as it is to use a different representation in which we assign each of the three possible *p* states

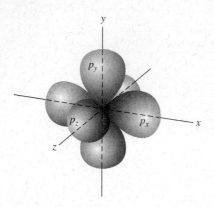

FIGURE 9.7 Probability distributions of three different p electrons.

a label that gives the direction in space corresponding to the lobes of maximum probability. Thus p_z is the state with regions of large electron probability along the z axis, and similarly for p_x and p_y. Figure 9.7 shows a schematic representation of these probability distributions. (The p_z state corresponds exactly to $m_l = 0$; p_x and p_y correspond to *mixtures* of $m_l = +1$ and $m_l = -1$.) Just as the uncertainty principle does not allow us to observe the two lobes of probability in the xy plane, it also forbids us from observing the separate p_x and p_y probability distributions. However, the distributions do exist (even through we can't observe them), and two atoms can interact with one another by means of these electron clouds.

We consider the structure of molecules containing p electrons based on this model of the three mutually perpendicular p states p_x, p_y, and p_z. We discuss three applications of this type of covalent bonding: pp bonds, sp directed bonds, and sp hybrid states.

pp Covalent Bonds

Consider what happens when we bring together two p-shell atoms, whose probability distributions are each similar to Figure 9.7. We assume that the atoms approach along the x axis, as in Figure 9.8. As the atoms are brought together, the p_x states overlap (Figure 9.9a), giving (if the two wave functions add) an increased electron charge density between the two nuclei and contributing to the bonding of the atoms in the molecule. There is a much weaker overlap between the p_y states (Figure 9.9b) and also between the p_z states (which are not shown in the figure). Because the overlap of the p_y states is not along the line connecting the nuclei, there are components to the binding force that oppose one another, and only a much smaller resultant force acts along the line connecting the nuclei (Figure 9.9b). In addition, there is less overlap of the p_y states. The net result is that the p_y states (and also the p_z states) are less effective in binding the molecule than the p_x states.

This somewhat oversimplified model suggests that the p_x state should have a much greater bonding effect (and also a greater antibonding effect) than the p_y and p_z states. It also suggests that the bonding and antibonding effects of p_y states should be the same as those of p_z states.

Now we can consider the energies of the molecular states as a function of the nuclear separation distance R. We assume that we are dealing with two atoms

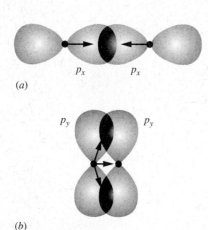

FIGURE 9.9 (*a*) Overlap of p_x probability distributions. The vectors indicate the force on the nuclei due to the overlap. (*b*) Overlap of p_y probability distributions. The off-axis forces give a smaller resultant force along the axis.

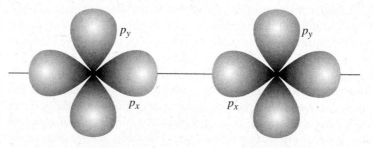

FIGURE 9.8 Two atoms with p electrons. The p_z probability distribution, which extends perpendicular to the page, is not shown.

having filled $1s$ and $2s$ states and valence electrons in the $2p$ shell. When the $1s$ states of the two atoms overlap, the result is $1s$ bonding and antibonding *molecular* states, just as in the case of H_2. There are altogether four $1s$ electrons in the molecule, and with two in each state the $1s$ bonding and antibonding molecular states are filled to capacity. The same is true of the $2s$ states. The *atomic $2s$* levels form bonding and antibonding *molecular* states; because each atom has a filled $2s$ shell, the four $2s$ electrons fill both the bonding and antibonding molecular states.

The atoms have partially filled $2p$ shells, so the final molecular bonding depends critically on the molecular $2p$ states. For each *atomic p* state (p_x, p_y, p_z) there are corresponding bonding and antibonding *molecular* states. However, the bonding and antibonding effects of these states are not equivalent, as Figure 9.9 illustrates. The p state that happens to lie along the line of approach (p_x) has an effect that is significantly greater than the p states that lie off the line of approach (p_y, p_z). The p_x bonding state must therefore lie lower in energy than the p_y and p_z bonding states, and the p_x antibonding state must lie higher in energy than the p_y and p_z antibonding states. Figure 9.10 illustrates the energies of the molecular states. The relative stability of a molecule can be determined based on the filling of the bonding and antibonding states with electrons (two per state, corresponding to spin up and spin down electrons). The following example illustrates how these states are filled.

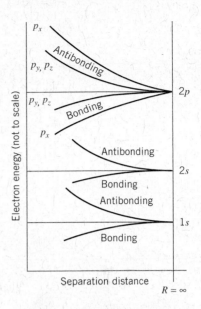

FIGURE 9.10 Bonding and antibonding $2p$ states.

Example 9.1

Based on the filling of the bonding and antibonding states, predict the relative stability of the molecules (a) N_2, (b) O_2, and (c) F_2.

Solution

(a) Nitrogen ($1s^2 2s^2 2p^3$) has seven electrons: two each in the filled $1s$ and $2s$ shells, and three electrons in the $2p$ shell. In the N_2 molecule, there are therefore 14 electrons. We start in Figure 9.10 with two in the bonding $1s$ state, then two in the antibonding $1s$, then two more in the bonding $2s$, followed by two in the antibonding $2s$ for a total of eight electrons in the s states. That leaves six $2p$ electrons for the $2p$ molecular states. We can place two each in the three lowest $2p$ bonding molecular states, thus filling those states. No electrons go into the $2p$ antibonding states. With

only bonding $2p$ electrons, N_2 forms a very stable diatomic molecule.

(b) Oxygen ($1s^2 2s^2 2p^4$) has eight electrons, so the O_2 molecule has a total of 16 electrons. As in N_2, the first eight electrons fill the $1s$ and $2s$ states, leaving eight additional electrons for the $2p$ states. The first six of those fill the three bonding states, and so the remaining two must go into $2p$ antibonding states. With six bonding and two antibonding valence electrons, we would expect that O_2 is less stable than N_2, which has only bonding valence electrons.

(c) Fluorine ($1s^2 2s^2 2p^5$) has nine electrons, so of the 18 electrons in the F_2 molecule 10 must be placed in the $2p$ states: six in the bonding states and four in the antibonding states. Thus F_2 should be less stable than O_2.

How well do the properties of these molecules agree with our predictions? N_2 has a dissociation energy of 9.8 eV and is not reactive under most circumstances. O_2 has a smaller dissociation energy (5.1 eV); the O_2 molecular bonds can be broken by relatively modest chemical reactions, as, for example, the oxidation of metals exposed to air. F_2 has an even smaller dissociation energy (1.6 eV);

fluorine gas reacts quite violently with many substances, and the F_2 molecule can be broken apart by exposure to visible light (which has photon energies of $2-4\,eV$) in a process known as *photodissociation*. The properties of these $2p$ molecules are thus quite consistent with expectations based on the filling of the bonding and antibonding states. Similar relationships occur for the $3p$, $4p$, $5p$, and $6p$ homonuclear molecules.

sp Molecular Bonds

It is often the case that a stable molecule is formed from two atoms, one with an s-state valence electron and the other with one or more p-state valence electrons. Consider, for example, the HF molecule. The F atom has five electrons in the p shell, so of the three $2p$ atomic states, two will each have their capacity of two electrons, and the third will have a single electron. We ignore the four paired p electrons, which do not significantly affect the molecular bonding, and concentrate instead on the single unpaired p electron. The two-lobed probability distribution corresponds to a two-lobed $p-$state wave function, in which the signs of ψ are opposite for the two lobes. The $1s$ wave function of H has only one sign (Figure 9.11). As the H and F atoms approach each other from a large distance, the H wave function and the F wave function can combine to give an increased electron probability in the region between the nuclei, and hence a *bonding sp* state is formed. It is also possible to have *antibonding sp* states, which result from the H and F wave functions having opposite signs and producing a reduced electron probability density between the nuclei.

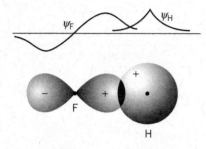

FIGURE 9.11   Overlap of s and p wave functions.

Table 9.2 gives dissociation energies and nuclear separation distances for some sp-bonded diatomic molecules.

Consider now the structure of the water molecule, H_2O. Oxygen has eight electrons, four of which occupy the $2p$ shell. When we place these electrons in the $2p$ *atomic* states, we begin with one electron each in the p_x, p_y, and p_z states, and then the fourth $2p$ electron must pair with one of the first three. An oxygen atom therefore has two unpaired $2p$ electrons, each of which can form a bond with the

TABLE 9.2 Properties of *sp*-Bonded Molecules

Molecule	Dissociation Energy (eV)	Equilibrium Separation (nm)
HF	5.90	0.092
HCl	4.48	0.128
HBr	3.79	0.141
HI	3.10	0.160
LiF	5.98	0.156
LiCl	4.86	0.202
NaF	4.99	0.193
NaCl	4.26	0.236
KF	5.15	0.217
KCl	4.43	0.267

$1s$ electron of H to form a molecule of H_2O. Figure 9.12 shows a representation of the electron probability distributions we might expect for an oxygen atom and for a molecule of H_2O. Such a molecule has *directed* bonds, which have a fixed, measurable relative direction in space. The expected angle between the two bonds is 90°; this angle can be measured experimentally by, for example, measuring the electric dipole moment of the atom, and the result, 104.5°, is somewhat larger than we expect. This discrepancy can be interpreted as arising from the Coulomb repulsion of the two H atoms, which tends to spread the bond angle somewhat.

As another example, consider the NH_3 (ammonia) molecule. With $Z = 7$, the nitrogen atom has three unpaired p electrons, one each in the p_x, p_y, and p_z atomic states. Each of these can form a bond with a H atom to form the NH_3 molecule, and we expect to find three mutually perpendicular sp bonds (Figure 9.13). The measured bond angle is 107.3°, again indicating some repulsion between the H atoms.

Table 9.3 lists some bond angles measured for other molecules that have sp directed bonds. As you can see, the bond angle does indeed approach 90° in many cases. Based on the discussion given above, you should be able to explain why this happens as the Z of the central atom increases.

sp Hybrid States

One example of a $2p$ atom we have so far not considered is carbon, and for a special reason: carbon forms a great variety of molecular bonds, with a resulting diversity in the type and complexity of molecules containing carbon. It is this diversity that is the basis for the many kinds of *organic molecules* that can form, based on various kinds of carbon molecular bonds, and so an understanding of the physics of carbon molecular bonds is essential to the understanding of many fundamental questions of structure and processes in molecular biology.

Carbon, with six electrons, has the configuration $1s^2 2s^2 2p^2$, so we expect carbon under ordinary circumstances to show a valence of 2, with the two $2p$ electrons contributing to the structure, and we might therefore expect to form stable molecules such as CH_2, with directed sp bonding (similar to H_2O) and a bond angle of roughly 90°. Instead, what forms is CH_4 (methane) in a tetrahedral structure (Figure 9.14), with four equivalent bonds. For another example, the elements of the third column of the periodic table (boron, aluminum, gallium, . . .) have the outer configuration $ns^2 np$ ($n = 2$ for boron, $n = 3$ for aluminum, etc.), and we expect these elements to form compounds as if they had a single valence p electron. We therefore expect halides such as BCl or GaF, oxides such as B_2O or Al_2O, nitrides such as B_3N or Al_3N, hydrides such as BH or GaH, and so forth. Instead we find that boron, aluminum, and gallium generally behave as if they had three valence electrons, and form compounds such as BCl_3, Al_2O_3, AlN, and B_2H_6. Furthermore, the three valence electrons seem to be equivalent; there seems to be no way, for example, to associate two of the valence electrons with s states and one with a p state. The bonds formed by the three electrons make equivalent angles of 120° with one another.

It is the effect of *sp hybridization* that is responsible for the valence of three (rather than one) in boron and four (rather than two) in carbon. The four bonds in CH_4 are *equivalent* and *identical*, which would *not* be expected if we had two ss bonds and two sp bonds; similarly, in BF_3 or BCl_3, the three bonds are identical and are clearly *not* identified with two sp bonds and one pp bond.

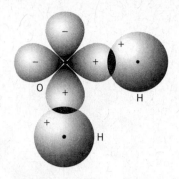

FIGURE 9.12   Overlap of electronic wave functions in H_2O.

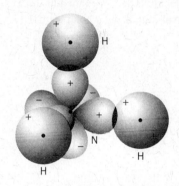

FIGURE 9.13   Overlap of electronic wave functions in NH_3.

TABLE 9.3 Bond Angles of *sp* Directed Bonds

Molecule	Bond Angle
H_2O	104.5°
H_2S	93.3°
H_2Se	91.0°
H_2Te	89.5°
NH_3	107.3°
PH_3	93.3°
AsH_3	91.8°
SbH_3	91.3°

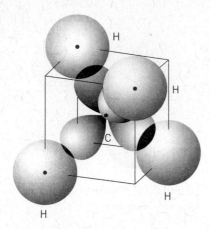

FIGURE 9.14   Tetrahedral arrangement of molecular bonds in CH_4.

The normal meaning of *hybrid* is an offspring resulting from the union of parents of different types, in which the offspring is not exactly like either parent, but retains some of the attributes of each. In the case of molecules, hybridization refers to a process by which the states can no longer be identified as either *s* or *p* states, but rather are mixtures of *s* and *p* states. The formation of *sp* hybrids is normally as follows:

1. In an atom with the configuration $2s^2 2p^n$, one of the $2s$ electrons is excited to the $2p$ shell, giving a configuration $2s^1 2p^{n+1}$.
2. The hybrid states are formed by taking equal mixtures of the wave functions representing the $2s$ state and each of the $2p$ states. For example, in the case of boron, the configuration of $2s^2 2p^1$ is converted to $2s^1 2p^2$. Assuming the $2p$ states to be $2p_x$ and $2p_y$, the resultant *hybrid* wave functions can be represented as different combinations of ψ_{2s}, ψ_{2p_x}, and ψ_{2p_y}, such as

$$\psi = \psi_{2s} + \psi_{2p_x} + \psi_{2p_y} \quad \text{or} \quad \psi = \psi_{2s} - \psi_{2p_x} + \psi_{2p_y} \quad \text{or} \quad \psi = \psi_{2s} + \psi_{2p_x} - \psi_{2p_y}$$

Illustrations of the probability distributions expected for sp, sp^2, and sp^3 hybrids are shown in Figure 9.15. Keep in mind that these do not yet represent *molecular* states—they are merely the contribution of one of the atoms to the bonding electronic distributions of the molecule.

The tetrahedral structure of CH_4 is therefore merely the result of the symmetrical spatial arrangement of the four sp^3 hybrid states of C, with each hybrid state bonded to one H. The bond angle of such a symmetrical tetrahedron is $109.5°$, in good agreement with the measured bond angles of CH_4 and other sp^3 hybrids, shown in Table 9.4.

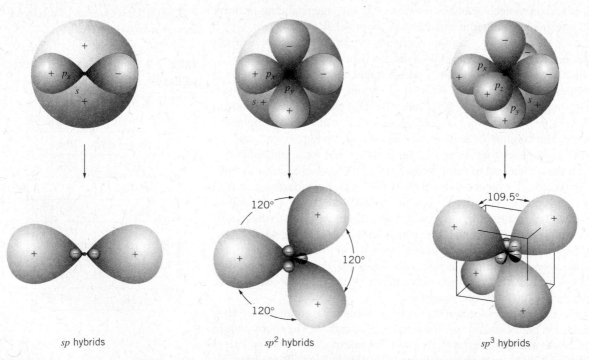

FIGURE 9.15   Probability distributions in sp, sp^2, and sp^3 hybrids.

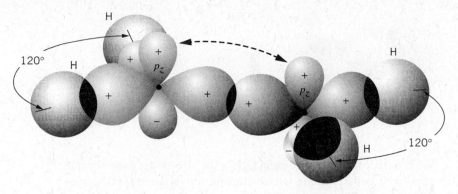

FIGURE 9.16 Molecular bonding in C_2H_4. For clarity the p_z overlap is not shown; the dashed arrows indicate the bond formed by the unhybridized p_z states.

It is also possible for only two $2p$ electrons in the $2s^1 2p^3$ configuration of carbon to become involved in hybrid states; the third $2p$ electron then is available, for example, to form ordinary pp molecular states. The ethylene molecule C_2H_4 is an example of such a structure. The three sp^2 hybrids form bond angles of $120°$; each carbon atom has two of its three sp^2 hybrid states joined to a H atom, and the third is joined to the other carbon. The unhybridized p electron also forms a bond between the two carbons, in a manner similar to that shown in Figure 9.9b for the "off-axis" pp molecular states. Figure 9.16 shows a representation of the bonding in C_2H_4. There are two bonds between the pair of carbons, one from the sp^2 hybrid and the other from the unhybridized p_z state. Another example of sp^2 hybrids in carbon is benzene (C_6H_6), in which each carbon is joined to one H and two other carbons by the sp^2 hybrids, with again one unhybridized p orbital available to bond the carbons. The basic structure of benzene is a ring of carbon atoms, as shown in Figure 9.17, with the expected angle of $120°$ between the hybrid states, which gives the molecule its characteristic hexagonal shape.

TABLE 9.4   Bond Angles of sp^3 Hybrids

Molecule	Bond Angle
CCl_4	$109.5°$
C_2H_6	$109.3°$
C_2Cl_6	$109.3°$
$CClF_3$	$108.6°$
CH_3Cl	$110.5°$
$SiHF_3$	$108.2°$
SiH_3Cl	$110.2°$
$GeHCl_3$	$108.3°$
GeH_3Cl	$110.9°$

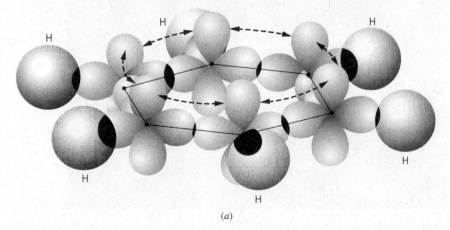
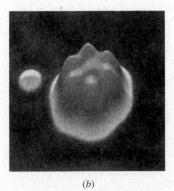

(a)   (b)

FIGURE 9.17 (a) Molecular bonding in benzene. As in Figure 9.16, the overlap of the unhybridized p_z states is not shown, but the dashed arrows indicate the p_z bonds. (b) Scanning electron microscope image showing electron distribution of the benzene molecular structure on a copper surface.

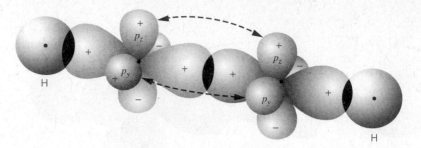

FIGURE 9.18 Molecular bonding in C_2H_2. The carbons are joined by three bonds, one from the p_x hybrid and two from the unhybridized p_y and p_z states (dashed arrows).

Finally, it is also possible for carbon to form sp hybrids, leaving two unhybridized p states. Acetylene (C_2H_2) is an example of such a molecule, with the two carbons now joined by three bonds, one from the sp hybrid and two from the unhybridized p states. Figure 9.18 shows the bonding arrangement in C_2H_2.

This variety of bonds entered into by carbon is the basis for the varied properties of organic molecules, from the simple ones we have studied here, to the complex ones that form the basis of living things. However, it is not only carbon that shows sp hybridization, but other atoms as well. (Indeed, the failure of NH_3 to show the expected $90°$ bond angle could be blamed on sp hybridization rather than on the repulsion of the H atoms.) It is also possible to have $3s-3p$ hybrids (silicon) and $4s-4p$ hybrids (germanium). It is this hybridization that gives these materials, like carbon, a valence of 4 and a symmetrical bonding arrangement, which are partly responsible for the usefulness of Si and Ge as semiconducting materials, as we will discuss in Chapter 11. It is also interesting to speculate on the possibility of a new type of organic chemistry, including new life forms, based on Si or Ge, rather than C (Figure 9.19).

FIGURE 9.19 Artist's image of an alien landscape populated by life forms based on silicon. Because silicon is similar to carbon in its chemical behavior, it can bond with hydrogen and oxygen to form complex molecules. Creatures based on silicon would look crystalline in appearance and be able to survive in high-temperature environments.

9.3 IONIC BONDING

In covalent bonding, as we have seen, the bonding electrons do not belong to any particular atom in the molecule, but rather are shared among the atoms. It is also possible to form a molecule that results from the extreme opposite case in which valence electrons are not shared but instead spend all of their time in the neighborhood of only one of the atoms of the molecule. Consider the ionic molecule NaCl. Suppose we have a neutral sodium atom ($1s^2 2s^2 2p^6 3s^1$), with one $3s$ electron outside of a filled shell, and a neutral chlorine atom ($1s^2 2s^2 2p^6 3s^2 3p^5$), which lacks one electron from a filled $3p$ shell. To remove the outer electron from Na requires 5.14 eV, the *ionization energy* of Na, and we are left with a positively charged Na$^+$ ion. If we then attach that electron to the Cl atom, creating a negatively charged Cl$^-$ ion, the energy *released* is 3.61 eV, the *electron affinity* of Cl. The energy is released because the filled $3p$ shell is an especially stable configuration, which is energetically very favorable. Thus, if we borrow 5.14 eV to ionize the Na, we get back immediately 3.61 eV by attaching the electron to the Cl. We can get back the remaining 1.53 eV (= 5.14 eV − 3.61 eV) by moving the Na$^+$ and Cl$^-$ close enough together that their Coulomb potential energy is −1.53 eV. The separation distance corresponding to this potential energy is found from the potential energy equation, $U = q_1 q_2 / 4\pi\varepsilon_0 R$:

$$ R = -\frac{e^2}{4\pi\varepsilon_0}\frac{1}{U} = \frac{1.44\,\text{eV}\cdot\text{nm}}{1.53\,\text{eV}} = 0.941\,\text{nm} $$

That is, as long as the Cl$^-$ and Na$^+$ are closer together than 0.941 nm, the Coulomb attraction will supply enough energy to overcome the difference between the ionization energy of Na and the electron affinity of Cl. Put another way, Na$^+$ and Cl$^-$ *ions* separated by less than 0.941 nm have a more stable configuration than neutral Na and Cl *atoms*.

However, we cannot push the Na$^+$ and Cl$^-$ ions too close together, because eventually the filled p shells begin to overlap, which causes the ions to repel one another. The Pauli principle forbids additional electrons in either filled p shell. Forcing the two ions closer together therefore requires that some of the electrons be pushed from the $2p$ or $3p$ shells into a higher shell, in order to "make room" for the overlapping electrons. Because this takes energy, we must therefore add energy to the Na$^+$ + Cl$^-$ system in order to reduce the separation of the ions beyond a certain point. We can imagine this energy as a sort of "potential energy" of repulsion, which increases rapidly as we try to force the ions close together.

In summary, when the ions are far apart, they attract one another, and when they are too close together, they repel one another; in between there must be an equilibrium position where the attractive and repulsive forces are balanced. It is this equilibrium position that determines the size of an ionic molecule.

Figure 9.20 shows the energy of the NaCl molecule as a function of the separation distance between the nuclei. Taking the energy zero point to refer to the neutral atoms, we can separate the molecular energy into three terms: the constant $\Delta E = 1.53$ eV, which represents the difference in energy between the ions and the neutral atoms; the Coulomb attraction U_C between the ions; and the Pauli "repulsion" U_R, which is represented in Figure 9.20 as a potential energy that rises rapidly for small R and falls to 0 for values of R that exceed the sum of

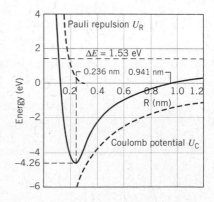

FIGURE 9.20 Molecular energy of NaCl. The zero of the energy scale represents neutral Na and Cl atoms. The solid curve is the sum of the three contributions to the molecular energy.

TABLE 9.5 Properties of Some Ionic Diatomic Molecules

Molecule	Dissociation Energy (eV)	Equilibrium Separation (nm)
NaCl	4.26	0.236
NaF	4.99	0.193
NaH	2.08	0.189
LiCl	4.86	0.202
LiH	2.47	0.239
KCl	4.43	0.267
KBr	3.97	0.282
RbF	5.12	0.227
RbCl	4.64	0.279

the ionic radii, at which point the electron shells no longer overlap and the Pauli repulsion does not occur. The sum of these three terms gives the molecular energy of NaCl, which shows a minimum at an equilibrium separation R_{eq} of 0.236 nm. At that distance the energy is -4.26 eV, so the binding energy or dissociation energy (the energy needed to split the molecule into neutral atoms) is 4.26 eV.

Table 9.5 shows the equilibrium separation and dissociation energy of several ionic molecules.

Remember that we are concerned here with isolated molecules and not with collections of molecules in solids. When we speak of NaCl in this section, we mean not the solid salt but rather a gas of NaCl molecules. The spacing between atoms in a solid can be very different from the spacing in a molecule.

Example 9.2

(a) What is the value of the Pauli repulsion energy of NaCl at the equilibrium separation? (b) Estimate the value of the Pauli repulsion energy at a separation of 0.1 nm.

Solution

(a) At the equilibrium separation distance, the Coulomb energy is

$$U_C = -\frac{1}{4\pi\varepsilon_0}\frac{e^2}{R_{eq}} = -\frac{1.44\,\text{eV}\cdot\text{nm}}{0.236\,\text{nm}} = -6.10\,\text{eV}$$

and the Pauli repulsion energy can be found from the molecular energy E:

$$U_R = E - U_C - \Delta E$$
$$= -4.26\,\text{eV} - (-6.10\,\text{eV}) - 1.53\,\text{eV} = 0.31\,\text{eV}$$

(b) From Figure 9.20 we estimate $E = +4.0$ eV at $R = 0.1$ nm. The Coulomb energy at $R = 0.1$ nm is -14.4 eV, and the repulsion energy is

$$U_R = E - U_C - \Delta E$$
$$= +4.0\,\text{eV} - (-14.4\,\text{eV}) - 1.53\,\text{eV} = 16.9\,\text{eV}$$

Note how rapidly the repulsion energy increases at small R.

Covalent and ionic bonding represent two opposite extreme cases, one in which electrons are shared between two atoms and the other in which the electrons always are associated with one of the atoms. How can we decide whether two atoms will join by covalent or ionic bonding? The answer depends on the willingness of the atoms to share their electrons, or equivalently the degree to which one atom dominates the other in its desire to have the valence electrons all to itself.

For homonuclear diatomic molecules, we expect a purely covalent bond; because the two atoms are exactly alike, neither can dominate and the electron is completely shared between them. For heteronuclear molecules, however, the situation is quite different. There can be no purely ionic or purely covalent bond. The two atoms have different atomic numbers and different electronic configurations, so the wave function of a valence or bonding electron, even if shared, will not be exactly the same near one atom as it is near the other. The electron must therefore spend more time near one atom than near the other; this means that one atom may have a slight excess of negative charge and the bond, even if we think of it as being covalent, will also have a small ionic character. Conversely, a "pure" ionic bond has a small covalent character; we learned in Chapter 7 that electron wave functions do not suddenly drop off to zero amplitude, but instead fall exponentially to zero. Therefore, even in an ionic molecule like NaCl, the wave function of the electron that was transferred to the Cl^- ion is not zero at the location of the Na^+ ion; the wave function may indeed have a very small amplitude at the Na^+ ion, but it is not zero. The electron thus spends some of its time, even if very little, being shared between the atoms, and the bonding, while mostly ionic, has a small covalent character as well.

An empirical way to characterize the relative ionic character of a molecule is based on the *electric dipole moment* of the molecule. An electric dipole, as illustrated in Figure 9.21, consists of two charges $+q$ and $-q$ separated by a distance r. The electric dipole moment of this arrangement is defined as

$$p = qr \qquad (9.4)$$

In a purely covalent molecule, there is no excess charge on either atom, so in effect $q = 0$ and the dipole moment is expected to be zero. In an ionic molecule, on the other hand, a net positive charge resides on one atom and a net negative charge on the other, so the dipole moment is not zero.

If NaCl were a purely ionic molecule, we would expect a dipole moment of

$$p = qR_{eq} = (1.60 \times 10^{-19} \text{ C})(0.236 \times 10^{-9} \text{ m}) = 3.78 \times 10^{-29} \text{ C} \cdot \text{m}$$

The measured electric dipole moment of NaCl is 3.00×10^{-29} C·m, which is 0.79 or 79% of the maximum value corresponding to a purely ionic bond. Because NaCl is only partially ionic, the measured dipole moment is smaller than the full ionic value. By taking the ratio between the measured electric dipole moment and the maximum value calculated from Eq. 9.4, we can determine a measure of the fractional ionic character of the molecular bond:

$$\text{fractional ionic character} = \frac{p_{measured}}{qR_{eq}} \qquad (9.5)$$

Linus Pauling (1901–1994, United States). Known mostly as a chemist, his work on molecular bonds transcends the traditional chemistry-physics boundary. He is the only recipient of two unshared Nobel Prizes, the chemistry prize in 1954 for his work on molecular bonds and the peace prize in 1963 for his efforts toward a ban on the testing of nuclear weapons.

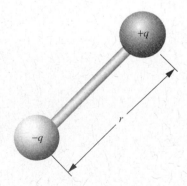

FIGURE 9.21 An electric dipole.

TABLE 9.6 Electronegativities of Some Elements

H	2.20										
Li	0.98	Be	1.57	C	2.55	N	3.04	O	3.44	F	3.98
Na	0.93	Mg	1.31	Si	1.90	P	2.19	S	2.58	Cl	3.16
K	0.82	Ca	1.00	Ge	2.01	As	2.18	Se	2.55	Br	2.96
Rb	0.82	Sr	0.95	Sn	1.96	Sb	2.05	Te	2.10	I	2.66
Cs	0.79	Ba	0.89								

Is there a property of atoms that allows us to predict whether they will join with one another in ionic or covalent bonds? One property that has some predictive value is the *electronegativity*, which can be roughly defined as the capability of an atom to attract electrons when it forms chemical bonds. The electronegativity of an atom can be computed from the sum of the energy cost of removing an electron (the ionization energy) and the energy gain in adding an electron (the electron affinity). Table 9.6 shows the electronegativity values for some elements.

If the two atoms in a molecule have equal electronegativities, they have equal tendencies to attract electrons and should therefore form covalent bonds. If the two electronegativities are very different, one atom will have a greater tendency to attract electrons; as a result, there will be a net negative charge on one atom and a net positive charge of equal magnitude on the other. The molecular bonding will then have at least a partial ionic character.

We therefore expect a rough relationship between the fractional ionic character (determined by comparing the measured electric dipole moment with its maximum expected value) and the magnitude of the *difference* between the electronegativity values. Figure 9.22 shows the fractional ionic characters plotted against the electronegativity differences. Although the points are scattered, you can see that there is indeed a direct relationship between the quantities: molecules with small

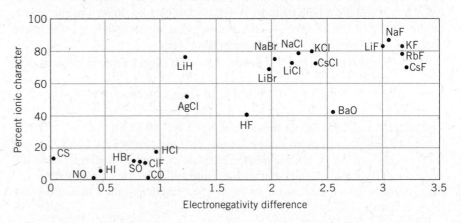

FIGURE 9.22 The fractional ionic character of bonds in diatomic molecules.

electronegativity differences have only a small ionic character and therefore form covalent bonds, while molecules with large electronegativity differences form mostly ionic bonds.

9.4 MOLECULAR VIBRATIONS

So far we have considered the properties of molecules based on the configurations of their electrons. The electrons largely determine the strength of the molecular bonds and their geometry. Molecules can absorb or emit energy by changing the configuration of their electrons, just as atoms can do. However, molecules can also absorb or emit energy, generally in much smaller quantities, in other ways. One way is for the atoms in a molecule to vibrate about their equilibrium positions, just like a mass on a spring. Another way is for the atoms to rotate about the center of mass of the molecule. We consider molecular vibrations in this section and rotations in the next section. The vibrational and rotational energies are often unique to particular molecules and can serve as a kind of "fingerprint" in identifying simple molecules.

Vibrations in Quantum Mechanics

A classical vibrating system such as a mass m on a spring of force constant k has an oscillation frequency $\omega = \sqrt{k/m}$. (Here we are dealing with the angular or circular frequency $\omega = 2\pi f$, measured in radians/second.) The oscillator has potential energy $U = \frac{1}{2}kx^2$, and with maximum amplitude x_m its (constant) total energy is $E = \frac{1}{2}kx_m^2$. For the classical oscillator, there are no restrictions on the total energy or the oscillator frequency—any value of the energy is allowed, and the frequency and energy can be varied independently.

As we learned in Chapter 5, the quantum oscillator behaves very differently. Only certain values of the energy are allowed. The allowed energy values, for an oscillator that can move in only one dimension, are

$$E_N = (N + \tfrac{1}{2})\hbar\omega = (N + \tfrac{1}{2})hf \qquad N = 0, 1, 2, 3, \cdots \qquad (9.6)$$

where ω is the classical oscillator frequency. The ground-state energy is $\frac{1}{2}\hbar\omega = \frac{1}{2}hf$, and the excited states are equally spaced with energy differences of $\hbar\omega = hf$. Figure 9.23 shows the excited states of a one-dimensional quantum oscillator. Note that the ground state does not have $E = 0$; this is a consequence of the uncertainty principle—an oscillator with $E = 0$ would have a displacement of exactly zero and also a momentum of exactly zero, which would violate the position-momentum form of the uncertainty principle. The minimum energy of $\frac{1}{2}hf$ is often called the *zero-point energy*, and it can have observable consequences (see Problem 31).

The oscillator can emit electromagnetic radiation in jumps to lower states, or it can absorb radiation in jumps to higher states. However, the jumps cannot occur

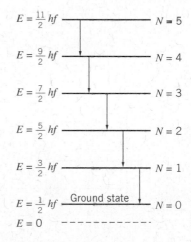

FIGURE 9.23   Some energy levels and transitions in a one-dimensional quantum oscillator.

in arbitrary steps. *The allowed transitions are those that change N by one unit.* This restriction is called a *selection rule* and applies only to transitions that occur through the emission or absorption of electromagnetic radiation.

$$\text{Vibrational selection rule:} \quad \Delta N = \pm 1 \quad (9.7)$$

Other causes of jumps, such as collisions, are not subject to this restriction.

Although this model of a quantum oscillator is oversimplified, we do find in nature many systems that do exhibit this type of behavior. In a molecule, the individual atoms can vibrate relative to the center of mass. A nucleus, as we discuss in Chapter 12, is sometimes modeled as a vibrating liquid drop. Both of these systems have some properties in common with the quantum oscillator, although the effective force constant k is not quite constant in those systems, which can lose stiffness as the excitation energy increases. In a solid, the atoms can often behave like three-dimensional oscillators, the properties of which are different from the one-dimensional oscillator we are considering here. We will discuss the oscillations of the atoms in a three-dimensional solid from a statistical viewpoint in Chapter 11.

Vibrating Diatomic Molecules

A molecule with more than two atoms can vibrate in many different ways, so for simplicity we'll consider only molecules with two atoms. Figure 9.24 shows a representation of the two atoms in a diatomic molecule, which oscillate so that the center of mass remains fixed.

To calculate the frequencies of the emitted transitions, we must know the mass m and the effective force constant k. Because both atoms in the molecule are participating in the vibration, the vibrating mass is not the mass of either atom alone but instead is a combination of both masses. Let the masses of the two atoms be m_1 and m_2. As they both pass through their equilibrium positions, the total energy of the molecule is only kinetic energy, so

$$E_T = \frac{1}{2}m_1 v_1^2 + \frac{1}{2}m_2 v_2^2 = \frac{p_1^2}{2m_1} + \frac{p_2^2}{2m_2} \quad (9.8)$$

In a frame of reference in which the center of mass of the molecule is fixed, the total momentum is zero, so $p_1 = p_2$ and the energy can be written (with $p = p_1 = p_2$) as

$$E_T = \frac{1}{2}p^2\left(\frac{1}{m_1} + \frac{1}{m_2}\right) = \frac{1}{2}p^2\left(\frac{m_1 + m_2}{m_1 m_2}\right) = \frac{p^2}{2m} \quad (9.9)$$

where

$$m = \frac{m_1 m_2}{m_1 + m_2} \quad (9.10)$$

That is, the energy of the system is the same as if it were a single mass m, moving with momentum p. This m is a sort of effective mass of the whole molecule and is known as the *reduced mass*. It is the mass we should use in calculating the vibrational frequency. (Previously we used the reduced mass of the electron and proton in our analysis of the hydrogen atom—see Section 6.8.)

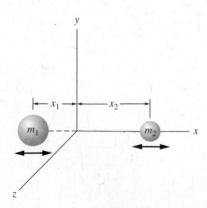

FIGURE 9.24 The two atoms in a vibrating diatomic molecule. The center of mass of the molecule remains fixed at the origin of the coordinate system.

Notice that $m = m_1/2$ when $m_1 = m_2$, as in a homonuclear molecule—the effective mass is half the mass of an individual atom. Whenever one mass is much greater than the other, the reduced mass has a value nearly equal to that of the lighter mass. This is consistent with our expectation, because the inertia of the heavier mass reduces its tendency to move, and most of the vibrational motion is done by the lighter mass.

The calculation of the effective force constant for vibrational motion is illustrated in the following example.

Example 9.3

Find the vibrational frequency and photon energy for H_2.

Solution

To find the frequency, we must know the force constant k. To estimate k, we consider the molecule to behave like a simple harmonic oscillator in the vicinity of its equilibrium separation R_{eq}, and we fit a parabolic oscillator potential energy $U = \frac{1}{2}kx^2$ to the molecular energy in that region. Figure 9.25 shows the region of the energy minimum and a parabola that approximates the curve near the minimum. The equation of the parabola is

$$E - E_{min} = \tfrac{1}{2}k(R - R_{eq})^2 \qquad (9.11)$$

The constant k can be estimated from the graph by finding the value of $R - R_{eq}$ that is necessary for a certain value of $E - E_{min}$. As shown in the figure, when $E - E_{min} = 0.50$ eV, the value of $R - R_{eq}$ is $\frac{1}{2}(0.034 \text{ nm}) = 0.017$ nm. Solving for k, we obtain

$$k = \frac{2(E - E_{min})}{(R - R_{eq})^2} = \frac{2(0.50 \text{ eV})}{(0.017 \text{ nm})^2}$$

$$= 3.5 \times 10^3 \text{ eV/nm}^2 = 3.5 \times 10^{21} \text{ eV/m}^2$$

The reduced mass of molecular hydrogen is half the mass of a hydrogen atom. We can now calculate the vibrational frequency:

$$f = \frac{1}{2\pi}\sqrt{\frac{k}{m}} = \frac{1}{2\pi}\sqrt{\frac{kc^2}{mc^2}}$$

$$= \frac{1}{2\pi}\sqrt{\frac{(3.5 \times 10^{21} \text{ eV/m}^2)(3.00 \times 10^8 \text{ m/s})^2}{(0.5)(1.008 \text{ u})(931.5 \times 10^6 \text{ eV/u})}}$$

$$= 1.3 \times 10^{14} \text{ Hz}$$

The corresponding photon energy is

$$E = hf = (4.14 \times 10^{-15} \text{ eV} \cdot \text{s})(1.3 \times 10^{14} \text{ Hz}) = 0.54 \text{ eV}$$

The wavelength of this radiation is 2.3 μm, which is in the infrared region of the spectrum. Molecular vibrations typically give photons in the infrared.

Note that the parabola gives a reasonable approximation to the actual energy curve only for energies up to about 1 eV above the minimum, which corresponds to the excited state with $N = 2$. If we were to excite the H_2 molecule to energies greater than 1 eV above its ground state, we would expect to see deviations from the behavior predicted by the simple harmonic oscillator. In particular, all the transitions would no longer have the same energy, and changes other than $\Delta N = \pm 1$ may occur. In other molecules, the harmonic oscillator approximation may remain valid for larger vibrational quantum numbers, but eventually it will fail in all cases at sufficiently large N.

Vibrational energies can often be found in tabulations of molecular properties. Often these tabulations give the energy in units of cm^{-1}. To convert to eV, multiply by the conversion factor $hc = 1.24 \times 10^{-4}$ eV $\cdot$ cm.

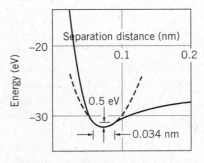

FIGURE 9.25 Fitting a parabola (dashed line) to the energy minimum of H_2.

Summarizing our conclusions from this section for the vibrational motion of molecules, we expect the simple harmonic oscillator to give a reasonable description of the motion near the energy minimum, with a sequence of emission or absorption photons all of energy hf corresponding to changes of one unit in the vibrational quantum number. The emitted or absorbed radiations are generally in the infrared region of the spectrum.

The result of Example 9.3 shows that in a gas of hydrogen molecules at room temperature, where the mean translational kinetic energy is about 0.025 eV, it is highly unlikely that a collision between molecules can deliver 0.54 eV to one molecule and set it vibrating. This explains why the vibrational degrees of freedom are "frozen out" at room temperature and thus why the heat capacity of H_2 does not agree with the value expected for 7 degrees of freedom, as we discussed in Chapter 1.

The vibrational energy of F_2 is 0.11 eV, so even at room temperature (see Table 1.1) the heat capacity of fluorine gas is starting to increase above the value that is characteristic of 5 degrees of freedom (which would include only translational and rotational motions). By 1000 K, where the mean thermal energy is around 0.075 eV, the vibrational states of F_2 can be easily excited and so the heat capacity of F_2 at 1000 K is characteristic of 7 degrees of freedom. Oxygen has a larger vibrational energy (0.20 eV); it is difficult to add this much energy in a gas in which the mean energy is only 0.025 eV, and so at 300 K its heat capacity remains at the value characteristic of 5 degrees of freedom (translation + rotation); the heat capacity for oxygen increases at 1000 K but not quite to the value for 7 degrees of freedom. Nitrogen has an even greater vibrational energy (0.29 eV), and so its heat capacity is smaller than that of F_2 or O_2 at 1000 K.

9.5 MOLECULAR ROTATIONS

A second way that a molecule can absorb or emit energy is through rotations about its center of mass. The state of the rotational motion of the molecule is described by specifying its angular momentum. In this section we discuss the special treatment of angular momentum in quantum mechanics and the resulting rotational states in diatomic molecules.

Rotations in Quantum Mechanics

A classical rigid rotor with rotational inertia I has a rotational kinetic energy given by $K = \frac{1}{2}I\omega^2 = L^2/2I$ where $L = I\omega$ is the rotational angular momentum. (Compare this formula with the more familiar expression for translational kinetic energy, $K = \frac{1}{2}mv^2 = p^2/2m$, and note that we can get from the translational expression to the rotational expression by replacing the linear momentum p with the angular momentum L and the mass m with the rotational inertia I.)

There is no general result for the quantization of linear momentum or translational kinetic energy in quantum mechanics, so we have no obvious clues about how to quantize the rotational motion. We can imagine a rotor to be a particle of mass m rotating on the end of a "massless" rod of length r, but we know that in quantum mechanics it is not possible to have rotations in a fixed plane (see the discussion in Section 7.2 about the rotational form of the uncertainty relationship). The quantum version of the rigid rotor is thus more properly described as a particle confined to move in three dimensions on a sphere of radius r rather than in two dimensions in a circle of radius r.

The general mathematical solutions for the wave functions of a three-dimensional rotor are beyond the level of this text (some of the solutions can be obtained from the angular parts of the wave functions of the hydrogen atom), but the resulting energy levels have a particularly simple form:

$$E_L = \frac{L(L+1)\hbar^2}{2I} \qquad L = 0, 1, 2, 3, \cdots \qquad (9.12)$$

where L is the angular momentum quantum number. The rotational levels have energies of $0, 2(\hbar^2/2I), 6(\hbar^2/2I), 12(\hbar^2/2I), 20(\hbar^2/2I), \ldots$, as shown in Figure 9.26. Note that, in contrast to the vibrational energy levels, the rotational energy levels are *not* equally spaced—the spacing increases as the energy increases.

We can obtain Eq. 9.12 from the classical expression for the rotational kinetic energy by replacing the classical angular momentum with the quantum result given by Eq. 7.5, $|\mathbf{L}| = \sqrt{L(L+1)}\,\hbar$. (Here we are using an upper-case L to represent the angular momentum quantum number of a *system* such as a molecule instead of the lower-case l, which represents the angular momentum quantum number of a *particle*.)

Molecules and nuclei are again good examples of quantum rotors. Because molecules and nuclei don't have rigidly fixed bond lengths, they don't quite follow the rigid rotor spacing exactly. As the angular momentum increases and the system rotates at a greater rate, the bonds can stretch somewhat, increasing the rotational inertia and slightly decreasing the rotational spacing. Nevertheless, the rotational spacing is followed very closely.

The quantum state of the rotor can be changed through the emission or absorption of electromagnetic radiation. As with the vibrational states, there is a *selection rule* that limits the jumps that can occur when radiation is emitted or absorbed:

$$\text{Rotational selection rule:} \qquad \Delta L = \pm 1 \qquad (9.13)$$

The permitted rotational jumps can occur only through a change of one unit in the angular momentum quantum number. In practice, we find that this selection rule is not absolute, but that electromagnetic transitions that change the angular momentum quantum number by one unit are often several orders of magnitude more probable than those that change L by other amounts. Some of the transitions that change L by one unit are shown in Figure 9.26. Note that the photons emitted in these transitions all have different energies, in contrast to the vibrational excitations in which all emitted photons have the same energy.

Rotating Diatomic Molecules

Consider the diatomic molecule shown in Figure 9.27. The origin of the coordinate system is at the center of mass of the molecule, so that $x_1 m_1 = x_2 m_2$. The rotational inertia of the molecule is $I = m_1 x_1^2 + m_2 x_2^2$. We can write this in terms of the reduced mass $m = m_1 m_2/(m_1 + m_2)$ and the equilibrium separation $R_{eq} = x_1 + x_2$ as

$$I = m R_{eq}^2 \qquad (9.14)$$

and the energies of the rotational states are

$$E_L = \frac{L(L+1)\hbar^2}{2m R_{eq}^2} = BL(L+1) \qquad (9.15)$$

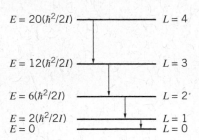

$E = 20(\hbar^2/2I)$ —— $L = 4$

$E = 12(\hbar^2/2I)$ —— $L = 3$

$E = 6(\hbar^2/2I)$ —— $L = 2$

$E = 2(\hbar^2/2I)$ —— $L = 1$
$E = 0$ —— $L = 0$

FIGURE 9.26   Some energy levels and transitions of a quantum rotor.

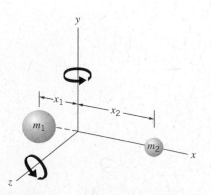

FIGURE 9.27   A rotating diatomic molecule.

where the rotational parameter B is defined for diatomic molecules as

$$B = \frac{\hbar^2}{2mR_{eq}^2} \tag{9.16}$$

The emitted or absorbed photons must follow the rotational selection rule $\Delta L = \pm 1$, so the energy of an emitted photon is the energy difference between two adjacent levels:

$$\Delta E = E_{L+1} - E_L = B(L+1)(L+2) - BL(L+1) = 2B(L+1) \tag{9.17}$$

In contrast to the transitions among the *vibrational* excitations, which all had the same energy, the energies of the *rotational* transitions depend on L, as you can see in Figure 9.26. The emitted photons have energies $2B, 4B, 6B, \ldots$.

Example 9.4

Calculate the energies and wavelengths of the three lowest radiations emitted by molecular H_2.

Solution
The photon energy equals the energy difference between the levels, which is given by Eq. 9.17. To evaluate this energy, we must first determine the rotational quantity B. For H_2, the reduced mass m is half the mass of a hydrogen atom. Then

$$B = \frac{\hbar^2}{2mR_{eq}^2} = \frac{\hbar^2 c^2}{2mc^2 R_{eq}^2}$$

$$= \frac{(197.3 \, \text{eV} \cdot \text{nm})^2}{2(0.5 \times 1.008 \, \text{u} \times 931.5 \, \text{MeV/u})(0.074 \, \text{nm})^2}$$

$$= 0.0076 \, \text{eV}$$

Equation 9.17 now gives the energies directly, and the corresponding wavelengths are found from $\lambda = hc/\Delta E$.

$$L = 1 \text{ to } L = 0: \quad \Delta E = 2B = 0.0152 \, \text{eV} \quad \lambda = 81.6 \, \mu\text{m}$$

$$L = 2 \text{ to } L = 1: \quad \Delta E = 4B = 0.0304 \, \text{eV} \quad \lambda = 40.8 \, \mu\text{m}$$

$$L = 3 \text{ to } L = 2: \quad \Delta E = 6B = 0.0456 \, \text{eV} \quad \lambda = 27.2 \, \mu\text{m}$$

The energies of the emitted photons form an arithmetic sequence with relative values $1, 2, 3, \ldots$, and the emitted wavelengths have the inverse sequence $1, 1/2, 1/3, \ldots$.

The emitted rotational transitions are (like the vibrational transitions) in the infrared region, but the wavelengths are 1–2 orders of magnitude larger than those of the vibrational transitions. This region of the spectrum corresponds to the far infrared and microwave radiations.

In contrast to the vibrational motion, for which the minimum energy was 0.54 eV, the minimum energy for the rotational motion in H_2 is only 0.015 eV. In hydrogen gas at room temperature, where the mean translational kinetic energy is 0.025 eV, it is easily possible for a collision between two molecules to cause one of them to start rotating. Thus the rotational motion can occur at room temperature, and H_2 behaves as if it has 2 rotational degrees of freedom (see Section 1.3). Gases whose molecules have greater mass than hydrogen have smaller rotational energies and their rotational states are even more easily excited in collisions, so for most gases the rotational degrees of freedom are present at ordinary temperatures.

Example 9.5

Figure 9.28 shows a portion of the rotational absorption spectrum of a molecule. Determine the rotational inertia of the molecule.

Solution

Each peak in the absorption spectrum corresponds to the molecule moving from one level of Figure 9.26 to the next higher level after absorbing a photon of the proper energy or frequency. The frequencies can be found from Eq. 9.17:

$$f = \frac{\Delta E}{h} = (L+1)\frac{\hbar}{2\pi I}$$

where Eq. 9.14 has been used to replace mR_{eq}^2 with the rotational inertia I. This gives the frequency of each peak in Figure 9.28, but doesn't allow us to find I because we don't know the L value for the peaks. We can avoid this problem by calculating the difference in frequency Δf between adjacent peaks L and $L+1$:

$$\Delta f = (L+2)\frac{\hbar}{2\pi I} - (L+1)\frac{\hbar}{2\pi I} = \frac{\hbar}{2\pi I}$$

or, estimating the spacing Δf between the peaks of Figure 9.28 as 6.2×10^{11} Hz,

$$I = \frac{\hbar}{2\pi \Delta f} = \frac{1.05 \times 10^{-34} \text{ J} \cdot \text{s}}{2\pi(6.2 \times 10^{11} \text{ Hz})}$$
$$= 2.7 \times 10^{-47} \text{ kg} \cdot \text{m}^2 = 0.016 \text{ u} \cdot \text{nm}^2$$

This value would correspond to, for example, a molecule with a reduced mass of 1 u (such as one consisting of hydrogen combined with a much heavier atom) and an equilibrium separation of 0.13 nm.

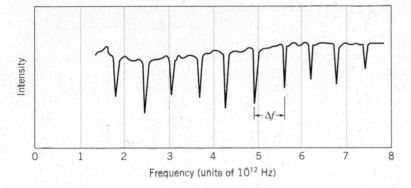

FIGURE 9.28 A molecular absorption spectrum.

Summarizing this section, we have seen that the rotational motion of a molecule results in a sequence of energy levels that are not equally spaced. The rotational energy levels are spaced 1–2 orders of magnitude closer than the vibrational levels. The emitted radiations form an ascending series of energies (or a descending series of wavelengths) in the far infrared or microwave region of the spectrum. The radiations are restricted by the selection rule that permits the rotational quantum-number L to change by only one unit.

9.6 MOLECULAR SPECTRA

A complex molecule can absorb or emit energy in a variety of ways, as indicated schematically by the energy-level diagram of Figure 9.29. The emitted or absorbed energy can change the electronic state (in analogy with the change of energy levels of electrons in atoms). The energy necessary to make this change is of the

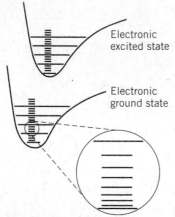

FIGURE 9.29 Electronic, vibrational, and rotational energy levels of a molecule.

order of eV, corresponding to photons in the visible range of the spectrum. Within the energy minimum of any electronic state, there are vibrational and rotational states. The vibrational states are equally spaced and have energy separations of typically 0.1–1 eV. The rotational states are not equally spaced, and they have a smaller energy separation of typically 0.01–0.1 eV. Because the rotational spacing is much smaller than the vibrational spacing, it is convenient to consider each vibrational state as providing the basis on which a sequence of rotational states is built. We discuss here only the vibrational and rotational structure, not the electronic excited states.

Figure 9.30 shows a detail of the rotational and vibrational structure. The states are labeled with the vibrational quantum number N and the rotational quantum number L. Depending on their properties, some molecules can show a purely vibrational structure (such as Figure 9.23) and others can show a purely rotational structure (such as in Figure 9.26). In many molecules, however, the transitions between the levels must *simultaneously* satisfy *both* the rotational and vibrational selection rules:

$$|\Delta N| = 1 \quad and \quad |\Delta L| = 1 \tag{9.18}$$

Consider the state with quantum numbers $N = 1$ and $L = 4$, as shown in Figure 9.30. The molecule cannot make a transition to the next lowest rotational state ($N = 1, L = 3$), because it would violate the vibrational selection rule $|\Delta N| = 1$. All transitions must simultaneously satisfy *both* selection rules. The molecule can make a transition to the state with $N = 0$ and $L = 3$ or to the state with $N = 0$ and $L = 5$. Absorption transitions must also satisfy these selection rules.

Let's now obtain a general expression for the energies of the transitions that satisfy both selection rules. The energy of the state with quantum numbers N and L can be written as the sum of the vibrational and rotational terms:

$$E_{NL} = (N + \tfrac{1}{2})hf + BL(L + 1) \tag{9.19}$$

where

$$N = 0, 1, 2, \ldots \quad and \quad L = 0, 1, 2, \ldots$$

The vibrational term is usually much larger than the rotational term, so the *emission* wavelengths in the spectrum will usually correspond to $N \to N - 1$ with $L \to L \pm 1$; the *absorption* wavelengths will be those for transitions in which N increases by one unit.

For absorption from initial state N, L to a final state $N + 1, L \pm 1$ the possible photon energies are

$$\Delta E = E_{N+1, L\pm1} - E_{NL}$$
$$= [(N + \tfrac{3}{2})hf + B(L \pm 1)(L \pm 1 + 1)] - [(N + \tfrac{1}{2})hf + BL(L + 1)]$$

$$\Delta E = hf + 2B(L + 1) \quad \text{for } L \to L + 1 \tag{9.20}$$

$$\Delta E = hf - 2BL \quad \text{for } L \to L - 1 \tag{9.21}$$

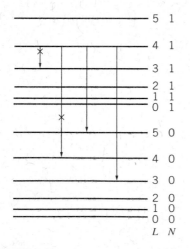

FIGURE 9.30 Combined rotational and vibrational energy levels of a molecule. Transitions marked with an X violate the selection rules and so are forbidden or strongly suppressed.

Figure 9.31 shows the expected spectrum of absorption photons. Starting from the center and increasing to the right is a series of photons whose energies are given by Eq. 9.20

$$hf + 2B, hf + 4B, hf + 6B, \ldots$$

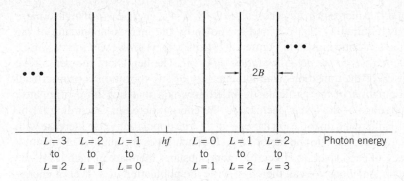

FIGURE 9.31 Expected sequences of absorption transitions between combined rotational and vibrational states. Each vertical line represents the absorption of a photon with a sharply defined energy.

Also starting from the center but decreasing to the left is a series of photons whose energies are given by Eq. 9.21:

$$hf - 2B, hf - 4B, hf - 6B, \ldots$$

Note that the photon of energy hf is missing at the center of the spectrum; such a photon would correspond to a "pure" vibrational transition, which would violate the rotational selection rule.

We can compare this ideal spectrum with the real spectrum of Figure 9.32, which shows the absorption transitions of the molecule HCl. Although the basic structure of Figure 9.31 is present, including the missing pure vibrational transition at the center of the spectrum, there are a number of differences, which we can explain based on the structure of the HCl molecule.

1. *The transitions are not equally spaced.* We expect all of the transitions of the spectrum to be separated by a constant energy $2B$ but, as you can see, this is not the case. The spacing of the lines appears to decrease as we move away from the center to the right and to increase as we move to the left. The explanation for this effect lies with our assumption that the molecule is a rigid structure with a constant rotational inertia. As the angular velocity and angular momentum of the rotating molecule increase, the somewhat nonrigid bond can stretch slightly, increasing the value of R_{eq}. As Eq. 9.20 shows, the photon energies increase less rapidly if R_{eq} grows, as the $L \to L + 1$ branch in Figure 9.32 illustrates. On the

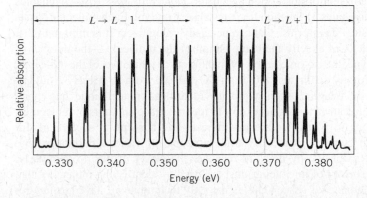

FIGURE 9.32 The absorption spectrum of molecular HCl.

other hand, an increasing R_{eq} causes the $L \to L - 1$ energies to increase more rapidly, as required by Eq. 9.21 and as shown by the increasing spacing of the lines of the left-hand branch in Figure 9.32. Both effects grow with increasing L.

2. *The heights of the peaks are quite different.* The heights of the peaks give the intensity of the transitions, and the intensity of any transition is proportional to the population of the particular level from which that transition originates. The populations of the levels decrease as the energy increases, according to the Maxwell-Boltzmann distribution factor $e^{-E/kT}$. The populations also increase with increasing L according to the factor $2L + 1$, which gives the angular momentum degeneracy of each level; in effect, the more substates there are in each level, the greater its population. We can therefore write the population of a level of energy E_{NL} as

$$p(E_{NL}) = (2L + 1)e^{-E_{NL}/kT} = (2L + 1)e^{-[(N+1/2)hf+BL(L+1)]/kT} \qquad (9.22)$$

where we have used E_{NL} from Eq. 9.19.

For the first few levels, the populations increase as L increases, because the exponential factor does not differ much from 1. However, as the energy continues to increase, the rapidly decreasing exponential factor begins to dominate and the populations decrease accordingly, becoming negligible for $L > 10$.

The level with maximum population, which corresponds to the most intense peak in the spectrum, can be found from Eq. 9.22 by locating its maximum, where $dp/dL = 0$. The result is

$$2L + 1 = \sqrt{\frac{2kT}{B}} \qquad (9.23)$$

For a measurement at room temperature ($kT \cong 0.025$ eV) and for $2B \cong 0.0026$ eV (estimated from the peak spacing in Figure 9.32), we obtain $L = 3$ for the peak of maximum intensity, in agreement with Figure 9.32.

3. *Each peak appears to be two very closely spaced peaks.* Chlorine consists of two types of atoms (isotopes) whose masses are approximately 35 u and 37 u. The differing masses result in slightly different vibrational and rotational energies for the two isotopes, with the heavier atom having the smaller energies. Thus the peaks for the Cl atoms with mass 37 u appear at slightly lower energies than the peaks for the Cl atoms with mass 35 u. The lighter atom occurs with about three times the abundance of the heavier atom, so the peaks for the lighter atom have the greater intensity.

Just as atomic spectroscopy gives us a way to identify atoms from their characteristic emission or absorption spectrum, *molecular spectroscopy* enables us to identify molecules by the radiation they absorb or emit. Each molecule has its own characteristic "fingerprint" that can be easily recognized. It is important that this technique tells us exactly the composition of the molecule—the number of atoms of each type, the isotopic ratios, even the state of ionization of the molecule. Thus we could easily distinguish CO from CO_2, $H^{35}Cl$ from $H^{37}Cl$, H_2^+ from H_2.

As you might imagine, such a precise identification technique has many applications to areas where it is necessary to identify trace amounts of molecules. Two applications in particular are of interest. The absorption spectra of our atmosphere can be used to identify trace amounts of various pollutants, and thus molecular spectroscopy helps to measure the purity of our air. Similarly, the absorption spectra of interstellar dust are used to identify the molecules that are present in interstellar space. This is the only technique we have (so far) for learning about the formation of complex molecules in our galaxy, because stars are too hot to permit molecules to exist.

Unfortunately, our atmosphere absorbs much of the infrared and microwave radiation that characterizes the spectra of these molecules, but spectrometers carried on satellites beyond the atmosphere permit the observation of those radiations and the identification of many varieties of molecules, including some relatively complex organic molecules. As an added benefit, when those spectrometers are aimed toward the Earth, they can measure how the infrared radiation emitted by the Earth is absorbed in its atmosphere, and thus detect the presence of various atmospheric pollutants.

Example 9.6

(a) From the absorption spectrum of HCl (Figure 9.32), determine the vibrational force constant. (b) Determine the rotational energy spacing of the peaks and compare with the expected value for HCl.

Solution
(a) The vibrational frequency can be found from the energy of the "missing" central transition in Figure 9.32. This energy can be read from the figure to be about $0.358\,\text{eV}$, and so

$$f = \frac{\Delta E}{h} = \frac{0.358\,\text{eV}}{4.14 \times 10^{-15}\,\text{eV}\cdot\text{s}} = 8.65 \times 10^{13}\,\text{Hz}$$

The reduced mass of HCl can be found from Eq. 9.10 to be $m = 0.98$ u. We can now find the force constant:

$$k = 4\pi^2 m f^2 = 4\pi^2 (0.98\,\text{u})(8.65 \times 10^{13}\,\text{Hz})^2$$
$$= 2.89 \times 10^{29}\,\text{u}\cdot\text{Hz}^2 = 2.99 \times 10^{21}\,\text{eV/m}^2$$

(b) From the figure, the energy spacing of the peaks is estimated to be $0.0026\,\text{eV}$. The expected spacing (see Figure 9.31) is $2B$, or

$$2B = \frac{\hbar^2}{mR_{eq}^2} = \frac{(hc)^2/4\pi^2}{(mc^2)R_{eq}^2}$$
$$= \frac{(1240\,\text{eV}\cdot\text{nm})^2/4\pi^2}{(0.98\,\text{u} \times 931.5\,\text{MeV/u})(0.127\,\text{nm})^2} = 0.00265\,\text{u}$$

This calculation gives excellent agreement with the value estimated from the spectrum.

Chapter Summary

		Section				Section
Covalent bonding	Bonding involving the sharing or overlap of electron wave functions, forming bonding and antibonding states	9.2	Reduced mass	$m = \dfrac{m_1 m_2}{m_1 + m_2}$		9.4
Ionic bonding	Bonding involving the electrostatic (Coulomb) attraction of oppositely charged ions	9.3	Rotational energies	$E_L = \dfrac{L(L+1)\hbar^2}{2mR_{eq}^2} = BL(L+1)$ $L = 0, 1, 2, 3, \cdots$		9.5
Fractional ionic character	$p_{\text{measured}}/qR_{eq}$	9.3	Rotational selection rule	$\Delta L = \pm 1$		9.5
Vibrational energies	$E_N = (N + \frac{1}{2})\hbar\omega = (N + \frac{1}{2})hf$ $N = 0, 1, 2, 3, \cdots$	9.4	Rotation-vibration energies	$E_{NL} = (N + \frac{1}{2})hf + BL(L+1)$		9.6
Vibrational selection rule	$\Delta N = \pm 1$	9.4	Population of energy level	$p(E_{NL}) = (2L+1)e^{-E_{NL}/kT}$ $= (2L+1)e^{-[(N+1/2)hf + BL(L+1)]/kT}$		9.6

Questions

1. Why does H_2 have a smaller radius and a greater binding energy than H_2^+?

2. The molecule LiH has a simple electronic structure. The H atom would like to gain an electron to fill its $1s$ subshell, but the Li atom would similarly like to fill its $2s$ subshell. Based on the atomic structure of H and Li, which would you expect to dominate in the desire for an additional electron? Are the electronegativity values of Table 9.6 consistent with this?

3. Consider molecules in which an alkali element X (X = Li, Na, K, Rb) is paired with either F or Cl to make XF or XCl. The molecule XF always has a greater dissociation energy and a smaller separation distance than the molecule XCl. Explain.

4. H_2^+ has a greater bond length (equilibrium separation) than H_2, but O_2^+ has a *smaller* bond length than O_2. Explain.

5. In general, would you expect ss bonds or pp bonds to be stronger? Why?

6. Explain why the bond angles of the sp directed bonds (Table 9.3) approach $90°$ as the atomic number of the central atom increases.

7. How do the molecular force constants k compare with those of ordinary springs? What do you conclude from this comparison?

8. Why is it unnecessary to consider rotations of the diatomic molecule of Figure 9.27 about the x axis? Estimate the rotational inertia for rotations about the x axis; compare the typical rotational energies with corresponding values for rotations about the y or z axis.

9. Explain how the equilibrium separation in a molecule can be determined by measuring the absorption or emission lines for rotational states.

10. How would the rotational energy spacing of D_2 (a form of "heavy hydrogen"; the mass of D is twice the mass of H) compare with the rotational spacing of H_2? How would their vibrational energy spacings compare? Their equilibrium separation distances?

11. For a molecule like HCl, estimate the number of rotational levels between the first two vibrational levels.

12. Why does an atom generally absorb radiation only from the ground state, while a molecule can absorb from many excited rotational or vibrational states?

13. If a collection of molecules were all in the $N = 0, L = 0$ ground state, how many lines would there be in the absorption spectrum?

Problems

9.1 The Hydrogen Molecule

1. Calculate the ionization energy of H_2.

2. Suppose a spherically symmetric negative charge of diameter R_{eq} were centered between two protons at the H_2^+ equilibrium configuration, so that the two protons just touch the surface of the sphere. What charge on the sphere is required to give the system a binding energy of 2.7 eV?

9.2 Covalent Bonding in Molecules

3. Bond strengths are frequently given in units of kilojoules per mole. Find the molecular dissociation energy (in eV) from the following bond strengths: (a) NaCl, 410 kJ/mole; (b) Li_2, 106 kJ/mole; (c) N_2, 945 kJ/mole.

4. Following the method of Example 9.1, predict the relative dissociation energies of the molecules NO, NF, and OF. Check your prediction by obtaining the bond strengths from tabulations.

5. For each of the following pairs of molecules, predict which will have the larger dissociation energy: (a) Li_2 and Be_2; (b) B_2 and C_2; (c) CO and O_2. Check your prediction against tabulated values of the bond strengths or dissociation energies.

6. (a) Which molecule would you expect to have the greater dissociation energy, F_2 or F_2^+? (b) N_2 or N_2^+? (c) NO or NO^+? (d) CN or CN^+? Check your predictions against tabulated dissociation energies.

9.3 Ionic Bonding

7. Calculate the Coulomb energy of KBr at the equilibrium separation distance.

8. The ionization energy of potassium is 4.34 eV; the electron affinity of iodine is 3.06 eV. At what separation distance will the KI molecule gain enough Coulomb energy to overcome the energy needed to form the K^+ and I^- ions?

9. (a) Assuming a separation of 0.193 nm, compute the expected electric dipole moment of NaF. (b) The measured dipole moment is 27.2×10^{-30} C·m. What is the fractional ionic character of NaF?

10. The equilibrium separation of HI is 0.160 nm and its measured electric dipole moment is 1.47×10^{-30} C·m. Find the fractional ionic character of HI.

11. The equilibrium separation of BaO is 0.194 nm and the measured electric dipole moment is 26.5×10^{-30} C·m. Calculate the fractional ionic character, assuming two valence electrons.

9.4 Molecular Vibrations

12. The effective vibrational force constant of CN is 1.017×10^4 eV/nm^2. (a) Relative to the ground state, find the energies of the first and second excited vibrational states. (b) In ionized CN$^+$ the force constant is weaker by 3.42%. What would be the energies of the first and second excited states in CN$^+$?

13. The vibrational transition in BeO is observed at a wavelength of 6.724 μm. What is the effective force constant of BeO?

14. Figure 9.33 shows the energy minimum of molecular NaCl, through which a parabola has been drawn. Following the methods of Example 9.3, find the effective vibrational force constant, the vibrational frequency and wavelength, and the vibrational photon energy for NaCl. In what range of the electromagnetic spectrum would such radiations be found? What is the maximum vibrational quantum number for which the parabolic approximation remains valid for NaCl?

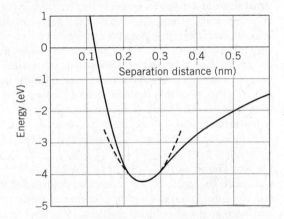

FIGURE 9.33 Problem 14.

15. The vibrational energy of CO is 0.2691 eV when the constituents of the molecule are the most abundant isotopes of carbon ($m = 12.00$ u) and oxygen ($m = 16.00$ u). (a) What would be the vibrational energy if the oxygen were replaced by the less abundant isotope with $m = 18.00$ u? (b) What would be the vibrational energy if the carbon in the original CO were replaced with radioactive carbon (used in radiocarbon dating) with mass 14.00 u?

9.5 Molecular Rotations

16. Derive Eq. 9.14 from the definition of the rotational inertia of a particle.

17. Following the method of Example 9.4, calculate the energies and wavelengths of the three lowest rotational transitions emitted by molecular NaCl. In what region of the electromagnetic spectrum would these radiations be found?

18. The atoms of a carbon dioxide molecule (CO_2) are arranged in a line, with the carbon in the center and the oxygens on either side at a distance of 0.116 nm. In the absence of vibrations, calculate the energies of the first five rotational states of CO_2 for rotations about an axis through the center of mass of the molecule.

19. Find the difference in the rotational parameter B of HCl for the two different masses of Cl (35 u and 37 u).

20. In observing the rotational transitions of a certain molecule, two photons are observed at wavelengths of 526.6 μm and 658.3 μm. There are no photons emitted with wavelengths between these two values. (a) Determine the rotational parameter B for this molecule, and find where these wavelengths occur in the sequence of rotational transitions. (b) From other experiments with this molecule, it is known that the equilibrium separation is 0.1172 nm. It is also known that this molecule contains carbon. Determine the identity of the other atom.

9.6 Molecular Spectra

21. Figure 9.30 illustrates the combined rotational-vibrational structure when the vibrational energy is much larger than the rotational energy. Make a sketch showing the reverse situation, in which the rotational energy is much larger than the vibrational energy. Use a scale in which $B = 10$ units and $hf = 2$ units; show rotational levels up to $L = 3$ and vibrational levels to $N = 3$.

22. (a) Sketch a diagram, similar to Figure 9.30, showing all possible absorption transitions from the $N = 0$ to the $N = 1$ states. Include rotational states up to $L = 5$. (b) Using the values $hf = 10$ units and $B = 0.25$ unit, show the energy spectrum of the absorption, including all transitions from part (a). Label each transition with the initial and final quantum numbers.

23. Based on the energy levels given by Eq. 9.19, calculate the energies of the *emitted* photons.

24. (a) What is the reduced mass of the KCl molecule? (b) With an equilibrium separation of 0.267 nm, find the spacing of the transitions in the combined rotational-vibrational spectrum.

25. Derive Eq. 9.23 from Eq. 9.22.

26. In a collection of HCl molecules, at what temperature would the number in the first excited vibrational state be 1/3 the number in the ground state? (Ignore the rotational structure.)

27. The most intense absorption line in the rotational-vibrational spectrum of CO at room temperature occurs for $L = 7$. Justify this value with a calculation. (The equilibrium separation of CO is 0.113 nm.)

General Problems

28. The energy of a diatomic molecule can be represented by the empirical function

$$E = \frac{A}{R^9} - \frac{B}{R}$$

where A and B are positive constants. Find A and B in terms of the equilibrium separation R_{eq} and the dissociation energy

E_0. Sketch the resulting function, choosing values for the constants appropriate for H_2.

29. Show that the angle between bonds in the tetrahedral carbon structure (Figure 9.14) is 109.5°.

30. Complete the following table, which compares properties of H_2 molecules when one or both of the H is replaced with an atom of deuterium (D), which is a "heavy hydrogen" atom with twice the mass of ordinary hydrogen.

Molecule	Vibrational frequency	R_{eq}	B
H_2	1.32×10^{14} Hz	0.074 nm	0.0076 eV
HD			
D_2			

31. Because of the "zero-point" energy (see Figures 9.23 and 9.29), the ground state of a molecule is not at the minimum of the molecular energy curve but at an energy of $hf/2$ above the minimum. The dissociation energy is the energy needed from that state to break apart the molecule. (a) Given that the vibrational energy of H_2 is 0.54 eV, find the "zero-point" energies of H_2, HD, and D_2. Assume the mass of deuterium ("heavy hydrogen") is twice the mass of hydrogen. (b) Given that the dissociation energy of H_2 is 4.52 eV, find the dissociation energies of HD and D_2. Assume that the added neutron does not change the molecular energy curve or the equilibrium separation.

32. Estimate the number of rotational states between the vibrational states for (a) H_2 (see Example 9.3); (b) HCl (see Example 9.6); (c) NaCl (see Problems 14 and 17).

33. Figure 9.34 shows the absorption spectrum of the molecule HBr. Following the basic procedures of Section 9.6, find: (a) the energy of the "missing" transition; (b) the effective force constant k; (c) the rotational spacing $2B$. Estimate the value of the rotational spacing expected for HBr and compare with the value deduced from the spectrum. Why are there only single lines and not double lines as in the case of HCl?

34. (a) In a collection of H_2 molecules at room temperature, what is the ratio of the number of molecules in the $N = 1$ vibrational state to the number in the $N = 0$ vibrational state? (Ignore the rotational structure for this problem.) (b) What is the ratio of the number in the $N = 2$ state to the number in the ground state?

35. Repeat Problem 34 for NaCl molecules.

36. (a) In a collection of H_2 molecules at room temperature in the $N = 0$ vibrational state, find the relative number of molecules in the first three rotational excited states compared with the number in the ground state. (b) Repeat part (a) if the temperature is 30 K. (Hint: Don't forget the degeneracy of the levels.)

37. Repeat Problem 36 for NaCl.

38. (a) Predict the fractional ionic content of KBr based on Figure 9.22. (b) Calculate the fractional ionic content and compare with your prediction.

39. In a 1987 study of emissions from molecules in the interstellar medium, the following frequencies were reported: 93.98 GHz, 140.97 GHz, and 234.94 GHz. (a) Assuming these radiations to be emitted in transitions between rotational states of a diatomic molecule (but not necessarily to be consecutive in sequence), what is the largest possible value of the rotational spacing $2B$ that would be consistent with these emissions? (b) Where do these transitions fit in the sequence of rotational emissions from this molecule? Give the corresponding angular momentum quantum numbers. (c) The paper that reported these emissions claimed them to be the first identification of phosphorus in the interstellar medium and that the molecule observed was PN. Verify that the observed rotational emissions are consistent with PN.

40. The star known as IRC +10216 or CW Leonis is a red giant in the late stages of its evolution. It is surrounded by a dust cloud that is rich in a variety of chemical elements that have been produced in fusion reactions in the star. Among the molecules observed in the dust cloud is AlCl, which has been identified through two close-lying strong rotational emissions with frequencies 160.312 GHz and 156.547 GHz. (a) Locate these transitions among the rotational states of AlCl. (b) What temperature is necessary to excite AlCl in order to produce these rotations?

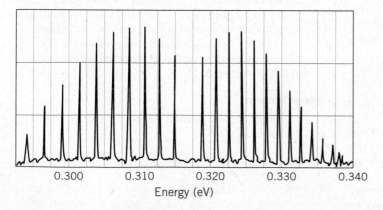

FIGURE 9.34 Problem 33.

Chapter 10

STATISTICAL PHYSICS

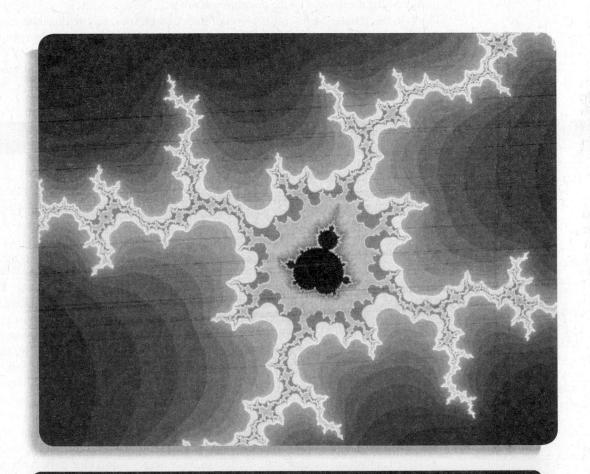

Some processes in nature are determined by the statistical distribution of random events. When the number of particles in a system is very large, the range of likely outcomes becomes very narrow, and the processes may seem deterministic. Other processes may result from deterministic laws but seem chaotic in their outcomes. Such processes can be represented by apparently irregular geometric patterns called fractals.

Many physics experiments are analyzed as if the interactions take place in single, isolated events. The emission of light from a gas at low density, Rutherford scattering, and Compton scattering are examples of experiments that can be analyzed in this way. On the other hand, consider the addition of energy to a gas in a container by raising its temperature. We know how the average energy of the atoms will change, but we can't analyze the behavior of individual atoms.

This sharing of energy among the many parts of a system cannot be simply discussed in terms of single isolated events. The analysis of such *cooperative* phenomena requires the techniques of *statistical physics*, in which we are concerned not with calculating the *exact* outcome of single, isolated events, but with predicting the *average* outcome of many cooperative events, based on the *statistical distribution* of the possible outcomes.

In this chapter we discuss the laws of statistical physics, and we illustrate some systems that are governed by *classical* statistics and some others that require *quantum* statistics. These statistical concepts are necessary to understand the bulk properties of matter, which are discussed briefly in this chapter and more extensively in Chapter 11.

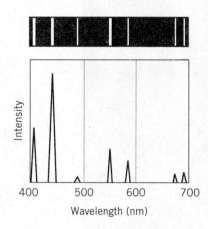

FIGURE 10.1 (top) The line spectrum of mercury in the visible region as it might be analyzed using a diffraction grating. (bottom) The intensity spectrum of the transitions.

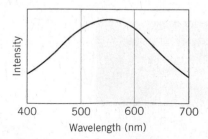

FIGURE 10.2 The continuous spectrum of an incandescent source in the visible region.

10.1 STATISTICAL ANALYSIS

When we pass an electric current through a tube containing a low-density gas, such as mercury vapor, light is emitted. An electron pushed to an excited state of an atom returns to the ground state with the emission of one or more photons. In the case of mercury vapor, we see individual photons corresponding to green light, blue light, orange light, and so forth. Each photon has a definite wavelength and corresponds to a transition between two levels of definite energies. Aside from the effect of the uncertainty principle, the wavelengths are "sharp." If we analyze the light with a high-resolution device such as a diffraction grating, the resulting spectrum (Figure 10.1) shows the sharpness of the spectral lines. We can understand this spectrum based on our knowledge of the excited states of a single mercury atom; as long as the density of the gas is low, the number of atoms in the tube does not affect the observed spectrum. We treat the light emitted by this collection of atoms as if the individual emissions occur singly and in isolation.

Consider now the contrasting case of the tungsten filament of an ordinary incandescent light bulb. Figure 10.2 shows the spectrum in this case, which has the continuous distribution of wavelengths that we call "white" light. All wavelengths are present, not just a finite number. Isolated tungsten atoms, like those of mercury, emit light at a finite number of discrete, well-defined wavelengths, but in a solid tungsten filament the cooperative effect of other nearby atoms changes the energy levels and makes the spectrum continuous. Even though the mercury vapor and the tungsten filament may contain roughly the same number of atoms, in one case we can ignore the presence of the other atoms, while in the other case we must consider the mutual influence of many or all of the atoms in the sample.

There are two ways to approach the analysis of a complex system. The first is to specify a set of *microscopic* properties, such as the position and velocity of each atom. For even a small system containing perhaps 10^{15} atoms, this is obviously a hopeless task. The second way is to recognize that such a description is not only impossibly complex, it is also unnecessary because it provides far more detail than is useful. We can understand and predict the behavior of systems containing many particles in terms of a few *macroscopic* properties, such as the temperature or the pressure of a gas. The development of relationships between microscopic and macroscopic properties was one of the great triumphs of 19th-century physics. In the case of a gas confined to a container, for example, kinetic theory gives the relationship between the microscopic motion of the molecules and the macroscopic temperature and pressure.

More generally, we can make a *statistical* analysis by counting the number of different arrangements of the microscopic properties of a system. For example, consider the distribution of 2 units of energy to a "gas" of four identical but distinguishable particles. Each particle can acquire energy only in integral units (in analogy with the simple harmonic oscillator). How can these four particles share the 2 units of energy? One way is for one particle to have the entire 2 units. There are four different ways to accomplish this distribution, corresponding to choosing each of the four particles to take the 2 units of energy. Another way to distribute the energy is to give two different particles 1 unit each. There are six ways to accomplish this distribution (Table 10.1). Each possible energy distribution is called a *macrostate*—a state of the system that can be observed through the measurement of a macroscopic property such as the temperature. In our simple system, there are two macrostates: macrostate A in which one particle has 2 units of energy, and macrostate B in which two particles each have 1 unit of energy. The different arrangements of microscopic variables corresponding to a single macrostate are called *microstates*. In our system, there are four microstates corresponding to macrostate A and six microstates corresponding to macrostate B. The number of microstates corresponding to a given macrostate is called the *multiplicity W*. For our system, $W_A = 4$ and $W_B = 6$.

One application of these statistical principles is to determine the direction of the natural evolution of a system. The *second law of thermodynamics* says, in effect, that isolated systems evolve in a direction such that the multiplicity increases.*

TABLE 10.1 Macrostates of a Simple System

Macrostate	Microstate (Energy of Particle)			
	Particle 1	Particle 2	Particle 3	Particle 4
A	2	0	0	0
	0	2	0	0
	0	0	2	0
	0	0	0	2
B	1	1	0	0
	1	0	1	0
	1	0	0	1
	0	1	1	0
	0	1	0	1
	0	0	1	1

---

*The second law is often expressed in terms of the *entropy S* by saying that an isolated system must evolve such that $\Delta S \geq 0$. The Austrian physicist Ludwig Boltzmann (1844–1906) developed the relationship between the entropy of a system and the multiplicity of its macrostate: $S = k \ln W$, where k is the Boltzmann constant. The increase in the entropy of a system as it changes from one macrostate to another is thus equivalent to an increase in the multiplicity.

That is, if we started our system in macrostate A and allowed the four particles to interact with one another, later we might find it in macrostate B; however, if the system began in macrostate B, we would be less likely to find it later in macrostate A, because a change from B to A involves a less probable decrease in multiplicity. As the number of particles in the system increases, differences in multiplicity become greater, and changes involving decreases in multiplicity become so improbable as to be unobservable.

Implicit in this statistical analysis is the following postulate:

All microstates of a system are equally probable.

Our system can be found with equal probability in any of the 10 microstates listed in Table 10.1. It is this postulate that allows us to assign a greater statistical weight to macrostate B; the system can be found with equal probability in any of the six microstates of B or the four microstates of A, so the relative weight of B is $6/4 = 3/2$. In a large number of identical systems, we expect to find 40% in macrostate A and 60% in macrostate B.

For another example, in the card game of poker a royal flush consisting of the A♥, K♥, Q♥, J♥, 10♥ is just as probable as the *particular* worthless hand 10♠, 8♣, 5♦, 4♥, 2♠. What makes a royal flush so special is that there are only four possible royal flushes among the 2,598,960 possible poker hands, most of which are worthless. Even though a royal flush is just as likely as a *particular* worthless hand, it is much less likely than *any* worthless hand. In the language of statistical physics, the macrostate "royal flush" has fewer microstates (and is therefore less probable) than the macrostate "worthless hand" or the macrostate "one pair."

The assumption of equal probabilities of the microstates allows us to do calculations on the system. For example, suppose we have a large number of identical systems like that of Table 10.1 whose microstates are distributed randomly. Let us reach into each system and measure the energy of a particle. What is the probability that the particle will have 2 units of energy? Among the 10 *equally probable* microstates consisting of a total of 40 particles, there are four particles with 2 units of energy. Thus $p(2) = 4/40 = 0.10$, and we expect that 10% of our measurements would find $E = 2$. Similarly, $p(1) = 0.30$ and $p(0) = 0.60$.

The statistical analysis of a complex system gives us a way of describing the state of the system, its average properties, and its evolution in time. Physics is often concerned with exactly these details, so the statistical analysis is very useful indeed. It can be applied to systems in which the number of particles is very small, such as a nucleus with the order of 10^2 particles, or very large, such as a star like the Sun with 10^{57} particles. Our next task is to determine whether there are differences between the statistical behaviors of classical and quantum systems.

10.2 CLASSICAL AND QUANTUM STATISTICS

To illustrate the differences between classical and quantum statistics, we first consider another example similar to the one of the previous section: the distribution of a total of 6 energy units to a collection of five identical but distinguishable oscillator-like particles, each of which can absorb energy in equal increments of 1 unit. Instead of a tabulation similar to Table 10.1, the energy distribution is illustrated in Figure 10.3. There are ten macrostates, labeled A through J. Each

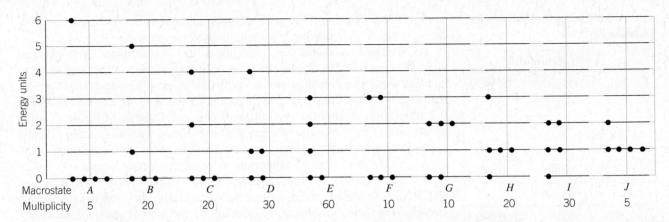

FIGURE 10.3 The macrostates of a system in which five identical particles share 6 units of energy.

dot indicates a particle with a certain energy; for instance, in macrostate B there are three particles with energy $E = 0$, one with $E = 1$, and one with $E = 5$. The multiplicity of each macrostate (the number of microstates) can be worked out in tabular form, or it can be calculated directly using standard methods from permutation theory:

$$W = \frac{N!}{N_0! N_1! N_2! N_3! N_4! N_5! N_6!} \quad (10.1)$$

where N is the total number of particles and N_E is the number with energy E.

The results of this calculation are an example of a more general approach that gives the total number of microstates when N distinguishable particles share Q integral units of energy:

$$W_{\text{total}} = \frac{(N + Q - 1)!}{Q!(N - 1)!} \quad (10.2)$$

so for our case we expect the total number of microstates to be $10!/6!4! = 210$, in agreement with the total of the values of W given in Figure 10.3.

As we did previously, let us calculate the probability to measure any particular value of the energy of a particle. This can be done by considering a collection of all 210 possible microstates and counting the number of times each value of the energy appears. Symbolically,

$$p(E) = \frac{\sum_i N_i W_i}{N \sum_i W_i} \quad (10.3)$$

where N_i represents the number of particles with energy E in each particular macrostate. The sums are carried out over all of the macrostates (10 in the case of Figure 10.3).

Table 10.2 gives the probabilities associated with measuring each of the possible energies. Note that the probability decreases with increasing energy. The probabilities are plotted in Figure 10.4. The smooth curve is an exponential of the

TABLE 10.2 Energy Probabilities for the System of Figure 10.3

Energy	Probability
0	0.400
1	0.267
2	0.167
3	0.095
4	0.048
5	0.019
6	0.005

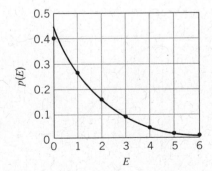

FIGURE 10.4 The probabilities from Table 10.2. The curve is an exponential function that closely fits the points.

TABLE 10.3 Energy Probabilities for Quantum Particles

Energy	Probability	
	Integral Spin	Spin ½
0	0.420	0.333
1	0.260	0.333
2	0.160	0.200
3	0.080	0.067
4	0.040	0.067
5	0.020	0.000
6	0.020	0.000

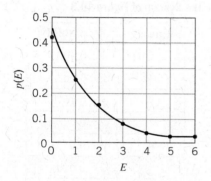

FIGURE 10.5 The probabilities from Table 10.3 for particles with integral spin. The curve that characterizes the points rises more steeply than the exponential at low energy.

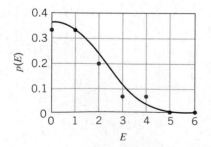

FIGURE 10.6 The probabilities from Table 10.3 for spin ½. The curve that characterizes the points becomes approximately flat at low energies.

form $p \propto e^{-\beta E}$, where β is a constant chosen to fit the data. You can see that the decrease of p with increasing E is approximately exponential.

This example represents the application of classical statistics. Although real systems composed of many particles have too many macrostates and microstates to tabulate, we can analyze their properties by determining the distribution function that describes how the energy is shared among the parts of a system. Later in this chapter we discuss the application of classical statistics to the distribution of energies of the molecules of a gas. Our experience with this simple example serves as a useful guide: The true classical distribution function turns out to be an exponential of the form $e^{-\beta E}$. However, there are other phenomena, such as the electrical conductivity and heat capacity of metals, the behavior of liquid helium, and thermal radiation, which cannot be successfully analyzed using classical statistics. For these phenomena, we must use the methods of *quantum statistics*.

Why should classical statistics differ from quantum statistics? Two of the assumptions made in the example of this section are inconsistent with basic principles of quantum physics:

1. *In quantum physics, identical particles must be treated as indistinguishable.* In calculating the multiplicity of the macrostates, we assumed the classical particles to be identical but distinguishable. That is, the particles are numbered 1 through 5 (or carry other distinguishing marks). In macrostate A, for instance, it is possible to distinguish the microstate in which particle 1 has $E = 6$ from the microstate in which particle 2 has $E = 6$, so we count each of these as separate microstates in determining the multiplicity of 5. If we regard the particles as indistinguishable quantum particles (such as electrons or photons), we can't tell the difference between these microstates. If we can't *observe* the five microstates of A as separate arrangements, we can't *count* them as separate arrangements. For identical quantum particles, the multiplicity of each macrostate becomes exactly 1.

2. *Quantum mechanics can impose limits on the maximum number of particles that can be assigned to any particular state.* For example, suppose the particles in our example are electrons. The Pauli principle forbids two electrons in a system from being in the same state of motion (or having the same set of quantum numbers). Electrons can have spin up or spin down, so if there are no other quantum numbers associated with these energy states, there can be no more than two electrons in any energy state. For electrons, macrostates A, B, C, F, G, H, and J are forbidden, because in each one there are more than two particles in the same energy state.

We can repeat the calculation of the probability $p(E)$ for two cases of quantum particles: photons or alpha particles, which have integral spins and are not restricted by the Pauli principle, and electrons or protons, which have a spin of ½ and *are* restricted by the Pauli principle to no more than two per energy state. In the first case, we have ten macrostates with equal multiplicities of 1, and in the second case we have three macrostates (D, E, I) again with equal multiplicities of 1. Table 10.3 gives the resulting values of $p(E)$. The probabilities for the particles with integral spins are plotted in Figure 10.5; the curve is approximately exponential but rises a bit more steeply at low energies. The values for the particles with $s = $ ½ are plotted in Figure 10.6; the behavior is no longer even approximately exponential; instead, it is rather flat near $E = 0$ and then drops rapidly to 0 for the higher energies.

The example we have analyzed in this section should be regarded only as indicating some differences between classical and quantum statistics. The true energy distribution functions for many-particle systems look different from the ones obtained for this simple five-particle system.

Particles with integral spin are described by a quantum distribution function that is approximately exponential but that rises more sharply at low energies; this difference is responsible for a number of interesting cooperative effects. Particles that obey the Pauli principle are described by a very different quantum distribution that becomes flat at low energies, like Figure 10.6. The quantum distribution functions for these two cases are discussed later in this chapter.

Example 10.1

For the case in which five particles share 6 units of energy, calculate the probability to observe a particle with 2 units of energy for (a) distinguishable classical particles, (b) indistinguishable quantum particles with integral spin, and (c) indistinguishable quantum particles with half-integral spin.

Solution

(a) For distinguishable classical particles, we can use Equation 10.3 with the multiplicities for the 10 macrostates given in Figure 10.3 to sum the probabilities to find 2 units of energy in all of the macrostates. The numerator of Eq. 10.3 evaluates to:

$$\sum N_i W_i = 0 \times 5 + 0 \times 20 + 1 \times 20 + 0 \times 30 + 1 \times 60$$
$$+ 0 \times 10 + 3 \times 10 + 0 \times 20 + 2 \times 30 + 1 \times 5$$
$$= 175$$

and so

$$p(2) = \frac{\sum N_i W_i}{N \sum W_i} = \frac{175}{5 \times 210} = 0.167$$

(b) For indistinguishable quantum particles with integral spin, each of the macrostates has multiplicity 1, so the numerator of Eq. 3 becomes

$$\sum N_i W_i = 0 \times 1 + 0 \times 1 + 1 \times 1 + 0 \times 1 + 1 \times 1$$
$$+ 0 \times 1 + 3 \times 1 + 0 \times 1 + 2 \times 1 + 1 \times 1 = 8$$

and

$$p(2) = \frac{\sum N_i W_i}{N \sum W_i} = \frac{8}{5 \times 10} = 0.160$$

(c) For indistinguishable quantum particles with half-integral spin, the states with more than 2 particles having the same value of the energy are not permitted and so have multiplicity 0 (states $A, B, C, F, G, H,$ and J). The remaining allowed states have multiplicity 1. The sum is now

$$p(2) = \frac{0 \times 1 + 1 \times 1 + 2 \times 1}{5 \times 3} = 0.200$$

Example 10.2

A system consists of two particles, each of which has a spin of 1. What are the possible values of the z component of the total spin and what is the multiplicity associated with each value, assuming the particles to be (a) distinguishable and (b) indistinguishable?

Solution

For each particle, we can have $m_s = +1, 0, -1$ and for their combination we have $M_S = m_{s1} + m_{s2}$ where $M_S \hbar$ gives the z component of the total spin S. Clearly M_S runs

from a maximum of $+2$ to a minimum of -2. (a) For distinguishable particles, there are 3 possible m_s values for each particle and so a total of $3 \times 3 = 9$ combinations. None of these combinations is prohibited. (b) For indistinguishable particles, we cannot count $m_{s1} = +1, m_{s2} = 0$ and $m_{s1} = 0, m_{s2} = +1$ as separate microstates, because we cannot distinguish particle 1 from particle 2. So there are 3 fewer microstates when the particles are indistinguishable. The multiplicities are:

Macrostate	Microstates		
M_S	m_{s1}, m_{s2}	**Distinguishable**	**Indistinguishable**
+2	+1, +1	1	1
+1	+1, 0	2	1
	0, +1		
0	+1, −1	3	2
	−1, +1		
	0, 0		
−1	0, −1	2	1
	−1, 0		
−2	−1, −1	1	1

For distinguishable particles, the possible total spins are $S = 2$ ($M_S = +2, +1, 0, −1, −2$ for a multiplicity of 5), $S = 1$ ($M_S = +1, 0, −1$ for multiplicity 3), and $S = 0$ ($M_S = 0$, multiplicity 1). The total multiplicity is $5 + 3 + 1 = 9$.

For indistinguishable particles, because we have states with $M_S = \pm 2$ we clearly must have $S = 2$. Assigning 5 of the microstates ($M_S = +2, +1, 0, −1, −2$) to $S = 2$,

there is only 1 microstate left, with $M_S = 0$, which must be assigned to $S = 0$. Note that for distinguishable particles we can have $S = 2, 1,$ or 0, while the indistinguishable quantum particles with integral spin can combine only to $S = 2$ or 0. This suggests another important difference between the statistics of classical and quantum systems—for quantum systems, the statistics will allow some spin combinations and forbid others.

10.3 THE DENSITY OF STATES

We have seen in the previous sections how the multiplicity of a state determines the probability to find a system in that state. We are now going to extend that concept to more complicated systems.

Let's consider a system (classical or quantum) composed of a large number of particles that can exist in many different energy states. The relative probability for the particles to have energy E is given by the *distribution function $f(E)$*. This function might tell us, for example, that it is less likely to find a particle in the system with larger amounts of energy than with smaller amounts. For example, if our system is a gas at temperature T, it is increasingly less probable to find molecules with energy far above kT (where k is the Boltzmann constant).

Our ultimate goal is to calculate the number of particles with any given value of the energy E. This number is in part determined by the distribution function $f(E)$. However, the calculation must also take into account that there may be many states available at the energy E. This additional factor is related to the multiplicity of microstates that we have considered in systems with small numbers of particles. In effect, to find the number of occupied states we combine the number of available states with the probability that each state is occupied.

There are two different ways of calculating the number of available states at the energy E. If the energy states are discrete and individually observable, the number of states available at the energy E is just the degeneracy of the state at that energy. For example, if our system is a dilute gas of hydrogen atoms, the number of states available at the energy E_n for a state with principal quantum number n is the degeneracy of that state, which is $2n^2$. As we go to higher excited states, the number of atoms in each state might decrease due to the distribution function $f(E)$ and might increase due to the degeneracy factor $2n^2$. The actual number of atoms that might be found in the excited state with a particular energy represents the combined effect of both factors.

For another example, consider the rotational excited states in a molecule such as HCl. The degeneracy of each level of energy E_L with angular momentum quantum number L is $2L + 1$. As we go to higher energies corresponding to larger values of the angular momentum, the distribution function decreases and the degeneracy factor increases, so that the net population of any level is determined by the combined effect of both factors. In Section 9.6, we saw an example of how these two factors produced a maximum intensity in the absorption spectrum for the state with a certain value of L.

The degeneracy factor and the distribution function can be combined to enable us to calculate the number of particles in the system having a certain value of the discrete energy E_n:

$$N_n = d_n f(E_n) \tag{10.4}$$

where d_n is the degeneracy of the level at energy E_n. The distribution function $f(E_n)$ is properly normalized to ensure that the total number of particles in our system is N:

$$N = \sum N_n = \sum d_n f(E_n) \tag{10.5}$$

where the sum is carried out over all the states (although often in practice the distribution function f decreases so rapidly for increasing E that it is sufficiently accurate to include only a few terms in the summation).

On the other hand, we might be analyzing a system in which the energy states are so numerous and spaced so closely together that we cannot observe them as separate and individual states. As a result, we may consider only the number of states in a small interval dE at the energy E, or equivalently the number of states between energies E and $E + dE$. This method brings to mind the analysis of the kinetic energy of the molecules in a gas (see Section 1.3) and the analysis of the probability distribution corresponding to solutions of the Schrödinger equation (see Section 5.3).

For this analysis we treat E as a continuous variable. (Of course, we know the states are discrete, but there are so many of them and they are so close together that this is really a very good approximation.) We define the *density of states* $g(E)$ so that $g(E)dE$ is the number of available states per unit volume in the interval dE at energy E (or between E and $E + dE$). The number of *populated* states in that interval is determined by the density of states factor and the distribution function:

$$dN = N(E)\,dE = Vg(E)f(E)\,dE \tag{10.6}$$

where V is the volume of the system (which must be included because the density of states is the number of states *per unit volume*). Again we'll assume that the distribution function is properly normalized so that the total number of particles in the system is fixed at N:

$$N = \int dN = \int_0^\infty N(E)\,dE = V \int_0^\infty g(E)f(E)\,dE \tag{10.7}$$

Be sure to remember the difference between N and $N(E)$. The symbol N represents the total number of particles in the system, while $N(E)$ represents the number *per unit energy interval* at energy E. That is, N is a pure number, while $N(E)$ is a function of E and has units of (energy)$^{-1}$.

In this chapter we discuss the application of classical statistics to the molecules of a gas and the application of quantum statistics to the electrons in a metal, which we can consider to be a "gas" of electrons. We also discuss the application of quantum statistics to the problem of cavity radiation that we first introduced in Chapter 3, so we'll need to understand the density of states of a gas of photons.

Density of States in a Gas of Particles

We'll first discuss the density of states in a "gas" of particles, such as electrons or molecules. This turns out to be a useful concept in numerous applications. For example, the space around a hot metal filament is filled with a cloud of emitted electrons. In a solid metal conductor such as copper, each atom contributes one loosely bound electron to the electric current; because these electrons belong to the material as a whole rather than to individual atoms, it is appropriate to treat the electrons as a gas in the metal. The same calculation applies, as we shall see, to an ordinary molecular gas such as nitrogen.

The basis of our calculation is the particle trapped in a three-dimensional region of space, which we assume to be a cubical box in which each side has length L. The cubical box represents a three-dimensional potential energy well. (See, for example, the analysis of the two-dimensional potential energy well in Section 5.4.) By an obvious extension of Eqs. 5.41 and 5.45, the wave function and energy levels for the infinite three-dimensional potential energy well can be written as

$$\psi(x,y,z) = A' \sin \frac{n_x \pi x}{L} \sin \frac{n_y \pi y}{L} \sin \frac{n_z \pi z}{L} \tag{10.8}$$

and

$$E = \frac{p^2}{2m} = \frac{1}{2m}(p_x^2 + p_y^2 + p_z^2) = \frac{h^2}{8mL^2}(n_x^2 + n_y^2 + n_z^2) = \frac{h^2}{8mL^2}n^2 \tag{10.9}$$

where $n^2 = n_x^2 + n_y^2 + n_z^2$.

So now the problem that confronts us in finding the density of states is: how many different combinations of the quantum numbers n_x, n_y, and n_z give us a particular value of the energy E? If the energy levels were well separated and could be observed individually, we could find their degeneracy and thus obtain the density of states. However, in our case the cube is so large compared with the typical de Broglie wavelength of the electrons that there is a huge number of energy levels in any interval dE, certainly far too many for us to attempt to count.

One way to find the number of energy states is to imagine a coordinate system in which the axes are labeled n_x, n_y, and n_z (Figure 10.7). The points with a common value of E lie on a spherical surface of radius $n = \sqrt{n_x^2 + n_y^2 + n_z^2}$, and the points with energies between E and $E + dE$ lie within a spherical shell between radius n and radius $n + dn$. Only positive values of n_x, n_y, and n_z may occur, and thus we have only one-eighth of the complete shell. The entire shell has surface area $4\pi n^2$ and thickness dn, so the number of points (each point representing a different combination of n_x, n_y, n_z leading to the same value of E) in this portion of the shell is $\frac{1}{8} \times 4\pi n^2 dn$.

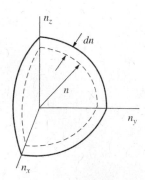

FIGURE 10.7 One-eighth of a spherical shell of radius n and thickness dn in the n_x, n_y, n_z coordinate system.

Finally, if the particles have spin s, each point in this coordinate system is really $2s + 1$ points, which represent each of the different possible spin orientations (because, in the absence of a magnetic field, each of these different spin orientations represents a different state having the same energy E). For example, if the particles are electrons, there are two spin orientations (one with $m_s = +1/2$ and one with $m_s = -1/2$), and for electrons $2s + 1 = 2$. The density of states must include an additional factor of $2s + 1$. Thus the number of states per unit volume is

$$g(n)\, dn = \frac{1}{8}\frac{2s + 1}{V} 4\pi n^2\, dn \qquad (10.10)$$

The number of states contained in this interval is the same whether we count the states in terms of n or E. That is, $g(n)dn = g(E)dE$. With $E = h^2 n^2/8mL^2$, we have $dE = 2h^2 n\, dn/8mL^2$ and so

$$g(E)\, dE = g(n)\, dn = \frac{4\pi(2s + 1)\sqrt{2}m^{3/2}}{h^3}E^{1/2}\, dE \qquad (10.11)$$

using $V = L^3$ for the volume of the cube. The density of states for the gas of particles is then

$$g(E) = \frac{4\pi(2s + 1)\sqrt{2}m^{3/2}}{h^3}E^{1/2} = \frac{4\pi(2s + 1)\sqrt{2}(mc^2)^{3/2}}{(hc)^3}E^{1/2} \qquad (10.12)$$

Density of States in a Gas of Photons

Let's consider now a gas of photons. Imagine a cavity such as we discussed in Section 3.3: a hollow metal box of volume V with walls at temperature T. The box (in the form of a cube with each side of length L) is filled with photons having a range of energies from 0 to ∞ (with an equivalent range of frequencies or wavelengths).

Let's imagine the box to be filled with electromagnetic standing waves. Any particular wave has energy $E = pc = c\sqrt{p_x^2 + p_y^2 + p_z^2}$. Each momentum component must obey the standing wave condition for that particular wavelength, and applying the boundary conditions that the electric field must vanish at the locations of the walls gives exactly the same restrictions on the wavelengths or wave numbers as the analysis of the three-dimensional infinite potential energy well (for which $\psi = 0$ at $x = 0$ and $x = L$, and similarly for y and z). Thus, for example, for the x direction we have the wave number k_x subject to the condition $k_x L = n_x \pi$, and so $p_x = \hbar k_x = \hbar \pi n_x/L$, with analogous results for the y and z directions. The indices n_x, n_y, and n_z are independent of one another and take positive values from 1 to ∞. We can therefore write the energy as

$$E = c\sqrt{p_x^2 + p_y^2 + p_z^2} = c\sqrt{\left(\frac{\hbar\pi n_x}{L}\right)^2 + \left(\frac{\hbar\pi n_y}{L}\right)^2 + \left(\frac{\hbar\pi n_z}{L}\right)^2}$$

$$= \frac{\hbar c\pi}{L}\sqrt{n_x^2 + n_y^2 + n_z^2} \qquad (10.13)$$

We now have exactly the same situation that we encountered for the electron gas shown in Figure 10.7. The number of points in the portion of the spherical shell is again proportional to its volume $\frac{1}{8} \times 4\pi n^2\, dn$.

Finally, we must account for the possible spin states of the photon. Because a photon has $s = 1$, we might expect that the density of states should include the factor $2s + 1 = 3$. However, the $s_z = 0$ state is not permitted for photons. This is equivalent to requiring that an electromagnetic wave have only one of two possible polarizations, either right circular or left circular. The factor that accounts for the spin multiplicity in the density of states is thus 2, and the number of states per unit volume is given by Eq. 10.10 with $2s + 1$ replaced by 2. With $E = \hbar c n / L$, we obtain $dE = \hbar c \, dn / L$ and finally

$$g(E) \, dE = g(n) \, dn = \frac{1}{\pi^2 (\hbar c)^3} E^2 \, dE \qquad (10.14)$$

with again $V = L^3$ for the volume of the cavity. The density of states for the photon gas is then

$$g(E) = \frac{1}{\pi^2 (\hbar c)^3} E^2 = \frac{8\pi}{(hc)^3} E^2 \qquad (10.15)$$

Note that, in contrast to the gas of particles, in which the density of states depends on $E^{1/2}$, the density of states for a photon gas depends on E^2.

Example 10.3

What is the number of available states per cubic meter in helium gas within a narrow interval of 0.0002 eV at the most probable molecular energy of 0.0086 eV at a temperature of 200 K?

Solution
The density of available states is $g(E)dE$, with $g(E)$ for particles given by Eq. 10.12, using $s = 0$ and $mc^2 = 3727$ MeV for helium atoms and $dE = 0.0002$ eV for the energy interval:

$$g(E) \, dE = \frac{4\pi\sqrt{2}(mc^2)^{3/2}}{(hc)^3} E^{1/2} \, dE$$

$$= \frac{4\pi\sqrt{2}(3727 \times 10^6 \text{ eV})^{3/2}}{(1240 \text{ eV} \cdot \text{nm})^3 (1\text{m}/10^9 \text{ nm})^3}$$

$$\times (0.0086 \text{ eV})^{1/2}(0.0002 \text{ eV})$$

$$= 3.9 \times 10^{28} \text{ m}^{-3}$$

There are about 10^{25} gas atoms per m^3, which suggests that only a very small fraction of the available states is populated in an ordinary gas.

Example 10.4

The interior of a star can be considered to be filled with photons of energies in the range of 1 MeV. What is the density of available states for a container of photons of energies in an interval of 10 keV at 1 MeV?

Solution
From Eq. 10.15 we have

$$g(E) \, dE = \frac{8\pi}{(hc)^3} E^2 \, dE = \frac{8\pi (10^9 \text{ nm/m})^3}{(1240 \text{ eV} \cdot \text{nm})^3} (10^6 \text{ eV})^2 (10^4 \text{ eV})$$

$$= 1.3 \times 10^{35} \text{ m}^{-3}$$

Such a large number of available states can accommodate a large number of photons. Note that this result depends critically on the energy of the photons. For visible light (photons of order 1 eV), a similarly narrow interval could accommodate about 10^{19} energy states per m^3.

10.4 THE MAXWELL-BOLTZMANN DISTRIBUTION

Next we must consider the possible forms of the distribution function $f(E)$, which describes how the available states at energy E are populated. We'll begin by considering a classical system, in which the density is relatively low. In practice this means that the average spacing between the particles is large compared with their de Broglie wavelength. The individual particles may have quantized energy levels, such as the excited states of atoms or molecules, but the overall system does not show quantum behavior. This limit works perfectly well to describe gases under ordinary conditions of temperature and pressure.

The distribution function that applies in this case is the *Maxwell-Boltzmann distribution*, which is given by

$$f_{MB}(E) = A^{-1}e^{-E/kT} \qquad (10.16)$$

for a system in thermal equilibrium at temperature T. (Here k is the Boltzmann constant.) As the energy increases, the occupation probability drops off exponentially. We have included a normalization constant A^{-1}, which helps to ensure that the total number of particles in the system is fixed at N. (The reason for writing the normalization constant as A^{-1} rather than simply A will become apparent later in this chapter.)

Let's apply the Maxwell-Boltzmann distribution to find the number of molecules having energy between E and $E + dE$ in a gas at temperature T that fills a container of volume V with a total of N molecules. From Eq. 10.5 we have (using the density of states from Eq. 10.12)

$$dN = N(E)\,dE = Vg(E)f_{MB}(E)\,dE = A^{-1}\frac{4\pi V(2s+1)\sqrt{2}(mc^2)^{3/2}}{(hc)^3}E^{1/2}e^{-E/kT}\,dE$$
$$(10.17)$$

The normalization constant is determined by the condition that the total number of molecules, obtained as in Eq. 10.7 by integrating over all energies, must be N:

$$N = \int dN = \int_0^\infty N(E)\,dE = A^{-1}\frac{4\pi V(2s+1)\sqrt{2}(mc^2)^{3/2}}{(hc)^3}\int_0^\infty E^{1/2}e^{-E/kT}dE$$
$$(10.18)$$

The definite integral is a standard form, and carrying out the integration gives $A^{-1} = N(hc)^3/V(2s+1)(2\pi mc^2 kT)^{3/2}$. We obtain for the energy distribution of gas molecules

$$N(E) = Vg(E)f_{MB}(E) = \frac{2N}{\sqrt{\pi}(kT)^{3/2}}E^{1/2}e^{-E/kT} \qquad (10.19)$$

which is identical to Eq. 1.22. This equation is the *Maxwell-Boltzmann energy distribution*.

This distribution is shown in Figure 10.8. Note that it rises to a maximum at the most probable energy $E_p = \frac{1}{2}kT$ and then falls gradually to zero as the energy increases, which suggests that it is increasingly rare to find a molecule with energy much larger than kT. The shaded strip of width dE represents the number $dN = N(E)dE$ given by Eq. 10.17, the number of molecules between E and $E + dE$. The mean or average energy is $E_m = \frac{3}{2}kT$, as we showed in Eq. 1.26.

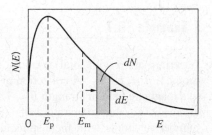

FIGURE 10.8 The Maxwell-Boltzmann energy distribution for molecules, showing the most probable energy $E_p = \frac{1}{2}kT$ and the mean energy $E_m = \frac{3}{2}kT$. The shaded strip represents the number of molecules dN with energies between E and $E + dE$.

In normalizing Eq. 10.19, Planck's constant has disappeared. We can therefore regard the Maxwell-Boltzmann energy distribution as describing the behavior of classical particles, such as an ordinary molecular gas, in which quantum effects do not contribute.

Example 10.5

A container holds one mole of argon gas at a temperature of 375 K. Calculate the fraction of the molecules in the container with energies between 0.025 eV and 0.026 eV.

Solution

We'll assume that $N(E)$ is varying sufficiently smoothly between 0.025 and 0.026 eV that we can use $dN = N(E)dE$ to find dN. This avoids having to use an integral to find the number. From Eq. 10.19 we then have, with

$kT = (8.617 \times 10^{-5} \text{ eV})(375 \text{ K}) = 0.0323 \text{ eV}$,

$$\frac{dN}{N} = \frac{N(E)\, dE}{N} = \frac{2}{\sqrt{\pi}\,(kT)^{3/2}} E^{1/2} e^{-E/kT}\, dE$$

$$= \frac{2}{\sqrt{\pi}\,(0.0323 \text{ eV})^{3/2}} (0.0255 \text{ eV})^{1/2}$$

$$\times e^{-0.0255 \text{ eV}/0.0323 \text{ eV}} (0.001 \text{ eV})$$

$$= 0.014 = 1.4\%$$

Example 10.6

(a) In a gas of atomic hydrogen at room temperature, what is the number of atoms in the first excited state at $E = 10.2$ eV, expressed as a ratio to the number in the ground state? (b) At what temperature would we expect to find 1/10 as many atoms in the first excited state as in the ground state?

Solution

(a) For this case we use the discrete form of the expression for the number of atoms given by Eq. 10.4. The degeneracy factor d_n is $2n^2$ for the state with principal quantum number n. At room temperature ($T = 293$ K), $kT = 0.0252$ eV. The ratio of the number in the excited state ($n = 2$) to the number in the ground state ($n = 1$) is then

$$\frac{N_2}{N_1} = \frac{d_2 e^{-E_2/kT}}{d_1 e^{-E_1/kT}} = \frac{8}{2} e^{-(E_2 - E_1)/kT} = 4 e^{-10.2 \text{ eV}/0.0252 \text{ eV}}$$

$$= 4 e^{-405} = 0.6 \times 10^{-175}$$

In order to have one atom in the excited state, we therefore require about 1.7×10^{175} atoms of hydrogen, about 3×10^{148} kg, a quantity greater than the mass of the universe!

(b) We now require $N_2/N_1 = 0.1$ and solve for T:

$$0.1 = 4 e^{-10.2 \text{ eV}/kT}$$

This gives $kT = 2.77$ eV or $T = 3.21 \times 10^4$ K.

Example 10.7

A certain atom with total atomic spin $\frac{1}{2}$ has a magnetic moment $\mu = 1.2\ \mu_B$. A collection of such atoms is placed in a magnetic field of strength $B = 7.5$ T. What is the ratio, at $T = 77$ K (the temperature of liquid nitrogen), of the number of atoms with their spins aligned parallel to the field to those with their spins aligned opposite to the field?

Solution

The energy of interaction with the magnetic field is $E = -\vec{\mu} \cdot \vec{B}$, so the energy of those aligned parallel to the field is $E_1 = -\mu B$ while those aligned antiparallel have energy $E_2 = +\mu B$. The degeneracies of the

states with $m_s = +\frac{1}{2}$ and $m_s = -\frac{1}{2}$ are identical, so $d_1 = d_2$. With $\mu B = 1.2(9.27 \times 10^{-24} \text{ J/T})(7.5 \text{ T}) = 8.34 \times 10^{-23}$ J $= 5.21 \times 10^{-4}$ eV and $kT = 0.00664$ eV, we have:

$$\frac{N_1}{N_2} = \frac{d_1 e^{-E_1/kT}}{d_2 e^{-E_2/kT}} = e^{-(E_1 - E_2)/kT} = e^{2\mu B/kT}$$

$$= e^{2(5.21 \times 10^{-4} \text{ eV})/0.00664 \text{ eV}} = 1.17$$

That is, about 54% of the atoms are aligned parallel to the magnetic field and 46% are aligned antiparallel to the field.

The Distribution of Molecular Speeds

From an experimental standpoint, it is often easier to measure the distribution of speeds than the distribution of energies. So let's use the Maxwell-Boltzmann distribution for the distribution of kinetic energies in a gas to obtain the distribution of molecular speeds, which then can be tested in the laboratory. That is, we wish to obtain an expression for the number of molecules with speeds in the interval dv at v (between v and $v + dv$), represented by $dN = N(v)dv$. The number of molecules dN in the interval is the same, whether we count them in terms of energy or speed, and so $dN = N(E)dE = N(v)dv$, or $N(v) = N(E)dE/dv$. With $E = \frac{1}{2}mv^2$ and $dE/dv = mv$, we have

$$N(v) = N(E)\frac{dE}{dv} = \frac{2N}{\sqrt{\pi}(kT)^{3/2}}\sqrt{\frac{mv^2}{2}}e^{-mv^2/2kT}mv = N\sqrt{\frac{2}{\pi}}\left(\frac{m}{kT}\right)^{3/2}v^2 e^{-mv^2/2kT}$$

(10.20)

This equation, which is known as the *Maxwell speed distribution*, is graphed in Figure 10.9. The shaded strip shows the number dN in the interval dv at v. The most probable speed occurs at the value $v_p = (2kT/m)^{1/2}$.

An example of an experiment to measure the distribution of molecular speeds is shown in Figure 10.10. A small hole in the side of an oven allows a stream of molecules to escape; we assume the hole to be small enough so that the distribution of speeds inside the oven is not changed. The beam of molecules is made to pass through a slot in a disc attached to an axle rotating at angular velocity ω. At the other end of the axle is a second slotted disc, but the slot is displaced from the first by an angle θ. In order for a molecule to pass through both slots and strike the detector, it must travel the length L of the axle in the same time that it takes the axle to rotate by the angle θ, and thus $L/v = \theta/\omega$. Keeping L and θ fixed, we can vary ω; measuring the number of molecules striking the detector for each different value of ω enables us to measure the Maxwell speed distribution. A set of such experimental results is shown in Figure 10.11, and the agreement between the measured speed distribution and that predicted according to Eq. 10.20 is very impressive.

From this example you can also see the importance of the interval dv. What we measure is always the product $N(v)\,dv$, and in this case the range of velocities is determined primarily by the width of the slots in the discs. To make dv very small,

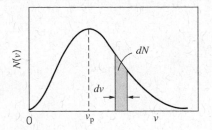

FIGURE 10.9   The Maxwell speed distribution for gas molecules. The shaded strip represents the number of molecules with speeds between v and $v + dv$.

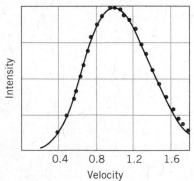

FIGURE 10.11   Result of measurement of the distribution of atomic speeds in thallium vapor. The solid line is obtained from the Maxwell speed distribution for an oven temperature of 870 K.

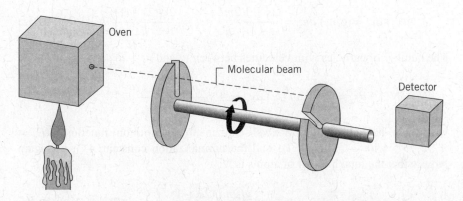

FIGURE 10.10   Apparatus to measure the distribution of molecular speeds.

and thereby measure v "exactly," we would need to make the slots very small, so that very few molecules could get through. For the "perfect" experiment, we make $dv \to 0$ by making the slot width equal to 0, and no molecules get through the apparatus!

Doppler Broadening of Spectral Lines

Atoms at rest are expected to emit very sharp, narrow spectral lines; this "natural line width" is usually determined by the uncertainty principle $\Delta E \Delta t \sim \hbar$, which makes the energy of the atomic excited states somewhat uncertain due to their finite lifetimes. Because the atoms of a gas are in motion, the frequency or wavelength of the light emitted by the atoms often can be Doppler shifted (see Eq. 2.22). Even if the container of gas is at rest relative to the observer, some of the molecules are moving toward the observer and will show an increased frequency, while others are moving away and will show a decreased frequency. As a result of the thermal motion, the spectral lines are broadened and show a width that can be much larger than the natural width.

For thermal motion, which involves velocities much smaller than the velocity of light, we can ignore motion transverse to the line of sight and consider only motion parallel to the line of sight from the observer to the source. Instead of the *speed* distribution, for this purpose we need to know how one component of the *velocity* of the gas atoms, let's assume it to be v_x, is distributed as a function of the temperature.

To find the velocity distribution, let's start with the density of states appropriate for a one-dimensional gas of particles. In Figure 10.7 we constructed a three-dimensional coordinate system with the axes labeled n_x, n_y, and n_z. To obtain the distribution function for v_x, we need only a one-dimensional coordinate system with the n_x axis (Figure 10.12). The "volume" available in one dimension for particles with energies in the range from E to $E + dE$ is just the range dn_x at n_x. The density of states (per unit length) in one dimension is then

$$g(n_x)dn_x = \frac{2s+1}{L}\, dn_x \tag{10.21}$$

With $E = p_x^2/2m = n_x^2 h^2/8mL^2$ we can find $v_x = n_x h/2mL$ and $dv_x = (h/2mL)\, dn_x$, so

$$g(v_x)\, dv_x = g(n_x)\, dn_x = \frac{2s+1}{L}\frac{2mL}{h}\, dv_x = \frac{2m(2s+1)}{h}\, dv_x \tag{10.22}$$

The number of particles with velocities between v_x and $v_x + dv_x$ is

$$dN = N(v_x)\, dv_x = Lg(v_x)f(v_x)\, dv_x = A^{-1}\frac{2mL(2s+1)}{h}e^{-mv_x^2/2kT}dv_x \tag{10.23}$$

where we have used the Maxwell-Boltzmann distribution function $f(E) = A^{-1}e^{-E/kT}$ with $E = \frac{1}{2}mv_x^2$. To find the normalization constant A^{-1}, we again require that the total number of atoms is N:

$$N = \int dN = \int_{-\infty}^{+\infty} N(v_x)dv_x = A^{-1}\frac{2mL(2s+1)}{h}\int_{-\infty}^{+\infty}e^{-mv_x^2/2kT}dv_x \tag{10.24}$$

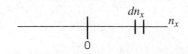

FIGURE 10.12 In one dimension, the velocity component is restricted to the interval between v_x and $v_x + dv_x$, which corresponds to the interval dn_x on the n_x axis.

Note that the integration limits are now $-\infty$ to $+\infty$, in contrast to the speed distribution in which the limits were 0 to ∞. We must include all atoms in the integral—those that are moving toward the observer as well as those that are moving away. Evaluating the integral, we find $A^{-1} = Nh\sqrt{m/2\pi kT}/2mL(2s + 1)$ and thus we have the *Maxwell velocity distribution*

$$N(v_x) = N \left(\frac{m}{2\pi kT}\right)^{1/2} e^{-mv_x^2/2kT} \tag{10.25}$$

This function is plotted in Figure 10.13. The shaded strip shows the number $dN = N(v_x)dv_x$ having velocity components between v_x and $v_x + dv_x$. This is a familiar curve known as the *Gaussian distribution* or the *normal distribution* (also called the "bell-shaped curve"), which has applications in many areas of probability and statistics.

To complete the analysis of the Doppler broadening, we must change the distribution function so it describes frequency or wavelength. That is, let's find out how many atoms will emit frequencies in the range from f to $f + df$, which we can write as $dN = N(f)df$. The relationship between frequency and velocity is given by Eq. 2.22. We'll change notation somewhat and let f_0 represent the unshifted frequency while f represents the observed (Doppler-shifted) frequency. We then have

$$f = f_0\sqrt{\frac{1 - v_x/c}{1 + v_x/c}} = f_0 \frac{1 - v_x/c}{\sqrt{1 - v_x^2/c^2}} \cong f_0(1 - v_x/c) \tag{10.26}$$

where we have replaced the square root in the denominator with 1 because $v_x \ll c$ for thermal motions. Solving for v_x, we obtain $v_x = c(1 - f/f_0)$ and taking the magnitude of the differentials we then have $|dv_x| = c\,df/f_0$. With these substitutions the number of atoms in the small interval becomes

$$dN = N(f)\,df = N(v_x)dv_x = N \left(\frac{m}{2\pi kT}\right)^{1/2} e^{-mc^2(1-f/f_0)^2/2kT} \frac{c\,df}{f_0} \tag{10.27}$$

and so the frequency distribution function is

$$N(f) = \frac{Nc}{f_0} \left(\frac{m}{2\pi kT}\right)^{1/2} e^{-mc^2(1-f/f_0)^2/2kT} \tag{10.28}$$

The frequency distribution function is shown in Figure 10.14. The broadening is usually characterized by giving the width of the spectral line Δf, defined as the range over which the intensity falls to half the maximum value on either side. (This is known as the "full width at half maximum" or FWHM.) The only factor in Eq. 10.28 that depends on frequency is the exponential, and so the width is determined by the frequencies at which the exponential (which is equal to 1 at $f = f_0$) falls by half: $e^{-mc^2(1-f/f_0)^2/2kT} = 1/2$ or, taking the logarithm of both sides,

$$-\frac{mc^2}{2kT}\left(1 - \frac{f}{f_0}\right)^2 = \ln(1/2) \tag{10.29}$$

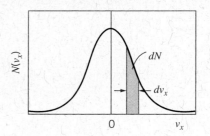

FIGURE 10.13 The Maxwell velocity distribution for gas molecules. The distribution is centered on $v_x = 0$. The shaded strip represents the number of molecules with velocity components between v_x and $v_x + dv_x$.

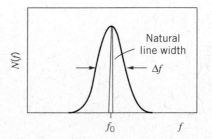

FIGURE 10.14 Doppler-broadened spectral line. The natural linewidth has been exaggerated for the drawing—typically the natural linewidth is no more than 10^{-5} to 10^{-4} of the broadened linewidth.

which can be solved to give $f = f_0(1 \pm \sqrt{(2 \ln 2)kT/mc^2})$. The two solutions (corresponding to the $+$ and $-$ signs) give the two points at which the distribution falls to half its maximum. The interval between those two points is

$$\Delta f = 2f_0 \sqrt{(2 \ln 2)kT/mc^2} \tag{10.30}$$

We can write a similar expression for the wavelength broadening because $\Delta f/f_0 = \Delta \lambda/\lambda_0$. The Doppler broadening is directly related to the temperature, and so a measurement of the width of spectral lines provides a way to determine the temperature of the emitting atoms. This is a powerful means of determining the temperatures of stars from observing the widths of their spectral lines.

10.5 QUANTUM STATISTICS

As we discussed in Section 10.2, the distribution functions for the indistinguishable particles of quantum physics are different from those of classical physics. Because of the unusual behavior of quantum systems, we must have separate distribution functions for particles that obey the Pauli exclusion principle (such as electrons) and particles that do not obey the Pauli principle (such as photons). We will not derive these distribution functions, but merely state them and discuss some of their properties and their applications.

Particles that do not obey the Pauli principle are those with integral spins (0, 1, 2, ..., in units of $\hbar$). Their statistical properties are determined by the *Bose-Einstein distribution function*:

$$f_{BE}(E) = \frac{1}{A_{BE}e^{E/kT} - 1} \tag{10.31}$$

Particles described by this distribution are known collectively as *bosons*. The constant A_{BE} serves as a kind of normalization constant, in analogy with the factor A in the Maxwell-Boltzmann distribution (and the comparison shows why we included this factor as A^{-1} in Eq. 10.16).

Particles of half-integral spin ($\frac{1}{2}, \frac{3}{2}, \ldots$) that obey the Pauli principle, such as electrons or nucleons, are described by the *Fermi-Dirac distribution function*:

$$f_{FD}(E) = \frac{1}{A_{FD}e^{E/kT} + 1} \tag{10.32}$$

These particles are known collectively as *fermions*.

How the minor change in sign in the denominator between f_{BE} and f_{FD} gives such a radical change in the form of the distribution function is not immediately obvious, and to show the differences we need to know more about the normalization coefficient A_{FD}, which is not a constant but depends on T. For the Bose-Einstein distribution, in most cases of practical interest A_{BE} is either independent of T or depends so weakly on T that the exponential term $e^{E/kT}$ dominates the temperature dependence. However, for the Fermi-Dirac distribution, A_{FD} is strongly dependent on T, and the dependence is usually approximately exponential, so A_{FD} is written as

$$A_{\text{FD}} = e^{-E_{\text{F}}/kT} \tag{10.33}$$

and the Fermi-Dirac distribution becomes

$$f_{\text{FD}}(E) = \frac{1}{e^{(E-E_{\text{F}})/kT} + 1} \tag{10.34}$$

where E_{F} is called the *Fermi energy*.

Let us look qualitatively at the differences between f_{BE} and f_{FD} at low temperatures. For the Bose-Einstein distribution, assuming for the moment $A_{\text{BE}} = 1$, in the limit of small T the exponential factor becomes large for large E, and so $f_{\text{BE}} \to 0$ for states with large energies. The only energy levels that have any real chance of being populated are those with $E = 0$, for which the exponential factor approaches 1, the denominator becomes very small, and $f_{\text{BE}} \to \infty$. Thus when T is small, all of the particles in the system try to occupy the lowest energy state. This effect is known as "Bose-Einstein condensation," and we will see that it has some rather startling consequences.

This effect is not possible for fermions, such as electrons. We know that the electrons in an atom, for example, do not all occupy the lowest energy state, no matter what the temperature. Let us see how the Fermi-Dirac distribution function prevents this. The exponential factor in the denominator of f_{FD} is $e^{(E-E_{\text{F}})/kT}$. For values of $E > E_{\text{F}}$, when T is small the exponential factor becomes large and f_{FD} goes to zero, just like f_{BE}. When $E < E_{\text{F}}$, however, the story is very different, for then $E - E_{\text{F}}$ is negative, and $e^{(E-E_{\text{F}})/kT}$ goes to zero for small T, so $f_{\text{FD}} = 1$. *The occupation probability is therefore only one per quantum state,* just as required by the Pauli principle. Even at very low temperatures, fermions do not "condense" into the lowest energy level.

In Figures 10.15 to 10.17, the three distributions f_{MB}, f_{BE}, and f_{FD} are plotted as functions of the energy E. (Note the qualitative similarity with Figures 10.4 to 10.6.) You can see, by comparing these figures, that all of the distribution functions fall to zero at large values of E; when $E \gg kT$, the occupation probability is very small, as we calculated for f_{MB} in the case of the first excited state of the hydrogen atom in Example 10.3. Notice also that, even though f_{MB} becomes large for small E, it remains finite. The Bose-Einstein distribution, f_{BE}, on the other hand, becomes infinite as $E \to 0$; this is the "condensation" effect referred to earlier, in which all of the particles try to occupy the lowest quantum state.

You can see that f_{FD} never becomes larger than 1.0, just as we expect for particles that obey the Pauli principle. The function f_{FD} has the value 1.0 for states with low energy (all states are filled), and it falls quickly to zero at high energy (all states are empty). The Fermi energy E_{F} gives the point at which the distribution function has the value 1/2. At absolute zero, all states below E_{F} are filled and all states above E_{F} are empty.

The normalization constant ultimately depends on the number of particles in the system, determined by integrating the distribution function $f(E)$ after multiplying it by the density of states $g(E)$. Note how changing the number of particles changes the normalizations of the different distributions. For the Maxwell-Boltzmann distribution, increasing N increases the area under the curve by raising the intercept, thus raising the entire curve. For the Fermi-Dirac

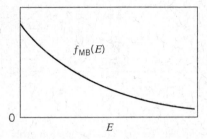

FIGURE 10.15 The Maxwell-Boltzmann distribution function.

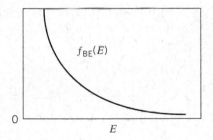

FIGURE 10.16 The Bose-Einstein distribution function.

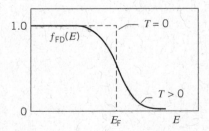

FIGURE 10.17 The Fermi-Dirac distribution function.

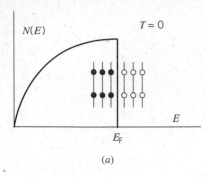

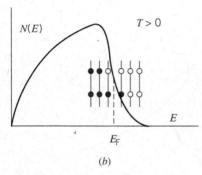

FIGURE 10.18 The occupation probability of electrons in an electron gas (a) at $T = 0$ and (b) at $T > 0$. The solid dots represent filled states and the open dots represent empty states. Each energy level can hold a maximum of 2 electrons (spin up and spin down).

distribution, on the other hand, increasing N increases the area by widening the curve to the right (increasing E_F) while keeping the intercept at 1.0.

Let's consider a gas of electrons described by the density of states function $g(E)$ given in Eq. 10.12 and thus with $N(E) = Vg(E)f_{FD}(E)$. Figure 10.18a shows a hypothetical set of energy levels and how they would be populated at $T = 0$ (with 2 electrons in each quantum state). As T increases, some levels above E_F are partially occupied ($f_{FD} > 0$), while some levels below E_F are partially empty ($f_{FD} < 1$). Figure 10.18b shows how the energy levels of a system might be populated at $T > 0$. The higher the temperature, the more the distribution spreads, but notice that only states in the vicinity of E_F are affected. The states at much lower energies remain filled, and those at much higher energies remain empty.

The Fermi energy varies only slightly with temperature for most materials, and we can regard it as constant for many applications. As we will see in Section 10.7, for electrons in a metal E_F depends on the electron density of the material, which doesn't change much with temperature. For some materials, notably semiconductors, the density of *conduction* electrons can change significantly with temperature, and thus E_F in these materials is temperature dependent.

Limit of Classical Statistics

Under what circumstances can we treat a system classically rather then according to the laws of quantum mechanics? The quantum behavior can be neglected if the de Broglie wavelength of a particle is much smaller than the physical separation between the particles. That is, no particle lies within the wave packet of its neighbors. If we take kT as a representative measure of the kinetic energy of a particle in a collection of particles at temperature T, then with $p^2/2m = kT$ we obtain the de Broglie wavelength as

$$\lambda = \frac{h}{p} = \frac{h}{\sqrt{2mkT}} = \frac{hc}{\sqrt{2mc^2kT}} \tag{10.35}$$

The density N/V gives the number of particles per unit volume, and so the average spacing d between particles is about $(N/V)^{-1/3}$. The condition for the applicability of classical physics is then $\lambda \ll d$ or $\lambda/d \ll 1$, which gives

$$\frac{\lambda}{d} = \frac{hc/\sqrt{2mc^2kT}}{(N/V)^{-1/3}} = \frac{hc(N/V)^{1/3}}{\sqrt{2mc^2kT}} \ll 1 \tag{10.36}$$

The normalization constant we found from Eq. 10.18 for the Maxwell-Boltzmann distribution can be written

$$A^{-1} = \frac{N(hc)^3}{V(2s+1)(2\pi mc^2kT)^{3/2}} = \frac{1}{(2s+1)\pi^{3/2}}\left[\frac{hc(N/V)^{1/3}}{\sqrt{2mc^2kT}}\right]^3 \tag{10.37}$$

The quantity in brackets on the right is just λ/d from Eq. 10.36. Apart from some small factors of order unity, Eq. 10.36 is equivalent to requiring that the normalization constant of the Maxwell-Boltzmann distribution is small: $A^{-1} \ll 1$, or that the number of *occupied* states in the gas is much smaller than the number of *available* states.

10.6 APPLICATIONS OF BOSE-EINSTEIN STATISTICS

Thermal Radiation

As we did in our discussion of thermal radiation in Chapter 3, we consider a cavity filled with electromagnetic radiation. For this calculation, we assume the box to be filled with a "gas" of photons. Photons have spin 1, so they are bosons and obey Bose-Einstein statistics.

The normalization parameter A_{BE} of the Bose-Einstein distribution, Eq. 10.34, depends on the total number of particles described by the distribution. Because photons are continuously created and destroyed as radiation is emitted or absorbed by the walls of the cavity, the total number of particles is not constant, and the parameter A_{BE} loses its significance. We can eliminate this factor from the Bose-Einstein distribution by setting $A_{BE} = 1$ in Eq. 10.34.

The number of photons having energy in the range E to $E + dE$ is, according to the Bose-Einstein distribution and using the density states for the photon gas from Eq. 10.15,

$$dN = N(E)\, dE = Vg(E)f_{BE}(E)\, dE = V\frac{8\pi}{(hc)^3}E^2\frac{1}{e^{E/kT} - 1}\, dE \qquad (10.38)$$

The radiant energy carried by photons with energy between E and $E + dE$ is $E\, dN = EN(E)dE$, and the contribution to the energy density (energy per unit volume) of photons with energy E is

$$u(E)\, dE = \frac{EN(E)\, dE}{V} = \frac{8\pi E^3}{(hc)^3}\frac{1}{e^{E/kT} - 1}\, dE \qquad (10.39)$$

The total energy density over all photon energies is

$$U = \int_0^\infty u(E)\, dE = \frac{8\pi}{(hc)^3}\int_0^\infty \frac{E^3\, dE}{e^{E/kT} - 1} = \frac{8\pi(kT)^4}{(hc)^3}\int_0^\infty \frac{x^3\, dx}{e^x - 1} \qquad (10.40)$$

where $x = E/kT$. The integral is a standard form and has the value $\pi^4/15$, so

$$U = \frac{8\pi^5 k^4}{15(hc)^3}T^4 \qquad (10.41)$$

This is identical with Stefan's law (Eq. 3.26) using the Stefan-Boltzmann constant from Eq. 3.42 and accounting for the factor $c/4$ that takes us from energy density of the radiation to radiant intensity I.

We can show that our expression for the energy density leads to Planck's equation for the intensity of cavity radiation by changing variables from E to λ. Substituting $E = hc/\lambda$ and $|dE| = (hc/\lambda^2)d\lambda$, we find

$$u(\lambda)\, d\lambda = u(E)\, dE = \frac{8\pi hc}{\lambda^5}\frac{1}{e^{hc/\lambda kT} - 1}\, d\lambda \qquad (10.42)$$

Multiplying by $c/4$ to convert from the energy density of the radiation to the intensity, we obtain the result that was given in Eq. 3.41.

Thus the Planck theory of blackbody radiation, which was so successful in accounting for experimental results, can be derived from the Bose-Einstein distribution for photons (but Planck's original work was done two decades before the development of Bose-Einstein statistics).

Liquid Helium

One of the most remarkable substances we can study in the laboratory is liquid helium. Here are some of its properties:

1. Helium gas is the most inert of the inert gases. Under normal conditions, it forms no compounds, and it has the lowest boiling point, 4.18 K, of any material.
2. Just below its boiling point of 4.18 K, helium behaves much like an ordinary liquid. As the helium boils, the escaping gas forms bubbles, like a boiling pot of water. As the liquid is cooled further, a sudden transition occurs at a temperature of 2.17 K: the violent boiling stops, and the liquid becomes absolutely still. (Evaporation continues, but only from the surface.)
3. As the liquid is cooled below 2.17 K, the specific heat and the thermal conductivity both increase suddenly and discontinuously. Figure 10.19 shows the specific heat as a function of temperature. The form of the figure looks rather like the Greek letter λ, and so the transition point at 2.17 K has become known as the *lambda point*. The thermal conductivity rises at the λ point by a factor of perhaps 10^6.
4. Above 2.17 K, liquid helium can be held in a vessel with a porous plug in the bottom. As soon as the liquid is cooled below 2.17 K, the liquid begins to flow easily through the plug.
5. Below the lambda point, liquid helium has the power to seemingly defy gravity, flowing up and over the walls of its container. The helium forms a thin film, which lines the walls of the container; the remaining liquid is then drawn up by the film like a siphon, and the helium can be seen dripping from the bottom of the container, as in the photograph of Figure 10.20.

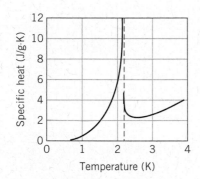

FIGURE 10.19 The specific heat of liquid helium. The discontinuity at 2.17 K is called the lambda point.

All of these strange properties occur because liquid helium obeys Bose-Einstein statistics. Ordinary helium has two electrons filling the 1s shell, so the total angular momentum of the electrons is zero. It happens that the helium nucleus (alpha particle) also has a spin of zero. Therefore the total spin of the atom (electron spin + nuclear spin) is zero, and a helium atom behaves like a boson. At 2.17 K, a *change of phase* occurs in the helium liquid. Above the lambda point, helium behaves like an ordinary liquid; below the lambda point, liquid helium begins to become a *superfluid*. As the temperature is decreased from the lambda point toward absolute zero, the relative concentration of the normal fluid decreases and that of the superfluid increases. The unusual properties of liquid helium are all caused by the superfluid component, which is also known as a *quantum liquid*. Because the helium atoms obey Bose-Einstein statistics, the Pauli principle does not prevent all of the atoms from being in the same quantum state. This begins to happen at the lambda point. We can think of the superfluid as being a single quantum state made up of a very large number of atoms; the atoms behave in a cooperative way, giving the superfluid its unusual properties.

FIGURE 10.20 Liquid helium can be seen dripping from the bottom of this container, as a result of a thin film flowing up and over the walls from the liquid inside the container.

By way of comparison, if we try the same kinds of experiments with the rarer isotope of helium, ^{3}He, the behavior is very different. Although ^{3}He has zero electronic spin, just like ^{4}He, it has only three particles rather than four in its nucleus, and its *nuclear* spin is $\frac{1}{2}$. The total atomic (electronic + nuclear) spin is therefore $\frac{1}{2}$, and ^{3}He behaves like a fermion and obeys the Fermi-Dirac distribution. Because the Pauli principle prevents more than one fermion from occupying any quantum state, no superfluidity is expected for ^{3}He, and indeed none is observed until ^{3}He is cooled to about 0.002 K. At this point the weak coupling of two ^{3}He to form a boson occurs, and the ^{3}He pairs can display the effects of Bose-Einstein statistics. (A related effect involving the pairing of electrons is responsible for superconductivity; see Section 11.5.)

Bose-Einstein Condensation

Let's consider the expression for the total number of particles of a system of bosons in a volume V. We can treat the bosons as a quantum system similar to the electron gas—particles with wave functions that vanish at the boundaries of the volume. The density of states is then given by Eq. 10.12 and the total number of particles in the volume V is then

$$N = \int dN = \int_0^\infty N(E)\, dE = \int_0^\infty V g(E) f_{BE}(E)\, dE$$

$$= \frac{(2s+1)4\pi\sqrt{2}Vm^{3/2}}{h^3} \int_0^\infty \frac{E^{1/2}}{A_{BE}e^{E/kT} - 1}\, dE \qquad (10.43)$$

Previously our approach to an equation of this type was to evaluate the integral and solve the resulting equation for the constant A, which provides the normalization to make the total number of particles equal to N. That procedure poses some difficulties for this integral, so we'll try a different approach: we'll see what Eq. 10.43 tells us about the *maximum* number of particles that can be accommodated in the volume V. Because A_{BE} is a pure number that is always $\geq +1$ (otherwise the denominator of f_{BE} could become negative, which makes no sense for a distribution function), we can find this maximum value by making the denominator in the integral as small as possible, that is, by putting $A_{BE} = 1$. The resulting integral has the value $1.306\pi^{1/2}(kT)^{3/2}$, and the maximum number of particles is then

$$N = 2.612V(2s+1)\left(\frac{2\pi mkT}{h^2}\right)^{3/2} \qquad (10.44)$$

It appears that we can violate this maximum limit by either (1) putting more particles into the volume V than Eq. 10.44 permits, or (2) lowering the temperature (and thus reducing the maximum N) so that the actual number of particles in the system becomes greater than the maximum limit given by Eq. 10.44 for that temperature. How can we still refer to a "maximum" number of particles in these cases?

To resolve this apparent difficulty, we must look more carefully at what happens for $E = 0$. Clearly the Bose-Einstein distribution function $f_{BE}(E)$ becomes infinite for $E = 0$ when $A_{BE} = 1$. The integral in Eq. 10.43 doesn't blow up at $E = 0$ because the numerator $E^{1/2}$ is zero at $E = 0$. But there is something very wrong with that restriction, which requires that the density of states (Eq. 10.12) be zero at $E = 0$. Our system must have a ground state, so there is at least one state at $E = 0$. This contradicts the calculation which puts $g(0) = 0$.

If we try to put more particles into the system than the maximum given by Eq. 10.44 (or equivalently if we try to reduce the temperature below its limit for a given N), the additional particles all can go into the $E = 0$ ground state, which is not subject to the restriction on the maximum value of N. This is Bose-Einstein condensation—all excess particles are "condensing" into the ground state. Note that the use of the word "condensation" to describe this effect does *not* refer to a gas condensing into a liquid. The particles are "condensed" into the same quantum state, where they are all described by the same wave function, but they are all still in the gaseous state.

What we are thus actually calculating in Eq. 10.44 is the number of particles in all states *except* the ground state, that is, the number in all the excited states. Let's call that value N_{ex}. It is this value that is limited by the restriction on the maximum number of particles that can be accommodated by the Bose-Einstein distribution. The number of particles in the ground state, N_0, has no restriction. The total number of particles is $N_{total} = N_0 + N_{ex}$.

Let's solve Eq. 10.44 for the critical temperature at which we expect this condensation to occur:

$$T_{BEC} = \left(\frac{N_{total}/V}{2.612(2s + 1)} \right)^{2/3} \frac{h^2}{2\pi mk} \tag{10.45}$$

Above this temperature, all of the particles can be in excited states of the system without restriction. When the temperature is reduced to T_{BEC}, the excited states are all fully populated, and any further reduction of the temperature below this value means that particles must be transforming from excited states into the ground state. With N_{ex} as the number of particles calculated in Eq. 10.44, we can combine that equation with Eq. 10.45 to give $N_{ex}/N_{total} = (T/T_{BEC})^{3/2}$, or

$$\frac{N_0}{N_{total}} = 1 - \left(\frac{T}{T_{BEC}} \right)^{3/2} \tag{10.46}$$

This applies only to temperatures at or below T_{BEC}. At $T = T_{BEC}, N_0 = 0$—all of the particles are in the excited states. As T is reduced below T_{BEC}, the fraction N_0/N_{total} increases, approaching 1 (all particles in the ground state) as $T \to 0$. This is the Bose-Einstein condensation.

Einstein first predicted this effect in dilute gases in 1925, but it took 70 years until the first experiments were done in 1995. The reason for this is apparent from considering the temperature necessary to observe the condensation. If we start with an ordinary gas at room temperature (with a density of around 2.4×10^{25} molecules/m^3), then Eq. 10.45 gives a temperature of around 0.001 K = 1 mK. This is the temperature at which the condensate begins to form, and to have a significant number of particles in the condensate we must be well below this temperature. It is clear that *very* low temperatures are required to observe the condensate. However, even if the gas molecules were only weakly interacting, a gas at ordinary densities would become a liquid at these temperatures. It is therefore necessary to work with gases at extremely low densities, and from Eq. 10.45 you can see that as we reduce the density, the temperature necessary to observe the Bose-Einstein condensation becomes even smaller.

To avoid the gas condensing into a liquid, we want to molecules to be very far apart (corresponding to a very low density). How far apart must they be? In an ordinary gas, the mean free path (average distance between collisions) is the order of 100 molecular diameters. To avoid molecules from colliding and therefore sticking together (which might trigger the formation of the liquid), let's assume the spacing between molecules must be the order of 100 times larger than it is in an ordinary gas, which means the density must be smaller by a factor of $(10^{-2})^3 = 10^{-6}$. From Eq. 10.45 we see that if the density is smaller by 10^{-6}, then T_{BEC} will be smaller by a factor of $(10^{-6})^{2/3} = 10^{-4}$. Thus instead of a temperature of 1 mK, the observation of a Bose-Einstein condensation requires temperatures of the order of 100 nK.

Such incredibly low temperatures require extraordinary means to produce, and that is why it took 70 years to observe the first Bose-Einstein condensate. Several successful experiments have been done since 1995, and the experiments generally use a combination of *laser cooling* and *magnetic cooling* to achieve these temperatures. In laser cooling, a collection of gas atoms is illuminated with a laser beam that is tuned to one of the atomic absorption frequencies. A gas atom that happens to be moving toward the laser will absorb a photon and slow down. However, a gas atom that is moving away from the laser will absorb a photon and speed up. How can this result in an overall slowing of the gas atoms?

The trick in laser cooling is to take advantage of the Doppler broadening of the absorption due to the distribution in the velocities of the gas atoms. The laser beam is tuned so that its frequency is a bit below the central frequency of the broadened peak. Frequencies smaller than the central frequency correspond to atoms moving toward the laser beam; such atoms can absorb at the frequency of the laser and thus slow down. Atoms moving in the opposite direction cannot absorb at that frequency (because their Doppler shifts are opposite) and so are not affected. If a second laser beam, also tuned below the central frequency, illuminates the atoms from the opposite side, then atoms moving in either direction will be slowed and therefore cooled. In practice, the gas is illuminated by lasers in all six directions to achieve cooling by slowing the atoms in three dimensions. The excited state formed by absorbing the photon will decay back to the ground state by emitting a photon, but the emission occurs in random directions and so doesn't change the velocity distribution of the atoms.

Laser cooling by itself is insufficient to achieve the temperatures necessary for Bose-Einstein condensation. It is also necessary to use some form of magnetic cooling. For example, suppose we confine the atoms in a region in which there is a magnetic field that is produced by a set of coils. The atoms are moving very slowly and do not have enough energy to escape the potential energy barrier that is established by the magnetic field. If the magnetic field strength is then reduced slightly, the more energetic atoms can escape. The remaining atoms still confined by the (weaker) magnetic field are the ones with smaller kinetic energies, and thus they have a lower temperature. With each lowering of the field strength, the more energetic atoms escape and the remaining gas cools. This is similar to evaporative cooling of a cup of hot coffee—the faster-moving atoms are the most likely to leave the liquid, and the remaining atoms have a smaller average kinetic energy and thus a lower temperature.

The first observations of a Bose-Einstein condensation were reported in 1995 by Eric Cornell and Carl Wieman using Rb vapor and by Wolfgang Ketterle using Na vapor. Figure 10.21 shows the velocity distribution illustrating the formation of the condensate in Rb vapor from the original work of Cornell and Wieman. At a temperature of 400 nK, the distribution shows a broad peak corresponding to a Maxwell velocity distribution centered at $v = 0$. At 200 nK, a narrow peak is superimposed on top of the Maxwell distribution at $v = 0$. This represents the atoms all moving with the same speed, as would be expected for a condensate with a large number of atoms in the same state of motion. At still lower temperatures (50 nK), the Maxwell distribution has disappeared, so that nearly all of the atoms are in the condensed ground state, consistent with Eq. 10.45.

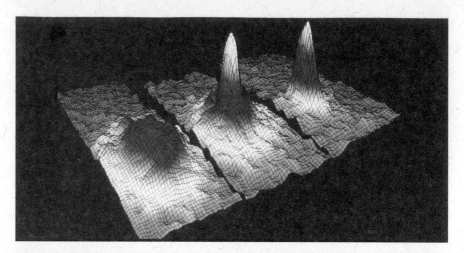

FIGURE 10.21 Bose-Einstein condensation in Rb atoms. The graphs show a representation of the velocity distribution at 400 nK (left), 200 nK (center), and 50 nK (right). At 400 nK, there is a broad Maxwellian distribution, but as the temperature is reduced, the molecules condense into a single quantum state characterized by a much narrower velocity profile.

For the experimental observations of the Bose-Einstein condensation, Cornell, Wieman, and Ketterle shared the 2001 Nobel Prize in physics.

10.7 APPLICATIONS OF FERMI-DIRAC STATISTICS

Now let's consider some applications of Fermi-Dirac statistics. We'll discuss several different systems consisting of spin-$\frac{1}{2}$ particles: electrons in metals, electrons and neutrons in stars, and ^{3}He in liquid ^{4}He.

The Free Electron Model of Metals

In a metal, the valence electrons are not very strongly bound to individual atoms, and consequently they travel rather freely throughout the volume of the metal. We can treat these electrons as a "gas" that obeys the Fermi-Dirac distribution, with a density of states given by Eq. 10.12. The number of electrons with energies between E and $E + dE$ is then

$$dN = N(E)\,dE = Vg(E)f_{FD}(E)\,dE = V\frac{8\pi\sqrt{2}m^{3/2}}{h^3}E^{1/2}\frac{1}{e^{(E-E_F)/kT}+1}\,dE \tag{10.47}$$

Figure 10.22 shows a graph of $N(E)$. Note that the energy kT is only a small interval compared with the range of occupied energy states. When the temperature of the metal is increased from 0 K to 300 K (room temperature), only a very small fraction of the electrons is affected—a small number move from filled states just below E_F to formerly empty states just above E_F.

We can find a numerical value for E_F at $T = 0$ by normalizing Eq. 10.47 so that the sample contains a total number N of these free electrons:

$$N = \int dN = \int_0^\infty N(E)\,dE = \frac{8\pi V\sqrt{2}m^{3/2}}{h^3}\int_0^\infty \frac{E^{1/2}}{e^{(E-E_F)/kT}+1}\,dE \tag{10.48}$$

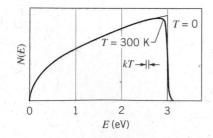

FIGURE 10.22 The number of occupied energy levels for electrons at $T = 0$ and $T = 300$ K, according to the Fermi-Dirac distribution. The Fermi energy E_F is chosen to be 3.0 eV.

At $T = 0$, the Fermi-Dirac distribution function has the value 1 for $E < E_F$ and 0 for $E > E_F$, so the integral reduces to

$$N = \frac{8\pi V \sqrt{2} m^{3/2}}{h^3} \int_0^{E_F} E^{1/2}\, dE = \frac{8\pi V \sqrt{2} m^{3/2}}{h^3} \frac{2}{3} E_F^{3/2} \qquad (10.49)$$

Solving for E_F, we obtain

$$E_F = \frac{h^2}{2m} \left(\frac{3N}{8\pi V} \right)^{2/3} \qquad (10.50)$$

We can also find the mean or average energy of the electrons

$$E_m = \frac{1}{N} \int_0^\infty E N(E)\, dE \qquad (10.51)$$

and it is left as an exercise to show that

$$E_m = \tfrac{3}{5} E_F \qquad (10.52)$$

Example 10.8

Compute the Fermi energy E_F for sodium.

Solution

Each sodium atom contributes one valence electron to the metal, and so the number of electrons per unit volume, N/V, is equal to the number of sodium atoms per unit volume. This in turn can be found from the density ρ and the molar mass M of sodium:

$$\frac{N}{V} = \frac{\rho N_A}{M} = \frac{(0.971 \times 10^3 \text{ kg/m}^3)(6.02 \times 10^{23} \text{ atoms/mole})}{0.023 \text{ kg/mole}}$$

$$= 2.54 \times 10^{28} \text{ m}^{-3}$$

The Fermi energy now can be found from Eq. 10.50:

$$E_F = \frac{h^2}{2m} \left(\frac{3}{8\pi} \frac{N}{V} \right)^{2/3} = \frac{h^2 c^2}{2mc^2} \left(\frac{3}{8\pi} \frac{N}{V} \right)^{2/3}$$

$$= \frac{(1240 \text{ eV} \cdot \text{nm})^2}{2(0.511 \times 10^6 \text{ eV})} \left(\frac{3}{8\pi} 2.54 \times 10^{28} \text{ m}^{-3} \right)^{2/3}$$

$$= 3.15 \text{ eV}$$

The average energy of the valence electrons is $\tfrac{3}{5} E_F$ or 1.89 eV. *Even at the absolute zero of temperature*, the electrons still have quite a large average energy.

From Figure 10.22 we see that the change in $N(E)$ between $T = 0$ and $T = 300$ K (room temperature) is relatively small, and so these values for E_F and E_m are approximately correct at room temperature.

The meaning of these numbers is as follows. Instead of isolated atoms with individual energy levels, we consider the metal to be a single system with a very large number of energy levels (at least as far as the valence electrons are concerned). Electrons fill these energy levels, in accordance with the Pauli principle, beginning at $E = 0$. By the time we add 2.54×10^{22} valence electrons to 1 cm^3 of sodium, we have filled energy levels up to $E_F = 3.15$ eV; all levels below E_F are filled and all levels above E_F are empty. Electrons have an almost continuous energy distribution (the levels are discrete, but they are very close together) from $E = 0$ to $E = E_F$, with an average energy of 1.89 eV. At $T = 300$ K, a relatively small number of electrons is excited from below E_F to above E_F;

the range over which electrons are excited is of order $kT \cong 0.025$ eV, so that only electrons within about 0.025 eV of E_F are affected by the change from $T = 0$ to $T = 300$ K.

In a similar fashion, if we apply a modest electric field to a metal, the only effect is to change the state of motion of a relatively small number of electrons near the Fermi energy. Most of the electrons can't be affected by the electric field, because all of the nearby states are already filled. In Chapter 11, we'll discuss the heat capacity and the electrical conductivity of metals based on the Fermi-Dirac distribution of the electrons.

White Dwarf Stars

A star like the Sun has a constant radius because the outward pressure due to the radiation traveling from the center (where the fusion reactions take place) balances the inward gravitational force that tends toward collapse. Eventually the hydrogen fuel will be converted to helium, the rate of fusion reactions will decrease, and gravity will take over. The Sun will collapse to a smaller and smaller radius, until further contraction is stopped by the Pauli principle. This is the *white dwarf* stage of stellar evolution.

Let's consider a star of mass M to be composed originally of hydrogen, with equal numbers of protons (hydrogen nuclei) and electrons. (The star is too hot for atomic hydrogen to form, so we consider the star to be a "gas" of protons and a "gas" of electrons occupying the same spherical volume.) After the hydrogen has been converted into helium, the star will contain N electrons and $N/2$ helium nuclei (alpha particles). The helium nuclei are bosons, so the Pauli principle does not apply to them during the collapse. The collapse ends when the electrons cannot be forced closer together because the Pauli principle would be violated. At that point, all of the electron energy levels are filled from 0 to the Fermi energy. The average energy E_m of the electrons is $\frac{3}{5}E_F$, as given by Eq. 10.52, and so the total energy of N electrons is

$$E_{elec} = NE_m = \frac{3}{5}NE_F = \frac{3}{5}N\frac{h^2}{2m_e}\left(\frac{3N}{8\pi V}\right)^{2/3} = \frac{3Nh^2}{10m_e}\frac{1}{R^2}\left(\frac{9N}{32\pi^2}\right)^{2/3} \quad (10.53)$$

assuming the electrons to be distributed uniformly throughout a sphere of radius R and volume $V = \frac{4}{3}\pi R^3$.

The total gravitational energy of the star can be found from the mass distribution of the helium (the electron mass is negligible in comparison with the helium mass). For simplicity, we assume the star to be of uniform density. The result (see Problem 38) is

$$E_{grav} = -\frac{3}{5}\frac{GM^2}{R} = -\frac{3}{5}\frac{GN^2m_\alpha^2}{4R} \quad (10.54)$$

with $M = (N/2)m_\alpha$. The total energy is

$$E = E_{elec} + E_{grav} = \frac{3Nh^2}{10m_e}\frac{1}{R^2}\left(\frac{9N}{32\pi^2}\right)^{2/3} - \frac{3GN^2m_\alpha^2}{20R} \quad (10.55)$$

The star collapses until its energy reaches a minimum value, at which point we can find its radius by setting dE/dR equal to 0 and solving the resulting equation, which gives

$$R = \frac{h^2}{GN^{1/3}m_e m_\alpha^2} \left(\frac{9}{4\pi^2}\right)^{2/3} \qquad (10.56)$$

Let's consider the white dwarf star Sirius B, which has a mass of 2.09×10^{30} kg (about 5% greater than the mass of the Sun). Equation 10.56 gives a radius of 7.2×10^6 m for Sirius B. The measured radius is about 5.6×10^6 m. The difference between the calculated and observed values is probably due mostly to the relativistic motion of the electrons. Our calculation of the Fermi energy assumed the electrons to move nonrelativistically. The Fermi energy of Sirius B is about 200 keV, which means that the kinetic energy of electrons near the Fermi energy is not small compared with the rest energy (511 keV). We also oversimplified the structure of the star by assuming it to be of uniform density (which was necessary to obtain Eq. 10.54 for the gravitational energy). Nevertheless, even this very rough calculation gives us a good approximation to the properties of white dwarf stars and demonstrates another system in which Fermi-Dirac statistics can be applied.

Note that the radius of the white dwarf is comparable to the radius of the Earth; that is, the white dwarf has the mass of the Sun but the radius of the Earth. The average density of Sirius B is about 10^9 kg/m^3, which is about one million times the average density of objects on Earth. The white dwarf is indeed an extreme state of matter!

If we treat the electrons relativistically, we can obtain an estimate for the mass of the star at which the Pauli principle applied to the electrons is not able to prevent gravitational collapse. This value, which is called the *Chandrasekhar limit*, is about 1.4 solar masses. For stars with greater masses, the extreme density forces the protons and electrons to combine into neutrons until the star collapses into a *neutron star*, composed entirely of neutrons. Because neutrons obey the Pauli principle, we can apply Fermi-Dirac statistics to analyze the properties of neutron stars (see Problem 27). From a calculation similar to that for the white dwarf, we can find the radius at which the energy of a neutron star is at a minimum:

$$R = \frac{h^2}{GN^{1/3}m_n^3} \left(\frac{9}{32\pi^2}\right)^{2/3} \qquad (10.57)$$

In this equation, N refers to the number of neutrons in the star. For a star of 1.5 solar masses, the radius would be about 11 km with a density of about 5×10^{17} kg/m^3.

Neutron stars are commonly observed as *pulsars* in which the magnetic field of the neutrons traps electrons outside the neutron star, and the rapid rotation of the neutron star causes a beam of electromagnetic radiation from the accelerated electrons to sweep past the Earth somewhat like the rotating light in a lighthouse. For stars heavier than about 6 solar masses, not even the Pauli principle applied to the neutrons can prevent further gravitational collapse, and the star will either explode as a supernova or collapse to a black hole.

The Heat Capacity of Dilute Solutions of ^{3}He in ^{4}He

Helium has two stable isotopes, ^{3}He and ^{4}He. The isotope ^{3}He is very rare (about 10^{-6} in abundance relative to ^{4}He) in natural He gas. The two isotopes are chemically identical and have the same electronic structure, but differ in their atomic masses (^{3}He has a mass of about 3 u and ^{4}He is about 4 u). The difference comes about as a result of their nuclei: the nucleus of ^{3}He contains 2 protons and 1 neutron, while the nucleus of ^{4}He contains 2 protons and 2 neutrons. Protons, neutrons, and electrons all have a spin of $\frac{1}{2}$. In ^{4}He, the 2 electrons combine to a total spin of 0, as do the 2 protons and the 2 neutrons. The total spin of ^{4}He is therefore 0. In ^{3}He, the 2 electrons combine to give a spin of 0 as do the 2 protons, but with only 1 neutron the total spin of ^{3}He is $\frac{1}{2}$. As a result, ^{4}He behaves like a boson and ^{3}He like a fermion.

As discussed above, ^{4}He becomes a superfluid at temperatures below 2.17 K, while ^{3}He does not. In a dilute mixture of liquid ^{3}He and ^{4}He below 2.17 K, the ^{4}He serves as a mostly inert background medium for the ^{3}He, and so we can treat the ^{3}He as a dilute "gas" of fermions, just as we treated the electron gas in analyzing metals.

The heat capacity of a dilute mixture of ^{3}He in ^{4}He is relatively straightforward to measure, so let's try to calculate the heat capacity using the Fermi-Dirac distribution to describe the ^{3}He. Starting with fermions at $T = 0$, we add energy until the collection is at temperature T. Because all of the energy states below E_F are filled at $T = 0$, most of the particles are not able to absorb any additional energy. A particle with energy far below E_F cannot absorb energy of the order of kT and move to an empty state, because there are no empty states nearby. The only particles that can change their state are those within a small energy range of about kT at E_F, as shown in Figure 10.22. In going from temperature 0 to temperature T, only a relatively small number of particles moves from just below E_F to just above E_F, with the rest of the particles remaining in their same energy states.

Figure 10.23 shows a greatly magnified view of the region around E_F. As the temperature is raised from 0 to T, the particles in the region just below E_F move to fill states just above E_F. In particular, consider the small number of particles dN in the narrow region of width dE located a small energy $-\varepsilon = E - E_F$ below E_F. The particles fill states in that region at $T = 0$, but when the temperature is raised to T they move to fill states in the corresponding region at an energy ε above E_F. Each particle in that narrow interval thus gains an energy of 2ε, and the total energy gained by all the particles in that narrow strip is $dE_{ex} = 2\varepsilon dN$. The strip has height N_{ex} given by the difference between $N(E)$ at $T = 0$ and $N(E)$ at temperature T:

$$N_{ex} = N(E, T = 0) - N(E, T) = Vg(E)[1 - f_{FD}(E)] \qquad (10.58)$$

where we have used $N(E) = Vg(E)f_{FD}(E)$ and $f_{FD} = 1$ for $T = 0$. The width of the strip is dE, so the number of particles in the strip is $dN = N_{ex}dE$. With $-\varepsilon = E - E_F$ and $|d\varepsilon| = |dE|$, we obtain the energy gained by the particles in the narrow strip:

$$dE_{ex} = 2\varepsilon\, dN = 2\varepsilon N_{ex}\, d\varepsilon = 2\varepsilon Vg(E_F - \varepsilon)\left(1 - \frac{1}{e^{-\varepsilon/kT} + 1}\right) d\varepsilon \qquad (10.59)$$

The energy difference ε, which we defined as the energy of the strip below E_F, runs from E_F (where $E = 0$) to 0 (where $E = E_F$). The total excitation energy of all of the particles that are excited from below E_F to above E_F is

$$E_{ex} = \int dE_{ex} = 2V \int_{E_F}^{0} \varepsilon g(E_F - \varepsilon)\left(1 - \frac{1}{e^{-\varepsilon/kT} + 1}\right) d\varepsilon \qquad (10.60)$$

We can simplify the calculation by noting that $g(E)$ is a very slowly varying function compared with f_{FD} in the region near E_F, and even though we are

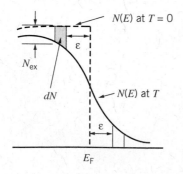

FIGURE 10.23 The region near E_F, showing $N(E)$ at $T = 0$ and at temperature T. When the temperature is raised from 0 to T, the dN particles in the shaded region move to the corresponding region above E_F, increasing their energy by 2ε in the process.

integrating from E_F to 0 the integral is nonzero only in a region very close to E_F. We can therefore take $g(E_F - \varepsilon) \cong g(E_F)$ and bring it out of the integral. Again because the integrand is nonzero only in a very small region, we can replace the lower limit on the integral by ∞.

$$E_{ex} = 2Vg(E_F) \int_\infty^0 \varepsilon \left(1 - \frac{1}{e^{-\varepsilon/kT} + 1}\right) d\varepsilon \tag{10.61}$$

The heat capacity is defined as $C = dE_{ex}/dT$. The derivative with respect to T can be moved inside the integral, and so

$$C = \frac{dE_{ex}}{dT} = 2Vg(E_F) \int_\infty^0 \varepsilon \left[\frac{-e^{-\varepsilon/kT}}{(e^{-\varepsilon/kT} + 1)^2}\left(\frac{\varepsilon}{kT^2}\right)\right] d\varepsilon$$

$$= \frac{2Vg(E_F)}{kT^2}(kT)^3 \int_0^\infty \frac{x^2 e^x}{(e^x + 1)^2} dx \tag{10.62}$$

where $x = \varepsilon/kT$. The integral is a standard form that has the value $\pi^2/6$. Putting in the value for $g(E_F)$ from Eqs. 10.12 and 10.50, we finally obtain

$$C = \frac{\pi^2 k^2 NT}{2E_F} \tag{10.63}$$

This equation predicts that the heat capacity of a dilute gas of fermions at low temperature should be proportional to T. Figure 10.24 shows the low-temperature heat capacity of a dilute mixture of 5% ^{3}He in ^{4}He, and you can see how well the data agree with the prediction. The relationship is indeed linear in T at low temperature. This same behavior also describes the low-temperature heat capacity of metals, in which the electrons can also be treated as a dilute gas of fermions, but as we will see in the next chapter the contribution of the atoms to the heat capacity can often be much larger than the contribution of the electrons.

Equation 10.63 predicts that the slope of the plot of C against T should be $\pi^2 k^2 N/2E_F$, which works out to be $1.24\,\text{J/K}^2$ for the experiment that obtained the results shown in Figure 10.24 (a 5% mixture of ^{3}He in ^{4}He and a total of 0.5 mole of liquid). The measured slope is $3.11\,\text{J/K}^2$, which differs from the expected slope by a factor of 2.5. The discrepancy comes about because we treated the ^{3}He atoms like those of a gas, in which the particles move freely. However, when ^{3}He moves through ^{4}He, there are viscous and other forces that act on the atoms. We can account for the difference by assigning the ^{3}He an "effective mass," which is greater than its actual mass; the greater mass simulates the sluggish behavior of the ^{3}He atoms as they move through ^{4}He. This same factor of about 2.5 appears in experiments with very different concentrations of ^{3}He, so it is not related to any interaction of ^{3}He atoms with other ^{3}He atoms. It also arises in other experiments, such as the study of heat conduction in ^{3}He-^{4}He mixtures, so it does indeed seem to describe the properties of the mixture itself rather than any particular experiment.

The success of the Fermi-Dirac distribution function in accounting for the properties of such a diverse array of systems—metals, white dwarf stars, and dilute mixtures of ^{3}He in ^{4}He—is truly impressive. In the next chapter, we shall explore in more detail how both Bose-Einstein and Fermi-Dirac statistics can be used to help us understand various properties of solids.

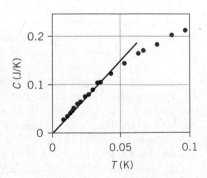

FIGURE 10.24 The heat capacity of 0.5 mole of a dilute (5%) mixture of ^{3}He in ^{4}He. The straight line is a fit to the linear portion of the plot below about 40 mK.

Chapter Summary

		Section			Section
Probability of energy observation	$p(E) = \dfrac{\sum N_i W_i}{N \sum W_i}$	10.2	Maxwell velocity distribution	$N(v_x) = N\left(\dfrac{m}{2\pi kT}\right)^{1/2} e^{-mv_x^2/2kT}$	10.4
Number of particles with discrete energy	$N_n = d_n f(E_n)$	10.3	Doppler broadening of spectral line	$\Delta f = 2f_0\sqrt{(2\ln 2)kT/mc^2}$	10.4
Density of states in gas of particles	$g(E) = \dfrac{4\pi(2s+1)\sqrt{2}(mc^2)^{3/2}}{(hc)^3}E^{1/2}$	10.3	Bose-Einstein distribution function	$f_{BE}(E) = \dfrac{1}{A_{BE}e^{E/kT} - 1}$	10.5
Density of states in gas of photons	$g(E) = \dfrac{1}{\pi^2(\hbar c)^3}E^2 = \dfrac{8\pi}{(hc)^3}E^2$	10.3	Fermi-Dirac distribution function	$f_{FD}(E) = \dfrac{1}{e^{(E-E_F)/kT} + 1}$	10.5
Maxwell-Boltzmann energy distribution	$N(E) = \dfrac{2N}{\sqrt{\pi}(kT)^{3/2}}E^{1/2}e^{-E/kT}$	10.4	Fermi energy	$E_F = \dfrac{h^2}{2m}\left(\dfrac{3N}{8\pi V}\right)^{2/3}$	10.7
Maxwell speed distribution	$N(v) = N\sqrt{\dfrac{2}{\pi}}\left(\dfrac{m}{kT}\right)^{3/2}v^2 e^{-mv^2/2kT}$	10.4	Radius of white dwarf star	$R = \dfrac{h^2}{GN^{1/3}m_e m_\alpha^2}\left(\dfrac{9}{4\pi^2}\right)^{2/3}$	10.7

Questions

1. Suppose a container filled with a gas moves with constant velocity v. How is the Maxwell velocity distribution different for such a gas, compared with the same container of gas at rest?

2. The population inversion necessary for the operation of a laser is sometimes called a "negative temperature." What is the meaning of a negative temperature? Does it have a physical interpretation?

3. Figure 10.25 shows two different experimental arrangements used to measure the distribution of molecular speeds. Based on the figures, explain how each apparatus might operate, and try to guess how the observed distribution of molecules might appear. Where do the fastest molecules land? The slowest?

4. How would Figure 10.13 change if the temperature of the gas were increased?

5. How is the speed distribution of a gas at temperature T different from that at temperature $2T$? The energy distribution? Sketch the speed and energy distributions at the two temperatures.

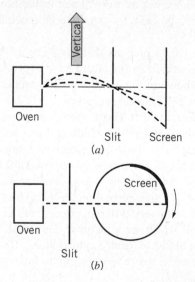

FIGURE 10.25 Question 3.

6. Consider a mixture of two gases of molecular masses m_1 and $m_2 = 2m_1$ in thermal equilibrium at temperature T. How do their speed distributions differ? How do their energy distributions differ?

7. It is generally more convenient, wherever possible, to use Maxwell-Boltzmann statistics rather than quantum statistics. Under what circumstances can a quantum system be described by Maxwell-Boltzmann statistics?

8. Suppose we had a gas of hydrogen *atoms* at relatively high density. Do the atoms behave as fermions or as bosons? Would a gas of deuterium (heavy hydrogen) atoms behave

any differently? (*Hint:* The nuclear spin is ½ for hydrogen and 1 for deuterium.)

9. The early universe contained a large density of neutrinos (massless spin-½ particles that travel at the speed of light). Which statistical distribution would be needed to describe the properties of the neutrinos?

10. Would you expect the photoelectric effect to depend on the temperature of the surface of the metal? Explain.

11. Estimate the mean kinetic energy of the "free" electrons in a metal if they obeyed Maxwell-Boltzmann statistics. How does this compare with the result of applying Fermi-Dirac statistics? Why is there such a difference?

Problems

10.1 Statistical Analysis

1. A collection of three noninteracting particles shares 3 units of energy. Each particle is restricted to having an integral number of units of energy. (*a*) How many macrostates are there? (*b*) How many microstates are there in each of the macrostates? (*c*) What is the probability of finding one of the particles with 2 units of energy? With 0 units of energy?

2. (*a*) Considering the numbers of heads and tails, how many macrostates are there when 5 coins are tossed? (*b*) What is the total number of possible microstates in tossing 5 coins? (*c*) Find the number of microstates for each macrostate, and be sure the total agrees with your answer to part (*b*).

3. Consider a system consisting of two particles, one with spin $s = 1$ and another with spin $s = \frac{1}{2}$. (*a*) Considering a microstate to be an assignment of the z component of the spins of each of the particles, what is the total number of microstates of the two-particle system? (*b*) How many macrostates are there for the total spin of the two-particle system? (*c*) Find the number of microstates for each macrostate, and be sure the total number agrees with your answer to part (*a*).

10.2 Classical and Quantum Statistics

4. Calculate the probabilities for $E = 0, 3$ and 5 listed in Table 10.2.

5. Calculate the probabilities given in Table 10.3 for $E = 0$ and $E = 3$ for (*a*) integral spin and (*b*) spin 1/2.

6. A system of four oscillator-like particles shares 8 units of energy. (That is, the particles can accept energy only in equal units, in which the oscillator spacing is 1 unit.) (*a*) List the macrostates, and for each macrostate give the number of microstates for distinguishable classical particles, indistinguishable quantum particles with integral spin, and indistinguishable quantum particles with half-integral spin. (*b*) Calculate the probability to find a particle with exactly 2 units of energy for each of the three different types of particles.

7. A system consists of two particles, each of which has a spin of $\frac{3}{2}$. (*a*) Assuming the particles to be distinguishable, what are the macrostates of the z component of the total spin, and what is the multiplicity of each? (*b*) What are the possible values of the total spin S and what is the multiplicity of each value? Verify that the total multiplicity matches that of part (*a*). (*c*) Now suppose the particles behave like indistinguishable quantum particles. What is the multiplicity of each of the macrostates of the z component of the total spin? (*d*) Show that for these quantum particles it is possible to have only combinations with total spin $S = 3$ or 1.

10.3 The Density of States

8. The universe is filled with photons left over from the Big Bang that today have an average energy of about 2×10^{-4} eV (corresponding to a temperature of 2.7 K). What is the number of available energy states per unit volume for these photons in an interval of 10^{-5} eV?

9. In certain semiconductors, the conducting regions are grown in very thin layers, which can be regarded as two-dimensional regions holding an electron gas. Calculate the density of states (per unit area) for a gas of particles of mass m and spin s confined to move in two dimensions in a square region of length L on each side.

10. Calculate the density of states (per unit area) for a collection of photons confined to a two-dimensional region in the shape of a square a length L on each side.

11. In a conductor like copper, each atom provides one electron that is available to conduct electric currents. If we assume that the electrons behave like a gas of particles at room temperature with a most probable energy of 0.0252 eV, what is the density of states in an interval of 1% about the most probable energy?

10.4 The Maxwell-Boltzmann Distribution

12. A system consists of N particles that can occupy two energy levels: a nondegenerate ground state and a three-fold

degenerate excited state, which is at an energy of 0.25 eV above the ground state. At a temperature of 960 K, find the number of particles in the ground state and in the excited state.

13. A system with nondegenerate energy levels has three energy states: a ground state at $E = 0$ and excited states at energies of 0.045 eV and 0.135 eV. At a temperature of 650 K, find the relative numbers of particles in the three states.

14. Show that the most probable speed v_p of the Maxwell speed distribution is $(2kT/m)^{1/2}$.

15. A container holds one mole of helium gas at a temperature of 293 K. (a) Show that the mean energy E_m of the molecules is 0.0379 eV. (b) How many molecules have energies in an interval of width $0.01E_m$ centered on E_m?

16. A cubic container holds one mole of argon gas at a temperature of 293 K. (a) How many molecules have speeds between 500 and 510 m/s? (b) How many molecules have velocities between 500 and 510 m/s in one particular direction? Explain any differences between the answers to (a) and (b).

17. The photosphere of the Sun has a temperature of 5800 K. (a) Calculate the energy linewidth of the first transition in the Lyman series of hydrogen in the Sun's photosphere. (b) For comparison, calculate the natural linewidth, assuming a lifetime of 10^{-8} s.

10.5 Quantum Statistics

18. Do we expect to be able to use Maxwell-Boltzmann statistics to analyze (a) nitrogen gas at standard conditions (room temperature, 1 atmosphere pressure); (b) liquid water at room temperature; (c) liquid helium at 4 K; (d) conduction electrons in copper at room temperature?

19. (a) What pressure must be applied to nitrogen gas at room temperature before Maxwell-Boltzmann statistics begins to fail? (b) To what temperature must we cool nitrogen gas at 1 atmosphere before Maxwell-Boltzmann statistics begins to fail?

10.6 Applications of Bose-Einstein Statistics

20. (a) Show that the total number of photons per unit volume at temperature T is $N/V = 8\pi(kT/hc)^3 \int_0^\infty x^2 dx/(e^x - 1)$. (b) The value of the integral is about 2.404. How many photons per cubic centimeter are there in a cavity filled with radiation at $T = 300$ K? At $T = 3$ K?

21. A blackbody is radiating at a temperature of 2.50×10^3 K. (a) What is the total energy density of the radiation? (b) What fraction of the energy is emitted in the interval between 1.00 and 1.05 eV? (c) What fraction is emitted between 10.00 and 10.05 eV?

22. Find the photon energy at which the blackbody energy spectrum $u(E)$ is a maximum. Compare this result with Wien's displacement law (see Chapter 3) and account for any differences.

10.7 Applications of Fermi-Dirac Statistics

23. Compute the Fermi energy and the average electron energy for copper.

24. Calculate the Fermi energy for magnesium, assuming two free electrons per atom.

25. A certain metal has a Fermi energy of 3.00 eV. Find the number of electrons per unit volume with energy between 5.00 eV and 5.10 eV for (a) $T = 295$ K; (b) $T = 2500$ K.

26. Derive Eq. 10.52 from Eq. 10.51.

27. Assume a neutron star consists of N neutrons (fermions with spin $1/2$) in a sphere of radius R and uniform density. The star is in equilibrium because the inward gravitational force, which tends to collapse the star, is opposed by a repulsion due to the Pauli principle, which prevents the neutrons from moving closer together. (a) Find an expression for the radius of a neutron star. (b) Evaluate the radius for a star of mass equal to 3 solar masses. (c) What is the density of the star?

28. Consider a neutron star of mass equal to twice the mass of the Sun. (a) Evaluate the Fermi energy and determine whether classical or relativistic kinematics should be used in the analysis. (b) Find the de Broglie wavelength of a neutron at the Fermi energy and compare with the average distance between neutrons.

29. For a 5.0% mixture of ^{3}He in 0.50 mole of ^{4}He, calculate the heat capacity in J/K at a temperature of 0.025 mK and compare with the data shown in Figure 10.24. The density of liquid ^{4}He at this temperature is 2.2×10^{28} atoms/m^3. Assume an effective mass of ^{3}He that is 2.5 times its ordinary mass.

General Problems

30. Show that a system of 2 indistinguishable quantum particles with spin 2 can combine only to a total spin of 0, 2, or 4.

31. Consider a collection of N noninteracting atoms with a single excited state at energy E. Assume the atoms obey Maxwell-Boltzmann statistics, and take both the ground state and the excited state to be nondegenerate. (a) At temperature T, what is the ratio of the number of atoms in the excited state to the number in the ground state? (b) What is the average energy of an atom in this system? (c) What is the total energy of the system? (d) What is the heat capacity of this system?

32. Suppose we have a gas in thermal equilibrium at temperature T. Each molecule of the gas has mass m. (a) What is the ratio of the number of molecules at the Earth's surface to the number at height h (with potential energy mgh)? (b) What is the ratio of the density of the gas at height h to the density ρ_0 at the surface? (c) Would you expect this simple model to give an adequate description of the Earth's atmosphere?

33. A collection of noninteracting hydrogen atoms is maintained in the $2p$ state in a magnetic field of strength 5.0 T. (a) At room temperature (293 K), find the fraction of the atoms in the $m_l = +1, 0,$ and -1 states. (b) If the $2p$ state made a

transition to the $1s$ state, what would be the relative intensites of the three normal Zeeman components? Ignore any effects of electron spin.

34. The following method is used to measure the molecular weight of very heavy molecules. A liquid containing the molecules is spun rapidly in a centrifuge, which establishes a variation in the density of the liquid. The density is measured, such as by absorption of light, to determine the molecular weight. Assign a fictitious "centrifugal" force to act on the molecules and show that the density varies as $\rho = \rho_0 e^{m\omega^2 x^2/2kT}$ where ω is the angular velocity of the centrifuge and x measures the distance along the centrifuge tube.

35. In sodium metal at room temperature, compute the energy difference between the points at which the Fermi-Dirac distribution function has the values 0.1 and 0.9. What do you conclude about the "sharpness" of the distribution?

36. In sodium metal (see Example 10.8), calculate the number of electrons per unit volume at room temperature in an interval of width $0.01E_F$ at the mean energy E_m.

37. Protons and neutrons are spin-$1/2$ particles in the nucleus. Find the average energy of the protons and neutrons in the nucleus of a uranium atom, which contains 92 protons and 143 neutrons and has the shape of a sphere of radius of 7.4×10^{-15} m.

38. Consider a uniform spherical distribution of matter of radius r and density ρ. (a) Imagine that a small increment of mass dm is brought from infinity to radius r. What is the change in the gravitational potential energy of the system consisting of the sphere and this mass increment? (b) Suppose we bring in from infinity a series of small mass increments that eventually form a thin spherical shell of radius r and thickness dr about the central sphere. What is the change in the potential energy of the system? (c) What is the total change in potential energy involved with creating a sphere of mass M and radius R?

39. (a) For a white dwarf star of mass equal to the mass of the Sun, find the de Broglie wavelength for electrons at the Fermi energy. Use nonrelativistic kinematics and assume the star to be composed of helium nuclei (alpha particles) and of uniform density. (b) Estimate the average distance between the electrons and compare with their de Broglie wavelength. What can you conclude from this comparison?

40. Measuring the relative population of magnetic substates of nuclei provides a direct way of determining the temperature for very cold systems, using a thermometer that is absolute and needs no calibration. The nucleus ^{60}Co behaves as if it has a spin of 5 and a magnetic moment of $\vec{\mu} = \gamma \vec{S}$, where $\vec{S}$ represents the nuclear spin and γ is a constant equal to 3.64×10^7 T^{-1}s^{-1}. When Co atoms are imbedded in a piece of magnetized iron, the Co nuclei experience a magnetic field of $\vec{B} = -B\hat{k}$, where $B = 29.0$ T and $\hat{k}$ is the unit vector in the z direction. (a) In a certain experiment using a Co in Fe thermometer, the ratio r of the population of the second lowest substate to that population of the lowest substate was observed to be $r = 0.419$. What is the corresponding temperature? (b) At that temperature, what is the ratio of the population of the $m = 0$ substate to that of the lowest substate?

41. The molecule cyanogen (CN), which is commonly found in interstellar gas clouds, has rotational excited states that can absorb visible light. These rotational states are populated by the warming effect of the cosmic background radiation that is a remnant of the creation of the universe. The energy of the first excited rotational state $(L = 1)$ is 4.71×10^{-4} eV above the ground state. (a) The ratio of the intensity of the radiation absorbed in the first excited rotational state to the intensity of the radiation absorbed in the rotational ground state is 0.421 ± 0.017. Assuming this factor represents the relative populations of the two states, calculate the temperature of the cyanogen molecules and its uncertainty. (b) Based on your deduced temperature, calculate the expected ratio of the absorption intensity from the *second* rotational state to that of the ground state and compare with the observed relative intensity (0.0121 ± 0.0014). Direct observation of this background radiation shows it to have the expected thermal radiation spectrum at this temperature (see Chapter 15).

Chapter 11

SOLID-STATE PHYSICS

This scanning electron microscope image shows crystals of the metallic element tungsten. The shape of the crystal is determined by the geometrical arrangement of tungsten atoms, which are bound together in the body-centered cubic structure.

In this chapter we study the way atoms or molecules combine to make solids. In particular, we discuss how the principles of quantum mechanics are essential in understanding the properties of solids.

At first thought, there seem to be so many different solids that to classify them and form some general rules for their properties would appear to be a hopeless task. The book you are reading is made of paper and cloth, held together by a glue made from resins, once liquid and now solid. Your desk might be made of wood, metal, or plastic; your chair might be made of similar materials, and might perhaps be covered with cloth, leather, or plastic fabric and contain fiber or synthetic foam padding. Around you, there might be many books and papers, pencils made of wood and metal and graphite, rubber erasers, pens of metal and plastic. A plastic body and a liquid crystal display surround the semiconductors that lie at the heart of your calculator or computer, your cell phone, and your portable multimedia player. Looking out through a glass window, you see structures made of wood, bricks, concrete, or metal, selected for strength, utility, or attractiveness. Each of these solids has a characteristic color, texture, strength, hardness, or ductility; it has a certain measurable electrical conductivity, heat capacity, thermal conductivity, magnetic susceptibility, and melting point; it has certain characteristic emission or absorption spectra in the visible, infrared, ultraviolet, or other regions of the electromagnetic spectrum.

It is a fair generalization to say that all of these properties depend on two features of the structure of the material: the type of atoms or molecules of which the substance is made, and the way those atoms or molecules are joined or stacked together to make the solid. It is the formidable task of the *solid-state* (or *condensed-matter*) *physicist* or *physical chemist* to try to relate the structure of materials to their observed physical or chemical properties.

Quantum mechanics plays a fundamental role in determining properties of the solid: mechanical, electrical, thermal, magnetic, optical, and so forth. In this chapter we illustrate the application of quantum mechanics to the study of solids by studying some of their thermal, electrical, and magnetic properties.

11.1 CRYSTAL STRUCTURES

Our discussion will concentrate on materials in which the atoms or molecules occupy regular or periodic sites; this structure is called a *lattice,* and materials with this structure are called *crystals*. Crystalline materials include many metals, chemical salts, and semiconductors. One property that distinguishes crystals is their *long-range order*—once we begin constructing the lattice in one location, we determine the placement of atoms that are quite far away. In this respect, the crystal is like a brick wall, in which the bricks are stacked in a periodic array and the placement of a brick is predetermined by the original arrangement of bricks far away compared with the size of a brick. (By contrast, *amorphous* materials such as glass or paper have no long-range order, and their structure is more similar to a pile of bricks than to a brick wall.)

Solid crystals can be classified by the cohesive forces that are responsible for holding the lattice together, as well as by the shape of the arrangement of the atoms in the lattice. We'll look at a few different ways that atoms can be bound together in solids.

Ionic Solids

As we learned in Chapter 9, the cohesive forces in ionic molecules originate from the electrostatic attraction between a closed-shell ion, such as Na^+, and another closed-shell ion, such as Cl^-. Ionic materials can also form solids readily, because a Na^+ ion can simultaneously attract many Cl^- ions to itself, thereby building up a solid structure. The ions are held together by electrostatic forces, so we might suppose that the more negative ions there are around a positive ion, the more stable and strong the solid will be. (Covalent bonds, on the other hand, involve specific electron wave functions and so are limited in the number of near neighbors that can participate in the bonding.)

Ionic solids are crystalline, rather than amorphous, because we can pack ions together more efficiently in a regular array than in a random arrangement. (The same is true for bricks: In a regular array, there are more bricks per unit volume than in a random pile, and their average separation is smaller.)

The simplest type of crystal lattice is the *cubic* lattice, in which we imagine the atoms to be placed at the corners of a succession of cubes that cover the volume of the crystal. Figure 11.1 shows the basic cubic structure. This type of stacking is not the most efficient, because there are large gaps at the center of each face of the cube, and also in the middle of the cube itself. We get a better stacking arrangement, which has more atoms per unit volume, if we place another atom either at the center of each face of the cube or at the center of the body of the cube. These two lattices are known as *face-centered cubic* (fcc) and *body-centered cubic* (bcc) and are illustrated in Figures 11.2 and 11.3.

The fcc lattice gives a slightly more efficient packing (more atoms per unit volume) and so it is usually the most stable structure. However, atoms do not stack like hard spheres, and often the bcc structure is preferred. These two crystal types, fcc and bcc, also occur for materials other than ionic solids, such as certain metals.

A common material that has the fcc lattice structure is NaCl, and for that reason the fcc lattice is often called the *NaCl structure*. In order to have the atoms attract one another, we must alternate Na^+ and Cl^- ions, as is shown in Figure 11.4.

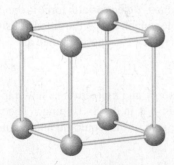

FIGURE 11.1   The simple cubic crystal. The atoms are shown as small spheres for clarity; in an actual solid, the atoms should be imagined as spheres in contact in this cubic geometry.

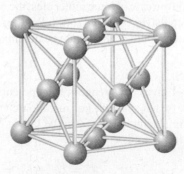

FIGURE 11.2 The face-centered cubic structure.

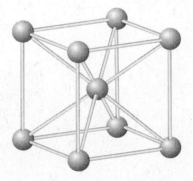

FIGURE 11.3 The body-centered cubic structure.

FIGURE 11.4 Packing of Na (small spheres) and Cl (large spheres) in a fcc crystal of NaCl.

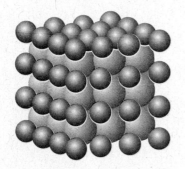

FIGURE 11.5 Packing of Cs (large spheres) and Cl (small spheres) in a bcc crystal of CsCl.

In this illustration you can see how the atoms pack together just like hard spheres in contact. Notice that a given Na^+ ion is attracted by 6 close Cl^- neighbors, and does not "belong to" any single Cl^- ion. *It is therefore wrong to consider ionic solids as being composed of molecules.*

A typical bcc structure is CsCl, as shown in Figure 11.5, and so the bcc lattice is often known as the *CsCl structure*. In this case each ion is surrounded by 8 neighbors of the opposite charge.

Each Na^+ ion in the NaCl structure is surrounded at a distance R by 6 Cl^- ions exerting attractive electrostatic forces. At the slightly larger distance of $R\sqrt{2}$ from each Na^+ ion are 12 Na^+ ions exerting repulsive forces, and at the still greater distance of $R\sqrt{3}$ there are 8 Cl^- ions exerting attractive forces. To find the total Coulomb potential energy U_C, we can continue in this way to add the alternating attractive and repulsive contributions:

$$U_C = \sum \frac{q_1 q_2}{4\pi\varepsilon_0 r} = \frac{e^2}{4\pi\varepsilon_0}\left(-6\frac{1}{R} + 12\frac{1}{R\sqrt{2}} - 8\frac{1}{R\sqrt{3}} + \cdots\right)$$

(11.1)

$$= -\frac{e^2}{4\pi\varepsilon_0}\frac{1}{R}\left(6 - \frac{12}{\sqrt{2}} + \frac{8}{\sqrt{3}} - \cdots\right) = -\alpha\frac{e^2}{4\pi\varepsilon_0}\frac{1}{R}$$

where α, called the *Madelung constant*, is the factor in parentheses in Eq. 11.1:

$$\alpha = 6 - \frac{12}{\sqrt{2}} + \frac{8}{\sqrt{3}} - \cdots$$

(11.2)

This quantity depends only on the geometry of the lattice and is evaluated by summing the slowly converging series of alternating positive and negative terms. The result is

$$\alpha = 1.7476 \qquad \text{(fcc or NaCl lattice)}$$

For the bcc lattice, a similar calculation gives

$$\alpha = 1.7627 \qquad \text{(bcc or CsCl lattice)}$$

As in the case of ionic molecules, the net attractive electrostatic force is opposed by a repulsive force due to the Pauli principle, which keeps the filled subshells from overlapping. The repulsive potential energy can be approximated as

$$U_R = AR^{-n}$$

(11.3)

where A gives the strength of the potential energy and n determines how rapidly it increases at small R. For most ionic crystals, n is in the 8–10 range. The total potential energy of an ion in the lattice is the sum of the Coulomb and repulsive potential energies:

$$U = U_C + U_R = -\alpha\frac{e^2}{4\pi\varepsilon_0}\frac{1}{R} + \frac{A}{R^n}$$

(11.4)

The energies are illustrated in Figure 11.6. There is a stable minimum in the energy, that determines both the equilibrium separation R_0 and the binding energy. To find this minimum, we set dU/dR to zero, which gives

$$A = \frac{\alpha e^2 R_0^{n-1}}{4\pi\varepsilon_0 n}$$

(11.5)

TABLE 11.1 Properties of Ionic Crystals

	Nearest-Neighbor Separation	Cohesive Energy (kJ/mol)	n	Structure
LiF	0.201	1030	6	fcc
LiCl	0.257	834	7	fcc
NaCl	0.281	769	9	fcc
NaI	0.324	682	9.5	fcc
KCl	0.315	701	9	fcc
KBr	0.330	671	9.5	fcc
RbF	0.282	774	8.5	fcc
RbCl	0.329	680	9.5	fcc
CsCl	0.356	657	10.5	bcc
CsI	0.395	600	12	bcc
MgO	0.210	3795	7	fcc
BaO	0.275	3029	9.5	fcc

The binding energy B of an ion in the crystal is the depth of the energy well at $R = R_0$, the equilibrium separation between nearest-neighbor ions. Substituting Eq. 11.5 into Eq. 11.4 and evaluating the resulting equation at $R = R_0$, we obtain

$$B = -U(R_0) = \frac{\alpha e^2}{4\pi\varepsilon_0 R_0}\left(1 - \frac{1}{n}\right) \qquad (11.6)$$

From thermodynamic measurements, it is possible to determine the bulk *cohesive energy* of a solid. In effect, the cohesive energy of an ionic solid is defined as the energy necessary to dismantle the solid into individual ions. Some measured values of the cohesive energies and nearest-neighbor spacings are given in Table 11.1. The value of the exponent n is determined from compressibility data.

The cohesive energy of a bulk sample can be calculated by multiplying the binding energy for a single ion, determined from Eq. 11.6, by the number of ions in the sample, except that such a calculation would count each ion twice.* In one mole of an ionic solid, there are Avogadro's number N_A of positive ions and also N_A negative ions, for a total of $2N_A$ ions per mole. The relationship between the molar cohesive energy E_{coh} and the ionic binding energy B is then

$$E_{coh} = \tfrac{1}{2}(B)(2N_A) = BN_A \qquad (11.7)$$

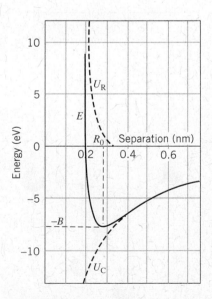

FIGURE 11.6 Contributions to the energy of an ionic crystal. Numerical values are for NaCl.

---

*Consider ions A and B. If we use Eq. 11.6 to compute the binding energy of ion A, the result includes the interaction of ion A with *all* the ions of the solid, including ion B. Similarly, the binding energy of ion B calculated from Eq. 11.6 includes the interaction of B with A. If we calculated the total binding energy of the solid by adding together the binding energies of all ions A and B, we would be including the interaction between A and B twice.

where the factor of 1/2 corrects for the problem of double counting of the ions.

The following examples illustrate the relationship between cohesive energy of the bulk solid and the binding energy per ion pair.

Example 11.1

(a) Determine the experimental value of the binding energy of an ion pair in the NaCl lattice from the cohesive energy. (b) Find the expected value of the binding energy based on the lattice parameters.

Solution

(a) From Eq. 11.7, we have

$$B = \frac{E_{coh}}{N_A} = \frac{769 \times 10^3 \text{ J/mol}}{(6.02 \times 10^{23} \text{ ions/mol})(1.60 \times 10^{-19} \text{ J/eV})}$$

$$= 7.98 \text{ eV}$$

(b) The calculated value of the ionic binding energy is obtained from Eq. 11.6:

$$B = \frac{\alpha e^2}{4\pi\varepsilon_0 R_0}\left(1 - \frac{1}{n}\right)$$

$$= \frac{(1.7476)(1.44 \text{ eV}\cdot\text{nm})}{0.281 \text{ nm}}(0.889) = 7.96 \text{ eV}$$

The agreement between the experimental and calculated values is very good.

Example 11.2

How much energy *per neutral atom* would be needed to take apart a crystal of NaCl?

Solution

If we supply an energy of E_{coh} to a mole of NaCl, we obtain N_A Na^+ ions and N_A Cl^- ions. To convert these to neutral atoms, we must remove an electron from each Cl^-, which costs us the electron affinity of Cl (3.61 eV), and

then we must attach that electron to the Na^+, which returns the ionization energy of Na (5.14 eV). The net cost per pair of Na and Cl atoms is

$$7.98 \text{ eV} + 3.61 \text{ eV} - 5.14 \text{ eV} = 6.45 \text{ eV}$$

Expending this much energy gives two neutral atoms (Na and Cl), so the net cost per atom is half that amount, or 3.23 eV.

The large cohesive energies of ionic solids such as NaCl gives them a common set of properties: They are hard, with high melting and vaporization temperatures (because it takes a lot of thermal energy to break the bonds). They are soluble in polar liquids such as water, in which the dipole moment of the water molecule can supply the electrostatic force necessary to break the ionic bonds. There are no free or valence electrons, so they are poor electrical conductors and not strongly magnetic. They are transparent to visible light (because light rays have too little energy to excite electrons from the filled shells), but absorb strongly in the infrared (corresponding to the vibrational frequencies of the atoms in their lattice sites).

Covalent Solids

As we discussed in Chapter 9, carbon forms molecules by covalent bonding of its four outer electrons in sp^3 hybrid orbits. Such bonds are highly *directional*, and we have seen how it is possible to calculate the angle between the bonds based on the symmetry of the bonding configuration. Solid carbon, in the form of diamond, is an example of a solid in which the interatomic forces are also of a covalent nature. As in a molecule, the four equivalent sp^3 hybrid states participate in covalent bonds, and because they are equivalent they must make equal angles with one another. The manner in which this is done is shown in Figure 11.7. A central carbon atom is covalently bound to four other carbons that occupy four

FIGURE 11.7   The tetrahedral structure of carbon.

TABLE 11.2 Some Covalent Solids

Crystal	Nearest-Neighbor Distance (nm)	Cohesive Energy (kJ/mol)
ZnS	0.235	609
C (diamond)	0.154	710
Si	0.234	447
Ge	0.244	372
Sn	0.280	303
CuCl	0.236	921
GaSb	0.265	580
InAs	0.262	549
SiC	0.189	1185

corners of a cube as shown. The angle between the bonds is $109.5°$, as it was in the covalently bonded molecules.

Figure 11.8 illustrates how the solid structure characteristic of diamond is constructed of such bonds. Each carbon has four close neighbors with which it shares electrons in covalent bonds. The basic structure is known as *tetrahedral*, and many compounds have a similar structure as a result of covalent bonding. Table 11.2 shows some of these compounds. The cohesive energy is the energy required to dismantle the solid into individual atoms. The structure is also known as the *zinc sulfide* or *zinc blende* structure.

Some of the covalent solids listed in Table 11.2 have bond energies larger than those of ionic solids. Substances such as diamond and silicon carbide are particularly hard. Other covalent solids with structures similar to carbon are silicon and germanium; the structure of these solids is responsible for their behavior as semiconductors.

The covalent solids do not have the same similarity of characteristics that ionic solids do, and so we cannot make the same generalizations. Carbon, in the diamond structure, has a large bond energy and is therefore very hard and transparent to visible light; germanium and tin have similar structures, but are metallic in appearance and highly reflective. Carbon (as diamond) has an extremely high melting point (4000 K); germanium and tin melt at much lower temperatures more characteristic of ordinary metals. Some (like diamond) are extremely poor electrical conductors, while others (like Si, Ge, and Sn) can conduct electricity but not nearly as well as most metals. Of course, these differences depend on the actual bond energy in the solid, which in turn depends on the type of atoms of which the solid is made. Those solids with large bond energies are hard, have high melting points, are poor electrical and thermal conductors, and are transparent to visible light. Those solids with small bond energies may have very different properties.

Metallic Bonds

The valence electrons in a metal are usually rather loosely bound, and frequently the electronic shells are only partially filled, so that metals tend not to form covalent bonds. The basic structure of metals is a "sea" or "gas" of approximately free electrons surrounding a lattice of positive ions. The metal is held together by the attractive force between each individual metal ion and the electron gas.

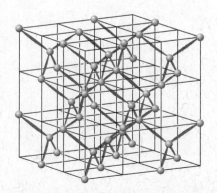

FIGURE 11.8 The lattice structure of diamond.

TABLE 11.3 Structure of Metallic Crystals

Metal	Crystal Type	Nearest-Neighbor Distance (nm)	Cohesive Energy (kJ/mol)
Fe	bcc	0.248	418
Li	bcc	0.304	158
Na	bcc	0.372	107
Cu	fcc	0.256	337
Ag	fcc	0.289	285
Pb	fcc	0.350	196
Co	hcp	0.251	424
Zn	hcp	0.266	130
Cd	hcp	0.298	112

The most common crystal structures of metallic solids are fcc, bcc, or a third type known as *hexagonal close-packed (hcp)*. The hcp structure is shown in Figure 11.9; like the fcc structure, it is a particularly efficient way of packing atoms together. Some metals and their characteristics are shown in Table 11.3. The cohesive energy of metal bonds tends to fall in the range 100–400 kJ/mol (1–4 eV/atom), making the metals less strongly bound than ionic or covalent solids. As a result, many metals have relatively low melting points (some below a few hundred °C). The relatively free electrons in the metal interact readily with photons of visible light, so metals are not transparent. The free electrons are responsible for the high electrical and thermal conductivity of metals. Because metallic bonds don't depend on any particular sharing or exchange of electrons between specific atoms, the exact nature of the atoms of the metal is not as important as it is in the case of ionic or covalent solids; as a result we can make many kinds of metallic alloys by mixing together different metals in varying proportions.

Molecular Solids

None of the solids we have discussed so far can be considered as composed of individual molecules. It is, however, possible for molecules to exert forces on one another and to bind together in solids. The electrons in a molecule are already shared in *molecular* bonds, so there are no available electrons to participate in ionic, covalent, or metallic bonds with other molecules. Moreover, molecules are electrically neutral, so there are no Coulomb forces involved. Molecular solids are held together by much weaker forces, which generally depend on the *electric dipole moments* of the molecules. Because these forces are much weaker than the internal forces that hold a molecule together, a molecule *can* retain its identity in a molecular solid.

The electric dipole moment of one molecule can exert an attractive force on the dipole moment of another. The dipole cohesive force (which is proportional to $1/R^3$) in molecular solids is generally weaker than the $1/R^2$ Coulomb force that is responsible for the cohesive energies of other solids. Molecular solids are therefore more weakly bound and have lower melting points than ionic, covalent, or metallic solids, because it takes less thermal energy to break the bonds of a molecular solid.

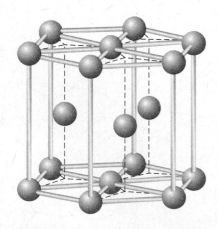

FIGURE 11.9 Arrangement of atoms in a hexagonal close-packed crystal.

Some molecules (called *polar molecules*) have permanent electric dipole moments consisting of a positive charge on one end of the molecule and an equal negative charge on the opposite end. For example, in a water molecule, the oxygen atom tends to attract all of the electrons of the molecule and so looks like the negative end of the dipole; the two "bare" protons are the positive ends of the dipole. The dipole forces between water molecules are responsible for the beautiful hexagonal patterns of snowflakes. When bonding of this sort involves hydrogen atoms, as it does in water, it is known as *hydrogen bonding*.

It is also possible to have dipole forces exerted between atoms or molecules that have no permanent dipole moments. Quantum mechanical fluctuations* can produce an instantaneous electric dipole moment in one atom, which then induces a dipole moment by polarizing a neighboring atom. The result is an attractive dipole-dipole force known as the *van der Waals force*, which is responsible for the bonding in certain molecular solids (as well as for such physical effects as surface tension and friction). Examples of solids that are bound by the van der Waals force include those composed of the inert gases (Ne, Ar, Kr, and Xe), symmetric molecules such as CH_4 and $GeCl_4$, halogens, and other gases such as H_2, N_2, and O_2.

The van der Waals force is extremely weak; it falls off with separation distance like R^{-7}. In inert gas crystals, the nearest neighbor distance is $0.3 - 0.4$ nm, but the cohesive energies are typically only 10 kJ/mol or 0.1 eV/atom. Solids bound by these weak forces have low melting points, because little thermal energy is required to break the bonds. In fact, because the induced dipole moment of an atom or molecule should be approximately proportional to its *total number* of electrons, we might expect that the melting points of nonpolar molecular solids should be roughly proportional to the number of electrons in each molecule. Figure 11.10 shows this relationship; although the properties of the individual solids cause

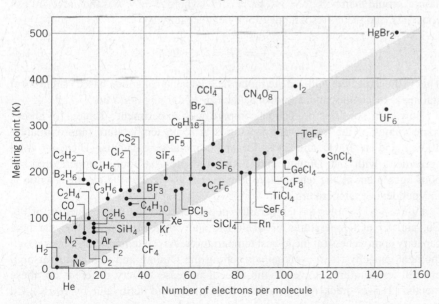

FIGURE 11.10 The melting points of molecular solids depend approximately on the number of electrons per molecule.

---

*These fluctuations are too rapid to be observed in the laboratory. Measurements give only the *average* value of this fluctuating dipole moment, which is zero.

considerable scatter of the points, the relationship is roughly as we expect it should be.

11.2 THE HEAT CAPACITY OF SOLIDS

Just as we discussed in Chapter 1 for the heat capacity of gases, the heat capacity of solids provides another example of the breakdown of classical statistical mechanics and the need for a more detailed theory based on quantum mechanics. (You might find it helpful to review the classical calculation of the heat capacity for gases in Section 1.3.)

Let's first consider what classical thermodynamics predicts for the heat capacity of a solid. In contrast to a gas, an atom in a solid occupies a specific position in the lattice, so no translational motion is possible. Thus there are no degrees of freedom corresponding to the translational motion. The atom can move only by vibrating about its equilibrium position in the lattice. We can imagine the atom to behave as if it were connected to all of its closest neighbors by springs. It can vibrate in any of the three coordinate directions independently of the other two—the initial displacements of the springs in the $x, y,$ and z directions can be chosen independently, and the initial velocities in each direction can be set independently of one another. Consequently there are 6 degrees of freedom in this situation—2 degrees of freedom (corresponding to the vibrational potential and kinetic energies) for each of the three directions. According to the equipartition theorem, the average energy for each degree of freedom is $\frac{1}{2}kT$, so the average energy per atom is $6 \times \frac{1}{2}kT = 3kT$. The total internal energy of one mole (N_A atoms) would then be $E_{int} = 3N_AkT = 3RT$ (where $R = N_Ak$ is the universal gas constant), and the corresponding molar heat capacity is

$$C = \frac{\Delta E_{int}}{\Delta T} = 3R = 24.9 \text{ J/mol} \cdot \text{K} \qquad (11.8)$$

This is the expected value of the molar heat capacity of solids based on classical statistical mechanics and is known as the *law of Dulong and Petit*.

How well does this prediction compare with experiment? Table 11.4 shows some values of the molar heat capacities at room temperature (approximately 300 K) and at 100 K and 25 K for some metallic elements. There is good agreement with the Dulong and Petit prediction at room temperature, but poor agreement as the temperature is reduced. (Note that the classical value is independent of temperature.)

Figure 11.11 shows the temperature dependence of the heat capacities of Pb, Cu, and Cr at temperatures between 1 K and 100 K. It appears that the heat capacity approaches 0 at the lowest temperatures. As the temperature is increased, the heat capacity rises, eventually reaching the Dulong and Petit value at high enough temperature. However, the rate of increase is very different for these metals: Pb rises quickly (approaching the Dulong and Petit value by 100 K), Cu rises more slowly, and Cr rises even more slowly.

Clearly the classical calculation fails to account for the heat capacities of these solids. One possible resolution of this problem would be to consider the application of quantum statistics to the electrons in these metals. In Chapter 10, we discussed the heat capacity of a gas of fermions. We applied the model to a dilute solution of ^3He in ^4He, but the result can apply equally as well to any system of particles governed by Fermi-Dirac statistics.

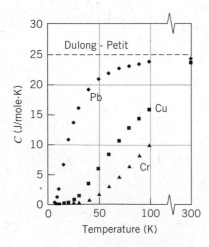

FIGURE 11.11 Molar heat capacity of Pb (diamonds), Cu (squares), and Cr (triangles) at temperatures below 100 K. The room temperature values are shown at the right.

TABLE 11.4 Heat Capacities of Common Metals*

Metal	T = 300 K J/kg·K	T = 300 K J/mole·K	T = 100 K J/mole·K	T = 25 K J/mole·K
Al	0.904	23.4	12.8	0.420
Ag	0.235	24.3	20.0	3.05
Au	0.129	23.4	21.1	5.11
Cr	0.461	23.8	10.0	0.199
Cu	0.387	23.9	16.0	0.971
Fe	0.450	24.6	12.0	0.398
Pb	0.128	24.7	23.8	14.0
Sn	0.222	23.8	22.0	6.80

*The value of the heat capacity depends on the circumstances under which it is determined. Usually measured values are observed at constant pressure (C_P), while calculated values are more easily obtained for constant volume (C_V). The first data column in this table is the experimental specific heat (heat capacity per unit mass) at constant pressure, and the remaining columns all give the molar heat capacity at constant volume. In most cases C_P is just a few percent larger than C_V.

We can treat the electrons in a metal as a Fermi gas. In deriving Eq. 10.63 for the heat capacity, the only assumption we made was that $kT \ll E_F$. For most metals, E_F is a few eV and even at room temperature kT is only 0.025 eV, so the approximation should be pretty good. Let's rewrite Eq. 10.63 for one mole of a substance ($N = N_A$) as follows:

$$C = \frac{\pi^2 k^2 N_A T}{2E_F} = \frac{\pi^2}{2}\frac{RkT}{E_F} \tag{11.9}$$

where $R = 8.31$ J/mole·K. Equation 11.9 is written as if each atom of the lattice contributes one electron to the electron gas, so that N (the number of electrons) is equal to N_A (for one mole of atoms). If, for example, the metal had a valence of 2, then we would have $N = 2N_A$.

For copper $E_F = 7.03$ eV and Eq. 11.9 gives $C = 0.146$ J/mole·K at room temperature. This value is far smaller than the experimental value, indicating that the electrons provide only a small contribution to the heat capacity, at least at room temperature. So the correct explanation for the behavior of the heat capacity must lie elsewhere than the electrons.

Einstein Theory of Heat Capacity

In an ordinary solid, most of the physical properties originate either with the valence electrons or with the latticework of atoms. Electrical conductivity, for example, originates with the valence electrons, while the propagation of mechanical waves is due to the lattice of atoms. The heat capacities of solids have contributions from both lattice and conduction electrons; at all but the lowest temperatures, the lattice contribution is dominant.

The explanation for the failure of classical physics to account for the heat capacity of solids was first given by Einstein, who assumed that the *oscillations* (not the atoms) of the solid obeyed Bose-Einstein statistics. Just as electromagnetic waves are analyzed as "particles" (quanta of electromagnetic energy, or photons) that obey Bose-Einstein statistics, so are mechanical or acoustic waves analyzed as "particles" (quanta of vibrational energy, called *phonons*) that also obey Bose-Einstein statistics. Einstein made the simplifying assumption that all of the phonons (oscillations) have the same frequency.

We have seen in Chapter 5 that a quantized oscillator has an energy of $\hbar\omega(n + \frac{1}{2})$. Each additional value of n represents an additional phonon; to go from a vibrational energy of $\frac{5}{2}\hbar\omega$ to $\frac{7}{2}\hbar\omega$ we must "create" a phonon of energy $\hbar\omega$. One mole of the solid contains N_A atoms and thus $3N_A$ oscillators. The density of states (number of states per unit volume) is thus $3N_A/V$, and the integral of Eq. 10.7 is evaluated only at the single energy $E = \hbar\omega$ (because all phonons have energy $\hbar\omega$). Using the Bose-Einstein distribution, the number of phonons is then $N = 3N_A/(e^{\hbar\omega/kT} - 1)$, and the total internal energy of the solid is the number of phonons times the energy of each phonon:

$$E_{\text{int}} = N\hbar\omega = 3N_A\hbar\omega\frac{1}{e^{\hbar\omega/kT} - 1} \tag{11.10}$$

The heat capacity can be found from dE_{int}/dT:

$$C = \frac{dE_{\text{int}}}{dT} = 3N_A\hbar\omega\frac{(e^{\hbar\omega/kT})(\hbar\omega/kT^2)}{(e^{\hbar\omega/kT} - 1)^2}$$

$$= 3R\left(\frac{\hbar\omega}{kT}\right)^2\frac{e^{\hbar\omega/kT}}{(e^{\hbar\omega/kT} - 1)^2} = 3R\left(\frac{T_E}{T}\right)^2\frac{e^{T_E/T}}{(e^{T_E/T} - 1)^2} \tag{11.11}$$

where we have replaced $\hbar\omega/k$ with the parameter T_E, called the *Einstein temperature*. The vibrational energy $\hbar\omega$ (or the Einstein temperature T_E) is an adjustable parameter of the theory and takes different values for different materials. Typically, T_E is of the order of several hundred kelvins.

When T is small, the exponential term in the denominator dominates, and $C \propto e^{-T_E/T}$, so indeed C approaches 0 for small T, in agreement with experiment. Figure 11.12 shows the molar heat capacity of Cu compared with the behavior predicted by Eq. 11.11, with $T_E = 225$ K giving the best fit to the data. As you can see, the agreement is reasonably good. However, even though the shape of the theoretical curve matches the overall trend of the data, it fails to do a good job at accounting for the behavior at the lowest temperatures (the data approach zero more slowly than the theory predicts).

In this calculation we have oversimplified by assuming all of the oscillations to have the same frequency. A better calculation, which was first done in 1912 by Peter Debye[*], assumes a distribution of frequencies with a density of states given by an expression of the same form as that for the "photon gas" of blackbody radiation; the predicted low temperature behavior is then

$$C = \frac{12\pi^4}{5}R\left(\frac{T}{T_D}\right)^3 \tag{11.12}$$

where T_D is a parameter of the theory known as the *Debye temperature*, which is different for different materials.

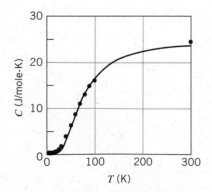

FIGURE 11.12 Molar heat capacity of Cu. The solid curve gives the temperature dependence expected according to the Einstein theory (Eq. 11.11).

[*]Peter Debye (1884–1966) was born in the Netherlands but spent most of his academic career in German universities (where at one point he served as Schrödinger's professor) and finally moved to the U.S. in 1940. He is perhaps best known for his analysis of X-ray diffraction patterns (such as Figure 3.8), for which he received the 1936 Nobel Prize in chemistry.

At the lowest temperatures, we therefore can identify two terms in the heat capacity: a term due to the electrons, which is linear in the temperature (Eq. 11.9), and another term due to the lattice vibrations of the atoms, which is proportional to T^3. Combining these two terms, we then expect the low-temperature heat capacity to be of the form $C = aT + bT^3$, where a is the coefficient of T in Eq. 11.9 and b is the coefficient of T^3 in Eq. 11.12. As T approaches 0, the T^3 term drops off more rapidly than the linear term, so at the very lowest temperatures we expect the electrons to have a more significant contribution. We can turn this equation into a linear graph and identify both contributions by writing $C/T = a + bT^2$ and plotting C/T as a function of T^2, which should give a straight line of slope b and y-intercept a. Figure 11.13 shows the results for copper. The data do indeed fall on a straight line, in excellent agreement with the Debye theory. If the electronic part if the heat capacity were not present (that is, if there were only the lattice contribution from Eq. 11.12), then the line would go through the origin and the intercept would be zero. So the intercept tells us about the electronic contribution to the heat capacity. The slope of the line tells us about the lattice contribution and depends on the Debye temperature.

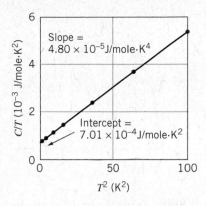

FIGURE 11.13 Molar heat capacity for Cu, plotted as C/T against T^2. The slope gives the lattice contribution and the intercept gives the contribution of the electrons.

Example 11.3

(a) From the slope of the line in Figure 11.13, determine the Debye temperature of copper. (b) Using the Fermi energy of copper (7.03 eV), determine the expected value of the intercept.

Solution

(a) The slope b is equal to the coefficient of T^3 in Eq. 11.12, so $b = 12\pi^4 R/5T_D^3$ and

$$T_D = \left(\frac{12\pi^4 R}{5b}\right)^{1/3} = \left[\frac{12\pi^4 (8.31 \text{ J/mole} \cdot \text{K})}{5(4.80 \times 10^{-5} \text{ J/mole} \cdot \text{K}^4)}\right]^{1/3}$$

$$= 343 \text{ K}$$

(b) The intercept a is the coefficient of T in the electronic contribution to the heat capacity (Eq. 11.9):

$$a = \frac{\pi^2 kR}{2E_F}$$

$$= \frac{\pi^2 (8.617 \times 10^{-5} \text{ eV/K})(8.31 \text{ J/mole} \cdot \text{K})}{2(7.03 \text{ eV})}$$

$$= 5.03 \times 10^{-4} \text{ J/mole} \cdot \text{K}^2$$

The expected value of the intercept based on the free-electron model doesn't quite agree with the experimental value from Figure 11.13 for exactly the same reason that our analysis of the ^{3}He data in Chapter 10 didn't agree with the predictions: an electron moving through a copper lattice doesn't behave as a free electron would in an electron gas. We can account for the forces exerted by the lattice on the electrons by assigning the electrons an "effective mass" that is larger than the mass of a free electron. The additional mass accounts for the "sluggish" behavior of the electrons in moving through the lattice. The mass enters the calculation through the Fermi energy (Eq. 10.50); making the mass larger results in a smaller value of E_F and thus a larger value of the intercept a. For copper, the effective mass of the electrons is about 1.4 times their free mass.

The Debye theory also gives good agreement with the experimental data at higher temperatures. Except at room temperature, the data of Table 11.4 seem to have little in common. At 100 K the heat capacities vary by more than a factor of 2, and they vary by two orders of magnitude at 25 K. In the Debye theory, the heat capacity for any substance can be written as a function of T/T_D. If we plot the heat

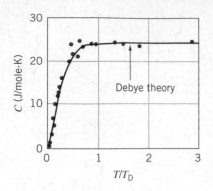

FIGURE 11.14 The heat capacity for eight different metals plotted against T/T_D. All values fall along the same curve calculated from the Debye theory.

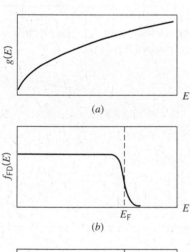

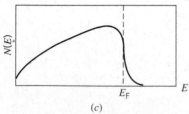

FIGURE 11.15 (*a*) The density of states factor for electrons (Eq. 10.12). (*b*) The Fermi-Dirac distribution function (Eq. 10.34). (*c*) The number of occupied states per unit energy interval, determined from the product of (*a*) and (*b*).

capacities of the eight metals listed in Table 11.4 against T/T_D, the large variation among the values for different metals disappears, as shown in Figure 11.14. The data for all substances fall along the same curve calculated from the Debye theory. Understanding these widely different materials in a common basis is a great triumph for the quantum theory and for the application of Bose-Einstein statistics.

11.3 ELECTRONS IN METALS

In metals, each atom contributes one or more loosely bound electrons to an "electron gas" of nearly free electrons that can easily move throughout the metal. In analogy with an ordinary molecular gas, these electrons move freely and experience forces only when they scatter from the ion cores in the lattice. For now, we'll assume the distribution of occupied electron states is determined by the density of states for the electron gas and the Fermi-Dirac distribution function. In the next section, we will see that a more detailed analysis of the properties of the interaction of the electrons with the atoms of the lattice forbids certain ranges of energy values, but we'll ignore that effect for this discussion. With these assumptions, we can use the electron gas model to study many of the properties of metals, such as electrical conduction, heat capacity, and heat conduction.

Figure 11.15 reviews the main details of the Fermi-Dirac energy distribution, which we discussed in Chapter 10, as it might be applied to electrons in metals. The distribution of occupied electron states is determined by the product of the density of states factor, Eq. 10.12, and the Fermi-Dirac distribution function, Eq. 10.34. At $T = 0$, all states above the Fermi energy E_F are empty and all states below E_F are occupied. For temperatures greater than 0, E_F identifies the point at which the Fermi-Dirac factor has the value $1/2$. The difference between $N(E)$ at $T = 0$ and at room temperature was illustrated in Figure 10.22; only a small number of electrons near E_F are affected by the temperature change.

We calculated the Fermi energy at $T = 0$ by using Eq. 10.49, in which the integral of $N(E)$ over all energies gives the total number of electrons N. The same procedure can be used to find E_F at any temperature:

$$N = \int_0^\infty N(E)\, dE = \frac{8\sqrt{2}\pi V m^{3/2}}{h^3} \int_0^\infty \frac{E^{1/2}\, dE}{e^{(E-E_F)/kT} + 1} \qquad (11.13)$$

In principle, we can evaluate the integral and solve for E_F, as we did to obtain Eq. 10.50. However, the integral cannot be evaluated in closed form. The solution can be approximated as

$$E_F(T) \approx E_F(0)\left[1 - \frac{\pi^2}{12}\left(\frac{kT}{E_F(0)}\right)^2\right] \qquad (11.14)$$

Here $E_F(T)$ represents the Fermi energy at temperature T and $E_F(0)$ represents the Fermi energy at $T = 0$ (Eq. 10.50). At room temperature, $kT = 0.025$ eV, and for most metals the Fermi energy is a few eV, so the change in the Fermi energy between 0 K and room temperature is only about 1 part in 10^4. We can therefore regard the Fermi energy as a constant for our applications, and we will represent it simply as E_F. Table 11.5 shows the Fermi energies of some metals.

Electrical Conduction

When an electric field $\vec{E}$ is applied to a metal, a current flows in the direction of the field. The flow of charges is described in terms of a *current density* $\vec{j}$, the current per unit cross-sectional area. In an ordinary metal, the current density is proportional to the applied electric field:

$$\vec{j} = \sigma \vec{E} \tag{11.15}$$

where the proportionality constant σ is the *electrical conductivity* of the material. We would like to understand the conductivity in terms of the properties of the metal.

The free electrons in our electron gas experience a force $\vec{F} = -e\vec{E}$ and a corresponding acceleration $-e\vec{E}/m$. We observe that in a conductor the current is constant in time, so the increase in velocity from the electric field must be opposed, in this case by collisions with the lattice. This model of conduction in metals views the electrons as accelerated by the field only for short intervals, following which they are slowed by collisions. The net result is that the electrons acquire on the average a steady *drift velocity* $\vec{v}_d$, given by the acceleration times the average time τ between collisions:

$$\vec{v}_d = \frac{-e\vec{E}}{m}\tau \tag{11.16}$$

The magnitude of the current density is determined by the number of charge carriers and their average speed:

$$\vec{j} = -ne\vec{v}_d \tag{11.17}$$

where n is the density of electrons available for conduction. Substituting for the drift velocity, we obtain

$$\vec{j} = \frac{ne^2\tau}{m}\vec{E} \tag{11.18}$$

and the conductivity is therefore

$$\sigma = \frac{ne^2\tau}{m} \tag{11.19}$$

The unknown factor in Eq. 11.19 is the time between collisions, which we can express as

$$\tau = \frac{l}{v_{av}} \tag{11.20}$$

where l is the *mean free path* of the electrons, the average distance the electrons travel between collisions, and v_{av} is their average speed through the lattice. (Note that this speed is *not* the same as the drift speed, which is the small increment of speed that comes about from applying the electric field.)

Let's see how this theory compares with experimentally observed conductivities.

TABLE 11.5　Fermi Energies of Some Metals

Metal	E_F (eV)
Ag	5.50
Au	5.53
Ba	3.65
Ca	4.72
Cs	1.52
Cu	7.03
Li	4.70
Mg	7.11
Na	3.15

Example 11.4

Assuming the average distance traveled between collisions is roughly the distance between atoms, estimate the electrical conductivity of copper at room temperature.

Solution

The nearest neighbor spacing between atoms in a copper lattice is 0.256 nm. If we treat the electron gas semiclassically, the average kinetic energy of an electron is $\frac{3}{2}kT$, and so the average speed of an electron at room temperature ($kT = 0.0252$ eV) is

$$v_{av} = \sqrt{\frac{3kT}{m}}$$

$$= \sqrt{\frac{3(0.0252 \text{ eV})}{0.511 \times 10^6 \text{ eV}/c^2}}$$

$$= 1.15 \times 10^5 \text{ m/s}$$

The average time between collisions is then

$$\tau = l/v_{av} = (0.256 \times 10^{-9} \text{ m})/(1.15 \times 10^5 \text{ m/s})$$
$$= 2.22 \times 10^{-15} \text{ s}$$

The density of copper atoms is

$$n = \frac{\rho N_A}{M} = \frac{(8.96 \times 10^3 \text{ kg/m}^3)(6.02 \times 10^{23} \text{ atoms/mole})}{0.0635 \text{ kg/mole}}$$

$$= 8.49 \times 10^{28} \text{ atoms/m}^3$$

and the conductivity is

$$\sigma = \frac{ne^2\tau}{m}$$

$$= \frac{(8.49 \times 10^{28} \text{ m}^{-3})(1.60 \times 10^{-19} \text{ C})^2(2.22 \times 10^{-15} \text{ s})}{9.11 \times 10^{-31} \text{ kg}}$$

$$= 5.30 \times 10^6 \ \Omega^{-1} \text{ m}^{-1}$$

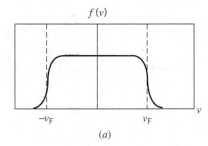

(a)

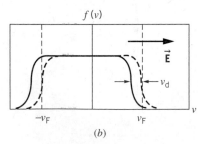

(b)

FIGURE 11.16   (a) The Fermi-Dirac velocity distribution function. (b) When an electric field is applied, the distribution shifts as the electrons are accelerated in a direction opposite to the field.

The measured conductivity of copper at room temperature is $5.96 \times 10^7 \ \Omega^{-1}\text{m}^{-1}$, so our calculation is off by more than an order of magnitude. Moreover, the temperature dependence is wrong: this calculation predicts that the conductivity should decrease as $T^{-1/2}$ in this temperature region, while the measured conductivity decreases like T^{-1}.

We have actually made a couple of errors in this calculation, both of which have to do with ignoring the effects of quantum mechanics in the conduction process. Let's see how we can remedy these defects.

Quantum Theory of Electrical Conduction

Figure 11.16a represents the Fermi distribution of electron velocities in the metal. Like the Fermi energy distribution, it is flat at $v = 0$ and it falls to zero near the Fermi velocity v_F, but (unlike the energy distribution) it has positive and negative branches because the electrons can move in either direction. When an electric field is applied, all electrons acquire on the average an additional velocity component equal to the drift velocity v_d, which shifts the entire velocity distribution to the left (opposite the field direction), as shown in Figure 11.16b. Even though the entire distribution shifts, the net effect of applying the electric field is centered on a small number of electrons in the vicinity of v_F. The electric field causes some electron states near v_F in the direction of $\vec{E}$ to become unoccupied, while an equal number of states near v_F in a direction opposite to $\vec{E}$ become occupied.

The only electrons affected by the field are those in a narrow interval near the Fermi energy. These electrons are moving with a speed $v_F = \sqrt{2E_F/m}$, which for copper with $E_F = 7.03$ eV works out to be 1.57×10^6 m/s. This speed is about an order of magnitude larger than the speed we found in Example 11.4, which would give us a shorter average time between collisions and hence a *smaller* conductivity. This seems to make the disagreement between theory and experiment even worse! Moreover, the Fermi energy is nearly independent of temperature, so the speed of electrons near the Fermi energy should likewise not change with temperature, nor should the conductivity.

That leaves the mean free path *l* as our last resort in fixing the calculation. However, in a perfectly arranged lattice, the mean free path should be infinite! Because atoms are mostly empty space, an electron should have a clear path through the material without scattering from the lattice ions. A perfect lattice should have an infinite conductivity! In practice, we find that the mean free path of an electron may be many hundreds of times the spacing between atoms, so that an encounter of an electron with a lattice ion is not very common.

In a real metallic lattice, two effects contribute to the scattering of electrons: (1) the atoms are in random thermal motion (oscillating about their equilibrium positions) and therefore do not occupy exactly the positions of a perfectly arranged lattice, and (2) lattice imperfections and impurities cause deviations from the ideal lattice. The first effect is temperature dependent and dominates at high temperatures; the second effect is independent of temperature and dominates at the lowest temperatures. In fact, because the average vibrational potential energy (which depends on the square of the vibrational amplitude) is proportional to the temperature, the average area that a vibrating atom presents to an electron moving through the lattice is also proportional to T. Therefore the conductivity decreases like T^{-1}, in agreement with observations.

Figure 11.17 shows the resistivity (the inverse of conductivity) of sodium metal as a function of the temperature. You can see the temperature-independent part at low temperature and the temperature-dependent part at higher temperatures (which increases linearly with T).

The same principles that govern electrical conductivity in a metal also govern *thermal conductivity*. Heat entering the material causes electrons in a small interval (of width kT) near the Fermi energy to move more rapidly, and those electrons can transfer their energy to the lattice in collisions with the ions. Assuming the mean free paths for electrical conduction and thermal conduction are the same, the ratio of the thermal conductivity to the electrical conductivity should be independent of the material but should depend only on the temperature (because the interval kT determines the number of electrons available for thermal conduction).

The ratio $K/\sigma T$ (where K is the thermal conductivity) should be the same for all materials and all temperatures. The proportionality of the thermal and electrical conductivities is known as the *Wiedemann-Franz law*. The ratio $K/\sigma T$ can be calculated from the parameters of the electron gas model to be

$$L = \pi^2 k^2 / 3e^2 = 2.44 \times 10^{-8} \text{ W} \cdot \Omega / \text{K}^2 \qquad (11.21)$$

which is called the *Lorenz number*. Figure 11.18 shows the ratio $K/\sigma T$ for a variety of metals at room temperature. In this region the thermal and electrical conductivities vary by nearly two orders of magnitude, but the ratio remains fairly constant and agrees with the Lorenz value. This agreement is another successful application of Fermi-Dirac statistics to the properties of electrons in solids.

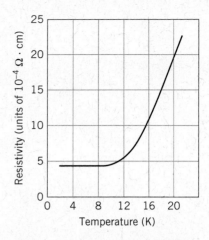

FIGURE 11.17 The electrical resistivity of sodium metal as a function of the temperature.

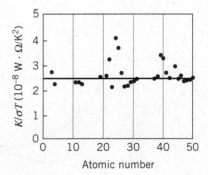

FIGURE 11.18 The Wiedemann-Franz ratio $K/\sigma T$ for various metals at room temperature. The solid line is the Lorenz number.

11.4 BAND THEORY OF SOLIDS

The model of treating a conductor as a gas of free electrons has taken us a long way toward understanding the properties of materials, but to gain a deeper understanding we must consider the interaction of the electrons with the atoms of the lattice. We'll see that this interaction leads to a profound difference between the electron energies of the free electron gas (continuous from zero up to the Fermi energy) and the electron energies in an interacting system (an alternating series of allowed and forbidden energy regions).

When two identical atoms, such as sodium, are very far apart, the electronic levels in one are not affected by the presence of the other. The $3s$ electron of each atom has a single energy with respect to its nucleus. As we bring the atoms closer together, the electron wave functions begin to overlap, and two different $3s$ levels form, depending on whether the two wave functions add or subtract. This effect is responsible for molecular binding, as discussed in Section 9.2. Figure 11.19 shows a representation of the energy levels.

As we bring together more atoms, the same sort of effect occurs. When the sodium atoms are far apart, all $3s$ electrons have the same energy, and as we begin to move them together, the energy levels begin to "split." The situation for five atoms is shown in Figure 11.20. There are now five energy levels that result from the five overlapping electron wave functions. As the number of atoms is increased to the very large numbers that characterize an ordinary piece of metal (perhaps 10^{22} atoms), the levels become so numerous and so close together that we can no longer distinguish the individual levels, as shown in Figure 11.21. We can regard the N atoms as forming an almost continuous *band* of energy levels. Because those levels were identified with the $3s$ atomic levels of sodium, we refer to the $3s$ *band*.

Each energy band in a solid with N atoms has a total of N individual levels. Each level can hold $2(2l + 1)$ electrons (corresponding to the two different orientations of the electron spin and the $2l+1$ orientations of the electron orbital angular momentum) so that the capacity of each band is $2(2l + 1)N$ electrons.

Figure 11.22 shows a more complete representation of the energy bands in sodium metal. The $1s$, $2s$, and $2p$ bands are each full; the $1s$ and $2s$ bands each contain $2N$ electrons and the $2p$ band contains $6N$ electrons. The $3s$ band *could*

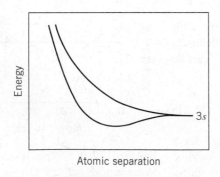

FIGURE 11.19 Splitting of $3s$ level when two atoms are brought together.

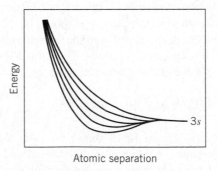

FIGURE 11.20 Splitting of $3s$ level when five atoms are brought together.

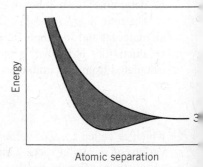

FIGURE 11.21 Formation of $3s$ band by a large number of atoms.

accommodate $2N$ electrons as well; however, each of the N atoms contributes only one $3s$ electron to the solid, and so there is a total of only N $3s$ electrons available. The $3s$ band is therefore half full. Above the $3s$ band is a $3p$ band, which could hold $6N$ electrons, but which is completely empty.

The situation we have described is the ground state of sodium metal. When we add energy to the system (thermal or electrical energy, for example), the electrons can move from the filled states to any of the empty states. In this case, electrons from the partially full $3s$ band can absorb a small amount of energy and move to empty $3s$ states within the $3s$ band, or they can absorb a larger amount of energy and move to the $3p$ band.

We can describe this situation in a more correct way using the Fermi-Dirac distribution. At a temperature of $T = 0\,\mathrm{K}$, all electron levels below the Fermi energy E_F are filled and all levels above the Fermi energy are empty. In the case of sodium, the Fermi energy is in the middle of the $3s$ band, because all electron levels below that energy are occupied (Figure 11.23). At higher temperatures the Fermi energy gives the level at which the occupation probability is 0.5; the Fermi energy does not change significantly as we increase the temperature, but the occupation probability of the levels above E_F is no longer zero. Figure 11.24 shows a situation in which the thermal excitation of electrons leads to a small population of the $3p$ band and some vacant states in the $2p$ band.

Sodium is an example of a substance that is a good electrical conductor. When we apply a very modest potential difference, of the order of 1 V, electrons can easily absorb energy because there are N unoccupied states within the $3s$ band, all within an energy of about 1 eV. Electrons absorb energy as they are accelerated by the applied voltage, and they are therefore free to move as long as there are many unoccupied states within the accessible energy range. In sodium there are N relatively free electrons that can easily move to N unoccupied energy states, and sodium is therefore a good conductor.

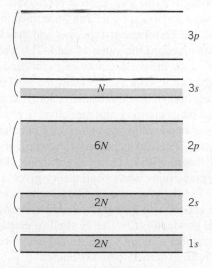

FIGURE 11.22 Energy bands in sodium metal.

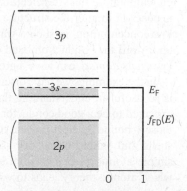

FIGURE 11.23 Energy bands in sodium at $T = 0$ (the filled $1s$ and $2s$ bands are not shown). The Fermi energy is at the center of the half-filled $3s$ band. Note that the Fermi-Dirac distribution function is drawn with the energy axis vertical.

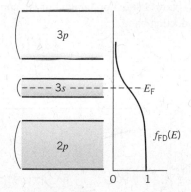

FIGURE 11.24 Energy bands in sodium at $T > 0$. The $2p$ band is no longer completely full (there are a few vacant states near the top), and the $3p$ band is no longer completely empty.

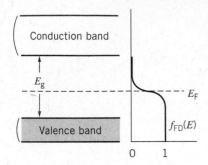

FIGURE 11.25　When $E_g \gg kT$, there are no electrons in the conduction band. This situation characterizes an insulator.

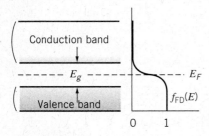

FIGURE 11.26　Band structure of a semiconductor. The gap is much smaller than in an insulator, so there is now a small population of the conduction band.

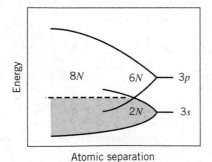

FIGURE 11.27　Band structure in magnesium. The filled 3s and empty 3p bands overlap, forming a single partially filled band.

The band structure of sodium, in which the Fermi level lies in the middle of a band, is characteristic of many good electrical conductors. A completely different situation occurs when the Fermi level lies in the gap between two bands, so that the band below E_F is completely full and the band above is completely empty. If the gap energy E_g is large compared with kT, then even though the Fermi-Dirac distribution spreads as the temperature is raised, it doesn't spread enough to result in a significant population of states in the upper band (called the *conduction band*) or a significant number of empty states in the lower band, called the *valence band*. This situation is shown in Figure 11.25. There are many electrons in the valence band available for electrical conduction, but there are few empty states for them to move through, so they do not contribute to the electrical conductivity. There are many empty states in the conduction band, but at ordinary temperatures there are so few electrons in that band that their contribution to the electrical conductivity is also very small. These substances are classified as *insulators* and in general they have two properties: a large energy gap (a few electron-volts) between the valence and conduction bands, and a Fermi level that is in the gap between the bands (i.e., a filled valence band and an empty conduction band).

A material with the same basic structure but a much smaller energy gap (1 eV or less) shows quite a different behavior. These materials are known as *semiconductors*. Figure 11.26 shows a representation of such a substance at ordinary temperatures. There are now many electrons in the conduction band, and of course many empty states accessible to them, so that they can conduct relatively easily. There are also many empty states in the valence band, so that some of the electrons in the valence band can also contribute to the electrical conductivity by moving about through those states. We consider these two mechanisms of electrical conduction in detail in Section 11.6. For now we note two characteristic properties of semiconductors that relate directly to the band structure as shown in Figure 11.26. (1) Because thermal excitation across the gap is relatively probable, the electrical conductivity of semiconductors depends more strongly on temperature than the electrical conductivity of insulators or conductors. (2) It is possible to alter the structure of these materials, by adding impurities in very low concentration, in such a way that the Fermi energy changes and may move up toward the conduction band or down toward the valence band. This process, known as *doping*, can have a great effect on the conductivity of a semiconductor.

In the examples we have discussed so far, it is not apparent why the band theory is so useful in understanding the properties of a solid. Sodium, for example, is expected to be a good conductor based on its atomic properties alone (a relatively loosely bound 3s electron); on the other hand, solid xenon has only filled atomic shells and should be a poor conductor. These conclusions follow either from simple atomic theory or from band theory. However, there are many cases in which atomic theory leads to wrong predictions while band theory gives correct results. We consider two examples. (1) Magnesium has a filled 3s shell, and on the basis of atomic theory alone we expect it to be a poor electrical conductor. It is, however, a very good electrical conductor. (2) The 2p shell of carbon has only two electrons of the maximum number of six. Carbon should therefore be a relatively good conductor; instead it is an extremely poor conductor.

We can understand both of these materials based on the unusual way the bands of these solids behave when the atoms are close enough so that the band gap disappears and the bands overlap. In magnesium (Figure 11.27), for example, the (filled) 3s and (empty) 3p bands overlap, and the result is a single band with a capacity of $2N + 6N = 8N$ levels. Only $2N$ of those are filled, and so magnesium behaves like a material with a single band filled only to one-fourth its capacity. Magnesium is therefore a very good conductor.

In carbon, the overlap of the electronic wave functions at close range first causes mixing of the $2s$ and $2p$ bands, in a way similar to magnesium; a single band is created with a capacity of $8N$ electrons (Figure 11.28). The $2s$ states contribute $2N$ electrons, and the $2p$ states contribute another $2N$ (out of a maximum capacity of $6N$). As the atoms approach still closer, the band divides into two separate bands, each with a capacity of $4N$ electrons. Because carbon has four valence electrons (two $2s$ and two $2p$), the lower $4N$ states are completely filled and the upper $4N$ states of the conduction band are completely empty. Carbon is therefore an insulator. Germanium and silicon have the same type of structure as carbon, but their equilibrium separation is greater, so the gap between the valence and conduction bands is smaller, about 1 eV; it is this feature that causes Ge and Si to be semiconductors.

*Justification of Band Theory

The band theory of solids has had great success in accounting for the properties of metals, insulators, and semiconductors. In this section we consider a different approach to band theory that is based on the quantum mechanics of an electron moving through a lattice of ions. In analogy with solutions to the Schrödinger equation discussed in Chapter 5, in which an electron in a potential energy well shows discrete energy levels, we will see that an electron in a periodic potential energy provided by a lattice of ions can show energy bands.

To simplify the problem we consider only a one-dimensional lattice of ions (Figure 11.29). The electron is represented by a de Broglie wave traveling through the lattice. The interaction between the electron and the lattice can be represented as a scattering problem, similar to Bragg scattering (Section 3.1). The Bragg condition for scattering is

$$2d \sin \theta = n\lambda \qquad (n = 1, 2, 3, \ldots) \qquad (11.22)$$

where d is the atomic spacing and θ is the angle of incidence measured from the plane of atoms (*not* from the normal). In a two-dimensional lattice, the incident wave can be scattered in many different directions, depending on the plane where we imagine the reflection to occur (recall Figure 3.6); in one dimension, however, only one possible reflection can occur—the incident wave can be reflected back in the opposite direction. We can use the Bragg condition for this case, with $d = a$ (the spacing between the ions or atoms of the lattice) and $\theta = 90°$ (the angle between the "reflecting plane" and the incident wave). With $2a \sin 90° = n\lambda$ and $\lambda = 2\pi/k$ (where k is the wave number), we find

$$k = n\frac{\pi}{a} \qquad (11.23)$$

For wave numbers that do not satisfy this condition, the electron propagates freely through the lattice and behaves like a free particle whose energy is only kinetic:

$$E = \frac{p^2}{2m} = \frac{\hbar^2 k^2}{2m} \qquad (11.24)$$

---

*This is an optional section that may be skipped without loss of continuity.

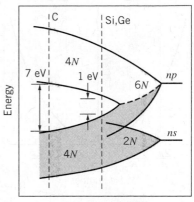

FIGURE 11.28 Band structure of carbon ($n = 2$), silicon ($n = 3$), and germanium ($n = 4$). The combined $ns + np$ band splits into two bands, each of which can hold $4N$ electrons. The atomic separation of carbon gives it a gap of about 7 eV and makes it an insulator, while the larger separation of silicon and germanium results in a smaller gap of about 1 eV and makes them semiconductors.

FIGURE 11.29 One-dimensional Bragg scattering. The only possible scattering is a reflection back in the opposite direction.

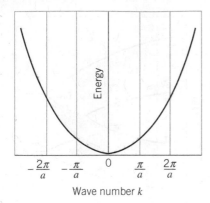

FIGURE 11.30 The parabolic relationship between energy and wave number for a free particle.

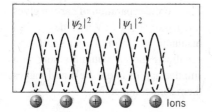

FIGURE 11.31 Probability densities for two different standing waves in the one-dimensional lattice.

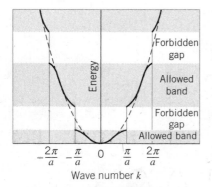

FIGURE 11.32 The relationship between energy and wave number for a one-dimensional lattice. The dashed curve is the parabola that represents the free particles. The solid curves represent waves scattered by the lattice.

There are no restrictions on k, so all values of E are allowed. This relationship between E and k for free electrons defines a parabola, as shown in Figure 11.30.

For wave numbers that satisfy the Bragg condition, the reflected and incident waves add to produce standing waves, which always result when we superpose two waves of equal wavelengths traveling in opposite directions. Depending on the phase difference between the waves, their amplitudes can add or subtract, so two different possible standing waves can result. Their probability densities are shown in Figure 11.31. For one of the waves (ψ_1), the electrons are more likely to be found close to the positive ions; these electrons are more tightly bound to the lattice—the energy of the electron is a bit lower than that of the free electron, due to the negative potential energy between the electron and the ions. Electrons represented by the other wave (ψ_2) are most likely to be found in the region between the ions; they are less tightly bound, so their energies are a bit above those of the unscattered electrons (for which the probability density is flat, so they are equally likely to be found at any location).

The resulting dependence of the energy of the electrons on the wave number k is illustrated by the S-shaped curve segments in Figure 11.32. For wave numbers that are far from satisfying the Bragg condition (that is, values of k that are not close to $n\pi/a$), the curve segments overlap the dashed parabola representing the free particle. Close to the wave numbers that satisfy the Bragg condition, however, the energy deviates from that of the free particle, a bit below the parabola for the more tightly bound electrons that spend more time near the ions and a bit above the parabola for the less tightly bound electrons that are more likely found between the ions.

Notice from Figure 11.32 that, even though all values of k are permitted, there are certain allowed bands of energy values separated by forbidden gaps. An electron traveling in this lattice is permitted to have energies only in the regions corresponding to the allowed bands. This indicates how a periodic array of atoms results in energy bands.

A more detailed calculation in three dimensions gives a better representation of the allowed and forbidden bands, but you can see that even this basic one-dimensional model shows how the bands can arise from the interactions of the electrons with a periodic lattice.

11.5 SUPERCONDUCTIVITY

At low temperatures, the resistivity of a metal (the inverse of its conductivity) is nearly constant. As the temperature of a material is lowered, the lattice contribution to the resistivity decreases while the impurity contribution remains approximately constant, and as we approach $T = 0$ K the resistivity should approach a constant value. Many metals, known as *normal* metals, behave in this way, as illustrated in Figure 11.17.

The behavior of another class of metals is quite different. These metals behave normally as the temperature is decreased, but at some critical temperature T_c (which depends on the properties of the metal), the resistivity drops suddenly to zero, as shown in Figure 11.33. These materials are known as *superconductors*. The resistivity of a superconductor is not merely very small at temperatures below T_c; it vanishes! Such materials can conduct electric currents even in the absence of an applied voltage, and the conduction occurs with no i^2R (joule heating) losses.

Superconductivity has been observed in 28 elements at ordinary pressures, in several additional elements at high pressure, and in hundreds of compounds and alloys. Since the original discovery of superconductivity in Hg in 1911, the focus of research has been to search for materials with the highest possible critical temperature, because many possible large-scale applications of superconductivity are presently impractical owing to the high cost of keeping materials below their critical temperatures. Table 11.6 summarizes some superconducting materials and their critical temperatures. You can see that before 1986, progress in raising the critical temperature was very slow, but since 1986 dramatic and rapid increases in T_c have been achieved.

Conspicuously absent from the list of superconductors are the best metallic conductors (Cu, Ag, Au), which suggests that superconductivity is *not* caused by a good conductor getting better but instead must involve some fundamental change in the material. In fact, superconductivity results from a kind of paradox: ordinary materials can be good conductors if the electrons have a relatively weak interaction with the lattice, but superconductivity results from a *strong* interaction between the electrons and the lattice.

Consider an electron moving through the lattice. As it moves, it attracts the positive ions and disturbs the lattice, much as a boat moving through water creates a wake. These disturbances propagate as lattice vibrations, which can then interact with another electron. In effect, two electrons interact with one another through the intermediary of the lattice; the electrons move in correlated pairs that do not lose energy by interacting with the lattice. (These pairs are not necessarily traveling together through the lattice; they may be separated by a large distance.) In the absence of a net current, the members of a pair have opposite momenta; when a net current is established, both members of the pair acquire a slight increase in momentum in the same direction, and this motion is responsible for the current.

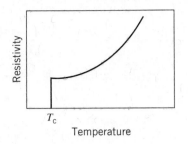

FIGURE 11.33   Resistivity of a superconductor.

TABLE 11.6 Some Superconducting Materials

Material	T_c (K)	Gap (meV)	Year
Zn	0.85	0.24	1933
Al	1.18	0.34	1937
Sn	3.72	1.15	1913
Hg	4.15	1.65	1911
Pb	7.19	2.73	1913
Nb	9.25	3.05	1930
Nb_3Sn	18.1		1954
Nb_3Ge	23.2		1973
$La_xBa_{2-x}CuO_4$	36		1986
$La_xSr_{2-x}CuO_5$	40		1986
$YBa_2Cu_3O_7$	93		1987
$Tl_2Ba_2Ca_2Cu_3O_{10}$	125		1988
$Hg_{12}Tl_3Ba_{30}Ca_{30}Cu_{45}O_{127}$	138		1994

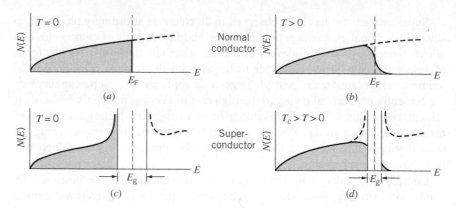

FIGURE 11.34 The number of filled electron states in (a) a normal conductor at $T = 0$; (b) a normal conductor at $T > 0$; (c) a superconductor at $T = 0$; (d) a superconductor at $T > 0$ but less than T_c. In (c), there is an energy gap of width E_g, and states displaced from within the gap pile up on either side of the gap. At higher temperatures, as in (d), the gap is narrower and there are empty states below the gap and filled states above the gap. As the temperature is increased to T_c, the gap width becomes 0 and the distribution of occupied states of the superconductor approaches that of the normal conductor. The gap width is exaggerated in the figure; generally, $E_g \sim 10^{-3}$ eV.

According to the successful BCS theory of superconductivity,* below the critical temperature there is a small energy gap E_g in the occupation probability of electrons in a superconductor (Figure 11.34). Below the gap, the electrons form pairs, which are known as *Cooper pairs*. Once a single Cooper pair forms, it is energetically favorable for other pairs to form, so the change from the normal state above T_c to the superconducting state below T_c is quite sudden. (As shown in Figure 11.34c, the population can *exceed* the limits imposed by the Fermi-Dirac distribution of at most one electron per quantum state. When the electrons are paired, they no longer behave like fermions and so it is possible to have more than one in each quantum state.)

When a superconductor is cooled below T_c, the gap opens and Cooper pairs begin to form. As the material is cooled further, the gap widens. Values of the energy gap listed in Table 11.6 correspond to the limiting case as $T \to 0$. The energy gaps, which can be regarded as representing the binding energy of a Cooper pair, are very small, of the order of 10^{-3} eV. At $T = 0$, all states below the gap are occupied. When $0 < T < T_c$, there are some unoccupied states below the gap, and some states above the gap are occupied by normal (unpaired) electrons.

It seems reasonable that there should be a direct relationship between the critical temperature and the energy gap: the larger the energy gap, the more thermal energy is required to break the Cooper pairs to destroy the superconductivity. The BCS theory gives this relationship:

$$E_g = 3.53kT_c \tag{11.25}$$

---

*The theory of superconductivity was developed in 1957 by John Bardeen, Leon N. Cooper, and J. Robert Schrieffer, who were awarded the 1972 Nobel Prize in physics for their work. Bardeen also shared the 1956 Nobel Prize for his research on semiconductors and his development of the transistor.

As the temperature is raised to T_c, the gap width decreases, and the superconductivity disappears above T_c where the gap width becomes zero.

Beginning in 1986, a new class of superconductors was discovered with unusually high values of T_c. In the 75 years from the discovery of superconductivity in 1911 until 1986, the highest T_c had gone from 4 K to about 23 K. In 1986, several materials were discovered with T_c in the range 30–40 K. By 1987 it was up to 93 K, and it rose to 138 K in 1994. Crossing the boundary at 77 K is important, because it means that cooling can be accomplished with liquid nitrogen instead of liquid helium, which costs nearly an order of magnitude more than liquid nitrogen. The rapid increase in T_c has led to the hope that it might be possible to develop materials that are superconductors at room temperature. Such materials could enable the transmission of electric power over long distances without resistive losses.

The high-T_c superconductors are oxides of copper in combination with other elements. They are ceramics, which means that they are rather brittle and not easily formed into wires to carry current. The crystal structure is characterized by planes of copper and oxygen between planes of the other elements. It seems likely that the superconductivity occurs in the copper oxide planes, but it is not yet clear that the complete explanation for these new superconductors is given by the BCS theory.

Superconducting materials have many applications that take advantage of their abilities to carry electrical currents without resistive losses. Electromagnets can be constructed that carry large currents and therefore produce large magnetic fields (of order 5 to 10 T). Currents as large as 100 A can be carried by very fine superconducting wires, of order 0.1 mm diameter, and thus such magnets can be constructed in a smaller space, using less material, than would be possible with ordinary conductors. Once started, a current in a superconducting loop of wire can circulate for years with no external source to drive it. Superconducting wires are also used to produce magnetic fields in magnetically levitated trains and in magnetic resonance imaging, and also to bend beams of particles in high-energy accelerators, such as the Large Hadron Collider, which began operation in 2008.

Josephson Effect

Imagine a thin layer of insulating material sandwiched between two identical superconductors. The insulating layer is thin enough that the electron pairs can tunnel through from one superconductor to another. This is a typical case of one-dimensional barrier penetration, such as we discussed in Chapter 5.

Figure 11.35 represents the arrangement. In the superconductors (regions 1 and 3), the electron pairs move freely, so the wave functions are of the form of the free particle: $\psi_1(x) = Ae^{i(kx+\alpha_1)}$ and $\psi_3(x) = Ae^{i(kx+\alpha_3)}$, where α_1 and α_3 are arbitrary phase angles (the amplitude A is taken to be real). At the two boundaries on either side of the barrier, the waves are $\psi_1 = Ae^{i\phi_1}$ and $\psi_3 = Ae^{i\phi_3}$, where ϕ_1 and ϕ_3 represent the values of the exponents at the boundaries. Inside the barrier of height U_0, the wave function is of the form $\psi_2(x) = Be^{k'x} + Ce^{-k'x}$, where $k' = \sqrt{2mU_0/\hbar^2}$. Applying the two boundary conditions on ψ, we ultimately obtain a current of the form

$$i = i_0 \sin(\phi_1 - \phi_3) \qquad (11.26)$$

The existence of this "supercurrent" through the junction was first predicted by British physicist Brian Josephson in 1969, and it is now known as the *Josephson effect*. Josephson shared the 1973 Nobel Prize in physics for this discovery.

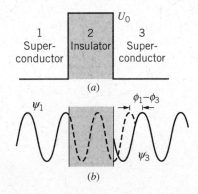

FIGURE 11.35 (*a*) A thin insulator sandwiched between two superconductors provides a potential energy barrier of height U_0 to the flow of current. (*b*) There is a phase difference of $\phi_1 - \phi_3$ between the wave functions ψ_1 and ψ_3 in the superconductors.

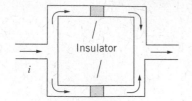

FIGURE 11.36 Two Josephson junctions can be combined to form a quantum interference device.

One important application of the Josephson effect is in the measurement of very weak magnetic fields. Consider a device with two Josephson junctions, as shown in Figure 11.36. Under ordinary circumstances the current divides equally in the two branches, and the phase difference is the same across both junctions. Now suppose a magnetic field is applied perpendicular to the plane of the loop. An additional current is induced around the loop. This additional current, either clockwise or counterclockwise depending on the direction of the magnetic field, will add to the current in one of the Josephson junctions and subtract from the current in the other, causing a relative change in the phase differences across the two junctions. When the currents combine, this phase difference causes maxima and minima in the net current leaving the loop in a manner that is very similar to the maxima and minima in double-slit interference. Observing the maxima and minima serves as a sensitive measurement of the magnetic field in the loop. This device is called a SQUID (Superconducting QUantum Interference Device) and allows measurements of magnetic fields of less than 10^{-17} T. Such sensitive devices allow precise mapping of the magnetic fields inside the brain and also find use in other medical procedures including magnetic resonance imaging (MRI).

In another application, a DC voltage ΔV is applied across a single Josephson junction. This voltage changes the energy of an electron pair on one side of the junction by $2e\Delta V$. For example, if region 1 is made more positive than region 3, the wave function at the boundary in region 1 would become $\psi_1 = Ae^{i(\varphi_1 + 2e\Delta Vt/\hbar)}$ using the usual procedure (see Eq. 5.6) for including the time dependence of the wave function (the factor $e^{-i\omega t}$ with $\omega = E/\hbar = -2e\Delta V/\hbar$). The current through the junction then becomes

$$i = i_0 \sin\left(\phi_1 - \phi_3 + \frac{2e\Delta V}{\hbar}t\right) \tag{11.27}$$

Applying a DC voltage to the junction produces an AC current! Because frequencies can be measured very precisely, a measurement of the frequency of this AC current can be used to determine ΔV. As a result, since 1990 the AC Josephson effect has been accepted by the General Conference on Weights and Measures as the international standard for the volt.

11.6 INTRINSIC AND IMPURITY SEMICONDUCTORS

A semiconductor is a material with an energy gap E_g of order 1 eV between the valence band and the conduction band. At $T = 0$, all states in the valence band are full and all states in the conduction band are empty; recall that the Fermi-Dirac distribution is a step function at $T = 0$ and gives an occupation probability of exactly 1 for all states below E_F and exactly 0 for all states above E_F. As the temperature is raised, however, some states above E_F are occupied and some states below E_F are empty. At room temperature, the relationship between the Fermi energy, the valence and conduction bands, and the electron energy distribution might be as shown in Figure 11.26.

Although the value of the room-temperature Fermi-Dirac distribution function is nearly zero in the conduction band, it is not *exactly* zero; Figure 11.37 shows a greatly magnified view of $f_{FD}(E)$ near the bottom of the conduction band. The value of $E - E_F$ is about 0.5 eV if E_F lies near the middle of the 1 eV energy gap, and therefore $E - E_F \gg kT$, because at room temperature $kT \sim 0.025\,\text{eV}$. The 1 in the denominator of the Fermi-Dirac distribution is therefore negligible, and $f_{FD}(E)$ is approximately exponential, as shown in Figure 11.37.

Assuming the Fermi energy to lie near the middle of the gap, the occupation probability near the bottom of the conduction band is of order $e^{-E_g/2kT} \cong 10^{-9}$. Thus one atom in 10^9 contributes an electron to the electrical conductivity; compare this with a metal in which essentially *every* atom contributes an electron to the conductivity. (On the other hand, consider an insulator, which has a band structure very similar to that of a semiconductor, except the energy gap is perhaps 5 eV instead of 1 eV. This small difference in the size of the energy gap has an enormous effect on the occupation probability of the conduction band at room temperature: $e^{-E_g/2kT} \cong 10^{-44}$. Thus in a sample containing of order 10^{20} atoms, there may be 10^{11} conduction electrons in a semiconductor, 10^{20} in a conductor, and none in an insulator.)

Figure 11.38 shows the corresponding region near the top of the valence band. If there are a few filled states in the conduction band, there must be a few *empty* states in the valence band, and the Fermi-Dirac distribution is just a tiny bit smaller than 1; in fact it is approximately $1 - e^{(E-E_F)/kT}$. This number is about $1 - 10^{-9}$, based on our discussion for the electrons in the conduction band. (Because all the electrons in the conduction band came originally from the valence band, the number of electrons in the conduction band is *exactly* equal to the number of vacancies in the valence band. The Fermi-Dirac distribution is therefore symmetric in the conduction and valence bands, so the Fermi energy must lie at the center of the gap.)

In practice it is much easier to analyze the behavior of a relatively small number of vacancies in the valence band rather than the large number of electrons in that band. When we apply an electric field to the semiconductor, the electrons in the *conduction* band can move easily, because there are many empty states to move into. There are few vacancies in the *valence* band, however. Under the influence of the electric field, an electron in the valence band can move only if there is a vacancy nearby for it to move into. When that electron moves into the vacancy, it creates another vacancy, which can in turn be filled by another electron. In this way, electrons moving in one direction cause an apparent motion of the vacancy in the opposite direction. The situation is similar to the motion of cars in a parking lot with one vacancy (Figure 11.39).

These vacancies in the valence band are known as *holes*, and they behave as if they have positive charges. In an electric field, electrons in the conduction band acquire a drift velocity in a direction opposite to the field (see Eq. 11.16) but (because electrons carry negative charge) they give a current opposite to the velocity and thus in the same direction as the field (see Eq. 11.17). The holes in the valence band acquire a velocity in the same direction as the field and give a current in the same direction as their velocity (that is, in the direction of the field). The current due to the electrons in the conduction band is therefore in the same direction as the current due to the holes in the valence band.

The current in a semiconductor therefore consists of two parts: the negatively charged electrons in the conduction band and the positively charged holes in the valence band. Although the number of electrons in the conduction band is equal to the number of holes in the valence band, the two contributions to the current are in general not equal, because the electrons in the conduction band move more easily than the electrons in the valence band. Typically, the contribution of the electrons to the current at room temperature is about two to four times the contribution of the holes.

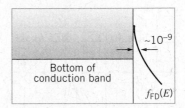

FIGURE 11.37 The tail of the Fermi-Dirac distribution function near the bottom of the conduction band. On the scale of this greatly magnified drawing, the 1 of $f_{FD}(E)$ would be about 1000 km off to the right, and E_F is about 1 m below the edge of this page.

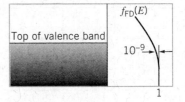

FIGURE 11.38 The Fermi-Dirac distribution function near the top of the valence band, showing the small fraction of empty states.

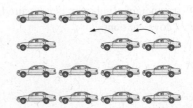

FIGURE 11.39 One car moves to the left, filling the vacancy but creating a new vacancy, which is then filled by the next car moving to the left. The motion of cars to the left, each filling a vacant space, is equivalent to the motion of the vacant space to the right.

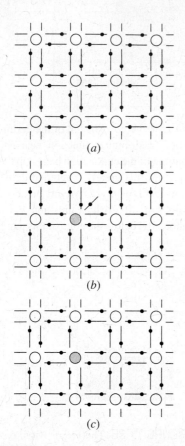

FIGURE 11.40 (a) Covalent bonding in Si or Ge. Each atom provides four electrons for covalent bonds with its neighbors. (b) When a Si or Ge atom is replaced with a valence-5 atom (shaded), there is an extra electron that does not participate in covalent bonds. (c) If a Si or Ge atom is replaced by a valence-3 atom (shaded), one electron from a neighboring atom is not paired in a covalent bond.

The material we have been describing thus far is an *intrinsic* semiconductor and is characterized by several features: (1) the number of electrons in the conduction band is equal to the number of holes in the valence band; (2) the Fermi energy lies at the middle of the gap; (3) the electrons contribute most to the current, but the holes are important also; (4) about 1 electron in 10^9 contributes to the conduction.

Because only 1 electron in 10^9 contributes to the conductivity of an intrinsic semiconductor, the presence of impurities can significantly alter the conductivity of the semiconductor in a way that might not be easily controllable. However, if impurities with known properties are *deliberately* introduced into the semiconductor in carefully controlled amounts, their contribution to the conductivity can be precisely determined. At impurity levels of only 1 part in 10^6 or 10^7, the impurity contribution to the conductivity dominates the intrinsic contribution.

Such materials are known as *impurity* semiconductors, and the process of introducing the impurity is known as *doping*. Impurity semiconductors can be of two varieties: those in which the impurity contributes additional electrons to the conduction band and those in which the impurity contributes additional holes to the valence band.

Let us consider a material such as silicon or germanium, in which there are four valence electrons in hybrid orbitals. In the band theory view, these fill the $4N$ states of the valence band; in the atomic view the lattice is constructed so that each Ge or Si atom has four neighbors with which it shares an electron, and so all electrons participate in covalent bonding (Figure 11.40a). Now suppose we replace one of the Si or Ge atoms with an atom that has five valence electrons, such as phosphorus, arsenic, or antimony. Four of the five electrons form covalent bonds with the neighboring Si or Ge atoms, but the fifth electron is relatively weakly bound to the impurity atom and can be easily detached to contribute to the conductivity (Figure 11.40b). Alternatively, we could replace one of the Si or Ge atoms with an atom that has three valence electrons, such as boron, aluminum, gallium, or indium. Its three valence electrons form covalent bonds with the neighboring Si or Ge (Figure 11.40c), but one of the surrounding atoms has an unpaired electron. Completing the four pairs of covalent bonds is energetically very favorable, so an electron is easily captured to complete the symmetry of the lattice. This creates a hole in the valence band and therefore contributes to the conductivity.

On an energy-level diagram, the electron energies of these impurity atoms appear as discrete levels in the energy gap, either just below the conduction band (as in Figure 11.41a), or just above the valence band (as in Figure 11.41b). The energy needed for these electrons to enter the conduction band, or for electrons from the valence band to fill the low-lying empty states, is relatively small, about 0.01 eV in Ge and 0.05 eV in Si. As a result, even at room temperature ($kT \sim 0.025$ eV) these excitations can occur easily.

The energy levels formed by valence-5 impurities are known as *donor states* and the impurity is known as a *donor*, because electrons are "donated" to the conduction band. A semiconductor that has been doped with donor impurities is known as an *n-type* semiconductor, because the conductivity is due mostly to the negative electrons.

The energy levels formed by valence-3 impurities are known as *acceptor states*, because they can "accept" electrons from the valence band. A material that has been doped with acceptor impurities is known as a *p-type* semiconductor, because the conductivity is due mostly to the positively charged holes. (Remember that *n*-type and *p*-type materials are both *electrically neutral* because they are made from neutral atoms. The designations *n* and *p* refer only to the charge carriers, not

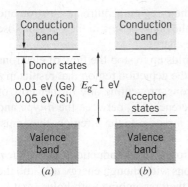

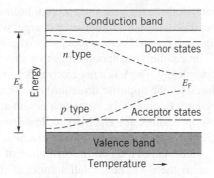

FIGURE 11.41 (a) Energy levels of donor states. (b) Energy levels of acceptor states.

FIGURE 11.42 In a semiconductor, the Fermi energy moves toward the middle of the gap as the temperature increases.

to the material itself. Depending on whether we add or remove electrons, n-type and p-type materials can become either negatively or positively charged).

At $T = 0$ the Fermi energy in n-type semiconductors lies between the donor states and the conduction band (remember, all states below E_F are full and all above E_F are empty; at $T = 0$ the donor states are all occupied). In p-type semiconductors, the Fermi energy at $T = 0$ lies between the valence band and the acceptor states. As the temperature is raised, the thermal excitation of electrons from the valence band to the conduction band (as in an intrinsic semiconductor) causes the Fermi energy to move toward the center of the energy gap, as shown in Figure 11.42. For low doping levels and at a high enough temperature, the material may behave like an intrinsic semiconductor.

11.7 SEMICONDUCTOR DEVICES

The *p-n* Junction

When a p-type semiconductor is placed in contact with an n-type semiconductor (Figure 11.43) electrons flow from the n-type material into the p-type material, until equilibrium is established. This equilibrium occurs when the Fermi energies in the two substances become identical.

The resulting energy level diagram is shown in Figure 11.44. The region between the two materials is known as the *depletion* region, because it has been somewhat depleted of charge carriers. Electrons from the donor states of the n-type material fill the holes of the acceptor states of the p-type material. In this region the donor states *do not* provide electrons for the conduction band and the acceptor states *do not* provide holes in the valence band.

Actually, these devices are not made by bringing two different materials into contact, but rather by doping one side of a material so that it becomes n type and the other side so that it becomes p type. The doping is carefully controlled, and typically depletion layers have a thickness of the order of $1\ \mu m$.

The excess electrons that have entered the p-type material give that side of the depletion region a negative charge, which tends to repel additional electrons from the n region. There is a corresponding positive charge in the n region (because it has lost electrons to the p region). These charges are associated with the fixed

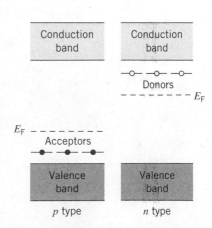

FIGURE 11.43 n-type and p-type semiconductors before contact.

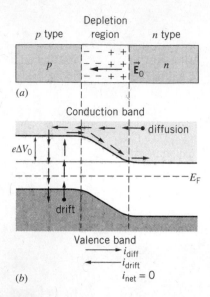

(a)

(b)

FIGURE 11.44 (a) A p-n junction. The electric field in the depletion region inhibits the additional flow of electrons. (b) Energy levels in p-n junction. Energetic electrons in the tail of the Fermi-Dirac distribution contribute to the diffusion current, while thermal excitation gives an equal and opposite drift current.

ions in those regions; the acceptor atoms in the p region acquire an electron and become fixed sites of negative charge, while the donor atoms in the n region lose an electron and become fixed sites of positive charge.

In equilibrium, enough negative charge builds up to stop the flow of electrons completely. There is a net electric field $\vec{E}_0$ in the depletion region that results in a force (in the opposite direction) on the electrons, preventing any further flow of charge. Equivalently, there is a potential difference ΔV_0 between the n-type and p-type regions; for electrons to flow from the n region to the p region, they must climb the energy barrier of height $e\Delta V_0$.

In the tail of the Fermi distribution of electrons in the conduction band of the n region, there will be a small number of electrons with enough energy to climb the energy barrier and enter the p region, where they recombine with holes (that is, they "fall" from the conduction band of the p region into the valence band). This gives the *diffusion* or *recombination* contribution to the current, which is directed from the p region to the n region (the current direction always being opposite to the direction of electron flow).

Even though holes provide the dominant contribution to the conduction in the p region, there are also electrons that provide a smaller contribution to the current. Electrons are thermally excited from the valence band in the p region to the conduction band, where they are accelerated by the electric field and travel into the n region. This gives the *drift* or *thermal* contribution to the current. At equilibrium, the two contributions to the current cancel one another, so that the net current is zero, as shown in Figure 11.44.

Let us now apply an external voltage ΔV_{ext} across the junction so that the p-type material is made more positive than the n material; that is, we connect the + terminal of a battery to the p side of the junction and the − terminal of the battery to the n side (Figure 11.45). The effect of the battery is to *lower* the energy hill by an amount $e\Delta V_{ext}$. (The vertical axis shows electron energy, and a potential difference of ΔV_{ext} gives an electron energy of $-e\Delta V_{ext}$.) This situation is called a *forward voltage* or *forward biasing*. The forward bias causes the depletion region to become narrower, because the battery pulls electrons out of the p region and injects them back into the n region. Because the energy hill is lower, more electrons can diffuse from the n region into the p region, so the diffusion current is considerably increased. (That is, there are more electrons in the n region in the tail of the Fermi distribution with energies above the bottom of the conduction band of the p region.) The drift current, however, is unaffected by the presence of the battery or the height of the hill. There is now a net current through the junction in the forward direction.

Now we reverse the battery connections (Figure 11.46), a situation known as *reverse voltage* or *reverse biasing*. This *raises* the hill by the amount $e\Delta V_{ext}$, *widens* the depletion region (because the battery pulls more electrons from the n region and injects them into the p region), and *decreases* the diffusion current. The drift current is again unchanged, so now there is a relatively small net current in the reverse direction.

Figure 11.47 shows the upper tail of the Fermi-Dirac distribution of the electrons extending into the conduction band of the n-type region. Only those electrons in the portion of the tail above the energy E_c of the bottom of the conduction band of the p-type region can flow back across into the p-type region and it is these electrons that produce the diffusion current. The number of electrons in that tail above the energy E_c is approximately

$$N_1 = ne^{-(E_c - E_F)/kT} \tag{11.28}$$

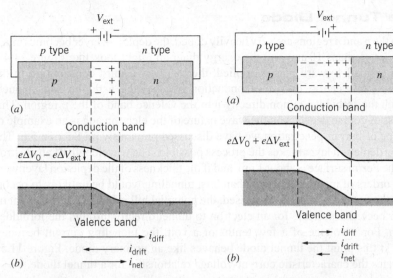

FIGURE 11.45 (*a*) A forward-biased *p-n* junction. (*b*) The energy level diagram. The potential energy hill is smaller, the diffusion current is larger, and there is a net forward current.

FIGURE 11.46 (*a*) A reversed-biased *p-n* junction. (*b*) The energy level diagram. The potential energy hill is larger, the diffusion current is smaller, and there is a small reverse current.

where n is some proportionality factor, and where we have approximated the Fermi-Dirac function as an exponential by neglecting the 1 in the denominator. (Because $E_c - E_F \geqslant 1\text{eV}$ and $kT = 0.025$ eV, this is an excellent approximation.) The diffusion current is proportional to N_1, and because the drift and diffusion currents are equal, the drift current is also proportional to N_1. Applying ΔV_{ext} changes the level E_c to $E_c - e\Delta V_{ext}$, and the number of electrons in the tail above $E_c - e\Delta V_{ext}$ is

$$N_2 = ne^{-(E_c - e\Delta V_{ext} - E_F)/kT} \tag{11.29}$$

The diffusion current is now proportional to N_2; applying the bias did not change the drift current, so it is still proportional to N_1. The net current is given by the difference:

$$i \propto N_2 - N_1 = ne^{-(E_c - E_F)/kT}(e^{e\Delta V_{ext}/kT} - 1) \tag{11.30}$$

We can rewrite this expression as

$$i = i_0(e^{e\Delta V_{ext}/kT} - 1) \tag{11.31}$$

This function is plotted in Figure 11.48, and it is immediately obvious why such *p-n* junctions, also known as *diodes*, have the property of *rectifying* varying currents. When the applied voltage is such that the junction is forward biased, a large forward current can flow. (When $\Delta V_{ext} = 1\text{V}$, $i = 2 \times 10^{17}i_0$.) When the applied voltage is such that the junction is reverse biased, only a very small current can flow. (When $\Delta V_{ext} = -1\text{V}$, $i \cong -i_0$.) Even very small forward voltages can produce large forward currents; even very large reverse voltages can produce only small reverse currents.

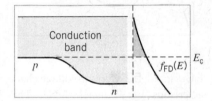

FIGURE 11.47 The diffusion current depends on the number of electron states in the tail of the Fermi-Dirac distribution above the energy E_c of the bottom of the conduction band in the *p*-type material.

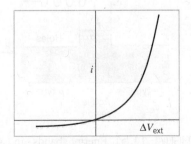

FIGURE 11.48 Current-voltage characteristics of an ideal *p-n* junction.

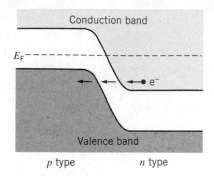

FIGURE 11.49 Energy level diagram of a *p-n* junction under heavy doping. Electrons can tunnel across the narrow gap.

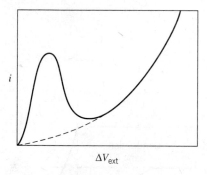

FIGURE 11.50 Current-voltage characteristics of a tunnel diode. The dashed curve shows the characteristics of an ordinary *p-n* junction diode.

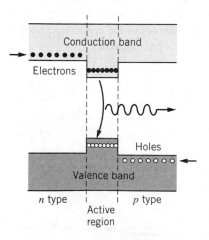

FIGURE 11.51 Energy bands in a diode laser. The active region has a smaller gap than the *n*-type and *p*-type regions on either side.

The Tunnel Diode

When the *p* and *n* regions are very heavily doped, the depletion layer becomes much narrower, perhaps 10 nm, and the energy diagram might look like Figure 11.49. When a small forward bias is applied, there is now a third contribution to the current—an electron from the conduction band of the *n* region can "tunnel" through the forbidden region directly into the valence band of the *p* region. This process of course depends on the wave nature of the electron and is an example of the type of barrier penetration we have discussed previously, in Section 5.6. The narrow depletion layer makes the process possible. The wavelength of an electron near the Fermi surface is about 1 nm, and if the thickness of the depletion layer were many orders of magnitude larger than this, tunneling would be unlikely to occur.

As the forward voltage is increased, the potential hill is lowered, and soon it no longer becomes possible for an electron to tunnel directly through the forbidden region. For a voltage of a few tenths of a volt, the tunneling current becomes zero. At this point the tunnel diode behaves like an ordinary diode. Figure 11.50 illustrates the characteristic current-voltage relationship for a tunnel diode.

Tunnel diodes are useful in electric circuits as high-speed elements, because the characteristics of the device can change as rapidly as the bias voltage can be changed. They can also be used as switches. If we were to pass current through the tunnel diode so that we were on the peak of the characteristic curve, a small increase in the current would cause the voltage to jump suddenly to a much larger value.

Photodiodes

A photodiode is a *p-n* junction whose operation involves the emission or absorption of light. These devices operate on principles similar to ordinary atoms. An electron in the valence band can absorb a photon and make a transition to the conduction band. Photons of visible light have energies of order 2 to 3 eV, so a semiconductor with its gap of order 1 eV is just right for such a transition. Conversely, an excited electron from the conduction band can drop back down to the valence band, emitting a photon in the process.

A common device that emits visible light is the LED, or light-emitting diode. An external current supplies the energy necessary to excite electrons to the conduction band, and when the electrons fall back down to recombine with holes, a photon is emitted. The energy is of course equal to the difference in energy of the electronic states. By varying the chemical composition, it is possible to produce LEDs emitting any color of visible light. LEDs find wide use as indicator lights and in video displays, including televisions and computer monitors. Broad-spectrum LEDs emitting white light are used in environmental lighting.

It is possible for photodiodes to operate in reverse, in which an incoming photon is absorbed in the depletion region and produces an electron-hole pair. An electric field sweeps up the electron-hole pairs and produces an electric signal. Such devices are used to produce electric current (as in a silicon solar cell) or to count photons (as in light meters for cameras or detectors of X rays or gamma rays in space probes).

Figure 11.51 shows another application of the emission of light by a semiconductor, in this case a *diode laser* or *semiconductor laser*. A thin layer of semiconducting material is sandwiched between *n*-type and *p*-type regions having a slightly larger energy gap. Electrons are injected from an external circuit into the *n*-type material, from which they diffuse into the middle layer. The electrons are prevented from diffusing into the *p*-type layer by a potential barrier, so they tend to concentrate in the middle layer. In a similar fashion, holes are injected into the *p*-type layer and again concentrate in the middle layer. This creates a population inversion similar to that described in our discussion of the laser in Section 8.7. An electron drops into the valence band accompanied by the emission of a photon; that photon then induces other transitions leading to the avalanche of photons that gives the lasing action.

The physical construction of a typical diode laser is illustrated in Figure 11.52. The lasing material is a narrow (0.2 μm) layer of GaAs, and the *p*-type and *n*-type layers are GaAlAs a few μm in thickness. The ends of the material are cleaved to create mirrorlike surfaces that reflect a portion of the light wave, enhancing stimulated emission in the active region. This device emits at a wavelength of 840 nm, in the near infrared region. Diode lasers at this wavelength are commonly used in communication to send signals along optical fibers. By varying the materials of the laser, it is possible to obtain visible radiation in almost any color. Diode lasers find common use in bar-code scanners and in players for CDs and DVDs. There are also many medical uses, for example in laser surgery.

Diode lasers are of small size, and they consume very little power (typically 10 mW, compared with the standard HeNe laser that may consume several watts). As a result, diode lasers can be powered by ordinary batteries. The light signal can be turned on or off in switching times that are characteristic of semiconductors (< 100 ps), and thus we have a device that can rapidly modulate the beam. Even though conventional lasers are capable of performing many of the same functions as diode lasers, the small size, low cost, low power consumption, and rapid switching times make diode lasers the superior choice for most low-power applications.

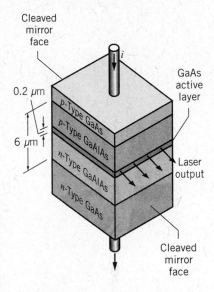

FIGURE 11.52 A diode laser. The lasing action occurs in the thin GaAs layer.

11.8 MAGNETIC MATERIALS

Our final example of the application of quantum physics to solids concerns the magnetic behavior of materials. The magnetic susceptibility of atoms was discussed briefly in Section 8.4 (see Figure 8.10). Most atoms have permanent magnetic dipole moments, due either to the spin or orbital angular momentum of the electrons (or both). Ordinarily, these magnetic moments point in random directions, so the net total magnetic dipole moment of a sample of the material is zero. However, when a magnetic field is applied, the magnetic moments rotate into partial or full alignment with the applied field, and the vector sum of the dipole moments gives the material a net magnetization. Specifically, the total magnetization $\vec{\mathbf{M}}$ is defined as the sum of all the individual atomic magnetic dipole moments $\vec{\mu}_i$ per unit volume:

$$\vec{\mathbf{M}} = \frac{\sum_{i=1}^{N} \vec{\mu}_i}{V} \qquad (11.32)$$

where the sum is carried out over all the N individual particles (atoms, for example) in the material.

Over a fairly wide range of applied magnetic fields, in many materials we find that the net magnetization is directly proportional to the applied field $\vec{\mathbf{B}}_{app}$. That is, the stronger the applied field, the more the individual magnetic moments rotate into alignment with the field. (Clearly this proportionality cannot continue indefinitely as the field strength increases, because eventually all dipoles will be aligned with the field and further increases in the field will have little or no effect.)

The proportionality constant between the magnetization and the applied field is called the *magnetic susceptibility* χ:

$$\mu_0 \vec{\mathbf{M}} = \chi \vec{\mathbf{B}}_{app} \tag{11.33}$$

For many materials, χ is small (10^{-5} to 10^{-1}) and positive. These materials are called *paramagnetic*. In other materials, χ is observed to be negative (that is, the direction of the net magnetization is opposite to the direction of the applied field). These materials, which are called *diamagnetic*, usually have no permanent atomic magnetic moments, often because they have only paired electrons (as for example the inert gases); their magnetic behavior is due to a slight change in the orbital motion of the atomic electrons in response to the applied field. Diamagnetism is ordinarily a very weak effect, with susceptibilities in the range of -10^{-5} to -10^{-4}. (The effect responsible for diamagnetism, the alteration of the orbital motion of the electrons, can also occur in paramagnetic materials, but it is generally much weaker than the paramagnetism. However in some materials, such as copper, the paramagnetism is so weak that the diamagnetism is dominant.) In yet other materials, the magnetization remains after the applied field is removed. These include the *ferromagnetic* substances, and the susceptibility is undefined for ferromagnets.

Magnetic effects are often strongly temperature dependent. In fact materials can change from ferromagnetic to paramagnetic as the temperature is increased.

Paramagnetism of Electron Gas

Let's begin by investigating the magnetic behavior of an electron gas. When we apply a magnetic field to an electron gas, the energy of an electron in the field is $E = -\vec{\mu}_s \cdot \vec{\mathbf{B}}_{app}$, where $\vec{\mu}_s = -(e/m)\vec{\mathbf{s}}$ is the spin magnetic moment of the electron. The energy of the interaction of an electron with the field is then (assuming the field is in the z direction)

$$E = -\vec{\mu}_s \cdot \vec{\mathbf{B}}_{app} = \frac{e}{m}\vec{\mathbf{s}} \cdot \vec{\mathbf{B}}_{app} = \frac{e}{m}s_z B_{app} = \frac{e}{m}m_s \hbar B_{app} = \pm \mu_B B_{app} \quad (11.34)$$

with $m_s = \pm 1/2$. The symbol μ_B represents the Bohr magneton $e\hbar/2m$ (see Eq. 7.23). The electrons with $m_s = +1/2$ gain energy $\mu_B B_{app}$ and those with $m_s = -1/2$ lose an equal amount of energy.

Figure 11.53 shows the Fermi-Dirac distribution of populated electron states separately for the spin-up and spin-down electrons (half of the electrons are in each group). All of the spin-up electrons move up in energy by $\mu_B B_{app}$ and all of the spin-down electrons move down. Because the two groups of electrons are in contact, the higher-energy electrons in the shaded region with spin up flip their spins and fill the vacant energy states in the spin-down group until the two groups equalize their Fermi energies. As a result, there is an excess of electrons with spin down in a strip of width $\Delta E = 2\mu_B B_{app}$. The number of electrons in the strip is $\Delta N = \frac{1}{2}Vg(E)f_{FD}(E)\Delta E$, where the factor of $1/2$ comes from the fact that the spin-up and spin-down distributions each have $1/2$ of the total number of electrons. The drawing of Figure 11.53 has been exaggerated; the strips are very narrow compared with E_F, and we assume that we are at a reasonably low temperature in which the Fermi-Dirac distribution is fairly sharp, so we can evaluate the density of states at $E = E_F$ and take $f_{FD} = 1$:

$$\Delta N = N(\downarrow) - N(\uparrow) = Vg(E_F)\mu_B B_{app} \tag{11.35}$$

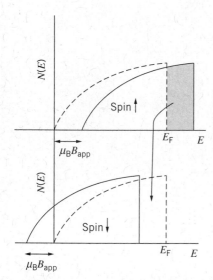

FIGURE 11.53   The dashed lines show the number of electrons in the Fermi-Dirac distribution drawn separately for spin up (top) and spin down (bottom). In an applied magnetic field, the energies of the spin-up electrons increase and the energies of the spin-down electrons decrease. To maintain the same Fermi energy with the field on, the spin-up electrons in the shaded strip flip their spins and move into the equivalent area with spin down.

From Eq. 11.32, the z component of the magnetization is $V^{-1} \sum \mu_{iz} = V^{-1} \mu_B \Delta N$, so

$$\chi = \frac{\mu_0 M}{B_{app}} = \mu_0 \mu_B^2 \, g(E_F) = \frac{3 \mu_0 \mu_B^2}{2 E_F} \frac{N}{V} \qquad (11.36)$$

using $g(E_F) = 3N/2VE_F$. This calculation was first done by Wolfgang Pauli, and the result is often called the *Pauli paramagnetic susceptibility*. Note that N/V in this equation is the number of free electrons per unit volume, which might differ from the number of atoms per unit volume in materials in which there is more than one valence electron per atom.

Equation 11.36 can be used to calculate values for the susceptibility of materials in which the electrons behave like a gas of fermions. However, comparing calculated susceptibility values with experimental values can often be challenging because of the many different units that are used. On the surface, it appears that the susceptibility is a dimensionless quantity, but in fact it can be calculated in many different ways: susceptibility per unit volume (as in Eq. 11.36) or per unit mass or per mole. Furthermore, it can be expressed in either cgs units or SI units (our choice). The tabulated values in the literature are often given as the molar susceptibility in cgs units. The conversion procedure is $\chi_{volume}^{SI} = 4\pi \chi_{volume}^{cgs} = 4\pi (\rho/M) \chi_{molar}^{cgs}$, where ρ is the density of the solid and M is its molar mass (don't confuse this with the magnetization!). With the conversion written in this way, ρ and M must be in cgs units.

With that warning in mind, let's look at how Eq. 11.36 compares with experiment. Table 11.7 shows a few calculated and measured values. The agreement is surprisingly good, particularly in view of our omitting a number of important effects from the calculation, including a diamagnetic correction. For many metals, the diamagnetic contribution is larger than the Pauli paramagnetism, and as a result their susceptibility is negative (copper and gold are examples). The diamagnetic contribution is important because the Pauli susceptibility is so small (only a small fraction of the electrons near the Fermi energy contribute). Metal ions in salts typically have paramagnetic susceptibilities that can be several orders of magnitude larger, as we discuss next.

It is also interesting to note that Eq. 11.36 predicts that the susceptibility of these substances should be independent of temperature. Raising the temperature should broaden the distribution of both the spin-up and spin-down distributions (Figure 11.53) by about the same amount, making a negligible contribution to the susceptibility. Indeed, the susceptibility of these metals has only a very weak dependence on temperature. In sodium, for example, the susceptibility changes by only a few percent between room temperature (300 K) and liquid helium temperature (4 K).

TABLE 11.7 Pauli Magnetic Susceptibility of Some Solids

Element	Experimental susceptibility ($\times 10^{-6}$)		Calculated susceptibility ($\times 10^{-6}$)
	cgs, per mole	SI, per volume	
Al	16.5	20.7	15.7
K	20.8	5.8	6.2
Mg	13.1	11.8	12.2
Na	16	8.5	8.2

Example 11.5

(a) Compute the Pauli paramagnetic susceptibility of Mg. (b) The experimental value of the molar susceptibility is 13.1×10^{-6} in cgs units. What is the value of the volume susceptibility in SI units?

Solution

(a) For Mg (which has valence 2),

$$\frac{N}{V} = \frac{2\rho N_A}{M}$$

$$= \frac{2(1.74 \times 10^3 \text{ kg/m}^3)(6.02 \times 10^{23} \text{ atoms/mole})}{0.0243 \text{ kg/mole}}$$

$$= 8.62 \times 10^{28} \text{ m}^{-3}$$

The susceptibility is

$$\chi = \frac{3\mu_0 \mu_B^2}{2E_F} \frac{N}{V}$$

$$= \frac{3(4\pi \times 10^{-7} \text{ T} \cdot \text{m/A})(9.27 \times 10^{-24} \text{ J/T})^2(8.62 \times 10^{28} \text{ m}^{-3})}{2(7.13 \text{ eV})(1.602 \times 10^{-19} \text{ J/eV})}$$

$$= 12.2 \times 10^{-6}$$

To see how the units cancel, it is helpful to realize that the magnetic moment has units of either J/T or $A \cdot m^2$ (the latter coming from the definition $\mu = iA$ for the magnetic moment of a current i in a loop of area A).

(b)

$$\chi_{\text{volume}}^{\text{SI}} = 4\pi \frac{\rho}{M} \chi_{\text{molar}}^{\text{cgs}}$$

$$= 4\pi \frac{1.74 \text{ g/cm}^3}{24.3 \text{ g/mole}}(13.1 \times 10^{-6})$$

$$= 11.8 \times 10^{-6}$$

Note that the density and molar mass are in cgs units for the conversion.

Paramagnetism of Atoms and Ions

Instead of the free electrons, let's consider the contribution of the atoms or ions to the paramagnetism. We'll represent the effective electronic spin of each atom by J, which might represent the total intrinsic spin S of the electrons in the atom, their total orbital angular momentum L, or a combination of both (the nuclear spin is excluded, because nuclear magnetic effects are negligible compared with electronic magnetic effects). Depending on the number of electrons in the atom, J might be integral or half-integral.

The angular momentum J has all of the usual properties of quantum angular momentum. It has z component $J_z = m_J \hbar$, where m_J runs from $-J$ to $+J$ in integer steps. For any J, there are $2J + 1$ possible values of m_J. For example, if $J = 3/2$, then $m_J = -3/2, -1/2, +1/2, +3/2$. Associated with this angular momentum there is an effective magnetic moment $\vec{\mu} = -g_J \mu_B \vec{J}$, where g_J is a dimensionless factor of order unity that describes how the spin and orbital angular momenta combine to give J. (The minus sign is present because electrons have negative charge.) Assuming the magnetic field defines the z direction, the energy of interaction of this magnetic moment with the applied field is $E = -\vec{\mu} \cdot \vec{B}_{\text{app}} = g_J \mu_B m_J B_{\text{app}}$.

We'll treat the atoms or ions as if they are independent of one another, so that they can be described by Maxwell-Boltzmann statistics. Because the magnetic substates m_J are nondegenerate, we can write the number of atoms in each magnetic substate in the form of Eq. 10.4: $N_{m_J} = A^{-1}e^{-E/kT} = A^{-1}e^{-g_J \mu_B m_J B_{\text{app}}/kT}$, where A^{-1} is the normalization constant for the Maxwell-Boltzmann distribution.

The normalization condition requires that N be the total number of atoms in all substates: $N = \sum_{m_J=-J}^{+J} N_{m_J} = A^{-1} \sum_{m_J=-J}^{+J} e^{-g_J \mu_B m_J B_{app}/kT}$. We therefore have:

$$N_{m_J} = \frac{N e^{-g_J \mu_B m_J B_{app}/kT}}{\sum_{m_J=-J}^{+J} e^{-g_J \mu_B m_J B_{app}/kT}} \tag{11.37}$$

The z component of the magnetization is then

$$M = V^{-1} \sum_{\text{all atoms}} \mu_z = -V^{-1} \sum_{m_J=-J}^{+J} N_{m_J} g_J \mu_B m_J$$

$$= -\frac{N V^{-1} g_J \mu_B \sum_{m_J=-J}^{+J} m_J e^{-g_J \mu_B m_J B_{app}/kT}}{\sum_{m_J=-J}^{+J} e^{-g_J \mu_B m_J B_{app}/kT}} \tag{11.38}$$

The magnetic susceptibility is

$$\chi = \frac{\mu_0 M}{B_{app}} = -\frac{\mu_0 N V^{-1} g_J \mu_B \sum_{m_J=-J}^{+J} m_J e^{-g_J \mu_B m_J B_{app}/kT}}{B_{app} \sum_{m_J=-J}^{+J} e^{-g_J \mu_B m_J B_{app}/kT}} \tag{11.39}$$

Let's examine the form of χ for the simplest case in which $J = 1/2$, so that we have only two terms in the sums ($m_J = -1/2$ and $+1/2$). Equation 11.39 becomes

$$\chi = \frac{-\mu_0 N g_J \mu_B [(-1/2) e^{g_J \mu_B B_{app}/2kT} + (+1/2) e^{-g_J \mu_B B_{app}/2kT}]}{V B_{app} (e^{g_J \mu_B B_{app}/2kT} + e^{-g_J \mu_B B_{app}/2kT})}$$

$$= \frac{\mu_0 N g_J \mu_B}{2 V B_{app}} \tanh(g_J \mu_B B_{app}/kT) \tag{11.40}$$

Figure 11.54 shows the susceptibility of a spin-$1/2$ atom as a function of $1/T$, so the high-temperature region is on the left side of the graph and the low-temperature region is on the right side. In the low-temperature region, the magnetic moments become fully aligned, and neither additional cooling nor an increase in the applied field can change the magnetization. In the high-temperature region, the graph is nearly a straight line, indicating that the susceptibility is linear in $1/T$. (How high must the temperature be to observe this linear behavior? For a fairly large field of 1 T the quantity $\mu_B B_{app}$ is about 0.060 meV. When T is 10 K, kT is 0.862 meV. Thus even with moderately large fields, "high temperature" for this discussion means anything above about 10 K. At low fields the "high temperature region" might reach all the way down to 1 K.)

In the high-temperature region, the exponents in Eq. 11.39 are small and we can use the approximation $e^x \cong 1 + x$ for small x (with $x = g_J \mu_B m_J B_{app}/kT$):

$$\chi = -\frac{\mu_0 N V^{-1} g_J \mu_B \sum_{m_J=-J}^{+J} m_J (1 - g_J \mu_B m_J B_{app}/kT)}{B_{app} \sum_{m_J=-J}^{+J} (1 - g_J \mu_B m_J B_{app}/kT)} \tag{11.41}$$

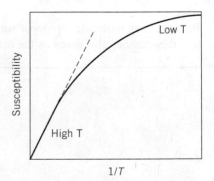

FIGURE 11.54 Magnetic susceptibility as a function of inverse temperature for a spin-$1/2$ material. Note the linear behavior at high temperature.

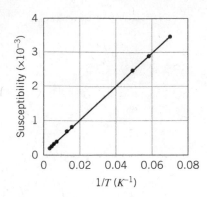

FIGURE 11.55 Susceptibility of paramagnetic copper sulfate ($CuSO_4 \cdot 5H_2O$) between room temperature and 14 K.

All quantities other than m_J come out of the sums, leaving us three summations to evaluate:

$$\sum_{m_J=-J}^{+J} 1 = 2J+1 \qquad \sum_{m_J=-J}^{+J} m_J = 0 \qquad \sum_{m_J=-J}^{+J} m_J^2 = \tfrac{1}{3}J(J+1)(2J+1) \quad (11.42)$$

After making these substitutions and clearing terms, the result is

$$\chi = \frac{\mu_0 N g_J^2 \mu_B^2 J(J+1)}{3VkT} \qquad (11.43)$$

This shows the linear dependence of the susceptibility on $1/T$, as we observed in the high-temperature region of Figure 11.54. This result is known as *Curie's law**, which is often expressed in the form $\chi = C/T$ where C (the combination of factors in Eq. 11.43) is called the *Curie constant*. (It is not a true constant, as it takes different values for different substances, but it is constant for any single substance.) Figure 11.55 shows the susceptibility of the paramagnetic salt copper sulfate, which clearly shows the expected linear behavior down to a temperature of about 14 K.

Example 11.6

The slope of the graph in Figure 11.55 is 0.0499 K. Assuming the paramagnetism resides with the copper ions, find the value of the quantity $g_J^2 J(J+1)$ for copper sulfate.

Solution

We first need the value of N/V, which requires the density (2.28×10^3 kg/m^3) and molar mass (0.250 kg/mole) of copper sulfate:

$$\frac{N}{V} = \frac{\rho N_A}{M}$$

$$= \frac{(2.28 \times 10^3 \text{ kg/m}^3)(6.02 \times 10^{23} \text{ atoms/mole})}{0.250 \text{ kg/mole}}$$

$$= 5.49 \times 10^{27} \text{ atoms/m}^3$$

Because each molecule of copper sulfate has only one Cu ion, this number also gives us the density of Cu ions. We

then have

$$C = \frac{\mu_0 N \mu_B^2}{3Vk} g_J^2 J(J+1)$$

$$= (4\pi \times 10^{-7} \text{ T} \cdot \text{m/A})(5.49 \times 10^{27} \text{ m}^{-3})$$

$$\times (9.27 \times 10^{-24} \text{ J/T})^2 g_J^2 J(J+1)$$

$$\times [3(1.38 \times 10^{-23} \text{ J/K})]^{-1}$$

$$= (0.0143 \text{ K}) g_J^2 J(J+1) = 0.0499 \text{ K}$$

As a result, $g_J^2 J(J+1) = 3.49$. Such experiments give us important information about the behavior of ions in crystals that is otherwise difficult to obtain. In this case we can learn about the value of the effective spin J that describes the copper ion.

In copper sulfate, the divalent copper ion Cu^{++} has the outer electronic configuration $3d^9$ (recall that neutral Cu atoms have the configuration $3d^{10}4s^1$). According to the rules for finding the total S and L for atoms (see Section 8.6), the nine $3d$ electrons in Cu^{++} have $S = 1/2$ (the maximum possible value of the total M_S) and $L = 2$ (the maximum possible resulting value of the total M_L). However, the measured value of $g_J^2 J(J+1)$ of 3.49 from Example 11.6 is more consistent with the configuration $S = 1/2, L = 0$ (for which its value would be 3.00), than

---

*Physicist Pierre Curie (1859–1906) was the husband of Marie Curie. Together they shared the 1903 Nobel Prize in physics for their research on radioactivity. Pierre Curie also contributed to the study of magnetism.

it is with any configuration involving $S = 1/2$ and $L = 2$ (which give values of either 2.40 or 12.60).

This is a common observation for paramagnetic crystals involving the transition metals (those in which the $3d$ shell is filling). In a magnetic field, which has a single preferred direction (usually taken to be the z direction), the component of the orbital angular momentum L_z along that direction is fixed, while the other components (L_x and L_y) average to zero. In a crystal, however, there is a strong electric field that may have three equivalent directions, and all three components of $\vec{L}$ might average to zero. As a result, the ion behaves as if it has $L = 0$ (we say that the orbital angular momentum is "quenched"), and so only the total S contributes to the magnetic moment. In contrast, the rare earth elements also form many paramagnetic crystals, but both L and S contribute to J. In the rare earths, the electrons in the unfilled $4f$ shell are shielded by the filled $5s$ and $5p$ shells, which have the strongest interactions with the electric field of the crystal (because the average radii of their orbits are larger). The "inner" $4f$ electrons are not much affected by the electric field of the crystal and so they contribute their large orbital angular momentum to the total J of the ion.

Ferromagnetism

In some materials, the individual magnetic dipoles of the ions can align with one another even in the absence of an external magnetic field. In this case a net magnetization M is present when B_{app} is zero, so clearly the susceptibility is undefined for these materials. The most familiar example of this behavior is *ferromagnetism*, in which the neighboring dipoles all align in the same direction. (It is also possible to have *antiferromagnetic* materials, in which the neighboring dipoles orient in opposite directions.)

It is perhaps tempting at first to think of the interaction responsible for ferromagnetism as a result of the magnetic field due to one dipole exerting a magnetic force to align the neighboring dipole. However, this dipole-dipole interaction is far too weak to account for the strong coupling between neighbors that produces ferromagnetic alignment. Instead, the effect results from the overlap of the wave functions of the electrons in neighboring atoms, in a manner similar to covalent bonding but depending on the spins of the electrons. This effect is very sensitive to the interaction between the spins and also to the distance between the neighboring ions.

In Fe, Co, and Ni the value of the atomic spacing results in an energy minimum for the parallel orientation of neighboring spins, and so these materials are ferromagnetic at room temperature (but they become paramagnetic at a sufficiently high temperature, where the thermal energy kT exceeds the interaction energy). In other materials (such as some of the rare earth elements) the interaction is weaker, so they may not exhibit their ferromagnetic behavior until they are cooled to a point where the thermal energy kT is smaller than the interaction energy. In still other cases, the atomic spacing of the pure element may not permit ferromagnetism, but a different atomic spacing in certain compounds containing that element may allow the overlap interaction that causes the spins to align. For example, Cr is weakly paramagnetic at room temperature, but CrO_2 (which is used to make magnetic recording tape) is ferromagnetic.

The band theory provides a framework for understanding ferromagnetic behavior. Consider iron, which has the electronic configuration $3d^6 4s^2$. The partially filled $3d$ band can be split into two subbands, one with spin up and one with spin down. In the absence of the overlap interaction, the bands are at the same

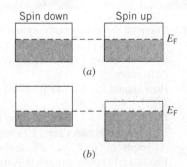

FIGURE 11.56 The $3d$ band in Fe. (a) In the absence of the interaction, there are 3 electrons per atom in the spin-up and spin-down subbands. (b) The interaction lowers the relative energy of the spin-up band, so there are now 4 electrons per atom with spin up and 2 with spin down.

energy and each band has 3 electrons per atom (out of its maximum capacity of 5) as in Figure 11.56a. The effect of the overlap interaction is to raise the energy of one band relative to the Fermi energy and lower the energy of the other (as in Figure 11.56b). Now there are roughly 4 electrons per atom in the spin-up subband and 2 in the spin-down subband, and this difference of approximately 2 electrons per atom in the 3d band is responsible for the net magnetization of iron.

Chapter Summary

		Section
Binding energy of ion in crystal	$B = \dfrac{\alpha e^2}{4\pi\varepsilon_0 R_0}\left(1 - \dfrac{1}{n}\right)$	11.1
Cohesive energy of crystal	$E_{\mathrm{coh}} = BN_{\mathrm{A}}$	11.1
Electron contribution to heat capacity	$C = \dfrac{\pi^2 k^2 N_{\mathrm{A}} T}{2E_{\mathrm{F}}} = \dfrac{\pi^2}{2}\dfrac{RkT}{E_{\mathrm{F}}}$	11.2
Einstein heat capacity	$C = 3R\left(\dfrac{T_{\mathrm{E}}}{T}\right)^2 \dfrac{e^{T_{\mathrm{E}}/T}}{(e^{T_{\mathrm{E}}/T}-1)^2}$	11.2
Debye heat capacity	$C = \dfrac{12\pi^4}{5}R\left(\dfrac{T}{T_{\mathrm{D}}}\right)^3$	11.2
Conductivity of free electron gas	$\sigma = ne^2\tau/m$	11.3
Lorenz number	$L = \pi^2 k^2/3e^2$ $= 2.44 \times 10^{-8}\ \mathrm{W}\cdot\Omega/\mathrm{K}^2$	11.3

		Section
BCS gap energy in superconductor	$E_{\mathrm{g}} = 3.53kT_{\mathrm{c}}$	11.5
Current in p-n junction diode	$i = i_0(e^{e\Delta V_{\mathrm{ext}}/kT} - 1)$	11.7
Pauli paramagnetic susceptibility	$\chi = \dfrac{3\mu_0\mu_{\mathrm{B}}^2}{2E_{\mathrm{F}}}\dfrac{N}{V}$	11.8
Paramagnetic susceptibility of atoms	$\chi = -\mu_0 N V^{-1} g_J \mu_{\mathrm{B}}$ $\times \displaystyle\sum_{m_J=-J}^{+J} m_J e^{-g_J \mu_{\mathrm{B}} m_J B_{\mathrm{app}}/kT}$ $\times \left(B_{\mathrm{app}} \displaystyle\sum_{m_J=-J}^{+J} e^{-g_J \mu_{\mathrm{B}} m_J B_{\mathrm{app}} kT}\right)^{-1}$	11.8
Curie's law	$\chi = \dfrac{\mu_0 N g_J^2\, \mu_{\mathrm{B}}^2 J(J+1)}{3VkT} = \dfrac{C}{T}$	11.8

Questions

1. Compare the equilibrium separations and binding energies of ionic *solids* (Table 11.1) with those of the corresponding ionic *molecules* (Table 9.5). Account for any systematic differences.

2. How should Eq. 11.7 be modified to be valid for MgO and BaO?

3. From Figure 11.11, estimate the Einstein temperature for lead. (*Hint:* Consider Eq. 11.11 when $T = T_{\mathrm{E}}$.)

4. A graph of C/T vs. T^2 for solid argon (similar to Figure 11.13) goes through the origin; that is, its y-intercept is zero. Explain.

5. Assuming that its other properties don't also change with temperature, at what temperature would you expect carbon to begin to behave like a semiconductor?

6. Would you expect the Wiedemann-Franz law to apply to semiconductors? To insulators?

7. (*a*) Why does the electrical conductivity of a metal decrease as the temperature is increased? (*b*) How would you expect the conductivity of a semiconductor to change with temperature?

8. Why is it that only the electrons near E_{F} contribute to the electrical conductivity?

9. Would you expect silicon to behave like an insulator at a low enough temperature? Would it behave like a conductor at a high enough temperature?

10. What determines the drift speed of an electron in a metal? What determines the Fermi speed?

11. Do the superconducting elements have any particular electronic structure or configuration in common?

12. Three different materials have filled valence bands and empty conduction bands, and the Fermi energy lies in the middle of the gap. The gap energies are 10 eV, 1 eV, and 0.01 eV. Classify the electrical properties of these materials at room temperature and at 3 K.

13. In what way does a p-n junction behave as a capacitor?

14. Semiconductors are sometimes called "nonohmic" materials. Why?

15. If a semiconductor is doped at a level of one impurity atom per 10^9 host atoms, what is the average spacing between the impurity atoms?

16. Explain the processes that contribute to the current in a forward-biased *p-n* junction. Do the same for a reverse-biased junction.

17. What limits the response time of a *p-n* junction when the external voltage is varied? Why does a tunnel diode not have the same limits?

18. The energy gap E_g is 0.72 eV for Ge and 1.10 eV for Si. At what wavelengths will Ge and Si be transparent to radiation? At what wavelengths will they begin to absorb significantly?

19. Why is a semiconductor better than a conductor for applications as a solar cell or photon detector? Would an insulator be even better?

20. Why is the magnetic susceptibility of an electron gas almost independent of temperature?

21. Would a sample of paramagnetic material be attracted or repelled by the N pole of a magnet? By the S pole? How would a sample of diamagnetic material behave? How would ferromagnetic material behave?

22. Is it possible for a material that has a positive magnetic susceptibility at room temperature to have a negative susceptibility at higher temperature?

Problems

11.1 Crystal Structures

1. Consider the packing of hard spheres in the simple cubic geometry of Figure 11.1. Imagine eight spheres in contact with their nearest neighbors, with their centers at the corners of the basic cube. (*a*) What fraction of the volume of each sphere is inside the volume of the basic cube? (*b*) Let *r* be the radius of each sphere and let *a* be the length of a side of the cube. Express *a* in terms of *r*. (*c*) What fraction of the volume of the cube is taken up by the portions of the spheres? This fraction is called the *packing fraction*.

2. Compute the packing fractions (see Problem 1) of (*a*) the fcc structure (Figure 11.2) and (*b*) the bcc structure (Figure 11.3). Which structure fills the space most efficiently?

3. Derive Eqs. 11.5 and 11.6.

4. Calculate the first three contributions to the electrostatic potential energy of an ion in the CsCl lattice.

5. (*a*) Find the binding energy per ion pair in CsCl from the cohesive energy. (*b*) Find the binding energy per ion pair in CsCl from Eq. 11.7. (*c*) Find the binding energy per atom for CsCl. The ionization energy of Cs is 3.89 eV.

6. (*a*) Find the binding energy per ion pair in LiF from the cohesive energy. (*b*) Find the binding energy per ion pair in LiF from Eq. 11.7. (*c*) Find the binding energy per atom for LiF. The ionization energy of Li is 5.39 eV, and the electron affinity of F is 3.45 eV.

7. Calculate the Coulomb energy and the repulsion energy for NaCl at its equilibrium separation.

8. The density of sodium is 0.971 g/cm^3 and its molar mass is 23.0 g. In the bcc structure, what is the distance between sodium atoms?

9. Copper has a density of 8.96 g/cm^3 and molar mass of 63.5 g. Calculate the center-to-center distance between copper atoms in the fcc structure.

10. Calculate the binding energy per atom for metallic Na and Cu.

11.2 The Heat Capacity of Solids

11. At what temperature do the lattice and electronic heat capacities of copper become equal to each other? Take $T_D = 343$ K and $E_F = 7.03$ eV. Which contribution is larger above this temperature? Below this temperature?

12. (*a*) The heat capacity of solid argon at a temperature of 2.00 K is 2.00×10^{-2} J/mole·K. What is the Debye temperature of solid argon? (See Question 4.) (*b*) What value do you expect for the heat capacity at a temperature of 3.00 K?

13. When C/T is plotted against T^2 for potassium, the graph gives a straight line with a slope of 2.57×10^{-3} J/mole·K^4. What is the Debye temperature for potassium?

14. At a temperature of 4 K, the heat capacity of silver is 0.0134 J/mole·K. The Debye temperature of silver is 225 K. (*a*) What is the electronic contribution to the heat capacity at 4 K? (*b*) What are the lattice and electronic contributions and the total heat capacity at 2 K?

11.3 Electrons in Metals

15. (*a*) In copper at room temperature, what is the electron energy at which the Fermi-Dirac distribution function has the value 0.1? (*b*) Over what energy range does the Fermi-Dirac distribution function for copper drop from 0.9 to 0.1?

16. Calculate the de Broglie wavelength of an electron with energy E_F in copper, and compare the value with the atomic separation in copper.

17. What is the number of occupied energy states per unit volume in sodium for electrons with energies between 0.10

and 0.11 eV above the Fermi energy at room temperature (293 K)?

18. The electrical conductivity of copper at room temperature is $5.96 \times 10^7 \, \Omega^{-1} m^{-1}$. Evaluate the average distance between electron scatterings. How many lattice spacings does this amount to?

19. At what temperature would the Fermi energy of Au be reduced by 1%? Compare this temperature with the melting point of Au (1337 K). Is it reasonable to assume the Fermi energy is a constant, independent of temperature?

20. A copper wire of diameter 0.50 mm carries a current of 2.5 mA. What fraction of the copper electrons contributes to the electrical conduction?

21. Use the Wiedemann-Franz ratio to calculate the thermal conductivity of copper at room temperature. The electrical conductivity is $5.96 \times 10^7 \, \Omega^{-1} \cdot m^{-1}$.

11.4 Band Theory of Solids

22. Estimate the ratio of the concentration of electrons in the conduction band of carbon (an insulator) and silicon (a semiconductor) at room temperature (293 K). The energy gaps are 5.5 eV for carbon and 1.1 eV for silicon. Assume the Fermi energy lies at the center of the gap.

23. Estimate the ratio of the number of electrons in the conduction bands of germanium ($E_g = 0.66$ eV) and silicon ($E_g = 1.12$ eV) at a temperature of 400 K. Assume the Fermi energy is at the center of the gap.

24. The valence band in Si has a width of 12 eV. In a cube of Si that measures 1.0 mm on each side, calculate (a) the total number of states in the valence band, and (b) the average spacing between the states. The density of sodium is 2.33 g/cm^3.

11.5 Superconductivity

25. (a) Zirconium metal has a superconducting transition temperature of 0.61 K. Assuming the validity of the BCS theory, what is the energy gap for Zr? (b) If a beam of photons were incident on superconducting Zr, what wavelength of photons would be sufficient to break up the Cooper pairs? In what region of the electromagnetic spectrum are these photons?

26. When superconducting tantalum metal is illuminated with a beam of photons, it is found that photon wavelengths of up to 0.91 mm are sufficient to destroy the superconducting state. According to the BCS theory, find the energy gap and critical temperature for Ta.

27. Find the frequency of the current that results when a voltage difference of 1.25 μV is applied across a Josephson junction.

11.6 Intrinsic and Impurity Semiconductors

28. The temperature of a sample of intrinsic silicon is increased by 100 K over room temperature (293 K). Estimate the increase in conductivity that we would expect from this increase in temperature. The gap energy in silicon is 1.1 eV.

29. (a) Estimate the fraction of the electrons in the valence band of intrinsic silicon that can be excited to the conduction band at a temperature of 100 K and at room temperature. Take the energy gap in silicon to be 1.1 eV. (b) By what factor would the conductivity of a metal change over this same temperature interval?

30. Estimate the temperature at which the density of conduction electrons in intrinsic germanium equals that of intrinsic silicon at room temperature (293 K). The gap energies are 1.12 eV for Si and 0.66 eV for Ge.

31. (a) When we replace an atom of silicon with an atom of phosphorus, the outer electron of phosphorus is screened so that the atom behaves like a single-electron atom with $Z_{eff} \cong 1$. Compute the energy of the electron, assuming that silicon has a dielectric constant of 12 that effectively reduces the electric field experienced by the electron. (b) The additional electron in Si has an effective mass that is 0.43 times the free electron mass. How does this correction factor change the electron energy?

32. Assuming the energy gap in intrinsic silicon is 1.1 eV and that the Fermi energy lies at the middle of the gap, calculate the occupation probability at 293 K of (a) a state at the bottom of the conduction band and (b) a state at the top of the valence band.

33. In a sample of germanium at room temperature (293 K), what fraction of the Ge atoms must be replaced with donor atoms in order to increase the population of the conduction band by a factor of 3? Assume all donor atoms are ionized, and take the energy gap in Ge to be 0.66 eV.

11.7 Semiconductor Devices

34. (a) In a p-n junction at room temperature, what is the ratio between the current with a forward bias of 2.00 V to the current with a forward bias of 1.00 V? (b) Evaluate the same ratio at a temperature of 400 K.

35. Under certain conditions, the current in a p-n junction at room temperature is observed to be 1.5 mA when a forward bias of 0.25 V is applied. If the same voltage were used to reverse bias the junction, what would be the current?

36. Gallium phosphide ($E_g = 2.26$ eV) and zinc selenide ($E_g = 2.87$ eV) are commonly used to make LEDs. What is the most prominent emission wavelength of these devices, and what color is the corresponding light?

37. LEDs of varying colors can be made by mixing GaN ($E_g = 3.4$ eV) and InN ($E_g = 0.7$ eV) in different proportions. Calculate the relative amounts of GaN and InN needed to produce an LED that emits (a) green light (550 nm) and (b) violet light (400 nm).

11.8 Magnetic Materials

38. Estimate the fraction of spin-up electrons in sodium that flip their spins (as in Figure 11.53) and thus contribute to the Pauli paramagnetism. Assume a magnetic field of 1 T.

39. In copper sulfate, the Cu^{++} ions behave as if they have $J = 1/2$, for which $g_J = 2$. In a magnetic field of 0.25 T, calculate the relative numbers of copper ions in the $m_J = +1/2$ and $m_J = -1/2$ states at (a) 300 K and (b) 4.2 K.

40. Compute the Pauli paramagnetic susceptibilities for (a) Li and (b) Ba. Compare with the experimental values, which are respectively 14.2×10^{-6} and 20.6×10^{-6} (in cgs units per mole).

41. (a) Calculate the expected paramagnetic susceptibility for gold, assuming it to behave like a free electron gas. (b) The observed cgs molar susceptibility at room temperature is -28.0×10^{-6}. Assuming the susceptibility consists only of the diamagnetic and paramagnetic parts, find the diamagnetic contribution to the susceptibility of gold.

42. (a) The experimental paramagnetic susceptibility of $MnCl_2$ at room temperature (293 K) in cgs units per mole is 14350×10^{-6}. What is the value of $g_J^2 J(J+1)$ for the Mn^{++} ion? (b) What is the electronic configuration expected for the Mn^{++} ion? For this configuration, and assuming that the electronic S corresponds to maximizing the total M_S, what is the value of S for Mn^{++}? For the arrangement with that S, what is the corresponding total L? (c) With $J = L + S$, what is the value of g_J that follows from the experimental susceptibility?

General Problems

43. By summing the contributions for the attractive and repulsive Coulomb potential energies, show that the Madelung constant has the value 2 ln 2 for a one-dimensional "lattice" of alternating positive and negative ions.

44. Plot the cohesive energies of ionic crystals (see Table 11.1 and other data that you may find) against their melting points (see, for example, the *Handbook of Chemistry and Physics*). Is there a correlation between cohesive energies and melting points?

45. Plot the cohesive energies of metallic crystals (see Table 11.3 and other data that you may find) against their melting points (see, for example, the *Handbook of Chemistry and Physics*). Is there a correlation between cohesive energies and melting points?

46. (a) By taking the derivative of the total potential energy of an ion in a lattice, find an expression for the force on the ion. (b) Suppose an ion is displaced from its equilibrium position by a small distance x, so that $R = R_0 + x$. Show that for small values of x, the force can be written as $F = -kx$. Express k in terms of the other parameters of the crystal. (c) Find the value of k for NaCl and evaluate the oscillation frequency for a sodium ion. (d) Suppose a sodium ion in the lattice absorbed a photon of this frequency and began to oscillate. Find the wavelength of the photon. In what region of the electromagnetic spectrum is this photon?

47. The electric field of a dipole is proportional to $1/r^3$. Assuming that the induced dipole moment of molecule B is proportional to the electric dipole field of molecule A,

show that the van der Waals force is proportional to r^{-7}. (*Hint:* Show that the potential energy of dipole B in the electric field caused by dipole A is proportional to r^{-6}.)

48. (a) Obtain the data for the heat capacity of aluminum between 1 K and 100 K (see, for example, the *Handbook of Chemistry and Physics*). Plot the data, and by trial and error find the value of the Einstein temperature that gives the best fit to the data. (b) Plot the data for temperatures below 10 K as C/T vs. T^2, determine the slope and intercept, and deduce the Debye temperature and effective mass for Al (using $E_F = 11.7$ eV).

49. (a) Obtain the data for the heat capacity of gold between 1 K and 100 K (see, for example, the *Handbook of Chemistry and Physics*). Plot the data, and by trial and error find the value of the Einstein temperature that gives the best fit to the data. (b) Plot the data for temperatures below 10 K as C/T vs. T^2, determine the slope and intercept, and deduce the Debye temperature and effective mass for Au (using $E_F = 5.53$ eV).

50. (a) A Cooper pair in a superconductor can be considered to be a bound state with an energy uncertainty that is of the order of the gap energy E_g. Assuming these pairs exist close to the Fermi energy, find the uncertainty in the location of a Cooper pair, which is a good estimate of its size. (b) Estimate the size of a Cooper pair for aluminum ($E_F = 11.7$ eV) and compare with the lattice spacing in aluminum (0.286 nm).

51. When a material such as germanium is used as a photon detector, an incoming photon makes many interactions and excites many electrons across the gap between the valence and the conduction band. (a) ^{137}Cs emits a 662-keV gamma ray. How many electrons are excited across the 0.66-eV gap of germanium by the absorption of this gamma ray? (b) The number N calculated in part (a) is subject to statistical fluctuations of $\sqrt{N}$. Compute the variation in N and the fractional variation in N. (c) What is the corresponding variation in the measured energy of the gamma ray? This result is the experimental resolution of the detector.

52. On a single graph, sketch the atomic paramagnetic susceptibility as a function of inverse temperature for $J = 1/2, 1$, and $3/2$. Assume all other coefficients ($N/V, g_J, B_{app}$) to be the same for these three cases. Plot the ratio of the susceptibility to its maximum for that spin, so you can compare the variation in the approach to saturation for these three spins.

53. The magnetic field at a distance r from a magnetic dipole μ is $B = \mu_0 \mu / 2\pi r^3$. Show that the dipole-dipole interaction energy is too small to account for the ferromagnetism of iron at all but the lowest temperatures. Assume an effective magnetic dipole moment of 2.2 μ_B per atom.

54. (a) Oxygen gas is observed to be paramagnetic at room temperature, but nitrogen gas is diamagnetic. Explain this based on the filling of the bonding and antibonding $2p$ orbitals in O_2 and N_2. (*Hint:* See Example 9.1.) (b) Would you expect NO gas to be paramagnetic or diamagnetic?

Chapter 12

NUCLEAR STRUCTURE AND RADIOACTIVITY

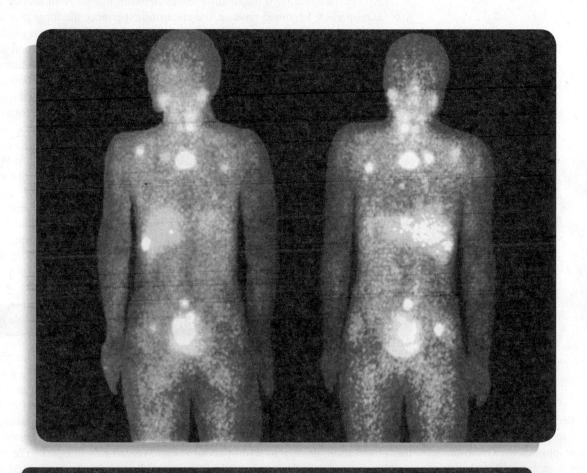

Radioactive isotopes have proven to be valuable tools for medical diagnosis. The photo shows gamma-ray emission from a man who has been treated with a radioactive element. The radioactivity concentrates in locations where there are active cancer tumors, which show as bright areas in the gamma-ray scan. This patient's cancer has spread from his prostate gland to several other locations in his body.

The nucleus lies at the center of the atom, occupying only 10^{-15} of its volume but providing the electrical force that holds the atom together. Within the nucleus there are Z positive charges. To keep these charges from flying apart, the nuclear force must supply an attraction that overcomes their electrical repulsion. This nuclear force is the strongest of the known forces; it provides nuclear binding energies that are millions of times stronger than atomic binding energies.

There are many similarities between atomic structure and nuclear structure, which will make our study of the properties of the nucleus somewhat easier. Nuclei are subject to the laws of quantum physics. They have ground and excited states and emit photons in transitions between the excited states. Just like atomic states, nuclear states can be labeled by their angular momentum.

There are, however, two major differences between the study of atomic and nuclear properties. In atomic physics, the electrons experience the force provided by an external agent, the nucleus; in nuclear physics, there is no such external agent. In contrast to atomic physics, in which we can often consider the interactions among the electrons as a perturbation to the primary interaction between electrons and nucleus, in nuclear physics the mutual interaction of the nuclear constituents is just what provides the nuclear force, so we cannot treat this complicated many-body problem as a correction to a single-body problem. We therefore cannot avoid the mathematical difficulties in the nuclear case, as we did in the atomic case.

The second difference between atomic and nuclear physics is that we cannot write the nuclear force in a simple form like the Coulomb force. There is no closed-form analytical expression that can be written to describe the mutual forces of the nuclear constituents.

In spite of these difficulties, we can learn a great deal about the properties of the nucleus by studying the interactions between different nuclei, the radioactive decay of nuclei, and the properties of some nuclear constituents. In this chapter and the next we describe these studies and how we learn about the nucleus from them.

12.1 NUCLEAR CONSTITUENTS

The work of Rutherford, Bohr, and their contemporaries in the years between 1911 and 1920 showed that the positive charge of the atom is confined in a very small nuclear region at the center of the atom, that the nucleus in an atom of atomic number Z has a charge of $+Ze$, and that the nucleus provides most (99.9%) of the atomic mass. It was also known that the masses of the atoms (measured in atomic mass units) were very nearly integers; a glance at Appendix D confirms this observation, usually to within about 0.1%. We call this integer A the *mass number*. It was therefore reasonable to suppose that nuclei are composed of a number A of more fundamental units whose mass is very close to 1 u. Because the only particle known at that time with a mass close to 1 u was the proton (the nucleus of hydrogen, with a mass of 1.0073 u and a charge of $+e$), it was postulated (incorrectly, as we shall see) that the nucleus of an atom of mass number A contained A protons.

Such a nucleus would have a nuclear charge of Ae rather than Ze; because $A > Z$ for all atoms heavier than hydrogen, this model gives too much positive charge to the nucleus. This difficulty was removed by the *proton-electron model*, in which it was postulated (again incorrectly) that the nucleus also contained $(A - Z)$ electrons. Under these assumptions, the nuclear mass would be about

Enrico Fermi (1901–1954, Italy-United States). There is hardly a field of modern physics to which he did not make contributions in theory or experiment. He developed the statistical laws for spin-½ particles, and in the 1930s he proposed a theory of beta decay that is still used today. He was the first to demonstrate the transmutation of elements by neutron bombardment (for which he received the 1938 Nobel Prize), and he directed the construction of the first nuclear reactor.

A times the mass of the proton (because the mass of the electrons is negligible) and the nuclear electric charge would be $A(+e) + (A - Z)(-e) = Ze$, in agreement with experiment. However, this model leads to several difficulties. First, as we discovered in Chapter 4 (see Example 4.7), the presence of electrons in the nucleus is not consistent with the uncertainty principle, which would require those electrons to have unreasonably large (~ 19 MeV) kinetic energies.

A more serious problem concerns the total *intrinsic spin* of the nucleus. From measurements of the *very* small effect of the nuclear magnetic moment on the atomic transitions (called the *hyperfine splitting*), we know that the proton has an intrinsic spin of $1/2$, just like the electron. Consider an atom of deuterium, sometimes known as "heavy hydrogen." It has a nuclear charge of $+e$, just like ordinary hydrogen, but a mass of two units, twice that of ordinary hydrogen. The proton-electron nuclear model would then require that the deuterium *nucleus* contain two protons and one electron, giving a net mass of two units and a net charge of one. Each of these three particles has a spin of $1/2$, and the rules for adding angular momenta in quantum mechanics would lead to a spin of deuterium of either $1/2$ or $3/2$. However, the measured total spin of deuterium is 1. For these and other reasons, the hypothesis that electrons are a nuclear constituent must be discarded.

The resolution of this dilemma came in 1932 with the discovery of the *neutron*, a particle of roughly the same mass as the proton (actually about 0.1% more massive) but having no electric charge. According to the *proton-neutron* model, a nucleus consists of Z protons and $(A - Z)$ neutrons, giving a total charge of Ze and a total mass of roughly A times the mass of the proton, because the proton and neutron masses are roughly the same.

The proton and neutron are, except for their electric charges, very similar to one another, and so they are classified together as *nucleons*. Some properties of the two nucleons are listed in Table 12.1.

The chemical properties of any element depend on its atomic number Z, but not on its mass number A. It is possible to have two different nuclei with the same Z but with different A (that is, with the same number of protons but different numbers of neutrons). Atoms of these nuclei are identical in all their chemical properties, differing only in mass and in those properties that depend on mass. Nuclei with the same Z but different A are called *isotopes*. Hydrogen, for example, has three isotopes: ordinary hydrogen ($Z = 1, A = 1$), deuterium ($Z = 1, A = 2$), and tritium ($Z = 1, A = 3$). All of these are indicated by the chemical symbol H. When we discuss nuclear properties it is important to distinguish among the different isotopes. We do this by indicating, along with the chemical symbol, the atomic number Z, the mass number A, and the *neutron number $N = A - Z$* in the following format:

$$_Z^A X_N$$

where X is any chemical symbol. The chemical symbol and the atomic number Z give the same information, so it is not necessary to include both of them in the isotope label. Also, if we specify Z then we don't need to specify *both* N and

TABLE 12.1 Properties of the Nucleons

Name	Symbol	Charge	Mass	Rest Energy	Spin
Proton	p	$+e$	1.007276 u	938.28 MeV	$1/2$
Neutron	n	0	1.008665 u	939.57 MeV	$1/2$

A. It is sufficient to give only the chemical symbol and *A*. The three isotopes of hydrogen would be indicated as $^1_1H_0$, $^2_1H_1$, and $^3_1H_2$, or more compactly as ^1H, ^2H, and ^3H. In Appendix D you will find a list of isotopes and some of their properties.

Example 12.1

Give the symbol for the following: (*a*) the isotope of helium with mass number 4; (*b*) the isotope of tin with 66 neutrons; (*c*) an isotope with mass number 235 that contains 143 neutrons.

Solution

(*a*) From the periodic table, we find that helium has $Z = 2$. With $A = 4$, we have $N = A - Z = 2$. Thus the symbol would be $^4_2He_2$ or ^4He.

(*b*) Again from the periodic table, we know that for tin (Sn), $Z = 50$. We are given $N = 66$, so $A = Z + N = 116$. The symbol is $^{116}_{50}Sn_{66}$ or ^{116}Sn.

(*c*) Given that $A = 235$ and $N = 143$, we know that $Z = A - N = 92$. From the periodic table, we find that this element is uranium, and so the proper symbol for this isotope is $^{235}_{92}U_{143}$ or ^{235}U.

12.2 NUCLEAR SIZES AND SHAPES

Like atoms, nuclei lack a hard surface or an easily definable radius. In fact, different types of experiments can often reveal different values of the radius for the same nucleus.

From a variety of experiments, we know some general features of the nuclear density. Its variation with the nuclear radius is shown in Figure 12.1. Because the nuclear force is the strongest of the forces, we might expect that this strong force would cause the protons and neutrons to congregate at the center of the nucleus, giving an increasing density in the central region. However, Figure 12.1 shows that this is not the case—the density remains quite uniform. This gives some important clues about the short range of the nuclear force, as we discuss in Section 12.4.

Another interesting feature of Figure 12.1 is that the density of a nucleus seems not to depend on the mass number *A*; very light nuclei, such as ^{12}C, have roughly the same central density as very heavy nuclei, such as ^{209}Bi. Stated another way, the number of protons and neutrons per unit volume is approximately constant over the entire range of nuclei:

$$\frac{\text{number of neutrons and protons}}{\text{volume of nucleus}} = \frac{A}{\frac{4}{3}\pi R^3} \cong \text{constant}$$

assuming the nucleus to be a sphere of radius *R*. Thus $A \propto R^3$, which suggests a proportionality between the nuclear radius *R* and the cube root of the mass number: $R \propto A^{1/3}$ or, defining a constant of proportionality R_0,

$$R = R_0 A^{1/3} \tag{12.1}$$

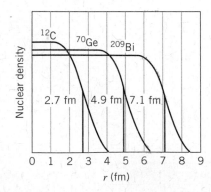

FIGURE 12.1 The radial dependence of the nuclear charge density.

The constant R_0 must be determined by experiment, and a typical experiment might be to scatter charged particles (alpha particles or electrons, for example) from the nucleus and to infer the radius of the nucleus from the distribution of scattered particles. From such experiments, we know the value of R_0 is approximately 1.2×10^{-15} m. (The exact value depends, as in the case of atomic physics, on exactly how we define the radius, and values of R_0 usually range from 1.0×10^{-15} m to 1.5×10^{-15} m.) The length 10^{-15} m is 1 femtometer (fm), but physicists often refer to this length as one fermi, in honor of the Italian-American physicist Enrico Fermi.

Example 12.2

Compute the approximate nuclear radius of carbon ($A = 12$), germanium ($A = 70$), and bismuth ($A = 209$).

Solution
Using Eq. 12.1, we obtain:

Carbon: $R = R_0 A^{1/3} = (1.2 \text{ fm})(12)^{1/3} = 2.7 \text{ fm}$

Germanium: $R = R_0 A^{1/3} = (1.2 \text{ fm})(70)^{1/3} = 4.9 \text{ fm}$

Bismuth: $R = R_0 A^{1/3} = (1.2 \text{ fm})(209)^{1/3} = 7.1 \text{ fm}$

As you can see from Figure 12.1, these values define the mean radius, the point at which the density falls to half the central value.

Example 12.3

Compute the density of a typical nucleus, and find the resultant mass if we could produce a nucleus with a radius of 1 cm.

Solution
Making a rough estimate of the nuclear mass m as A times the proton mass, we have

$$\rho = \frac{m}{V} = \frac{A m_p}{\frac{4}{3}\pi R^3} = \frac{A m_p}{\frac{4}{3}\pi R_0^3 A}$$

$$= \frac{1.67 \times 10^{-27} \text{ kg}}{\frac{4}{3}\pi (1.2 \times 10^{-15} \text{ m})^3} = 2 \times 10^{17} \text{ kg/m}^3$$

The mass of a hypothetical nucleus with a 1-cm radius would be

$$m = \rho V = \rho (\tfrac{4}{3}\pi R^3)$$

$$= (2 \times 10^{17} \text{ kg/m}^3)(\tfrac{4}{3}\pi)(0.01 \text{ m})^3$$

$$= 8 \times 10^{11} \text{ kg}$$

about the mass of a 1-km sphere of ordinary matter!

The result of Example 12.3 shows the great density of what physicists call *nuclear matter*. Although examples of such nuclear matter in bulk are not found on Earth (a sample of nuclear matter the size of a large building would have a mass as great as that of the entire Earth), they are found in certain massive stars, in which the gravitational force causes protons and electrons to merge into neutrons, creating a neutron star (see Section 10.7) that is in effect a giant atomic nucleus!

One way of measuring the size of a nucleus is to scatter charged particles, such as alpha particles, as in Rutherford scattering experiments. As long as the alpha particle is outside the nucleus, the Rutherford scattering formula holds, but when the distance of closest approach is less than the nuclear radius, deviations from the Rutherford formula occur. Figure 12.2 shows the results of a Rutherford scattering experiment in which such deviations are observed. (Problem 33 suggests how a value for the nuclear radius can be inferred from these data.)

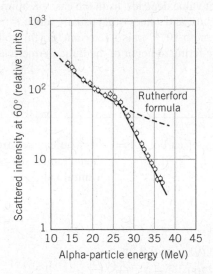

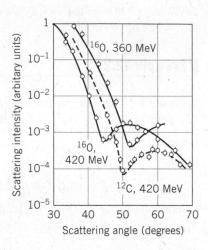

FIGURE 12.2   Deviations from the Rutherford formula in scattering from ^{208}Pb are observed for alpha-particle energies above 27 MeV.

FIGURE 12.3   Diffraction of 360-MeV and 420-MeV electrons by ^{12}C and ^{16}O nuclei.

Other scattering experiments can also be used to measure the nuclear radius. Figure 12.3 shows a sort of "diffraction pattern" that results from the scattering of energetic electrons by a nucleus. In each case the first diffraction minimum is clearly visible. (The intensity at the minimum doesn't fall to zero because the nuclear density doesn't have a sharp edge, as illustrated in Figure 12.1.) For scattering of radiation of wavelength λ by a circular disc of diameter D, the first diffraction minimum should appear at an angle of $\theta = \sin^{-1}(1.22\lambda/D)$. (Review Example 4.2 for another example of this calculation.) At an electron energy of 420 MeV, the observed minima for ^{16}O and ^{12}C give a radius of 2.6 fm for ^{16}O and 2.3 fm for ^{12}C (see Problem 34), in agreement with the values 3.0 fm and 2.7 fm computed from Equation 12.1.

12.3 NUCLEAR MASSES AND BINDING ENERGIES

Suppose we have a proton and an electron at rest separated by a large distance. The total energy of this system is the total rest energy of the two particles, $m_p c^2 + m_e c^2$. Now we let the two particles come together to form a hydrogen atom in its ground state. In the process, several photons are emitted, the *total* energy of which is 13.6 eV. The total energy of this system is the rest energy of the hydrogen atom, $m(H)c^2$, plus the total photon energy, 13.6 eV. Conservation of energy demands that the total energy of the system of isolated particles must equal the total energy of atom plus photons: $m_e c^2 + m_p c^2 = m(H)c^2 + 13.6$ eV, which we write as

$$m_e c^2 + m_p c^2 - m(H)c^2 = 13.6 \text{ eV}$$

That is, the rest energy of the combined system (the hydrogen atom) is less than the rest energy of its constituents (an electron and a proton) by 13.6 eV. This energy difference is the *binding energy* of the atom. We can regard the binding energy as either the "extra" energy we obtain when we assemble an atom from its components or else the energy we must supply to disassemble the atom into its components.

Nuclear binding energies are calculated in a similar way. Consider, for example, the nucleus of deuterium, $^2_1\text{H}_1$, which is composed of one proton and one neutron. The nuclear binding energy of deuterium is the difference between the total rest energy of the constituents and the rest energy of their combination:

$$B = m_n c^2 + m_p c^2 - m_D c^2 \tag{12.2}$$

where m_D is the mass of the deuterium nucleus. To finish the calculation, we replace the nuclear masses m_p and m_D with the corresponding *atomic* masses: $m(^1\text{H})c^2 = m_p c^2 + m_e c^2 - 13.6\,\text{eV}$ and $m(^2\text{H})c^2 = m_D c^2 + m_e c^2 - 13.6\,\text{eV}$. Substituting into Eq. 12.2, we obtain

$$B = m_n c^2 + [m(^1\text{H})c^2 - m_e c^2 + 13.6\,\text{eV}] - [m(^2\text{H})c^2 - m_e c^2 + 13.6\,\text{eV}]$$
$$= [m_n + m(^1\text{H}) - m(^2\text{H})]c^2$$

Notice that the electron mass cancels in this calculation. For deuterium, we then have

$$B = (1.008665\,\text{u} + 1.007825\,\text{u} - 2.014102\,\text{u})(931.5\,\text{MeV/u}) = 2.224\,\text{MeV}$$

Here we use $c^2 = 931.5\,\text{MeV/u}$ to convert mass units to energy units.

Let's generalize this process to calculate the binding energy of a nucleus X of mass number A with Z protons and N neutrons. Let m_X represent the mass of this nucleus. Then the binding energy of the nucleus is, by analogy with Eq. 12.2, the difference between the nuclear rest energy and the total rest energy of its constituents (N neutrons and Z protons):

$$B = N m_n c^2 + Z m_p c^2 - m_X c^2 \tag{12.3}$$

In order to use tabulated atomic masses to do this calculation, we must replace the nuclear mass m_X with its corresponding atomic mass: $m(^A_Z\text{X}_N)c^2 = m_X c^2 + Z m_e c^2 - B_e$, where B_e represents the total binding energy of all the electrons in this atom. Nuclear rest energies are of the order of 10^9 to 10^{11} eV, total electron rest energies are of the order of 10^6 to 10^8 eV, and electron binding energies are of the order of 1 to 10^5 eV. Thus B_e is very small compared with the other two terms, and we can safely neglect it to the accuracy we need for these calculations.

Substituting atomic masses for the nuclear masses m_p and m_X, we obtain an expression for the total binding energy of any nucleus $^A_Z\text{X}_N$:

$$B = [N m_n + Z m(^1_1\text{H}_0) - m(^A_Z\text{X}_N)]c^2 \tag{12.4}$$

The electron masses cancel in this equation, because on the right side we have the difference between the mass of Z hydrogen atoms (with a total of Z electrons) and the atom of atomic number Z (also with Z electrons). The masses that appear in Eq. 12.4 are *atomic* masses.

Example 12.4

Find the total binding energy B and also the average binding energy per nucleon B/A for $^{56}_{26}Fe_{30}$ and $^{238}_{92}U_{146}$.

Solution

From Eq. 12.4, for $^{56}_{26}Fe_{30}$ with $N = 30$ and $Z = 26$,

$$B = [30(1.008665\,u) + 26(1.007825\,u)$$
$$- 55.934937\,u](931.5\,MeV/u)$$
$$= 492.3\,MeV$$

$$\frac{B}{A} = \frac{492.3\,MeV}{56} = 8.790\,MeV \text{ per nucleon}$$

For $^{238}_{92}U_{146}$,

$$B = [146(1.008665\,u) + 92(1.007825\,u)$$
$$- 238.050788\,u](931.5\,MeV/u)$$
$$= 1802\,MeV$$

$$\frac{B}{A} = \frac{1802\,MeV}{238} = 7.570\,MeV \text{ per nucleon}$$

Example 12.4 gives us insight into an important aspect of nuclear structure. The values of B/A show that the nucleus ^{56}Fe is *relatively* more tightly bound than the nucleus ^{238}U; the average binding energy *per nucleon* is greater for ^{56}Fe than for ^{238}U. Alternatively, this calculation shows that, given a large supply of protons and neutrons, we would release more energy by assembling those nucleons into nuclei of ^{56}Fe than we would by assembling them into nuclei of ^{238}U.

Repeating this calculation for the entire range of nuclei, we obtain the results shown in Figure 12.4. The binding energy per nucleon starts at small values

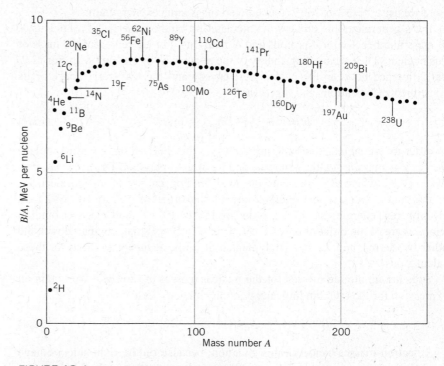

FIGURE 12.4 The binding energy per nucleon.

(0 for the proton and neutron, 1.11 MeV for deuterium), rises to a maximum of 8.795 MeV for ^{62}Ni, and then falls to values of around 7.5 MeV for the heavy nuclei.

The binding energy per nucleon is roughly constant over a fairly wide range of nuclei. From the region around $A = 60$ there is a sharp decrease for light nuclei, which is caused by their having an increasing relative fraction of loosely bound protons and neutrons on the surface. There is a gradual decrease for the more massive nuclei, due to the increasing Coulomb repulsion of the protons.

Figure 12.4 suggests that we can liberate energy from the nucleus in two different ways. If we split a massive nucleus (say, $A > 200$) into two lighter nuclei, energy is released, because the binding energy per nucleon is greater for the two lighter fragments than it is for the original nucleus. This process is known as *nuclear fission*. Alternatively, we could combine two light nuclei ($A < 10$, for example) into a more massive nucleus; again, energy is released when the binding energy per nucleon is greater in the final nucleus than it is in the two original nuclei. This process is known as *nuclear fusion*. We consider fission and fusion in greater detail in Chapter 13.

Proton and Neutron Separation Energies

If we add the ionization energy E_i (13.6 eV) to a hydrogen atom, we obtain a hydrogen ion H$^+$ and a free electron. In terms of the rest energies of the particles, we can write this process as $E_i + m(\text{H})c^2 = m(\text{H}^+)c^2 + m_e c^2$. If we generalize to an arbitrary element X, this becomes $E_i + m(\text{X})c^2 = m(\text{X}^+)c^2 + m_e c^2$, or

$$\text{X} \rightarrow \text{X}^+ + e^-: \quad E_i = m(\text{X}^+)c^2 + m_e c^2 - m(\text{X})c^2 = [m(\text{X}^+) + m_e - m(\text{X})]c^2$$

In the case of element X, the ionization energy gives the smallest amount of energy necessary to remove an electron from an atom, and we saw in Figure 8.8 how the ionization energy provides important information about the properties of atoms.

For nuclei, a process similar to ionization consists of removing the least tightly bound proton or neutron from the nucleus. The energy required to remove the least tightly bound proton is called the *proton separation energy* S_p. If we add energy S_p to a nucleus $^A_Z X_N$, we obtain the nucleus $^{A-1}_{Z-1} X'_N$ and a free proton. In analogy with the atomic case, we can write the separation energy as

$$^A_Z X_N \rightarrow {}^{A-1}_{Z-1} X'_N + \text{p}: \quad S_p = [m(^{A-1}_{Z-1}X'_N) + m(^1\text{H}) - m(^A_Z X_N)]c^2 \quad (12.5)$$

using atomic masses. Similarly, if we add the *neutron separation energy* S_n to nucleus $^A_Z X_N$, we obtain the nucleus $^{A-1}_{Z} X_{N-1}$ and a free neutron:

$$^A_Z X_N \rightarrow {}^{A-1}_{Z} X_{N-1} + \text{n}: \quad S_n = [m(^{A-1}_{Z}X_{N-1}) + m_n - m(^A_Z X_N)]c^2 \quad (12.6)$$

Proton and neutron separation energies are typically in the range of 5–10 MeV. It is no coincidence that this energy is about the same as the average binding energy per nucleon. The total binding energy B of a nucleus is the energy needed to take it apart into Z free protons and N free neutrons. This energy is the sum of A proton and neutron separation energies.

Example 12.5

Find the proton separation energy and the neutron separation energy of ^{125}Te.

Solution

To separate a proton, $^{125}\text{Te} \rightarrow {}^{124}\text{Sb} + p$. Using Eq. 12.5, the proton separation energy is

$$S_p = [m(^{124}\text{Sb}) + m(^1\text{H}) - m(^{125}\text{Te})]c^2$$
$$= (123.905936 \text{ u} + 1.007825 \text{ u}$$
$$- 124.904431 \text{ u})(931.50 \text{ MeV/u})$$
$$= 8.691 \text{ MeV}$$

For neutron separation, $^{125}\text{Te} \rightarrow {}^{124}\text{Te} + n$. The neutron separation energy is (from Eq. 12.6)

$$S_n = [m(^{124}\text{Te}) + m_n - m(^{125}\text{Te})]c^2$$
$$= (123.902818 \text{ u} + 1.008665 \text{ u}$$
$$- 124.904431 \text{ u})(931.50 \text{ MeV/u})$$
$$= 6.569 \text{ MeV}$$

The proton and neutron separation energies play a role in nuclei similar to that of the ionization energy in atoms. Figure 12.5 shows a plot of the neutron separation energies of nuclei with a "valence" neutron (and no valence proton) from $Z = 36$ to $Z = 62$. As we add neutrons, the neutron separation energy decreases smoothly except near $N = 50$ and $N = 82$, where there are more sudden decreases in the separation energy. In analogy with atomic physics (see Figure 8.8), these sudden decreases are associated with the filling of shells. The motions of neutrons and protons in the nucleus are described in terms of a shell structure that is similar to that of atomic shells, and when a neutron or proton is placed into a new shell it is less tightly bound and its separation energy decreases. The neutron separation data indicate that there are closed neutron shells at $N = 50$ and $N = 82$. Relationships such as Figure 12.5 provide important information about the shell structure of nuclei.

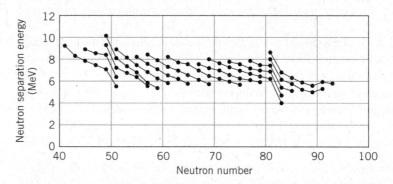

FIGURE 12.5 The neutron separation energy. The lines connect isotopes of the same element that have an odd neutron, starting on the left at $Z = 36$ and ending on the right at $Z = 62$.

12.4 THE NUCLEAR FORCE

Our successful experience with using the simplest atom, hydrogen, to gain insights into atomic structure suggests that we should begin our study of the nuclear force by looking at the simplest system in which that force operates—the deuterium nucleus, which consists of one proton and one neutron. For example, we might hope to learn something about the nuclear force from the photons emitted in transitions between the excited states of this nucleus. Unfortunately, this strategy does not work—deuterium has *no nuclear excited states*. When we bring a proton and an electron together to form a hydrogen atom, many photons may be emitted as the electron drops into its ground state; from this spectrum we learn the energies of the excited states. When we bring a proton and a neutron together to form a deuterium nucleus, only one photon (of energy 2.224 MeV) is emitted as the system drops directly into its ground state.

Even though we can't use the excited states of deuterium, we can learn about the nuclear force in the proton-neutron system by scattering neutrons from protons as well as by doing a variety of different experiments with heavier nuclei. From these experiments we have learned the following characteristics of the nuclear force:

1. The nuclear force is the strongest of the known forces, and so it is sometimes called the *strong* force. For two adjacent protons in a nucleus, the nuclear interaction is 10–100 times stronger than the electromagnetic interaction.

2. The strong nuclear force has a very short range—the distance over which the force acts is limited to about 10^{-15} m. This conclusion follows from the constant central density of nuclear matter (Figure 12.1). As we add nucleons to a nucleus, each added nucleon feels a force only from its nearest neighbors, and *not* from all the other nucleons in the nucleus. In this respect, a nucleus behaves somewhat like a crystal, in which each atom interacts primarily with its nearest neighbors, and additional atoms make the crystal larger but don't change its density. Another piece of evidence for the short range comes from Figure 12.4. Because the binding energy per nucleon is roughly constant, total nuclear binding energies are roughly proportional to A. For a force with long range (such as the gravitational and electrostatic forces, which have infinite range) the binding energy is roughly proportional to the square of the number of interacting particles. (For example, because each of the Z protons in a nucleus feels the repulsion of the other $Z - 1$ protons, the total electrostatic energy of the nucleus is proportional to $Z(Z - 1)$, which is roughly Z^2 for large Z.)

   Figure 12.6 illustrates the dependence of the nuclear binding energy on the separation distance between the nucleons. The binding energy is relatively constant for separation distances less than about 1 fm, and it is zero for separation distances much greater than 1 fm.

3. The nuclear force between any two nucleons does not depend on whether the nucleons are protons or neutron—the n-p nuclear force is the same as the n-n nuclear force, which is in turn the same as the nuclear portion of the p-p force.

A successful model for the origin of this short-range force is the *exchange force*. Suppose we have a neutron and a proton next to one another in the nucleus. The neutron emits a particle, on which it exerts a strong attractive force. The proton also exerts a strong force on the particle, perhaps strong enough to absorb the particle. The proton then emits a particle that can be absorbed by the neutron. The proton and neutron each exert a strong force on the exchanged particle, and thus they appear to exert a strong force on each other. The situation is similar to that shown in Figure 12.7, in which two people play catch with a ball to which each is attached by a spring. Each player exerts a force on the ball, and the effect is as if each exerted a force on the other.

How can a neutron of rest energy $m_n c^2$ emit a particle of rest energy mc^2 and still remain a neutron, without violating conservation of energy? The answer to this question can be found from the uncertainty principle, $\Delta E \Delta t \sim \hbar$.

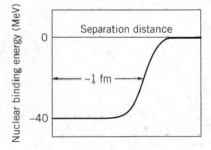

FIGURE 12.6 Dependence of nuclear binding energy on the separation distance of nucleons.

FIGURE 12.7 An attractive exchange force.

We don't know that energy has been conserved unless we measure it, and we can't measure it more accurately than the uncertainty ΔE in a time interval Δt. We can therefore "violate" energy conservation by an amount ΔE for a time interval of at most $\Delta t = \hbar/\Delta E$. The amount by which energy conservation is violated in our exchange force model is mc^2, the rest energy of the exchanged particle. This particle can thus exist only for a time interval (in the laboratory frame) of at most

$$\Delta t = \frac{\hbar}{mc^2} \qquad (12.7)$$

The longest distance this particle can possibly travel in the time Δt is $x = c\Delta t$, since it can't move faster than the speed of light. With $x = c\Delta t = c\hbar/mc^2$, we then have a relationship between the maximum range of the exchange force and the rest energy of the exchanged particle:

$$mc^2 = \frac{\hbar c}{x} \qquad (12.8)$$

Inserting into this expression an estimate for the range of the nuclear force of 10^{-15} m or 1 fm, we can estimate the rest energy of the exchanged particle:

$$mc^2 = \frac{\hbar c}{x} = \frac{200 \text{ MeV} \cdot \text{fm}}{1 \text{ fm}} = 200 \text{ MeV}$$

The exchanged particle cannot be observed in the laboratory during the exchange, for to do so would violate energy conservation. However, if we provide energy to the nucleons from an external source (for example, by causing a nucleus to absorb a photon), the "borrowed" energy can be repaid and the particle can be observed. When we carry out this experiment, the nucleus is found to emit pi mesons (pions), which have a rest energy of 140 MeV, remarkably close to our estimate of 200 MeV. Many observable properties of the nuclear force have been successfully explained by a model based on the exchange of pions. We discuss the properties of pions in Chapter 14. Other exchanged particles contribute to different aspects of the nuclear force. For example, an exchanged particle is responsible for the repulsive part of the force at very short range, which keeps the nucleons from all collapsing toward the center of the nucleus (see Problem 10).

12.5 QUANTUM STATES IN NUCLEI

Ideally we would like to solve the Schrödinger equation using the nuclear potential energy. This process, if it were possible, would give us a set of energy levels for the protons and neutrons that we could then compare with experiment (just as we did for the energy levels of electrons in atoms). Unfortunately, we cannot carry through with this program for several reasons: the nuclear potential energy cannot be expressed in a convenient analytical form, and it is not possible to solve the nuclear many-body problem except by approximation.

Nevertheless, we can make some simplifications that allow us to analyze the structure and properties of nuclei by using techniques already introduced in this book. We'll represent the nuclear potential energy as a finite potential well of radius R equal to the nuclear radius. That confines the nucleons to a nucleus-sized region and allows them to move freely inside that region.

Let's consider ^{125}Te (a nucleus very close to the center of the range of nuclei), which we analyzed in Example 12.5. The width of the potential energy well is equal to the nuclear radius, which we find from Eq. 12.1: $R = (1.2 \text{ fm})(125)^{1/3} = 6.0 \text{ fm}$. The second quantity we need to know is the depth of the potential energy well. We'll consider the neutrons and protons separately. The 73 neutrons in ^{125}Te will fill a series of energy levels in the potential energy well. The top of the well is at $E = 0$ (above which the neutrons would become free). The bottom of the well is at a negative energy $-U_0$. The neutrons fill the levels in the well starting at $-U_0$ and ending *not* at energy zero but at energy -6.6 MeV, as we found in Example 12.5. That is, we must add a minimum of 6.6 MeV to raise the least tightly bound neutron out of the well and turn it into a free neutron.

To find the energy difference between the bottom of the well and the highest filled state, we can consider the nucleus to be a "gas" of neutrons and protons whose energies are described by the Fermi-Dirac distribution (Chapter 10). A statistical distribution is intended to describe systems with large numbers of particles, but it should be a reasonable rough approximation for our "gas" of 73 neutrons. To find the energy of the highest filled state, we need the Fermi energy of the neutrons (using Eq. 10.50 with $V = \frac{4}{3}\pi R^3 = 900 \text{ fm}^3$ as the volume of the ^{125}Te nucleus):

$$E_F = \frac{h^2}{2m}\left(\frac{3N}{8\pi V}\right)^{2/3} = \frac{h^2 c^2}{2mc^2}\left(\frac{3N}{8\pi V}\right)^{2/3}$$
$$= \frac{(1240 \text{ MeV} \cdot \text{fm})^2}{2(940 \text{ MeV})}\left[\frac{3(73)}{8\pi(900 \text{ fm}^3)}\right]^{2/3} = 37.0 \text{ MeV}$$

Figure 12.8 shows the resulting potential energy well for neutrons. The depth of the well is the sum of the neutron separation energy S_n and the Fermi energy: $U_0 = S_n + E_F = 6.6 \text{ MeV} + 37.0 \text{ MeV} = 43.6 \text{ MeV}$.

A similar calculation for the 52 protons in ^{125}Te gives $E_F = 29.5 \text{ MeV}$. For the protons, $S_p + E_F = 8.7 \text{ MeV} + 29.5 \text{ MeV} = 38.2 \text{ MeV}$, much less than the depth we determined for the neutron well. The difference between the depths of the neutron and proton wells is due to the Coulomb repulsion energy of the protons, which makes the protons less tightly bound than the neutrons. Figure 12.9 gives a representation of the proton states in their potential energy well.

Quantum States and Radioactive Decay

Figure 12.10 shows the protons and neutrons near the top of their potential energy wells. Note that we can add energy to the nucleus that is less than the proton or neutron separation energies. In the region between $E = -S_n$ or $-S_p$ and $E = 0$ are the nuclear excited states in which a proton or a neutron can absorb energy and move from its ground state to one of the unoccupied higher states. As was the case with atoms, the nucleus can make transitions from excited states to lower excited states or to the ground state by photon emission. In the case of nuclei, those photons are called *gamma rays* and have typical energies of 0.1 MeV to a few MeV.

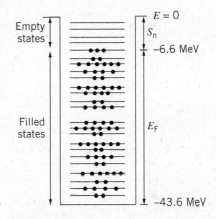

FIGURE 12.8   Neutron states in a potential energy well for the 73 neutrons of ^{125}Te.

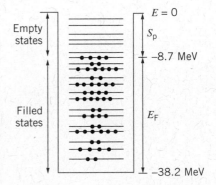

FIGURE 12.9 Proton states in a potential energy well for the 52 protons of ^{125}Te.

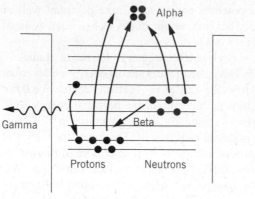

FIGURE 12.10 Proton and neutron states near the top of the well for ^{125}Te. Alpha decay is represented by two protons and two neutrons being boosted from negative energy bound states to positive energy free states and forming an alpha particle. Beta decay is represented by a neutron transforming into a proton. Gamma decay can occur among the empty states above the highest occupied proton and neutron states.

It is also possible to have other nuclear transformations that can be represented in Figure 12.10. It is clearly not possible for this nucleus spontaneously to emit a proton or a neutron—we have seen that it takes many MeV to boost a bound proton or neutron to a free state. However, it is possible simultaneously to boost *two* protons and *two* neutrons and form them into an alpha particle ($^4_2$He$_2$). If the energy gained in the formation of the alpha particle (its binding energy, 28.3 MeV) is greater than the sum of the four separation energies, there will be a net energy gain in the process; this energy can appear as the kinetic energy of the alpha particle that is emitted by the nucleus. This process is called *nuclear alpha decay*. You can see from the neutron and proton separation energies that this process does not occur for ^{125}Te.

Another type of transformation occurs under certain circumstances when a neutron changes into a proton and drops into one of the empty proton states at lower energy. Under other circumstances, in which the proton levels are higher and the neutron levels are lower, a proton can transform into a neutron and drop into one of the empty neutron states. This process is called *nuclear beta decay*. It is not always obvious from diagrams such as Figure 12.10 whether this type of transformation will occur, because changing a neutron to a proton increases the net Coulomb energy of the nucleus and thus increases the energy of all of the proton states. Neither the neutron-to-proton nor the proton-to-neutron transformation can occur for ^{125}Te.

12.6 RADIOACTIVE DECAY

Figure 12.11 shows a plot of all the known nuclei, with stable nuclei indicated by dark shading. For the lighter stable nuclei, the neutron and proton numbers are roughly equal. However, for the heavy stable nuclei, the factor $Z(Z - 1)$ in the Coulomb repulsion energy grows rapidly, so extra neutrons are required to supply the additional binding energy needed for stability. For this reason, all heavy stable nuclei have $N > Z$.

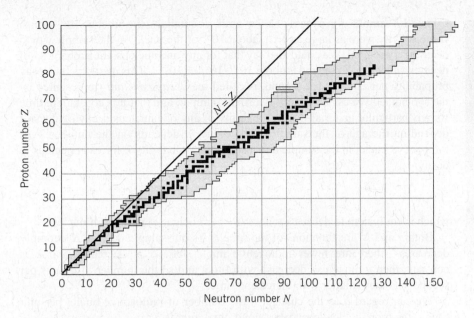

FIGURE 12.11 Stable nuclei are shown in color; known radioactive nuclei are in light shading.

Most of the nuclei represented in Figure 12.11 are unstable, which means that they transform themselves into more stable nuclei by changing their Z and N through *alpha decay* (emission of ^{4}He) or *beta decay* (changing a neutron to a proton or a proton to a neutron). Nuclei are unstable in excited states, which can transition to ground states through *gamma decay* (emission of photons). The three decay processes (alpha, beta, and gamma decay) are examples of the general subject of *radioactive decay*. In the remainder of this section, we establish some of the basic properties of radioactive decay, and in the following sections we treat alpha, beta, and gamma decay separately.

The rate at which unstable radioactive nuclei decay in a sample of material is called the *activity* of the sample. The greater the activity, the more nuclear decays per second. (The activity has nothing to do with the *kind* of decays or of radiations emitted by the sample, or with the *energy* of the emitted radiations. The activity is determined only by the *number* of decays per second.)

The basic unit for measuring activity is the *curie*.* Originally, the curie was defined as the activity of one gram of radium; that definition has since been replaced by a more convenient one:

$$1 \text{ curie (Ci)} = 3.7 \times 10^{10} \text{ decays/s}$$

One curie is quite a large activity, and so we work more often with units of millicurie (mCi), equal to 10^{-3} Ci, and microcurie (μCi), equal to 10^{-6} Ci.

---

*The SI unit of activity is the becquerel (Bq), named for Henri Becquerel, the French scientist who discovered radioactivity in 1896. One becquerel equals one decay/s, so 1 Ci = 3.7×10^{10}Bq.

Marie Curie (1867–1934, Poland-France). Her pioneering studies of the natural radioactivity of radium and other elements earned her two Nobel Prizes, the physics prize in 1903 for the discovery of radioactivity (shared with Henri Becquerel and with her husband, Pierre) and the unshared chemistry prize in 1911 for the isolation of pure radium. She established the Institute of Radium at the University of Paris, where she continued to pursue research in the medical applications of radioactive materials. Her daughter Irene was awarded the 1935 Nobel Prize in chemistry for the discovery of artificial radioactivity.

Consider a sample with a mass of a few grams, containing the order of 10^{23} atoms. If the activity were as large as 1 Ci, about 10^{10} of the nuclei in the sample would decay every second. We could also say that for any one nucleus, the probability of decaying during each second is about $10^{10}/10^{23}$ or 10^{-13}. This quantity, the decay probability per nucleus per second, is called the *decay constant* (represented by the symbol λ). We assume that λ is a small number, and that it is constant in time for any particular material—the probability of any one nucleus decaying doesn't depend on the age of the sample. The activity a depends on the number N of radioactive nuclei in the sample and also on the probability λ for each nucleus to decay:

$$a = \lambda N \qquad (12.9)$$

which is equivalent to decays/s = decays/s per nucleus × number of nuclei.

Both a and N are functions of the time t. As our sample decays, N certainly decreases—there are fewer radioactive nuclei left. If N decreases and λ is constant, then a must also decrease with time, and so the number of decays per second becomes smaller with increasing time.

We can regard a as the change in the number of radioactive nuclei per unit time—the more nuclei decay per second, the larger is a.

$$a = -\frac{dN}{dt} \qquad (12.10)$$

A minus sign must be present because dN/dt is negative (N is decreasing with time), and we want a to be a positive number.) Combining Eqs. 12.9 and 12.10 we have $dN/dt = -\lambda N$, or

$$\frac{dN}{N} = -\lambda \, dt \qquad (12.11)$$

This equation can be integrated directly to yield

$$N = N_0 e^{-\lambda t} \qquad (12.12)$$

where N_0 represents the number of radioactive nuclei originally present at $t = 0$. Equation 12.12 is the *exponential law of radioactive decay*, which tells us how the number of radioactive nuclei in a sample decreases with time. We can't easily measure N, but we can put this equation in a more useful form by multiplying on both sides by λ, which gives

$$a = a_0 e^{-\lambda t} \qquad (12.13)$$

where a_0 is the original activity ($a_0 = \lambda N_0$).

Suppose we count the number of decays of our sample in one second (by counting for one second the radiations resulting from the decays). Repeating the measurement, we could then plot the activity a as a function of time, as shown in Figure 12.12. This plot shows the exponential dependence expected on the basis of Eq. 12.13.

It is often more useful to plot a as a function of t on a semilogarithmic scale, as shown in Figure 12.13. On this kind of plot, Eq. 12.13 gives a straight line of slope $-\lambda$.

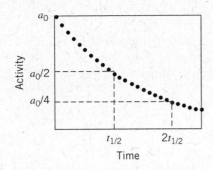

FIGURE 12.12 Activity of a radioactive sample as a function of time.

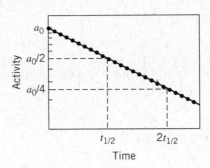

FIGURE 12.13 Semilog plot of activity versus time.

The *half-life*, $t_{1/2}$, of the decay is the time that it takes for the activity to be reduced by half, as shown in Figure 12.12. That is, when $t = t_{1/2}$, $a = \frac{1}{2}a_0 = a_0 e^{-\lambda t_{1/2}}$, from which we find

$$t_{1/2} = \frac{1}{\lambda}\ln 2 = \frac{0.693}{\lambda} \tag{12.14}$$

Another useful parameter is the *mean lifetime* τ (see Problem 37):

$$\tau = \frac{1}{\lambda} \tag{12.15}$$

When $t = \tau$, $a = a_0 e^{-1} = 0.37 a_0$.

Example 12.6

The half-life of ^{198}Au is 2.70 days. (*a*) What is the decay constant of ^{198}Au? (*b*) What is the probability that any ^{198}Au nucleus will decay in one second? (*c*) Suppose we had a 1.00-μg sample of ^{198}Au. What is its activity? (*d*) How many decays per second occur when the sample is one week old?

Solution

(*a*)

$$\lambda = \frac{0.693}{t_{1/2}} = \frac{0.693}{2.70\text{ d}}\frac{1\text{ d}}{24\text{ h}}\frac{1\text{ h}}{3600\text{ s}}$$

$$= 2.97 \times 10^{-6}\text{ s}^{-1}$$

(*b*) The decay probability per second is just the decay constant, so the probability of any ^{198}Au nucleus decaying in one second is 2.97×10^{-6}.

(*c*) The number of atoms in the sample is determined from the Avogadro constant N_A and the molar mass M:

$$N = \frac{mN_A}{M}$$

$$= \frac{(1.00 \times 10^{-6}\text{ g})(6.02 \times 10^{23}\text{ atoms/mole})}{198\text{ g/mole}}$$

$$= 3.04 \times 10^{15}\text{ atoms}$$

$$a = \lambda N = (2.97 \times 10^{-6}\text{ s}^{-1})(3.04 \times 10^{15})$$

$$= 9.03 \times 10^{9}\text{ Bq} = 0.244\text{ Ci}$$

(*d*) The activity decays according to Eq. 12.13:

$$a = a_0 e^{-\lambda t}$$

$$= (9.03 \times 10^{9}\text{ Bq})e^{-(2.97 \times 10^{-6}\text{ s}^{-1})(7\text{ d})(3600\text{ s/d})}$$

$$= 1.50 \times 10^{9}\text{ Bq}$$

Example 12.7

The half-life of ^{235}U is 7.04×10^8 y. A sample of rock, which solidified with the Earth 4.55×10^9 years ago, contains N atoms of ^{235}U. How many ^{235}U atoms did the same rock have at the time it solidified?

Solution

The age of the rock corresponds to

$$\frac{4.55 \times 10^9 \text{ y}}{7.04 \times 10^8 \text{ y}} = 6.46 \text{ half-lives}$$

Each half-life reduces N by a factor of 2, so the overall reduction in N has been $2^{6.46} = 88.2$. The original rock therefore contained $88.2N$ atoms of ^{235}U.

Conservation Laws in Radioactive Decays

Our study of radioactive decays and nuclear reactions reveals that nature is not arbitrary in selecting the outcome of decays or reactions, but rather that certain laws limit the possible outcomes. We call these laws *conservation laws*, and we believe these laws give us important insight into the fundamental workings of nature. Several of these conservation laws are applied to radioactive decay processes.

1. Conservation of Energy   Perhaps the most important of the conservation laws, conservation of energy tells us which decays are energetically possible and enables us to calculate rest energies or kinetic energies of decay products. A nucleus X will decay into a lighter nucleus X′, with the emission of one or more particles we call collectively x only if the rest energy of X is greater than the total rest energy of X′ + x The excess rest energy is known as the Q *value* of the decay X → X′ + x:

$$Q = [m_X - (m_{X'} + m_x)]c^2 \tag{12.16}$$

where the m's represent the *nuclear masses*. The decay is possible only if this Q value is positive. The excess energy Q appears as kinetic energy of the decay products (assuming X is initially at rest):

$$Q = K_{X'} + K_x \tag{12.17}$$

2. Conservation of Linear Momentum   If the initially decaying nucleus is at rest, then the total linear momentum of all of the decay products must sum to zero:

$$\vec{p}_{X'} + \vec{p}_x = 0 \tag{12.18}$$

Usually the emitted particle or particles x are much less massive than the residual nucleus X', and the *recoil momentum* $p_{X'}$ yields a very small kinetic energy $K_{X'}$.

If there is only one emitted particle x, Eqs. 12.17 and 12.18 can be solved simultaneously for $K_{X'}$ and K_x. If x represents two or more particles, we have more unknowns than we have equations, and no unique solution is possible. In this case, a range of values from some minimum to some maximum is permitted for the decay products.

3. Conservation of Angular Momentum   The total spin angular momentum of the initial particle before the decay must equal the total angular momentum (spin plus orbital) of all of the product particles after the decay. For example, the decay of a neutron (spin angular momentum = $1/2$) into a proton plus an electron is forbidden by conservation of angular momentum, because the spins of the proton and electron, each equal to $1/2$, can be combined to give a total of either 0 or 1, neither of which is equal to the initial angular momentum of the neutron. Adding integer units of orbital angular momentum to the electron does not restore angular momentum conservation in this decay process.

4. Conservation of Electric Charge   This is such a fundamental part of all decay and reaction processes that it hardly needs elaborating. The total net electric charge before the decay must equal the net electric charge after the decay.

5. Conservation of Nucleon Number   In some decay processes, we can create particles (photons or electrons, for example) which did not exist before the decay occurred. (This of course must be done out of the available energy—that is, it takes 0.511 MeV of energy to create an electron.) However, nature does *not* permit us to create or destroy protons and neutrons, although in certain decay processes we can convert neutrons into protons or protons into neutrons, *The total nucleon number A does not change in decay or reaction processes*. In some decay processes, A remains constant because both Z and N remain unchanged; in other processes Z and N both change in such a way as to keep their sum constant.

12.7 ALPHA DECAY

In alpha decay, an unstable nucleus disintegrates into a lighter nucleus and an alpha particle (a nucleus of ^{4}He), according to

$$^A_Z X_N \rightarrow ^{A-4}_{Z-2} X'_{N-2} + ^4_2 He_2 \qquad (12.19)$$

where X and X' represent different nuclei. For example, $^{226}_{88}Ra_{138} \rightarrow ^{222}_{86}Rn_{136} + ^4_2He_2$.

Decay processes release energy, because the decay products are more tightly bound than the initial nucleus. The energy released, which appears as the kinetic

FIGURE 12.14 A nucleus X alpha decays, resulting in a nucleus X' and an alpha particle.

energy of the alpha particle and the "daughter" nucleus X', can be found from the masses of the nuclei involved according to Eq. 12.16:

$$Q = [m(\text{X}) - m(\text{X}') - m(^4\text{He})]c^2 \qquad (12.20)$$

As we did in our calculations of binding energy, we can show that the electron masses cancel in Eq. 12.20, and so we can use *atomic masses*. This energy Q appears as kinetic energy of the decay products:

$$Q = K_{\text{X}'} + K_\alpha \qquad (12.21)$$

assuming we choose a reference frame in which the original atom X is at rest. Linear momentum is also conserved in the decay process, as shown in Figure 12.14, so that

$$p_\alpha = p_{\text{X}'} \qquad (12.22)$$

From Eqs. 12.21 and 12.22 we eliminate $p_{\text{X}'}$ and $K_{\text{X}'}$, because we normally don't observe the daughter nucleus in the laboratory. Typical alpha decay energies are a few MeV; thus the kinetic energies of the alpha particle and the nucleus are much smaller than their corresponding rest energies, and so we can use nonrelativistic mechanics to find

$$K_\alpha \cong \frac{A - 4}{A} Q \qquad (12.23)$$

Example 12.8

Find the kinetic energy of the alpha particle emitted in the alpha decay process $^{226}\text{Ra} \rightarrow {}^{222}\text{Rn} + {}^4\text{He}$.

Solution

From Eq. 12.20 the Q value is

$$Q = [m(^{226}\text{Ra}) - m(^{222}\text{Rn}) - m(^4\text{He})]c^2$$
$$= (226.025410 \text{ u} - 222.017578 \text{ u}$$
$$\qquad -4.002603 \text{ u})(931.5 \text{ MeV/u})$$
$$= 4.871 \text{ MeV}$$

The kinetic energy is given by Eq. 12.23:

$$K_\alpha = \frac{A - 4}{4} Q = \left(\frac{222}{226}\right)(4.871 \text{ MeV})$$
$$= 4.785 \text{ MeV}$$

Table 12.2 shows some sample alpha decays and their half-lives. You can see from the table that small changes in the decay energy (about a factor of 2) result in enormous changes in the half-life (24 orders of magnitude)! For example, for the isotopes ^{232}Th and ^{230}Th (which have the same Z and therefore the same Coulomb interaction between the alpha particle and the product nucleus) the kinetic energy changes by only 0.68 MeV (about 15%), while the half-life changes by about five orders of magnitude. Any successful calculation of the alpha decay probabilities must account for this sensitivity to the decay energy.

TABLE 12.2 Some Alpha Decay Energies and Half-Lives

Isotope	K_α (MeV)	$t_{1/2}$	λ (s^{-1})
^{232}Th	4.01	1.4×10^{10} y	1.6×10^{-18}
^{238}U	4.19	4.5×10^{9} y	4.9×10^{-18}
^{230}Th	4.69	7.5×10^{4} y	2.9×10^{-13}
^{241}Am	5.64	432 y	5.1×10^{-11}
^{230}U	5.89	20.8 d	3.9×10^{-7}
^{210}Rn	6.16	2.4 h	8.0×10^{-5}
^{220}Rn	6.29	56 s	1.2×10^{-2}
^{222}Ac	7.01	5 s	0.14
^{215}Po	7.53	1.8 ms	3.9×10^{2}
^{218}Th	9.85	0.11 μs	6.3×10^{6}

Quantum Theory of Alpha Decay

Alpha decay is an example of quantum-mechanical barrier penetration, as we discussed in Chapter 5. Suppose it is energetically possible for two neutrons and two protons to form an alpha particle, as represented in Figure 12.10. The alpha particle is trapped inside the nucleus by a barrier due to the Coulomb energy. The height of this barrier U_B is the Coulomb potential energy of the alpha particle and daughter nucleus at the radius R:

$$U_B = \frac{1}{4\pi\varepsilon_0}\frac{q_1 q_2}{r} = \frac{2(Z-2)e^2}{4\pi\varepsilon_0 R} \qquad (12.24)$$

which gives 30 to 40 MeV for a typical heavy nucleus. Here $q_1 = 2e$ is the electric charge of the alpha particle, and $q_2 = (Z-2)e$ is the electric charge of the nucleus after the decay, which is responsible for the Coulomb force.

Figure 12.15 shows the potential energy barrier encountered by the alpha particle as it tries to leave the interior of the nucleus ($r < R$). The energy of the alpha particle is typically in the range of 4–8 MeV, and so it is impossible for the alpha particle to surmount the barrier; the only way the alpha particle can escape is to "tunnel" through the barrier. A representation of the alpha particle wave function as it tunnels through the barrier is shown in Figure 12.15b.

The probability per unit time λ for the alpha particle to appear in the laboratory is the probability of its penetrating the barrier multiplied by the number of times per second the alpha particle strikes the barrier in its attempt to escape. If the alpha particle is moving at speed v inside a nucleus of radius R, it will strike the barrier as it bounces back and forth inside the nucleus at time intervals of $2R/v$. In a heavy nucleus with $R \sim 6$ fm, the α particle strikes the "wall" of the nucleus about 10^{22} times per second!

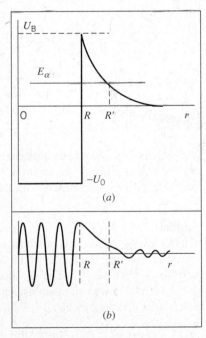

FIGURE 12.15   (a) The potential energy barrier for an alpha particle. (b) A representation of the wave function of the alpha particle.

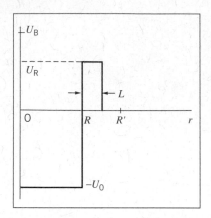

FIGURE 12.16 Replacing the Coulomb barrier for alpha decay with a flat barrier of height U_R.

The probability for the alpha particle to penetrate the barrier can be found by solving the Schrödinger equation for the potential energy shown in Figure 12.15. To simplify this calculation, we can replace the Coulomb barrier with a "flat" barrier, as shown in Figure 12.16. As we discussed in Chapter 5, the probability to penetrate a potential energy barrier is determined by the exponential factor e^{-2kL}, where L is the thickness of the barrier and where $k = \sqrt{(2m/\hbar^2)(U_0 - E)}$ for a barrier of height U_0 and a particle of energy E. The decay probability can then be estimated as

$$\lambda = \frac{v}{2R}e^{-2kL} \tag{12.25}$$

which includes both the rate at which the particle strikes the barrier and its probability to penetrate it. By making suitable rough estimates for the thickness and height of the barrier (see Problem 41), you should be able roughly to reproduce the range of values for the decay probabilities given in Table 12.2.

An exact calculation of the decay probability can be done by replacing the Coulomb barrier with a series of thin, flat barriers that are chosen to fit the Coulomb barrier as closely as possible. This calculation was first done in 1928 by George Gamow and was one of the first successful applications of the quantum theory.

Some nuclei can be unstable to the emission of other particles or collections of particles. Nuclei that have a large abundance of protons (those at the left-hand boundary of the light shaded region of Figure 12.11) may emit protons in a rare process similar to alpha decay. In this way they reduce their proton excess and move closer to stability. An example of this process is $^{151}_{71}\text{Lu}_{80} \rightarrow \,^{150}_{70}\text{Yb}_{80} + \text{p}$.

Other nuclei have recently been shown to emit clusters of particles such as ^{12}C, ^{14}C, or ^{20}Ne. The following example illustrates this process.

Example 12.9

The nucleus ^{226}Ra decays by alpha emission with a half-life of 1600 y. It also decays by emitting ^{14}C. Find the Q value for ^{14}C emission and compare with that for alpha emission (see Example 12.8).

Solution

If ^{226}Ra emits ^{14}C, which contains 6 protons and 8 neutrons, the resulting nucleus is ^{212}Pb, so the decay process is $^{226}\text{Ra} \rightarrow \,^{212}\text{Pb} + \,^{14}\text{C}$. The Q value can be found from Eq. 12.16, where we can again use atomic masses because the electron masses cancel.

$$\begin{aligned} Q &= [m(^{226}\text{Ra}) - m(^{212}\text{Pb}) - m(^{14}\text{C})]c^2 \\ &= (226.025410 \text{ u} - 211.991898 \text{ u} \\ &\quad -14.003242 \text{ u})(931.5 \text{ MeV/u}) \\ &= 28.197 \text{ MeV} \end{aligned}$$

Even though the Q value far exceeds the Q value for alpha decay (4.871 MeV), the Coulomb barrier for ^{14}C decay is roughly 3 times higher and thicker than it is for alpha decay [change $q_1 q_2$ to $6(Z - 6)e^2$ in Eq. 12.24]. As a result, the probability for ^{14}C decay turns out to be only about 10^{-9} of the probability for alpha decay; that is, ^{226}Ra emits one ^{14}C for every 10^9 alpha particles. See Problem 42 for a calculation of the relative decay probabilities.

12.8 BETA DECAY

In beta decay a neutron in the nucleus changes into a proton (or a proton into a neutron); Z and N each change by one unit, but A doesn't change. The emitted particles, which were called beta particles when first observed in 1898, were soon identified as electrons. In the most basic beta decay process, a free neutron decays into a proton and an electron: $n \rightarrow p + e$ (plus a third particle, as we discuss later).

The emitted electron is *not* one of the orbital electrons of the atom. It also is not an electron that was previously present within the nucleus, for as we have seen (Example 4.7) the uncertainty principle forbids electrons of the observed energies to exist inside the nucleus. The electron is "manufactured" by the nucleus out of the available energy. If the rest energy *difference* between the nuclei is at least $m_e c^2$, this will be possible.

In the 1910s and 1920s, beta decay experiments revealed two difficulties. First, the decay $n \rightarrow p + e^-$ appears to violate the law of conservation of angular momentum, as we discussed in Section 12.6. Second, measurements of the energy of the emitted electrons showed that the energy spectrum of the electrons is continuous, from zero up to some maximum value K_{max}, as shown in Figure 12.17. This implies an apparent violation of conservation of energy, because all electrons should emerge from the decay $n \rightarrow p + e^-$ with precisely the same energy. Instead, all electrons emerge with less energy, but in varying amounts.

For example, in the decay $n \rightarrow p + e^-$, the Q value is

$$Q = (m_n - m_p - m_e)c^2 = 0.782 \, \text{MeV} \tag{12.26}$$

Except for a very small correction, which accounts for the recoil energy of the proton, all of this energy should appear as kinetic energy of the electron, and all emitted electrons should have *exactly* this energy. However, experiments in the 1920s showed that all the emitted electrons have less than this energy—they have a continuous range of energies from 0 to 0.782 MeV.

The problem of this "missing" energy was very puzzling until 1930 when Wolfgang Pauli found the ingenious solution to *both* the apparent violations of conservation of angular momentum and energy—he suggested that there is a *third* particle emitted in beta decay. Electric charge is already conserved by the proton and electron, so this new particle cannot have electric charge. If it has spin $1/2$, it will satisfy conservation of angular momentum, because we can combine the spins of the three decay particles to give $1/2$, which matches the spin of the original decaying neutron. The "missing" energy is the energy carried away by this third particle, and the observed fact that the energy spectrum extends all the way to the value 0.782 MeV suggests that this particle has a very small mass.

This new particle is called the *neutrino* ("little neutral one" in Italian) and has the symbol ν. As we discuss in Chapter 14, every particle has an *antiparticle,* and the antiparticle of the neutrino is the *antineutrino* $\overline{\nu}$. It is, in fact, the antineutrino that is emitted in neutron beta decay. The complete decay process is thus

$$n \rightarrow p + e^- + \overline{\nu} \tag{12.27}$$

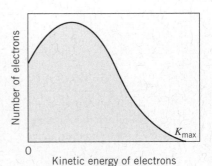

FIGURE 12.17 Spectrum of electrons emitted in beta decay.

(vertical axis) Number of electrons

(horizontal axis) Kinetic energy of electrons

K_{max}

0

Neutron decay can also occur in a nucleus, in which a nucleus with Z protons and N neutrons decays to a nucleus with $Z + 1$ protons and $N - 1$ neutrons:

$$^A_Z X_N \rightarrow \,^{A}_{Z+1} X'_{N-1} + e^- + \bar{\nu} \qquad (12.28)$$

The Q value for this decay is

$$Q = [m(^A X) - m(^A X')]c^2 \qquad (12.29)$$

It can be shown (Problem 23) that the electron masses cancel in calculating Q, so it is *atomic masses* that appear in Eq. 12.29. The antineutrino does not appear in the calculation of the Q value because its mass is negligibly small (of the order of eV/c^2, compared with the atomic masses measured in units of 10^3 MeV/c^2).

The energy released in the decay (the Q value) appears as the energy E_ν of the antineutrino, the kinetic energy K_e of the electron, and a small (usually negligible) recoil kinetic energy of the nucleus X':

$$Q = E_\nu + K_e + K_{X'} \cong E_\nu + K_e \qquad (12.30)$$

The electron (which must be treated relativistically, because its kinetic energy is *not* small compared with its rest energy) has its maximum kinetic energy when the antineutrino has a negligibly small energy. Figure 12.17 shows the energy distribution of electrons emitted in a typical negative beta decay. The electron and neutrino share the decay energy Q; the kinetic energy of the electron (equal to $Q - K_e$) ranges from 0 (when the neutrino has its maximum energy, $E_\nu = Q$) to Q (when $E_\nu = 0$).

Another beta decay process is

$$p \rightarrow n + e^+ + \nu \qquad (12.31)$$

in which a *positive electron,* or *positron*, is emitted. The positron is the antiparticle of the electron; it has the same mass as the electron but the opposite electric charge. The neutrino emitted in this process is similarly the antiparticle of the antineutrino that is emitted in neutron beta decay.

Proton beta decay has a negative Q value, and so it is never observed in nature for free protons. (This is indeed fortunate—if the free proton were unstable to beta decay, stable hydrogen atoms, the basic material of the universe, could not exist!) However, protons in some nuclei can undergo this decay process:

$$^A_Z X_N \rightarrow \,^{A}_{Z-1} X'_{N+1} + e^+ + \nu \qquad (12.32)$$

The Q value for this process is (Problem 23)

$$Q = [m(^A X) - m(^A X') - 2m_e]c^2 \qquad (12.33)$$

in which the masses are *atomic masses*. In this case, the positron and neutrino share the decay energy Q (again neglecting the small recoil energy of the nucleus X′). Figure 12.18 shows the energy distribution of positrons emitted in a typical positive beta decay.

A nuclear decay process that competes with positron emission is *electron capture*; the basic electron capture process is

$$p + e^- \rightarrow n + \nu \qquad (12.34)$$

in which a proton captures an atomic electron from its orbit and converts into a neutron plus a neutrino. The electron necessary for this process is one of the inner orbital electrons in an atom, and we identify the capture process by the shell from which the captured electron comes: K-shell capture, L-shell capture, and so forth. (The electronic orbits that come closest to, or even penetrate, the nucleus have the higher probability to be captured.) In nuclei the process is

$$_Z^A X_N + e^- \rightarrow {}_{Z-1}^{A} X'_{N+1} + \nu \qquad (12.35)$$

and the Q value, using atomic masses, is

$$Q = [m(^A X) - m(^A X')]c^2 \qquad (12.36)$$

In this case, neglecting the small initial kinetic energy of the electron and the recoil energy of the nucleus, the neutrino takes all of the available final energy:

$$E_\nu = Q \qquad (12.37)$$

In contrast to other beta-decay processes, a *monoenergetic* neutrino is emitted in electron capture.

Table 12.3 gives some typical beta decay processes, along with their Q values and half-lives.

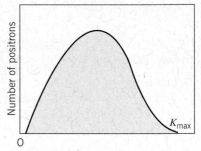

FIGURE 12.18   Spectrum of positrons emitted in beta decay.

TABLE 12.3 Typical Beta Decay Processes

Decay	Type	Q (MeV)	$t_{1/2}$
$^{19}O \rightarrow {}^{19}F + e^- + \bar{\nu}$	β^-	4.82	27 s
$^{176}Lu \rightarrow {}^{176}Hf + e^- + \bar{\nu}$	β^-	1.19	3.6×10^{10} y
$^{25}Al \rightarrow {}^{25}Mg + e^+ + \nu$	β^+	3.26	7.2 s
$^{124}I \rightarrow {}^{124}Te + e^+ + \nu$	β^+	2.14	4.2 d
$^{15}O + e^- \rightarrow {}^{15}N + \nu$	EC	2.75	122 s
$^{170}Tm + e^- \rightarrow {}^{170}Er + \nu$	EC	0.31	129 d

Example 12.10

^{23}Ne decays to ^{23}Na by negative beta emission. What is the maximum kinetic energy of the emitted electrons?

Solution

This decay is of the form given by Eq. 12.28, ^{23}Ne → ^{23}Na + e$^-$ + $\bar{\nu}$, and the Q value is found from Eq. 12.29, using *atomic* masses:

$$Q = [m(^{23}\text{Ne}) - m(^{23}\text{Na})]c^2$$
$$= (22.994467 \text{ u} - 22.989769 \text{ u})(931.5 \text{ MeV/u})$$
$$= 4.376 \text{ MeV}$$

Neglecting the small correction for the kinetic energy of the recoiling nucleus, the maximum kinetic energy of the electrons is equal to this value (which occurs when the neutrino has a negligible energy).

Example 12.11

^{40}K is an unusual isotope, in that it decays by negative beta emission, positive beta emission, and electron capture. Find the Q values for these decays.

Solution

The process for negative beta decay is given by Eq. 12.28, ^{40}K → ^{40}Ca + e$^-$ + $\bar{\nu}$, and the Q value is found from Eq. 12.29 using atomic masses:

$$Q_{\beta^-} = [m(^{40}\text{K}) - m(^{40}\text{Ca})]c^2$$
$$= (39.963998 \text{ u} - 39.962591 \text{ u})(931.5 \text{ MeV/u})$$
$$= 1.311 \text{ MeV}$$

Equation 12.32 gives the decay process for positive beta emission, ^{40}K → ^{40}Ar + e$^+$ + ν, and the Q value is given

by Eq. 12.33:

$$Q_{\beta^+} = [m(^{40}\text{K}) - m(^{40}\text{Ar}) - 2m_e]c^2$$
$$= [39.963998 \text{ u} - 39.962383 \text{ u} - 2(0.000549 \text{ u})]$$
$$\times (931.5 \text{ MeV/u})$$
$$= 0.482 \text{ MeV}$$

For electron capture, ^{40}K + e$^-$ → ^{40}Ar + ν, and from Eq. 12.36:

$$Q_{ec} = [m(^{40}\text{K}) - m(^{40}\text{Ar})]c^2$$
$$= (39.963998 \text{ u} - 39.962383 \text{ u})(931.5 \text{ MeV/u})$$
$$= 1.504 \text{ MeV}$$

12.9 GAMMA DECAY AND NUCLEAR EXCITED STATES

Following alpha or beta decay, the final nucleus may be left in an excited state. Just as an atom does, the nucleus will reach its ground state after emitting one or more photons, known as nuclear gamma rays. The energy of each photon is the energy difference between the initial and final nuclear states, less a negligibly small correction for the recoil kinetic energy of the nucleus. The energies of emitted gamma rays are typically in the range of 100 keV to a few MeV. Nuclei can likewise be excited from the ground state to an excited state by absorbing a photon of the appropriate energy, in a process similar to the resonant absorption by atomic states.

Figure 12.19 shows a typical energy-level diagram of excited nuclear states and some of the gamma-ray transitions that can be emitted. Typical values for the half-lives of the excited states are 10^{-9} to 10^{-12} s, although there are occasional cases of excited states with half-lives of hours, days, or even years.

When a gamma-ray photon is emitted, the nucleus must recoil to conserve momentum. The photon has energy E_γ and momentum $p_\gamma = E_\gamma/c$. The nucleus recoils with momentum p_R. If the nucleus is initially at rest, then momentum conservation requires that $p_R = p_\gamma$ in magnitude (and that the nucleus recoil in a direction opposite to that of the gamma ray). The recoil kinetic energy K_R is small, so that nonrelativistic equations can be used for the nucleus (of mass M):

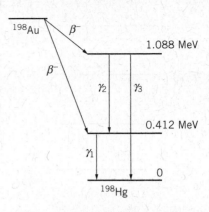

FIGURE 12.19   Some gamma rays emitted following beta decay.

$$K_R = \frac{p_R^2}{2M} = \frac{p_\gamma^2}{2M} = \frac{E_\gamma^2}{2Mc^2} \qquad (12.38)$$

For a medium-mass nucleus of $A = 100$ and a large gamma-ray energy of 1 MeV, the recoil kinetic energy is only 5 eV. Suppose the gamma ray is emitted when the nucleus jumps from an initial state with energy E_i to a final state with energy E_f. Conservation of energy then gives $E_i = E_f + E_\gamma + K_R$, so the energy of the emitted gamma ray is

$$E_\gamma = E_i - E_f - K_R \cong E_i - E_f \qquad (12.39)$$

The gamma-ray energy is equal to the difference between the initial and final energy states, because the recoil kinetic energy of the nucleus is negligibly small.

In calculating the energies of alpha and beta particles emitted in radioactive decays, we have assumed that no gamma rays are emitted. If there are gamma rays emitted, the available energy (Q value) must be shared between the other particles and the gamma ray, as the following example shows.

Example 12.12

^{12}N beta decays to an excited state of ^{12}C, which subsequently decays to the ground state with the emission of a 4.43-MeV gamma ray. What is the maximum kinetic energy of the emitted beta particle?

Solution

To determine the Q value for this decay, we first need to find the mass of the product nucleus ^{12}C *in its excited state*. In the ground state, ^{12}C has a mass of 12.000000 u, so its mass in the excited state (indicated by ^{12}C*) is

$$m(^{12}C^*) = 12.000000 \text{ u} + \frac{4.43 \text{ MeV}}{931.5 \text{ MeV/u}}$$
$$= 12.004756 \text{ u}$$

In this decay, a proton is converted to a neutron, so it must be an example of positron decay. The Q value is, according to Eq. 12.33,

$$Q = [m(^{12}N) - m(^{12}C^*) - 2m_e]c^2$$
$$= [12.018613 \text{ u} - 12.004756 \text{ u} - 2(0.000549 \text{ u})]$$
$$\times (931.5 \text{ MeV/u})$$
$$= 11.89 \text{ MeV}$$

(Notice that we could have just as easily found the Q value by first finding the Q value for decay to the *ground state*, 16.32 MeV, and then subtracting the excitation energy of 4.43 MeV, because the decay to the excited state has that much less available energy.)

Neglecting the small correction for the recoil kinetic energy of the ^{12}C nucleus, the maximum electron kinetic energy is 11.89 MeV.

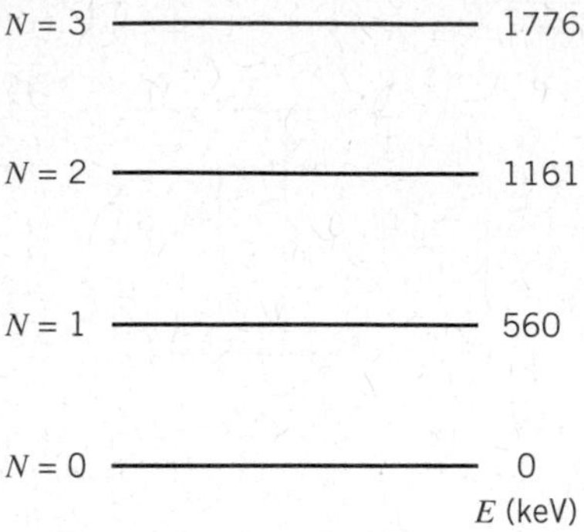

$N = 3$ ———————— 1776

$N = 2$ ———————— 1161

$N = 1$ ———————— 560

$N = 0$ ———————— 0

E (keV)

FIGURE 12.20 Nuclear vibrational states in the nucleus ^{120}Te. The states are labeled with the vibrational quantum number N. Note that the states are nearly equally spaced, as is expected for vibrations.

Nuclear Excited States

The study of nuclear gamma emission is an important tool of the nuclear physicist; the energies of the gamma rays can be measured with great precision, and they provide a powerful means of deducing the energies of the excited states of nuclei. This type of *nuclear spectroscopy* is very similar to the methods of molecular spectroscopy discussed in Chapter 9. In fact, the nuclear excited states can be formed in ways that are similar to molecular excited states:

1. Proton or Neutron Excitation   Nuclear excited states can be formed when a proton or a neutron is excited from a filled state to one of the empty states shown in Figure 12.8 or 12.9, just as in molecules we can form an excited state by promoting an electron from a lower state to one of the empty molecular orbitals. When a proton or neutron drops from an excited state to a lower state, a gamma-ray photon is emitted. The energy of the photon is equal to the energy difference between the states (neglecting the small recoil kinetic energy of the nucleus). To estimate the average energy for this type of excitation, we note from Figure 12.8 that 73 neutrons occupy an energy of 37.0 MeV, so the average spacing of the filled levels is (37.0 MeV)/73 = 0.5 MeV. The spacing between the empty states, among which gamma rays are emitted, should be about the same.

2. Nuclear Vibrations   The nucleus can vibrate like a jiggling water droplet. The vibrational excited states are equally spaced, just like the molecular vibrational states shown in Figure 9.22. Unlike a molecule, a nucleus vibrates like an incompressible fluid—for example, if the "equator" bulges outward, the "poles" must move inward to keep the density constant. The separation of the equally spaced vibrational states is about 0.5–1 MeV. Figure 12.20 shows an example of some vibrational nuclear excited states. Although the selection rules for photon emission in nuclei are not as strongly restrictive as they are in molecules, nuclei in higher vibrating states usually jump to lower vibrating states by changing the vibrational quantum number by one unit and emitting a gamma-ray photon in the process.

3. Nuclear Rotations   The nucleus can rotate, showing the same $L(L+1)$ spacing as a molecule (see Figure 9.25). Figure 12.21 shows an example of rotational nuclear excited states. The spacing between the rotational ground state and the first rotational excited state is typically 0.05–0.1 MeV. (Note that in nuclei, as in molecules, the rotational spacing is generally much smaller than the vibrational spacing.) Nuclei in higher rotational states can jump to lower rotational states by emitting gamma-ray photons; in the case of nuclei, the selection rule restricting the change in the rotational quantum number to one unit, which was strongly followed by molecules, does not strongly apply to nuclei. In nuclei, the rotational quantum number generally changes by one or two units when gamma-ray photons are emitted in transitions between the nuclear rotational states.

*Nuclear Resonance

One way of studying atomic systems is to do *resonance* experiments. In such experiments, radiation from a collection of atoms in an excited state is incident

---

*This is an optional section that may be skipped without loss of continuity.

on a collection of identical atoms in their ground state. The ground-state atoms can absorb the photons and jump to the corresponding excited state. However, as we have seen, the emitted photon energy is less than the transition energy by the recoil kinetic energy K_R; moreover, it is less than the photon energy required for resonance by $2K_R$, because the *absorbing* atom must recoil also. The absorption experiment is still possible, because the excited states don't have "exact" energies—a state with a mean lifetime τ has an energy uncertainty ΔE that is given by the uncertainty relationship: $\Delta E \tau \sim \hbar$. That is, the state lives on the average for a time τ, and during that time we can't determine its energy to an accuracy less than ΔE. For typical atomic states, $\tau \sim 10^{-8}$ s, so $\Delta E \sim 10^{-7}$ eV. Because K_R, which is of the order of 10^{-10} eV, is much less than the width ΔE, the "shift" caused by the recoil is not large, and the widths of the emitting and absorbing atomic states cause sufficient overlap for the absorption process to occur. Figure 12.22 illustrates this case.

The situation is different for nuclear gamma rays. A typical lifetime might be 10^{-10} s, and so the widths are the order of $\Delta E \sim 10^{-5}$ eV. The photon energies are typically 100 keV = 10^5 eV, and so K_R is of order 1 eV. This situation is depicted in Figure 12.23, and you can immediately see that because K_R is so much larger than the width ΔE, no overlap of emitter and absorber is possible, so resonance absorption cannot occur.

In 1958, it was discovered that the overlap of the emitter and absorber can be restored by placing the radioactive nuclei and the absorbing nuclei in crystals. The crystalline binding energies are large compared with K_R, so the individual atoms are held tightly to their positions in the crystal lattice and are not free to recoil; if any recoil is to occur, it must be the whole crystal that recoils. This effect is to make the mass M that appears in Eq. 12.38 not the mass of an atom, but the mass of the entire crystal, perhaps 10^{20} times larger than an atomic mass. (As an analogy, imagine the difference between striking a brick with a baseball bat, and striking a brick wall!) Once again the recoil kinetic energy is made small, and resonant absorption can occur (Figure 12.24). For this discovery, Rudolf Mössbauer was awarded the 1961 Nobel Prize in physics, and the process of achieving nuclear resonance by embedding the emitting and absorbing nuclei in crystal lattices is now known as the *Mössbauer effect*.

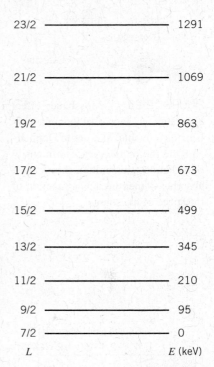

L	E (keV)
23/2	1291
21/2	1069
19/2	863
17/2	673
15/2	499
13/2	345
11/2	210
9/2	95
7/2	0

FIGURE 12.21   An example of nuclear rotational states in the nucleus ^{165}Ho. The states are labeled with the rotational quantum number L. The energies closely follow the expected $L(L+1)$ spacing.

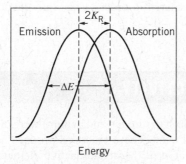

FIGURE 12.22   Representative emission and absorption energies in an atomic system.

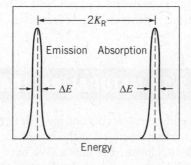

FIGURE 12.23   Representative emission and absorption energies in a nuclear system.

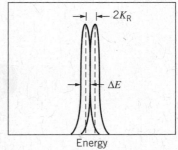

FIGURE 12.24   Emission and absorption energies for nuclei bound in a crystal lattice.

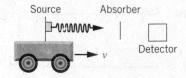

FIGURE 12.25   Mössbauer effect apparatus. A source of gamma rays is made movable, in order to Doppler-shift the photon energies. The intensity of radiations transmitted through the absorber is measured as a function of the speed of the source.

The small remaining difference between the emission and absorption energies can be eliminated to obtain complete overlap by Doppler-shifting either the emission or absorption energies. The Doppler-shifted frequency when a source moves toward the observer at speed v is given by Eq. 2.22, $f' = f(1 + v/c)$, where we ignore the $\sqrt{1 - v^2/c^2}$ term because $v \ll c$. Using $E = hf$ for the photon energy, we have

$$E' = E(1 + v/c) \tag{12.40}$$

If we take the width ΔE as a representative estimate of how far we would like to Doppler-shift the photon energy, then $E' \cong E + \Delta E$, and so $E + \Delta E \cong E + E(v/c)$. Solving for v, we find

$$v \cong c\frac{\Delta E}{E} \tag{12.41}$$

Estimating $\Delta E \sim 10^{-5}$ eV (the width of the state) and $E \sim 100$ keV (the energy of the photon), we have

$$v \cong (3 \times 10^8 \text{ m/s})\frac{10^{-5} \text{ eV}}{10^5 \text{ eV}} = 3 \text{ cm/s}$$

Such low speeds can be easily and accurately produced in the laboratory.

Figure 12.25 shows a diagram of the apparatus to measure the Mössbauer effect. The resonant absorption is observed by looking for decreases in the number of gamma rays that are transmitted through the absorber. At resonance, more gamma rays are absorbed and so the transmitted intensity decreases. Typical results are shown in Figure 12.26.

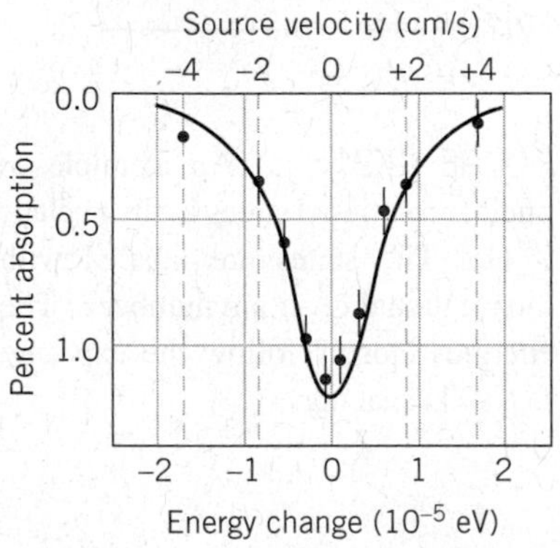

The Mössbauer effect is an extremely precise method for measuring small changes in the energies of photons. In one particular application, the Zeeman splitting of *nuclear* (not atomic) states can be observed. When a nucleus is placed in a magnetic field, the Zeeman effect causes an energy splitting of the nuclear m states, similar to the atomic case. However, nuclear magnetic moments are about 2000 times smaller than atomic magnetic moments, and a typical energy splitting would be about 10^{-6} eV. To observe such an effect directly we would need to measure photon energies to 1 part in 10^{11} (a photon energy of 10^5 eV is shifted by 10^{-6} eV), but using the Mössbauer effect, this is not difficult.

In Chapter 15 we discuss another application of this extremely precise technique, in which the energy gained when a photon "falls" through several meters of the Earth's gravitational field is measured in order to test one prediction of Einstein's general theory of relativity.

FIGURE 12.26 Typical results in a Mössbauer effect experiment. A velocity of 2 cm/s Doppler shifts the gamma rays enough to move the emission and absorption energies off resonance.

12.10 NATURAL RADIOACTIVITY

All of the elements beyond the very lightest (hydrogen and helium) were produced by nuclear reactions in the interiors of stars. These reactions produce not only stable elements, but radioactive ones as well. Most radioactive elements have half-lives that are much smaller than the age of the Earth (about 4.5×10^9 y), so those radioactive elements that may have been present when the Earth was formed have decayed to stable elements. However, a few of the radioactive elements created long ago have half-lives that are as large as or even greater than the age of the Earth. These elements can still be observed to undergo radioactive decay and account for part of the background of *natural radioactivity* that surrounds us.

Radioactive decay processes either change the mass number A of a nucleus by four units (alpha decay) or don't change A at all (beta or gamma decay). A radioactive decay process can be part of a sequence or series of decays if a *radioactive* element of mass number A decays to another *radioactive* element of mass number A or $A - 4$. Such a series of processes will continue until a stable element is reached. A hypothetical such series is illustrated in Figure 12.27. Because gamma decays don't change Z or A, they are not shown; however, most of the alpha and beta decays are accompanied by gamma-ray emissions.

The A values of the members of such decay chains differ by a multiple of 4 (including zero as a possible multiple) and so we expect four possible decay chains, with A values that can be expressed as $4n, 4n + 1, 4n + 2$, and $4n + 3$, where n is an integer. One of the four naturally occurring radioactive series is illustrated in Figure 12.28. Each series begins with a relatively long-lived member, proceeds through many α and β decays, which may have very short half-lives, and finally ends with a stable isotope. Three of these series begin with isotopes having half-lives comparable to the age of the Earth, and so are still observed today. The neptunium series $(4n + 1)$ begins with ^{237}Np, which has a half-life of "only" 2.1×10^6 y, much less than the 4.5×10^9 y since the formation of the Earth. Thus all of the ^{237}Np that was originally present has long since decayed to ^{209}Bi.

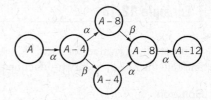

FIGURE 12.27   An example of a hypothetical radioactive decay chain.

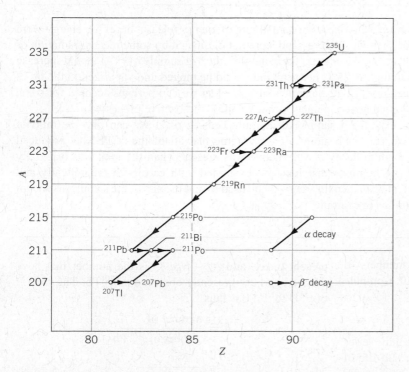

FIGURE 12.28 The ^{235}U decay chain. The diagonal lines represent α decays, and the horizontal lines show β decays.

Example 12.13

Compute the Q value for the $^{238}U \rightarrow \, ^{206}Pb$ decay chain, and find the rate of energy production per gram of uranium.

Solution

Because A changes by 32, there must be 8 alpha decays in the chain. These 8 alpha decays would decrease Z by 16 units, from 92 to 76. However, the final Z must be 82, so there must also be 6 beta decays in the chain. We recall that for β^- decays, the electron masses combine with the nuclear masses in the computation of the Q value and we can therefore use atomic masses. Thus for the entire decay chain,

$$Q = [m(^{238}U) - m(^{206}Pb) - 8m(^4He)]c^2$$
$$= [238.050788 \, u - 205.974465 \, u - 8(4.002603 \, u)]$$
$$\times (931.5 \, \text{MeV/u})$$
$$= 51.7 \, \text{MeV}$$

The half-life of the decay is 4.5×10^9 y, so λ, the decay probability per atom, is

$$\lambda = \frac{\ln 2}{t_{1/2}} = \frac{0.693}{(4.5 \times 10^9 \, \text{y})(3.16 \times 10^7 \, \text{s/y})}$$
$$= 4.9 \times 10^{-18} \, \text{s}^{-1}$$

One gram of ^{238}U is $\frac{1}{238}$ mole and therefore contains $\frac{1}{238} \times 6 \times 10^{23}$ atoms. The decay rate (activity) of the ^{238}U is given by the decay probability per atom per unit time multiplied by the number of atoms:

$$a = \lambda N$$
$$= \left(4.9 \times 10^{-18} \, \frac{\text{decays}}{\text{atom} \cdot \text{s}} \right) \left(\frac{1}{238} \times 6 \times 10^{23} \, \text{atoms} \right)$$
$$= 12,000 \, \text{decays/s}$$

Each decay releases 51.7 MeV, and so the rate of energy production is

$$12,000 \frac{\text{decays}}{\text{s}} \times 51.7 \frac{\text{MeV}}{\text{decay}} \times 10^6 \frac{\text{eV}}{\text{MeV}} \times 1.6 \times 10^{-19} \frac{\text{J}}{\text{eV}}$$
$$= 1.0 \times 10^{-7} \, \text{W}$$

This may seem like a very small rate of energy release, but if the energy were to appear as thermal energy and were not dissipated by some means (radiation or conduction to other matter, for example), the 1-g sample of ^{238}U would increase in temperature by 25°C per year and would be melted and vaporized in the order of one century! This calculation suggests that we can perhaps account for some of the internal heat of planets through natural radioactive processes.

If we examine a sample of uranium-bearing rock, we can find the ratio of ^{238}U atoms to ^{206}Pb atoms. If we assume that all of the ^{206}Pb was produced by the uranium decay and that none was present when the rock was originally formed (assumptions that must be examined with care both theoretically and experimentally), then this ratio can be used to find the age of the sample, as shown in the following example.

Example 12.14

Three different rock samples have ratios of numbers of ^{238}U atoms to ^{206}Pb atoms of 0.5, 1.0, and 2.0. Compute the ages of the three rocks.

Solution

Because all of the other members of the uranium series have half-lives that are much shorter than the half-life of ^{238}U (4.5×10^9 y), we ignore the intervening decays and consider only the ^{238}U decay. Let N_0 be the original number of ^{238}U atoms, so that $N_0 e^{-\lambda t}$ is the number that are still

present today, and $N_0 - N_0 e^{-\lambda t}$ is the number that have decayed and are presently observed as ^{206}Pb. The ratio R of ^{238}U to ^{206}Pb is thus

$$R = \frac{\text{number of } ^{238}U}{\text{number of } ^{206}Pb}$$
$$= \frac{N_0 e^{-\lambda t}}{N_0 - N_0 e^{-\lambda t}}$$
$$= \frac{1}{e^{\lambda t} - 1}$$

Solving for t and recalling that $\lambda = 0.693/t_{1/2}$, we find:

$$t = \frac{t_{1/2}}{0.693} \ln\left(\frac{1}{R} + 1\right) \qquad (12.42)$$

We can then obtain the values of t corresponding to the three values of R,

$$R = 0.5 \qquad t = 7.1 \times 10^9 \text{ y}$$
$$R = 1.0 \qquad t = 4.5 \times 10^9 \text{ y}$$
$$R = 2.0 \qquad t = 2.6 \times 10^9 \text{ y}$$

The oldest rocks on Earth, dated by similar means, have ages of about 4.5×10^9 y. The age of the first rock analyzed above, 7.1×10^9 y, suggests either that the rock had an extraterrestrial origin, or else that our assumption of no initial ^{206}Pb was incorrect. The age of the third rock suggests that it solidified only 2.6×10^9 y ago; previous to that time it was molten and the decay product ^{206}Pb may have "boiled away" from the ^{238}U.

There are a number of other naturally occurring radioactive isotopes that are not part of the decay chain of the heavy elements. A partial list is given in Table 12.4; some of these can also be used for radioactive dating.

Other radioactive elements are being produced continuously in the Earth's atmosphere as a result of nuclear reactions between air molecules and the high-energy particles known as "cosmic rays." The most notable and useful of these is ^{14}C, which beta decays with a half-life of 5730 y. When a living plant absorbs CO_2 from the atmosphere, a small fraction (about 1 in 10^{12}) of the carbon atoms is ^{14}C, and the remainder is stable ^{12}C (99%), and ^{13}C (1%). When the plant dies, its intake of ^{14}C stops, and the ^{14}C decays. If we assume that the composition of the Earth's atmosphere and the flux of cosmic rays have not changed significantly in the last few thousand years, we can find the age of specimens of organic material by comparing their ^{14}C/^{12}C ratios to those of living plants. The following example shows how this *radiocarbon dating* technique is used.

TABLE 12.4 Some Naturally Occurring Radioactive Isotopes

Isotope	$t_{1/2}$
^{40}K	1.25×10^9 y
^{87}Rb	4.8×10^{10} y
^{92}Nb	3.2×10^7 y
^{113}Cd	9×10^{15} y
^{115}In	5.1×10^{14} y
^{138}La	1.1×10^{11} y
^{176}Lu	3.6×10^{10} y
^{187}Re	4×10^{10} y
^{232}Th	1.41×10^{10} y

Example 12.15

(a) A sample of carbon dioxide gas from the atmosphere fills a vessel of volume 200.0 cm^3 to a pressure of 2.00×10^4 Pa (1 Pa = 1 N/m^2, about 10^{-5} atm) at a temperature of 295 K. Assuming that all of the ^{14}C beta decays were counted, how many counts would be accumulated in one week? (b) An old sample of wood is burned, and the resulting carbon dioxide is placed in an identical vessel at the same pressure and temperature. After one week, 1420 counts have been accumulated. What is the age of the sample?

Solution

(a) We first find the number of atoms present in the vessel, using the ideal gas law:

$$N = \frac{PV}{kT} = \frac{(2.00 \times 10^4 \text{ N/m}^2)(2.00 \times 10^{-4} \text{ m}^3)}{(1.38 \times 10^{-23} \text{ J/K})(295 \text{ K})}$$
$$= 9.82 \times 10^{20} \text{ atoms}$$

If the fraction of ^{14}C atoms is 10^{-12}, there are 9.82×10^8 atoms of ^{14}C present. The activity is

$$a = \lambda N = \frac{0.693}{(5730 \text{ y})(3.16 \times 10^7 \text{ s/y})} 9.82 \times 10^8$$
$$= 3.76 \times 10^{-3} \text{ decays/s}$$

In one week the number of decays is 2280.

(b) An identical sample that gives only 1420 counts must be old enough for only 1420/2280 of its original activity to remain. With $1420 = 2280e^{-\lambda t}$, we have

$$t = \frac{1}{\lambda} \ln\left(\frac{2280}{1420}\right) = \frac{5730 \text{ y}}{0.693} \ln\left(\frac{2280}{1420}\right) = 3920 \text{ y}$$

Chapter Summary

		Section			Section
Nuclear radius	$R = R_0 A^{1/3}, R_0 = 1.2$ fm	12.2	Q value of decay $X \rightarrow X' + x$	$Q = [m_X - (m_{X'} + m_x)]c^2$	12.6
Nuclear binding energy	$B = [Nm_n + Zm(^1_1H_0) - m(^A_ZX_N)]c^2$	12.3	Q value of alpha decay	$Q = [m(X) - m(X') - m(^4He)]c^2$	12.7
Proton separation energy	$S_p = [m(^{A-1}_{Z-1}X'_N) + m(^1H) - m(^A_ZX_N)]c^2$	12.3	Kinetic energy of alpha particle	$K_\alpha \cong Q(A-4)/A$	12.7
Neutron separation energy	$S_n = [m(^{A-1}_ZX_{N-1}) + m_n - m(^A_ZX_N)]c^2$	12.3	Q values of beta decay	$Q_{\beta^-} = [m(^AX) - m(^AX')]c^2,$ $Q_{\beta^+} = [m(^AX) - m(^AX') - 2m_e]c^2$	12.8
Range of exchanged particle	$mc^2 = \hbar c/x$	12.4	Recoil in gamma decay	$K_R = E_\gamma^2/2Mc^2$	12.9
Activity	$a = \lambda N, \lambda = \ln 2/t_{1/2} = 0.693/t_{1/2}$	12.5			
Radioactive decay law	$N = N_0 e^{-\lambda t}, a = a_0 e^{-\lambda t}$	12.5			

Questions

1. The magnetic dipole moment of a deuterium nucleus is about 0.0005 Bohr magneton. What does this imply about the presence of an electron in the nucleus, as the proton-electron model requires?

2. Suppose we have a supply of 20 protons and 20 neutrons. Do we liberate more energy if we assemble them into a single ^{40}Ca nucleus or into two ^{20}Ne nuclei?

3. Atomic masses are usually given to a precision of about the sixth decimal place in atomic mass units (u). This is true for both stable and radioactive nuclei, even though the uncertainty principle requires that an atom with a lifetime Δt has a rest energy uncertain by $\hbar/\Delta t$. Based on the typical lifetimes given for nuclear decays, are we justified in expressing atomic masses to such precision? At what lifetimes would such precision not be justified?

4. Only two stable nuclei have $Z > N$. (a) What are these nuclei? (b) Why don't more nuclei have $Z > N$?

5. In a deuterium nucleus, the proton and neutron spins can be either parallel or antiparallel. What are the possible values of the total spin of the deuterium nucleus? (It is not necessary to consider any orbital angular momentum.) The magnetic dipole moment of the deuterium nucleus is measured to be nonzero. Which of the possible spins is eliminated by this measured value?

6. Why is the binding energy per nucleon relatively constant? Why does it deviate from a constant value for low mass numbers? For high mass numbers?

7. A neutron, which has no electric charge, has a magnetic dipole moment. How is this possible?

8. The electromagnetic interaction can be interpreted as an exchange force, in which photons are the exchanged particle. What does Eq. 12.8 imply about the range of such a force? Is this consistent with the conventional interpretation of the electromagnetic force? What would you expect for the rest energy of the exchanged particle that carries the gravitational force?

9. What is meant by assuming that the decay constant λ is a constant, independent of time? Is this a requirement of theory, an axiom, or an experimental conclusion? Under what circumstances might λ change with time?

10. If we focus our attention on a specific nucleus in a radioactive sample, can we know exactly how long that nucleus will live before it decays? Can we predict which half of the nuclei in a sample will decay during one half-life? What part of quantum physics is responsible for this?

11. A certain radioactive sample is observed to undergo 10,000 decays in 10 s. Can we conclude that $a = 1000$ decays/s if (a) $t_{1/2} \gg 10$ s; (b) $t_{1/2} = 10$ s; (c) $t_{1/2} \ll 10$ s?

12. Suppose we wish to do radioactive dating of a sample whose age we guess to be t. Should we choose an isotope whose half-life is (a) $\gg t$; (b) $\sim t$; or (c) $\ll t$?

13. The alpha particle is a particularly tightly bound nucleus. Based on this fact, explain why heavy nuclei alpha decay and light nuclei don't.

14. Can you suggest a possible origin for the helium gas that is part of the Earth's atmosphere?

15. Estimate the recoil kinetic energy of the residual nucleus following alpha decay. (This energy is large enough to drive the residual nucleus out of certain radioactive sources; if the residual nucleus is itself radioactive, there is the chance of spread of radioactive material. A thin coating over the source is necessary to prevent this.)

16. Why does the electron energy spectrum (Figure 12.17) look different from the positron energy spectrum (Figure 12.18) at low energies?

17. Will electron capture always be energetically possible when positron beta decay is possible? Will positron beta decay always be energetically possible when electron capture is possible?

18. All three beta decay processes involve the emission of neutrinos (or antineutrinos). In which processes do the neutrinos have a continuous energy spectrum? In which is the neutrino monoenergetic?

19. Neutrinos always accompany electron capture decays. What other kind of radiation always accompanies electron capture? (*Hint:* It is not nuclear radiation.) What other kind of nonnuclear radiation might accompany β^- or β^+ decays in bulk samples?

20. The positron decay of ^{15}O goes directly to the ground state of ^{15}N; no excited states of ^{15}N are populated and no γ rays follow the beta decay. Yet a source of ^{15}O is found to emit γ rays of energy 0.51 MeV. Explain the origin of these γ rays.

21. Would ^{92}Nb be a convenient isotope to use for determining the age of the Earth by radioactive dating? (See Table 12.4.) What about ^{113}Cd?

22. The natural decay chain $^{238}_{92}$U $\rightarrow$ $^{206}_{82}$Pb consists of several alpha decays, which decrease A by 4 and Z by 2, and negative beta decays, which increase Z by 1. (See Example 12.13.) As shown in Figures 12.27 and 12.28, sometimes a decay chain can proceed through different branches. Does the number of alpha decays and beta decays in the chain depend on this branching?

23. It has been observed that there is an increased level of radon gas ($Z = 86$) in the air just before an earthquake. Where does the radon come from? How is it produced? How is it released? How is it detected?

24. Which of the decay processes discussed in this chapter would you expect to be most sensitive to the chemical state of the radioactive sample?

25. In Figure 12.26, only 1% of the gamma intensity is absorbed, even at resonance. For complete resonance, we would expect 100% absorption. What factors might contribute to this small absorption?

Problems

12.1 Nuclear Constituents

1. Give the proper isotopic symbols for: (*a*) the isotope of fluorine with mass number 19; (*b*) an isotope of gold with 120 neutrons; (*c*) an isotope of mass number 107 with 60 neutrons.

2. Tin has more stable isotopes than any other element; they have mass numbers 114, 115, 116, 117, 118, 119, 120, 122, 124. Give the symbols for these isotopes.

12.2 Nuclear Sizes and Shapes

3. (*a*) Compute the Coulomb repulsion energy between two nuclei of ^{16}O that just touch at their surfaces. (*b*) Do the same for two nuclei of ^{238}U.

4. Find the nuclear radius of (*a*) ^{197}Au; (*b*) ^{4}He; (*c*) ^{20}Ne.

12.3 Nuclear Masses and Binding Energies

5. Find the total binding energy, and the binding energy per nucleon, for (*a*) ^{208}Pb; (*b*) ^{133}Cs; (*c*) ^{90}Zr; (*d*) ^{59}Co.

6. Find the total binding energy, and the binding energy per nucleon, for (*a*) ^{4}He; (*b*) ^{20}Ne; (*c*) ^{40}Ca; (*d*) ^{55}Mn.

7. Calculate the total nuclear binding energy of ^{3}He and ^{3}H. Account for any difference by considering the Coulomb interaction of the extra proton of ^{3}He.

8. Find the neutron separation energy of: (*a*) ^{17}O; (*b*) ^{7}Li; (*c*) ^{57}Fe.

9. Find the proton separation energy of: (*a*) ^{4}He; (*b*) ^{12}C; (*c*) ^{40}Ca.

12.4 The Nuclear Force

10. The nuclear attractive force must turn into a repulsion at very small distances to keep the nucleons from crowding too close together. What is the mass of an exchanged particle that will contribute to the repulsion at separations of 0.25 fm?

11. The weak interaction (the force responsible for beta decay) is produced by an exchanged particle with a mass of roughly 80 GeV. What is the range of this force?

12.5 Quantum States in Nuclei

12. Determine the depth of the proton and neutron potential energy wells for (*a*) ^{16}O; (*b*) ^{235}U.

13. The two-neutron separation energies of ^{160}Dy and ^{164}Dy are, respectively, 15.4 MeV and 13.9 MeV, and the two-proton separation energies of ^{158}Dy and ^{162}Dy are, respectively, 12.4 MeV and 14.8 MeV. From these data alone, determine whether alpha decay is energetically allowed for ^{160}Dy and ^{164}Dy.

12.6 Radioactive Decay

14. What fraction of the original number of nuclei present in a sample will remain after (a) 2 half-lives; (b) 4 half-lives; (c) 10 half-lives?

15. A certain sample of a radioactive material decays at a rate of 548 per second at $t = 0$. At $t = 48$ minutes, the counting rate has fallen to 213 per second. (a) What is the half-life of the radioactivity? (b) What is its decay constant? (c) What will be the decay rate at $t = 125$ minutes?

16. What is the decay probability per second per nucleus of a substance with a half-life of 5.0 hours?

17. Tritium, the hydrogen isotope of mass 3, has a half-life of 12.3 y. What fraction of the tritium atoms remains in a sample after 50.0 y?

18. Suppose we have a sample containing 2.00 mCi of radioactive ^{131}I ($t_{1/2} = 8.04$ d). (a) How many decays per second occur in the sample? (b) How many decays per second will occur in the sample after four weeks?

19. Ordinary potassium contains 0.012 percent of the naturally occurring radioactive isotope ^{40}K, which has a half-life of 1.3×10^9 y. (a) What is the activity of 1.0 kg of potassium? (b) What would have been the fraction of ^{40}K in natural potassium 4.5×10^9 y ago?

12.7 Alpha Decay

20. Derive Eq. 12.23 from Eqs. 12.21 and 12.22.

21. For which of the following nuclei is alpha decay permitted? (a) ^{210}Bi; (b) ^{203}Hg; (c) ^{211}At.

22. Find the kinetic energy of the alpha particle emitted in the decay of ^{234}U.

12.8 Beta Decay

23. Derive Eqs. 12.29, 12.33, and 12.36.

24. Find the maximum kinetic energy of the electrons emitted in the negative beta decay of ^{11}Be.

25. ^{75}Se decays by electron capture to ^{75}As. Find the energy of the emitted neutrino.

26. ^{15}O decays to ^{15}N by positron beta decay. (a) What is the Q value for this decay? (b) What is the maximum kinetic energy of the positrons?

12.9 Gamma Decay and Nuclear Excited States

27. The nucleus ^{198}Hg has excited states at 0.412 and 1.088 MeV. Following the beta decay of ^{198}Au to ^{198}Hg, three gamma rays are emitted. Find the energies of these three gamma rays.

28. Compare the recoil energy of a nucleus of mass 200 that emits (a) a 5.0-MeV alpha particle, and (b) a 5.0-MeV gamma ray.

29. A certain nucleus has the following sequence of rotational states E_L (energies in keV): $E_0 = 0$, $E_1 = 100.1$, $E_2 = 300.9$, $E_3 = 603.6$, and $E_4 = 1010.0$. Assuming that the emitted gamma rays occur only from changes of one or two units in the rotational quantum number, find all possible photon energies that can be emitted from these states. Sketch the excited states, showing the allowed transitions.

12.10 Natural Radioactivity

30. The radioactive decay of ^{232}Th leads eventually to stable ^{208}Pb. A certain rock is examined and found to contain 3.65 g of ^{232}Th and 0.75 g of ^{208}Pb. Assuming all of the Pb was produced in the decay of Th, what is the age of the rock?

31. The 4n radioactive decay series begins with $^{232}_{90}$Th and ends with $^{208}_{82}$Pb. (a) How many alpha decays are in the chain? (See Question 22.) (b) How many beta decays? (c) How much energy is released in the complete chain? (d) What is the radioactive power produced by 1.00 kg of ^{232}Th ($t_{1/2} = 1.40 \times 10^{10}$ y)?

32. A piece of wood from a recently cut tree shows 12.4 ^{14}C decays per minute. A sample of the same size from a tree cut thousands of years ago shows 3.5 decays per minute. What is the age of this sample?

General Problems

33. Figure 12.2 suggests that the Rutherford scattering formula fails for $60°$ scattering when K is about 28 MeV. Use the results derived in Chapter 6 to find the closest distance between alpha particle and nucleus for this case, and compare with the nuclear radius of ^{208}Pb. Suggest a possible reason for any discrepancy.

34. Assuming the nucleus to diffract like a circular disk, use the data shown in Figure 12.3 to find the nuclear radius for ^{12}C and ^{16}O. How does changing the electron energy from 360 MeV to 420 MeV affect the deduced radius for ^{16}O? (*Hint:* Use the extreme relativistic approximation from Chapter 2 to relate the electron's energy and momentum to find its de Broglie wavelength.)

35. A radiation detector is in the form of a circular disc of diameter 3.0 cm. It is held 25 cm from a source of radiation, where it records 1250 counts per second. Assuming that the detector records every radiation incident upon it, find the activity of the sample (in curies).

36. What is the activity of a container holding 125 cm^3 of tritium (^{3}H, $t_{1/2} = 12.3$ y) at a pressure of 5.0×10^5 Pa (about 5 atm) at $T = 300$ K?

37. With a radioactive sample originally of N_0 atoms, we could measure the mean, or average, lifetime τ of a nucleus by measuring the number N_1 that live for a time t_1 and then decay, the number N_2 that decay after t_2 and so on:

$$\tau = \frac{1}{N_0}(N_1 t_1 + N_2 t_2 + \cdots)$$

(a) Show that this is equivalent to $\tau = \lambda \int_0^\infty e^{-\lambda t} t \, dt$. (b) Show that $\tau = 1/\lambda$. (c) Is τ longer or shorter than $t_{1/2}$?

38. Complete the following decays:
    (a) ^{27}Si $\rightarrow$ ^{27}Al $+$
    (b) ^{74}As $\rightarrow$ ^{74}Se $+$
    (c) ^{228}U $\rightarrow$ α $+$
    (d) ^{93}Mo $+$ e$^-$ $\rightarrow$
    (e) ^{131}I $\rightarrow$ ^{131}Xe $+$

39. ^{239}Pu decays by alpha emission with a half-life of 2.41×10^4 y. Compute the power output, in watts, that could be obtained from 1.00 gram of ^{239}Pu.

40. ^{228}Th alpha decays to an excited state of ^{224}Ra, which in turn decays to the ground state with the emission of a 217-keV photon. Find the kinetic energy of the alpha particle.

41. By replacing the Coulomb barrier in alpha decay with a flat barrier (see Figure 12.16) of thickness $L = \frac{1}{2}(R' - R)$, equal to half the thickness of the Coulomb barrier that the alpha particle must penetrate, and height $U_0 = \frac{1}{2}(U_B + K_\alpha)$, equal to half the height of the Coulomb barrier above the energy of the alpha particle, estimate the decay half-lives for ^{232}Th and ^{218}Th and compare with the measured values given in Table 12.2. (*Hint:* In calculating the speed of the alpha particle inside the nucleus, assume that the well depth is 30 MeV.) Although the results of this rough calculation do not agree well with the measured values, the calculation does indicate how barrier penetration is responsible for the enormous range of observed half-lives. How would you refine the calculation to obtain better agreement with the measured values?

42. (a) Using the same replacements described in Problem 41, estimate the decay probability of ^{226}Ra for alpha emission and for ^{14}C emission. (See Examples 12.8 and 12.9.) (b) Using the results of part (a), estimate the number of ^{14}C emitted relative to the number of alpha particles emitted by a source of ^{226}Ra.

43. Compute the recoil proton kinetic energy in neutron beta decay (a) when the electron has its maximum energy; (b) when the neutrino has its maximum energy.

44. In the beta decay of ^{24}Na, an electron is observed with a kinetic energy of 2.15 MeV. What is the energy of the accompanying neutrino?

45. The first excited state of ^{57}Fe decays to the ground state with the emission of a 14.4-keV photon in a mean lifetime of 141 ns. (a) What is the width ΔE of the state? (b) What is the recoil kinetic energy of an atom of ^{57}Fe that emits a 14.4-keV photon? (c) If the kinetic energy of recoil is made negligible by placing the atoms in a solid lattice, resonant absorptions will occur. What velocity is required to Doppler-shift the emitted photon so that resonance does not occur?

46. What is the probability of a ^{14}C atom in atmospheric CO_2 decaying in your lungs during a single breath? The atmosphere is about 0.03% CO_2. Assume you take in about 0.5 L of air in each breath and exhale it 3.5 s later.

Chapter 13

NUCLEAR REACTIONS AND APPLICATIONS

Nuclear reactors produce intense beams of neutrons that can be used to measure how radiation exposure affects various materials. They also produce rare radioisotopes that can be used for medicine and applications in industry. The photo shows the core of a nuclear reactor, which is submerged in water that acts as a neutron moderator. The glow comes from Cerenkov radiation, which is emitted when electrons from radioactive decays move at speeds greater than the speed of light in water.

The knowledge of the nucleus that we can obtain from studying radioactive decays is limited, because only certain radioactive processes occur in nature, only certain isotopes are made in those processes, and only certain excited states of nuclei (those that happen to follow radioactive decays) can be studied. Nuclear reactions, however, give us a controllable way to study *any* nuclear species, and to select any excited states of that species.

In this chapter we discuss some of the different nuclear reactions that can occur, and we study the properties of those reactions. Two nuclear reactions are of particular importance: fission and fusion. We pay special attention to those processes and we discuss how they are useful as sources of energy (or, more correctly, as *converters* of nuclear energy into thermal or electrical energy).

We conclude our study of nuclear physics with an introduction to some of the ways that methods of nuclear physics can be applied to problems in a variety of different areas.

13.1 TYPES OF NUCLEAR REACTIONS

In a typical nuclear reaction laboratory experiment, a beam of particles of type x is incident on a target containing nuclei of type X. After the reaction, an outgoing particle y is observed in the laboratory, leaving a residual nucleus Y. Symbolically, we write the reaction as

$$x + X \rightarrow y + Y$$

For example,

$$^2_1\text{H}_1 + ^{63}_{29}\text{Cu}_{34} \rightarrow \text{n} + ^{64}_{30}\text{Zn}_{34}$$

Like a chemical reaction, a nuclear reaction must be balanced—the total number of protons must be the same before and after the reaction, and also the total number of neutrons must remain the same. In the example above, there are 30 protons on each side and 35 neutrons on each side. (The forces responsible for nuclear beta decay can change neutrons into protons or protons into neutrons, but these forces act on a typical time scale of at least 10^{-10} s. The projectile and target nuclei are within the range of one another's nuclear forces for an interval of at most 10^{-20} s, so there is not enough time for this type of proton-neutron conversion to take place.) The protons and neutrons can be rearranged among the reacting nuclei, but their numbers cannot change.

A nuclear reaction takes place under the influence of forces internal to the system of projectile and target. The absence of external forces means that the reaction conserves energy, linear momentum, and angular momentum.

In most experiments, we observe only the outgoing light particle y; the heavy residual nucleus Y usually loses all its kinetic energy (by collisions with other atoms) and therefore stops within the target.

We assume that we produce the reaction by bombarding target nuclei X, initially at rest, with projectiles x of kinetic energy K_x. The product particles then share this kinetic energy, plus or minus any additional energy from the rest energy difference of the initial and final nuclei. (We consider energy in nuclear reactions in Section 13.3.)

The bombarding particles x can be either charged particles, supplied by a suitable nuclear accelerator, or neutrons, whose source may be a nuclear reactor. Accelerators for charged particles, illustrated in Figures 13.1 and 13.2, are of two basic types. In a cyclotron, a particle is held in a circular orbit by a magnetic field and receives a small "kick" by an electric field twice each time it travels around the circle; a particle may make perhaps 100 orbits before finally emerging with a kinetic energy of the order of 10 to 20 MeV per unit of electric charge. In the Van de Graaff accelerator, a particle is accelerated only once from a single high-voltage terminal, which may be at a potential of as much as 25 million volts; the kinetic energy of the particle is then about 25 MeV per unit of charge.

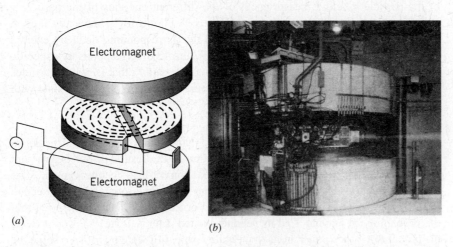

(a)

(b)

FIGURE 13.1 (a) Schematic diagram of a cyclotron accelerator. Charged particles are bent in a circular path by a magnetic field and are accelerated by an electric field each time they cross the gap. (b) A cyclotron accelerator. The magnets are in the large cylinders at the top and the bottom. The beam of particles is visible as it collides with air molecules after leaving the cyclotron.

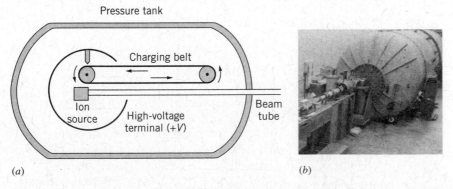

(a)

(b)

FIGURE 13.2 (a) Diagram of a Van de Graaff accelerator. Particles from the ion source are accelerated from the high-voltage terminal to ground. (b) A typical Van de Graaff accelerator laboratory. The ion source and high-voltage terminal are inside the large pressure tank.

In nuclear reaction experiments, we usually measure two basic properties of the particle y: its energy, and its probability to emerge at a certain angle with a certain energy. We look briefly at these two types of measurements.

1. Measuring the particle energy

If neither the residual nucleus Y nor the outgoing particle y had excited states, then by using conservation of energy and momentum, we could calculate exactly the energy of y when measured at a certain angle. If the nucleus Y is left in an excited state, then the kinetic energy of y is reduced by (approximately) the energy of the excited state above the ground state, because the two particles Y and y must still share the same amount of total energy. Each higher excited state of the nucleus Y corresponds to a certain reduced energy of the particle y, and a measurement of the different energies of the particle y tells us about the excited states of the nucleus Y. Figure 13.3 shows an example of a typical set of experimental results and the corresponding deduced excited states of the residual nucleus. Each peak in Figure 13.3 corresponds to a specific energy of y, and therefore to a specific excited state of Y; that is, when particles with energy 9.0 MeV are observed, the nucleus Y is left in the excited state with energy 1.0 MeV.

2. Measuring the reaction probability

Notice that the different peaks in Figure 13.3 have different heights. This feature of the results of our experiment tells us that it is more probable for the reaction to lead to one excited state than to another. This is an example of the reaction probability, the second of the properties of y that we can determine. For example, Figure 13.3 shows that the probability of leaving Y in its second excited state (1.0 MeV) is about twice the probability of leaving Y in its first excited state. If it were possible to solve the Schrödinger equation with the nuclear potential energy, we could calculate these reaction probabilities and compare them with experiment. Unfortunately we can't solve this many-body problem, so we must work backward by measuring the reaction probabilities and then trying to infer some properties of the nuclear force.

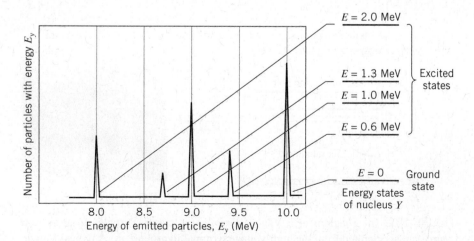

FIGURE 13.3 A sample spectrum of energies of the outgoing particle y, and the corresponding excited states of Y.

The Reaction Cross Section

Reaction probabilities are usually expressed in terms of the *cross section,* which is a sort of effective area presented by the target nucleus to that projectile for a specific reaction, for all possible energies and directions of travel of the outgoing particle y. The larger the reaction probability, the larger is the corresponding cross section. In general, the cross section depends on the energy of the incident particle, K_x.

The cross section σ is expressed in units of area, but the area is a very small one, of the order of 10^{-28} m^2. Nuclear physicists use this as a convenient unit of measure for cross sections, and it is known as one *barn* (b): 1 barn = 10^{-28} m^2. Notice that the area of the disc of a single nucleus of medium weight is about 1 barn; however, reaction cross sections often can be very much greater or less than one barn. For example, consider the cross section for these reactions involving certain isotopes of the neighboring elements iodine and xenon:

$$I + n \rightarrow I + n \text{ (inelastic scattering)} \qquad \sigma = 4\,b$$
$$Xe + n \rightarrow Xe + n \text{ (inelastic scattering)} \qquad \sigma = 4\,b$$
$$I + n \rightarrow I + \gamma \text{ (neutron capture)} \qquad \sigma = 7\,b$$
$$Xe + n \rightarrow Xe + \gamma \text{ (neutron capture)} \qquad \sigma = 10^6\,b$$

You can see that, although the neutron inelastic scattering cross sections of I and Xe are similar, the neutron capture cross sections are very different. These measurements are therefore telling us something interesting and unusual about the properties of the nucleus Xe.

Suppose a beam of particles is incident on a thin target of area S, which contains a total of N nuclei. The effective area of each nucleus is the cross section σ, and so the total effective area of all the nuclei in the target is (ignoring shadowing effects) σN. The fraction of the target area that this represents is $\sigma N / S$, and as long as this ratio is small, shadowing effects are negligible. This fraction is the probability for the reaction to occur.

Suppose the incident particles strike the target at a rate of I_0 particles per second, and suppose the outgoing particles y are emitted at a rate of R per second. (This is also the rate at which the product nucleus Y is formed.) Then the reaction probability can also be expressed as the rate of y divided by the rate of x, or R/I_0. Setting the two expressions for the reaction probability equal to each other, we obtain $\sigma N / S = R/I_0$, or

$$R = \frac{\sigma N}{S} I_0 \tag{13.1}$$

This gives a relationship between the reaction cross section and the rate of emission of y.

In a reactor, the intensity of neutrons is usually expressed in terms of the rate at which neutrons cross a unit area perpendicular to the beam, or *neutron flux* ϕ (neutrons/cm^2/s). The cross section is σ (square centimeter per nucleus per incident neutron). The rate R also depends on the number of target nuclei. Suppose the mass of the target is m; the number of target nuclei is then $N = (m/M)N_A$, where M is the molar mass, and N_A is Avogadro's constant (6.02×10^{23} atoms per mole). Thus, for neutron-induced reactions, using Eq. 13.1 we obtain

$$R = \phi \sigma N = \phi \sigma \frac{m}{M} N_A \tag{13.2}$$

Example 13.1

For a certain incident proton energy the reaction $p + {}^{56}Fe \rightarrow n + {}^{56}Co$ has a cross section of 0.40 b. If we bombard a target in the form of a 1.0-cm-square, 1.0-μm-thick iron foil with a beam of protons equivalent to a current of 3.0 μA, and if the beam is spread uniformly over the entire surface of the target, at what rate are the neutrons produced?

Solution

We first calculate the number of nuclei in the target. The volume of the target is $V = (1.0\,\text{cm})^2 (1.0\,\mu m) = 1.0 \times 10^{-4}\,\text{cm}^3$, and (using the density of iron of 7.9 g/cm³) its mass is $m = \rho V = (7.9\,\text{g/cm}^3)(1.0 \times 10^{-4}\,\text{cm}^3) = 7.9 \times 10^{-4}$ g. The number of atoms (or nuclei) is then

$$N = \frac{mN_A}{M} = \frac{(7.9 \times 10^{-4}\,\text{g})(6.02 \times 10^{23}\,\text{atoms/mole})}{56\,\text{g/mole}}$$

$$= 8.5 \times 10^{18}\,\text{atoms}$$

Next we need to find the number of particles per second in the incident beam. We are given that the current is $3.0 \times 10^{-6}\,\text{A} = 3.0 \times 10^{-6}$ C/s, and with each proton having a charge of 1.6×10^{-19} C, the beam intensity is

$$I_0 = \frac{3.0 \times 10^{-6}\,\text{C/s}}{1.6 \times 10^{-19}\,\text{C/particle}} = 1.9 \times 10^{13}\,\text{particles/s}$$

From Eq. 13.1 we can now find R:

$$R = \frac{N\sigma I_0}{S}$$

$$= (8.5 \times 10^{18}\,\text{nuclei})(0.40 \times 10^{-24}\,\text{cm}^2/\text{nucleus})$$
$$\times (1.9 \times 10^{13}\,\text{particles/s})\,(1\,\text{cm}^2)^{-1}$$

$$= 6.5 \times 10^7\,\text{particles/s}$$

About 10^8 neutrons per second are emitted from the target.

13.2 RADIOISOTOPE PRODUCTION IN NUCLEAR REACTIONS

Often we use nuclear reactions to produce radioactive isotopes. In this procedure, a stable (nonradioactive) isotope X is irradiated with the particle x to form the radioactive isotope Y; the outgoing particle y is of no interest and is not observed. In this case we don't observe the individual particles Y as they are produced in the reaction; instead, we irradiate the target to produce some number of radioactive Y nuclei that remain within the target. After the irradiation we observe the radioactive decay of the nuclei Y.

We would like now to calculate the activity of the isotope Y that is produced from a given exposure to a certain quantity of the particle x for a certain time t. Let R represent the constant rate at which Y is produced; this quantity is related to the cross section and to the intensity of the beam of x, as given in Eq. 13.1. In a time interval dt, the number of Y nuclei produced is $R\,dt$. The isotope Y is radioactive, so the number of nuclei of Y that decay in the interval dt is $\lambda N\,dt$, where λ is the decay constant ($\lambda = 0.693/t_{1/2}$) and N is the number of Y nuclei present. The net change dN in the number of Y nuclei is

$$dN = R\,dt - \lambda N\,dt \tag{13.3}$$

or

$$\frac{dN}{dt} = R - \lambda N \tag{13.4}$$

The solution to this differential equation is

$$N(t) = \frac{R}{\lambda}(1 - e^{-\lambda t}) \tag{13.5}$$

and the activity is

$$a(t) = \lambda N = R(1 - e^{-\lambda t}) \qquad (13.6)$$

Notice that, as expected, $a = 0$ at $t = 0$ (there are no nuclei of type Y present at the start). For large irradiation times $t \gg t_{1/2}$, this expression approaches the constant value R. When t is small compared with the half-life $t_{1/2}$, the activity increases linearly with time:

$$a(t) = R[1 - (1 - \lambda t + \cdots)] \cong R\lambda t \qquad (t \ll t_{1/2}) \qquad (13.7)$$

Figure 13.4 shows the relationship between $a(t)$ and t. As you can see, not much activity is gained by irradiating for more than about two half-lives.

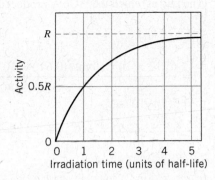

FIGURE 13.4 Formation of activity in a nuclear reaction.

Example 13.2

Thirty milligrams of gold are exposed to a neutron flux of 3.0×10^{12} neutrons/cm^2/s for 1.0 minute. The neutron capture cross section of gold is 99 b. Find the resultant activity of ^{198}Au.

Solution
From Appendix D we find that the stable isotope of gold has a mass number of $A = 197$, and that radioactive ^{198}Au has a half-life of 2.70 d $= 3.88 \times 10^3$ min. Thus, using Eq. 13.2,

$$R = \phi\sigma\frac{m}{M}N_A$$

$$= \left(3.0 \times 10^{12} \frac{\text{neutrons}}{\text{cm}^2 \cdot \text{s}}\right)\left(99 \times 10^{-24} \frac{\text{cm}^2}{\text{neutron} \cdot \text{nucleus}}\right)$$

$$\times \left(\frac{0.030 \text{ g}}{197 \text{ g/mole}}\right)(6.02 \times 10^{23} \text{ atoms/mole})$$

$$= 2.7 \times 10^{10} \text{ s}^{-1}$$

In this case $t \ll t_{1/2}$, so we can use Eq. 13.7:

$$a = R\lambda t = (2.7 \times 10^{10} \text{ s}^{-1})\left(\frac{0.693}{3.88 \times 10^3 \text{ min}}\right)(1.0 \text{ min})$$

$$= 4.8 \times 10^6 \text{ s}^{-1} = 130 \text{ } \mu\text{Ci}$$

Example 13.3

The radioactive isotope ^{61}Cu ($t_{1/2} = 3.41$ h) is to be produced by alpha particle reactions on a target of ^{59}Co. A foil of cobalt, measuring 1.5 cm $\times$ 1.5 cm in area and 2.5 μm in thickness, is placed in a 12.0-μA beam of alpha particles; the beam uniformly covers the target. For the alpha energy selected, the reaction has a cross section of 0.640 b. (a) At what rate is the ^{61}Cu produced? (b) What is the resulting activity of ^{61}Cu after 2.0 h of irradiation?

Solution
(a) The reaction is ^{59}Co $+ ^4$He $\rightarrow ^{61}$Cu $+ 2$n. The mass of the target is $m = \rho V = (8.9 \text{ g/cm}^3)(1.5 \text{ cm})^2(2.5 \times 10^{-4} \text{ cm}) = 5.0 \times 10^{-3} \text{ g}$ and the number of target atoms is

$$N = \frac{mN_A}{M}$$

$$= \frac{(5.0 \times 10^{-3} \text{ g})(6.02 \times 10^{23} \text{ atoms/mole})}{58.9 \text{ g/mole}}$$

$$= 5.12 \times 10^{19} \text{ atoms}$$

The rate at which the beam strikes the target is

$$I_0 = \frac{12.0 \times 10^{-6} \text{ A}}{2 \times 1.60 \times 10^{-19} \text{ C/particle}}$$

$$= 3.75 \times 10^{13} \text{ particles/s}$$

The rate at which the ^{61}Cu is produced is, using Eq. 13.1,

$$R = \frac{N\sigma I_0}{S}$$

$$= \frac{(5.12 \times 10^{19} \text{ atoms})(0.640 \times 10^{-24} \text{ cm}^2)(3.75 \times 10^{13} \text{ s}^{-1})}{(1.5 \text{ cm})^2}$$

$$= 5.5 \times 10^8 \text{ s}^{-1}$$

(b) The activity is determined from Eq. 13.6:

$$a = R(1 - e^{-\lambda t})$$

$$= (5.5 \times 10^8 \text{ s}^{-1})\,(1 - e^{-(0.693)(2.0 \text{ h})/(3.41\text{h})})$$

$$= 1.8 \times 10^8 \text{ s}^{-1} = 4.9 \text{ mCi}$$

13.3 LOW-ENERGY REACTION KINEMATICS

We assume for this discussion that the velocities of the nuclear particles are sufficiently small that we can use nonrelativistic kinematics. We consider a projectile x moving with momentum $\vec{\mathbf{p}}_x$ and kinetic energy K_x. The target is at rest, and the reaction products have momenta $\vec{\mathbf{p}}_y$ and $\vec{\mathbf{p}}_Y$ and kinetic energies K_y and K_Y. The particles y and Y are emitted at angles θ_y and θ_Y with respect to the direction of the incident beam. Figure 13.5 illustrates this reaction. We assume that the resultant nucleus Y is not observed in the laboratory (if it is a heavy nucleus, moving relatively slowly, it generally stops within the target).

As we did in the case of radioactive decay, we use energy conservation to compute the Q value for this reaction (assuming X is initially at rest):

initial energy = final energy

$$m_N(x)c^2 + K_x + m_N(X)c^2 = m_N(y)c^2 + K_y + m_N(Y)c^2 + K_Y \qquad (13.8)$$

The m's in Eq. 13.8 represent the *nuclear* masses of the reacting particles. However, as we have discussed, the number of protons must be balanced in a nuclear reaction:

$$Z_x + Z_X = Z_y + Z_Y \qquad (13.9)$$

We can therefore add equal numbers of electron masses to each side of Eq. 13.8 and, neglecting as usual the electron binding energy, the nuclear masses become atomic masses with no additional corrections needed. Rewriting Eq. 13.8, we obtain

$$[m(x) + m(X) - m(y) - m(Y)]c^2 = K_y + K_Y - K_x \qquad (13.10)$$

The rest energy difference between the initial particles and final particles is defined to be the *Q value* of the reaction

$$Q = (m_i - m_f)c^2 = [m(x) + m(X) - m(y) - m(Y)]c^2 \qquad (13.11)$$

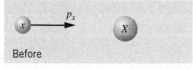

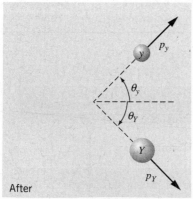

FIGURE 13.5 Momenta of particles before (top) and after (bottom) the reaction.

and, combining Eqs. 13.10 and 13.11, we see that the Q value is equal to the difference in kinetic energy between the final particles and initial particle:

$$Q = K_y + K_Y - K_x \qquad (13.12)$$

Example 13.4

(a) Compute the Q value for the reaction $^2\text{H} + {}^{63}\text{Cu} \rightarrow \text{n} + {}^{64}\text{Zn}$. (b) Deuterons of energy 12.00 MeV are incident on a ^{63}Cu target, and neutrons are observed with 16.85 MeV of kinetic energy. Find the kinetic energy of the ^{64}Zn.

Solution

(a) The Q value can be found using Eq. 13.11 with masses from Appendix D:

$$Q = [m(^2\text{H}) + m(^{63}\text{Cu}) - m(\text{n}) - m(^{64}\text{Zn})]c^2$$

$$= (2.014102\,\text{u} + 62.929597\,\text{u} - 1.008665\,\text{u}$$
$$- 63.929142\,\text{u})(931.5\,\text{MeV/u})$$
$$= 5.488\,\text{MeV}$$

(b) From Eq. 13.12, we find

$$K_Y = Q + K_x - K_y$$
$$= 5.488\,\text{MeV} + 12.00\,\text{MeV} - 16.85\,\text{MeV}$$
$$= 0.64\,\text{MeV}$$

Reactions for which $Q > 0$ convert nuclear energy to kinetic energy of y and Y. They are called *exothermic* or *exoergic* reactions. Reactions with $Q < 0$ require energy input, in the form of the kinetic energy of x, to be converted into nuclear binding energy. These are known as *endothermic* or *endoergic* reactions.

In an endoergic reaction, we must supply at least enough kinetic energy to provide the additional rest energy of the reaction products. There is thus some minimum, or *threshold*, kinetic energy of x, below which the reaction will not take place. This threshold kinetic energy not only must supply the additional rest energy of the products, but also must supply some kinetic energy of the products; even at the minimum energy, the products cannot be at rest, for that would violate conservation of linear momentum—the momentum p_x before the collision would not be equal to the momentum of the final products after the collision if they were formed at rest.

This problem is most easily analyzed in the center-of-mass reference frame. In the lab frame before the reaction, the center of mass moves with velocity $v = m(x)v_x/[m(x) + m(X)]$. If we travel with that velocity and observe the reaction, we would see x moving with velocity $v_x - v$ and X moving with velocity $-v$, as shown in Figure 13.6. If x has exactly the threshold kinetic energy, in this reference frame the reaction products y and Y would be at rest.

We must conserve total relativistic energy $K + mc^2$ in the reaction, and we restrict our discussion to small velocities $v \ll c$ so that the nonrelativistic expression for the kinetic energy can be used. Energy conservation in the center-of-mass frame gives:

$$\frac{1}{2}m(x)(v_x - v)^2 + \frac{1}{2}m(X)(-v)^2 + m(x)c^2 + m(X)c^2 = m(y)c^2 + m(Y)c^2 \tag{13.13}$$

where v_x represents the threshold velocity in the lab frame. Substituting the value of v and doing a bit of algebra, we can find the threshold kinetic energy (in the laboratory reference frame):

$$K_{\text{th}} = -Q\left(1 + \frac{m(x)}{m(X)}\right) \tag{13.14}$$

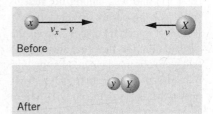

Before

After

FIGURE 13.6 Reaction at threshold in center-of-mass reference frame.

Example 13.5

Calculate the threshold kinetic energy for the reaction $p + {}^3H \rightarrow {}^2H + {}^2H$ (a) if protons are incident on ${}^3H$ at rest; (b) if ${}^3H$ (tritons) are incident on protons at rest.

Solution

The Q value is

$$Q = [m({}^1H) + m({}^3H) - 2m({}^2H)]c^2$$
$$= (1.007825\,u + 3.016049\,u - 2 \times 2.014102\,u)$$
$$\times (931.5\,MeV/u)$$
$$= -4.033\,MeV$$

(a) When protons are incident on ${}^3H$, the identification is $x = {}^1H, X = {}^3H$, so

$$K_{th} = -Q\left(1 + \frac{m({}^1H)}{m({}^3H)}\right)$$

$$= (4.033\,MeV)\left(1 + \frac{1.007825\,u}{3.016049\,u}\right) = 5.381\,MeV$$

(b) When ${}^3H$ is incident on protons, the identification of x and X is reversed, so

$$K_{th} = -Q\left(1 + \frac{m({}^3H)}{m({}^1H)}\right)$$

$$= (4.033\,MeV)\left(1 + \frac{3.016049\,u}{1.007825\,u}\right) = 16.10\,MeV$$

This calculation illustrates a general result: Less energy is required for a nuclear reaction if a light particle is incident on a heavy target than if a heavy particle is incident on a light target.

Lise Meitner (1878–1968, Germany-Sweden). Known for her research into radioactivity, Meitner discovered the radioactive element protactinium ($Z = 91$) and was among the first to study the properties of beta decay. Her most important discovery was the explanation for the puzzling results that were observed when uranium was bombarded with neutrons. She suggested that the uranium could split into two pieces, and she proposed the name "fission" for this process. Element 109 is named in her honor.

13.4 FISSION

The massive nucleus ${}^{254}Cf$ ($Z = 98$) can be produced in accelerators by collisions between suitably chosen projectiles and targets. This nucleus is of special interest because it is also produced in supernova explosions, and knowledge of its properties provides a key to understanding the formation of the elements in stars, as we discuss later in this chapter. ${}^{254}Cf$ is radioactive, decaying with a half-life of 60.5 d. The Q values for positive and negative beta decay of ${}^{254}Cf$ are both negative, so that mode of decay is not available. Alpha decay is energetically possible but the Coulomb barrier is very high, making that decay mode improbable. Instead, ${}^{254}Cf$ decays by splitting into two pieces of much smaller masses—for example,

$$^{254}_{98}Cf_{156} \rightarrow {}^{140}_{54}Xe_{86} + {}^{110}_{44}Ru_{66} + 4n$$

This mode of decay is known as *nuclear fission*. Fission can occur as a spontaneous radioactive decay process for a relatively small number of massive nuclei, and it can also be induced in other nuclei by adding energy to make the nucleus less stable. In addition to the two fission fragments, some neutrons are usually emitted in the fission process.

We can consider the nucleus to be a mixture of protons and neutrons moving about under the mutual attraction of their nuclear forces and (in the case of the protons) the repulsion of their Coulomb forces. For many nuclei, the result of these interactions is a spherical shape that has often been compared to a drop of liquid floating freely in a region where no external forces act. The equilibrium shape will be close to spherical, and if the nucleus is distorted (for example, by stretching it in one direction) it can vibrate about its equilibrium shape and eventually return to its spherical shape somewhat like a stretched spring or other elastic system returns to its original configuration. Figure 13.7 shows a schematic representation of the energy of the drop as a function of the distortion.

For other nuclei, the equilibrium shape is not spherical but is already distorted; their surfaces are like an ellipse rotated about its long axis. For these nuclei, the major axis might be 30–50% longer than the minor axis. If these nuclei are stretched by a small amount and released, they will usually revert to their distorted equilibrium shape. But if the stretching is sufficiently large, they may not return to equilibrium but instead may split in two, as represented in Figure 13.8.

This occurs because of the rather delicate balance between the nuclear force that keeps the nucleus together and the Coulomb repulsion force, which makes the nucleus less stable. When the stretching is sufficiently large, the total attractive nuclear force is reduced (because on the average the protons and neutrons have fewer "near neighbors" with which to interact), but the Coulomb force, because it has a long range, is not reduced significantly. The center of the distorted shape can be "pinched off" and the delicate balance between the nuclear and Coulomb forces is upset. The Coulomb force can then drive the two fragments apart.

Figure 13.8 shows a kind of "barrier" between the distorted equilibrium shape and the fissioned nucleus. This fission barrier has a height of roughly 6 MeV, but as we know it is possible to "tunnel" through the barrier. Thus nuclei can undergo fission with smaller amounts of excitation energy. It is very unlikely for a nucleus to tunnel through the barrier at its thickest, but as the excitation energy is increased the barrier becomes less thick and fission becomes more probable. For the radioactive decay of ^{254}Cf, the probability to penetrate the fission barrier (99.7%) is greater than the probability to penetrate the barrier to alpha decay (0.3%).

The splitting of a nucleus such as ^{254}Cf into two fragments does not always produce the same set of final nuclei. Many different processes are possible, with the actual outcome determined according to statistical probability. For ^{254}Cf it is most likely that one fragment will have a mass near $A = 110$ and the other near $A = 140$, but other mass distributions can occur. Figure 13.9 shows the mass distribution of the fragments in the fission of ^{254}Cf and ^{235}U.

The number of neutrons emitted in fission can also vary. For ^{254}Cf, the average is about 3.9 neutrons per fission.

Energy Released in Fission

According to Figure 12.4, the binding energy per nucleon of ^{254}Cf is about 7 MeV. If a nucleus of ^{254}Cf splits into two nuclei with $A = 127$, the binding energy per nucleon of the final nuclei would be about 8 MeV. Thus the binding energy of each of those 254 nucleons increases from about 7 MeV to about 8 MeV, which gives an increase in the total binding energy of the nucleus of about 250 MeV. If the final nucleons are more tightly bound, that means that an equivalent amount of energy has been released to some other form.

This energy release from a single nucleus is an enormous quantity. For comparison, chemical processes such as combustion usually release a few eV per atom. In fission we have an energy release at the atomic level that is 10^8 times larger than the energy released in chemical processes!

Where does this energy go? Let's imagine that ^{254}Cf suddenly breaks in half to form two nuclei with $Z = 49$ and $A = 127$ that are just touching at their surfaces. The radius of each of these nuclei is $1.2(127)^{1/3} = 6.0$ fm, and the Coulomb repulsion of these two nuclei would be $U = (49e)^2/4\pi\varepsilon_0(2R) = 286$ MeV, which is quite close to our estimate of the nuclear binding energy released. These two charged objects repel one another, so that the Coulomb energy quickly becomes kinetic energy. Most of the energy released in fission is thus produced as

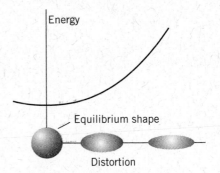

FIGURE 13.7 The energy of a nucleus with a spherical equilibrium shape increases as the distortion increases.

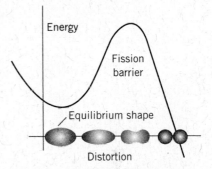

FIGURE 13.8 The energy of a nucleus with a nonspherical equilibrium shape. If enough energy is added, the nucleus can tunnel through the fission barrier and split into two pieces.

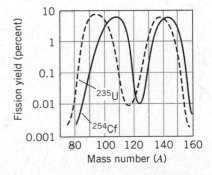

FIGURE 13.9 The mass distribution of fission fragments from ^{254}Cf (solid line) and ^{235}U (dashed line).

Nuclear binding energy

↓

Coulomb potential energy of nuclear fragments

↓

Kinetic energy of fragments

About 80% of the fission energy is released in this form. The fragments do not travel very far before dissipating their kinetic energy through atomic collisions, which are usually observed as a temperature increase of the material. In a power reactor, this temperature increase can be used to produce steam, which can drive a turbine to produce electricity. The remaining 20% of the energy released appears as decay products (betas and gammas) of the highly radioactive fragments and kinetic energies of the neutrons that may also be emitted during fission.

Induced Fission

The radioactive decay of ^{254}Cf is an example of the spontaneous fission of a nucleus that is sufficiently unstable that it can tunnel through the fission barrier with no additional energy needed. However, ^{254}Cf is an artificially produced nucleus that does not occur in nature and in which the energy released in fission is essentially stored in the nucleus by the nuclear reaction that was used to produce the ^{254}Cf. There are other examples of fissionable nuclei that may occur naturally or are produced artificially but that do not fission spontaneously. These nuclei can be made to fission by the addition of some energy, which might be in the form of an absorbed photon but more often occurs with the absorption of a neutron. In these cases the energy input is very small compared with the energy released in the fission process. One such nucleus is ^{235}U, which might absorb a neutron to make ^{236}U and then fission according to

$$^{235}_{92}U_{143} + n \rightarrow\, ^{93}_{37}Rb_{56} +\, ^{141}_{55}Cs_{86} + 2n$$

As in the case of spontaneous fission, many different outcomes are possible, with a statistical distribution of the masses of the fragments. In the fission of ^{235}U, the most probable outcome has fragments of mass numbers near $A = 90$ and $A = 140$ (as in Figure 13.9), and the average number of neutrons is about 2.5. Examples of easily produced fissionable nuclei include ^{239}Pu (obtained from the beta decay of ^{239}U, which is made when ^{238}U absorbs a neutron) and ^{233}U (obtained in a similar manner from ^{232}Th).

In a bulk sample of uranium, each of the neutrons emitted in fission can be absorbed by another nucleus of ^{235}U and thus induce another fission process, resulting in the emission of still more neutrons, followed by more fissions, and so forth. As long as the average number of neutrons available to produce new fissions is greater than 1 per reaction, the number of fissions grows with time. This avalanche or *chain reaction* of fission events, each with the release of about 200 MeV of energy, can either occur under very rapid and uncontrolled conditions, as in a nuclear weapon, or else under slower and carefully controlled conditions, as in a nuclear reactor.

Electrical Power from Fission

Electrical power can be generated using the thermal energy released in fission to boil water. Ordinary uranium by itself cannot serve as fuel for the reactor for several reasons, of which three in particular stand out: enrichment, moderation, and control.

Enrichment   To maintain a steady energy production from fission reactions, we would like for one neutron from each fission to be available to produce another fission. Generally the average number of neutrons produced is greater than one, but neutrons can be lost from the reaction in a variety of ways (for example, by non-fission absorption by ^{238}U in uranium fuel). Natural uranium consists of only about 0.7% of ^{235}U and 99.3% ^{238}U, which means that most of the available uranium nuclei generally do not participate in the fission process but instead remove neutrons from being able to produce other fissions. To overcome this problem it is necessary to use *enriched* uranium, in which the abundance of ^{235}U is increased beyond its natural value of 0.7%. Most power reactors use uranium enriched to 3–5% ^{235}U. Enrichment is a difficult process because ^{235}U and ^{238}U are chemically identical. It is achieved only by taking into account the small mass difference between the two isotopes (for example, by forcing gaseous uranium through a porous barrier in which the more massive ^{238}U atoms diffuse more slowly).

Moderation   The neutrons produced in fission typically have kinetic energies of a few MeV. Such energetic neutrons have a relatively low probability of inducing new fissions, because the fission cross section generally decreases rapidly with increasing neutron energy. We therefore must slow down, or *moderate*, these neutrons in order to increase their chances of initiating fission events. The fissionable material is surrounded by a *moderator*, and the neutrons lose energy in collisions with the atoms of the moderator. When a neutron is scattered from a heavy nucleus like uranium, the energy of the neutron is changed hardly at all, but in a collision with a very light nucleus, the neutron can lose substantial energy. The most effective moderator is one whose atoms have about the same mass as a neutron; hydrogen is therefore the first choice. Ordinary water is frequently used as a moderator, because collisions with the protons are very effective in slowing the neutrons; however, neutrons have a relatively high probability of being absorbed by the water according to the reaction $p + n \rightarrow {}_1^2H_1 + \gamma$. So-called "heavy water," in which the hydrogen is replaced by deuterium, is more useful as a moderator, because it has virtually no neutron absorption cross section. A heavy-water reactor, which has more available neutrons, can use ordinary (nonenriched) uranium as fuel; a reactor using ordinary water as moderator has fewer neutrons available to produce fission, and must therefore have more ^{235}U in its core.

Carbon is a light material that is solid, stable, and abundant, and that has a relatively small neutron absorption cross section. Enrico Fermi and his co-workers built the first nuclear reactor in 1942 at the University of Chicago; this reactor used carbon, in the form of graphite blocks, as moderator.

Control   To produce a stable nuclear reactor, the average number of neutrons in each fission reaction that is available to produce the next set of fission reactions must be exactly equal to 1. If it is even slightly greater than 1, the reaction rate will grow exponentially out of control. Control of the reaction rate is usually accomplished by inserting into the core of the reactor control rods made of cadmium, which has a very large cross section for absorbing neutrons and thus removing them from the fission process. However, small fluctuations in the reaction rate occur much too rapidly for any mechanical system to move the control rods in and out to control the number of neutrons emitted in the fission reaction.

Fortunately, nature has provided us with the solution to this problem. About 1% of the neutrons emitted in fission are *delayed neutrons*, produced not at the instant of fission but somewhat later, following the radioactive decays of the fission fragments. For example ^{93}Rb, which might be produced in the fission of

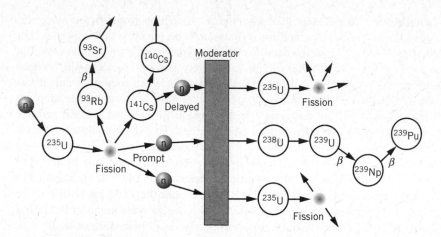

FIGURE 13.10 A typical sequence of processes in fission. A nucleus of ^{235}U absorbs a neutron and fissions; two prompt neutrons and one delayed neutron are emitted. Following moderation, two neutrons cause new fissions and the third is captured by ^{238}U, resulting finally in ^{239}Pu.

^{235}U, beta decays with a half-life of 6 s to ^{93}Sr, which is occasionally (in about 1% of decays) produced in a very high excited state that is unstable to neutron emission. The neutron appears to emerge with the 6-s half-life of the beta decay. This short delay time is enough to allow the control rods to be adjusted to maintain a constant reaction rate. The reactor is designed so that the neutron replication rate is just less than 1 for the prompt neutrons and exactly equal to 1 for prompt + delayed neutrons, which allows the control rods to work effectively.

Figure 13.10 summarizes some of the processes that can occur in fission. A nucleus of ^{235}U captures a neutron and fissions into two heavy fragments and two prompt neutrons; one of the fragments emits a delayed neutron. The three neutrons are slowed by passage through the moderator. Two of the neutrons cause new fissions, and the third is captured by ^{238}U, eventually to form fissionable ^{239}Pu, which can be recovered from the fuel by chemical means. Not shown in this diagram are other processes that can occur: escape of neutrons through the surface of the reactor, capture in the moderator, and fission of ^{238}U by fast (unmoderated) neutrons.

Fission Reactors

In a fission reactor, the heat produced in the fuel must be extracted to generate electrical power. It must also be extracted for reasons of safety, because enough heat is produced to melt the core and cause a serious accident. For this reason, reactors contain an emergency core cooling system that is designed to prevent the core from overheating if the heat extraction system should fail.

Extracting the fission energy from the reactor core can be accomplished through several different techniques. In one design, called the *pressurized water reactor* and illustrated in Figure 13.11, the heat is extracted in a two-step process. Water circulates through the core under great pressure, to prevent its turning to steam. This hot water then in turn heats a second water system, which actually delivers steam to the turbine. The steam never enters the reactor core, so it does not become radioactive, and thus there is no radioactive material in the vicinity of the turbine.

The power reactors in the United States are mostly the pressurized-water type using enriched uranium as fuel and ordinary water as moderator. Canada also

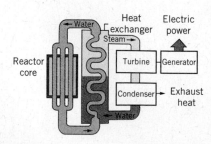

FIGURE 13.11 The components of a pressurized-water reactor.

uses pressurized water reactors, but heavy water and natural uranium are used. In a variation on this design, the pressurized water is replaced with a liquid metal such as sodium, which has the advantages of remaining liquid at much higher temperatures than water and of having a larger thermal conductivity than water. Yet another design uses gas flow through the core to extract the heat; the hot gas is then used to produce steam. Reactors in Great Britain are gas-cooled and graphite-moderated.

There are yet other technological problems associated with nuclear power that are the subjects of active debate and investigation. Some of the radioactive isotopes among the fission fragments have very long half-lives, of the order of many years. The radioactive waste from reactors must be stored in a manner that prevents leakage of radioactive material into the biological environment. Many people are concerned about the safety of nuclear reactors, not only regarding proper design and operation, but also about their resistance to natural disasters such as earthquakes or to acts of terrorism or sabotage.

In 1986, a graphite-moderated power reactor at Chernobyl in the former U.S.S.R. suffered a serious accident due to the disabling of the core cooling system, which is designed to extract the intense heat generated in the reactor core. The resulting temperature rise ignited the graphite moderator and caused an explosion of the reactor containment vessel, releasing radioactive fission products and exposing the inhabitants of the region to life-threatening radiation doses. The water-moderated power reactors used in the United States cannot suffer this kind of accident.

The vulnerability of reactors to natural disasters was dramatically revealed by the earthquake-triggered tsunami that struck Japan's Fukushima reactor complex in 2011. The flooding of the reactor buildings caused the pumps supplying cooling water to fail. As a result, the reactor core overheated due to the radioactive decay of the fission products, and a partial meltdown of the fuel rods occurred. The ensuing release of radioactivity contaminated a wide region of the Japanese countryside.

Finally, as in all heat engines, the disposal of the exhaust or waste heat (primarily from the steam recondensing to water) generates considerable thermal pollution. Nuclear power plants are generally less efficient at converting fuel to electrical power compared with plants that burn fossil fuels, because nuclear plants operate at lower temperatures; while fossil-fuel plants can have efficiencies as large as 40%, nuclear plants are generally in the range of 30 to 35%. A plant operating at 30% efficiency produces 50% more thermal pollution than one that generates the same amount of power at 40% efficiency.

A Naturally Occurring Fission Reactor

We conclude this section with a fascinating example of nature at work—the first sustained nuclear fission reactor on Earth was *not* the one constructed by Fermi in Chicago in 1942, but a *natural* fission reactor in Africa, which is believed to have operated two billion years ago for a period of perhaps several hundred thousand years. This reactor of course used naturally occurring uranium as a fuel and naturally occurring water as a moderator.

It would not be possible to build such a reactor today, because the capture of neutrons by the protons in water results in too few neutrons remaining to sustain a chain reaction in uranium with only 0.7% of ^{235}U. However, two billion years ago, naturally occurring uranium contained a much larger fraction of ^{235}U than does present-day uranium. Both ^{235}U and ^{238}U are radioactive, but the half-life of ^{235}U is only about one-sixth as great as the half-life of ^{238}U. If we go back in time about 2×10^9 y, which is half of one half-life of ^{238}U, there was about 40% more ^{238}U than there is today, but there was $2^3 = 8$ times as much ^{235}U. Naturally occurring uranium was then about 3% ^{235}U, and, at such enrichments, ordinary water can serve as an effective moderator.

A deposit of such uranium, in a large enough mass and with ground water present to act as moderator, could have "gone critical" and begun to react. The reaction could have been controlled by the boiling of the water—when enough heat had been generated to evaporate some of the water, the reaction would slow down and perhaps stop, because of the lack of a moderator. When the uranium had cooled sufficiently to allow more liquid water to collect, the reactor would have started up again. This cycle could in principle have continued indefinitely, until enough ^{235}U was used up or until geological changes resulted in the removal of the water.

The discovery of this reactor followed the observation that the uranium that was being mined from that region in Africa contained too little ^{235}U. The discrepancy was a very small one—the samples contained 0.7171% ^{235}U, compared with the usual 0.7202%—but it was enough to stimulate the curiosity of the researchers. They guessed that the only mechanism that could result in the consumption of ^{235}U was the nuclear fission process, and this guess was tested by searching in the ore for stable isotopes that result from the radioactive decay of fission products. When such isotopes were found, and in particular when they were found in abundances very different from what would be expected from "natural" mineral deposits, the existence of the natural reactor was confirmed.

13.5 FUSION

Energy may also be released in nuclear reactions in the process of fusion, in which two light nuclei combine to form a heavier nucleus. The energy released in this process is the excess binding energy of the heavy nucleus compared with the lighter nuclei; from Figure 12.4, we see that this process can release energy as long as the final nucleus is less massive than about $A = 60$.

For example, consider the reaction

$$^2_1\text{H}_1 + ^2_1\text{H}_1 \rightarrow ^3_1\text{H}_2 + ^1_1\text{H}_0$$

The Q value is 4.0 MeV, and so this nuclear reaction liberates about 1 MeV per nucleon, roughly the same as the fission reaction. This reaction can occur when a beam of deuterons is accelerated on to a deuterium target. In order to observe the reaction, we must get the incident and target deuterons close enough that the nuclear force can produce the reaction; that is, we must overcome the mutual Coulomb repulsion of the two particles. We can estimate this Coulomb repulsion by calculating the electrostatic repulsion of two deuterons when they are just touching. The radius of a deuteron is about 1.5 fm, and the electrostatic potential energy of the two charges separated by about 3 fm is about 0.5 MeV. A deuteron with 0.5 MeV of kinetic energy can overcome the Coulomb repulsion and initiate a reaction in which 4.5 MeV of energy (0.5 MeV of incident kinetic energy plus the 4-MeV Q value) is released.

Doing this reaction in a typical accelerator, in which the beam currents are typically in the microampere range, would produce only a small amount of energy (of the order of a few watts). To obtain significant amounts of energy from fusion, it is necessary to work with much larger quantities of deuterium. For example, the fusion energy from the deuterium in a liter of ordinary water (which contains 0.015% D_2O) would be equivalent to the chemical energy obtained from burning about 300 liters of gasoline.

A more promising approach consists of heating deuterium gas to a high enough temperature so that each atom of deuterium has about 0.25 MeV of thermal kinetic energy (hence the name *thermonuclear* fusion). Then in a collision between two deuterium atoms, the total of 0.5 MeV of kinetic energy would be sufficient to overcome the Coulomb repulsion.

The difficulty with this approach is in heating the deuterium gas to a sufficient temperature; from the expression $\frac{1}{2}kT$ for the thermal kinetic energy of a gas molecule, we can calculate that an energy of 0.25 MeV corresponds to a temperature of the order of 10^9 K. Even assuming that barrier penetration (Section 5.6) would allow a reasonable probability to penetrate the Coulomb barrier at lower kinetic energies (perhaps corresponding to one-tenth of the calculated temperature), it is hard to imagine conditions under which these temperatures can be created. However, such conditions do exist in the interiors of stars, which produce their energy through fusion reactions. Fusion processes thus support all life on Earth. Scientists and engineers who are seeking to develop fusion processes for electrical power generation face the challenge of duplicating, for a brief instant of time and on a much smaller scale, the conditions in the interior of stars.

Fusion Processes in Stars

In the basic fusion process that occurs in stars (including our Sun), four protons combine to make one ^{4}He. Stars are composed of ordinary hydrogen rather than deuterium, so it is first necessary to convert the hydrogen to deuterium. This is done according to the reaction

$$^1_1H_0 + {}^1_1H_0 \rightarrow {}^2_1H_1 + e^+ + \nu$$

This process involves converting a proton to a neutron and is analogous to the beta-decay processes discussed in Chapter 12. Once we have obtained ^{2}H (deuterium), the next reaction that can occur is

$$^2_1H_1 + {}^1_1H_0 \rightarrow {}^3_2He_1 + \gamma$$

followed by

$$^3_2He_1 + {}^3_2He_1 \rightarrow {}^4_2He_2 + 2{}^1_1H_0$$

Note that the first two reactions must occur *twice* in order to produce the two ^{3}He we need for the third reaction; see the schematic diagram of Figure 13.12. We can write the net process as

$$4{}^1_1H_0 \rightarrow {}^4_2He_2 + 2e^+ + 2\nu + 2\gamma$$

For the calculation of the Q value in terms of *atomic* masses, four electrons must be added to the left side to make four neutral hydrogen atoms. To balance the reaction we must also add four electrons to the right side; two of these are associated with the ^{4}He atom, and the other two can be combined with the two positrons according to the reaction $e^+ + e^- \rightarrow 2\gamma$, so that the additional gamma rays are available as energy from the reaction. The two positrons disappear in this

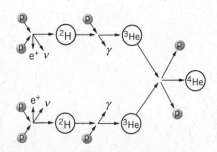

FIGURE 13.12 Schematic diagram of processes in the fusion of protons to form helium.

process; the only masses remaining are four hydrogen *atoms* and the one helium *atom,* and so

$$Q = (m_i - m_f)c^2 = [4m(^1\text{H}) - m(^4\text{He})]c^2$$
$$= (4 \times 1.007825\,\text{u} - 4.002603\,\text{u})(931.5\,\text{MeV/u}) = 26.7\,\text{MeV}$$

Each fusion reaction liberates about 26.7 MeV of energy.

At what rate do these reactions occur in the Sun? About 1.4×10^3 W of solar power is incident on each square meter of the Earth's surface. At our distance of about 1.5×10^{11} m from the Sun, its energy is spread over a sphere of area $4\pi r^2 = 28 \times 10^{22}\,\text{m}^2$, and thus the power output from the Sun is about 4×10^{26} W, which corresponds to about 2×10^{39} MeV/s. Each fusion reaction liberates about 26 MeV, and thus there must be about 10^{38} fusion reactions per second, consuming about 4×10^{38} protons per second. (Don't worry about running out of protons—the Sun's mass is about 2×10^{30} kg, which corresponds to about 10^{57} protons, enough to burn for the next few billion years.)

The sequence of reactions described above is called the *proton-proton* cycle and probably represents the source of the Sun's energy. However, it is probably *not* the primary source of fusion energy in many stars, because the first reaction (in which two protons combine to form a deuteron), which is similar to beta decay, takes place only on a very long time scale (as we discuss in the next chapter), and is therefore very unlikely to occur. A more likely sequence of reactions is the *carbon cycle*:

$$^{12}\text{C} + {}^1\text{H} \rightarrow {}^{13}\text{N} + \gamma$$
$$^{13}\text{N} \rightarrow {}^{13}\text{C} + e^+ + \nu$$
$$^{13}\text{C} + {}^1\text{H} \rightarrow {}^{14}\text{N} + \gamma$$
$$^{14}\text{N} + {}^1\text{H} \rightarrow {}^{15}\text{O} + \gamma$$
$$^{15}\text{O} \rightarrow {}^{15}\text{N} + e^+ + \nu$$
$$^{15}\text{N} + {}^1\text{H} \rightarrow {}^{12}\text{C} + {}^4\text{He}$$

A symbolic diagram of the process is shown in Figure 13.13. Notice that the ^{12}C plays the role of catalyst; we neither produce nor consume any ^{12}C in these reactions, but the presence of the carbon permits this sequence of reactions to take place at a much greater rate than the previously discussed proton-proton cycle. The net process is still described by $4{}^1\text{H} \rightarrow {}^4\text{He}$, and of course the Q value is the same. The Coulomb repulsion between H and C is larger than the Coulomb repulsion between two H nuclei, so more thermal energy and a correspondingly higher temperature are needed for the carbon cycle. The carbon cycle probably becomes important at a temperature of about 20×10^6 K, while the Sun's interior temperature is "only" 15×10^6 K.

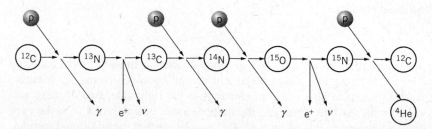

FIGURE 13.13 Sequence of events in the carbon cycle.

Fusion Reactors

For a controlled thermonuclear reactor, several reactions could be used, such as

$$^2\text{H} + {}^2\text{H} \rightarrow {}^3\text{H} + {}^1\text{H} \qquad Q = 4.0\,\text{MeV}$$
$$^2\text{H} + {}^2\text{H} \rightarrow {}^3\text{He} + \text{n} \qquad Q = 3.3\,\text{MeV}$$
$$^2\text{H} + {}^3\text{H} \rightarrow {}^4\text{He} + \text{n} \qquad Q = 17.6\,\text{MeV}$$

The third reaction, known as the D-T (deuterium-tritium) reaction, has the largest energy release and is perhaps the best candidate for a fusion reactor. When deuterium gas (or a deuterium-tritium mixture) is heated to a high temperature, the atoms become ionized; the resulting gas of hot, ionized particles is called a *plasma*. To increase the probability of collisions between the ions that would result in fusion, there are three requirements for the plasma: (1) *a high density n*, so that the particles have a high probability of collision; (2) *a high temperature T*, in the range of 10^8 K, which increases the probability for the particles to penetrate their mutual Coulomb barrier; and (3) *a long confinement time τ*, during which the high temperature and density must be maintained. The first and third of these parameters can be combined using some fairly general considerations based on the power needed to heat the plasma (which is proportional to the density n) and the power derived from fusions in the plasma (which is proportional to $n^2\tau$). For the fusion power to exceed the input power, the product $n\tau$ must exceed a certain minimum value; this condition is

$$n\tau \geq 10^{20}\,\text{s}\cdot\text{m}^{-3} \qquad (13.15)$$

which is known as *Lawson's criterion*. The capability of a plasma to produce energy through fusion can be characterized by the value of its Lawson's parameter $n\tau$ and its temperature T.

The electrical repulsion of the ionized particles in a plasma tends to force the ions away from one another and toward the walls of their container, where they would lose energy in collisions with the cooler atoms of the walls. To maintain the density and temperature, two techniques are under development. In *magnetic confinement*, intense magnetic fields are used to trap the motion of the particles, and in *inertial confinement*, the plasma is heated and compressed so quickly that fusion occurs before the fuel can expand and cool.

Magnetic confinement A magnetic field can confine a plasma because the charged particles spiral around the magnetic field lines. Figure 13.14 shows a toroidal magnetic confinement geometry. There are two contributions to the magnetic field: One is along the toroid axis and another is around the axis. The combination of these two fields gives a helical field along the toroid axis, and the charged particles are confined as they spiral about the field lines. This type of device is a called a *tokamak* (from the Russian acronym for "toroidal magnetic chamber"). A current passed through the plasma serves both to heat the plasma and to create one of the magnetic field components. Figure 13.15 shows the Tokamak Fusion Test Reactor at Princeton University, which operated from 1982 to 1997 and achieved an ion temperature of 5.1×10^8 K and a fusion power level of 10.7 MW. This device came very close to reaching Lawson's criterion with a plasma density of $n = 10^{20}$ particles/m^3 (five orders of magnitude smaller than an ordinary gas) and a confinement time of $\tau = 0.2$ s.

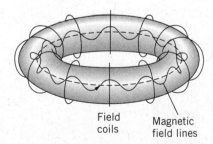

Field coils   Magnetic field lines

FIGURE 13.14 The toroidal geometry of plasma confinement. The ionized atoms circulate around the ring, trapped by the magnetic field lines. The coils produce a magnetic field along the axis of the toroid (dashed line). Another field component is produced by a current along the axis in the plasma. The two components of the field produce the helical field lines shown.

FIGURE 13.15 The inside of the Tokamak Fusion Test Reactor. The technician at the left gives a measure of the size of the toroidal chamber. (Dietmar Krause/Princeton Plasma Physics Laboratory.)

The development of magnetic confinement devices has produced a steady march toward achieving a self-sustaining fusion reactor by increasing the values of both Lawson's parameter $n\tau$ and the temperature, as illustrated in Figure 13.16. Devices have closely approached "breakeven," where the power produced by fusion reactions equals the power necessary to heat the plasma. A true self-sustaining reactor requires the attainment of "ignition," where the power produced by fusion reactions can maintain the reactor with no external source of energy. The next generation of fusion reactor development is the ITER (originally, the International Thermonuclear Experimental Reactor), currently under construction in France as a collaboration among many nations and expected to be operational in the year 2016. The ITER is planned to produce fusion power levels that are 5–10 times what is necessary to heat the plasma (that is, 5–10 times the "breakeven" condition).

Inertial confinement   Inertial confinement takes the opposite approach by compressing the fuel to high densities for very short confinement times. In one method, which is illustrated in Figure 13.17, a small pellet of D-T fuel is struck simultaneously from many directions by intense laser beams that first vaporize the pellet and convert it to a plasma, and then heat and compress it to the point at which fusion can occur. The laser pulses are very short, typically lasting only about 1 ns, and thus according to Lawson's criterion the density must exceed 10^{29} particles/m^3. However, because of inefficiencies of the lasers and other losses a self-sustaining laser fusion reactor must exceed this minimum by perhaps 2–3 orders of magnitude. Figure 13.18 shows the target chamber of the National Ignition Facility at the Lawrence Livermore National Laboratory. Operating for the first time in 2010, it is designed so that a 2-mm diameter pellet of D-T is struck simultaneously by 192 laser beams that deliver an energy of 1 MJ in a pulse lasting a few ns, which is expected to compress the pellet to a central density that is 100 times that of lead.

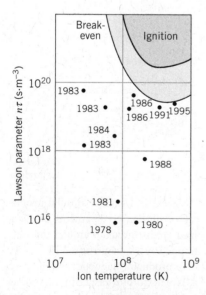

FIGURE 13.16   The approach to breakeven and ignition in fusion reactors, shown as a plot of Lawson's parameter against temperature.

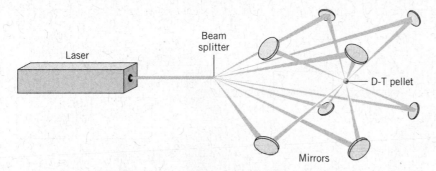

FIGURE 13.17 Inertial confinement fusion initiated by a laser.

FIGURE 13.18 Workers inside the 10-m-diameter target chamber where 192 laser beams (which enter the chamber through the circular ports) strike the target that is held by the positioning arm at the right. (Courtesy Lawrence Livermore National Laboratory.)

In the D-T fusion reaction, most of the energy is carried by the neutrons (recall that in the fission reaction only a small fraction of the energy went to the neutrons). This presents some difficult problems for the recovery of the energy and its conversion into electrical power. One possibility for a fusion reactor design is shown in Figure 13.19. The reaction area is surrounded by lithium, which captures neutrons by the reaction

$$^6_3\text{Li}_3 + \text{n} \rightarrow ^4_2\text{He}_2 + ^3_1\text{H}_2$$

The kinetic energies of the reaction products are rapidly dissipated as heat, and the thermal energy of the liquid lithium can be used to convert water to steam in order to generate electricity. This reaction has the added advantage of producing tritium (^3H), which is needed as a fuel for the fusion reactor.

One difficulty with the D-T fusion process is the large number of neutrons released in the reactions. Although fusion reactors will not produce the radioactive wastes that fission reactors do, the neutrons are sure to make radioactive the immediate area surrounding the reactor, and the structural damage to materials resulting from exposure to large fluxes of neutrons may weaken critical parts of

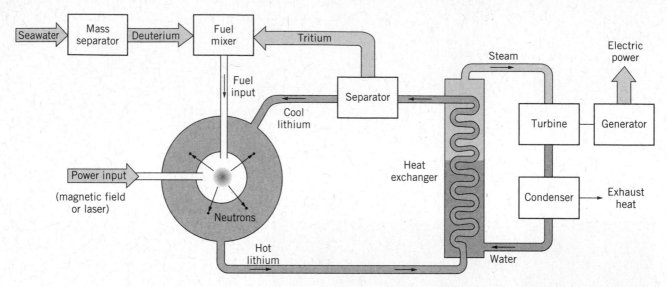

FIGURE 13.19 Design of a fusion reactor.

the reactor vessel. Here once again the lithium is helpful, because a 1-m thickness of lithium should be sufficient to stop essentially all of the neutrons.

Fusion energy is the subject of vigorous research in many laboratories in the United States and around the world; the technological problems are being attacked with a variety of methods, and researchers are hopeful that solutions can be found during the next 20 years so that fusion can help to supply our electrical power needs.

13.6 NUCLEOSYNTHESIS

After a star's hydrogen has been converted to helium through fusion reactions, gravitational collapse can occur that raises the temperature of the core of the star from about 10^7 K to about 10^8 K. At this point there is enough thermal kinetic energy to overcome the Coulomb repulsion of the helium nuclei, and helium fusion can begin. In this process three ^{4}He are converted into ^{12}C by the two-step process

$$^4\text{He} + {}^4\text{He} \rightarrow {}^8\text{Be}$$
$$^8\text{Be} + {}^4\text{He} \rightarrow {}^{12}\text{C}$$

The first reaction is endothermic, with a Q value of 92 keV. The nucleus ^{8}Be is unstable and decays back into two alpha particles in a time of the order of 10^{-16} s. Even so, the Boltzmann factor $e^{-\Delta E/kT}$ suggests that at 10^8 K there will be a small concentration of ^{8}Be. The second reaction has a particularly large cross section; in spite of the rapid breakup of ^{8}Be, there is still a good chance to form ^{12}C. The net Q value for the process is 7.3 MeV, or about 0.6 MeV per nucleon, much less than the 6.7 MeV per nucleon produced by hydrogen burning.

Once enough ^{12}C has formed in the core, other alpha particle reactions become possible, such as

$$^{12}\text{C} + {}^4\text{He} \rightarrow {}^{16}\text{O}$$
$$^{16}\text{O} + {}^4\text{He} \rightarrow {}^{20}\text{Ne}$$
$$^{20}\text{Ne} + {}^4\text{He} \rightarrow {}^{24}\text{Mg}$$

Each of these reactions is exothermic, releasing a few MeV of energy and contributing to the star's energy production. At still higher temperatures (10^9 K) carbon burning and oxygen burning begin:

$$^{12}\text{C} + {}^{12}\text{C} \rightarrow {}^{20}\text{Ne} + {}^4\text{He}$$
$$^{16}\text{O} + {}^{16}\text{O} \rightarrow {}^{28}\text{Si} + {}^4\text{He}$$

Eventually ^{56}Fe is reached, at which point no further energy is gained by fusion (Figure 12.4).

If this explanation of the formation of elements is correct, we expect the abundances of the elements to have the following properties:

1. Large relative abundances of the light, even-Z elements; small relative abundances of odd-Z elements.
2. Little or none of the elements between He and C (Li, Be, B), which are not produced in these reactions.
3. Large relative abundance of Fe, the end product of the fusion cycle.

Figure 13.20 shows the relative abundances of the light elements in the solar system, and they are in agreement with all of the three above expectations. Each even-Z element is 10 to 100 times more abundant than its odd-Z neighbors; there is a prominent peak at Fe; the heavy elements with $Z > 30$ *combined* are less abundant than every element but one in the range C to Zn; and the three elements Li, Be, B are far less abundant than the elements in the range C to Zn.

The light odd-Z elements can be produced by alternative reactions among the fusion products, for example:

$$^{12}\text{C} + {}^{12}\text{C} \rightarrow {}^{23}\text{Na} + {}^1\text{H}$$
$$^{16}\text{O} + {}^{16}\text{O} \rightarrow {}^{31}\text{P} + {}^1\text{H}$$

The abundance of nitrogen is nearly equal to that of its neighbors C and O, which are the most abundant of elements beyond H and He; nitrogen has a greater abundance than any other odd-Z element shown, and greater than all even-Z elements with $Z > 8$. The formation of nitrogen must therefore be a relatively

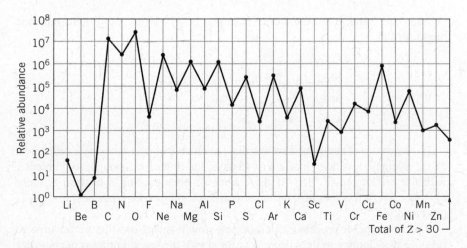

FIGURE 13.20 Relative abundances (by weight) of the elements beyond helium in the solar system.

common process in stars. The element B is rare, so alpha particle reactions are of no help in forming nitrogen. The most likely sources of N are

$$^{12}C + {}^1H \rightarrow {}^{13}N + \gamma \qquad\qquad {}^{16}O + {}^1H \rightarrow {}^{17}F + \gamma$$
$$^{13}N \rightarrow {}^{13}C + e^+ + \nu \qquad \text{and} \qquad {}^{17}F \rightarrow {}^{17}O + e^+ + \nu$$
$$^{13}C + {}^1H \rightarrow {}^{14}N + \gamma \qquad\qquad {}^{17}O + {}^1H \rightarrow {}^{14}N + {}^4He$$

The stable isotopes ^{13}C and ^{17}O are found in natural carbon and oxygen with abundances of 1.1% and 0.04%, which suggests that these reactions do indeed take place.

The production of the elements beyond iron requires the presence of neutrons, which are not produced in the reactions we have listed so far, because neutrons are likely to be emitted only in reactions with nuclei that have an excess of neutrons. If enough of the heavier isotopes, such as ^{13}C, ^{17}O, or ^{21}Ne, are formed, the following reactions can produce neutrons:

$$^{13}C + {}^4He \rightarrow {}^{16}O + n$$
$$^{17}O + {}^4He \rightarrow {}^{20}Ne + n$$
$$^{21}Ne + {}^4He \rightarrow {}^{24}Mg + n$$

How are the heavy elements built up by neutron capture? Consider the effect of neutron capture on ^{56}Fe:

$$^{56}Fe + n \rightarrow {}^{57}Fe \qquad \text{(stable)}$$
$$^{57}Fe + n \rightarrow {}^{58}Fe \qquad \text{(stable)}$$
$$^{58}Fe + n \rightarrow {}^{59}Fe \qquad (t_{1/2} = 45\,\text{d})$$

What happens next depends on the number of available neutrons. If that number is small, the chances of ^{59}Fe encountering a neutron *before* it decays to ^{59}Co are small, and the process might continue as follows:

$$^{59}Fe \rightarrow {}^{59}Co + e^- + \overline{\nu}$$
$$^{59}Co + n \rightarrow {}^{60}Co \qquad (t_{1/2} = 5\,\text{y})$$
$$^{60}Co \rightarrow {}^{60}Ni + e^- + \overline{\nu}$$

On the other hand, if the number of neutrons is very large, a different sequence might result:

$$^{59}Fe + n \rightarrow {}^{60}Fe \qquad (t_{1/2} = 3 \times 10^5\,\text{y})$$
$$^{60}Fe + n \rightarrow {}^{61}Fe \qquad (t_{1/2} = 6\,\text{m})$$
$$^{61}Fe \rightarrow {}^{61}Co + e^- + \overline{\nu}$$
$$^{61}Co \rightarrow {}^{61}Ni + e^- + \overline{\nu}$$

If the density of neutrons is so low that the chance of encountering a neutron is, on the average, less than once every 45 days, the first process ought to dominate, with the production of ^{60}Ni. If the chance of encountering a neutron is more like once every few minutes, the second process should dominate, and no ^{60}Ni is produced.

The first type of process, which occurs *slowly* and allows the nuclei time to beta decay, is known as the *s process* (*s* for slow); the second process occurs very *rapidly* and is known as the *r process* (*r* for rapid).

Figure 13.21 illustrates how the r and s processes can proceed from ^{56}Fe. The s process never strays very far from the region of the stable nuclei, while the r process can produce many nuclei that have a large excess of neutrons. The larger the excess of neutrons, the shorter is the half-life of these nuclei. Eventually the half-life becomes so short that no neutron is captured before the beta decay occurs to the next higher Z. All nuclei produced in the r process will eventually decay toward the stable nuclei, generally moving by beta decays along the diagonal line of constant mass number A.

Some stable nuclei are produced only through the s process, others are produced only through the r process, and some may be produced through both processes. Often the natural abundance of the isotopes of an element can suggest the relative roles of these two processes. In Figure 13.21, you can see that the stable isotope ^{70}Zn cannot be produced in the s process, because the half-life of ^{69}Zn is too short (56 min). Other isotopes for which the r process is important are ^{76}Ge, ^{82}Se, ^{86}Kr, ^{96}Zr, and ^{122}Sn. The isotope ^{64}Ni can be produced either through the s process (as shown in Figure 13.21) or through the r process (such as through beta decays beginning with ^{64}Fe). On the other hand, ^{64}Zn (the most abundant isotope of zinc) is produced *only* through the s process, because r-process beta decays proceeding along the $A = 64$ line are stopped at stable ^{64}Ni and cannot reach ^{64}Zn.

The heaviest element that can be built up out of s-process neutron captures is ^{209}Bi; the half-lives of the isotopes beyond ^{209}Bi are too short to allow the s process to continue. The presence in nature of heavier elements such as thorium or uranium suggests that the r process must operate in this region as well.

The r process most likely occurs during supernova explosions, following the breakdown and implosion of a star that has used up its fusion reserves. In a very short time, lasting of the order of seconds, the star implodes, produces an enormous flux of neutrons (perhaps 10^{32} n/cm^2/s), and builds up all elements to

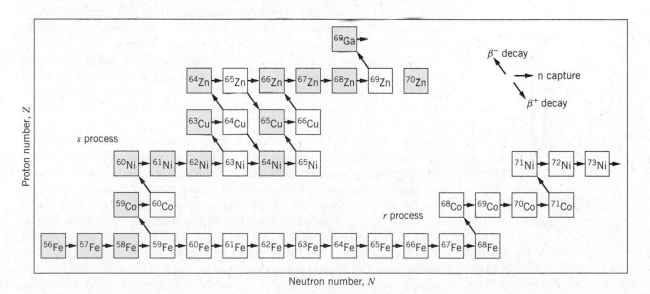

FIGURE 13.21 A section of the chart of the nuclides (Figure 12.11), showing the s- and r-process paths from ^{56}Fe. Shaded squares represent stable nuclei, and unshaded squares represent radioactive nuclei. Many r-process paths are possible, as the short-lived nuclei beta decay; only one of those possible paths is shown. All the nuclei in the r-process path are unstable and may beta decay toward the stable nuclei.

about $A = 260$. When the final explosion occurs, these elements are hurled out into space, to become part of new star systems. The heavy atoms of which the Earth is made may have been produced in such an explosion.

13.7 APPLICATIONS OF NUCLEAR PHYSICS

In this chapter, we have discussed how fission and fusion reactions can be used to generate electrical power, and in the previous chapter we discussed how the radioactive decay of various isotopes can be used to date the historical origin of material containing those isotopes. These are but a few of the many ways that nuclear decays and reactions can be applied to the solution of practical problems. In this section we discuss briefly some other applications of the techniques of nuclear physics.

Neutron Activation Analysis

Nearly every radioactive isotope emits characteristic gamma rays, and many chemical elements can be identified by their gamma ray spectra. For example, when ^{59}Co (the only stable isotope of cobalt) is placed in a flux of neutrons (such as is found near the core of a reactor), neutron absorption results in the production of the radioactive isotope ^{60}Co, which beta decays with a half-life of 5.27 years. Following the beta decay, ^{60}Ni emits two gamma rays of energies 1.17 MeV and 1.33 MeV and of equal intensity. If we place in a flux of neutrons a material of unknown composition, and if we observe, following the neutron bombardment, two gamma rays of equal intensity and energies 1.17 MeV and 1.33 MeV, it is a safe bet that the unknown sample contained cobalt. In fact, from the rate of gamma emission we could deduce exactly how much cobalt the material contains, assuming that we know the neutron flux and the neutron capture cross section of ^{59}Co. This technique is known as *neutron activation analysis*, and has been used in many applications in which the elements are present in such small quantities that chemical identification is not practical. Typically, neutron activation analysis can be used to identify elements in quantities of the order of 10^{-9} g, and sensitivity down to 10^{-12} g is often possible.

Such a sensitive and precise technique finds application in a variety of areas, in which the chemical composition must be determined for samples that are available only in microscopic quantities or that must be analyzed in a nondestructive manner. For example, the chemical composition of various types of pottery can help us trace the geographical origin of the clay from which they were made; such analyses of pottery shards can trace the trading routes of prehistoric people. Art forgeries can be detected by a knowledge of the chemical composition of paints, because techniques for producing pigments have changed over the last four centuries with corresponding changes in the level of impurities in paints. The chemical analysis of tiny quantities of material such as paint, gunshot residues, soil, or hair can provide important evidence in criminal investigations. Neutron activation analysis of samples of the hair of such historical figures as Napoleon or Newton has revealed the chemicals to which they were exposed centuries ago. An example of a neutron activation analysis study of a sample of hair is shown in Figure 13.22.

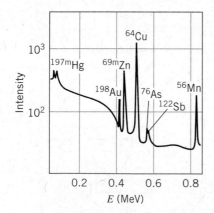

FIGURE 13.22 Gamma-ray spectrum following neutron activation of a sample of human hair. The sample shows traces of mercury, gold, zinc, copper, arsenic, antimony, and manganese. [From D. DeSoete et al., *Neutron Activation Analysis* (Wiley Interscience, 1972).]

Medical Radiation Physics

One of the most important applications of nuclear physics has been in medicine, for both diagnostic and therapeutic purposes. The use of X rays for producing images for medical diagnosis is well known, but X rays are of limited value. They show distinct and detailed images of bones, but they are generally less useful in making images of soft tissue. Radioactive isotopes can be introduced into the body in chemical forms that have an affinity for certain organs, such as bone or the thyroid gland. A sensitive detector (called a "gamma-ray camera") can observe the radiations from the isotopes that are concentrated in the organ and can produce an image that shows how the activity is distributed in the patient. These detectors are capable of determining where each gamma-ray photon originates in the patient. Figure 13.23 shows an image of the brain, taken after the patient was injected with the radioactive isotope ^{99}Tc ($t_{1/2} = 6$ h). The images clearly show an area of the brain where the activity has concentrated. Ordinarily the brain does not absorb impurities from the blood, so such concentrations often indicate a tumor or other abnormality.

Another technique that reveals a wealth of information is *positron emission tomography* (PET), in which the patient is injected with a positron-emitting isotope that is readily absorbed by the body. Examples of isotopes used are ^{15}O ($t_{1/2} = 2$ min), ^{13}N ($t_{1/2} = 10$ min), ^{11}C ($t_{1/2} = 20$ min), and ^{18}F ($t_{1/2} = 110$ min). These isotopes are produced with a cyclotron, and because of the short half-lives the cyclotron must be present at the site of the diagnostic facility. When a positron emitter decays, the positron quickly annihilates with an electron and produces two 511-keV gamma rays that travel in opposite directions. By surrounding the patient with a ring of detectors, it is possible to determine exactly where the decay occurred, and from a large number of such events, the physician can produce an image that reconstructs the distribution of the radioisotope in the patient. One advantage of the PET scan over X-ray techniques such as the CAT (computerized axial tomography) scan is that it can produce a dynamic image—changes in the patient during the measuring time can be observed. Figure 13.24 shows a brain scan of a patient who was injected with glucose labeled with ^{18}F. Active areas of the brain metabolize glucose more rapidly, and so they become more concentrated with ^{18}F, allowing medical workers to observe regions of the brain associated with different mental activities.

Radiation therapy takes advantage of the effect of radiations in destroying unwanted tissue in the body, such as a cancerous growth or an overactive thyroid gland. The effect of the passage of radiation through matter is often to ionize the

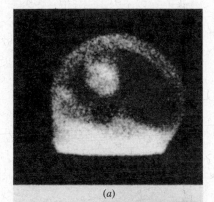

(a)

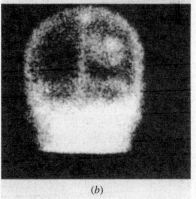

(b)

FIGURE 13.23 Scintillation camera image of the brain, following intravenous injection of 20 mCi of ^{99m}Tc. (a) Side view, with the patient's face to the left. (b) Back view. The bright circular spot shows concentration of blood in a lesion, possibly a tumor. Other bright areas show the scalp and the major veins.

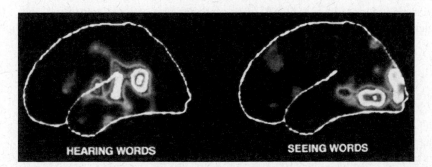

HEARING WORDS   SEEING WORDS

FIGURE 13.24 PET scan showing different areas of the brain that are active when either hearing words or seeing words.

Rosalyn Yalow (1921–2011, United States). After receiving her Ph.D. in nuclear physics, she researched the medical applications of radioactive isotopes. Yalow developed the technique of radioimmunoassay, which uses radioactive tracers to measure small amounts of substances in the blood or other fluids. Her development of this technique was recognized with the award of the Nobel Prize in medicine in 1977.

atoms. The ionized atoms can then participate in chemical reactions that lead to their incorporation into molecules and subsequent alteration of their biological function, possibly the destruction of a cell or the modification of its genetic material. For example, an overactive thyroid gland is often treated by giving the patient radioactive ^{131}I, which collects in the thyroid. The beta emissions from this isotope damage the thyroid cells and ultimately lead to their destruction. Certain cancers are treated by implanting needles or wires containing radium or other radioactive substances. The decays of these radioisotopes cause localized damage to the cancerous cells.

Other cancers can be treated using beams of particles that cause nuclear reactions within the body at the location of the tumor. Pions and neutrons are used for this purpose. The absorption of a pion or a neutron by a nucleus causes a nuclear reaction, and the subsequent emission of particles or decays by the reaction products again causes local damage that is concentrated at the site of the tumor, inflicting maximum damage to the tumor and minimum damage to the surrounding healthy tissue.

Alpha-Scattering Applications

Radioactive sources emitting alpha particles have been used in a variety of applications. Most of these take advantage of the persistence of radioactive decay—the decays can be depended upon to occur at a fixed rate in any location.

Alpha particles from radioactive decay can be absorbed and their energy converted into another form, such as electrical power obtained through thermoelectric conversion. The power levels are not large (of the order of 1 W per gram of material; see Problem 31), but they are sufficient to power many devices, from cardiac pacemakers to the Voyager spacecraft, which photographed Jupiter, Saturn, and Uranus.

Scattering of alpha particles emitted by a radioactive source is the basis of operation of ionization-type smoke detectors; alpha particles from the decay of ^{241}Am are scattered by the ionized atoms that result from combustion. When the smoke detector senses a decrease in the rate at which alphas are counted (due to some of them being scattered away from the detector), the alarm is triggered.

Other applications of alpha-particle scattering are used for materials analysis. In *Rutherford backscattering*, the analysis uses the reduction in energy of an alpha particle that is scattered through an angle of 180°. Although our discussion of Rutherford scattering in Chapter 6 assumed that the target nucleus was infinitely heavy and thus acquired no energy in the scattering, in practice a small amount of energy is given even to a heavy nucleus. By allowing the target nucleus to recoil, we can find the loss in energy ΔK of an alpha particle of kinetic energy K that scatters through 180° (see Problem 29):

$$\Delta K = K \left[\frac{4m/M}{(1 + m/M)^2} \right] \qquad (13.16)$$

where m is the mass of the alpha particle and M is the mass of the target nucleus. For a heavy nucleus ($m/M = 0.02$), the loss in energy is of order 0.5 MeV, which is easily measurable. Figure 13.25 shows a sample of the spectrum of alpha particles backscattered from a thin foil containing copper, silver, and gold. Note the Z^2 dependence of the scattering probability that characterizes Rutherford scattering (see Eq. 6.14), and also note the sensitivity of the technique even to the two naturally occurring isotopes of copper. (However, the two isotopes of silver cannot be resolved.) The Surveyor spacecraft that landed on the Moon and the

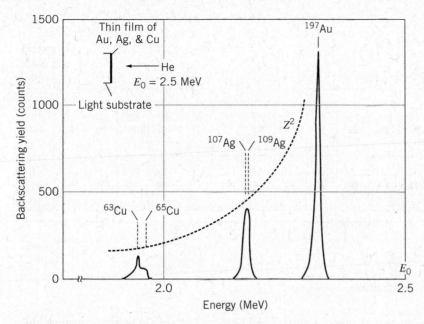

FIGURE 13.25 Backscattering spectrum of 2.5-MeV α particles from a thin film of copper, silver, and gold. The dashed line shows the Z^2 behavior of the cross section expected from the Rutherford formula. Note the appearance of the two isotopes of copper. [From M.-A. Nicolet, J. W. Mayer, and I. V. Mitchell, *Science* **177**, 841 (1972). Copyright ©1972 by the AAAS.]

Viking landers on Mars carried Rutherford backscattering experiments to analyze the chemical composition of the surface of those bodies.

Superheavy Elements

The known atoms beyond uranium ($Z = 92$) are *all* radioactive, with half-lives short compared with the age of the Earth. They are therefore not present in terrestrial matter, but they can be produced in the laboratory. The production process for the series of elements beginning with neptunium ($Z = 93$), called *transuranic* elements, follows the same process outlined in Section 13.6: neutron capture followed by beta decay. Using similar techniques researchers have produced elements up to $Z = 100$ (fermium). Beyond this point, too few atoms are produced for neutron capture to reveal the presence of the next element. Instead, reactions with accelerated charged particles are used.

Many of the elements in this series have half-lives of only minutes or seconds, and thus the production and identification of these elements requires painstaking experimental efforts—the isotopes are often produced in quantities of a few atoms! Although most of these elements have not been produced in sufficient quantity to study their chemical properties, it is expected that their place in the periodic table will be as shown in Figure 13.26, up to the inert gas with $Z = 118$. All elements up to 118 have been observed (although names for elements beyond $Z = 112$ have not yet been chosen—the symbols shown in Figure 13.26 represent placeholder names).

The extreme instability of these transuranic elements results from the increased Coulomb repulsion of the nuclear protons as Z increases; these elements decay by alpha decay or by spontaneous fission. However, strong theoretical evidence

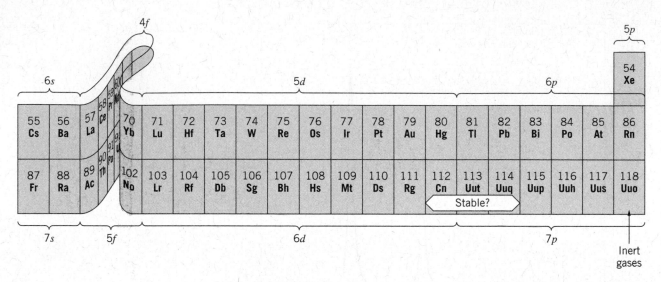

FIGURE 13.26 New massive elements in the periodic table.

suggests that elements around $Z = 114, N = 184$ should be stable against alpha decay, beta decay, and spontaneous fission. This region around element 114 is often called the "island of stability."

Although the production and observation of nuclei of such *superheavy* elements are not likely to have any immediate applications, their study would be of great interest to test our understanding of the ordering of the periodic table, and the comparison of their chemical and physical properties with those of the $5d$ and $6p$ elements would be a test of our ordering of the elements. Many of the artificially produced transuranic elements have already found applications in research and technology. The alpha decays of ^{238}Pu and ^{239}Pu have been used as power sources for spacecraft, and ^{241}Am serves as an alpha source for smoke detectors. The radioisotope ^{254}Cf decays by spontaneous fission; the neutrons released in the decay have many applications, including medical treatment and materials analysis.

Chapter Summary

		Section			Section
Reaction rate	$R = \sigma NI_0/S$ or $R = \phi\sigma N = \phi\sigma m N_A/M$	13.1	^{235}U fission reaction (sample)	$^{235}_{92}$U$_{143}$ + n $\rightarrow$ $^{93}_{37}$Rb$_{56}$ + $^{141}_{55}$Cs$_{86}$ + 2n	13.4
Production of activity in reaction	$a(t) = \lambda N = R(1 - e^{-\lambda t})$ $\cong R\lambda t \ (t \ll t_{1/2})$	13.2	D-T fusion reaction	^{2}H + ^{3}H $\rightarrow$ ^{4}He + n	13.5
Reaction Q value $(x + X \rightarrow y + Y)$	$Q = (m_i - m_f)c^2$ $= [m(x) + m(X) - m(y) - m(Y)]c^2$ $= K_y + K_Y - K_x$	13.3	Lawson's criterion	$n\tau \geq 10^{20}$ s$\cdot$m^{-3}	13.5
Threshold kinetic energy	$K_{th} = -Q[1 + m(x)/m(X)]$	13.3	Alpha backscattering	$\Delta K = K\left[\dfrac{4m/M}{(1 + m/M)^2}\right]$	13.7

Questions

1. The cross sections for reactions induced by protons generally increase as the kinetic energy of the proton increases, while cross sections for neutron-induced reactions generally decrease with increasing neutron energy. Explain this behavior.

2. Cross sections for reactions induced by thermal neutrons ($K \sim kT$, where T is room temperature) are often several orders of magnitude larger than cross sections for the same reaction induced by fast neutrons ($K \sim$ MeV). Justify this difference by comparing the time spent in the vicinity of a target nucleus by a thermal neutron and a fast neutron.

3. When two nuclei approach one another in a nuclear reaction, there is a Coulomb repulsion between them. Does this potential energy affect the kinematics of the reaction? Does it affect the cross section?

4. The most abundant component of our atmosphere is ^{14}N. Assuming that cosmic rays supply sufficient high-energy protons and neutrons, explain how the radioactive isotopes ^{14}C and ^{3}H can be formed.

5. Consider the photodisintegration reaction $A + \gamma \rightarrow B + C$. In terms of the binding energies of A, B, and C, what are the requirements for this reaction? Would you expect to observe photodisintegration more readily for light nuclei ($A < 56$) or heavy nuclei ($A > 56$)?

6. Would you expect to observe the radiative capture of an alpha particle $X + \alpha \rightarrow X' + \gamma$ for heavy nuclei?

7. In what sense is photodisintegration (see Question 5) the inverse of radiative capture (see Question 6)? How are the photon energies related?

8. Comment on the following statement: The fission reaction is useful for energy production because of the large kinetic energies given to the neutrons emitted following fission.

9. ^{238}U is fissionable, but only with neutrons in the MeV range. Explain why ^{238}U is not a suitable reactor fuel.

10. What is the difference between a slow neutron and a delayed neutron? Between a fast neutron and a prompt neutron?

11. In a typical fission reaction, which fragment (heavier or lighter) has the larger kinetic energy? The larger momentum? The larger speed?

12. When charged particles travel in a medium faster than light travels in that medium, Cerenkov radiation is emitted. This is the origin of the blue glow of the water that surrounds a reactor core. What might be the identity of these charged particles from a reactor? What velocities and kinetic energies do they need to produce Cerenkov radiation? (The index of refraction of water is 1.33.)

13. In general, would you expect fission fragments to decay by positive or negative beta decay? Why?

14. Among the fission products that build up in reactor fuel elements is xenon, which has an extremely large neutron capture cross section (see Section 13.1). What effect does this buildup have on the operation of the reactor?

15. Helium has virtually no neutron absorption cross section. Would helium be a better reactor moderator than carbon, which has a small but nonzero cross section?

16. Estimate the number of fissions per second that must occur in a 1000-MW power plant, assuming a 30% efficiency of energy conversion.

17. The fission cross section for ^{235}U for slow neutrons is about 10^6 times the fission cross section of ^{238}U for slow neutrons, yet for fast neutrons the fission cross sections of ^{235}U and ^{238}U are roughly the same. Explain this effect.

18. Consider two fragments of uranium fission with atomic numbers Z and $92 - Z$. Estimating their mass numbers as 2.5 times their atomic numbers, find an expression for the Coulomb potential energy of the two fragments when they are just touching, and show that this expression is maximized when the two fragments are identical. Why then is the fission fragment mass distribution, Figure 13.9, not a maximum at $A = 118$?

19. Assume that ^{235}U splits into two fragments with mass numbers of 90 and 145, with each fragment having roughly the same ratio of Z/A as ^{235}U. On this basis, explain why neutrons are emitted in fission.

20. Why is it necessary to convert a proton to a neutron in the first step of the proton-proton fusion cycle? Why can't two protons fuse directly?

21. Explain why a fusion reactor requires a high particle density, a high temperature, and a long confinement time.

22. In the argument leading to Lawson's criterion, Eq. 13.15, it was mentioned that the power necessary to heat the plasma is proportional to the particle density n, while the power obtained from fusion is proportional to n^2. Explain these two proportionalities.

23. Why do radioactive decay power sources use alpha emitters rather than beta emitters?

Problems

13.1 Types of Nuclear Reactions

1. Fill in the missing particle in these reactions:

    (a) ^{4}He + ^{14}N $\rightarrow$ ^{17}O +   (c) ^{27}Al + ^{4}He $\rightarrow$ n +

    (b) ^{9}Be + ^{4}He $\rightarrow$ ^{12}C +   (d) ^{12}C + $\rightarrow$ ^{13}N + n

2. In a certain nuclear reaction, outgoing protons are observed with energies 16.2 MeV, 14.8 MeV, 11.6 MeV, 8.9 MeV, and 6.7 MeV. No energies higher than 16.2 MeV are observed. Construct a level scheme of the product nucleus.

3. In order to determine the cross section for neutron capture, you are irradiating a thin gold foil, in the form of a circular disk of diameter 3.0 mm and thickness 1.81 μm, with neutrons to produce the reaction $n + {}^{197}Au \rightarrow {}^{198}Au + \gamma$. By observing the outgoing gamma-ray photons in a detector, you determine that the gold decays at a rate of 5.37×10^6 per second. From an independent measurement, you have determined the neutron flux to be 7.25×10^{10} neutrons/cm^2/s. What value do you deduce for the cross section for this reaction?

4. The element cobalt is commonly used for measuring the intensity of neutron beams through the reaction $n + {}^{59}Co \rightarrow {}^{60}Co + \gamma$. By observing the radioactive decay of ${}^{60}Co$, it is possible to deduce the rate at which it is produced in the reaction. The cross section for this reaction is 37.0 b. A thin disk of Co-Al alloy has a diameter of 1.00 cm and a mass of 46 mg; the alloy contains 0.44% Co by weight. With neutrons spread uniformly over the surface of the foil, it is concluded that ${}^{60}Co$ is produced at the rate of 1.07×10^{12} per second. What is the rate at which neutrons strike the target?

5. A beam of 20.0 μA of protons is incident on 2.0 cm^2 of a target of ${}^{107}Ag$ of thickness 4.5 μm producing the reaction $p + {}^{107}Ag \rightarrow {}^{105}Cd + 3n$. Neutrons are observed at a rate of 8.5×10^6 per second. What is the cross section for this reaction at this proton energy?

6. A beam of alpha particles is incident on a target of ${}^{63}Cu$, resulting in the reaction $\alpha + {}^{63}Cu \rightarrow {}^{66}Ga + n$. Assume the cross section for the particular alpha energy to be 1.25 b. The target is in the form of a foil, 2.5 μm thick. The beam has a circular cross section of diameter 0.50 cm and a current of 7.5 μA. Find the rate of neutron emission.

13.2 Radioisotope Production in Nuclear Reactions

7. A radioactive isotope of half-life $t_{1/2}$ is produced in a nuclear reaction. What fraction of the maximum possible activity is produced in an irradiation time of (a) $t_{1/2}$; (b) $2t_{1/2}$; (c) $4t_{1/2}$?

8. List five nuclear reactions, consisting of a light stable projectile nucleus (mass 4 or less) incident on a heavy stable target nucleus, that can produce the radioactive nucleus ${}^{56}Co$.

9. Show that Eq. 13.5 is a solution to Eq. 13.4.

10. The radioisotope ${}^{15}O$ ($t_{1/2} = 122$ s) is used to measure respiratory function. Patients inhale the gas, which is made by irradiating nitrogen gas with deuterons (${}^2H$). Consider a cubical cell measuring 1.24 cm on each edge, which holds nitrogen gas at a pressure of 2.25 atm and a temperature of 293 K. One face of the cube is uniformly irradiated with a deuteron beam having a current of 2.05 A. At the chosen deuteron energy, the reaction cross section is 0.21 b. (a) At what rate is ${}^{15}O$ produced in the cell? (b) After an irradiation lasting for 60.0 s, what is the activity of ${}^{15}O$ in the cell?

11. Neutron capture in sodium occurs with a cross section of 0.53 b and leads to radioactive ${}^{24}Na$ ($t_{1/2} = 15$ h). What is the activity that results when 1.0 μg of Na is placed in a neutron flux of 2.5×10^{13} neutrons/cm^2/s for 4.0 h?

13.3 Low-Energy Reaction Kinematics

12. Derive Eq. 13.14 from Eq. 13.13.

13. Find the Q value of the reactions:
    (a) $p + {}^{55}Mn \rightarrow {}^{54}Fe + 2n$
    (b) ${}^3He + {}^{40}Ar \rightarrow {}^{41}K + {}^2H$

14. Find the Q value of the reactions:
    (a) ${}^6Li + n \rightarrow {}^3H + {}^4He$
    (b) $p + {}^2H \rightarrow 2p + n$
    (c) ${}^7Li + {}^2H \rightarrow {}^8Be + n$

15. In the reaction ${}^2H + {}^3He \rightarrow p + {}^4He$, deuterons of energy 5.000 MeV are incident on ${}^3He$ at rest. Both the proton and the alpha particle are observed to travel along the same direction as the incident deuteron. Find the kinetic energies of the proton and the alpha particle.

16. (a) What is the Q value of the reaction $p + {}^4He \rightarrow {}^2H + {}^3He$? (b) What is the threshold energy for protons incident on ${}^4He$ at rest? (c) What is the threshold energy if ${}^4He$ are incident on protons at rest?

13.4 Fission

17. (a) Find the Q value of the fission decay ${}^{254}Cf \rightarrow {}^{127}In + {}^{127}In$, in which ${}^{254}Cf$ splits in half. (b) Find the Q value for the more probable fission process ${}^{254}_{98}Cf_{156} \rightarrow {}^{140}_{54}Xe_{86} + {}^{110}_{44}Ru_{66} + 4n$. Masses are: $m({}^{127}In) = 126.917353$ u, $m({}^{140}Xe) = 139.921641$ u, $m({}^{110}Ru) = 109.914136$ u.

18. Find the energy released in the fission of 1.00 kg of uranium that has been enriched to 3.0% in the isotope ${}^{235}U$.

19. We can understand why ${}^{235}U$ is readily fissionable, and ${}^{238}U$ is not, with the following calculation. (a) Find the energy difference between ${}^{235}U + n$ and ${}^{236}U$. We can regard this as the "excitation energy" of ${}^{236}U$. (b) Repeat for ${}^{238}U + n$ and ${}^{239}U$. (c) Comparing your results for (a) and (b), explain why ${}^{235}U$ will fission with very low energy neutrons, while ${}^{238}U$ requires fast neutrons of 1 to 2 MeV of energy to fission. (d) From a similar calculation, predict whether ${}^{239}Pu$ requires low-energy or higher-energy neutrons to fission.

20. Find the Q value (and therefore the energy released) in the fission reaction ${}^{235}U + n \rightarrow {}^{93}Rb + {}^{141}Cs + 2n$. Use $m({}^{93}Rb) = 92.922042$ u and $m({}^{141}Cs) = 140.920046$ u.

13.5 Fusion

21. (a) Calculate the Q value for the six reactions or decays of the carbon cycle of fusion. (b) By accounting for the electron masses, show that the total Q value for the carbon cycle is identical with that of the proton-proton cycle.

22. Show that the D-T fusion reaction releases 17.6 MeV of energy.

23. In the D-T fusion reaction, the kinetic energies of ${}^2H$ and ${}^3H$ are small, compared with typical nuclear binding energies. (Why?) Find the kinetic energy of the emitted neutron.

24. (a) If a tokamak fusion reactor were able to achieve a confinement time of 0.60 s, what minimum particle density is required? (b) If the reactor were able to achieve 10 times the density found in part (a), what is the minimum plasma temperature required for ignition of a self-sustaining fusion reaction?

13.6 Nucleosynthesis

25. Find the energy released when three alpha particles combine to form ^{12}C.

26. To what temperature must helium gas be heated before the Coulomb barrier is overcome and fusion reactions begin?

27. Trace the path of the s process from the stable isotope ^{63}Cu to the stable isotope ^{75}As, showing the neutron capture and beta decay processes.

28. Show how the s process proceeds from stable ^{81}Br to stable ^{95}Mo.

13.7 Applications of Nuclear Physics

29. An alpha particle of mass m makes an elastic head-on collision with an atom of mass M at rest. Show that the loss in kinetic energy of the alpha particle is given by Eq. 13.16.

30. (a) Calculate the energy loss of a 2.50-MeV alpha particle after backscattering from an atom of copper, silver, and gold. Compare your calculated values with the peak energies in Figure 13.25. (b) Calculate the expected energy difference between the peaks for the two isotopes of copper and also for the two isotopes of silver. Explain why the silver peaks are closer together than the copper peaks. Can you estimate the relative abundances of the two isotopes of copper from the figure?

31. A radioactive source is to be used to produce electrical power from the alpha decay of ^{238}Pu ($t_{1/2} = 88$ y). (a) What is the Q value for the decay? (b) Assuming 100% conversion efficiency, how much power could be obtained from the decay of 1.0 g of ^{238}Pu?

General Problems

32. A small sample of paint is placed in a neutron flux of 3.0×10^{12} neutrons/cm^2/s for a period of 2.5 min. At the end of that period the activity of the sample is found to include 105 decays/s of ^{51}Ti ($t_{1/2} = 5.8$ min) and 12 decays/s ^{60}Co ($t_{1/2} = 5.27$ y). Find the amount, in grams, of titanium and cobalt in the original sample. Use the following information: Cobalt is pure ^{59}Co, which has a cross section of 19 b; titanium is 5.25 percent ^{50}Ti, which has a cross section of 0.14 b.

33. A 2.0-mg sample of copper (69% ^{63}Cu, 31% ^{65}Cu) is placed in a reactor where it is exposed to a neutron flux of 5.0×10^{12} neutrons/cm^2/s. After 10.0 min the resulting activities are 72 μCi of ^{64}Cu ($t_{1/2} = 12.7$ h) and 1.30 mCi of ^{66}Cu ($t_{1/2} = 5.1$ min). Find the cross sections of ^{63}Cu and ^{65}Cu.

34. A beam of neutrons of intensity I is incident on a thin slab of material of area A, thickness dx, density ρ, and atomic weight M. The neutron absorption cross section is σ. (a) What is the loss in intensity dI of this beam in passing through the material? (b) A beam of original intensity I_0 passes through a thickness x of the material. Show that the intensity of the emerging beam is $I = I_0 e^{-n\sigma x}$, where n is the number of absorber nuclei per unit volume. (c) Assume that the total cross section for neutrons incident on copper is 5.0 b. What fraction of the intensity of a neutron beam is lost after traveling through copper of thickness 1.0 mm? 1.0 cm? 1.0 m?

35. A reaction in which two particles join to form a single excited nucleus, which then decays to its ground state by photon emission, is known as *radiative capture*. Find the energy of the gamma ray emitted in the radiative capture of an alpha particle by ^{7}Li. Assume alpha particles of very small kinetic energy are incident on ^{7}Li at rest.

36. How much energy is required (in the form of gamma-ray photons) to break up ^{7}Li into ^{3}H + ^{4}He? This reaction is known as *photodisintegration*.

37. The nucleus ^{113}Cd captures a thermal neutron ($K = 0.025$ eV), producing ^{114}Cd in an excited state; the excited state of ^{114}Cd decays to the ground state by emitting a photon. Find the energy of the photon.

38. When a neutron collides head-on with an atom at rest, the loss in its kinetic energy is given by Eq. 13.16. (a) What fraction of its energy will a neutron lose in a head-on collision with an atom of hydrogen, deuterium, or carbon? (b) Consider a neutron with an initial energy of 2.0 MeV. How many head-on collisions must it make with carbon atoms for its energy to be reduced to the thermal range (0.025 eV)? (c) Is the result of part (b) an underestimate or an overestimate of the actual number of collisions necessary to "thermalize" the neutrons? Explain.

39. Suppose we have 100.0 cm^3 of water, which is 0.015% D$_2$O. (a) Compute the energy that could be obtained if all the deuterium were consumed in the ^{2}H + ^{2}H → ^{3}H + p reaction. (b) As an alternative, compute the energy released if two-thirds of the deuterium were fused to form ^{3}H, which is then combined with the remaining one-third in the D-T reaction.

40. (a) Find the Q value of the reaction ^{4}He + ^{4}He → ^{8}Be. (b) In a gas of ^{4}He at a temperature of 10^8 K, estimate the relative amount of ^{8}Be present.

Chapter 14

ELEMENTARY PARTICLES

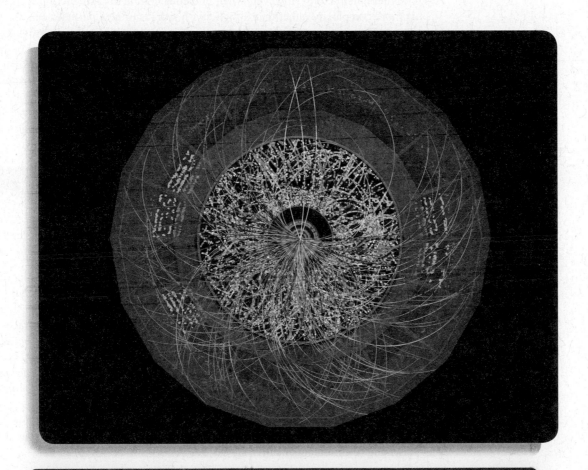

Particle tracks from a head-on collision of two lead ions at the Large Hadron Collider at CERN. Thousands of product particles are produced in each collision. As they travel outward from the collision site at the center, their energy loss and the curvature of their path in a magnetic field help to identify the particles. The goal of this experiment is to produce a "soup" of quarks and gluons, which is believed to characterize the universe just microseconds after the Big Bang.

The search for the basic building blocks of nature has occupied the thoughts of scientific investigators since the Greeks introduced the idea of atomism 2500 years ago. As we look carefully at complex structures, we find underlying symmetries and regularities, that help us to understand the laws that determine how they are put together. The regularities of crystal structure, for example, suggest to us that the atoms of which the crystal is composed must follow certain rules for arranging themselves and joining together. As we look more deeply, we find that although nature has constructed all material objects out of roughly 100 different kinds of atoms, we can understand these atoms in terms of only three particles: the electron, proton, and neutron. Our attempts to look further within the electron have been unsuccessful—the electron seems to be a fundamental particle, with no internal structure. However, when nucleons collide at high energy, the result is more complexity rather than simplicity; hundreds of new particles can emerge as products of these reactions. If there are hundreds of basic building blocks, it seems unlikely that we could ever uncover any fundamental dynamic laws of their behavior. However, experiments show a new, underlying regularity that can be explained in terms of a small number of truly fundamental particles called *quarks*.

In this chapter, we examine the properties of many of the particles of physics, the laws that govern their behavior, and the classifications of these particles. We also show how the quark model helps us to understand some properties of the particles.

14.1 THE FOUR BASIC FORCES

All of the known forces in the universe can be grouped into four basic types. In order of increasing strength, these are: *gravitation,* the *weak interaction, electromagnetism,* and the *strong interaction.*

1. The Gravitational Interaction   Gravity is of course exceedingly important in our daily lives, but on the scale of fundamental interactions between particles in the subatomic realm, it is of no importance at all. To give a relative figure, the gravitational force between two protons just touching at their surfaces is about 10^{-38} of the strong force between them. The principal difference between gravitation and the other interactions is that, on the practical scale, gravity is cumulative and infinite in range. Tiny gravitational interactions, such as the force exerted by one atom of the Earth on one atom of your body, combine to produce observable effects. The other forces, while much stronger than gravity at the microscopic level, do not affect objects on the large scale, either because they have a short range (the strong and weak forces) or their effect is negated by shielding (electromagnetism).

2. The Weak Interaction   The weak interaction is responsible for nuclear beta decay (see Section 12.8) and other similar decay processes involving fundamental particles. It does not play a major role in the binding of nuclei. The weak force between two neighboring protons is about 10^{-7} of the strong force between them, and the range of the weak force is on the scale of 0.001 fm. Nevertheless, the weak force is important in understanding the behavior of fundamental particles, and it is critical in understanding the evolution of the universe.

3. The Electromagnetic Interaction   Electromagnetism is important in the structure and the interactions of the fundamental particles. For example, some particles interact or decay primarily through this mechanism. Electromagnetic forces are of infinite range, but the shielding effect generally diminishes their effect for ordinary objects. Many common macroscopic forces (such as friction, air resistance, drag, and tension) are ultimately due to electromagnetic forces at the atomic level. Within the atom, electromagnetic forces dominate. The electromagnetic force between neighboring protons in a nucleus is about 10^{-2} of the strong force, but within the nucleus the electromagnetic forces can act cumulatively because there is no shielding. As a result, the electromagnetic force can compete with the strong force in determining the stability and the structure of nuclei.

4. The Strong Force   The strong force, which is responsible for the binding of nuclei, is the dominant one in the reactions and decays of most of the fundamental particles. However, as we shall see, some particles (such as the electron) do not feel this force at all. It has a relatively short range, on the order of 1 fm.

The relative strength of a force determines the time scale over which it acts. If we bring two particles close enough together for any of these forces to act, then a longer time is required for the weak force to cause a decay or reaction than for the strong force. As we shall see, the mean lifetime of a decay process is often a signal of the type of interaction responsible for the process, with strong forces being at the shortest end of the time scale (often down to 10^{-23} s). Table 14.1 summarizes the four forces and some of their properties.

Particles can interact with one another in decays and reactions through any of the basic forces. Table 14.1 indicates which particles can interact through each of the four forces. All particles can interact through the gravitational and weak forces. A subset of those can interact through the electromagnetic force (for example, the neutrinos are excluded from this category), and a still smaller subset can interact through the strong force. When two strongly interacting particles are within the range of each other's strong force, we can often neglect the effects of the weak and electromagnetic forces in decay and reaction processes; because their relative strengths are so much smaller than that of the strong force, their effects are much smaller than those of the strong force. (However, these forces are not always negligible—the weak interaction between protons is responsible for a critical step in one of the fusion processes that occurs in stars.)

Even though the proton is a strongly interacting particle, a proton and an electron will *never* interact through the strong force. The electron is able to ignore the strong force of the proton and respond only to its weak or electromagnetic force.

Each of the four forces can be represented in terms of the emission or absorption of particles that carry the interaction, just as we represent the force between nucleons in the nucleus in terms of the exchange of pions (see Section 12.4). Associated with each type of force is a field that is carried by its characteristic particle, as shown in Table 14.2.

TABLE 14.1 The Four Basic Forces

Type	Range	Relative Strength	Characteristic Time	Typical Particles
Strong	1 fm	1	$< 10^{-22}$ s	π, K, n, p
Electromagnetic	∞	10^{-2}	$10^{-14} - 10^{-20}$ s	e, μ, π, K, n, p
Weak	10^{-3} fm	10^{-7}	$10^{-8} - 10^{-13}$ s	All
Gravitational	∞	10^{-38}	Years	All

TABLE 14.2 The Field Particles

Force	Field Particle	Symbol	Charge (e)	Spin (h)	Rest Energy (GeV)
Strong	Gluon	g	0	1	0
Electromagnetic	Photon	γ	0	1	0
Weak	Weak boson	W^+, W^-	± 1	1	80.4
		Z^0	0	1	91.2
Gravitational	Graviton		0	2	0

- **The strong force** between quarks is carried by particles called *gluons*, which have been observed through indirect techniques.
- **The electromagnetic force** between particles can be represented in terms of the emission and absorption of *photons*.
- **The weak force** is carried by the *weak bosons* $W^\pm$ and Z^0, which are responsible for processes such as nuclear beta decay. For example, the beta decay of the neutron (a weak interaction) can be represented as

$$n \rightarrow p + W^- \qquad \text{followed by} \qquad W^- \rightarrow e^- + \bar{\nu}_e$$

Because the decay $n \rightarrow p + W^-$ would violate energy conservation, the existence of the W^- is restricted by the uncertainty principle, and its range can be determined in a manner similar to that of the pion (see Eq. 12.8).

- **The gravitational force** is carried by the *graviton*, which is expected to exist based on theories of gravitation but has not yet been observed.

14.2 CLASSIFYING PARTICLES

One way of studying the elementary particles is to classify them into different categories based on certain behaviors or properties and then to look for similarities or common characteristics among the classifications. We have already classified some particles in Table 14.1 according to the types of forces through which they interact. Another way of classifying them might be according to their masses. In the early days of particle physics, it was observed that the lightest particles (including electrons, muons, and neutrinos) showed one type of behavior, the heaviest group (including protons and neutrons) showed a different behavior, and a middle group (such as pions and kaons) showed a still different behavior. The names originally given to these groups are based on the Greek words for light, middle, and heavy: *leptons* for the light particles, *mesons* for the middle group, and *baryons* for the heavier particles. Even though the classification by mass is now obsolete (leptons and mesons have been discovered that are more massive than protons or neutrons), we keep the original names, which now describe instead a group or *family* of particles with similar properties. When we compare our first two ways of classifying particles, we find an interesting result: The leptons do not interact through the strong force, but the mesons and baryons do.

We can also classify particles by their intrinsic spins. Every particle has an intrinsic spin; you will recall that the electron has a spin of $\frac{1}{2}$, as do the proton and neutron. We find that the leptons all have spins of $\frac{1}{2}$, the mesons all have integral spins $(0, 1, 2, \ldots)$, and the baryons all have half-integral spins $(\frac{1}{2}, \frac{3}{2}, \frac{5}{2}, \ldots)$.

Antiparticles

One additional property that is used to classify a particle is the nature of its *antiparticle.** Every particle has an antiparticle, which is identical to the particle in such properties as mass and lifetime, but differs from the particle in the sign of its electric charge (and in the sign of certain other properties, as we discuss later). The antiparticle of the electron is the positron e^+, which was discovered in the 1930s through reactions initiated by cosmic rays. The positron has a charge of $+e$ (opposite to that of the electron) and a rest energy of 0.511 MeV (identical to that of the electron). The antiproton $\bar{p}$ was discovered in 1956 (see Example 2.18); it has a charge of $-e$ and a rest energy of 938 MeV. A stable atom of antihydrogen could be constructed from a positron and an antiproton; the properties of this atom would be identical to those of ordinary hydrogen.

Antiparticles of stable particles (such as the positron and the antiproton) are themselves stable. However, when a particle and its antiparticle meet, the *annihilation reaction* can occur: the particle and antiparticle both vanish, and instead two or more photons can be produced. Conservation of energy and momentum requires that, neglecting the kinetic energies of the particles, when two photons are emitted each must have an energy equal to the rest energy of the particle. Examples of annihilation reactions are:

$$e^- + e^+ \rightarrow \gamma_1 + \gamma_2 \quad (E_{\gamma_1} = E_{\gamma_2} = 0.511 \, \text{MeV})$$
$$p + \bar{p} \rightarrow \gamma_1 + \gamma_2 \quad (E_{\gamma_1} = E_{\gamma_2} = 938 \, \text{MeV})$$

We call the kind of stuff of which we are made *matter* and the other kind of stuff *antimatter.* There may indeed be galaxies composed of antimatter, but we cannot tell by the ordinary techniques of astronomy, because *light and antilight are identical*! To put it another way, the photon and antiphoton are the same particle, so matter and antimatter emit the same photons. The only way to tell the difference is by sending a chunk of our matter to the distant galaxy and seeing whether or not it is annihilated with the corresponding emission of a burst of photons. (It is indeed possible, but *highly unlikely,* that the first astronaut to travel to another galaxy may suffer such a fate! The first intergalactic handshake would indeed be quite an event!)

In our classification scheme it is usually easy to distinguish particles from antiparticles. We begin by defining *particles* to be the stuff of which ordinary matter is made—electrons, protons, and neutrons. Ordinary matter is not composed of neutrinos, so we have no basis for distinguishing a neutrino from an antineutrino, but the conservation laws in the beta decay process can be understood most easily if we define the *antineutrino* to be the particle that accompanies negative beta decay and the *neutrino* to be the particle that accompanies positron decay and electron capture. For a heavy baryon, such as the Λ (lambda), we take advantage of its radioactive decay, which leads eventually to ordinary protons and neutrons; that is, the Λ is the particle that decays to n, and the $\overline{\Lambda}$ ("anti-lambda") therefore decays to $\bar{n}$. Similarly, in the case of the leptons, the μ^- and the μ^+ are antiparticles of one another; because μ^- decays to ordinary e^- (and has many properties in common with the electron) it is the *particle,* while μ^+ is the antiparticle.

---

*We use two systems to indicate antiparticles. Sometimes the symbol for the particle will be written along with the electric charge to indicate particle or antiparticle, as, for example, e^+ and e^-, or μ^+ and μ^-. Other times the antiparticle will be written with a bar over the symbol—for example, ν and $\bar{\nu}$ or p and $\bar{p}$.

Three Families of Particles

Table 14.3 summarizes the three families of material particles.

Leptons   The leptons interact only through the weak or electromagnetic inter-actions. No experiment has yet been able to reveal any internal structure for the leptons; they appear to be truly fundamental particles that cannot be split into still smaller particles. All known leptons have spin $\frac{1}{2}$.

Table 14.4 shows the six known leptons, grouped as three pairs of particles. Each pair includes a charged particle (e^-, μ^-, τ^-) and an uncharged neutrino (ν_e, ν_μ, ν_τ). Each lepton has a corresponding antiparticle. We have already discussed the electron neutrino and antineutrino in connection with beta decay (Section 12.8), and the decay of cosmic-ray muons was discussed as confirming the time dilation effect in special relativity (Section 2.4). The neutrino masses are very small but nonzero. The rest-energy limits shown in Table 14.4 come from attempts at direct measurement, but indirect evidence from astrophysics and cosmology suggests that the rest energies of all three neutrinos are less than 1 eV.

Mesons   Mesons are strongly interacting particles having integral spin. A partial list of some mesons is given in Table 14.5. Mesons can be produced in reactions through the strong interaction; they decay to other mesons or leptons through the strong, electromagnetic, or weak interactions. For example, pions can be produced in reaction of nucleons, such as

$$p + n \rightarrow p + p + \pi^- \qquad \text{or} \qquad p + n \rightarrow p + n + \pi^0$$

TABLE 14.3 Families of Particles

Family	Structure	Interactions	Spin	Examples
Leptons	Fundamental	Weak, electromagnetic	Half integral	e, ν
Mesons	Composite	Weak, electromagnetic, strong	Integral	π, K
Baryons	Composite	Weak, electromagnetic, strong	Half integral	p, n

TABLE 14.4 The Lepton Family

Particle	Antiparticle	Particle Charge (e)	Spin ($\hbar$)	Rest Energy (MeV)	Mean Life (s)	Typical Decay Products
e^-	e^+	-1	$\frac{1}{2}$	0.511	∞	
ν_e	$\overline{\nu}_e$	0	$\frac{1}{2}$	$< 2\,\text{eV}$	∞	
μ^-	μ^+	-1	$\frac{1}{2}$	105.7	2.2×10^{-6}	$e^- + \overline{\nu}_e + \nu_\mu$
ν_μ	$\overline{\nu}_\mu$	0	$\frac{1}{2}$	< 0.19	∞	
τ^-	τ^+	-1	$\frac{1}{2}$	1777	2.9×10^{-13}	$\mu^- + \overline{\nu}_\mu + \nu_\tau$
ν_τ	$\overline{\nu}_\tau$	0	$\frac{1}{2}$	< 18	∞	

TABLE 14.5 Some Selected Mesons

Particle	Antiparticle	Charge* (e)	Spin (h)	Strangeness*	Rest Energy (MeV)	Mean Life (s)	Typical Decay Products
π^+	π^-	+1	0	0	140	2.6×10^{-8}	$\mu^+ + \nu_\mu$
π^0	π^0	0	0	0	135	8.4×10^{-17}	$\gamma + \gamma$
K^+	K^-	+1	0	+1	494	1.2×10^{-8}	$\mu^+ + \nu_\mu$
K^0	$\overline{K}^0$	0	0	+1	498	0.9×10^{-10}	$\pi^+ + \pi^-$
η	η	0	0	0	548	5.1×10^{-19}	$\gamma + \gamma$
ρ^+	ρ^-	+1	1	0	775	4.4×10^{-24}	$\pi^+ + \pi^0$
η'	η'	0	0	0	958	3.2×10^{-21}	$\eta + \pi^+ + \pi^-$
D^+	D^-	+1	0	0	1869	1.0×10^{-12}	$K^- + \pi^+ + \pi^+$
J/ψ	J/ψ	0	1	0	3097	7.1×10^{-21}	$e^+ + e^-$
B^+	B^-	+1	0	0	5279	1.6×10^{-12}	$D^- + \pi^+ + \pi^-$
Υ	Υ	0	1	0	9460	1.2×10^{-20}	$e^+ + e^-$

*The charge and strangeness are those of the particle. Values for the antiparticle have the opposite sign. The spin, rest energy, and mean life are the same for a particle and its antiparticle.

and the pions can decay according to

$$\pi^- \to \mu^- + \overline{\nu}_\mu \qquad \text{(mean life} = 2.6 \times 10^{-8} \text{ s)}$$
$$\pi^0 \to \gamma + \gamma \qquad \text{(mean life} = 8.4 \times 10^{-17} \text{ s)}$$

The first decay is caused by the weak interaction (indicated by the lifetime and by the presence of a neutrino among the decay products) and the second is caused by the electromagnetic interaction (indicated by the lifetime and the photons).

Because mesons are not observed in ordinary matter, the classification into particles and antiparticles is somewhat arbitrary. For the charged mesons such as π^+ and π^- or K^+ and K^-, which are not part of ordinary matter, the positive and negative particles are antiparticles of one another but there is no way to choose which is matter and which is antimatter. For some uncharged mesons (such as π^0 and η) the particle and antiparticle are identical, while for others (such as K^0 and $\overline{K}^0$) they may be distinct.

Baryons   The baryons are strongly interacting particles with half-integral spins ($1/2$, $3/2$, ...). A partial listing of some baryons is given in Table 14.6. Like the leptons, the baryons have distinct antiparticles. Like the mesons, the baryons can be produced in reactions with nucleons through the strong interaction; for example, the Λ^0 baryon can be produced in the following reaction:

$$p + p \to p + \Lambda^0 + K^+$$

The Λ^0 then decays through the weak interaction according to

$$\Lambda^0 \to p + \pi^- \qquad \text{(mean life} = 2.6 \times 10^{-10} \text{ s)}$$

TABLE 14.6 Some Selected Baryons

Particle	Antiparticle	Charge* (e)	Spin (ℏ)	Strangeness*	Rest Energy (MeV)	Mean Life (s)	Typical Decay Products
p	$\bar{\text{p}}$	+1	$\frac{1}{2}$	0	938	∞	
n	$\bar{\text{n}}$	0	$\frac{1}{2}$	0	940	886	$p + e^- + \bar{\nu}_e$
Λ^0	$\bar{\Lambda}^0$	0	$\frac{1}{2}$	−1	1116	2.6×10^{-10}	$p + \pi^-$
Σ^+	$\bar{\Sigma}^+$	+1	$\frac{1}{2}$	−1	1189	8.0×10^{-11}	$p + \pi^0$
Σ^0	$\bar{\Sigma}^0$	0	$\frac{1}{2}$	−1	1193	7.4×10^{-20}	$\Lambda^0 + \gamma$
Σ^-	$\bar{\Sigma}^-$	−1	$\frac{1}{2}$	−1	1197	1.5×10^{-10}	$n + \pi^-$
Ξ^0	$\bar{\Xi}^0$	0	$\frac{1}{2}$	−2	1315	2.9×10^{-10}	$\Lambda^0 + \pi^0$
Ξ^-	$\bar{\Xi}^-$	−1	$\frac{1}{2}$	−2	1322	1.6×10^{-10}	$\Lambda^0 + \pi^-$
Δ^*	$\bar{\Delta}^*$	+2, +1, 0, −1	$\frac{3}{2}$	0	1232	5.6×10^{-24}	$p + \pi$
Σ^*	$\bar{\Sigma}^*$	+1, 0, −1	$\frac{3}{2}$	−1	1385	1.8×10^{-23}	$\Lambda^0 + \pi$
Ξ^*	$\bar{\Xi}^*$	−1, 0	$\frac{3}{2}$	−2	1533	7.2×10^{-23}	$\Xi + \pi$
Ω^-	$\bar{\Omega}^-$	−1	$\frac{3}{2}$	−3	1672	8.2×10^{-11}	$\Lambda^0 + K^-$

*The charge and strangeness are those of the particle. Values for the antiparticle have the opposite sign. The spin, rest energy, and mean life are the same for a particle and its antiparticle.

Even though neutrinos are not produced in this decay process, the lifetime indicates that the decay proceeds through the weak interaction. Other baryons can be identified in Table 14.6 that decay through the strong, electromagnetic, or weak interactions.

14.3 CONSERVATION LAWS

In the decays and reactions of elementary particles, conservation laws provide a way to understand why some processes occur and others are not observed, even though they are expected on the basis of other considerations. We frequently use the conservation of energy, linear momentum, and angular momentum in our analysis of physical phenomena. These conservation laws are closely connected with the fundamental properties of space and time; we believe those laws to be absolute and inviolable.

We also use other kinds of conservation laws in analyzing various processes. For example, when we combine two elements in a chemical reaction, such as hydrogen + oxygen → water, we must balance the reaction in the following way:

$$2H_2 + O_2 \rightarrow 2H_2O$$

The process of balancing a reaction can also be regarded as a way of accounting for the electrons that participate in the process: A molecule of water contains 10 electrons, and so the atoms that combine to make up the molecule must likewise include 10 electrons.

In nuclear processes, we are concerned not with electrons but with protons and neutrons. In the alpha decay of a nucleus, such as

$$^{235}_{92}U_{143} \rightarrow\ ^{231}_{90}Th_{141} + ^{4}_{2}He_2$$

or in a reaction such as

$$p + ^{63}_{29}Cu_{34} \rightarrow\ ^{63}_{30}Zn_{33} + n$$

we balance the number of protons and also the number of neutrons. We might be tempted to conclude that nuclear processes conserve both proton number and neutron number, but the separate conservation laws are not satisfied in beta decays, for example

$$n \rightarrow p + e^- + \overline{\nu}_e$$

which does not conserve either neutron number or proton number. However, it does conserve the total neutron number plus proton number, which is equal to 1 both before and after the decay. (This conservation law of total nucleon number includes the separate laws of conservation of proton number and neutron number as a special case.)

Lepton Number Conservation

In negative beta decay we always find an antineutrino emitted, never a neutrino. Conversely, in positron beta decay, it is the neutrino that is always emitted. We account for these processes by assigning each particle a *lepton number L*. The electron and neutrino are assigned lepton numbers of $+1$, and the positron and antineutrino are assigned lepton numbers of -1; all mesons and baryons are assigned lepton numbers of zero. Lepton number conservation in positive and negative beta decay then works as follows:

$$\begin{array}{ccccccc} n & \rightarrow & p + & e^- & + & \overline{\nu}_e \\ L: 0 & \rightarrow & 0 + & 1 & + & (-1) \end{array}$$

$$\begin{array}{ccccccc} p & \rightarrow & n + & e^+ & + & \nu_e \\ L: 0 & \rightarrow & 0 + & (-1) & + & 1 \end{array}$$

You can see that the total lepton number is 0 both before and after these decays, which accounts for the appearance of the antineutrino in negative beta decay and the neutrino in positron decay.

According to the lepton conservation law, these processes are forbidden:

$$\begin{array}{ccccc} e^- & + p & \rightarrow & n + & \overline{\nu}_e \\ L: 1 & + 0 & \rightarrow & 0 + & (-1) \end{array}$$

$$\begin{array}{cccc} p & \rightarrow & e^+ & + \gamma \\ L: 0 & \rightarrow & -1 & + 0 \end{array}$$

In keeping track of leptons, we must count each type of lepton (e, μ, τ) separately. Evidence for this comes from a variety of experiments. For example, the distinction between electron-type and muon-type leptons is clear from an experiment in which a beam of muon-type antineutrinos is incident on a target of protons:

$$\overline{\nu}_\mu + p \rightarrow n + \mu^+$$

Emmy Noether (1882–1935, Germany-United States). Known both as a mathematician and a theoretical physicist, she explored the role of conservation laws in physics. In an important result now known as Noether's theorem, she discovered that each symmetry of the mathematical equations describing a phenomenon gives a conserved quantity. For example, the symmetry of equations to translations in time leads to conservation of energy, and the invariance to translations in space leads to conservation of linear momentum.

If there were no difference between electron-type and muon-type leptons, the following reaction would be possible: $\bar{\nu}_\mu + p \rightarrow n + e^+$. However, this outcome is never observed, which indicates the distinction between the two types of leptons and the need to account separately for each type.

Another example of the difference between the types of leptons comes from the failure to observe the decay $\mu^- \rightarrow e^- + \gamma$. If there were only one common type of lepton number, this decay would be possible. The failure to observe this decay (in comparison with the commonly observed decay $\mu^- \rightarrow e^- + \bar{\nu}_e + \nu_\mu$, which conserves both muon-type and electron-type lepton number) suggests the need for the different kinds of lepton numbers. We call these lepton numbers L_e, L_μ, and L_τ, and we have the following conservation law for leptons:

In any process, the lepton numbers for electron-type leptons, muon-type leptons, and tau-type leptons must each remain constant.

The following examples illustrate the conservation of these lepton numbers.

$$
\begin{array}{ccccccc}
 & \bar{\nu}_e & + & p & \rightarrow & e^+ & + & n \\
L_e: & -1 & + & 0 & \rightarrow & -1 & + & 0
\end{array}
$$

$$
\begin{array}{ccccccc}
 & \nu_\mu & + & n & \rightarrow & \mu^- & + & p \\
L_\mu: & 1 & + & 0 & \rightarrow & 1 & + & 0
\end{array}
$$

$$
\begin{array}{ccccccccc}
 & \mu^- & \rightarrow & e^- & + & \bar{\nu}_e & + & \nu_\mu \\
L_e: & 0 & \rightarrow & 1 & + & (-1) & + & 0 \\
L_\mu: & 1 & \rightarrow & 0 & + & 0 & + & 1
\end{array}
$$

$$
\begin{array}{ccccc}
 & \pi^- & \rightarrow & \mu^- & + & \bar{\nu}_\mu \\
L_\mu: & 0 & \rightarrow & 1 & + & (-1)
\end{array}
$$

Studying these examples, we can understand why sometimes neutrinos appear and sometimes antineutrinos appear.

Baryon Number Conservation

Baryons are subject to a similar conservation law. All baryons are assigned a baryon number $B = +1$, and all antibaryons are assigned $B = -1$. All nonbaryons (mesons and leptons) have $B = 0$. We then have the law of conservation of baryon number:

In any process, the total baryon number must remain constant.

(The conservation of nucleon number A is a special case of conservation of baryon number, in which all the baryons are nucleons. In particle physics, it is customary to use B instead of A to represent all baryons, including the nucleons.) No violation of the law of baryon conservation has ever been observed, although the Grand Unified Theories (see Section 14.8) suggest that the proton can decay in a way that would violate conservation of baryon number.

As an example of conservation of baryon number, consider the reaction that was responsible for the discovery of the antiproton:

$$p + p \rightarrow p + p + p + \bar{p}$$

On the left side, the total baryon number is $B = +2$. On the right side, we have three baryons with $B = +1$ and one antibaryon with $B = -1$, so the total baryon number is $B = +2$ on the right side also. On the other hand, the reaction $p + p \rightarrow p + p + \bar{n}$ violates baryon number conservation and is therefore forbidden.

Strangeness Conservation

The number of mesons that can be created or destroyed in decays or reactions is not subject to a conservation law like the number of leptons or baryons. For example, the following reactions can be used to produce pions:

$$p + p \rightarrow p + n + \pi^+ \qquad\qquad p + p \rightarrow p + n + \pi^+ + \pi^0$$
$$p + p \rightarrow p + p + \pi^0 \qquad\qquad p + p \rightarrow p + p + \pi^+ + \pi^-$$

As long as enough energy is available, any number of pions can be produced in these reactions.

If we try the same type of reaction to produce K mesons, a different type of behavior is observed. The reactions $p + p \rightarrow p + n + K^+$ and $p + p \rightarrow p + p + K^0$ never occur, even though the incident proton is given enough energy to produce this particle. We do, however, observe reactions such as $p + p \rightarrow p + n + K^+ + \bar{K}^0$ and $p + p \rightarrow p + p + K^+ + K^-$, which are very similar to the reactions that produce two pions. Why do reactions producing π mesons give any number (odd or even) but reactions producing K mesons give them only in pairs?

Here's another example of this unusual behavior. The reaction $\pi^- + p \rightarrow \pi^+ + \Sigma^-$ conserves electric charge and baryon number and so would be expected to occur, but it does not. Instead, the following reaction is easily observed: $\pi^- + p \rightarrow K^+ + \Sigma^-$. Usually when we fail to observe a reaction or decay process that is expected to occur, we look for the violation of some conservation law such as electric charge or baryon number. Is there a new conserved quantity whose violation prohibits the reaction from occurring?

There are also decay processes that suggest that our labeling of the particles is incomplete. The uncharged η and π^0 mesons decay very rapidly ($10^{-16} - 10^{-18}$ s) into two photons; on the basis of the systematic behavior of mesons, we would expect the K^0 to decay similarly to two photons in a comparable time. The observed decay of the K^0 takes place much more slowly (10^{-10} s); moreover, the decay products are not photons, but π mesons and leptons. Is a new conservation law responsible for restricting the decay of the K^0?

As a final example of the need for a new conservation law, the heavy charged mesons are all strongly interacting particles, and we expect them to decay into the lighter mesons through the strong interaction with very short lifetimes. For example, the decay $\rho^+ \rightarrow \pi^+ + \pi^0$ occurs in a lifetime of about 10^{-23} s. But the decay $K^+ \rightarrow \pi^+ + \pi^0$ occurs very slowly, in a time of the order of 10^{-8} s, and in fact the different decay mode $K^+ \rightarrow \mu^+ + \nu_\mu$ is more probable. What is responsible for slowing the decay of the K meson by 15 orders of magnitude?

These unusual behaviors are explained by the introduction of a new conserved quantity. This quantity is called the *strangeness S*, and we can use it to explain the properties of the K-meson decays. The K^0 and K^+ are assigned strangeness of $S = +1$; the π mesons and leptons are nonstrange particles ($S = 0$). The decay $K^0 \rightarrow \gamma + \gamma$, which is an electromagnetic decay (as indicated by the photons), is forbidden because the electromagnetic interaction conserves strangeness ($S = +1$ on the left, $S = 0$ on the right). The decay $K^+ \rightarrow \pi^+ + \pi^0$ does not occur in the typical strong interaction time of 10^{-23} s because the strong interaction cannot change strangeness. It occurs in the typical weak interaction time of 10^{-8} s (and the corresponding weak interaction decay $K^+ \rightarrow \mu^+ + \nu_\mu$ occurs as often) because the weak interaction *does not* conserve strangeness; decays that are caused by the weak interaction can change the strangeness by one unit.

We can summarize these results in the *law of conservation of strangeness*:

In processes governed by the strong or electromagnetic interactions, the total strangeness must remain constant. In processes governed by the weak interaction, the strangeness either remains constant or changes by one unit.

The strangeness quantum numbers of the mesons and baryons are given in Tables 14.5 and 14.6. The strangeness of an antiparticle has the opposite sign to that of the corresponding particle.

Conservation of strangeness in the strong interaction explains why the K mesons are always produced in pairs in proton-proton collisions. The protons and neutrons are non-strange particles ($S = 0$), so the only way to conserve strangeness in the collisions that produce K mesons is to produce them in pairs, always one with $S = +1$ and the other with $S = -1$.

The baryons also come in strange and nonstrange varieties. Looking at the lifetimes in Table 14.6, we see that the Λ^0 decays into $p + \pi^-$ with a lifetime of about 10^{-10} s, while we would expect a strongly interacting particle to decay to other strongly interacting particles with a lifetime of about 10^{-23} s. If the strangeness of the Λ^0 is assigned as -1, these decays change S and are forbidden to go by the strong interaction, and so must be due to the weak interaction, with the characteristic 10^{-10} s lifetime. The strangeness violation also tells us why the electromagnetic decay $\Lambda^0 \rightarrow n + \gamma$ does not occur (while the decay $\Sigma^0 \rightarrow \Lambda^0 + \gamma$ does occur, with a typical electromagnetic lifetime of 10^{-19} s). It also explains why the reaction $\pi^- + p \rightarrow \pi^+ + \Sigma^-$, which is permitted by all other conservation laws. is never observed—the initial state has $S = 0$ and the final state has $S = -1$, so it violates strangeness conservation.

The weak interaction can change the strangeness by at most *one* unit. As a result, processes such as $\Xi^0 \rightarrow n + \pi^0$ ($S = -2 \rightarrow S = 0$) are absolutely forbidden, even by the weak interaction.

Example 14.1

The Ω^- baryon has $S = -3$. (*a*) It is desired to produce the Ω^- using a beam of K^- incident on protons. What other particles are produced in this reaction? (*b*) How might the Ω^- decay?

Solution

(*a*) Reactions usually proceed only through the strong interaction, which conserves strangeness. We consider the reaction

$$K^- + p \rightarrow \Omega^- + ?$$

On the left side, we have $S = -1, B = +1$, and electric charge $Q = 0$. On the right side, we have $S = -3$, $B = +1$, and $Q = -1$. We must therefore add to the right side

particles with $S = +2$, $B = 0$, and $Q = +1$. Scanning through the tables of mesons and baryons, we find that we can satisfy these criteria with K^+ and K^0, so one possible reaction is

$$K^- + p \rightarrow \Omega^- + K^+ + K^0$$

(*b*) The Ω^- cannot decay by the strong interaction, because no $S = -3$ final states are available. It must therefore decay to particles having $S = -2$ through the weak interaction, which can change S by one unit. One of the product particles must be a baryon in order to conserve baryon number. Two possibilities are

$$\Omega^- \rightarrow \Lambda^0 + K^- \quad \text{and} \quad \Omega^- \rightarrow \Xi^0 + \pi^-$$

14.4 PARTICLE INTERACTIONS AND DECAYS

In this section we briefly summarize the properties of the elementary particles and how they are measured.

Atoms and molecules can be taken apart relatively easily and nonviolently, enabling us to study their structure. However, the elementary particles, most of which are unstable and do not exist in nature, must be created in violent collisions. (The particle theorist Richard Feynman once compared this process with studying fine Swiss watches by smashing them together and looking at the pieces that emerge from the collision.) For this purpose we need a high-energy beam of particles and a suitable target of elementary particles. The only strongly interacting, stable elementary particle is the proton, and thus a hydrogen target is a logical choice. To get a reasonable density of target atoms, researchers often use liquid, rather than gaseous, hydrogen.

For a suitable beam, we must be able to accelerate a particle to very high energies (so that the energy of the particle may be hundreds of times its rest energy mc^2). A stable charged particle is the logical choice for the beam; stability is required because of the relatively long time necessary to accelerate the particle to high energies, and a charged particle is required so that electromagnetic fields may be used to accelerate the particle. Once again the proton is a convenient choice, and thus many particle physics reactions are produced using beams of high-energy protons. For example, at the Fermi National Accelerator Laboratory (Fermilab) near Chicago protons are accelerated to 1000 GeV ($v/c = 0.9999996$) around a track of radius 1000 m (Figure 14.1).

One type of particle physics reaction can thus be represented as

$$p + p \rightarrow \text{product particles}$$

Richard P. Feynman (1918–1988, United States). Seldom is one person known for both exceptional insights into theoretical physics and exceptional methods of teaching first-year physics. He received the Nobel Prize for his work on the theory that couples quantum mechanics to electromagnetism, and his text and film *Lectures on Physics* give unusual perspectives to many areas of basic physics for undergraduates.

FIGURE 14.1 An aerial view of the Tevatron at the Fermi National Accelerator Laboratory. Beams of protons and antiprotons circulate in opposite directions around the 1-km ring and collide at two locations, at the upper center and lower left. (Courtesy Fermi National Accelerator Laboratory.)

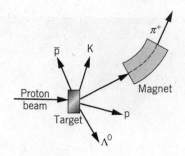

FIGURE 14.2 The production of secondary particle beams. The magnet helps to select the mass and momentum of the desired particle.

Among the product particles may be a variety of mesons or even heavier particles of the baryon family, of which the nucleons are the lightest members. The study of the nature and properties of these particles is the goal of particle physics.

In many cases, conservation laws restrict the nature of the product particles, and it would be desirable to have other types of beams available. One possibility is indicated in Figure 14.2. A proton beam is incident on a target—the nature of the target is not important. Like Feynman's Swiss watch parts, many different particles emerge. By suitable focusing and selection of the momentum, we can extract a beam of the *secondary* particles created in the reactions. The particle must live long enough to be delivered to a second target, which might be tens of meters away; even if the particle were traveling at the speed of light, it would need about 10^{-7} s to make its journey. Although this is a very short time interval by ordinary standards, on the time scale of elementary particles, it is a very long time—in fact, none of the unstable mesons or baryons (except the neutron) lives that long.

Although our efforts to make a secondary beam would seem to be in vain, we have forgotten one very important detail. The lifetime of the particle is measured in its rest frame, while we are observing its flight in the laboratory frame, in which the particle is moving at speeds extremely close to the speed of light. The *time dilation* factor results in a lifetime, observed in our frame of reference, which might be hundreds of times longer than the *proper lifetime*. This factor extends the range of available secondary beams to those particles with lifetimes as short as 10^{-10} s, and makes it possible to obtain secondary beams to study such reactions as

$$\pi + p \rightarrow \text{particles}$$

and

$$K + p \rightarrow \text{particles}$$

even though the proper lifetimes of the π and K are in the range of 10^{-10} to 10^{-8} s.

Detecting Particles

Observing the products of these reactions, which may involve dozens of high-energy charged and uncharged particles, poses a great technological problem for the experimenter. The detector must completely surround the reaction area, so that particles are recorded no matter what direction they travel after the reaction. The particles must produce visible tracks in the detector, so that their identity and direction of travel can be determined. The detector must provide sufficient mass to stop the particles and measure their energy. A magnetic field must be present, so that the resulting curved trajectory of a charged particle can be used to determine its momentum and the sign of its charge. Figure 14.3 shows tracks left in a *bubble chamber*, a large tank filled with liquid hydrogen in which the passage of a charged particle causes microscopic bubbles resulting from the ionization of the hydrogen atoms. The bubbles can be illuminated and photographed to reveal the tracks. Figure 14.4 shows a large detector system that is used both to display the tracks of particles and to measure their energies; Figure 14.5 shows a sample of the results that are obtained with this type of detector.

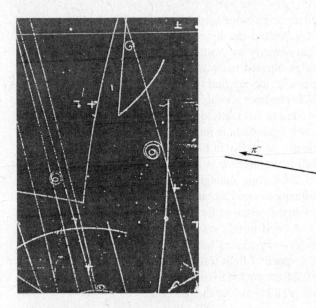

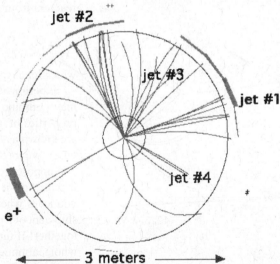

FIGURE 14.3 A bubble chamber photograph of a reaction between particles. At right is shown a diagram indicating the particles that participate in the reaction. An incident pion collides with a proton in the liquid hydrogen, producing a K^0 and a Λ^0, both of which subsequently decay. (Photo Courtesy Lawrence Berkeley National Laboratory.)

FIGURE 14.4 A large detector at the Tevatron at Fermilab. The proton and antiproton beams travel along the central axis of the detector and collide in its interior. The arches that have been removed on either side are the calorimeter detectors that in operation are pushed together so they can record the energies of all particles that leave the reaction. The inner detectors record the tracks of the particles. (Courtesy Fermi National Accelerator Laboratory.)

FIGURE 14.5 A sample of the tracks left by a multitude of particles from a single proton-antiproton collision recorded with the detector of Figure 14.4. A magnetic field causes the tracks to curve; the radius of curvature determines the momentum of the particle, and the direction of curvature determines the sign of the charge of the particle. The jets are showers of particles resulting from a quark or antiquark that is produced in the reaction. In this case the jets come from the top and antitop quarks. (Courtesy Fermi National Accelerator Laboratory.)

From a careful analysis of the paths of particles, such as those of Figures 14.3 or 14.5, we can deduce the desired quantities of mass, linear momentum, and energy. The other important property we would like to know is the lifetime of the decay of the product particles, because many of the products are often unstable. If we know the speed of a particle, we can find its lifetime by simply observing the length of its track in a bubble chamber photograph. (Even for uncharged particles, which leave no tracks, we can use this method to deduce the lifetime, because the subsequent decay of the uncharged particle into two charged particles defines the length of its path rather clearly, as shown in Figure 14.3.)

This method works well if the lifetime is of the order of 10^{-10} s or so, such that the particle leaves a track long enough to be measured (millimeters to centimeters). With careful experimental technique and clever data analysis, this can be extended to track lengths of the order of 10^{-6} m, and so lifetimes down to about 10^{-16} s can be measured in this way (with a little help from the time dilation factor). But many of our particles have lifetimes of only 10^{-23} s, and a particle moving at even the speed of light travels only the diameter of a nucleus in that time! How can we measure such a lifetime? Furthermore, how do we even know such a particle exists at all? Consider the reaction

$$\pi + p \rightarrow \pi + p + x$$

where x is an unknown particle with a lifetime of about 10^{-23} s, which decays into two π mesons according to $x \rightarrow \pi + \pi$. How do we distinguish the above reaction from the reaction

$$\pi + p \rightarrow \pi + p + \pi + \pi$$

which leads to the same particles as actually observed in the laboratory?

Experimental evidence suggests that the two π mesons in this type of reaction may combine for an instant (10^{-23} s) to form an entity with all of the usual properties of a particle—a definite mass, charge, spin, lifetime, etc. These states are known as *resonance particles*, and we now look at the indirect evidence from which we infer their existence.

Suppose you receive a package in the mail from a friend. When you open it, you find it contains many small, irregular pieces of broken glass. How do you learn whether your friend sent you a beautiful glass vase that was broken in shipment or a package of broken glass as a practical joke? You try to put the pieces together! If the pieces fit together, it is a good assumption that the vase was once whole, although the mere fact that they fit together doesn't *prove* that it was once whole. It's just the simplest possible assumption *consistent with our experience*. (An alternative assumption that the pieces were manufactured separately and just happen by chance to fit together is highly improbable.)

How then do we detect a "particle" that lives for only 10^{-23} s? We look at its decay products (which live long enough to be seen in the laboratory), and putting the pieces back together, we infer that they once may have been a whole particle.

For example, suppose in the laboratory we observe two π mesons emitted as shown in Figure 14.6. We measure the direction of travel and the linear momentum of the π mesons as shown. A second and a third event each produces two π mesons as also shown in the figure. Are these three events consistent with the existence of the same resonance particle?

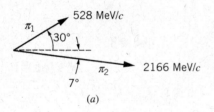

(a)

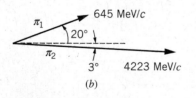

(b)

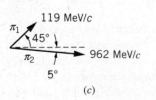

(c)

FIGURE 14.6 Three possible decays of an unknown particle into two π mesons. The direction and momentum of each π meson are indicated.

Let us assume that in each case, a particle moving at an unknown speed decayed into the two particles as shown, one with energy E_1 and momentum $\vec{\mathbf{p}}_1$ and the other with energy E_2 and momentum $\vec{\mathbf{p}}_2$. Each decay must conserve energy and momentum, so we can use the decay information to find the energy $E = E_1 + E_2$ and momentum $\vec{\mathbf{p}} = \vec{\mathbf{p}}_1 + \vec{\mathbf{p}}_2$ of the decaying particle, and then we can find its rest energy according to $mc^2 = \sqrt{E^2 - c^2\vec{\mathbf{p}}^2} = \sqrt{(E_1 + E_2)^2 - c^2(\vec{\mathbf{p}}_1 + \vec{\mathbf{p}}_2)^2}$. Carrying out the calculation, we find that, for the decay shown in part (a) of Figure 14.6, $mc^2 = 764\,\text{MeV}$, while for part (b), $mc^2 = 775\,\text{MeV}$. It is therefore possible that these two events result from the decays of identical particles. Part (c) of the figure gives $mc^2 = 498\,\text{MeV}$, which differs considerably from parts (a) and (b).

Of course, these two events are not sufficient to identify conclusively the existence of a resonance particle with a rest energy in the range of 770 MeV. It could be a mere accident, just like the chance fitting together of two pieces of broken glass. What is needed is a large (statistically significant) number of events, in which we can combine the energy and momenta of the two emitted π mesons in such a way that the deduced mass of the resonance particle is always the same. Figure 14.7 is an example of such a result. There is a background of events with a continuous distribution of energies, like beta decay electrons; these come from events like part (c) of Figure 14.6. There is also present a very prominent peak at 775 MeV. We identify this energy as the rest energy of the resonance particle, which is known as the ρ (rho) meson. (How do we know it is a meson? It must be a strongly interacting particle, because it decays so rapidly. Therefore the only possibilities are mesons, with integral spin, or baryons, with half-integral spin. Pi mesons have integral spin, and two integral spins can combine to give only another integral spin, so it must be a meson.)

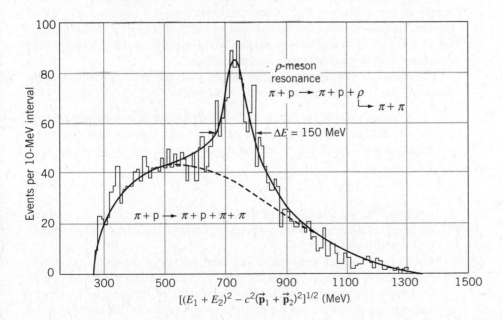

FIGURE 14.7 The resonance identified as the ρ meson. The horizontal axis shows the energy and momentum of the two decay π mesons, combined to be equivalent to the mass of the resonance particle.

We can also infer the lifetime of the particle from Figure 14.7. The particle lives only for about 10^{-23} s, and so if we are to measure its rest energy we have only 10^{-23} s in which to do it. But the uncertainty principle requires that an energy measurement made in a time interval Δt be uncertain by an amount roughly $\Delta E \cong \hbar / \Delta t$. This energy uncertainty ΔE is observed as the *width* of the peak in Figure 14.7. We don't always deduce the same value 775 MeV for the rest energy of the ρ meson; sometimes our value is a bit larger and sometimes a bit smaller. *The width of the resonance peak tells us the lifetime of the particle.* (The width is not really precisely defined, but physicists usually take as the width the interval between the two points where the height of the resonance is one-half its maximum value above the background, as shown in Figure 14.7.) The width of $\Delta E = 150$ MeV leads to a value of $\Delta t = \hbar / \Delta E = 4.4 \times 10^{-24}$ s for the lifetime of the ρ meson.

14.5 ENERGY AND MOMENTUM IN PARTICLE DECAYS

In analyzing the decays and reactions of elementary particles, we apply many of the same laws that we used for nuclear decays and reactions: energy, linear momentum, and total angular momentum must be conserved, and the total value of the quantum numbers associated with electric charge, lepton number, and baryon number (which we previously called nucleon number) must be the same before and after the decay or reaction. In reactions of elementary particles, we are often concerned with the production of new varieties of particles. The energy necessary to manufacture these particles comes from the kinetic energy of the reaction constituents (usually the incident particle); this energy is usually quite large (hence the name *high-energy physics* for this type of research), so *relativistic equations* must be used for energy and momentum.

The decays of elementary particles can be analyzed in a way similar to the decays of nuclei, following the same two basic rules:

1. The energy available for the decay is the difference in rest energy between the initial decaying particle and the particles that are produced in the decay. By analogy with our study of nuclear decays, we call this the Q value:

$$Q = (m_i - m_f)c^2 \qquad (14.1)$$

where $m_i c^2$ is the rest energy of the initial particle and $m_f c^2$ is the total rest energy of all the final product particles. (Of course, the decay will occur only if Q is positive.)

2. The available energy Q is shared as kinetic energy of the decay products in such a way as to conserve linear momentum. As in the case of nuclear decays, for a decay of a particle at rest into two final particles, the particles have equal and opposite momenta, and we can find unique values for the energies of the two final particles. For decays into three or more particles, each particle has a spectrum or distribution of energies from zero up to some maximum value (as was the case with nuclear beta decay).

Example 14.2

Compute the energies of the proton and π meson that result from the decay of a Λ^0 at rest.

Solution

The decay process is $\Lambda^0 \rightarrow p + \pi^-$. Using the rest energies from Tables 14.5 and 14.6, we have:

$$Q = (m_{\Lambda^0} - m_p - m_{\pi^-})c^2$$
$$= 1116\,\text{MeV} - 938\,\text{MeV} - 140\,\text{MeV}$$
$$= 38\,\text{MeV}$$

and so the total kinetic energy of the decay products must be:

$$K_p + K_\pi = 38\,\text{MeV}$$

Using the relativistic formula for kinetic energy, we can write this as

$$K_p + K_\pi = \left(\sqrt{c^2 p_p^2 + m_p^2 c^4} - m_p c^2\right)$$
$$+ \left(\sqrt{c^2 p_\pi^2 + m_\pi^2 c^4} - m_\pi c^2\right) = 38\,\text{MeV}$$

Conservation of momentum requires $p_p = p_\pi$. Substituting for one of the unknown momenta in the above equation and solving algebraically for the other, we obtain

$$p_\pi = p_p = 101\,\text{MeV}/c$$

The kinetic energies can be found by substituting these momenta into the relativistic formula:

$$K_\pi = 33\,\text{MeV} \qquad \text{and} \qquad K_p = 5\,\text{MeV}$$

Example 14.3

What is the maximum kinetic energy of the electron emitted in the decay $\mu^- \rightarrow e^- + \bar{\nu}_e + \nu_\mu$?

Solution

The Q value for this decay is $Q = m_\mu c^2 - m_e c^2 = 105.2\,\text{MeV}$, because the neutrinos have negligible rest energy. If the μ^- is at rest, this energy is shared by the electron and the neutrinos: $Q = K_e + E_{\bar{\nu}_e} + E_{\nu_\mu}$. When the electron has its maximum kinetic energy, the two neutrinos carry away the minimum energy. This minimum cannot be zero, because that would violate momentum conservation: the electron would be carrying momentum that would not be balanced by the neutrino momenta to give a net of zero (the μ^- is at rest, so $\sum \vec{p}_i = \sum \vec{p}_f = 0$). We assume that the electron has its maximum energy when the neutrinos are emitted in exactly the opposite direction to the electron; otherwise some of the decay energy is "wasted" by providing transverse momentum components for the neutrinos, and not as much energy will be available for the electron. It does not matter which of the neutrinos carry the energy and momentum (they may even share it in any proportion), so we let E_ν and p_ν be the total neutrino energy and momentum; these are of course related by $E_\nu \cong c p_\nu$,

because neutrinos are presumed to be of negligible mass and thus to travel at nearly the speed of light. If we let E_e and p_e represent the energy and momentum of the electron, then linear momentum conservation gives

$$p_e - p_\nu = 0$$

For the electron, $E_e = \sqrt{c^2 p_e^2 + m_e^2 c^4}$. Together, these equations give:

$$Q = K_e + E_\nu = E_e - m_e c^2 + c p_\nu = E_e - m_e c^2 + c p_e$$
$$= E_e - m_e c^2 + \sqrt{E_e^2 - m_e^2 c^4}$$

Solving, we find $E_e = Q/2m_\mu c^2 + m_e c^2$ and so

$$K_e = E_e - m_e c^2 = Q^2/2m_\mu c^2 = 52.3\,\text{MeV}$$

The original rest energy of the μ^- is shared essentially equally by the electron and the two neutrinos in this case: $(K_e)_{\text{max}} = (E_\nu)_{\text{max}} \cong Q/2$. Note how different this is from the case of the beta decay of the neutron, where the heavy proton resulting from the decay could absorb considerable recoil momentum at a cost of very little energy, so nearly all of the available energy could be given to the electron, and in that case $(K_e)_{\text{max}} \cong Q$.

Example 14.4

Find the maximum energy of the positrons and of the π mesons produced in the decay $K^+ \rightarrow \pi^0 + e^+ + \nu_e$.

Solution

The Q value for this decay is

$$Q = (m_K - m_\pi - m_e)c^2 = 494\,\text{MeV} - 135\,\text{MeV} - 0.5\,\text{MeV}$$
$$= 358.5\,\text{MeV}$$

This energy must be shared among the three products:

$$Q = K_\pi + K_e + E_\nu$$

The electron and π meson have their maximum energies when the neutrino has negligible energy: $Q = K_\pi + K_e$, and conservation of momentum in this case (if the neutrino has negligible momentum) requires $p_\pi = p_e$. Using relativistic kinetic energy, we have

$$Q = K_\pi + K_e = \sqrt{(pc)^2 + (m_\pi c^2)^2} - m_\pi c^2$$
$$+ \sqrt{(pc)^2 + (m_e c^2)^2} - m_e c^2$$

where $p = p_e = p_\nu$. Inserting the numbers, we obtain

$$494\,\text{MeV} = \sqrt{(pc)^2 + (135\,\text{MeV})^2} + \sqrt{(pc)^2 + (0.5\,\text{MeV})^2}$$

Clearing the two radicals involves quite a bit of algebra, but we can simplify the problem if we inspect this expression and notice that the solution must have a large value of pc, certainly greater than 100 MeV. (Otherwise the two terms could not sum to nearly 500 MeV.) Thus $(pc)^2 \gg (0.5\,\text{MeV})^2$, and we can neglect the electron rest energy term in the second radical, which simplifies the equation somewhat:

$$494\,\text{MeV} = \sqrt{(pc)^2 + (135\,\text{MeV})^2} + pc$$

Solving, we find $pc = 229\,\text{MeV}$, which gives $(E_e)_{max} = 229\,\text{MeV}$ and $(E_\pi)_{max} = 266\,\text{MeV}$. Figure 14.8 shows the observed energy spectra of e^+ and π^0 from the K^+ decay, and the energy maxima are in agreement with the calculated values. (The shapes of the energy distributions are determined by statistical factors, as in the case of nuclear beta decay. The statistical factors are different for e^+ and π^0, because the π^0 also has its maximum energy when the e^+ appears at rest and the ν carries the recoil momentum.)

You should repeat this calculation and convince yourself that (1) the π^0 has its maximum energy also when $K_e = 0$ $(E_e = m_e c^2)$ and (2) the e^+ does *not* have its maximum energy when $K_\pi = 0$.

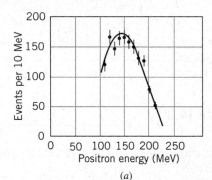

(a)

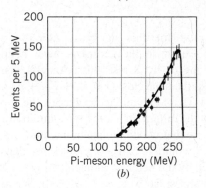

(b)

FIGURE 14.8 The spectrum of positrons and π mesons from the decay of the K^+ meson.

14.6 ENERGY AND MOMENTUM IN PARTICLE REACTIONS

The basic experimental technique of particle physics consists of studying the product particles that result from a collision between an incident particle (accelerated to high energies) and a target particle (often at rest). The kinematics of the reaction process must be analyzed using relativistic formulas, because the kinetic energies of the particles are usually comparable to or greater than their rest energies. In this section we derive some of the relationships that are needed to analyze these reactions, using the formulas for relativistic kinematics we obtained in Chapter 2. An important purpose of these reactions is the production of new varieties of particles, so we concentrate on calculating the threshold energy needed to produce these particles. (You might find it helpful to review the discussion in Chapter 13 on *nonrelativistic* reaction thresholds.)

Consider the following reaction:

$$m_1 + m_2 \rightarrow m_3 + m_4 + m_5 + \cdots$$

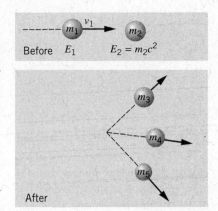

Before E_1 $E_2 = m_2 c^2$

After

FIGURE 14.9   A reaction between particles in the laboratory reference frame.

where the m's represent both the particles and their masses. Any number of particles can be produced in the final state. Here m_1 is the incident particle, which has total energy E_1, kinetic energy $K_1 = E_1 - m_1 c^2$, and momentum $cp_1 = \sqrt{E_1^2 - m_1^2 c^4}$ in the *laboratory* frame of reference. The target particle m_2 is at rest in the laboratory. Figure 14.9 illustrates this reaction in the laboratory frame of reference.

Just as we did for nuclear reactions, we define the Q value to be the difference between the initial and final rest energies:

$$Q = (m_i - m_f)c^2 = [m_1 + m_2 - (m_3 + m_4 + m_5 + \cdots)]c^2 \qquad (14.2)$$

If Q is positive, rest energy is turned into kinetic energy, so that the product particles $m_3, m_4, m_5, \ldots$ have more combined kinetic energy than the initial particles m_1 and m_2. If Q is negative, some of the initial kinetic energy of m_1 is turned into rest energy.

Example 14.5

Compute the Q values for the reactions (a) $\pi^- + p \rightarrow K^0 + \Lambda^0$; (b) $K^- + p \rightarrow \Lambda^0 + \pi^0$.

Solution

(a) Using rest energies from Tables 14.5 and 14.6,

$$Q = [m_{\pi^-} + m_p - (m_{K^0} + m_{\Lambda^0})]c^2$$
$$= 140\,\text{MeV} + 938\,\text{MeV} - 498\,\text{MeV} - 1116\,\text{MeV}$$
$$= -536\,\text{MeV}$$

This reaction has a negative Q value, and energy must be supplied in the form of initial kinetic energy to produce the additional rest energy of the products.

(b)

$$Q = [m_{K^-} + m_p - (m_{\Lambda^0} + m_{\pi^0})]c^2$$
$$= 494\,\text{MeV} + 938\,\text{MeV} - 1116\,\text{MeV} - 135\,\text{MeV}$$
$$= 181\,\text{MeV}$$

A positive Q value indicates that there is enough rest energy in the initial particles to produce the final particles; in fact there is 181 MeV of energy (plus the kinetic energy of the incident particle) left over for kinetic energy of the Λ^0 and π^0.

Threshold Energy

When the Q value is negative, there is a minimum kinetic energy that m_1 must have in order to initiate the reaction. As in the non-relativistic nuclear physics case, this *threshold kinetic energy* K_{th} is larger than the magnitude of Q. The Q value is the energy necessary to create the additional mass of the product particles,

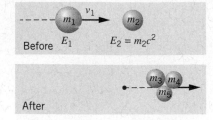

FIGURE 14.10   The reaction of Figure 14.9 when m_1 has the threshold kinetic energy. The product particles move together as a unit in the direction of the original momentum.

but to conserve momentum the product particles cannot be formed at rest, so the threshold energy must not only create the additional particles but must also give them sufficient kinetic energy so that linear momentum is conserved in the reaction.

If Figure 14.9 represents a reaction with a negative Q value, clearly the reaction is not being done at the threshold kinetic energy. In the reaction as it is drawn, not only have the new particles been created, they have been given both forward momentum (to the right in the figure), which is necessary to conserve the initial momentum of m_1, as well as transverse momentum. This transverse momentum, which must sum to zero in order to conserve momentum, is not necessary either to create the particles or to satisfy conservation of momentum. At the minimum or threshold condition, this transverse momentum is zero.

Also at threshold, the most efficient way to provide momentum to the final particles is to have them all moving together with the same speed, as in Figure 14.10. (This is equivalent to creating the particles at rest if we examine the collision from a reference frame in which the total initial momentum is zero, such as in a head-on collision of two particles.) Let's represent the bundle of final particles, all moving as a unit, as a total mass M. Then conservation of momentum ($p_{\text{initial}} = p_{\text{final}}$) gives $p_1 = p_M$ and conservation of total relativistic energy ($E_{\text{initial}} = E_{\text{final}}$) gives $E_1 + E_2 = E_M$, where p_M and E_M represent the momentum and total relativistic energy of the final bundle of particles. Then

$$\sqrt{(p_1 c)^2 + (m_1 c^2)^2} + m_2 c^2 = \sqrt{(p_M c)^2 + (m_M c^2)^2} = \sqrt{(p_1 c)^2 + (m_M c^2)^2}$$

(14.3)

Squaring both sides and solving, we obtain

$$\sqrt{(p_1 c)^2 + (m_1 c^2)^2} = \frac{(Mc^2)^2 - (m_1 c^2)^2 - (m_2 c^2)^2}{2m_2 c^2}$$

(14.4)

The threshold kinetic energy of m_1 is then

$$\begin{aligned} K_{\text{th}} = E_1 - m_1 c^2 &= \sqrt{(p_1 c)^2 + (m_1 c^2)^2} - m_1 c^2 \\ &= \frac{(Mc^2)^2 - (m_1 c^2)^2 - (m_2 c^2)^2}{2m_2 c^2} - m_1 c^2 \\ &= \frac{(Mc^2 - m_1 c^2 - m_2 c^2)(Mc^2 + m_1 c^2 + m_2 c^2)}{2m_2 c^2} \end{aligned}$$

(14.5)

With $Q = m_1 c^2 + m_2 c^2 - Mc^2$ and $M = m_3 + m_4 + m_5 + \cdots$, this becomes

$$K_{\text{th}} = (-Q) \frac{m_1 + m_2 + m_3 + m_4 + m_5 + \cdots}{2m_2}$$

(14.6)

This can also be written as

$$K_{\text{th}} = (-Q) \frac{\text{total mass of all particles involved in reaction}}{2 \times \text{mass of target particle}}$$

(14.7)

In the limit of low speeds, the relativistic threshold formula reduces to the non-relativistic formula for nuclear reactions derived in Chapter 13 (see Problem 20).

Example 14.6

Calculate the threshold kinetic energy to produce π mesons from the reaction $p + p \rightarrow p + p + \pi^0$.

Solution
The Q value is

$$Q = m_p c^2 + m_p c^2 - (m_p c^2 + m_p c^2 + m_\pi c^2)$$

$$= -m_\pi c^2 = -135 \, \text{MeV}$$

Using Eq. 14.7 we can find the threshold kinetic energy:

$$K_{th} = (-Q)\frac{4m_p + m_\pi}{2m_p}$$

$$= (135 \, \text{MeV})\frac{4(938 \, \text{MeV}) + 135 \, \text{MeV}}{2(938 \, \text{MeV})} = 280 \, \text{MeV}$$

Such energetic protons are produced at many accelerators throughout the world, and as a result the properties of the π mesons can be carefully investigated.

Example 14.7

In 1956 an experiment was performed at Berkeley to search for the antiproton, which could be produced in the reaction $p + p \rightarrow p + p + p + \bar{p}$. What is the threshold energy for this reaction?

Solution
The rest energy of the antiproton is identical to the rest energy of the proton (938 MeV), so the Q value is

$$Q = m_p c^2 + m_p c^2 - 4(m_p c^2) = -2m_p c^2$$

Thus

$$K_{th} = (2m_p c^2)\frac{6m_p c^2}{2m_p c^2} = 6m_p c^2 = 5628 \, \text{MeV}$$

$$= 5.628 \, \text{GeV}$$

For the discovery of the antiproton produced in this reaction, Owen Chamberlain and Emilio Segrè were awarded the Nobel Prize in physics in 1959.

It is interesting to compute the "efficiency" of these reactions—that is, how much of the initial kinetic energy we supply actually goes into producing the final particles, and how much is "wasted" in the laboratory kinetic energies of the reaction products. In the first example, we supply 280 MeV of kinetic energy to produce 135 MeV of rest energy, for an efficiency of about 50%. In the second example, $6m_p c^2$ of kinetic energy produces only $2m_p c^2$ of rest energy, for an efficiency of only 33%. As the rest energies of the product particles become larger, the efficiency decreases, and relatively more energy must be supplied. For example, to produce a particle with a rest energy of 50 GeV in a proton-proton collision, we need to supply about 1250 GeV of initial kinetic energy. Only 4% of the energy supplied actually goes into producing the new particles; the remaining 96% must go to kinetic energy of the products in order to balance the large initial momentum of the incident particle. To produce a 100-GeV particle requires not twice as much energy, but four times as much. This is obviously not a pleasant situation for particle physicists, who must build increasingly more powerful accelerators to accomplish their goals of producing more massive particles.

One way out of this difficulty would be to do an experiment in which two particles with equal and opposite momentum collide head-on. In effect, we would be doing this experiment in the center-of-mass (CM) frame, where at threshold the production of new particles is 100% efficient—*none* of the initial kinetic energy

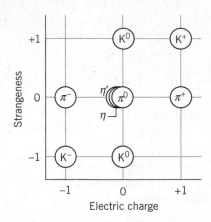

FIGURE 14.11 The relationship between electric charge and strangeness for the spin-0 mesons.

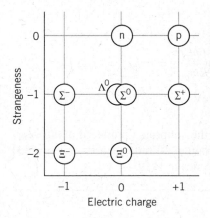

FIGURE 14.12 The relationship between electric charge and strangeness for the spin-$\frac{1}{2}$ baryons.

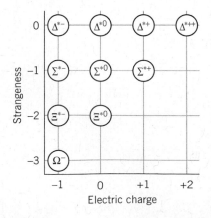

FIGURE 14.13 The relationship between electric charge and strangeness for the spin-$\frac{3}{2}$ baryons.

goes into kinetic energy of the products, which are produced at rest in the CM frame. Thus a 50-GeV particle could be produced by a head-on collision between two protons with as little as 25 GeV of kinetic energy. Of course, this great gain in efficiency is at a cost of the technological difficulty of making such collisions occur.

There are now *colliding beam* accelerators in operation, in which beams of particles (such as electrons or protons) can occasionally be made to collide. For example, in the Fermilab accelerator (Figure 14.1), beams of protons and antiprotons (each of energy 1 TeV = 1000 GeV) circulate around the ring in opposite directions and collide twice during each revolution. The Large Hadron Collider (hadron meaning strongly interacting particles), which is on the border between Switzerland and France, became operational in 2009; it collides two beams of protons each at an energy of 7 GeV in order to search for new particles in an even higher range of rest energies. Other colliding beam accelerators bring together electrons and positrons at energies of 50 to 100 GeV. In each case, all of the available energy can go into the production of new particles.

14.7 THE QUARK STRUCTURE OF MESONS AND BARYONS

Although the classifications and properties of the elementary particles seem like a complicated and disordered collection, there is an underlying order that suggests that a scheme of remarkable simplicity is at work. We can illustrate this order if we plot a diagram that has strangeness along the y axis and electric charge along the x axis. If the families of particles are placed in their proper locations on the graphs, regular geometrical patterns begin to emerge. Figures 14.11 to 14.13 show such plots for the lower mass spin-0 mesons, the spin-$\frac{1}{2}$ baryons, and the spin-$\frac{3}{2}$ baryons. In 1964, Murray Gell-Mann and George Zweig independently and simultaneously recognized that such regular patterns are evidence of an underlying structure in the particles. They showed that they could duplicate these patterns if the mesons and baryons were composed of three fundamental particles, which soon became known as *quarks*. These three quarks, known as up (u), down (d), and strange (s), have the properties listed in Table 14.7. We will shortly see that we now believe that six quarks are necessary to account for all known mesons and baryons.

Let us see how the quark model works in the case of the spin-0 mesons. The quarks have spin $\frac{1}{2}$, so the simplest scheme to form a spin-0 meson would be to combine two quarks, with their spins directed oppositely. However the mesons have baryon number $B = 0$, while a combination of two quarks would have $B = \frac{1}{3} + \frac{1}{3} = \frac{2}{3}$. A combination of a quark and an antiquark, on the other hand, would have $B = 0$, because the antiquark has $B = -\frac{1}{3}$.

For example, suppose we combine a u quark with a $\overline{d}$ ("antidown") quark, obtaining the combination $u\overline{d}$. This combination has spin zero and electric charge $\frac{2}{3}e + \frac{1}{3}e = +e$. (A d quark has charge $-\frac{1}{3}e$, so $\overline{d}$ has charge $+\frac{1}{3}e$.) The properties of this combination are identical with the π^+ meson, and so we identify the π^+ with the combination $u\overline{d}$. Continuing in this way, we find nine possible combinations of one of the three original quarks from Table 14.7 with an antiquark, as listed in Table 14.8, and plotting those nine combinations on a graph of strangeness against electric charge, we obtain Figure 14.14, which looks identical to Figure 14.11.

TABLE 14.7 Properties of the Three Original Quarks

Name	Symbol	Charge (e)	Spin (ℏ)	Baryon Number	Strangeness	Antiquark
Up	u	$+\frac{2}{3}$	$\frac{1}{2}$	$+\frac{1}{3}$	0	$\bar{u}$
Down	d	$-\frac{1}{3}$	$\frac{1}{2}$	$+\frac{1}{3}$	0	$\bar{d}$
Strange	s	$-\frac{1}{3}$	$\frac{1}{2}$	$+\frac{1}{3}$	-1	$\bar{s}$

Values shown for the charge, baryon number, and strangeness are those for the quark; values for the antiquark have the opposite sign.

TABLE 14.8 Possible Quark-Antiquark Combinations

Combination	Charge (e)	Spin (ℏ)	Baryon Number	Strangeness
$u\bar{u}$	0	0, 1	0	0
$u\bar{d}$	$+1$	0, 1	0	0
$u\bar{s}$	$+1$	0, 1	0	$+1$
$d\bar{u}$	-1	0, 1	0	0
$d\bar{d}$	0	0, 1	0	0
$d\bar{s}$	0	0, 1	0	$+1$
$s\bar{u}$	-1	0, 1	0	-1
$s\bar{d}$	0	0, 1	0	-1
$s\bar{s}$	0	0, 1	0	0

The baryons have $B = +1$ and spin $\frac{1}{2}$ or $\frac{3}{2}$, which suggests immediately that three quarks make a baryon. The 10 possible combinations of the three original quarks are listed in Table 14.9, and we can arrange them into two patterns as shown in Figures 14.15 and 14.16, which are identical to those for the spin-$\frac{1}{2}$ and spin-$\frac{3}{2}$ baryons.

Using the quark model, we can analyze the decays and reactions of the elementary particles, based on two rules:

1. Quark-antiquark pairs can be created from energy quanta, and conversely can annihilate into energy quanta. For example,

$$\text{energy} \rightarrow u + \bar{u} \quad \text{or} \quad d + \bar{d} \rightarrow \text{energy}$$

This energy can be in the form of gamma rays (as in electron-positron annihilation), or else it can be transferred to or from other particles in the decay or reaction.

2. The weak interaction can change one type of quark into another through emission or absorption of a W^+ or W^-, for example $s \rightarrow u + W^-$. The W then decays by the weak interaction, such as $W^- \rightarrow \mu^- + \bar{\nu}_\mu$ or $W^- \rightarrow d + \bar{u}$. The strong and electromagnetic interactions cannot change one type of quark into another.

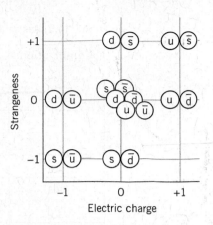

FIGURE 14.14   Spin-0 quark-antiquark combinations; compare with Figure 14.11.

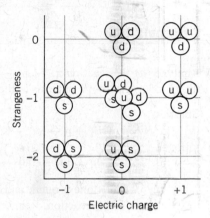

FIGURE 14.15   Spin-$\frac{1}{2}$ three-quark combinations; compare with Figure 14.12.

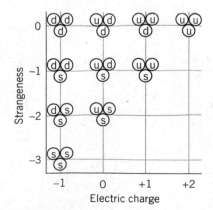

FIGURE 14.16   Spin-$\frac{3}{2}$ three-quark combinations; compare with Figure 14.13.

TABLE 14.9 Possible Three-Quark Combinations

Combination	Charge (e)	Spin ($\hbar$)	Baryon Number	Strangeness
uuu	+2	$\frac{3}{2}$	+1	0
uud	+1	$\frac{1}{2}, \frac{3}{2}$	+1	0
udd	0	$\frac{1}{2}, \frac{3}{2}$	+1	0
uus	+1	$\frac{1}{2}, \frac{3}{2}$	+1	−1
uss	0	$\frac{1}{2}, \frac{3}{2}$	+1	−2
uds	0	$\frac{1}{2}, \frac{3}{2}$	+1	−1
ddd	−1	$\frac{3}{2}$	+1	0
dds	−1	$\frac{1}{2}, \frac{3}{2}$	+1	−1
dss	−1	$\frac{1}{2}, \frac{3}{2}$	+1	−2
sss	−1	$\frac{3}{2}$	+1	−3

Example 14.8

Analyze (a) the reaction $\pi^- + p \to \Lambda^0 + K^0$ and (b) the decay $\pi^+ \to \mu^+ + \nu_\mu$ in terms of the constituent quarks.

Solution

(a) The reaction $\pi^- + p \to \Lambda^0 + K^0$ can be rewritten as follows:

$$d\bar{u} + uud \to uds + d\bar{s}$$

Each side contains one u quark and two d quarks, which don't change in the reaction. Removing these "spectator" quarks from each side of the reaction, we are left with the remaining transformation:

$$\bar{u} + u \to s + \bar{s}$$

The u and $\bar{u}$ annihilate, and from the resulting energy s and $\bar{s}$ are created.

(b) The π^+ has the quark composition $u\bar{d}$. There are no quarks in the final state ($\mu^+ + \nu_\mu$), so we must find a way to get rid of the quarks. One possible way is to change the u quark into a d quark: $u \to d + W^+$. The net process can thus be written as

$$u\bar{d} \to d + \bar{d} + W^+$$

with the products then undergoing the following processes to produce the final observed particles:

$$d + \bar{d} \to \text{energy} \quad \text{and} \quad W^+ \to \mu^+ + \nu_\mu$$

You may have noticed that some of the heavier mesons listed in Table 14.5 were not included in Figure 14.11, and they cannot be accounted for among the quark-antiquark combinations listed in Table 14.8. Where do these particles fit in our scheme?

In 1974, a new meson called J/ψ was discovered at a rest energy of 3.1 GeV. (It was given different names J and ψ by the two competing experimental groups that first reported its discovery.) This new meson was expected to decay to lighter mesons in a characteristic strong interaction time of around 10^{-23} s. Instead, its lifetime was stretched by 3 orders of magnitude to about 10^{-20} s, and its decay products were e^+ and e^-, which are more characteristic of an electromagnetic process. Why is the rapid, strong interaction decay path blocked for this particle? This was soon explained by assuming the J/ψ to be composed of a new quark c, called the *charm* quark, and its antiquark $\bar{c}$. The existence of the c quark had been predicted 4 years earlier as a way to explain the failure to observe the decay $K^0 \to \mu^+ + \mu^-$, which violates no previously known law but is nevertheless not observed.

The c quark, which carries a charge of $+\frac{2}{3}e$, has a property, charm, that operates somewhat like strangeness. We assign a charm quantum number $C = +1$ to the c quark (and assign $C = -1$ to its antiquark $\bar{c}$). All other quarks are assigned $C = 0$. We can now construct a new set of mesons by combining the c quark with the $\bar{u}, \bar{d}$, and $\bar{s}$ antiquarks and by combining the $\bar{c}$ antiquark with the u, d, and s quarks. Instead of nine spin-0 mesons, there are now 16, and the two-dimensional graphs of Figures 14.11 and 14.14 must be extended to a third dimension to show the C axis (Figure 14.17). All of these new mesons, called D, have been observed in high-energy collision experiments. Baryons containing this new quark have also been discovered, analogous to the Λ, Σ, Ξ, and Ω particles but with an s quark replaced by a c quark.

In 1977, the same sequence of events was repeated with another meson, Υ (upsilon). The rest energy was determined to be about 9.5 GeV, and again its decay was slowed to about 10^{-20} s and occurred into $e^+ + e^-$ rather than into mesons. Once again, a new quark was postulated: the b (bottom) quark with a new quantum "bottomness" number $B = -1$ and a charge of $-\frac{1}{3}e$. (The letter B is used to represent baryon number as well as bottomness. It should always be apparent from the discussion which one is meant.) The Υ is assigned as the combination $b\bar{b}$. Many new particles containing the b quark have been discovered, including B mesons (in which a b quark is paired with a different antiquark) and baryons similar to Λ, Σ, and Ξ with a b quark replacing one of the s quarks.

A sixth quark was discovered in 1994 in proton-antiproton collisions at Fermilab. These collisions created this new quark and its antiquark, both of which decayed into a shower of secondary particles (as in Figure 14.5). By measuring the energy and momentum of the secondary particles, the experimenters were able to determine the mass of the new quark to be 172 GeV (roughly the mass of a tungsten atom). This new quark is known as t (top) and has a new associated property of "topness" with a quantum number $T = +1$.

It may now seem that we are losing sight of our goal to achieve simplicity (to add the "bottomness" axis to Figure 14.17 we would need to depict a four-dimensional space!) and that we are moving toward replacing a complicated array of particles with an equivalently complicated array of quarks. However, there is good reason to believe that there are no more than six fundamental quarks. In the next section, we discuss how we are indeed on the path to a simple explanation of the fundamental particles.

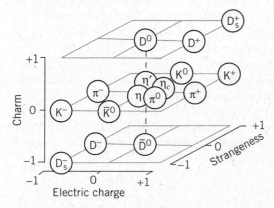

FIGURE 14.17 The relationship among electric charge, strangeness, and charm for the spin-0 mesons.

TABLE 14.10 Properties of the Quarks

Type	Symbol	Antiparticle	Charge (e)	Spin ($\hbar$)	Baryon Number	Rest Energy (MeV)	C	S	T	B
Up	u	$\bar{u}$	$+\frac{2}{3}$	$\frac{1}{2}$	$+\frac{1}{3}$	330	0	0	0	0
Down	d	$\bar{d}$	$-\frac{1}{3}$	$\frac{1}{2}$	$+\frac{1}{3}$	330	0	0	0	0
Charm	c	$\bar{c}$	$+\frac{2}{3}$	$\frac{1}{2}$	$+\frac{1}{3}$	1500	+1	0	0	0
Strange	s	$\bar{s}$	$-\frac{1}{3}$	$\frac{1}{2}$	$+\frac{1}{3}$	500	0	−1	0	0
Top	t	$\bar{t}$	$+\frac{2}{3}$	$\frac{1}{2}$	$+\frac{1}{3}$	172,000	0	0	+1	0
Bottom	b	$\bar{b}$	$-\frac{1}{3}$	$\frac{1}{2}$	$+\frac{1}{3}$	4700	0	0	0	−1

Table 14.10 shows the six quarks and their properties. The masses of the quarks cannot be directly determined, because a free quark has yet to be observed. The rest energies shown in Table 14.10 are estimates based on the "apparent" masses that quarks have when bound in various particles. For example, the observed rest energy of the proton is the sum of the rest energies of its three quark constituents less the binding energy of the quarks. Since we don't know the binding energy, we can't determine the rest energy of a free quark. The rest energies shown in Table 14.10 are often called those of *constituent* quarks.

The quark model does a great deal more than allow us to make geometrical arrangements of particles such as Figure 14.17. It can be used to explain many observed properties of the particles, such as their masses and magnetic moments, and to account for their decay lifetimes and reaction probabilities. Nevertheless, a free quark has never been observed, despite heroic experiments to search for them. How can we be sure that they exist? In experiments that scatter high-energy electrons from protons, we observe more particles scattered at large angles than we would expect if the electric charge of the proton were uniformly distributed throughout its volume, and from the analysis of the distribution of the scattered electrons we conclude that inside the proton are three point-like objects that are responsible for the scattering. This experiment is exactly analogous to Rutherford scattering, in which the presence of the nucleus as a compact object inside the atom was revealed by the distribution of scattered alpha particles at angles larger than expected. Like Rutherford's experiment, the observed cross section depends on the electric charge of the object doing the scattering, and from these experiments we can deduce charges of magnitude $\frac{1}{3}e$ and $\frac{2}{3}e$ for these point-like objects. These experiments give clear evidence for the presence of quarks inside the proton.

We don't yet know why free quarks have not been observed. Perhaps they are so massive that no accelerator yet built has enough energy to liberate one. Perhaps the force between quarks increases with distance (in contrast with electromagnetism or gravitation, which *decrease* with separation distance), so that an infinite amount of energy would be required to separate a quark from a nucleon. Or (as is now widely believed) perhaps the basic theory of quark structure forbids the existence of free quarks.

Quarkonium

The theoretical analysis of the structure of baryons poses mathematical difficulties that are characteristic of all three-body mechanical or quantum-mechanical systems. Instead, we can learn a bit about the interactions of quarks from examining the properties of two-body systems, especially the quark-antiquark combinations in mesons.

We'll look briefly at the combination of a quark with its own antiquark. The binding energies of quarks in mesons are very large (hundreds of MeV), so the quark-antiquark pairs of the light quarks (u, d, s) must be treated relativistically, because the binding energies and thus the kinetic energies in the bound state are roughly the same as the rest energies. However, for the more massive quarks (c, b, t), the binding energies are small compared with the rest energies and we can use nonrelativistic methods (such as the Schrödinger equation) for the analysis.

There is a well-studied analogy for the properties of a quark-antiquark combination. When a positron travels through matter, before it annihilates it forms an atom-like bound state with an electron. This bound electron-positron system is called *positronium*. The positron and the electron each orbit about their center of mass, in states that are similar to atomic states in hydrogen. We can label these states by the values of their total spin angular momentum S and orbital angular momentum L, as we did for atomic states in Chapter 8. (Don't confuse these labels with strangeness and lepton number.) The spins of the two particles can be parallel (hence with total spin $S = 1$) or antiparallel ($S = 0$). The total orbital angular momentum of the two particles about the center of mass can be $L = 0$ for s states, $L = 1$ for p states, etc. Finally, the system can exist with different radial wave functions that we can label with principal quantum number $n = 1, 2, 3, \ldots$, exactly as we did in the hydrogen atom. Figure 14.18a shows some of the bound states in positronium.

The positronium structure is very similar to the bound states in a quark-antiquark system, which is correspondingly known as *quarkonium*. Figure 14.18b shows the quarkonium excited states for the $c\bar{c}$ system, and the states for the $b\bar{b}$ system are shown in Figure 14.18c. There is a great similarity between the excited states of the two quarkonium systems. Note in particular the states with $S = 1$ and $L = 0$ in $c\bar{c}$ and $b\bar{b}$. (These would be equivalent to the $1s$, $2s$, $3s$, and $4s$ states

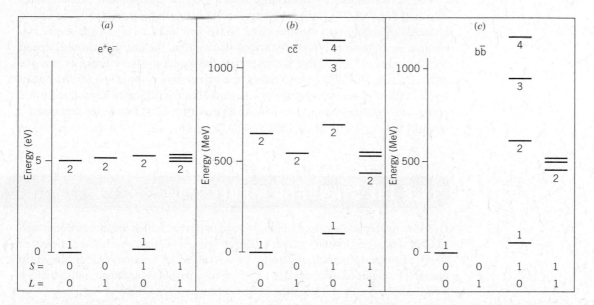

FIGURE 14.18 The energy levels of (a) positronium (e^+e^-), (b) $c\bar{c}$ quarkonium, and (c) $b\bar{b}$ quarkonium. The atom-like states are labeled with the value of the principal quantum number n. The zero of the energy scale is at 2980 MeV for $c\bar{c}$ and 9389 MeV for $b\bar{b}$. The $n = 2$ states with $S = 0, L = 0$ and $S = 0, L = 1$ in $b\bar{b}$ have not yet been discovered.

in a hydrogenic system. The lowest $S = 1, L = 0$ $c\bar{c}$ state is the J/ψ meson, and the corresponding $b\bar{b}$ state is the Υ meson.) Moreover, the overall structure of the excited states of the quarkonium systems is very similar to that of positronium (especially the $n = 2$ states of positronium and $c\bar{c}$).

Knowledge of the excited states of quarkonium allows us to guess at an effective potential energy for which we can solve the Schrödinger equation to try to calculate the energies of the states. The similarity with positronium certainly suggests that we try a Coulomb-like potential energy that depends on $1/r$. However, it cannot be an electromagnetic interaction—the electromagnetic interaction between two quarks separated by a distance on 0.5 fm (a typical size of a meson) is at most about 1 MeV, which is less than 1% of the energy differences of the quarkonium excited states shown in Figure 14.18. (Moreover, the $c\bar{c}$ electromagnetic interaction should be 4 times stronger than the $b\bar{b}$ electromagnetic interaction, but there is no evidence of this in Figure 14.18.) The $1/r$ interaction grows weaker with increasing separation, so we must add another term that grows stronger with separation, which accounts for the failure to produce a free quark. Several different potential energies have been tried for this additional term, the simplest being a term that is linear in the separation r. The net effective potential energy is then of the form

$$U(r) = -\frac{a}{r} + br \tag{14.8}$$

The Schrödinger equation can be solved numerically for this potential energy, with the constants a and b adjusted to give best agreement with experiment. The constant b turns out to have a value of about 1 GeV/fm. This large value is consistent with the failure to observe a free quark—to separate the quarks in a meson even to an atom-sized distance would require about 10^5 GeV, far greater than the beam energy of any accelerator.

One of the especially interesting features of the quarkonium excited states shown in Figure 14.18 is the rough agreement between the energies of the $c\bar{c}$ and $b\bar{b}$ states. (See especially the four states with $S = 1$ and $L = 0$.) This is surprising, because in a simple two-body hydrogen-like system, the energies should depend on the masses of the orbiting particles, and the b quark is three times as massive as the c quark. It would be interesting to continue this comparison for the bound states of the $t\bar{t}$ system involving the top quark, but the difficulty of producing this particle in significant quantities (owing to its large mass) has so far prevented a study of the excited states of the $t\bar{t}$ system.

14.8 THE STANDARD MODEL

Ordinary matter is composed of protons and neutrons, which are in turn composed of u and d quarks. Ordinary matter is also composed of electrons. In the radioactive decay of ordinary matter, electron-type neutrinos are emitted. Our entire world can thus be regarded as composed of four spin-$\frac{1}{2}$ particles (and their antiparticles), which can be grouped into a pair of leptons and a pair of quarks:

$$(e, \nu_e) \quad \text{and} \quad (u, d)$$

Within each pair, the charges of the two particles differ by one unit: -1 and 0, $+\frac{2}{3}$ and $-\frac{1}{3}$.

When we do experiments with high-energy accelerators, we find new types of particles: muons and muon neutrinos, plus mesons and baryons with the new properties of strangeness and charm. We can account for the structure of these particles with another pair of leptons and another pair of quarks:

$$(\mu, \nu_\mu) \quad \text{and} \quad (c, s)$$

Once again, the particles come in pairs differing by one unit of charge.

At even higher energies, we find a new generation of particles consisting of another pair of leptons (tau and its neutrino) and a new pair of quarks (top and bottom), which permits us to continue the symmetric arrangement of the fundamental particles in pairs:

$$(\tau, \nu_\tau) \quad \text{and} \quad (t, b)$$

Is it possible that there are more pairs of leptons and quarks that have not yet been discovered? At this point, we strongly believe the answer to be "No." Every particle so far discovered can be fit into this scheme of 6 leptons and 6 quarks. Furthermore, the number of lepton generations can be determined by the decay rates of the heaviest particles, and a limit of 3 emerges from these experiments. Finally, according to present theories the evolution of the universe itself would have proceeded differently if there had been more than three types of neutrinos. For these reasons, it is generally believed that there are no more than three generations of particles.

The strong force between quarks is carried by an exchanged particle, called the *gluon*, which provides the "glue" that binds quarks together in mesons and baryons. (There are actually eight different gluons in the model.) A theory known as *quantum chromodynamics* describes the interactions of quarks and the exchange of gluons. In this theory, the internal structure of the proton consists of three quarks "swimming in a sea" of exchanged gluons. Like the quarks, the gluons cannot be observed directly, but there is indirect evidence of their existence from a variety of experiments.

The theory of the structure of the elementary particles we have described so far is known as the *Standard Model*. It consists of 6 leptons and 6 quarks (and their antiparticles), plus the field particles (photon, 3 weak bosons, 8 gluons) that carry the various forces. It is remarkably successful in accounting for the properties of the fundamental particles, but it lacks the unified treatment of forces we would expect from a complete theory.

The first step toward unification was taken in 1967 with the development of the *electroweak* theory by Stephen Weinberg and Abdus Salam. In this theory, the weak and electromagnetic interactions are regarded as separate aspects of the same basic force (the electroweak force), just as electric and magnetic forces are distinct but part of a single phenomenon, electromagnetism. The theory predicted the existence of the W and Z particles; their discovery in 1983 provided a dramatic confirmation of the theory.

The next-higher level of unification would be to combine the strong and electroweak forces into a single interaction. Theories that attempt to do this are called *Grand Unified Theories* (GUTs). By incorporating leptons and quarks into a single theory, the GUTs explain many observed phenomena: the fractional electric charge of the quarks and the difference of one unit of charge between the members of the quark and lepton pairs within each generation. The GUTs also predict new phenomena, such as the conversion of quarks into leptons, which would permit the proton (which we have so far assumed to be an absolutely stable particle) to decay into lighter particles with a lifetime of at least 10^{31} y. Searches for photon decay (by looking for evidence of decays in a large volume of matter; see Figure 14.19) have so far been unsuccessful and have placed lower limits on the proton lifetime of at least 10^{32} y.

A missing part of the Standard Model is an explanation of why the particles have the masses that we observe. A complete theory ought to be able to calculate the masses of the particles. It has been proposed that there is a field pervading the entire universe and that particles acquire their particular masses as a result of the strength of their interactions with this field, somewhat like a particle moving through a viscous medium seems to have more inertia and thus a greater "effective" mass. This field is known as the *Higgs field* and the particle that carries the field interaction is called the *Higgs boson*. This particle has been searched for but not yet found; estimates of its expected mass are in excess of $100 \, \text{GeV}/c^2$. The Large Hadron Collider, the world's most powerful particle accelerator, is currently searching for evidence of the Higgs boson by colliding beams of protons at an energy of 7000 GeV.

Another shortcoming of the Standard Model is that it is based on massless neutrinos. Although the upper limit (see Table 14.4) on the mass of the electron neutrino is very small (2 eV), the limits on the other neutrino masses are much larger. Measurements of the flux of neutrinos reaching Earth from the Sun, produced in the fusion reactions discussed in Chapter 13, have consistently revealed a large deficit—the intensity of electron neutrinos observed on Earth is only about 1/3 of what is predicted based on models of how fusion reactions occur in the Sun's interior. Recent measurements at the Sudbury Neutrino Observatory in Canada have revealed that, although the intensity of electron neutrinos from the Sun is only 1/3 of the expected value, the total intensity

FIGURE 14.19 The Superkamiokande detector system in Japan was designed to search for proton decay. The water tank, 40 m in diameter and located 1000 m underground, holds 50,000 tons of water. The tank is lined with more than 10,000 photomultiplier detectors that respond to flashes of light that would be emitted when one of the protons in the water decayed. Here the tank has been partly emptied so that the technicians (in the boat) can service the photomultipliers.

of *all* neutrinos (including muon and tau neutrinos) reaching us from the Sun agrees with the predicted rate. This is very puzzling, because the fusion reactions in the Sun should produce only electron neutrinos; the reacting particles in the solar interior are not sufficiently energetic to produce mu and tau leptons. This mystery has been explained by proposing that the electron neutrinos are produced in the solar interior at the expected rate, but that during their journey from the Sun to Earth, the purely electron neutrinos become a mixture of roughly equal parts electron, muon, and tau neutrinos. This nicely explains why the rate of electron neutrinos from the Sun appears to be only about 1/3 of what is expected (the other 2/3 of the electron neutrinos having been converted into muon or tau neutrinos). This phenomenon of *neutrino oscillation* (which refers to neutrinos oscillating from one type to another) can occur only if the neutrinos have mass. The required masses are very small, well within the experimental limits, but the neutrino masses are definitely not zero. The Standard Model must be extended to include nonzero neutrino masses, and the rules for conservation of lepton number must be modified to allow one type of neutrino to transform into another.

The search for a consistent explanation of the elementary particles has led physicists to work with exotic theories. In *string theory* the particles are replaced by tiny (10^{-33} cm) strings, whose vibrations give rise to the properties we observe as particles. These theories exist in spacetimes with 10 or more dimensions, and at present seem to be far beyond any possible experimental test. Another extension of the Standard Model is called *supersymmetry*; this theory proposes that there is a higher symmetry between the spin-$\frac{1}{2}$ particles (such as the quarks and leptons) and particles with integral spin, so that under this theory there would be electrons and quarks with a spin of 0 and W and Z particles and photons with a spin of $\frac{1}{2}$. The masses of these supersymmetric particles are estimated to be very much larger than their ordinary partners, perhaps in the range of $100\ \text{GeV}/c^2$, but even in this range they should be observable through experiments currently planned at the Large Hadron Collider.

There is so far no conclusive verification for any of the GUTs, nor is there a successful theory that incorporates the remaining force, gravity, into a unified theory. The quest for unification and its experimental tests remains an active area of research in particle physics.

Chapter Summary

		Section			Section
Forces	Strong, electromagnetic, weak, gravitational	14.1	Conservation of baryon number B	*In any process, B remains constant.*	14.3
Field particles	Gluon (g), photon (γ), weak boson ($W^{\pm}, Z^0$), graviton	14.1	Conservation of strangeness S	*In strong and electromagnetic processes, S remains constant; in weak processes, $\Delta S = 0$ or ± 1.*	14.3
Leptons	$e^-, \nu_e, \mu^-, \nu_\mu, \tau^-, \nu_\tau$	14.2			
Mesons	$\pi^{\pm}, \pi^0, K^{\pm}, K^0, \overline{K}^0, \eta, \rho^{\pm}, \eta',$ $D^{\pm}, \psi, B^{\pm}, \Upsilon, \ldots$	14.2	Q value in decays or reactions	$Q = (m_i - m_f)c^2$	14.5, 14.6
Baryons	p, n, $\Lambda^0, \Sigma^{\pm,0}, \Xi^{-,0}, \Omega^-, \ldots$	14.2	Threshold energy in reactions	$K_{th} = -Q(m_1 + m_2 + m_3 + m_4 + m_5 + \cdots)/2m_2$	14.6
Conservation of lepton number L	*In any process, L_e, L_μ, and L_τ remain constant.*	14.3	Quarks	u, d, c, s, t, b	14.7

Questions

1. Some conservation laws are based on fundamental properties of nature, while others are based on systematics of decays and reactions and have as yet no fundamental basis. Give the basis for the following conservation laws: energy, linear momentum, angular momentum, electric charge, baryon number, lepton number, strangeness.

2. Does the presence of neutrinos among the decay products of a particle always indicate that the weak interaction is responsible for the decay? Do all weak interaction decays have neutrinos among the decay products? Which decay product indicates an electromagnetic decay?

3. Do all strongly interacting particles also feel the weak interaction?

4. In what ways would physics be different if there were another member of the lepton family less massive than the electron? What if there were another lepton more massive than the tau?

5. Suppose a proton is moving with high speed, so that $E \gg mc^2$. Is it possible for the proton to decay, such as into $n + \pi^+$ or $p + \pi^0$?

6. On planet anti-Earth, antineutrons beta decay into antiprotons. Is a neutrino or an antineutrino emitted in this decay?

7. List some experiments that might distinguish antineutrons from neutrons. Among others, you might consider (a) neutron capture by a nucleus; (b) beta decay; (c) the effect of a magnetic field on a beam of neutrons.

8. The Σ^0 can decay to Λ^0 without changing strangeness, so it goes by the electromagnetic interaction; the charged $\Sigma^\pm$ decay to p or n by the weak interaction in characteristic lifetimes of 10^{-10} s. Why can't $\Sigma^\pm$ decay to Λ^0 by the strong interaction in a much shorter time?

9. The Ω^- particle decays to $\Lambda^0 + K^-$. Why doesn't it also decay to $\Lambda^0 + \pi^-$?

10. Explain why we do not account for the number of mesons in decays or reactions with a "meson number" in analogy with lepton number or baryon number.

11. Consider that leptons and baryons both obey conservation laws and are both fermions; mesons do not obey a conservation law and are bosons. Can you think of another particle (other than a meson) that has integral spin and can be emitted or absorbed in unlimited numbers?

12. Can antibaryons be produced in reactions between baryons and mesons?

13. List some similarities and differences between the properties of photons and neutrinos.

14. Is it reasonable to describe a resonance as a definite particle, when its mass is uncertain (and therefore variable) by 20%?

15. Why are most particle physics reactions endothermic $(Q < 0)$?

16. Although doubly charged baryons have been found, no doubly charged mesons have yet been found. What would be the effect on the quark model if a meson with charge $+2e$ were found? How could such a meson be interpreted within the quark model?

17. All direct quark transformations must involve a change of charge; for example, $u \rightarrow d$ is allowed (accompanied by the emission of a W^-), but $s \rightarrow d$ is not. Can you suggest a two-step process that might permit the transformation of an s quark into a d?

18. The decay $K^+ \rightarrow \pi^+ + e^+ + e^-$ is at least five orders of magnitude less probable than the decay $K^+ \rightarrow \pi^0 + e^+ + \nu_e$. Based on Question 17, can you explain why?

19. The D mesons decay to π and K mesons with a lifetime of 10^{-13} s. (a) Why is the lifetime so much slower than a typical strong interaction lifetime? Is a quantum number not conserved in the decay? (b) What interaction is responsible for the decay?

20. The Δ^* baryons are found with electric charges $+2, +1, 0$, and -1. Based on the quark model, why do we expect no Δ^* with charge -2?

21. Although we cannot observe quarks directly, indirect evidence for quarks in nucleons comes from the scattering of high-energy particles, such as electrons. When the de Broglie wavelength of the electrons is small compared with the size of a nucleon (~ 1 fm), the electrons appear to be scattered from massive, compact objects much smaller than a nucleon. To which phenomenon discussed previously in this text is this similar? Can the scattering be used to deduce the mass of the struck object? How does the scattering depend on the electric charge of the struck object? What would be the difference between scattering from a particle of charge e and one of charge $\frac{2}{3}e$?

Problems

14.1 The Four Basic Forces

1. Identify the interaction responsible for the following decays (approximate half-lives are given in parentheses):

   (a) $\Delta^* \rightarrow p + \pi$ (10^{-23} s)
   (b) $\eta \rightarrow \gamma + \gamma$ (10^{-18} s)
   (c) $K^+ \rightarrow \mu^+ + \nu_\mu$ (10^{-8} s)
   (d) $\Lambda^0 \rightarrow p + \pi^-$ (10^{-10} s)
   (e) $\eta' \rightarrow \eta + 2\pi$ (10^{-21} s)
   (f) $K^0 \rightarrow \pi^+ + \pi^-$ (10^{-10} s)

2. What is the range of the W^- particle that is responsible for the weak interaction of a proton and a neutron?

14.2 Classifying Particles

3. Give one possible decay mode of the following mesons:
   (a) π^- (b) ρ^- (c) D^- (d) $\overline{K}^0$

4. Give one possible decay mode of the following antibaryons:
   (a) $\overline{n}$ (b) $\overline{\Lambda}^0$ (c) $\overline{\Omega}^-$ (d) $\overline{\Sigma}^0$

5. Suggest a possible decay mode for the K^0 meson that involves the emission of:
   (a) ν_e (b) $\overline{\nu}_e$ (c) ν_μ (d) $\overline{\nu}_\mu$
   Is it possible to have a decay mode of the K^0 that involves the emission of ν_τ or $\overline{\nu}_\tau$?

14.3 Conservation Laws

6. Name the conservation law that would be violated in each of the following decays:
   (a) $\pi^+ \rightarrow e^+ + \gamma$ (e) $\Lambda^0 \rightarrow n + \gamma$
   (b) $\Lambda^0 \rightarrow p + K^-$ (f) $\Omega^- \rightarrow \Xi^0 + K^-$
   (c) $\Omega^- \rightarrow \Sigma^- + \pi^0$ (g) $\Xi^0 \rightarrow \Sigma^0 + \pi^0$
   (d) $\Lambda^0 \rightarrow \pi^- + \pi^+$ (h) $\mu^- \rightarrow e^- + \gamma$

7. Each of the following reactions violates one (or more) of the conservation laws. Name the conservation law violated in each case:
   (a) $\nu_e + p \rightarrow n + e^+$
   (b) $p + p \rightarrow p + n + K^+$
   (c) $p + p \rightarrow p + p + \Lambda^0 + K^0$
   (d) $\pi^- + n \rightarrow K^- + \Lambda^0$
   (e) $K^- + p \rightarrow n + \Lambda^0$

8. Supply the missing particle in each of the following decays:
   (a) $K^- \rightarrow \pi^0 + e^- + $
   (b) $K^0 \rightarrow \pi^0 + \pi^0 + $
   (c) $\eta \rightarrow \pi^+ + \pi^- + $

9. Each of the reactions below is missing a single particle. Supply the missing particle in each case.
   (a) $p + p \rightarrow p + \Lambda^0 + $ (d) $K^- + n \rightarrow \Lambda^0 + $
   (b) $p + \overline{p} \rightarrow n + $ (e) $\overline{\nu}_\mu + p \rightarrow n + $
   (c) $\pi^- + p \rightarrow \Xi^0 + K^0 + $ (f) $K^- + p \rightarrow K^+ + $

14.4 Particle Interactions and Decays

10. Carry out the calculations of mc^2 for the three decays of Figure 14.6.

11. Determine the energy uncertainty or width of (a) η; (b) η'; (c) Σ^0; (d) Δ^*.

14.5 Energy and Momentum in Particle Decays

12. A Σ^- baryon is produced in a certain reaction with a kinetic energy of 3642 MeV. If the particle decays after one mean lifetime, what is the longest possible track this particle could leave in a detector?

13. Repeat the calculation of Example 14.4 for the case in which the π meson has zero kinetic energy, and show that the electron energy in this case is less than the maximum value.

14. Find the Q values of the following decays:
    (a) $\pi^0 \rightarrow \gamma + \gamma$ (c) $D^+ \rightarrow K^- + \pi^+ + \pi^+$
    (b) $\Sigma^+ \rightarrow p + \pi^0$

15. Find the Q values of the following decays:
    (a) $\pi^- \rightarrow \mu^- + \overline{\nu}_\mu$ (c) $\Sigma^0 \rightarrow \Lambda^0 + \gamma$
    (b) $K^0 \rightarrow \pi^+ + \pi^-$

16. Find the kinetic energies of each of the two product particles in the following decays (assume the decaying particle is at rest):
    (a) $K^0 \rightarrow \pi^+ + \pi^-$ (b) $\Sigma^- \rightarrow n + \pi^-$

17. Find the kinetic energies of each of the two product particles in the following decays (assume the decaying particle is at rest):
    (a) $\Omega^- \rightarrow \Lambda^0 + K^-$ (b) $\pi^+ \rightarrow \mu^+ + \nu_\mu$

18. A Σ^- with a kinetic energy of 0.250 GeV decays into $\pi^- + n$. The π^- moves at 90° to the original direction of travel of the Σ^-. Find the kinetic energies of π^- and n and the direction of travel of n.

19. A K^0 with a kinetic energy of 276 MeV decays in flight into π^+ and π^-, which move off at equal angles with the original direction of the K^0. Find the energies and directions of motion of the π^+ and π^-.

14.6 Energy and Momentum in Particle Reactions

20. Show Eq. 14.6 reduces to Eq. 13.14 in the nonrelativistic limit.

21. Determine the Q values of the following reactions:
    (a) $K^- + p \rightarrow \Lambda^0 + \pi^0$
    (b) $\pi^+ + p \rightarrow \Sigma^+ + K^+$
    (c) $p + p \rightarrow p + \pi^+ + \Lambda^0 + K^0$

22. Determine the Q values of the following reactions:
    (a) $\gamma + n \rightarrow \pi^- + p$
    (b) $K^- + p \rightarrow \Omega^- + K^+ + K^0$
    (c) $p + p \rightarrow p + \Sigma^+ + K^0$

23. Find the threshold kinetic energy for the following reactions. In each case the first particle is in motion and the second is at rest.
    (a) $p + p \rightarrow n + \Sigma^+ + K^0 + \pi^+$
    (b) $\pi^- + p \rightarrow \Sigma^0 + K^0$

24. Find the threshold kinetic energy for the following reactions. In each case the first particle is in motion and the second is at rest.
    (a) $p + n \rightarrow p + \Sigma^- + K^+$
    (b) $\pi^+ + p \rightarrow p + p + \overline{n}$

14.7 The Quark Structure of Mesons and Baryons

25. Analyze the following reactions in terms of the quark content of the particles and reduce them to fundamental processes involving the quarks:
    (a) $K^- + p \rightarrow \Omega^- + K^+ + K^0$
    (b) $\pi^+ + p \rightarrow \Sigma^+ + K^+$
    (c) $\gamma + n \rightarrow \pi^- + p$

26. Analyze the following reactions in terms of the quark content of the particles and reduce them to fundamental processes involving the quarks:

    (a) $K^- + p \rightarrow \Lambda^0 + \pi^0$
    (b) $p + p \rightarrow p + \pi^+ + \Lambda^0 + K^0$
    (c) $\gamma + p \rightarrow D^+ + \overline{D}^0 + n$

27. Analyze the following decays in terms of the quark content of the particles and reduce them to fundamental processes involving the quarks:

    (a) $\Omega^- \rightarrow \Lambda^0 + K^-$ (c) $\pi^0 \rightarrow \gamma + \gamma$
    (b) $n \rightarrow p + e^- + \overline{\nu}_e$ (d) $D^+ \rightarrow K^- + \pi^+ + \pi^+$

28. Analyze the following decays in terms of the quark content of the particles and reduce them to fundamental processes involving the quarks:

    (a) $K^0 \rightarrow \pi^+ + \pi^-$ (c) $\Sigma^- \rightarrow n + \pi^-$
    (b) $\Delta^{*++} \rightarrow p + \pi^+$ (d) $\overline{D}^0 \rightarrow K^+ + \pi^-$

29. Based on Figure 14.17, give the quark content of the six D mesons.

General Problems

30. Table 14.5 lists the most likely decay mode of the K^+ meson; Example 14.4 gives another possible decay. List four other possible decays that are allowed by the conservation laws.

31. It is desired to form a beam of Λ^0 particles to use for the study of reactions with protons. The Λ^0 are produced by reactions at one target and must be transported to another target 2.0 m away so that at least half of the original Λ^0 remain in the beam. Find the speed and the kinetic energy of the Λ^0 for this to occur.

32. Find a decay mode, other than that listed in Table 14.6, for (a) Ω^-; (b) Λ^0; (c) Σ^+ that satisfies the applicable conservation laws.

33. Consider the reaction $p + p \rightarrow p + p + \pi^0$ discussed in Example 14.6, but viewed instead from a frame of reference in which the two protons collide head-on with equal velocities. (a) At threshold in this frame of reference, the product particles are formed at rest. Find the proton velocities in this case. (b) Use the Lorentz velocity transformation to switch to the laboratory frame of reference in which one of the protons is at rest, and find the velocity of the other proton. (c) Find the kinetic energy of the incident proton in the laboratory frame and compare with the value found in Example 14.6.

34. The D_s^+ meson (rest energy = 1969 MeV, $S = +1, C = +1$; see Figure 14.17) has a lifetime of 0.5×10^{-12} s. (a) Which interaction is responsible for the decay? (b) Among the possible decay modes are $\phi + \pi^+$, $\mu^+ + \nu_\mu$, and $K^+ + \overline{K}^0$. How do the S and C quantum numbers change in these three decays? (The ϕ meson has a spin of 1, a rest energy of 1020 MeV, and a quark content of $s\overline{s}$.) (c) Analyze the three decay modes according to the quark content of the initial and final particles. (d) Why is the decay into $K^+ + \pi^+ + \pi^-$ allowed, while the decay into $K^- + \pi^+ + \pi^+$ is forbidden?

35. In the decay $K^+ \rightarrow \pi^+ + \pi^+ + \pi^-$ with the initial K meson at rest, what is the maximum kinetic energy of the pi mesons?

36. A beam of π^- mesons with a speed of $0.9980c$ is incident on a target of protons at rest. The reaction produces two particles, one of which is a K^0 meson that is observed to travel with momentum 1561 MeV/c in a direction that makes an angle of $20.6°$ with the direction of the incident pions. (a) Find the momentum and the direction of the second product particle. (b) Find the energy of that particle. (c) Find the rest energy of the second particle and deduce its identity.

Chapter 15

COSMOLOGY: THE ORIGIN AND FATE OF THE UNIVERSE

Today we scan the skies at all wavelengths from the very short (X rays and gamma rays) to the very long (radio waves). New and unexpected discoveries have occurred at these wavelengths: quasars, pulsars, supernovas, black holes — all of which suggest that the universe is not at all static and eternal, as was once believed, but instead is active, evolving, and teeming with radiation. Van Gogh's painting *The Starry Night* suggests exactly that view, even though it was painted in 1889, long before any of these discoveries were made.

In the short time of a few hundred years, developments in astronomy have taken us from the belief that the Earth and its human population were the center of the universe, to a role that approaches insignificance. Before the 16th century, it was widely believed that the planets, Sun, Moon, and stars revolved about a central Earth. By the early 20th century, astronomers had discovered that we inhabit one minor star of a vast number in our galaxy, and that the universe contains an equally vast number of other galaxies.

Gravity is the dominant force that determines the structure of the present universe, but Newton's theory is insufficient to explain a number of observations of the motion of celestial objects. For this purpose we need a different theory, the *general theory of relativity,* which was proposed by Albert Einstein in 1916. Although the mathematics of this theory is beyond the level of this text, we can summarize some of its features and discuss its experimental predictions and their verification. Like the special theory, the general theory of relativity offers us a new way of thinking about space and time.

In this chapter, we briefly survey the field of *cosmology*, the study of the universe on the large scale, including its origin, evolution, and future. For this study we must rely not only on relativity (special and general) and quantum theory, but also on fundamental results from atomic and molecular physics, statistical physics, thermodynamics, nuclear physics, and particle physics.

We begin with three discoveries that fundamentally altered our concept of the universe: it is expanding, it is filled with electromagnetic radiation, and most of its mass is mysteriously hidden from our view. We show how these discoveries have been incorporated, using results from general relativity, into a theory of the origin of the universe known as the Big Bang theory. We then consider other measurements that support this theory, and we conclude with some speculations on the future of the universe.

15.1 THE EXPANSION OF THE UNIVERSE

The evidence for the expansion of the universe comes from the change in wavelength of the light emitted by distant galaxies. In Chapter 2, we analyzed a similar effect as the relativistic Doppler shift (Eq. 2.22), which we can write in terms of wavelength as

$$\lambda' = \lambda \sqrt{\frac{1 + v/c}{1 - v/c}} = \lambda \frac{1 + v/c}{\sqrt{1 - v^2/c^2}} \qquad (15.1)$$

where v represents the relative velocity between the source of the light and the observer. Here λ' is the wavelength we measure on Earth and λ is the wavelength emitted by the moving star or galaxy in its own rest frame.

The light emitted by a star such as the Sun has a continuous spectrum. As light passes through the star's atmosphere, some of it is absorbed by the gases in the atmosphere, so the continuous *emission* spectrum has a few dark *absorption* lines superimposed (see Figure 6.15). Comparison between the known wavelengths of these lines (measured on Earth for sources at rest relative to the observer) and the Doppler-shifted wavelengths allows the speed of the star to be deduced from Eq. 15.1.

Of the stars in our galaxy, some are found to be moving toward us, with their light shifted toward the shorter wavelengths (blue), and others are moving away from us, with their light shifted toward the longer wavelengths (red). The average speed of these stars relative to us is about 30 km/s ($10^{-4}c$). The change in wavelength for these stars is very small. Light from nearby galaxies, those of our "local" group, again shows either small blue shifts or small red shifts.

However, when we look at the light from distant galaxies, we find it to be systematically red shifted, and by a large amount. Some examples of these measurements are shown in Figure 15.1. We do *not* see a comparable number of red and blue shifts, as we would expect if the galaxies were in random motion. All of the galaxies beyond our local group seem to be moving away from us.

The *cosmological principle* asserts that the universe must look the same from any vantage point, and so we must conclude that any other observer in the universe would draw the same conclusion: *The galaxies would be observed to recede from every point in the universe.*

Hubble's Law

In the 1920s, astronomer Edwin Hubble was using the 100-inch telescope on Mount Wilson in California to study the wispy nebulae. By resolving individual stars in the nebulae, Hubble was able to show that they are galaxies like the Milky Way, composed of hundreds of billions of stars. When Hubble measured the wavelength shifts of the light from the galaxies and deduced their speeds, he made two remarkable conclusions: the galaxies are moving away from us, and *the farther away a galaxy is from us, the faster it is moving*. This proportionality between the speed of the galaxy and its distance d is known as *Hubble's law*:

$$v = H_0 d \qquad (15.2)$$

The proportionality constant H_0 is known as the *Hubble parameter*.

Figure 15.2a shows a plot of Hubble's data for the deduced speeds against the distance. Although the points scatter quite a bit (due primarily to uncertainties in the distance measurements), there is a definite indication of a linear relationship. (Hubble's distance calibration was incorrect, so the labels on the horizontal axis do not correspond to the actual distances to the galaxies.) More modern data based on observing supernovas in distant galaxies are shown in Figure 15.2b. There is again clear evidence for a linear relationship, and the slope of the line gives a value of the Hubble parameter of about 72 km/s/Mpc*, within a range of about $\pm 10\%$. The Hubble parameter can also be determined from a variety of other cosmological experiments. These agree with the supernova data, and the best current value is

$$H_0 = 72 \, \frac{\text{km/s}}{\text{Mpc}}$$

The uncertainty in this value is on the order of $\pm 4\%$.

The Hubble parameter has the dimension of inverse time. As we show later, H_0^{-1} is a rough measure of the age of the universe. The best value of H_0 gives an age of 14×10^9 y. If the speed of recession has been changing, the true age can be less than H_0^{-1}.

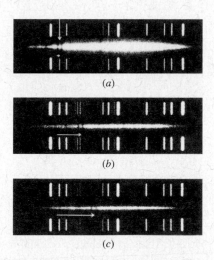

FIGURE 15.1 Red shifts of galaxies. (*a*) The horizontal band in the center shows a continuous spectrum with two dark lines superimposed (vertical arrow near left side), which represent absorption by calcium. Above and below the absorption spectrum are emission spectra for calibration. The recessional speed of this galaxy (which is in our local group) is 1200 km/s, so the red shift is very small. (*b*) The absorption spectrum of a galaxy with a recessional speed of 15,000 km/s. The red shift of the two calcium lines (indicated by the horizontal arrow) is significant. (*c*) The red shift of a galaxy with a recessional speed of 22,000 km/s. (Data courtesy Hale Observatories.)

---

*A parsec, pc, is a measure of distance on the cosmic scale; it is the distance that corresponds to one angular second of parallax. Because parallax is due to the Earth's motion around the Sun, the parallax angle 2α is the diameter $2R$ of the Earth's orbit divided by the distance d to the star or galaxy. Thus $\alpha = R/d$ radians, which gives 1 pc = 3.26 light-years = 3.084×10^{13} km. One megaparsec, Mpc, is 10^6 pc.

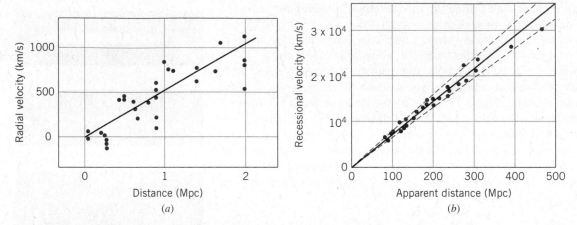

FIGURE 15.2 (*a*) Hubble's original data showing the linear relationship between the recessional speed of a galaxy and its distance from Earth. (*b*) Modern data showing Hubble's law. Data are based on observations of supernovas in distant galaxies using the Hubble Space Telescope. The solid line represents a Hubble parameter of 72 km/s/Mpc, and the dashed lines show the limits corresponding to ±7 km/s/Mpc. [Data from W. L. Freedman et al., *Astrophysical Journal* **553**, 47 (2001).]

Edwin Hubble (1889–1953, United States). His observational work with large telescopes revealed the existence of galaxies, and he was the first to measure their size and distance. Hubble's discovery of the recessional motion of the galaxies was one of the most exciting and important in the history of astronomy.

How does the Hubble law show that the universe is expanding? Consider the unusual universe represented by the three-dimensional coordinate system shown in Figure 15.3*a*, where each point represents a galaxy. With the Earth at the origin, we can determine the distance d to each galaxy. If this universe were to expand, with all the points becoming further apart, as in Figure 15.3*b*, the distance to each galaxy would be increased to d'. Suppose the expansion were such that every dimension increased by a constant ratio k in a time t; that is, $x' = kx$, and so forth. Then $d' = kd$, and a given galaxy moves away from us by a distance $d' - d$ in a time t, so its apparent recessional speed is

$$v = \frac{d' - d}{t} = d\frac{k - 1}{t} \qquad (15.3)$$

If we compare two galaxies 1 and 2,

$$\frac{v_1}{v_2} = \frac{d_1}{d_2} \qquad (15.4)$$

a relationship identical with Hubble's law, Eq. 15.2. Thus, in an expanding universe, it is perfectly natural that the further away from us a galaxy might be, the faster we observe it to be receding.

Notice also from Figure 15.3 that this is true no matter which point we happen to choose as our origin. From *any* point in the "universe" of Figure 15.3, the other points would be observed to satisfy Eq. 15.4 and thus also Hubble's law. We can further demonstrate this with two analogies. If we glue some spots to a balloon (Figure 15.4) and then inflate it, *every* spot observes all other spots to be moving away from it, and the farther away a spot is from any point, the faster its separation grows. For a three-dimensional analogy, consider the loaf of raisin bread shown in Figure 15.5 rising in an oven. As the bread rises, every raisin observes all the others to be moving away from it, and the speed of recession increases with the separation.

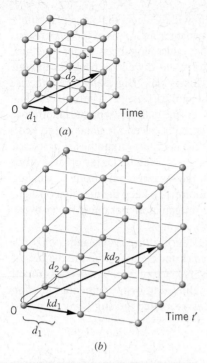

(a)

(b)

FIGURE 15.3 The expansion of a coordinate space, showing that the apparent speed of recession depends on the distance; d_2 is greater than d_1, and d_2 increases faster than d_1.

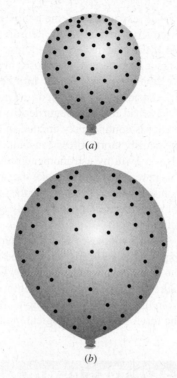

(a)

(b)

FIGURE 15.4 As a balloon is inflated, every observer on the surface experiences a velocity-distance relationship of the form of the Hubble law.

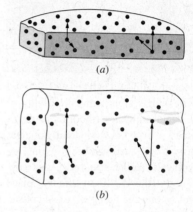

(a)

(b)

FIGURE 15.5   Another system in which the Hubble law is valid.

The correct interpretation of the cosmological redshifts requires the techniques of *general* relativity, which we discuss later in this chapter. According to general relativity, the shift in wavelength is caused by a stretching of the entire fabric of spacetime. Imagine small photos of galaxies glued to a rubber sheet. As the sheet is stretched, the distance between the galaxies increases, but they are not "in motion" according to the terms we usually use in physics to describe motion. However, the stretching of the space between the galaxies causes the wavelength of a light signal from one galaxy to increase by the total amount of the stretching before it is received at another galaxy. This is very different from the usual interpretation of the Doppler formula (Eq. 15.1). (In fact, for some galaxies the wavelength shift is so large that the special relativity formula would imply a recessional speed greater than the speed of light!) At low speeds, the Doppler interpretation of the redshift (that is, calculating a speed from the Doppler formula and using that speed in Hubble's law) gives results that correspond with those based on an expansion of spacetime. However, for very large cosmological redshifts, a more correct analysis must be based on the stretching model:

$$\frac{\lambda'}{\lambda} = \frac{R_0}{R} \qquad (15.5)$$

where R_0 represents a "size" or distance scale factor of the universe at the present time and R represents a similar factor at the time the light was emitted.

George Gamow (1904–1969, United States). His significant contributions to nuclear physics (theories of alpha decay, beta decay, and nuclear structure), astrophysics (nucleosynthesis, stellar structure), and cosmology (the Big Bang theory) place him among the first rank of scientists. Gamow was also one of the most successful writers of popular science, to which he brought unusual and amusing perspectives.

The expansion of the universe has been widely accepted since Hubble's discoveries in the 1920s. There are, however, two interpretations of this expansion. (1) If the galaxies are separating, long ago they must have been closer together. The universe was much denser in its past history, and if we look back far enough we find a single point of infinite density. This is the "Big Bang" hypothesis, developed in 1948 by George Gamow and his colleagues. (2) The universe has always had about the same density it does now. As the galaxies separate, additional matter is continuously created in the empty space between the galaxies, to keep the density more or less constant. This is the "Steady State" hypothesis, proposed also in 1948 by astronomer Fred Hoyle and others. New galaxies created from this new matter would make the universe look the same not only from all vantage points, but also *at all times* in the present and future. (To keep the density constant, the rate of creation need be only about one hydrogen atom per cubic meter every billion years.)

Both hypotheses had their supporters, and during the 1940s and 1950s the experimental evidence did not seem to favor either one over the other. In the 1960s, the new field of radio astronomy revealed the presence of a universal background radiation in the microwave region, which is believed to be the remnant radiation from the Big Bang. This single observation has propelled the Big Bang theory to the forefront of cosmological models.

15.2 THE COSMIC MICROWAVE BACKGROUND RADIATION

When a gas expands adiabatically, it cools. The same is true for the universe: the expansion is accompanied by cooling. As we go back in time, we find a hotter, denser universe. Far enough back in time, the universe would have been too hot for stable matter to form. Its composition was then a "gas" of particles and photons. The unstable particles eventually decayed to stable ones, and the stable particles eventually clumped together to form matter. The photons that filled the universe remained, but their wavelengths were stretched by the continuing expansion. Today those photons have a much lower temperature, but they still uniformly fill the universe.

The wavelength spectrum of those photons is that of an isolated object (blackbody) emitting thermal radiation at the temperature T that characterizes the universe at a particular time. The wavelengths change as the universe expands, but the radiation retains an ideal thermal spectrum at a temperature that decreases with time. In the 1940s, the Big Bang cosmologists (Gamow and others) predicted that this "fireball" would today be at a temperature of the order of 5 to 10 K; such photons would have a typical energy kT of the order of 10^{-3} eV or a wavelength of order 1 mm, in the microwave region of the spectrum.

The properties of this background radiation can be described using the formulas for thermal radiation we developed in Section 10.6. The number of photons dN in the energy interval dE at E (that is, with energies between E and $E + dE$) was given by Eq. 10.38. Writing that equation as the number per unit volume (number density), we get

$$\frac{N(E)\,dE}{V} = \frac{8\pi E^2}{(hc)^3}\frac{1}{e^{E/kT}-1}\,dE \qquad (15.6)$$

To find the *total* number of photons of all energies per unit volume we integrate Eq. 15.6 over energy:

$$\frac{N}{V} = \frac{1}{V} \int_0^\infty N(E)\, dE = \frac{8\pi}{(hc)^3} \int_0^\infty \frac{E^2\, dE}{e^{E/kT} - 1} = \frac{8\pi}{(hc)^3} (kT)^3 \int_0^\infty \frac{x^2\, dx}{e^x - 1}$$

(15.7)

where we have substituted $x = E/kT$. The definite integral is a standard form with a value approximately equal to 2.404. Equation 15.7 shows that the total number of photons per unit volume is proportional to the cube of the temperature, and evaluating the constants we find

$$N/V = (2.03 \times 10^7 \text{ photons/m}^3 \cdot \text{K}^3) T^3$$

(15.8)

We can write Eq. 10.41 for the energy density (energy per unit volume) in the same form by evaluating the constants:

$$U = \frac{8\pi^5 k^4}{15(hc)^3} T^4 = (4.72 \times 10^3 \text{ eV/m}^3 \cdot \text{K}^4) T^4$$

(15.9)

and the mean (average) energy per photon at temperature T is obtained from the ratio of Eqs. 15.9 and 15.8:

$$E_m = \frac{U}{N/V} = (2.33 \times 10^{-4} \text{ eV/K}) T$$

(15.10)

We now look at the experimental evidence for the existence of this microwave radiation and the determination of its temperature. From Eq. 10.42 we see that the measurement of the radiant energy density at *any* wavelength is enough for a determination of the temperature T, although to demonstrate that the radiation actually has an ideal thermal spectrum requires measurement over a range of wavelengths.

The first experimental evidence for this radiation was obtained in a 1965 experiment by Arno Penzias and Robert Wilson, who used a microwave antenna tuned to a wavelength of 7.35 cm. At this wavelength they recorded an annoying "hiss" from their antenna that could not be eliminated, no matter how much care they took in refining the measurement. After painstaking efforts to eliminate the "noise," they concluded that it was coming from no identifiable source and was striking their antenna from all directions, day and night, summer and winter. From the radiant energy at that wavelength they deduced a temperature of 3.1 ± 1.0 K, and it was later concluded that the radiation was the present remnant of the Big Bang "fireball." For this experiment, Penzias and Wilson shared the 1978 Nobel Prize in physics.

Since that original experiment there have been many additional studies, at various wavelengths in the range 0.05 to 100 cm, all giving about the same temperature. The most recent measurements were made with the Cosmic Background Explorer (COBE) satellite, which was launched into Earth orbit 1989, and the Wilkinson Microwave Anisotropy Probe (WMAP) satellite, which was launched into solar orbit in 2001. Previously, no precise data from Earth-bound observations were available below a wavelength of 1 cm because of atmospheric absorption. The COBE and WMAP satellites were able to obtain very precise data on the intensity of the background radiation in the wavelength range between 1 cm and 0.05 cm (0.0001 eV and 0.0025 eV).

Vera Rubin (1928–, United States). An observational astronomer, she has made pioneering discoveries about the motion of stars and galaxies. By observing the Doppler shifts of stars in galaxies, she deduced that their rotational velocities are not consistent with attraction only by a large concentration of mass at the galactic center. Rubin's work has been among the leading contributions to understanding the existence and amount of dark matter in the universe. She was awarded the National Medal of Science in 1993.

clusters than we can account for from the galaxies alone. We conclude that dark matter also surrounds clusters of galaxies.

Support for the existence of dark matter comes from the observation of light from distant galaxies that passes by a cluster of galaxies on its way to Earth. This light is deflected by the gravitational field of the cluster in a process known as "gravitational lensing." From the amount of the deflection of the light, it is possible to deduce the quantity of matter in the cluster. The results of these observations show that there is much more matter in the cluster than we would deduce from the luminous matter alone, suggesting the presence of relatively large amounts of dark matter. A dramatic illustration of this effect occurs in the colliding galaxies of the "Bullet Cluster," first analyzed in 2006, in which the distribution of ordinary matter (revealed by the glowing interstellar gas) is different from the distribution of dark matter (revealed by the gravitational lensing effect).

What kinds of objects make up this dark matter? Speculations about its nature are divided loosely into two categories: MACHOs (Massive Compact Halo Objects) and WIMPs (Weakly Interacting Massive Particles). Possible MACHOs include massive black holes, neutron stars, burnt-out white dwarf stars, or brown dwarf stars (Jupiter-sized objects of too small a mass to become a star). The WIMPs include neutrinos, magnetic monopoles, and other exotic types of stable elementary particles produced during the Big Bang. The major difference between the two types of objects is that MACHOs are made from baryons (protons and neutrons, like ordinary matter), while WIMPs are made from some other, more exotic type of non-baryonic matter. Current theories suggest that most of the dark matter is of the non-baryonic form, but (aside from neutrinos) no examples of this type of matter have yet been produced in any laboratory on Earth.

15.4 THE GENERAL THEORY OF RELATIVITY

The interpretation of the observational data describing our universe must be done using methods from the general theory of relativity, which is in essence a theory of gravitation developed by Albert Einstein between 1911 and 1915. The mathematical level of this theory is beyond the level of this text, but we will try to appreciate how the theory helps us to understand the structure and evolution of the universe.

The special theory of relativity arose from a thought experiment of Einstein's in which he imagined trying to catch up with a light beam. The general theory also arose from a thought experiment. Here are Einstein's words:

> *I was sitting in a chair in the patent office at Bern when all of a sudden a thought occurred to me: If a person falls freely he will not feel his own weight. This simple thought made a deep impression on me. It impelled me toward a theory of gravitation.*

Figure 15.10 illustrates a freely falling person in two situations: in the Earth's gravity and in interstellar space where the gravitational field is negligibly weak. In both cases the person is in an isolated chamber and therefore unable to use outside objects to deduce the motion of the chamber. From within the chamber

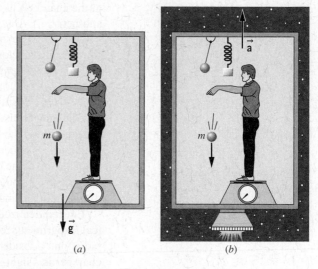

FIGURE 15.10 The effects of freely falling appear identical from within the chamber in (a) the Earth's gravity and (b) interstellar space.

FIGURE 15.11   The effects appear identical when the chamber is (a) at rest in a uniform gravitational field and (b) accelerating in interstellar space.

the two cases look exactly equivalent; no measuring instrument operating entirely within the chamber can distinguish between the two cases. An acceleration $\vec{a} = \vec{g}$ in a gravitational field $\vec{g}$ is equivalent to an acceleration of 0 in a negligible gravitational field.

It appears that an acceleration is able to "cancel out" the effects of a gravitational field. Let us go one step further and ask whether an acceleration can *produce* the effects of a gravitational field. Consider the situations illustrated in Figure 15.11. In one case the observer is at rest near the Earth, where the gravitational field is $\vec{g}$. In the other case, the observer is in empty space where the gravitational field is negligibly small, but the rocket engines are firing so that the chamber has an acceleration $\vec{a} = -\vec{g}$. There are various experiments in the chamber: a scale displays the weight of the observer (actually, the normal force exerted on the observer by the scale), a ball drops to the floor, a mass stretches a spring, and a pendulum oscillates. All of these experiments give identical results in the two chambers. Once again, there is no experiment that can be done within the chamber to distinguish the two cases.

This leads us to the *principle of equivalence:*

There is no local experiment that can be done to distinguish between the effects of a uniform gravitational field in a nonaccelerating inertial frame and the effects of a uniformly accelerating (noninertial) reference frame.

By "local" we mean that the experiments must be done within the chamber, and also that the chamber must be sufficiently small that the gravitational field is uniform. Near the surface of the Earth, for example, not all $\vec{g}$ vectors inside the chamber would be parallel; they point toward the center of the Earth, so there would be a slight angle between the $\vec{g}$ vectors on opposite sides of the chamber. If we make the chamber small, this effect is negligible and the $\vec{g}$ vectors everywhere in the chamber, like the $\vec{a}$ vectors in the accelerating chamber, are parallel to one another.

The principle of equivalence appears in a slightly different (and weaker) form in introductory physics, where it is stated in terms of the equivalence of *inertial* and *gravitational* mass. That is, the mass m that appears in the expression $F = ma$ (inertial mass) is identical to the mass m that appears in the expression $F = GMm/r^2$ (gravitational mass). It follows from this form of the principle of equivalence that all objects, regardless of their masses, fall with the same acceleration in the Earth's gravity. This was first tested by Galileo in the famous (and perhaps apocryphal) experiment in which he dropped two different masses from the top of the leaning tower of Pisa and observed them to fall at the same rate. In recent years other more precise experiments have established the equivalence of gravitational and inertial mass to about 1 part in 10^{11}.

Einstein realized that the principle of equivalence applied not only to mechanical experiments but to all experiments, even ones based on electromagnetic radiation. Consider the arrangement shown in Figure 15.12. At the top of the chamber is a light source that emits a wave of frequency f. At the bottom of the chamber and a distance H away is a detector that observes the wave and measures its frequency. When the light wave is emitted in the accelerating chamber, the source has speed v, which we assume to be small compared with the speed of light c. When it is detected, after a time of flight $t \approx H/c$, the floor is moving with a speed $v + at$. In effect there is a relative speed $\Delta v = at$ between the source and the detector, so there is a Doppler shift in the frequency given by Eq. 2.22:

$$f' = f\sqrt{\frac{1 + \Delta v/c}{1 - \Delta v/c}} \approx f(1 + \Delta v/c) \qquad (15.13)$$

or, in terms of the frequency difference $\Delta f = f' - f$,

$$\frac{\Delta f}{f} = \frac{\Delta v}{c} = \frac{at}{c} = \frac{aH}{c^2} \qquad (15.14)$$

Now let us compare the result of this experiment with a similar one done in a chamber at rest in a uniform gravitational field g. If the results of the two experiments are to be identical (as required by the principle of equivalence), there must be a frequency shift given by Eq. 15.14 with $a = g$:

$$\frac{\Delta f}{f} = \frac{gH}{c^2} \qquad (15.15)$$

The principle of equivalence thus predicts a change in frequency of a light wave falling in the Earth's gravity.

In 1959, R. V. Pound and G. A. Rebka allowed 14.4-keV photons from the radioactive decay of ^{57}Co to fall down the Harvard tower, a distance of 22.6 m. The expected fractional change in frequency, $\Delta f/f = gH/c^2$, was 2.46×10^{-15}; that is, to detect the effect, they had to measure the frequency or energy of the photon at the bottom of the tower to a precision of about 1 part in 10^{15}! The Mössbauer effect (Section 12.9) makes it possible to achieve such a level of precision, and the measured result was $\Delta f/f = (2.57 \pm 0.26) \times 10^{-15}$, consistent with the equivalence principle. Similar experiments based on comparisons between the

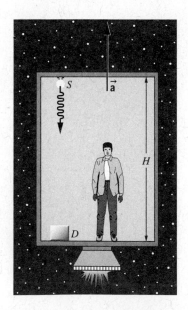

FIGURE 15.12   A source S emits a light wave that is recorded by a detector D in a chamber that is accelerating upward.

frequency of radiation emitted by satellites and received by ground stations have confirmed the predictions of the principle of equivalence to a precision of about 1 part in 10^4.

Because the Global Positioning System (GPS) relies on frequency measurements on the surface of the Earth from transmitters in orbiting satellites, its accuracy depends on applying a correction due to the gravitational frequency shift predicted by general relativity. Without this correction, errors in the GPS locating system of roughly 10 km per day would accumulate.

Note that in Eq. 15.15 the frequency shift depends on the difference in gravitational potential ΔV (potential energy per unit mass) between the source and the detector:

$$\Delta V = \frac{\Delta U}{m} = \frac{(mgH - 0)}{m} = gH \qquad (15.16)$$

For the satellite, even though the gravitational field through which the radiation travels is not uniform, the same conclusion holds: the frequency shift depends on the difference in gravitational potential between the source and the observer. Consider, for example, light leaving the surface of a star of mass M and radius R. The gravitational potential at the surface is $V = -GM/R$. If the light is observed on the Earth, where the gravitational potential is negligible compared with that of the star, the frequency shift is

$$\frac{\Delta f}{f} = \frac{\Delta V}{c^2} = -\frac{GM}{Rc^2} \qquad (15.17)$$

Photons climbing out of a star's gravitational field lose energy and are therefore shifted to smaller frequencies or longer wavelengths (red shifted). This effect is difficult to observe for two reasons: (1) the motion of the star causes a Doppler shift that is generally greater than the gravitational shift, and (2) the spectral lines are Doppler broadened by the thermal motion of the atoms near the surface of a star (see Section 10.4). Nevertheless, the effect has been confirmed for a few stars including the Sun.

Example 15.1

The Lyman α line in the hydrogen spectrum has a wavelength of 121.5 nm. Find the change in wavelength of this line in the solar spectrum due to the gravitational shift.

Solution

From Eq. 15.17, we have

$$\frac{\Delta \lambda}{\lambda} = -\frac{\Delta f}{f} = \frac{GM}{Rc^2}$$
$$= \frac{(6.67 \times 10^{-11} \, \text{N} \cdot \text{m}^2/\text{kg}^2)(1.99 \times 10^{30} \, \text{kg})}{(6.96 \times 10^8 \, \text{m})(3.00 \times 10^8 \, \text{m/s})^2}$$
$$= 2.12 \times 10^{-6}$$

The shift in wavelength is

$$\Delta \lambda = (2.12 \times 10^{-6})(121.5 \, \text{nm})$$
$$= 0.257 \, \text{pm}$$

This shift in wavelength is small in comparison with the Doppler shifts due to the Sun's rotation and the thermal broadening of its spectral lines (see Problem 6).

Space and Time in General Relativity

From *special* relativity we learn that the laws of physics must be the same in all inertial frames, and that there is no preferred inertial frame relative to which it is possible to determine the absolute velocity of an observer. From this point of view, it seems that an accelerated (noninertial) frame is a preferred frame, because it is possible to determine the absolute acceleration. In the *general* theory, Einstein sought to remove this restriction, so that motion would be relative for *all* observers, even accelerated ones. The principle of equivalence, which tells us that we can't distinguish between acceleration and a gravitational field, removes acceleration from its privileged role.

Ultimately, general relativity is a theory of geometry. The motion of a particle is determined by the properties of the space and time coordinates through which it moves. Because space and time are intimately coupled in relativity (see, for example, the Lorentz transformation in Section 2.5), we regard them as components of a combined coordinate system called *spacetime*.

The equivalence between accelerated motion and gravity suggests a relationship between spacetime coordinates and gravity. In the classical description, we would say that the presence of matter sets up a gravitational field, which then determines how objects move in response to that field. According to general relativity, we say that the presence of matter (and energy) causes spacetime to warp or curve; the motion of particles is then determined by the shape of the coordinate system. It is sometimes said that "Geometry tells matter how to move, and matter tells geometry how to curve." From the configuration of mass and energy, general relativity gives us a procedure for calculating the curvature of spacetime, and the motion of a particle or a light beam then follows directly.

Let's consider the simple case of a particle that moves in only one dimension—for example, a bead sliding without friction on a straight wire. We can plot the motion of the bead on an xt coordinate system, which is the two-dimensional spacetime of the bead. For example, if the bead moves with constant velocity along the wire from position x_A at time t_A to position x_B at time t_B, the motion in spacetime is represented by the straight line shown in Figure 15.13a.

Now suppose the wire is mounted vertically in an accelerating chamber in a gravity-free environment. To an observer in the chamber, the bead will appear to accelerate downward as the chamber is accelerated upward. The path in spacetime is now curved, as suggested by Figure 15.13b.

If the acceleration is replaced by the equivalent gravitational field, the motion of the bead, as observed from inside the chamber, is exactly the same—the bead appears to accelerate downward. General relativity describes this situation as a change in the shape of spacetime; the presence of matter (which classically we would describe as the source of the gravitational field) distorts the xt spacetime as indicated in Figure 15.13c. If we imagine the spacetime coordinate system as a grid laid out on a rubber sheet, the gravitating matter stretches the sheet, and the particle moves from A to B along the most direct path in the curved spacetime.

It is convenient to define the *spacetime interval ds*, in effect the separation between two events (such as the particle passing through successive points) in two-dimensional spacetime, as

$$(ds)^2 = (c\,dt)^2 - (dx)^2 \tag{15.18}$$

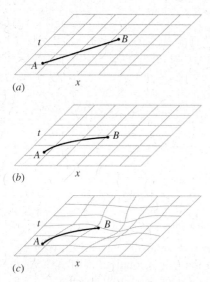

FIGURE 15.13 (a) The path through flat spacetime of a particle moving at constant velocity. (b) The path through flat spacetime of an accelerating particle. (c) The path of a particle through spacetime curved by matter.

This quantity is invariant under the Lorentz transformation, as you can prove by substituting dx' and dt' from Eq. 2.23. The trajectory of the particle in spacetime can be regarded as a collection of infinitesimal intervals. The particle is merely following the contour of spacetime, so the interval serves both to define the trajectory and to represent the shape of spacetime. The extension to three spatial dimensions gives an interval

$$(ds)^2 = (c\,dt)^2 - (dx)^2 - (dy)^2 - (dz)^2 \qquad (15.19)$$

To characterize a "curved" four-dimensional spacetime, we might write the interval as

$$(ds)^2 = g_0(c\,dt)^2 - g_1(dx)^2 - g_2(dy)^2 - g_3(dz)^2 \qquad (15.20)$$

where the four coefficients g_i describe the curvature of the spacetime and its deviation from a Euclidian nature (for which all $g_i = 1$).

The interval of Eq. 15.19 is characteristic of our familiar Euclidian space, which we call "flat." Figure 15.14 summarizes some of the characteristics of that space: a straight line is the shortest distance between two points, the sum of the angles of a triangle is $180°$, parallel lines never meet, the ratio between the circumference and the diameter of a circle is π, and so forth.

Figure 15.15 shows a curved non-Euclidian geometry, the surface of a sphere. Here the shortest distance between points is an arc of a great circle, the sum of the angles of a triangle is greater than $180°$, parallel lines can meet, and the ratio between the circumference and the diameter of a circle is less than π. The saddle-shaped geometry shown in Figure 15.16 has a different kind of curvature, in which the ratio between the circumference and the diameter of a circle is greater than π.

To appreciate the importance of curved spacetime, consider the experiment illustrated in Figure 15.17. A light beam is emitted by a source in the chamber and travels across the chamber to the opposite wall. If the chamber is in an inertial frame and free from gravitational fields, the beam travels horizontally across the chamber and strikes the opposite wall at the same height above the floor as the source. (This holds even if the chamber moves at constant velocity, as you can prove using special relativity.) Observers inside and outside the chamber agree on this conclusion.

If the chamber accelerates, the situation is different. Suppose the light is emitted when the chamber is at rest, relative to a particular inertial frame. The light beam then has no transverse component of velocity in this frame and moves horizontally. As the chamber accelerates, the light beam acquires no transverse velocity component, but the chamber's velocity increases. To the inertial observer, the beam travels along a horizontal straight line, but the chamber accelerates forward so that the beam strikes the opposite wall at a lower height than the source. To an observer in the chamber, the beam appears to follow the curved path shown in Figure 15.17a and strikes the wall at a lower height than the source.

According to the principle of equivalence, the observer in the chamber should record the same outcome if the chamber is at rest in a uniform gravitational field (Figure 15.17b). General relativity explains this observation through the curvature

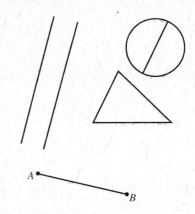

FIGURE 15.14 A flat space and its Euclidian geometrical properties.

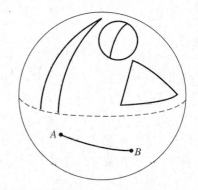

FIGURE 15.15 A curved space and its non-Euclidian geometrical properties.

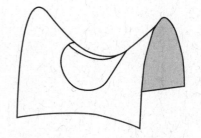

FIGURE 15.16   Another non-Euclidian curved space.

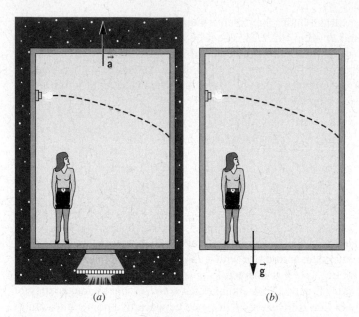

FIGURE 15.17 (*a*) According to an observer in an accelerating chamber, the light beam follows a curved path. (*b*) An observer at rest in a uniform gravitational field finds the same outcome.

of spacetime in the vicinity of the mass that is responsible for the gravitational field. The light beam is merely seeking out the shortest possible path in the curved spacetime, just like an ant crawling along a line on the spherical surface of Figure 15.15. All paths in the curved spacetime are curved.

It is tempting to seek an alternative explanation for the outcome shown in Figure 15.17*b*. For instance, we can assume each photon in the light beam to have an effective mass $m = E/c^2$ and then calculate its trajectory in the gravitational field as we would that of any classical particle of mass m. However, as we discuss in the next section, this method gives results that do not agree with observations for the path of photons in a gravitational field. The curvature of spacetime, which provides the correct explanation, is an inescapable consequence of the principle of equivalence.

General relativity gives a relationship between curvature and the density of mass and energy in space, which can be written symbolically as

$$\text{curvature of space} = \frac{8\pi G}{c^4}(\text{mass-energy density}) \qquad (15.21)$$

Note that this expression incorporates gravitation (Newton's constant G) and special relativity (the speed of light c). If no matter or energy is present, the right-hand side is zero; as a result, the curvature is zero and space is flat. In the limit of classical kinematics ($c \to \infty$) and in the limit of weak gravitational fields ($G \to 0$), space is nearly flat and we can safely use the Newtonian gravitational theory. This is equivalent to saying that if we take a small enough region of the sphere of Figure 15.15, or if we increase its radius to a sufficiently large value, the geometry is approximately Euclidian. Just as classical kinematics can be regarded as the limiting case of *special* relativity (for low speeds), so can classical gravitation be regarded as the limiting case of *general* relativity (for weak fields). In calculating the orbit of an Earth satellite or the trajectory of a space probe to Mars, Newton's theory gives entirely satisfactory results. Close to the Sun and to compact or massive stars, the curving of space can lead to observable effects, as we discuss in the next section.

15.5 TESTS OF GENERAL RELATIVITY

Newtonian gravitation and Einstein's general relativity each give predictions that can be tested against experiment, but under most circumstances the differences between the two predictions are extremely small. At the surface of the Earth, space is curved by only about 1 part in 10^9; even at the surface of the Sun, the curvature is only about 1 part in 10^6.

Nevertheless, there are experiments we can do that are precise enough to detect the difference between flat spacetime and curved spacetime. In this section we discuss several of these experiments.

Deflection of Starlight

When a beam of light from a star passes close to the Sun, it is deflected from its original direction, as shown in Figure 15.18. The star appears to be displaced from its true position by an angle θ.

It is possible to analyze this situation using Newtonian gravitation and special relativity by assigning the photons in the beam an effective mass $m = E/c^2$ and assuming them to be deflected by the Newtonian gravitational force. The experiment then looks very much like Rutherford scattering, and by analogy with the Rutherford scattering formula (see Problem 9) it is possible to calculate the deflection angle (in radians):

$$\theta = \frac{2GM}{Rc^2} \tag{15.22}$$

where M is the mass of the Sun and R is its radius. Substituting the numbers gives $\theta = 0.87''$ as the prediction of special relativity and Newtonian gravitation.

General relativity gives a different view. Spacetime in the vicinity of the Sun is curved, and the light beam is simply following the most direct path along the curved spacetime (Figure 15.19 is a two-dimensional representation of this effect). According to general relativity, the expected deflection is $1.74''$, exactly twice the value predicted by the Newtonian formula.

Measuring this effect requires the observation of a beam of light, such as from a star, that passes near the edge of the Sun. Starlight near the Sun can be observed only during a total solar eclipse. In 1919, just a few years after Einstein completed his general theory, two expeditions of British astronomers traveled to Africa and to South America to observe the solar eclipse and to measure the apparent changes in positions of stars whose light grazed the Sun. Their results for the deflection

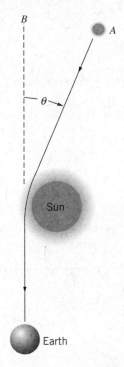

FIGURE 15.18 A light beam passing near the Sun is deflected. To an observer on Earth, the star at A appears to be at B.

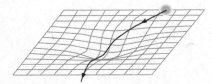

FIGURE 15.19 The path of a light beam from a star through curved spacetime.

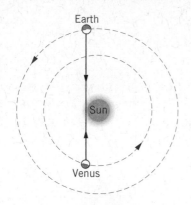

FIGURE 15.20   An electromagnetic signal travels between Earth and Venus at superior conjunction.

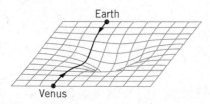

FIGURE 15.21   Path of signal between Earth and Venus in curved spacetime.

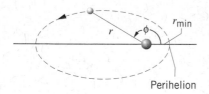

FIGURE 15.22   The elliptical orbit of a planet about a star.

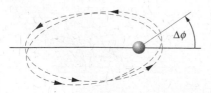

FIGURE 15.23   Precession of the perihelion (greatly exaggerated). After each orbit, the perihelion advances by an angle $\Delta\phi$.

angles, $1.98'' \pm 0.18''$ and $1.69'' \pm 0.45''$, gave strong support for the new general theory. In the years since those early results, this experiment has been repeated at nearly every total solar eclipse, and the overall agreement with general relativity is within 10%. Radio emission from quasars has also been used to confirm this effect, and here the agreement with general relativity is within 2%.

These experimental results give a clear distinction between Newtonian gravity (even with special relativity included) and general relativity.

Delay of Radar Echoes

When a line joining Earth and another planet (Venus, for example) passes through the Sun, the situation is known as "superior conjunction" and is illustrated in Figure 15.20. Based on the orbits of Earth and Venus, we can calculate how long it takes a radar signal sent from Earth to be reflected from Venus and return to Earth (about 20 minutes). Near superior conjunction, the signal passes close to the Sun, and therefore, according to general relativity, it does not travel in a Euclidian straight line, but instead follows a path through curved spacetime (Figure 15.21). It therefore takes the signal a bit longer than the expected time to make the round trip (think of the time intervals as extended as the beam passes close to the Sun, thus lengthening the time to travel the path). This time delay is expected to be about 10^{-4} s, and it has been confirmed to within a few percent (the limit on precision being uncertainties in the surface of Venus, since we don't know if the signal is being reflected from a mountain or a valley). More precise experiments were done in the late 1970s using a signal sent between Earth and the Viking landers on Mars. In this case, the result was consistent with general relativity to within about 0.1%. Signals from NASA's Cassini spacecraft, which entered orbit around Saturn in 2004, have provided the most sensitive test of the time delay, agreeing with the predictions of general relativity to within 0.002%.

Precession of Perihelion of Mercury

Consider a simple planetary system, shown in Figure 15.22, consisting of a single planet in orbit about a star of mass M such as the Sun. According to Newtonian gravitation, the orbit is a perfect ellipse with the star at one focus. The equation of the ellipse is

$$r = r_{\min}\frac{1+e}{1+e\cos\phi} \tag{15.23}$$

where $r_{\min}$ is the minimum distance between planet and star and e is the *eccentricity* of the orbit (the degree to which the ellipse is noncircular; $e = 0$ for a circle). When $r = r_{\min}$, the planet is said to be at *perihelion*; this occurs regularly, at exactly the same point in space, whenever $\phi = 0, 2\pi, 4\pi, \ldots$. According to general relativity, the orbit is not quite a closed ellipse; the curved spacetime near the star causes the perihelion direction to *precess* somewhat, as shown in Figure 15.23. After completing one orbit, the planet returns to $r_{\min}$, but at a slightly different ϕ. The difference $\Delta\phi$ can be computed from general relativity, according to which the orbit is

$$r = r_{\min}\frac{1+e}{1+e\cos(\phi - \Delta\phi)} \tag{15.24}$$

where

$$\Delta\phi = \frac{6\pi GM}{c^2 r_{min}(1 + e)} \tag{15.25}$$

For the Sun, $6\pi GM/c^2 = 27.80$ km, and thus even for the smallest value of r_{min} (for Mercury, 46×10^6 km) $\Delta\phi$ is of order 10^{-6} rad, an extremely small quantity. However, this effect is *cumulative*; that is, it builds up orbit after orbit, and after N orbits, the perihelion has advanced by $N\Delta\phi$. We usually express this precession in terms of the total precession per century (per 100 Earth years), and some representative values are shown in Table 15.1.

The expected precessions are very small, of the order of seconds of arc per century, but nevertheless have been measured with great accuracy; for the three planets closest to the Sun, and for the asteroid Icarus, the measured values are in agreement with the predictions of general relativity. In the best case, the agreement is within about 1%.

These experiments are very difficult to do because (except for Mercury and Icarus) the eccentricities are small and locating the perihelion is difficult. A more serious problem is that other effects, not associated with general relativity, also cause an apparent precession of the perihelion. In the case of Mercury, the observed precession is actually about $5601''$ per century; of that, $5026''$ are due to the precession of the Earth's equinox (a classical Newtonian effect of the spinning Earth) and $532''$ are due to the gravitational pull of the other planets on Mercury (also a classical Newtonian effect). Only the difference of $43''$ is due to general relativity.

Gravitational Radiation

Just as an accelerated charge emits electromagnetic radiation that travels with the speed of light, an accelerated mass emits gravitational radiation that also travels with the speed of light. In effect, gravity waves are ripples that travel through spacetime. Waves produced by such motions as the planets around the Sun are exceedingly weak and beyond any hope of detection. Cataclysmic events in the universe, such as supernova explosions, and highly accelerated systems, such as compact binary objects, may produce observable gravitational waves. Detection

TABLE 15.1 Precession of Perihelia

Planet	N (orbits per century)	e	r_{min} (10^6km)	$N\Delta\phi$ (arc seconds per century) General Relativity	$N\Delta\phi$ (arc seconds per century) Observed
Mercury	415.2	0.206	46.0	43.0	43.1 ± 0.5
Venus	162.5	0.0068	107.5	8.6	8.4 ± 4.8
Earth	100.0	0.017	147.1	3.8	5.0 ± 1.2
Mars	53.2	0.093	206.7	1.4	
Jupiter	8.43	0.048	740.9	0.06	
Icarus	89.3	0.827	27.9	10.0	9.8 ± 0.8

of these waves would provide another important confirmation of the general relativity theory.

In analogy with the effect of a passing electromagnetic wave on a charge, a passing gravitational wave could be detected by its effect on matter. Several antennas have been built to search for gravity waves, but no conclusive experimental evidence has yet been obtained. Indirect evidence has come from the observations of the change in the orbital period of a binary pulsar (see next section). Interferometric techniques are being used to build new detectors to search for gravity waves. The Laser Interferometer Gravitational-wave Observatory (LIGO), which began operation in 2001, consists of two installations (located in the states of Washington and Louisiana) whose interferometer arms are 4 km in length. A passing gravitational wave would cause a small change in the length of one arm relative to the other, which would be detected through a change in the fringe pattern similar to that of the Michelson interferometer (Section 2.2).

15.6 STELLAR EVOLUTION AND BLACK HOLES

Although the large gravitational field of the Sun has provided several good tests of general relativity, the most stringent tests will come from measurements in even larger gravitational fields, where the curvature of spacetime is significantly greater. Such large gravitational fields can occur following the collapse of a star into a more compact object: a white dwarf, a neutron star, or a black hole.

We considered the collapse of an ordinary star like the Sun to a white dwarf star as an example of the application of Fermi-Dirac statistics in Section 10.7. As the supply of hydrogen fuel in a star begins to be used up, the star will contract because the radiation pressure that opposes gravitational collapse is reduced. Eventually the stable white dwarf stage can be reached, at which the Pauli principle applied to the electrons prevents further collapse. The average density of a white dwarf star such as Sirius B is about 10^9 kg/m^3, which is about 10^6 times the average density of the Sun.

If the star has a mass greater than about 1.4 solar masses (the *Chandrasekhar limit*), the gravitational force is sufficient to overcome the Pauli repulsion of the electrons, and further collapse can occur. For a star of this mass, the Fermi energy of the electrons (Eq. 10.50) is 0.30 MeV. Higher-energy electrons in the tail of the Fermi-Dirac distribution will have sufficient energy to produce the inverse beta-decay reaction:

$$e^- + p \rightarrow n + \nu_e$$

for which the threshold energy is 0.782 MeV, not too far above E_F. This reaction removes some electrons from the star, reducing the effects of Pauli repulsion, and allowing the star to collapse a bit. The Fermi energy *increases*, pushing more electrons above the 0.782-MeV threshold, resulting in more electrons being lost, and so on, until all (or very nearly all) of the electrons vanish. The star is now composed of neutrons, instead of protons and electrons. The Pauli repulsion of the

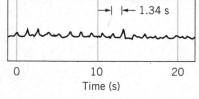

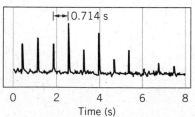

FIGURE 15.24   The radio signals from two different pulsars. The top signal is the record of the first pulsar discovered.

electrons no longer can oppose gravitational collapse, and so the star contracts until the Pauli principle applied to the *neutrons* (which also obey Fermi-Dirac statistics) prevents further collapse. As we calculated in Section 10.7, a neutron star of 1.5 solar masses would have a radius of 11 km and a density of about 5×10^{17} kg/m^3.

Are these neutron stars merely figments of the physicist's imagination or do they really exist? In 1967, radio astronomers at Cambridge University discovered an unusual signal among their observations—a regular pulsation, such as is shown in Figure 15.24, with a period of 1.34 s. No previously known astronomical object could produce such sharp and regular pulses, and at first the Cambridge group suspected that they might have discovered signals from an extraterrestrial intelligent civilization. (The object emitting the pulses was at first called LGM-1; LGM stands for "Little Green Men.") This notion was later discarded (unfortunately) and the object became known as a *pulsar*. Since 1967, hundreds of other pulsars have been discovered; all have extremely regular periods typically in the range 0.01–1 s.

The connection between pulsars and neutron stars was made soon after their discovery. The collapse of a rotating star to a neutron star causes the neutron star to rotate much more rapidly. Angular momentum is conserved during the collapse (see Problem 11), so the rotational angular velocity increases as the rotational inertia decreases. A relatively slow rotation rate of the original star can become a very rapid rotation rate for the neutron star.

The intense magnetic field of such a rapidly rotating object traps any emitted charged particles and accelerates them to high speeds, especially near the magnetic poles, where they give off radiation (Figure 15.25). As the star rotates, this beam of emitted radiation sweeps around like a searchlight or a lighthouse, and we see a pulse of radiation whenever the beam sweeps through the Earth. The observed interval between the pulses is, according to this interpretation, the rotational period of the neutron star.

If this explanation of a pulsar as a rotating neutron star is correct, we ought to see the pulsars slowing down somewhat, as the radiated energy is compensated by a decrease in the neutron star's rotational kinetic energy. This effect has been seen for nearly all pulsars, and amounts to about 1 part in 10^9 per day.

Pulsars have now been observed at many different wavelengths (optical, X ray, γ ray, radio) and with such great precision of timing that the slowing down of 10^{-9} per day is easily observable.

Although the exact mechanism of the collapse of a star to a neutron star is not yet understood, we suspect that the violent explosions known as *supernovas* leave a neutron star as a remnant. In 1054, Chinese astronomers observed a supernova explosion (which they called a "guest star") that was visible in the daytime over many days. Today we see the expanding shell of that explosion as the Crab Nebula (Figure 15.26). At the center of the Crab Nebula is a pulsar, rotating with a frequency of 30 Hz. It is remarkable that none of the many photographs of the Crab that were taken before 1967 revealed this pulsar blinking on and off every 0.033 s; all of these photographs were taken over long exposure times, and so the pulsations were not observable. When careful measures are taken, however, the blinking effect can be seen quite clearly (Figure 15.27). This suggests that, at least in this instance, pulsars may be identified as supernova remnants.

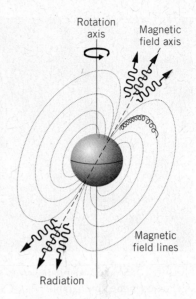

FIGURE 15.25   Charged particles trapped by the magnetic field lines of a neutron star are given large accelerations near the magnetic poles, from which a directional radiation beam emerges. If this radiation beam intercepts the Earth as the neutron star rotates, we see it as a pulse of radiation.

FIGURE 15.26   The Crab Nebula, remnant of a supernova observed in the year 1054.

 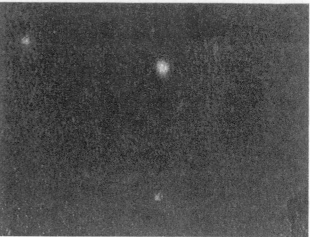

FIGURE 15.27 The visible pulsar of the Crab Nebula. The two exposures show the pulsar blinking on and off relative to the other stars in the photograph.

A Binary Pulsar

In 1974, an unusual pulsar was discovered. The period of the pulsar was measured to be 59 ms, making it among the fastest observed up to that time. More surprising, the pulse rate appeared to be slowing down by about 0.1% per hour, but later was observed to be *increasing* by about the same amount. It was quickly realized that the decrease and increase of the pulse rate could be explained as a Doppler shift if the neutron star were moving first away from and later toward the Earth. To move in this way, the pulsar must be in orbit around an unseen companion. Thus we have a pulsar as part of a binary star system, or a "binary pulsar."

The orbital period of the binary system was determined to be about 8 hours. This is an extremely short period; for example, it is more than 250 times shorter than the orbital period of Mercury, the fastest moving planet in our solar system. To have such a short orbital period, the pulsar must be orbiting very close to its companion (which is believed to be another neutron star). At such close distances, the curvature of spacetime is large and the effects of general relativity should be measurable. In effect, the binary pulsar provides us with a "general relativity laboratory."

Among the general relativity effects that have been observed in the binary pulsar system is the delay of the pulses due to the curvature of spacetime. This situation is similar to Figures 15.20 and 15.21, except that the pulsar (instead of Venus) is the origin of the signals and the curvature is due to the companion star (instead of the Sun). An effect analogous to the precession of Mercury's perihelion has also been observed (except that, in the case of a star, the point of closest approach is called *periastron* rather than perihelion). The periastron of the binary pulsar changes by $4.23°$ per year, about 35,000 times more rapidly than that of Mercury; the change of the periastron is known to an accuracy of about 1 part in 10^5, three orders of magnitude more precisely than Mercury's.

The most remarkable observation from the binary pulsar is the slowing of its orbital period, which general relativity explains as caused by the emission of gravitational radiation. Because of its large centripetal acceleration (its orbital speed is about $0.001c$), the binary pulsar should emit gravity waves, and the energy radiated away is compensated by a loss in the orbital energy. This loss amounts of $76\,\mu s$ per year or $67\,ns$ per orbit, and the measured change in the orbital rate confirmed the prediction of general relativity to within 1%. In the absence of direct observation, this is the strongest evidence yet obtained for the existence of gravitational radiation. For the discovery of the binary pulsar and its contributions to the study of gravitation, Joseph Taylor and Russell Hulse were awarded the 1993 Nobel Prize in physics.

Black Holes

A neutron star is not the ultimate fate of the collapse of massive stars. Stars with masses less than two or three solar masses probably do end up as white dwarf stars or neutron stars. For more massive stars, the gravitational force is strong enough to overcome even the Pauli principle applied to the neutrons, and there is nothing to prevent the material in the star from suffering complete collapse down to a single point in space. To understand gravitational collapse, we must turn again to general relativity.

Within a year after Einstein's 1916 publication of the general theory, Karl Schwarzschild worked out the solutions to the equations for the curvature of spacetime near a spherically symmetric mass M. In spherical coordinates (r, θ, ϕ), the spacetime interval for this solution is

$$(ds)^2 = c^2 \left[1 - \frac{2GM}{c^2 r} \right] (dt)^2 - \frac{(dr)^2}{\left[1 - \dfrac{2GM}{c^2 r} \right]} - r^2 (d\theta)^2 - r^2 \sin^2 \theta (d\phi)^2$$

$$(15.26)$$

(Note that in the classical and zero-gravity limits $c \to \infty$ and $G \to 0$, the two factors in square brackets disappear, leaving us with an interval in spherical coordinates that is the exact analog of the three-dimensional interval expressed in Cartesian coordinates, Eq. 15.19.) The radial part of this solution (the dr term) has what appears to be a serious problem: the factor in the denominator can become zero for a particular r, causing that term in the equation to "blow up." This occurs when r has the value

$$r_S = \frac{2GM}{c^2} \qquad (15.27)$$

which is known as the *Schwarzschild radius*. None of the physical coordinates actually "blows up" at $r = r_S$, and an object falling toward M would notice no change in its motion as it crossed the Schwarzschild radius.

For external observers watching the falling object, the situation is very different. As the object falls, general relativity predicts that its clocks would appear to run ever more slowly, stopping completely when the object reaches r_S. The object appears to be frozen forever at that location! While the object is falling, the light it emits becomes increasingly red shifted, and the red shift becomes infinite at $r = r_S$, so the object disappears from view! The outside observer can obtain no information about the object once it passes through the Schwarzschild radius. For that reason, the Schwarzschild radius is often called the *event horizon*; no external observer can see into that horizon.

The falling object does encounter one crisis at r_S. At any time before crossing r_S, the object could reverse its fall and escape from the gravitational pull of M — for example, by firing its rockets. Once it passes r_S, no escape is possible. Inside r_S, the escape speed exceeds the speed of light, and nothing (*not even light*) can escape. No travel or communication is permitted from inside r_S to the outside world. However, the object inside r_S can continue to exert a gravitational force on external objects or, in the language of general relativity, to curve spacetime beyond r_S.

An object whose mass M lies totally within the corresponding radius r_S is said to be a *black hole*. To form such an object requires that matter be compressed to exceptional densities. Table 15.2 shows some values of r_S for representative objects. For the Earth to become a black hole, we would need to compress it into a sphere of radius less than 1 cm, and the Sun would become a black hole only if compressed to a 3-km radius! Nevertheless, it has been speculated that black holes are the end products of the collapse of massive stars, and that tiny (atom-sized) black holes may have been formed by the extreme densities and pressures in the early universe.

Far from a black hole, the gravitational field is Newtonian in character; the effects of curved spacetime are small, and the black hole cannot be distinguished from any other gravitating object. Close to a black hole (or close to any massive, compact object), the effects of curved spacetime can become significant. The discovery of a massive black hole could therefore provide another "laboratory" for testing the predictions of general relativity where the effects may be substantially larger than in the vicinity of the Sun.

Many stars are members of binary systems, in which two stars orbit about their common center of mass. In many cases, a visible star appears to orbit with an invisible companion, and gases from the visible star emit intense X rays as they are accelerated toward the invisible companion. It is believed that these invisible companions in binary systems are black holes. Many such systems have been observed in our galaxy.

By observing the rotational motion of galaxies, it is possible to deduce the mass at the galactic center. For some galaxies, this mass turns out to be greater than 10^9 times the mass of the Sun. No known phenomenon other than a black hole permits so much mass to be concentrated in so small a region. Similar evidence derived from rotational motions suggests that even our near neighbor,

TABLE 15.2 Black Hole Event Horizons

Object	Mass (kg)	Ordinary Radius (m)	r_S (m)
^{238}U nucleus	4×10^{-25}	7×10^{-15}	6×10^{-52}
Physics book	1	0.1	1.5×10^{-27}
Earth	6×10^{24}	6×10^6	8.9×10^{-3}
Sun	2×10^{30}	7×10^8	3×10^3
Galaxy	$\sim 2 \times 10^{41}$	$\sim 10^{20}$	3×10^{14}
Universe	$\sim 10^{51}$	$10^{26}(?)$	$\sim 10^{24}$

the Andromeda galaxy, has a black hole of a few million solar masses at its center. Radio emissions from the center of our own Milky Way galaxy also suggest a black hole of a few million solar masses.

A surprising development in black hole theory occurred in 1974, when Steven Hawking showed that black holes could be sources of particle emission. According to quantum mechanics, particle-antiparticle pairs can spontaneously appear, as long as they exist for a short enough time that the uncertainty principle is not violated. That is, the particles can "borrow" an energy of $2mc^2$ as long as the loan is repaid (the particles vanish) within a time of at most $\Delta t \sim \hbar/2mc^2$. If a particle-antiparticle pair arises outside the event horizon of a black hole, its gravitational field can provide the energy necessary to repay the loan so that the particle and antiparticle can become real. Usually the particle and antiparticle fall back into the black hole, restoring the energy balance. However, one of the members of the pair may have enough energy to escape into the outside world. The black hole thus appears to be emitting particles. In the process, the black hole loses mass. The rate of mass loss is inversely proportional to the mass of the black hole; massive black holes that result from the collapse of stars emit particles at too low a rate to be observed. However, tiny black holes of atomic or nuclear size, which may have been formed in the early evolution of the universe, could be very bright sources of radiation.

Black holes provide both a fertile area for theoretical speculation and a challenge for the skill of experimenters. It has been suggested that material that falls into a black hole reappears at another time and place in the universe, or perhaps in another universe. Thus a black hole, if this speculation is correct, could be used for time travel or to travel between different universes. Other proposals suggest harnessing a black hole as an energy source. It has been estimated that, if black holes are indeed the end products of the evolution of massive stars, there could be as many as 10^9 massive black holes in our galaxy, which makes it likely that many black holes are within our observational reach. Or, perhaps we will someday observe a minihole ending its existence with a burst of radiation. As we continue to refine our ability to study the skies at visible, X-ray, and γ-ray wavelengths, black holes will figure prominently in our investigations.

15.7 COSMOLOGY AND GENERAL RELATIVITY

General relativity can be applied to calculate the properties of the universe as a whole. For this case, the mass-energy density term in Eq. 15.21 must describe the entire universe. We are not interested in the "local" variations in density on a scale of galactic size, but rather in the average density of the entire universe, evaluated over a distance that is large compared with the spacing between galaxies. (In a similar way, when we speak of the density of a solid we are interested not in the variations on the atomic scale but rather in the average density of the entire material, evaluated over a distance that is large compared with the spacing between atoms.) The density of the universe is not a constant; it changes with time as the universe expands.

Solving the equations of general relativity for the large-scale structure of the universe gives the following result, which is known as the Friedmann equation:

$$\left(\frac{dR}{dt}\right)^2 = \frac{8\pi}{3}G\rho R^2 - kc^2 \tag{15.28}$$

Here $R(t)$ represents the size or distance scale factor of the universe at time t, and ρ represents the total mass-energy density at the same time. (The density is expressed in mass units, such as kg/m^3, even if it represents radiation.)

The constant k that appears in Eq. 15.28 specifies the overall geometrical structure of the universe: $k = 0$ if the universe is flat, like Figure 15.14; $k = +1$ if the universe is curved and closed, like Figure 15.15; $k = -1$ if the universe is curved and open, like Figure 15.16. When $k = +1$, the distance factor $R(t)$ is directly related to the size or "radius" of the universe, but its meaning is not so apparent when $k = 0$ or $k = -1$, because in both of the latter cases the universe is infinite in extent. In these cases $R(t)$ should be regarded as a scale factor that represents the expansion of space; the absolute magnitude of R in this case is not significant, and only its variation with time is of interest, because any particular length (such as the distance between two galaxies) will vary with time just as R does.

To solve Eq. 15.28, we must therefore specify the constant k. On the large scale, our universe seems quite close to being flat (as we discuss in Section 15.10), and we therefore take $k = 0$. This simplifies the mathematics and gives results that are not too far different from what we obtain with $k = \pm 1$, so for rough estimates our calculation should be acceptable.

The density ρ in Eq. 15.28 must include both the matter and the radiation present in the universe. The present universe is dominated by matter; the contribution of radiation to the total density is negligible. As the universe expands, the amount of matter remains constant but the volume increases like R^3. Thus the matter density ρ_m decreases with increasing R according to $\rho_m \propto R^{-3}$. Putting this result into Eq. 15.28 and integrating, we find

$$R(t) = At^{2/3} \tag{15.29}$$

where A is a constant. Using this result to eliminate R from Eq. 15.28, we obtain

$$t = \frac{1}{\sqrt{6\pi G\rho_m}} \tag{15.30}$$

In contrast, the early universe was dominated by radiation; the mass density of the matter was negligible. From Eq. 10.42 we see that the energy density of the radiation depends on $d\lambda/\lambda^5$. All wavelengths scale with R, so we have $d\lambda \propto R$ and $\lambda^5 \propto R^5$. Thus the energy density of radiation ρ_r decreases with increasing R according to $\rho_r \propto R^{-4}$. Inserting this result into Eq. 15.28 and integrating, we obtain

$$R(t) = A't^{1/2} \tag{15.31}$$

where A' is a constant, and so

$$t = \sqrt{\frac{3}{32\pi G\rho_r}} \tag{15.32}$$

The Hubble parameter can be defined in terms of the time variation of the scale factor:

$$H = \frac{1}{R}\frac{dR}{dt} \qquad (15.33)$$

As the universe evolves, the value of H changes. Its present value H_0 is revealed in experiments with Hubble's law (Eq. 15.2).

If the universe has been expanding at a constant rate ($R \propto t$), then H^{-1} is the age of the universe. In the two cases we derived above, the age is less than H^{-1}. A matter-dominated universe expanding since $t = 0$ has an age of $\frac{2}{3}H^{-1}$, while a radiation-dominated universe has an age of $\frac{1}{2}H^{-1}$. In either case we can take H^{-1} as a rough measure of the age at any time.

We can therefore characterize the universe by several parameters: a shape parameter k, which describes whether it is flat or curved, open or closed; a radius or scale parameter $R(t)$, which measures the size of the universe as a function of time; the density ρ, which represents both matter and energy, and which is also a function of time; the Hubble parameter H, which is proportional to the rate of expansion; and also a deceleration parameter q, which tells us the rate at which the expansion is slowing down (see Problem 12). The challenge to the observational astronomer is to obtain data on the distribution and motion of the stars and galaxies that can be analyzed to obtain values for these parameters.

15.8 THE BIG BANG COSMOLOGY

The present universe is characterized by a relatively low temperature and a low density of particles. Its structure and evolution are dominated by the gravitational force. Because the universe has been expanding and cooling, in the distant past it must have been characterized by a higher temperature and a greater density of particles. Let us imagine we could run the cosmic clock backward and examine the universe at earlier times, even before the formation of stars and galaxies. At some point in its history, the temperature of the universe must have been high enough to ionize atoms; at that time the universe consisted of a plasma of electrons and positive ions, and the electromagnetic force was important in determining the structure of the universe. At still earlier times, the temperature was hot enough that collisions between the ions would have knocked loose individual nucleons, so the universe consisted of electrons, protons, and neutrons, along with radiation. In this era the strong nuclear force was important in determining the evolution of the universe. At still earlier times the weak interaction played a significant role.

If we try to go back still further, we reach a time when the matter of the universe consisted only of quarks and leptons. Because we have never observed a free quark, we don't know much about their individual interactions, and so we can't describe this very early state of the universe. If someday we are able to understand the interactions of free quarks, we can penetrate this barrier and look to still earlier times. Eventually we reach a fundamental barrier when the universe had an age of only 10^{-43} s, which is known as the *Planck time* (see Problem 26). Before this time, quantum theory and gravity are hopelessly intertwined, and none of our present theories gives us any clue about the structure of the universe.

Later than the Planck time, but still before the condensation of bulk matter, the universe consisted of particles, antiparticles, and radiation in approximate thermal equilibrium at temperature T. The universe at this time was radiation-dominated: the energy density of the radiation exceeded the energy density of the matter. In a radiation-dominated universe, we can use Eq. 15.32 to find a relationship between the temperature and the age. Inserting the radiation density from Eq. 15.9, remembering to convert to mass units such that $\rho_r = U/c^2$, and evaluating all numerical factors, we obtain

$$T = \frac{1.5 \times 10^{10} \text{ s}^{1/2} \cdot \text{K}}{t^{1/2}} \qquad (15.34)$$

where the temperature T is in K and the time t is in seconds. This equation relates the age of the early universe to its temperature.

The radiation of the early universe consisted of high-energy photons, whose average energy at the temperature T can be roughly estimated as kT, where k is the Boltzmann constant. The interactions between the radiation and the matter can be represented by two processes:

$$\text{photons} \rightarrow \text{particle} + \text{antiparticle}$$
$$\text{particle} + \text{antiparticle} \rightarrow \text{photons}$$

That is, photons can engage in pair production, in which their energy becomes the rest energy of a particle-antiparticle pair, or a particle and antiparticle can annihilate into photons. In each case, the energy of the photons must be at least as large as the rest energy of the particle and antiparticle.

Example 15.2

(a) At what temperature is the thermal radiation in the universe energetic enough to produce nucleons and anti-nucleons? (b) What is the age of the universe when it cools to that temperature?

Solution

(a) Let us consider the formation of proton-antiproton or neutron-antineutron pairs by photons:

$$\gamma + \gamma \rightarrow \text{p} + \bar{\text{p}} \quad \text{and} \quad \gamma + \gamma \rightarrow \text{n} + \bar{\text{n}}$$

To produce these reactions, the photons must have an energy at least as great as the nucleon rest energy, or about 940 MeV. The temperature of the photons must then be

$$T = \frac{E}{k} = \frac{mc^2}{k}$$

$$= \frac{940 \text{ MeV}}{8.6 \times 10^{-5} \text{ eV/K}} = 1.1 \times 10^{13} \text{ K}$$

(b) From Eq. 15.34 we can find the age of the universe when the photons have this temperature:

$$t = \left(\frac{1.5 \times 10^{10} \text{ s}^{1/2} \cdot \text{K}}{T} \right)^2$$

$$= \left(\frac{1.5 \times 10^{10} \text{ s}^{1/2} \cdot \text{K}}{1.1 \times 10^{13} \text{ K}} \right)^2 = 2 \times 10^{-6} \text{ s}$$

That is, at times earlier than $2 \, \mu\text{s}$, the universe was hot enough for the photons to produce nucleon-antinucleon pairs, but after $2 \, \mu\text{s}$ the photons were not energetic enough to produce nucleon-antinucleon pairs. The annihilation reaction continues to occur, but after this time nucleon-antinucleon pair production ceases.

In this calculation we are using average photon energies as estimates. Photons in the tail of the thermal spectrum are sufficiently energetic to produce nucleon-antinucleon pairs even after $2 \, \mu\text{s}$, but *on the average* the photons have too little energy. More precisely, we could state that the rate of nucleon-antinucleon pair production drops rapidly at around $2 \, \mu\text{s}$ and becomes negligible at times much greater than $2 \, \mu\text{s}$.

Let us now look at some of the major developments in the evolution of the universe.

$t = 10^{-6}$ s We begin the story at a time of $1\,\mu$s. From Eq. 15.34 we find $T = 1.5 \times 10^{13}$ K or $kT = 1300$ MeV. The scale factor is smaller than that of the present universe by the red shift, $2.7\,$K$/1.5 \times 10^{13}$ K $= 1.8 \times 10^{-13}$. If the universe were closed and finite, its radius would be smaller than the present observable radius (10^{26} m) by this factor, so the universe at that time is about the present size of the solar system (10^{13} m). At $1\,\mu$s, the universe consists of p, $\overline{\text{p}}$, n, $\overline{\text{n}}$, e$^-$, e$^+$, μ^-, μ^+, π^0, π^-, π^+, and perhaps other particles, plus photons, neutrinos, and antineutrinos. Because both pair production and annihilation can occur, the number of particles is roughly equal to the number of antiparticles for each species. Furthermore, the number of photons is roughly equal to the number of nucleons, which is in turn roughly equal to the number of electrons. The relative number of neutrons and protons is determined by three factors:

1. *The Boltzmann factor* $e^{-\Delta E/kT}$. Protons have less rest energy than neutrons, so there are more of them at any given temperature. The energy difference ΔE is $(m_{\text{n}} - m_{\text{p}})c^2 = 1.3$ MeV, so the neutron-to-proton ratio can be expressed as $e^{-1.5 \times 10^{10}/T}$ with T in Kelvins. For $T \sim 10^{13}$ K, this ratio is very nearly 1, but it becomes different from 1 as T approaches 10^{10} K.
2. *Nuclear reactions.* Reactions such as n $+ \nu_{\text{e}} \rightleftarrows$ p $+$ e$^-$ and n $+$ e$^+ \rightleftarrows$ p $+ \overline{\nu}_{\text{e}}$ can go in either direction and tend to make it easy for protons to turn into neutrons or neutrons into protons, as long as there are plenty of e$^-$, e$^+$, ν_{e}, and $\overline{\nu}_{\text{e}}$ around.
3. *Neutron decay.* The neutron half-life is about 10 min, which is going to be important only at later times. For $t < 1$ s, there has not yet been enough time for an appreciable number of neutrons to decay.

At $t = 1\,\mu$s, all three of these factors keep the neutron-to-proton ratio very close to 1.

$t = 10^{-2}$ s Between 10^{-6} s and 10^{-2} s, the temperature drops from 1.5×10^{13} K ($kT = 1300$ MeV) to 1.5×10^{11} K ($kT = 13$ MeV), and the distance scale factor increases by a factor of 100. The photons have on the average too little energy (13 MeV) to produce pions and muons, and because the pion and muon lifetimes are much shorter than 10^{-2} s, they have decayed into electrons, positrons, and neutrinos. Pair production of nucleons and antinucleons no longer occurs, but nucleon-antinucleon annihilation continues. As we discuss later, there is very slight imbalance of matter over antimatter of perhaps 1 part in 10^9. During this interval, all of the antimatter and most (99.9999999%) of the matter is annihilated. Pair production of electrons and positrons can still occur, so the universe consists of p, n, e$^-$, e$^+$, photons, and neutrinos. The neutron-to-proton ratio remains about 1.

$t = 1$ s Between 10^{-2} s and 1 s, the temperature drops to 1.5×10^{10} K ($kT = 1.3$ MeV). In this interval, the Boltzmann factor, which determines the neutron-to-proton ratio, becomes different from 1; by $t = 1$ s, the nucleons consist of about 73% protons and 27% neutrons. During this period, the influence of the neutrinos has been decreasing; to convert a proton to a neutron by capturing an antineutrino

$(\bar{\nu}_e + p \rightarrow n + e^+)$ requires an antineutrino of at least 1.8 MeV, above the mean neutrino energy (1.3 MeV) at this temperature. This begins the time of "neutrino decoupling," when the interactions of matter and primordial neutrinos no longer occur. From this time on, the neutrinos continue to fill the universe, cooling along with the expansion of the universe. These primordial neutrinos presently have roughly the same density as the microwave photons, but a slightly lower temperature (about 2 K).

$t = 6\,s$ Between 1 s and 6 s ($T = 6 \times 10^9$ K or $kT = 0.5$ MeV), the average photon energy decreases and becomes insufficient to produce electron-positron pairs. Electron-positron annihilation continues, and as a result all of the positrons and nearly all (99.9999999%) of the electrons are annihilated. The electrons have too little energy to convert protons to neutrons ($e^- + p \rightarrow n + \nu_e$ no longer occurs), and so the only remaining weak interaction process that influences the relative number of protons and neutrons is the radioactive decay of the neutron, which has a half-life of 10 minutes and so has not appreciably occurred by this time. The nucleons are now about 84% protons and 16% neutrons, or about 5 times as many protons as neutrons.

The composition of the universe after $t = 6$ s consists of some number N protons, the same number N electrons, and about $0.2N$ neutrons. There are no remaining positrons or antinucleons. Because particle-antiparticle annihilation has substantially reduced the number of nucleons while the number of photons remained stable, there are about $10^9 N$ photons (and about the same number of neutrinos).

| Example 15.3

Estimate the relative number of neutrons and protons among the nucleons at $t = 1$ s.

Solution

At this time, the temperature is 1.5×10^{10} K. The neutron-to-proton ratio is determined by the Boltzmann factor, $e^{-\Delta E/kT}$, where ΔE is the neutron-proton rest energy difference. The exponent in the Boltzmann factor is

$$\frac{\Delta E}{kT} = \frac{1.3\,\text{MeV}}{(8.62 \times 10^{-5}\,\text{eV/K})(1.5 \times 10^{10}\,\text{K})} = 1.0$$

so the ratio of neutrons to protons is

$$\frac{N_n}{N_p} = e^{-\Delta E/kT} = e^{-1.0} = 0.37$$

The relative number of protons is then

$$\frac{N_p}{N_p + N_n} = \frac{1}{1 + N_n/N_p} = \frac{1}{1 + 0.37} = 0.73$$

The nucleons consist of 73% protons and 27% neutrons.

15.9 THE FORMATION OF NUCLEI AND ATOMS

Let's review developments in the Big Bang cosmology up to $t = 6$ s. (1) A hot, dense universe, full of photons and elementary particles of all varieties, has cooled to below 10^{10} K. (2) Most of the unstable particles have decayed away. (3) All of the original antimatter and most of the original matter annihilated one another, leaving a small number of protons, an equal number of electrons, and about one-fifth as many neutrons. (4) Neutrinos, which have about the same density as photons, decoupled at about 1 s and will continue cooling as the universe expands.

As the neutrons and protons collide with one another, it is possible to form a deuteron (^{2}H nucleus):

$$n + p \rightarrow {}^2H + \gamma$$

but the high density of photons can also produce the inverse reaction:

$$\gamma + {}^2H \rightarrow n + p$$

We recall from Chapter 12 that the deuteron binding energy is 2.22 MeV. In order to have any appreciable buildup of deuterons, the photons present must first cool until their energies are below 2.22 MeV; otherwise the deuterons will be broken up as quickly as they can be formed. The energy 2.22 MeV corresponds to a temperature $T = 2.5 \times 10^{10}$ K, and we therefore might expect deuterons to be formed as soon as the temperature drops below 2.5×10^{10} K. However, this does not happen. The radiation does not have a single energy, but rather has a thermal spectrum. A small fraction of the photons has energies *above* 2.22 MeV, and these photons continue to break apart the deuterons (Figure 15.28).

Before matter-antimatter annihilation occurred, there were about as many photons as nucleons and antinucleons, but after $t = 0.01$ s, the ratio of nucleons to photons is about 10^{-9}; about $\frac{1}{6}$ of the nucleons are neutrons. If the fraction of photons above 2.22 MeV is greater than $\frac{1}{6} \times 10^{-9}$, there will be at least one energetic photon per neutron, which effectively prevents deuteron formation. Our next job is to calculate to what temperature the photons must cool before fewer than $\frac{1}{6} \times 10^{-9}$ of them are above 2.22 MeV.

The number density of thermal photons was given by Eq. 15.6. We expect that the temperature must be much less than 2.5×10^{10} K, and so we are interested in the distribution where $E \gg kT$, for which it is approximately

$$\frac{N(E)\,dE}{V} = \frac{8\pi E^2}{(hc)^3} e^{-E/kT}\,dE \tag{15.35}$$

and the total number density above some energy E_0 is determined by integrating the number density from E_0 to ∞: $N_{E>E_0}/V = \int_{E_0}^{\infty} N(E)dE/V$, which can be shown to be

$$\frac{N_{E>E_0}}{V} = \frac{8\pi}{(hc)^3}(kT)^3 e^{-E_0/kT}\left[\left(\frac{E_0}{kT}\right)^2 + 2\left(\frac{E_0}{kT}\right) + 2\right] \tag{15.36}$$

Equation 15.8 gives the *total* number density of photons, and thus the fraction f above E_0 is $(N_{E>E_0}/V)/(N/V)$, which can be evaluated to be

$$f = \frac{N_{E>E_0}/V}{N/V} = 0.42 e^{-E_0/kT}\left[\left(\frac{E_0}{kT}\right)^2 + 2\left(\frac{E_0}{kT}\right) + 2\right] \tag{15.37}$$

For $f = \frac{1}{6} \times 10^{-9}$, corresponding to the number needed to prevent deuteron formation, Eq. 15.37 gives $E_0/kT = 28$. With $E_0 = 2.22$ MeV, the required temperature is thus about 9×10^8 K; when $T > 9 \times 10^8$ K, the number of photons with $E > 2.22$ MeV is greater than the number of neutrons, and deuteron (^{2}H) formation is prevented. When T drops below 9×10^8 K (which occurs at about $t = 250$ s), deuterons can be produced.

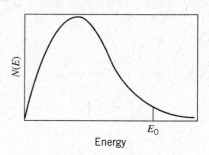

FIGURE 15.28 The thermal radiation spectrum. The photons above $E_0 = 2.22$ MeV are sufficiently energetic to break apart deuterium nuclei.

From 6 s to 250 s, very little (except expansion and the corresponding temperature decrease) happens in the universe, but after $t = 250$ s things happen very quickly. Deuterons form and then react with the many protons and neutrons available to give

$$^2\text{H} + \text{p} \rightarrow {}^3\text{He} + \gamma \quad \text{and} \quad {}^2\text{H} + \text{n} \rightarrow {}^3\text{H} + \gamma$$

The energies of formation of these nuclei are, respectively, 5.49 MeV and 6.26 MeV, well above the 2.22 MeV threshold of the deuteron formation. If the photons are not energetic enough to break apart the deuterons, they are certainly not energetic enough to break apart ^{3}He and ^{3}H. The final steps in the formation of the heavier nuclei are

$$^3\text{He} + \text{n} \rightarrow {}^4\text{He} + \gamma \quad \text{and} \quad {}^3\text{H} + \text{p} \rightarrow {}^4\text{He} + \gamma$$

There are no stable nuclei with $A = 5$, so no further reactions of this sort are possible. Nor is it possible to have $^4\text{He} + {}^4\text{He}$ reactions because ^{8}Be is highly unstable. (It would be possible to form stable ^{6}Li and ^{7}Li, but these are made in very small quantities relative to H and He; from Li further reactions are possible, such as $^7\text{Li} + {}^4\text{He} \rightarrow {}^{11}\text{B}$, and so forth, but these occur in still smaller quantities. The end products ^{2}H and He, along with the leftover original protons, make up about 99.9999% of the nuclei after the era of nuclear reactions.)

By $t = 250$ s, the original 16% neutrons present at $t = 6$ s had beta-decayed to about 12%, leaving 88% protons. Most of the ^{2}H, ^{3}H, and ^{3}He were "cooked" into heavier nuclei, so we can assume the universe to be composed mostly of ^{1}H and ^{4}He nuclei. Of the N nucleons present at $t = 250$ s, 12% ($0.12N$) were neutrons and $0.88N$ were protons. The $0.12N$ neutrons combined with $0.12N$ protons, forming $0.06N$ ^{4}He, and leaving $0.88N - 0.12N = 0.76N$ protons. The universe then consisted of $0.82N$ nuclei, of which $0.06N$ (7.3%) were ^{4}He and $0.76N$ (92.7%) were protons. Helium is about four times as massive as hydrogen, so by *mass* the universe is about 24% helium.

At this point the universe began a long and uneventful period of cooling, during which the *strong* interactions ceased to be of importance.

The final step in the evolution of the primitive universe is the formation of neutral hydrogen and helium atoms from the ^{1}H, ^{2}H, ^{3}He, and ^{4}He nuclei and the free electrons. In the case of hydrogen, this takes place when the photon energy drops below 13.6 eV; otherwise any atoms that might happen to form will be immediately ionized by the radiation. There are still about 10^9 photons for every proton, and so we must wait for the radiation to cool until the fraction of photons above 13.6 eV is less than about 10^{-9}. We can solve Eq. 15.37 for $f = 10^{-9}$ to obtain $E_0/kT = 26$. With $E_0 = 13.6$ eV, the corresponding temperature is $T = 6070$ K, which occurs at time $t = 6.1 \times 10^{12}$ s $= 190,000$ y. These final estimates are actually not quite correct. We have been considering only the energy density of radiation present in the universe. As the universe cools, the contribution of the matter to the total energy density becomes more significant, and so the temperature drops more slowly than we would estimate. This contribution may increase this time by about a factor of 2 to about 380,000 y, and the radiation temperature is decreased by about a factor of $\sqrt{2}$, to $T = 4300$ K.

After neutral atoms have formed, there are virtually no charged particles left in the universe, and the radiation field is not energetic enough to ionize the atoms. This is the time of the decoupling of the radiation field from the matter, and now

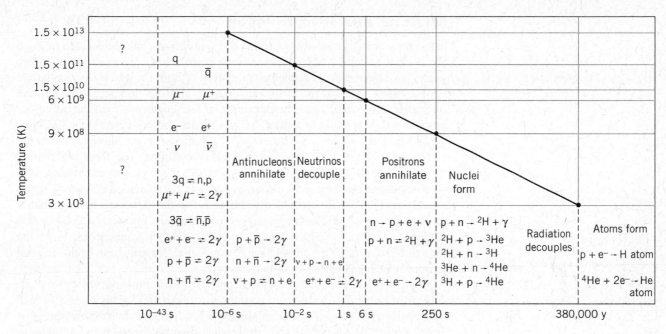

FIGURE 15.29 Evolution of the universe according to the Big Bang cosmology. The blue line shows the temperature and time in the radiation-dominated era before decoupling. The most important reactions in each era are shown.

electromagnetism, the third of the four basic forces, is no longer important in shaping the evolution of the universe. The large-scale development of the universe is from this point governed only by gravity.

The time after $t = 380,000$ y has been comparatively uneventful, at least from the point of view of cosmology. Density fluctuations of the hydrogen and helium triggered the condensation of galaxies, and then first-generation stars were born. Supernova explosions of the material from these stars permitted the formation of second generation systems, among which planets formed from the rocky debris.

Meanwhile, the decoupled radiation field, unaffected by the gravitational coming and going of matter, began the long journey that eventually took it, cooled again by a factor of 1600, to the radio telescopes of 20th-century Earth.

The details of the Big Bang cosmology are summarized in Figure 15.29. It is a remarkable story, all the more so because we can understand most of its details—with the possible exception of the first instant—with nothing more than some basic theories of modern physics, most of which we can study (on a much smaller scale!) in our laboratories on Earth.

15.10 EXPERIMENTAL COSMOLOGY

Far from being a science that involves only speculations about the distant past or the indefinite future, cosmology has in recent decades become a precise experimental science, involving observational results from high-resolution observatories on Earth and in space, as well as laboratory measurements of nuclear and particle properties that provide insight into cosmological phenomena. Here are a few of the observations and their implications.

Matter and Antimatter

In the early universe, there was roughly one nucleon and one antinucleon for each photon. If the numbers of nucleons and antinucleons had been exactly equal, there would have been either complete annihilation of both (in which case we would not be around to comment on the outcome) or else the clumping of matter and antimatter into galaxies and antigalaxies. Our telescopes can't tell the difference between galaxies made of matter and antimatter (because both emit the same light), but if there were large quantities of antimatter in the universe we should occasionally find a galaxy and an antigalaxy colliding, and their annihilation would light up the sky. We observe many galaxies in the process of colliding with their neighbors, but none show the intense annihilation radiation that would signal a matter-antimatter collision. Our conclusion is that the universe is made of matter and contains no significant concentrations of antimatter. For every 1,000,000,000 nucleons in the early universe there were 999,999,999 antinucleons; following the annihilation all of the antinucleons disappeared, leaving 1 out of the original 10^9 nucleons to make up the current universe.

After the matter-antimatter annihilation in the early universe, the ratio of the number of remaining nucleons to photons was about 10^{-9}. This number, which has remained constant since the annihilation era, is deduced from the measurement of the relative density of ^{2}H and ^{3}He in sites such as "first-generation" stars or interstellar gas, where no significant additional amounts of those atoms have been subsequently produced by fusion. The observed relative abundances of these atoms gives a nucleon to photon ratio in the range of $5-7 \times 10^{-10}$.

Where did the 10^{-9} excess of matter over antimatter in the early universe come from? We don't yet know the answer to this question, but evidence gathered in particle physics experiments may provide a clue. The first indication of an asymmetry between matter and antimatter was a 1964 experiment studying the decay of the neutral K meson, which shows a difference in behavior between K^0 and $\overline{K}^0$ only at the very low level of 1 part in 10^3 in the weak interaction or thus at a level of 10^{-10} relative to the strong interaction. (J. W. Cronin and V. L. Fitch received the 1980 Nobel Prize in physics for their work on this experiment.) Following the discovery of the b quark, it was hypothesized that the B^0 meson would show a similar effect, and so accelerators were built in the U. S. and Japan that produced the B^0 in large enough quantities to verify the asymmetry. Beginning in about 2001, scientists at these accelerators announced results that verified the matter-antimatter asymmetry previously observed only in the K^0 decay.

The distinction between matter and antimatter occurred at an early stage in the evolution of the universe, during the quark-antiquark era. The Grand Unified Theories (GUTs) include this asymmetry between quarks and antiquarks in a natural way, although there is as yet no accepted version of the GUTs that yields a convincing explanation for the K^0 and B^0 experiments.

Helium Abundance

Much of the matter in the universe has been formed and reformed, and so has lost its "memory" of the Big Bang. There is, however, "first generation" matter in stars and galaxies, that should show the roughly 24% helium abundance that characterized the formation of matter.

A variety of experiments suggests that the abundance of helium in the universe is 23 to 27% by mass, in excellent agreement with our rough estimate of 24%. These experiments include the emission of visible light from gas clouds near stars and the emission of radio waves by interstellar gas, both of which permit us to compare the amounts of hydrogen and helium present. In addition, the dynamics of stellar formation depends on the initial hydrogen and helium concentrations; present theories permit us to estimate their ratio from the observed properties of stars. The 24% abundance seems to be rather constant throughout the universe, as we would expect if it were predetermined by the Big Bang. (Not enough helium has been produced by nuclear fusion in stars in the last 14×10^9 y to change this ratio significantly.)

In fact (and here physics comes nearly full circle, from the very old and large to the very new and small), the early helium abundance is a function of the conditions before 10^{-6} s, when quarks and leptons filled the universe. The evolutionary rate in this era depends on the number of different kinds of quarks and leptons that can participate in reactions. It has been calculated that the helium abundance is probably not consistent with the existence of more than three generations of quarks and leptons. It is remarkable that extrapolations to an unobservable state of the universe can yield such insight into the fundamental structure of matter.

The Horizon Problem

Our telescopes permit us to look outward by about 10 billion light-years in any direction. No matter in what direction we look, the universe (which we are viewing as it looked 10 billion years ago) appears pretty much the same—the same types of galaxies and the same temperature of the background radiation. This is surprising, because regions that we observe in opposite directions are separated by 20 billion light years, while the universe is only 14 billion years old. If the universe had been expanding throughout its history at a uniform rate, those opposite regions of the sky could never have been connected by any signal and thus had no way to achieve the common characteristics that we now observe. (Imagine finding a block of copper that had been assembled from a random collection of copper atoms. If all parts of the block were at the same temperature, you would conclude that the block had been in existence for a long enough time for thermal energy to propagate throughout its volume. If we learned that the time since the block's assembly was less than that propagation time, it would be very puzzling to explain the achievement of thermal equilibrium in so short a time.)

This paradox is solved by a hypothesis called "inflation," which proposes that in the early universe rather than a constant expansion rate there was a sudden rapid growth (by perhaps 50 orders of magnitude) in a short interval of time between 10^{-35} s and 10^{-32} s. Before the time of inflation, the size of the universe was less than the distance through which distant parts could exchange energy since the time of the Big Bang. Thus all parts of the universe were able to achieve a common set of characteristics. After inflation, the size of the universe exceeded the maximum range of communication signals, but the homogeneous characteristics had already been achieved. The inflationary hypothesis thus neatly solves the horizon problem.

The Flatness Problem

For a flat universe ($k = 0$), we can combine Eqs. 15.28 and 15.33 to give

$$\rho_{cr} = \frac{3H^2}{8\pi G} = 0.97 \times 10^{-26} \, \text{kg/m}^3 \qquad (15.38)$$

This is the critical density corresponding to a flat universe. If the density is greater than this critical value, the universe is closed, and if it is less than this value, the universe is open. In discussing the density of the universe, it is useful to define the ratio between the actual density and this critical value:

$$\Omega = \frac{\rho}{\rho_{cr}} \tag{15.39}$$

As the universe expands, the gravitational interaction among its components slows the expansion rate. If $\Omega > 1$, the gravitational interaction will eventually halt and reverse the expansion. If $\Omega < 1$, the expansion will continue until the components are separated by infinite distances. If Ω is exactly equal to one, the expansion will also continue forever, but the components will arrive at their infinite separation just as they lose the last bit of kinetic energy.

By carefully measuring the variations in the temperature of regions of the microwave background, the WMAP satellite and other experiments have concluded that Ω is very close to 1, probably within 1%. It is of considerable interest to know whether Ω is *exactly* equal to 1 (for a flat universe), or just happens to be very close to 1 (for an open or closed universe).

By way of analogy, consider a projectile that is thrown upward from the surface of the Earth (ignore the gravity of the Sun and all other objects). The parameter Ω in effect measures the ratio between the gravitational potential energy and the kinetic energy: $\Omega = |U_{grav}|/K$. If the initial value of Ω is greater than 1, the gravitational energy exceeds the kinetic energy, so the projectile will rise to a maximum height and then fall back to Earth. When it reaches its maximum height, $K = 0$ and Ω becomes infinite. During the entire ascent, the value of Ω increases because the kinetic energy decreases more rapidly than the magnitude of the gravitational energy. If the projectile is launched so that $\Omega < 1$, there is more than enough kinetic energy to overcome the Earth's gravity, and the projectile will escape the pull of the Earth. When it reaches infinite separation, $\Omega = 0$ because $U_{grav} = 0$. During its entire outward journey, Ω decreases from its initial value and approaches zero. If we choose the initial velocity such that $\Omega = 1$, there is just enough energy to escape, and the projectile reaches infinite separation with $K = 0$. Throughout the entire journey, Ω remains exactly 1.

For the projectile as well as for the evolution of the universe, the conclusions are identical: If $\Omega = 1$ initially, it remains exactly 1 always, but if either $\Omega > 1$ or $\Omega < 1$, it grows further away from 1. If the early universe had $\Omega = 1.000001$, after the passage of 14 billion years Ω would have grown very large; similarly, if the initial value of Ω were 0.999999, by now it would be very close to 0. It has been calculated that for Ω to be within 1% of 1 today, it must have originally been in the range 1 ± 10^{-62}. Here again the inflation hypothesis is essential. Prior to the inflation era, the universe may have been open, flat, or closed, with any value of Ω. During inflation, the universe grew by so many orders of magnitude that the curvature became flat, much as the surface of a balloon becomes nearly flat when it is inflated by many orders of magnitude. As a result, Ω was indeed very close to 1 just after inflation, and we continue today to observe an Ω that remains close to 1.

The Composition and Age of the Universe

During the past 10 years, researchers have made enormous strides in determining the composition of the universe and its age. These determinations are based mostly on measurements of the properties of the cosmic microwave background radiation, in particular its geometrical distribution and its polarization properties. Among the most detailed experiments are those of the WMAP satellite (2001) and the

Boomerang balloon flights over Antarctica (1998 and 2003). These experiments, along with others, indicate that the universe is very nearly flat; that is, $\Omega = 1.00$ (so that $\rho = \rho_{cr}$) to within about 1%. They are also able to measure the relative densities of the various components of the mass-energy content of the universe. Ordinary baryonic matter contributes about 4.6% of the critical density, and dark (nonbaryonic) matter contributes 23%. Together, the two kinds of matter make up 28% of the critical density. If the density of the universe is equal to the critical density, what makes up the other 72%?

Beginning in about 1998, two teams of researchers were investigating the Hubble law by studying the exploding stars known as supernovas in the most distant galaxies (corresponding to large redshifts, with $\Delta\lambda/\lambda$ close to 0.9). Both teams found systematic and consistent departures from the Hubble law. The supernovas in these distant galaxies were fainter than expected, indicating that the galaxies were 10 to 15% farther from us than the Hubble law predicts. The research groups concluded that a mysterious force is accelerating the expansion of the universe. Generally we would expect that the expansion would be slowing down, owing to the gravitational interactions of its components. What could be responsible for increasing the rate of the expansion?

This unknown interaction is called "dark energy," and it represents the missing 72% of the composition of the universe. Although there are several theories about the nature of the dark energy, there is no convincing explanation of its origin or its role in the physical world. It has been suggested that the dark energy density does not diminish as the universe expands, so that as the densities of matter (both baryonic and non-baryonic) decrease with the expansion, eventually the dark energy begins to dominate and accelerates the expansion. We live at a time when this acceleration is dominant (which has occurred for about the past 5 billion years).

Another finding based on the observation of the background radiation is the age of the universe. There is general agreement that the age is 13.7×10^9 years, with an uncertainty of about 1%. The accelerated expansion solves a problem associated with the Hubble age. In a matter-dominated universe (which our universe has been for most of its existence), the present Hubble age should be $\frac{2}{3}H_0^{-1}$, which works out to be about 9×10^9 years. If the expansion has been accelerating, then the actual age can be greater than the Hubble age, which certainly seems to be the case.

It is fitting that this story ends where it began, with Einstein. When Einstein produced the general relativity theory in 1916 (a decade before Hubble's work), it was widely believed that the universe was static. In order for the equations of general relativity to allow a static solution, Einstein introduced into his equations an additional term, called the cosmological constant. After learning of Hubble's discovery of the expansion of the universe, Einstein called the introduction of the cosmological constant his "greatest blunder." It now appears that the cosmological constant is one possible explanation for the dark energy, and the term has been restored to the equations.

The increasing role of the dark energy suggests a sad fate for the universe. An open universe will expand forever, but in an accelerating open universe each observer's horizon will shrink increasingly rapidly. The most distant galaxies will separate from us at speeds greater than the speed of light (which is *not* a violation of special relativity, because no signals are being exchanged). The light from these distant galaxies will not be able to reach us, and the galaxies will gradually disappear. Future astronomers might be able to observe only local galaxies and might learn nothing of the expansion or the properties of the universe!

Chapter Summary

		Section
Hubble's law	$v = H_0 d$	15.1
Number density of photons	$N/V = (2.03 \times 10^7 \,\text{photons/m}^3 \cdot \text{K}^3) T^3$	15.2
Energy density of photons	$U = (4.72 \times 10^3 \,\text{eV/m}^3 \cdot \text{K}^4) T^4$	15.2
Gravitational frequency change	$\Delta f / f = gH/c^2$	15.4
Deflection of starlight	$\theta = 2GM/Rc^2$	15.5
Perihelion precession	$\Delta\phi = \dfrac{6\pi GM}{c^2 r_{min}(1 + e)}$	15.5
Schwarzschild radius	$r_S = 2GM/c^2$	15.6

		Section
Age of matter-dominated universe	$t = 1/\sqrt{6\pi G \rho_m}$	15.7
Age of radiation-dominated universe	$t = \sqrt{3/32\pi G \rho_r}$	15.7
Temperature of universe at age t	$T = \dfrac{1.5 \times 10^{10}\,\text{s}^{1/2} \cdot \text{K}}{t^{1/2}}$	15.8
Fraction of photons above E_0	$f = 0.42 e^{-E_0/kT}$ $\times \left[\left(\dfrac{E_0}{kT}\right)^2 + 2\left(\dfrac{E_0}{kT}\right) + 2 \right]$	15.9
Critical density of universe	$\rho_{cr} = \dfrac{3H^2}{8\pi G}$ $= 0.97 \times 10^{-26}\,\text{kg/m}^3$	15.10

Questions

1. If we were to measure the equivalence of gravitational and inertial mass, would we show that $m_{inertial} = m_{gravitational}$ or merely that $m_{inertial} \propto m_{gravitational}$?

2. Do tidal effects distinguish between Newtonian gravity and curved spacetime? What would be the shape of a drop of liquid following a path in a curved spacetime? Can such a drop distinguish between a uniform gravitational field and a uniform acceleration?

3. Suppose that the first measurement of deflection of starlight during a solar eclipse had been done after 1905, when the special theory of relativity was introduced, but before 1916, when the general theory was introduced. What would have been the effect of this measurement on the special theory?

4. If we could make a precise comparison of light from the Sun with light from the Moon, would the moonlight be red shifted, blue shifted, or unshifted relative to sunlight?

5. What difficulties might arise in the Pound and Rebka experiment on the gravitational red shift if the temperature of the source or the absorber varied?

6. Why are the abundances of Li, Be, and B so small?

7. Can we look out into the distant universe without also looking back into time?

8. Is Hubble's parameter a constant? Does it vary over large distances of space? Over long intervals of time?

9. Explain why the age of the universe must be less than H^{-1}.

10. Why is it difficult to obtain precise values for the Hubble parameter?

11. All natural processes are governed by the rule that the entropy must increase; the increase of entropy, as the universe "runs down," defines for us a direction of time. If the universe begins to contract and therefore to heat up, will the entropy of natural processes therefore decrease? Will the inhabitants of that universe observe time to be running backward?

12. The hydrogen in the universe contains a small fraction of deuterium. Assuming the deuterium originated in the Big Bang, what era of the Big Bang would we learn about by measuring the deuterium abundance? Can we accomplish this measurement using terrestrial hydrogen? What properties of deuterium could we use to determine its presence in distant regions of the galaxy?

13. Between $t = 1\,\text{s}$ and $t = 6\,\text{s}$, the neutron fraction should drop from 27 to 8%; instead it drops only to about 16%. Why don't more neutrons turn into protons during this era? Is it as difficult for protons to turn into neutrons?

14. If we were able to observe the neutrinos from the early universe, would they have a spectrum determined by the Planck distribution?

Problems

15.1 The Expansion of the Universe

1. Use Hubble's law to estimate the wavelength of the 590.0 nm sodium line as observed emitted from galaxies whose distance from us is (a) 1.0×10^6 light-years; (b) 1.0×10^9 light-years.

2. The light from a certain galaxy is red-shifted so that the wavelength of one of its characteristic spectral lines is doubled. Assuming the validity of Hubble's law, calculate the distance to this galaxy.

15.2 The Cosmic Microwave Background Radiation

3. (a) Taking $u(E) = EN(E)$ as the energy density of the thermal radiation, with $N(E)$ given in Eq. 15.6, differentiate to find the energy at which the maximum of the radiation energy spectrum occurs. (b) Evaluate the peak photon energy of the 2.7-K microwave background.

4. Starting with Eqs. 15.7 and 10.41, show how to evaluate the numerical constants that appear in Eqs. 15.8 and 15.9.

15.3 Dark Matter

5. Suppose an observer in a distant galaxy were observing the light from our Sun as the Sun moves directly toward the observer. Neglecting any net relative motion of the two galaxies, calculate the change in wavelength of the 121.5-nm Lyman series line due to the rotation of our galaxy.

15.4 The General Theory of Relativity

6. In Example 15.1 we calculated the change in wavelength of the Lyman α line due to the gravitational red shift. Compare this value with (a) the special relativistic Doppler shift due to the rotation of the Sun and (b) the thermal Doppler broadening (see Eq. 10.30). The Sun's radius is 6.96×10^8 m, its rotational period is 26 days, and it surface temperature is 6000 K.

7. A satellite is in orbit at an altitude of 150 km. We wish to communicate with it using a radio signal of frequency 10^9 Hz. What is the gravitational change in frequency between a ground station and the satellite? (Assume g doesn't change appreciably.)

8. According to the uncertainty principle, what is the minimum time interval necessary to measure a change in frequency of the magnitude observed in the Pound and Rebka experiment?

15.5 Tests of General Relativity

9. By drawing analogies between the Coulomb force law and the gravitational force law, use Eq. 6.8 for the deflection in Rutherford scattering to obtain Eq. 15.22 for the deflection of photons. Assume the photon behaves as if it has a mass $m = E/c^2$. (Hint: Write Eq. 6.8 in terms of the velocity of the particle instead of kinetic energy.)

10. In the binary star system known as PSR 1913 + 16, two neutron stars move about their common center of mass in highly elliptical orbits. Locate the orbital parameters for this motion, and add a row to Table 15.1 showing the precession angle expected from general relativity. (Hint: In Eq. 15.25, M is the total mass of the orbiting body and the central body.)

15.6 Stellar Evolution and Black Holes

11. (a) Show that Eq. 10.57 for the radius of a neutron star of mass M can be written $R = (12.3 \text{ km})(M/M_\odot)^{-1/3}$ where $M_\odot$ is the mass of the Sun. (b) Consider a star 1.5 times as massive as the Sun with a radius of 7×10^5 km (equal to the present radius of the Sun), rotating on its axis about once per year. (This is quite a slow rate of rotation—our Sun rotates about once per month.) If angular momentum is conserved in the collapse, what will be the final angular velocity? Assume the star can be represented as a sphere of uniform density, with rotational inertia $I = \frac{2}{5}MR^2$.

15.7 Cosmology and General Relativity

12. The rate of change of the cosmic expansion can be described in terms of a *deceleration parameter* $q = -R(d^2R/dt^2)/(dR/dt)^2$. (a) Evaluate q for the matter-dominated universe (Eq. 15.29) and the radiation-dominated universe (Eq. 15.31). (b) By differentiating Eq. 15.28, show that in a matter-dominated universe $q = 4\pi G\rho_m/3H^2$. (Hint: Use $\rho_m \propto R^{-3}$ to relate $d\rho_m/dt$ to dR/dt.)

15.8 The Big Bang Cosmology

13. Derive Eq. 15.34.

14. At what age did the universe cool below the threshold temperature for (a) nucleon production; (b) pi meson production?

15. (a) At what temperature was the universe hot enough to permit the photons to produce K mesons ($mc^2 = 500$ MeV)? (b) At what age did the universe have this temperature?

15.9 The Formation of Nuclei and Atoms

16. Derive Eqs. 15.36 and 15.37.

17. Suppose the difference between matter and antimatter in the early universe were 1 part in 10^8 instead of 1 part in 10^9. (a) Evaluate the temperature at which deuterium begins to form. (b) At what age does this occur? (c) Evaluate the temperature and the corresponding time of radiation decoupling (when hydrogen atoms form).

18. What was the age of the universe when the nucleons consisted of 60% protons and 40% neutrons?

15.10 Experimental Cosmology

19. Assuming that the density of the universe is equal to its critical value and that 4.6% of the universe is baryonic

matter, calculate the average number of baryons (nucleons) per cubic meter in the universe.

20. (a) Suppose the baryonic matter in the universe were composed of uniformly distributed stars of the mass of the Sun (2.0×10^{30} kg). What would be the average spacing between the stars? Express your answer in light-years. (b) Suppose instead that the baryonic matter were composed of uniformly distributed galaxies of the mass of the Milky Way (1.2×10^{42} kg). Expressed in light-years, what would be the average distance between the galaxies?

21. Suppose the non-baryonic dark matter consists entirely of neutrinos. What is the average rest energy of the neutrinos that could account for this part of the mass of the universe? As a rough estimate, assume that the neutrino density is the same as the present photon density.

General Problems

22. Photons of visible light have energies between about 2 and 3 eV. (a) Compute the number density of photons from the 2.73-K background radiation in that interval. (It is sufficient to characterize the visible region as $E = 2.5$ eV with $dE = 1.0$ eV.) (b) Assume the eye can detect about 100 photons/cm³. At what temperature would the background radiation be visible? At what age of the universe would this have occurred?

23. Consider the universe at a temperature of 5000 K. (a) At what age did this occur, and during which stage of the evolution of the universe? (b) Evaluate the average photon energy at that time. (c) If there are 10^9 photons per nucleon, evaluate the ratio between the radiation density and the mass density at that time.

24. The early universe was radiation dominated, and the present universe is matter dominated. (a) At what temperature were the radiation and matter densities equal? (b) What was the age of the universe when this occurred?

25. A neutron star of 2.00 solar masses is rotating at a rate of 1.00 revolutions per second. (a) What is the radius of the neutron star? (See Problem 11.) (b) Find its rotational kinetic energy. (c) If its rotational speed slows by 1 part in 10^9 per day, find the loss in rotational kinetic energy per day. (d) Assuming that the entire energy loss goes into radiation, find the radiative power. (e) If the star is 10^4 light-years from Earth, what would be the average power received by an antenna of area 10 m^2 if the star's energy were distributed uniformly in space instead of concentrated in a narrow beam?

26. Because we don't yet have a quantum theory of gravity, we cannot analyze the properties of the universe before the Planck time, about 10^{-43} s. If we assume that the properties of the universe during that era were determined by quantum theory, relativity, and gravity, the Planck time should be characterized by the fundamental constants of those three theories: h, c, and G. We can therefore write $t \propto h^i c^j G^k$, where i, j, and k are exponents to be determined. (a) Using a dimensional analysis, determine i, j, and k. (b) Assuming the proportionality parameter is of order unity, evaluate t. (c) What was the size of the observable universe at the Planck time?

27. Show that the spacetime interval given by Eq. 15.18 is invariant with respect to the Lorentz transformation. That is, show that $(ds)^2 = (ds')^2$, where $(ds')^2 = (c\,dt')^2 - (dx')^2$.

28. Light from star S in Figure 15.30 passes a distance b from a galaxy L, where it is deflected by an angle α and then reaches the observer O, who sees an image of the star at I. (For simplicity, assume the gravitational deflection takes place at a single point, and assume all angles in the figure are very small.) The galaxy (of mass M) is a distance d_L from the observer, and the star is a distance d_S from the observer. (a) For small angles, show that $\theta d_S = \beta d_S + (4GM/\theta d_L c^2)(d_S - d_L)$. (Hint: Use Eq. 15.22 for the deflection angle α when the impact parameter is b instead of R, but double the value to account for the difference between the special and general relativity predictions.) (b) Solve the resulting quadratic equation for θ and show that there are two image positions whose locations differ by $\Delta\theta = \sqrt{\beta^2 + 4\theta_E^2}$, where the Einstein angle θ_E is $\sqrt{4GM(d_S - d_L)/c^2 d_S d_L}$. This is an example of gravitational lensing, an effect of general relativity that has been observed for distant objects that appear in multiple images when their light travels a path through spacetime that is curved by an intervening galaxy. (c) When the star, lensing galaxy, and observer lie along a single line, something other than two images appears. Given the symmetry of the figure when $\beta = 0$, what do you expect to be observed in this case?

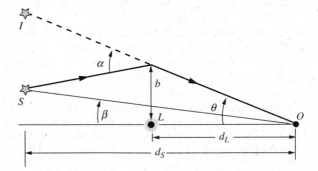

FIGURE 15.30 Problem 28.

Appendix

CONSTANTS AND CONVERSION FACTORS*

CONSTANTS

Speed of light	c	2.99792458×10^8 m/s
Charge of electron	e	$1.60217657 \times 10^{-19}$ C
Boltzmann constant	k	1.380649×10^{-23} J/K $= 8.617332 \times 10^{-5}$ eV/K
Planck's constant	h	$6.62606957 \times 10^{-34}$ J·s $= 4.13566752 \times 10^{-15}$ eV·s
	$\hbar = h/2\pi$	$1.054571726 \times 10^{-34}$ J·s $= 6.58211928 \times 10^{-16}$ eV·s
	hc	1239.8419 eV·nm (or MeV·fm)
	$\hbar c$	197.326972 eV·nm (or MeV·fm)
Gravitational constant	G	6.67384×10^{-11} N·m²/ kg²
Avogadro's constant	N_A	6.0221413×10^{23} mole^{-1}
Universal gas constant	R	8.314462 J/mole·K
Stefan-Boltzmann constant	σ	5.67037×10^{-8} W/m²·K⁴
Rydberg constant	R_∞	$1.097373156854 \times 10^7$ m^{-1}
Hydrogen ionization energy	$|E_1|$	13.6056925 eV
Bohr radius	a_0	$5.291772109 \times 10^{-11}$ m
Bohr magneton	μ_B	$9.2740097 \times 10^{-24}$ J/T $= 5.78838181 \times 10^{-5}$ eV/T
Nuclear magneton	μ_N	$5.0507835 \times 10^{-27}$ J/T $= 3.15245126 \times 10^{-8}$ eV/T
Fine structure constant	α	$1/137.03599907$
Electric constant	$e^2/4\pi\varepsilon_0$	1.4399645 eV·nm (or MeV·fm)

*The number of significant figures given for the numerical constants indicates the precision to which they have been determined; there is an experimental uncertainty, typically of a few parts in the last or next-to-last digit, except for the speed of light (which is exact).

SOME PARTICLE MASSES

	kg	u	MeV/c^2
Electron	$9.1093829 \times 10^{-31}$	$5.485799095 \times 10^{-4}$	0.51099893
Proton	$1.67262178 \times 10^{-27}$	1.0072764668	938.27205
Neutron	$1.67492735 \times 10^{-27}$	1.0086649160	939.56538
Deuteron	$3.3435835 \times 10^{-27}$	2.0135532127	1875.61286
Alpha	$6.6446568 \times 10^{-27}$	4.001506179	3727.3792

CONVERSION FACTORS

$1\,\text{eV} = 1.60217657 \times 10^{-19}\,\text{J}$

$1\,\text{u} = 931.49406\,\text{MeV}/c^2$

$\quad = 1.66053892 \times 10^{-27}\,\text{kg}$

$1\,\text{y} = 3.156 \times 10^7\,\text{s} \cong \pi \times 10^7\,\text{s}$

$1\,\text{barn (b)} = 10^{-28}\,\text{m}^2$

$1\,\text{curie (Ci)} = 3.7 \times 10^{10}\,\text{decays/s}$

$1\,\text{light-year} = 9.46 \times 10^{15}\,\text{m}$

$1\,\text{parsec} = 3.26\,\text{light-year}$

COMPLEX NUMBERS

The imaginary number i is defined as $\sqrt{-1}$. A *complex number or function* can be represented as having a real part, which does not depend on i, and an imaginary part, which depends on i. We can write a complex variable as $z = x + iy$, where the real part x and the imaginary part y are both real numbers or real functions. A complex wave function ψ can be written in terms of its real and imaginary parts as $\psi = \mathrm{Re}(\psi) + i\mathrm{Im}(\psi)$.

The complex conjugate of a complex number is obtained by substituting $-i$ for i, as in $z^* = x - iy$ or $\psi^* = \mathrm{Re}(\psi) - i\mathrm{Im}(\psi)$.

The squared magnitude of a complex number is defined as the product of the number and its complex conjugate, as in $|z|^2 = zz^*$ or $|\psi|^2 = \psi\psi^*$, and is equal to the sum of the squares of its real and imaginary parts:

$$|z|^2 = x^2 + y^2 \qquad \text{or} \qquad |\psi|^2 = [\mathrm{Re}(\psi)]^2 + [\mathrm{Im}(\psi)]^2$$

The complex exponential $e^{i\theta}$ can be represented in terms of real trigonometric functions as

$$e^{i\theta} = \cos\theta + i\sin\theta \qquad \text{and} \qquad e^{-i\theta} = \cos\theta - i\sin\theta$$

The squared magnitude of the complex exponential is equal to 1:

$$|e^{i\theta}|^2 = e^{i\theta}e^{-i\theta} = (\cos\theta + i\sin\theta)(\cos\theta - i\sin\theta) = \cos^2\theta + \sin^2\theta = 1$$

We can write the ordinary trigonometric functions in terms of these complex functions:

$$\sin\theta = \frac{1}{2i}(e^{i\theta} - e^{-i\theta}) \qquad \text{and} \qquad \cos\theta = \frac{1}{2}(e^{i\theta} + e^{-i\theta})$$

It is sometimes convenient to write the wave function in terms of a complex exponential as:

$$\psi = |\psi|e^{i\alpha}$$

where $|\psi|$ gives the magnitude of the wave function and α is its phase.

Appendix C

Periodic Table of the Elements

Key:
Symbol — Cl 17 — Atomic number
Atomic mass* — 35.453
$3p^5$ — Electron configuration

Transition elements

Group I	Group II											Group III	Group IV	Group V	Group VI	Group VII	Group 0
H 1 1.00794 $1s^1$																	**He** 2 4.00260 $1s^2$
Li 3 6.941 $2s^1$	**Be** 4 9.01218 $2s^2$											**B** 5 10.81 $2p^1$	**C** 6 12.011 $2p^2$	**N** 7 14.0067 $2p^3$	**O** 8 15.9994 $2p^4$	**F** 9 18.9984 $2p^5$	**Ne** 10 20.180 $2p^6$
Na 11 22.9998 $3s^1$	**Mg** 12 24.305 $3s^2$											**Al** 13 26.9815 $3p^1$	**Si** 14 28.0855 $3p^2$	**P** 15 30.9738 $3p^3$	**S** 16 32.065 $3p^4$	**Cl** 17 35.453 $3p^5$	**Ar** 18 39.948 $3p^6$
K 19 39.0983 $4s^1$	**Ca** 20 40.08 $4s^2$	**Sc** 21 44.9559 $3d^14s^2$	**Ti** 22 47.867 $3d^24s^2$	**V** 23 50.9415 $3d^34s^2$	**Cr** 24 51.996 $3d^54s^1$	**Mn** 25 54.9380 $3d^54s^2$	**Fe** 26 55.845 $3d^64s^2$	**Co** 27 58.9332 $3d^74s^2$	**Ni** 28 58.693 $3d^84s^2$	**Cu** 29 63.546 $3d^{10}4s^1$	**Zn** 30 65.38 $3d^{10}4s^2$	**Ga** 31 69.723 $4p^1$	**Ge** 32 72.64 $4p^2$	**As** 33 74.9216 $4p^3$	**Se** 34 78.96 $4p^4$	**Br** 35 79.904 $4p^5$	**Kr** 36 83.798 $4p^6$
Rb 37 85.4678 $5s^1$	**Sr** 38 87.62 $5s^2$	**Y** 39 88.9059 $4d^15s^2$	**Zr** 40 91.224 $4d^25s^2$	**Nb** 41 92.9064 $4d^45s^1$	**Mo** 42 95.96 $4d^55s^1$	**Tc** 43 (98) $4d^55s^2$	**Ru** 44 101.07 $4d^75s^1$	**Rh** 45 102.906 $4d^85s^1$	**Pd** 46 106.42 $4d^{10}5s^0$	**Ag** 47 107.868 $4d^{10}5s^1$	**Cd** 48 112.41 $4d^{10}5s^2$	**In** 49 114.82 $5p^1$	**Sn** 50 118.71 $5p^2$	**Sb** 51 121.76 $5p^3$	**Te** 52 127.60 $5p^4$	**I** 53 126.904 $5p^5$	**Xe** 54 131.29 $5p^6$
Cs 55 132.905 $6s^1$	**Ba** 56 137.33 $6s^2$	**La** 57 138.906 $5d^16s^2$	**Hf** 72 178.49 $5d^26s^2$	**Ta** 73 180.948 $5d^36s^2$	**W** 74 183.84 $5d^46s^2$	**Re** 75 186.207 $5d^56s^2$	**Os** 76 190.2 $5d^66s^2$	**Ir** 77 192.22 $5d^76s^2$	**Pt** 78 195.08 $5d^96s^1$	**Au** 79 196.967 $5d^{10}6s^1$	**Hg** 80 200.59 $5d^{10}6s^2$	**Tl** 81 204.383 $6p^1$	**Pb** 82 207.2 $6p^2$	**Bi** 83 208.980 $6p^3$	**Po** 84 (209) $6p^4$	**At** 85 (210) $6p^5$	**Rn** 86 (222) $6p^6$
Fr 87 (223) $7s^1$	**Ra** 88 (226) $7s^2$	**Ac** 89 (227) $6d^17s^2$	**Rf** 104 (265) $6d^27s^2$	**Db** 105 (268) $6d^37s^2$	**Sg** 106 (271) $6d^47s^2$	**Bh** 107 (272) $6d^57s^2$	**Hs** 108 (270) $6d^67s^2$	**Mt** 109 (276) $6d^77s^2$	**Ds** 110 (281) $6d^97s^1$	**Rg** 111 (280) $6d^{10}7s^1$	**Cn** 112 (285) $6d^{10}7s^2$	**Uut** 113 (284) $7p^1$	**Uuq** 114 (289) $7p^2$	**Uup** 115 (288) $7p^3$	**Uuh** 116 (293) $7p^4$	**Uus** 117 (292) $7p^5$	**Uuo** 118 (294) $7p^6$

Lanthanide series (57–71)

Ce 58 140.12 $4f^15d^16s^2$	**Pr** 59 140.908 $4f^36s^2$	**Nd** 60 144.24 $4f^46s^2$	**Pm** 61 (145) $4f^56s^2$	**Sm** 62 150.36 $4f^66s^2$	**Eu** 63 151.96 $4f^76s^2$	**Gd** 64 157.25 $5d^14f^76s^2$	**Tb** 65 158.925 $4f^96s^2$	**Dy** 66 162.50 $4f^{10}6s^2$	**Ho** 67 164.930 $4f^{11}6s^2$	**Er** 68 167.26 $4f^{12}6s^2$	**Tm** 69 168.934 $4f^{13}6s^2$	**Yb** 70 173.05 $4f^{14}6s^2$	**Lu** 71 174.967 $5d^14f^{14}6s^2$

Actinide series (89–103)

Th 90 232.038 $6d^27s^2$	**Pa** 91 231.036 $5f^26d^17s^2$	**U** 92 238.029 $5f^36d^17s^2$	**Np** 93 (237) $5f^46d^17s^2$	**Pu** 94 (244) $5f^67s^2$	**Am** 95 (243) $5f^77s^2$	**Cm** 96 (247) $5f^76d^17s^2$	**Bk** 97 (247) $5f^97s^2$	**Cf** 98 (251) $5f^{10}7s^2$	**Es** 99 (252) $5f^{11}7s^2$	**Fm** 100 (257) $5f^{12}7s^2$	**Md** 101 (258) $5f^{13}7s^2$	**No** 102 (259) $5f^{14}7s^2$	**Lr** 103 (262) $6d^15f^{14}7s^2$

* Atomic mass values are averaged over isotopes according to the percentages that occur on the earth's surface. For unstable elements, the mass number of the most stable known isotope is given in parentheses. Electron configurations of elements above 103 are tentative assignments based on the corresponding elements in the 6th row (period).

Source: IUPAC Commission on Atomic Weights and Isotopic Abundances, 2001.

Appendix D

TABLE OF ATOMIC MASSES

The table gives the atomic masses of some isotopes of each element. All naturally occurring stable isotopes are included (with their natural abundances shown in italics in the last column). Some of the longer-lived radioactive isotopes of each element are also included, with their half-lives. Each element has many other radioactive isotopes that are not included in this table. More complete listings can be found in the sources from which this table was derived: *Table of Isotopes* (8th Edition), edited by R. B. Firestone and V. S. Shirley (Wiley, 1999); G. Audi, A. H. Wapstra, and C. Thibault, "The 2003 Atomic Mass Evaluation," *Nuclear Physics* **A729,** 129 (2003).

In the half-life column, $My = 10^6 \, y$.

	Z	A	Atomic mass (u)	Abundance or Half-life		Z	A	Atomic mass (u)	Abundance or Half-life
H	1	1	1.0078250	99.985%	Na	11	21	20.997655	22.5 s
		2	2.014102	0.015%			22	21.994436	2.60 y
		3	3.016049	12.3 y			23	22.989769	100%
							24	23.990963	15.0 h
He	2	3	3.016029	0.000137%			25	24.989954	59 s
		4	4.002603	99.999863%			26	25.992633	1.1 s
Li	3	6	6.015123	7.59%					
		7	7.016005	92.41%	Mg	12	22	21.999574	3.88 s
		8	8.022487	0.84 s			23	22.994124	11.3 s
							24	23.985042	78.99%
Be	4	7	7.016930	53.2 d			25	24.985837	10.00%
		8	8.005305	0.07 fs			26	25.982593	11.01%
		9	9.012182	100%			27	26.984341	9.46 m
		10	10.013534	1.5 My			28	27.983877	20.9 h
		11	11.021658	13.8 s					
					Al	13	25	24.990428	7.18 s
B	5	8	8.024607	0.77 s			26	25.986892	0.72 My
		9	9.013329	0.85 as			27	26.981539	100%
		10	10.012937	19.8%			28	27.981910	2.24 m
		11	11.009305	80.2%			29	28.980445	6.56 m
		12	12.014352	20.2 ms					
					Si	14	26	25.992330	2.23 s
C	6	10	10.016853	19.3 s			27	26.986705	4.16 s
		11	11.011434	20.3 m			28	27.976927	92.23%
		12	12.000000	98.89%			29	28.976495	4.68%
		13	13.003355	1.11%			30	29.973770	3.09%
		14	14.003242	5730 y			31	30.975363	2.62 h
		15	15.010599	2.45 s			32	31.974148	132 y
N	7	13	13.005739	9.96 m	P	15	29	28.981801	4.14 s
		14	14.003074	99.63%			30	29.978314	2.50 m
		15	15.000109	0.37%			31	30.973762	100%
		16	16.006102	7.1 s			32	31.973907	14.3 d
		17	17.008450	4.2 s			33	32.971726	25.3 d
O	8	14	14.008596	70.6 s	S	16	30	29.984903	1.18 s
		15	15.003066	122 s			31	30.979555	2.57 s
		16	15.994915	99.76%			32	31.972071	95.02%
		17	16.999132	0. 038%			33	32.971459	0.75%
		18	17.999161	0.200%			34	33.967867	4.21%
		19	19.003580	26.9 s			35	34.969032	87.5 d
		20	20.004077	13.5 s			36	35.967081	0.02%
							37	36.971126	5.05 m
F	9	17	17.002095	64.5 s					
		18	18.000938	1.83 h	Cl	17	33	32.977452	2.51 s
		19	18.998403	100%			34	33.973763	1.53 s
		20	19.999981	11 s			35	34.968853	75.77%
		21	20.999949	4.2 s			36	35.968307	0.30 My
							37	36.965903	24.23%
Ne	10	18	18.005708	1.7 s			38	37.968010	37.2 m
		19	19.001880	17.2 s			39	38.968008	55.6 m
		20	19.992440	90.48%					
		21	20.993847	0.27%					
		22	21.991385	9.25%					
		23	22.994467	37.2 s					
		24	23.993611	3.4 m					

	Z	A	Atomic mass (u)	Abundance or Half-life		Z	A	Atomic mass (u)	Abundance or Half-life
Ar	18	34	33.980271	0.844 s	Cr	24	48	47.954032	21.6 h
		35	34.975258	1.78 s			49	48.951336	42.3 m
		36	35.967545	0.337%			50	49.946044	4.35%
		37	36.966776	35.0 d			51	50.944767	27.7 d
		38	37.962732	0.063%			52	51.940507	83.79%
		39	38.964313	269 y			53	52.940649	9.50%
		40	39.962383	99.60%			54	53.938880	2.36%
		41	40.964501	1.82 h			55	54.940840	3.50 m
		42	41.963046	32.9 y			56	55.940653	5.94 m
K	19	37	36.973376	1.23 s	Mn	25	52	51.945565	5.59 d
		38	37.969081	7.64 m			53	52.941290	3.7 My
		39	38.963707	93.26%			54	53.940359	312 d
		40	39.963998	1.25 Gy			55	54.938045	100%
		41	40.961826	6.73%			56	55.938905	2.58 h
		42	41.962403	12.3 h			57	56.938285	85.4 s
		43	42.960716	22.3 h					
Ca	20	38	37.976318	0.44 s	Fe	26	52	51.948114	8.27 h
		39	38.970720	0.86 s			53	52.945308	8.51 m
		40	39.962591	96.94%			54	53.939611	5.85%
		41	40.962278	0.103 My			55	54.938293	2.74 y
		42	41.958618	0.647%			56	55.934937	91.75%
		43	42.958767	0.135%			57	56.935394	2.12%
		44	43.955482	2.09%			58	57.933276	0.28%
		45	44.956187	163 d			59	58.934875	44.5 d
		46	45.953693	0.0035%			60	59.934072	1.5 My
		47	46.954546	4.54 d			61	60.936745	6.0 m
		48	47.952534	0.187%					
		49	48.955674	8.72 m	Co	27	57	56.936291	272 d
Sc	21	43	42.961151	3.89 h			58	57.935753	70.8 d
		44	43.959403	3.97 h			59	58.933195	100%
		45	44.955912	100%			60	59.933817	5.27 y
		46	45.955172	83.8 d			61	60.932476	1.65 h
		47	46.952408	3.35 d					
		48	47.952231	43.7 h	Ni	28	56	55.942132	6.08 d
Ti	22	44	43.959690	60 y			57	56.939794	35.6 h
		45	44.958126	3.08 h			58	57.935343	68.08%
		46	45.952632	8.25%			59	58.934347	0.076 My
		47	46.951763	7.44%			60	59.930786	26.22%
		48	47.947946	73.72%			61	60.931056	1.14%
		49	48.947870	5.41%			62	61.928345	3.63%
		50	49.944791	5.18%			63	62.929669	100 y
		51	50.946615	5.76 m			64	63.927966	0.93%
		52	51.946897	1.7 m			65	64.930084	2.52 h
V	23	48	47.952254	16.0 d	Cu	29	61	60.933458	3.33 h
		49	48.948516	329 d			62	61.932584	9.67 m
		50	49.947158	0.250%			63	62.929597	69.17%
		51	50.943960	99.750%			64	63.929764	12.7 h
		52	51.944775	3.74 m			65	64.927789	30.83%
		53	52.944338	1.60 m			66	65.928869	5.12 m
							67	66.927730	61.8 h

	Z	A	Atomic mass (u)	*Abundance* or Half-life		Z	A	Atomic mass (u)	*Abundance* or Half-life
Zn	30	62	61.934330	9.19 h	Br	35	77	76.921379	57.0 h
		63	62.933212	38.5 m			78	77.921146	6.46 m
		64	63.929142	*48.6%*			79	78.918337	*50.69%*
		65	64.929241	244 d			80	79.918529	17.7 m
		66	65.926033	*27.9%*			81	80.916291	*49.31%*
		67	66.927127	*4.1%*			82	81.916804	35.3 h
		68	67.924844	*18.8%*			83	82.915180	2.40 h
		69	68.926550	56 m	Kr	36	76	75.925910	14.8 h
		70	69.925319	*0.62%*			77	76.924670	74.4 m
		71	70.927722	2.45 m			78	77.920365	*0.35%*
							79	78.920082	35.0 h
Ga	31	67	66.928202	3.26 d			80	79.916379	*2.28%*
		68	67.927980	67.7 m			81	80.916592	0.229 My
		69	68.925574	*60.11%*			82	81.913484	*11.58%*
		70	69.926022	21.1 m			83	82.914136	*11.49%*
		71	70.924701	*39.89%*			84	83.911507	*57.00%*
		72	71.926366	14.1 h			85	84.912527	10.8 y
		73	72.925175	4.86 h			86	85.910611	*17.30%*
							87	86.913355	76.3 m
Ge	32	68	67.928094	271 d	Rb	37	83	82.915110	86.2 d
		69	68.927965	39.0 h			84	83.914385	33.1 d
		70	69.924247	*20.4%*			85	84.911790	*72.17%*
		71	70.924951	11.4 d			86	85.911167	18.6 d
		72	71.922076	*27.3%*			87	86.909181	*27.83%*
		73	72.923459	*7.8%*			88	87.911316	17.8 m
		74	73.921178	*36.7%*	Sr	38	82	81.918402	25.6 d
		75	74.922859	82.8 m			83	82.917557	32.4 h
		76	75.921403	*7.8%*			84	83.913425	*0.56%*
		77	76.923549	11.3 h			85	84.912933	64.8 d
							86	85.909260	*9.86%*
As	33	73	72.923825	80.3 d			87	86.908877	*7.00%*
		74	73.923929	17.8 d			88	87.905612	*82.58%*
		75	74.921596	*100%*			89	88.907451	50.6 d
		76	75.922394	26.3 h			90	89.907738	28.9 y
		77	76.920647	38.8 h	Y	39	87	86.910876	79.8 h
							88	87.909501	106.6 d
Se	34	72	71.927112	8.4 d			89	88.905848	*100%*
		73	72.926765	7.1 h			90	89.907152	64.1 h
		74	73.922476	*0.89%*			91	90.907305	58.5 d
		75	74.922523	120 d	Zr	40	88	87.910227	83.4 d
		76	75.919214	*9.4%*			89	88.908890	78.4 h
		77	76.919914	*7.6%*			90	89.904704	*51.45%*
		78	77.917309	*23.8%*			91	90.905646	*11.22%*
		79	78.918499	0.30 My			92	91.905041	*17.15%*
		80	79.916521	*49.6%*			93	92.906476	1.53 My
		81	80.917992	18.5 m			94	93.906315	*17.38%*
		82	81.916699	*8.7%*			95	94.908043	64.0 d
		83	82.919118	22.3 m			96	95.908273	*2.80%*
							97	96.910953	16.7 h

	Z	A	Atomic mass (u)	*Abundance or Half-life*		Z	A	Atomic mass (u)	*Abundance or Half-life*
Nb	41	91	90.906996	680 y	Pd	46	100	99.908506	3.63 d
		92	91.907194	35 My			101	100.908289	8.47 h
		93	92.906378	*100%*			102	101.905609	*1.02%*
		94	93.907284	20,300 y			103	102.906087	17.0 d
		95	94.906836	35.0 d			104	103.904036	*11.14%*
							105	104.905085	*22.33%*
Mo	42	90	89.913937	5.56 h			106	105.903486	*27.33%*
		91	90.911750	15.5 m			107	106.905133	6.5 My
		92	91.906811	*14.8%*			108	107.903892	*26.46%*
		93	92.906813	4000 y			109	108.905950	13.7 h
		94	93.905088	*9.3%*			110	109.905153	*11.72%*
		95	94.905842	*15.9%*			111	110.907671	23.4 m
		96	95.904679	*16.7%*					
		97	96.906021	*9.6%*	Ag	47	105	104.906529	41.3 d
		98	97.905408	*24.1%*			106	105.906669	24.0 m
		99	98.907712	65.9 h			107	106.905097	*51.84%*
		100	99.907477	*9.6%*			108	107.905956	2.37 m
		101	100.910347	14.6 m			109	108.904752	*48.16%*
							110	109.906107	24.6 s
Tc	43	95	94.907657	20.0 h					
		96	95.907871	4.3 d					
		97	96.906365	4.2 My	Cd	48	104	103.909849	57.7 m
		98	97.907216	4.2 My			105	104.909468	55.5 m
		99	98.906255	0.211 My			106	105.906459	*1.25%*
		100	99.907658	15.8 s			107	106.906618	6.50 h
							108	107.904184	*0.89%*
Ru	44	94	93.911360	51.8 m			109	108.904982	461 d
		95	94.910413	1.64 h			110	109.903002	*12.5%*
		96	95.907598	*5.5%*			111	110.904178	*12.8%*
		97	96.907555	2.79 d			112	111.902758	*24.1%*
		98	97.905287	*1.86%*			113	112.904402	*12.2%*
		99	98.905939	*12.8%*			114	113.903359	*28.7%*
		100	99.904219	*12.6%*			115	114.905431	53.5 h
		101	100.905582	*17.1%*			116	115.904756	*7.5%*
		102	101.904349	*31.6%*			117	116.907219	2.49 h
		103	102.906324	39.3 d					
		104	103.905433	*18.6%*					
		105	104.907753	4.44 h	In	49	111	110.905103	2.80 d
							112	111.905532	15.0 m
Rh	45	101	100.906164	3.3 y			113	112.904058	*4.29%*
		102	101.906843	207 d			114	113.904914	71.9 s
		103	102.905504	*100%*			115	114.903878	*95.71%*
		104	103.906656	42.3 s			116	115.905260	14.1 s
		105	104.905694	35.4 h					

	Z	A	Atomic mass (u)	Abundance or Half-life		Z	A	Atomic mass (u)	Abundance or Half-life
Sn	50	110	109.907843	4.11 h	Xe	54	122	121.908368	20.1 h
		111	110.907734	35.3 m			123	122.908482	2.08 h
		112	111.904818	*0.97%*			124	123.905893	*0.095%*
		113	112.905171	115.1 d			125	124.906395	16.9 h
		114	113.902779	*0.66%*			126	125.904274	*0.089%*
		115	114.903342	*0.34%*			127	126.905184	36.4 d
		116	115.901741	*14.54%*			128	127.903531	*1.91%*
		117	116.902952	*7.68%*			129	128.904779	*26.40%*
		118	117.901603	*24.22%*			130	129.903508	*4.07%*
		119	118.903308	*8.59%*			131	130.905082	*21.23%*
		120	119.902195	*32.58%*			132	131.904153	*26.91%*
		121	120.904235	27.0 h			133	132.905911	5.24 d
		122	121.903439	*4.63%*			134	133.905394	*10.44%*
		123	122.905721	129 d			135	134.907227	9.14 h
		124	123.905274	*5.79%*			136	135.907219	*8.86%*
		125	124.907784	9.64 d			137	136.911562	3.82 m
		126	125.907653	0.23 My					
Sb	51	119	118.903942	38.2 h	Cs	55	131	130.905464	9.69 d
		120	119.905072	15.9 m			132	131.906434	6.48 d
		121	120.903816	*57.21%*			133	132.905452	*100%*
		122	121.905174	2.72 d			134	133.906718	2.06 y
		123	122.904214	*42.79%*			135	134.905977	2.3 My
		124	123.905936	60.1 d			136	135.907312	13.0 d
		125	124.905254	2.76 y					
Te	52	118	117.905828	6.00 d	Ba	56	128	127.908318	2.43 d
		119	118.906404	16.1 h			129	128.908679	2.23 h
		120	119.904020	*0.09%*			130	129.906321	*0.106%*
		121	120.904936	19.2 d			131	130.906941	11.5 d
		122	121.903044	*2.55%*			132	131.905061	*0.101%*
		123	122.904270	*0.89%*			133	132.906007	10.5 y
		124	123.902818	*4.74%*			134	133.904508	*2.42%*
		125	124.904431	*7.07%*			135	134.905689	*6.59%*
		126	125.903312	*18.84%*			136	135.904576	*7.85%*
		127	126.905226	9.35 h			137	136.905827	*11.23%*
		128	127.904463	*31.74%*			138	137.905247	*71.70%*
		129	128.906598	69.6 m			139	138.908841	83.1 m
		130	129.906224	*34.08%*					
		131	130.908524	25.0 m	La	57	136	135.907636	9.87 m
I	53	125	124.904630	59.4 d			137	136.906494	60,000 y
		126	125.905624	12.9 d			138	137.907112	*0.090%*
		127	126.904473	*100%*			139	138.906353	*99.910%*
		128	127.905809	25.0 m			140	139.909478	1.68 d
		129	128.904988	15.7 My			141	140.910962	3.92 h
		130	129.906674	12.4 h					

	Z	A	Atomic mass (u)	*Abundance or Half-life*		Z	A	Atomic mass (u)	*Abundance or Half-life*
Ce	58	134	133.908925	76 h	Eu	63	149	148.917931	93.1 d
		135	134.909151	17.7 h			150	149.919702	36.9 y
		136	135.907172	*0.185%*			151	150.919850	*47.81%*
		137	136.907806	9.0 h			152	151.921745	13.5 y
		138	137.905991	*0.251%*			153	152.921230	*52.19%*
		139	138.906653	137.6 d			154	153.922979	8.59 y
		140	139.905439	*88.45%*			155	154.922893	4.75 y
		141	140.908276	32.5 d			156	155.924752	15.2 d
		142	141.909244	*11.11%*					
		143	142.912386	33.0 h	Gd	64	150	149.918659	1.79 My
		144	143.913647	285 d			151	150.920348	124 d
							152	151.919791	*0.20%*
Pr	59	139	138.908938	4.41 h			153	152.921750	240 d
		140	139.909076	3.39 m			154	153.920866	*2.18%*
		141	140.907653	*100%*			155	154.922622	*14.80%*
		142	141.910045	19.1 h			156	155.922123	*20.47%*
		143	142.910817	13.6 d			157	156.923960	*15.65%*
							158	157.924104	*24.84%*
Nd	60	140	139.909552	3.37 d			159	158.926389	18.5 h
		141	140.909610	2.49 h			160	159.927054	*21.9%*
		142	141.907723	*27.2%*			161	160.929669	3.66 m
		143	142.909814	*12.2%*					
		144	143.910087	*23.8%*	Tb	65	157	156.924025	71 y
		145	144.912574	*8.3%*			158	157.925413	180 y
		146	145.913117	*17.2%*			159	158.925347	*100%*
		147	146.916100	11.0 d			160	159.927168	72.3 d
		148	147.916893	*5.7%*			161	160.927570	6.91 d
		149	148.920149	1.73 h					
		150	149.920891	*5.6%*	Dy	66	154	153.924424	3.0 My
		151	150.923829	12.4 m			155	154.925754	9.9 h
							156	155.924283	*0.06%*
Pm	61	143	142.910933	265 d			157	156.925466	8.1 h
		144	143.912591	363 d			158	157.924409	*0.10%*
		145	144.912749	17.7 y			159	158.925739	144.4 d
		146	145.914696	5.53 y			160	159.925198	*2.3%*
		147	146.915139	2.62 y			161	160.926933	*18.9%*
		148	147.917475	5.37 d			162	161.926798	*25.5%*
		149	148.918334	53.1 h			163	162.928731	*24.9%*
							164	163.929175	*28.2%*
Sm	62	142	141.915198	72.5 m			165	164.931703	2.33 h
		143	142.914628	8.75 m					
		144	143.911999	*3.1%*	Ho	67	163	162.928734	4570 y
		145	144.913410	340 d			164	163.930234	29 m
		146	145.913041	103 My			165	164.930322	*100%*
		147	146.914898	*15.0%*			166	165.932284	26.8 h
		148	147.914823	*11.2%*			167	166.933133	3.0 h
		149	148.917185	*13.8%*					
		150	149.917276	*7.4%*					
		151	150.919932	90 y					
		152	151.919732	*26.7%*					
		153	152.922097	46.3 h					
		154	153.922209	*22.7%*					
		155	154.924640	22.3 m					

	Z	A	Atomic mass (u)	Abundance or Half-life		Z	A	Atomic mass (u)	Abundance or Half-life
Er	68	160	159.929083	28.6 h	W	74	178	177.945876	21.6 d
		161	160.929995	3.21 h			179	178.947070	37.0 m
		162	161.928778	0.14%			180	179.946704	0.12%
		163	162.930033	75.0 m			181	180.948197	121 d
		164	163.929200	1.60%			182	181.948204	26.5%
		165	164.930726	10.4 h			183	182.950223	14.3%
		166	165.930293	33.50%			184	183.950931	30.6%
		167	166.932048	22.87%			185	184.953419	75.1 d
		168	167.932370	26.98%			186	185.954364	28.4%
		169	168.934590	9.39 d			187	186.957160	23.7 h
		170	169.935464	14.91%					
		171	170.938030	7.52 h	Re	75	183	182.950820	70.0 d
							184	183.952521	38.0 d
Tm	69	167	166.932852	9.25 d			185	184.952955	37.40%
		168	167.934173	93.1 d			186	185.954986	3.72 d
		169	168.934213	100%			187	186.955753	62.60%
		170	169.935801	128.6 d			188	187.958114	17.0 h
		171	170.936429	1.92 y					
					Os	76	182	181.952110	22.1 h
Yb	70	166	165.933882	56.7 h			183	182.953126	13.0 h
		167	166.934950	17.5 m			184	183.952489	0.02%
		168	167.933897	0.13%			185	184.954042	93.6 d
		169	168.935190	32.0 d			186	185.953838	1.6%
		170	169.934762	3.0%			187	186.955750	1.6%
		171	170.936326	14.3%			188	187.955838	13.3%
		172	171.936381	21.8%			189	188.958147	16.2%
		173	172.938211	16.1%			190	189.958447	26.4%
		174	173.938862	31.8%			191	190.960930	15.4 d
		175	174.941276	4.19 d			192	191.961481	40.9%
		176	175.942572	12.8%			193	192.964152	30.1 h
		177	176.945261	1.9 h					
					Ir	77	189	188.958719	13.2 d
Lu	71	173	172.938931	1.37 y			190	189.960546	11.8 d
		174	173.940337	3.3 y			191	190.960594	37.3%
		175	174.940772	97.41%			192	191.962605	73.8 d
		176	175.942686	2.59%			193	192.962926	62.7%
		177	176.943758	6.65 d			194	193.965078	19.3 h
Hf	72	172	171.939448	1.87 y	Pt	78	188	187.959395	10.2 d
		173	172.940513	23.6 h			189	188.960834	10.9 h
		174	173.940046	0.16%			190	189.959932	0.014%
		175	174.941509	70 d			191	190.961677	2.86 d
		176	175.941409	5.26%			192	191.961038	0.78%
		177	176.943221	18.60%			193	192.962987	50 y
		178	177.943699	27.28%			194	193.962680	32.97%
		179	178.945816	13.62%			195	194.964791	33.83%
		180	179.946550	35.08%			196	195.964952	25.24%
		181	180.949101	42.4 d			197	196.967340	19.9 h
							198	197.967893	7.16%
Ta	73	179	178.945930	1.82 y			199	198.970593	30.8 m
		180	179.947465	0.012%					
		181	180.947996	99.988%					
		182	181.950152	114 d					

	Z	A	Atomic mass (u)	Abundance or Half-life		Z	A	Atomic mass (u)	Abundance or Half-life
Au	79	195	194.965035	186 d	Fr	87	212	211.996202	20.0 m
		196	195.966570	6.17 d			223	223.019736	22.0 m
		197	196.966569	*100%*	Ra	88	223	223.018502	11.43 d
		198	197.968242	2.696 d			224	224.020212	3.63 d
		199	198.968765	3.14 d			225	225.023612	14.9 d
Hg	80	194	193.965439	444 y			226	226.025410	1600 y
		195	194.966720	10.5 h	Ac	89	225	225.023230	10.0 d
		196	195.965833	*0.15%*			226	226.026098	29.4 h
		197	196.967213	64.1 h			227	227.027752	21.77 y
		198	197.966769	*10.0%*	Th	90	228	228.028741	1.91 y
		199	198.968280	*16.9%*			229	229.031762	7340 y
		200	199.968326	*23.1%*			230	230.033134	75,400 y
		201	200.970302	*13.2%*			231	231.036304	25.52 h
		202	201.970643	*29.9%*			232	232.038055	*100%*
		203	202.972872	46.6 d			233	233.041582	21.8 m
		204	203.973494	*6.9%*	Pa	91	230	230.034541	17.4 d
		205	204.976073	5.1 m			231	231.035884	32,800 y
Tl	81	201	200.970819	72.9 h			232	232.038592	1.31 d
		202	201.972106	12.2 d			234	234.043308	6.70 h
		203	202.972344	*29.52%*	U	92	233	233.039635	0.1592 My
		204	203.973864	3.78 y			234	234.040952	0.2455 My
		205	204.974428	*70.48%*			235	235.043930	*0.720%*
		206	205.976110	4.20 m			236	236.045568	23.42 My
Pb	82	202	201.972159	53,000 y			237	237.048730	6.75 d
		203	202.973391	51.9 h			238	238.050788	*99.274%*
		204	203.973044	*1.4%*			239	239.054293	23.5 m
		205	204.974482	17.3 My	Np	93	236	236.046570	0.154 My
		206	205.974465	*24.1%*			237	237.048173	2.14 My
		207	206.975897	*22.1%*			238	238.050946	2.117 d
		208	207.976652	*52.4%*	Pu	94	238	238.049560	87.74 y
		209	208.981090	3.25 h			239	239.052163	24,100 y
Bi	83	207	206.978471	32.9 y			240	240.053814	6561 y
		208	207.979742	0.368 My			241	241.056851	14.3 y
		209	208.980399	*100%*			242	242.058743	0.375 My
		210	209.984120	5.01 d	Am	95	241	241.056829	432 y
		211	210.987269	2.14 m			242	242.059549	16.0 h
Po	84	207	206.981593	5.80 h			243	243.061381	7370 y
		208	207.981246	2.90 y	Cm	96	246	246.067224	4760 y
		209	208.982430	102 y			247	247.070354	15.6 My
		210	209.982874	138.4 d			248	248.072349	0.348 My
At	85	209	208.986173	5.41 h	Bk	97	247	247.070307	1380 y
		210	209.987148	8.1 h					
		211	210.987496	7.21 h	Cf	98	251	251.079587	898 y
Rn	86	211	210.990601	14.6 h			254	254.087323	60.5 d
		222	222.017578	3.82 d					

	Z	A	Atomic mass (u)	*Abundance* or Half-life		Z	A	Atomic mass (u)	*Abundance* or Half-life
Es	99	252	252.082979	472 d	Sg	106	261	261.116117	0.23 s
Fm	100	257	257.095105	100.5 d	Bh	107	262	262.122892	0.10 s
Md	101	258	258.098431	51.5 d	Hs	108	264	264.128395	0.8 ms
No	102	259	259.101031	58 m	Mt	109	266	266.137299	1.7 ms
Lr	103	260	260.105504	3.0 m	Ds	110	270	270.144720	0.5 ms
Rf	104	261	261.108767	65 s	Rg	111	272	272.153615	4 ms
Db	105	262	262.114084	35 s					

ANSWERS TO ODD-NUMBERED PROBLEMS

Chapter 1

1. 4.527×10^6 m/s
3. (a) -7.79×10^5 m/s
   (b) 1.008×10^{-13} J, 3.995×10^{-13} J
5. (a) 2.13×10^6 m/s
   (b) 1.28×10^6 m/s
7. 4.34×10^{-5}
9. $6.1 \times 10^{-6} N$
11. $35.3°, v/\sqrt{2}, v/\sqrt{6}$
13. 2.47×10^6 m/s, -0.508×10^6 m/s
15. (a) 0.0104 eV   (b) 2550 m
17. 4.61×10^{12} rad/s

Chapter 2

1. 101 km/h at 62° east of south
3. 7×10^4 m/s
5. 2.6×10^8 m/s
7. (a) 357.1 ns   (b) 103 m   (c) 28.8 m
11. 0.402c
13. 5.0×10^7 m/s
17. $+0.937c, -0.572c$
19. (a) $+0.508\mu s$ (b) -81.5 m
21. (a) 2:00 P.M. (b) 3:00 P.M.
   (c) 1:00 P.M. and 3:00 P.M.
23. 8 y
25. (a) $K_i' = K_f' = 0.512mc^2$
   (b) $K_i = K_f = 0.458mc^2$
27. 0.958c
29. $v < 0.115c$
33. (a) 773 MeV/c^2 (b) 1227 MeV
35. 4.4×10^{-16} kg
37. $p_e = 10.5$ MeV/$c, K_e = 10.0$ MeV
   $p_p = 137.4$ MeV/$c, K_p = 10.0$ MeV
39. (a) 1268.1 MeV   (b) 298.8 MeV/c
   (c) 1232 MeV
41. 0.981c

43. $64.38 \mu s$
45. (a) $0.99875c$ (b) 400.5 y
47. 2.34 km, $1.07 \mu s$
49. (a) $0.648 \mu s$ (b) 335 m
51. (a) $E' = mc^2/\sqrt{(1 - u^2/c^2)(1 - v^2/c^2)}$

$$p' = \frac{m\sqrt{u^2 + v^2 - u^2v^2/c^2}}{\sqrt{(1 - u^2/c^2)(1 - v^2/c^2)}}$$

   (b) m^2c^4
53. (a) 3.1 MeV   (b) 7.8 MeV
55. 0.508c
57. 267.0 MeV, 28.9 MeV

Chapter 3

1. 1.32 mm
3. (a) 0.388 nm   (b) 7.2°
5. (a) 1.00×10^7 eV/$c, 5.33 \times 10^{-21}$ kg·m/s
   (b) 2.5×10^4 eV/$c, 1.3 \times 10^{-23}$ kg·m/s
   (c) 1.2 eV/$c, 6.6 \times 10^{-28}$ kg·m/s
   (d) 6.2×10^{-7} eV/$c, 3.3 \times 10^{-34}$ kg·m/s
7. (a) 0.124 nm   (b) 1.24×10^{-3} nm
   (c) 1.8 eV to 3.5 eV
9. 0.964 V
11. (a) 4.88 eV
19. 1.1 mm, 1.1×10^{-3} eV
21. 0.33 W
23. (a) 1.9×10^5 W/m^2 (b) 0.26%
25. (a) 10.33 keV   (b) 0.06 keV
29. 3.9×10^{-4} eV
31. 1.17 eV
33. 2.28 eV, 4.10×10^{-15} eV/s
35. (a) $2.52 \mu m$ (b) 0.405
37. 2.724 K
39. (a) 5.79×10^{-10} W/m^2, 1.91×10^{-11} W/m^2
   (b) 1.81×10^8/s, 2.97×10^7/s
41. 4.1 m/s

Chapter 4

1. (a) 13 fm   (b) 0.025 fm   (c) 0.73 nm
3. (a) $0.143c$ (b) $-9.72\,\text{MV}$
5. (a) $+0.010\,\text{V}$ (b) $+100\,\text{V}$ (c) $+1.0 \times 10^9\,\text{V}$
7. (a) 88 MeV   (b) 4.2 MeV   (c) 1.1 MeV
9. 33 nm
11. $15.2°(n=1), 31.8°(n=2), 52.2°(n=3)$
13. (a) 660 m   (b) 0.33 m
    (c) 0.017 mm   (d) 0.050 Hz
15. $8.4 \times 10^4\,\text{Hz}$
17. 5.8 nm
19. 33 MeV
21. $5.5 \times 10^{-7}\,\text{eV}$
23. 0.052 MeV
29. $v_{\text{group}} = 1.5 v_{\text{phase}}$
31. (b) 1.50 eV, 6.00 eV, 13.5 eV
33. (a) 0.0279 eV   (b) 3.3 nm   (c) 2.1 K
35. (a) 0.71 MeV   (b) 0.66 MeV
37. (a) 135 MeV   (b) $4.87 \times 10^{-24}\,\text{s}$ (c) 1.46 fm
39. $3 \times 10^{-9}\,\text{m}$

Chapter 5

1. (b) $-(B/m)\sqrt{2H/g}, H(1+B/mg)$
3. 1.1 eV
5. 150 eV, 600 eV, 1350 eV
7. (a) $c = a^2 b^2, d = a(b+1)$
    (b) $w = a(2b+1)$
9. $U(x) = -\hbar^2 b/mx, E = -\hbar^2 b^2/2m$
11. 10.1 eV, 18.9 eV
15. (a) 2.63×10^{-5} (b) 0.0106   (c) 5.42×10^{-3}
17. $5.00E_0, 10.00E_0$
19. $3E_0, 6E_0, 9E_0, 11E_0, 12E_0, \ldots$
21. (b) $\sqrt{3\hbar\omega_0/k}, \sqrt{5\hbar\omega_0/k}$
23. (a) 0   (b) $\hbar\omega_0 m/2$
25. $A^2 dx, 0.368 A^2 dx$
27. $B = D = -A\sqrt{E/(U_0 - E)}$
31. (a) 2160 eV   (b) $4.70 \times 10^4\,\text{eV}/c$
    (c) $4.2 \times 10^{-3}\,\text{nm}$
35. (b) $h^2 n^2/4L^2$
37. 0.157

Chapter 6

3. (a) $6.57 \times 10^{15}\,\text{Hz}, 45.7\,\text{nm}$
    (b) $3.48 \times 10^{15}\,\text{Hz}, 86.2\,\text{nm}$
5. (a) 22.8 fm   (b) 55.0 fm
    (c) 4.14 MeV, 0.86 MeV
7. 14 fm
9. (a) 8.4 MeV   (b) 1.61 fm
    (c) 6.00 fm   (d) 8.3×10^{-6}

11. 0.63 MeV
13. 28.2 fm, 19.9 fm, 38.9°
15. 121.51 nm, 102.52 nm, 97.21 nm
17. 2279 nm
19. 91.13 nm, 820.1 nm
21. $\Delta E = 0.306\,\text{eV}, 0.97\,\text{eV}, 2.86\,\text{eV}, 13.1\,\text{eV}$
23. 7.4 eV
27. $-54.40\,\text{eV}, -13.60\,\text{eV}, -6.04\,\text{eV}, -3.40\,\text{eV}$
31. 2.10 eV
33. (a) $6.58 \times 10^{12}\,\text{Hz}, 7.72 \times 10^{12}\,\text{Hz}$
    (b) $6.58 \times 10^9\,\text{Hz}, 6.68 \times 10^9\,\text{Hz}$
35. (a) 15   (b) 5   (c) 1
37. $7 \times 10^{-8}\,\text{eV}$
39. 48.23 nm
41. (a) 0.440 nm   (b) 11.3%

Chapter 7

1. $-me^4/32\pi^2\varepsilon_0^2\hbar^2$
3. 0.0108
5. $35°, 66°, 90°, 114°, 145°$
7. (a) 0, 1, 2, 3, 4, 5   (b) $+6$ to -6 in integer steps
    (c) 5   (d) 4
11. (a) 0, 0   (b) $0, 3.2 \times 10^{-11}$
    (c) $0, 3.2 \times 10^{-11}$
    (d) $1.1 \times 10^{-11}, 2.2 \times 10^{-11}$
13. $(3 \pm \sqrt{5})a_0$
15. 0.0054
17. Minima: $55°, 125°$ Maxima: $0°, 90°, 180°$
21. $3s, 2s, 1s, 3d$
23. (a) $7s, 7p, 7d, 7f, 7g, 7h, 7i$
    (b) $6p, 6f, 5p, 5f, 4p, 4f, 3p, 2p$
25. (a) 656.112 nm, 656.182 nm, 656.042 nm
27. $1.89\,\text{eV} \pm 2.55 \times 10^{-5}\,\text{eV}$,
    $1.89\,\text{eV} \pm 1.95 \times 10^{-5}\,\text{eV}$
29. 0.651, 0.440
31. $6a_0, 5a_0$
33. 3.3 mm

Chapter 8

1. (a) $(2, 1, +1, +1/2), (2, 1, +1, -1/2), \ldots$
    (b) 36   (c) 30   (d) 36
3. (a) 14   (b) $+3/2$ (c) $+8$
    (d) $+2$ (e) $+10$
5. (a) N, P, As, Sb, Bi
    (b) Co, Rh, Ir, Mt
7. (a) $[\text{Ar}]4s^2 3d^6$ (b) $+2$
    (c) $+2$ (d) $+3, +1$
9. $-1.51\,\text{eV}, -0.85\,\text{eV}$

11.   (a) 0.045 eV   (b) 3.374 eV, 4.373 eV, 4.750 eV
     (c) 3.54 eV, 2.02 eV
13.   3.68 eV, 15.5 eV, 63.7 eV
15.   (a) 5, 1   (b) 2, 4   (c) 2, 1
17.   0, 1, 2, 3, 4; 0,1
19.   (a) 1.12×10^{16} photons/s   (b) 763 V/m
21.   (a) 1.84   (b) 1.00
23. $1.40\,\text{eV}^{1/2}$, 6.4
25.   (a) 670.8 nm   (b) 58.4 nm
     (c) 230 nm, 50.6 nm

Chapter 9

1.   15.4 eV
3.   (a) 4.25 eV/molecule
     (c) 9.80 eV/molecule
5.   (a) Li_2 (c) CO
7.   5.11 eV
9.   (a) 30.9×10^{-30} C·m   (b) 88%
11.   42.7%
13. $4.689 \times 10^3\,\text{eV/nm}^2$
15.   (a) 0.2626 eV   (b) 0.2579 eV
17.   23.1 mm, 11.6 mm, 7.71 mm
19. 2.00×10^{-6} eV
23. $hf - 2B(L+1), hf + 2BL$
31.   (a) 0.27 eV, 0.23 eV, 0.19 eV
     (b) 4.56 eV, 4.60 eV
33.   (a) 0.317 eV   (b) $2.37 \times 10^{21}\,\text{eV/m}^2$
     (c) 2.1×10^{-3} eV
35.   (a) 8.2×10^{-2} (b) 6.8×10^{-3}
37.   (a) 2.99, 4.97, 6.91
     (b) 2.94, 4.70, 6.18
39.   (a) 194.4 μeV

Chapter 10

1.   (a) 3   (b) 3, 6, 1   (c) 20%, 40%
3.   (a) 6   (b) 2   (c) 4, 2
7.   (a) +3 (1), +2 (2), +1 (3), 0 (4), ...
     (b) 3 (7), 2 (5), 1 (3), 0 (1)
     (c) +3 (1), +2 (1), +1 (2), 0 (2), ...
9. $2\pi(2s+1)m/\hbar^2$
11. $2.7 \times 10^{23}\,\text{m}^{-3}$
13.   0.65, 0.29, 0.06
15.   (a) 0.0379 eV   (b) 2.78×10^{21}
17.   (a) 554 μeV   (b) 0.066 μeV
19.   (a) 10^3 atm   (b) 2.9 K
21.   (a) $1.84 \times 10^7\,\text{eV/m}^3$
     (b) 0.035   (c) 2.3×10^{-17}
23.   7.04 eV, 4.22 eV
25.   (a) $4.37 \times 10^{-36}\,\text{m}^{-3}$ (b) $6.00 \times 10^{-6}\,\text{m}^{-3}$

27.   (b) 8.6 km   (c) $2.2 \times 10^{18}\,\text{kg/m}^3$
29.   0.078 J/K
31.   (a) $e^{-E/kT}$ (b) $E/(e^{E/kT} + 1)$
     (c) $NE/(e^{E/kT} + 1)$
     (d) $R(E/kT)^2 e^{E/kT}/(e^{E/kT} + 1)^2$
33.   (a) 0.3295, 0.3333, 0.3372
35.   0.12 eV
37.   17.1 MeV, 22.9 MeV
39.   (a) 2.83×10^{-3} nm   (b) 1.39×10^{-3} nm
41.   (a) 2.78 ± 0.06 K   (b) 0.0137 ± 0.0017

Chapter 11

1.   (a) 1/8   (b) $a = 2r$ (c) 0.5236
5.   (a) 6.82 eV   (b) 6.45 eV   (c) 3.27 eV
7. -8.96 eV, 1.00 eV
9.   0.255 nm
11.   3.23 K
13.   91.1 K
15.   (a) 7.09 eV   (b) 0.11 eV
17. $2.3 \times 10^{24}\,\text{m}^{-3}$
19.   7080 K
21.   426 W/K·m
23.   790
25.   (a) 0.20 MeV   (b) 6.3 mm
27.   603 MHz
29.   (a) $1.9 \times 10^{-28}, 3.5 \times 10^{-10}$
31.   (a) -0.094 eV   (b) -0.040 eV
33. 4.2×10^{-6}
35. 6.8×10^{-5} mA
37.   (a) 57% and 43%   (b) 89% and 11%
39.   (a) 0.4997, 0.5003   (b) 0.480, 0.520
41.   (a) 10.8×10^{-6} (b) -45.3×10^{-6}
49.   (a) 135 K   (b) 165 K, $1.40 m_e$
51.   (a) 1.00×10^6 (b) $1.00 \times 10^3, 1.00 \times 10^{-3}$
     (c) 0.662 keV

Chapter 12

1.   (a) $^{19}_{9}\text{F}_{10}$ (b) $^{199}_{79}\text{Au}_{120}$ (c) $^{107}_{47}\text{Ag}_{60}$
3.   (a) 15 MeV   (b) 824 MeV
5.   (a) 1636.4 MeV, 7.868 MeV
     (b) 1118.5 MeV, 8.410 MeV
7.   7.718 MeV, 8.482 MeV
9.   (a) 19.814 MeV   (b) 15.958 MeV
11. 2.5×10^{-3} fm
13. ^{160}Dy: yes, ^{164}Dy: no
15.   (a) 35 min   (b) $0.020\,\text{min}^{-1}$ (c) $46\,\text{s}^{-1}$
17.   0.0598
19.   (a) 0.85 μCi   (b) 0.13%
21.   (a) Yes   (b) No   (c) Yes

25. 0.864 MeV
27. 0.412 MeV, 1.088 MeV, 0.676 MeV
29. 100.1 keV, 300.9 keV, 200.8 keV, ...
31. (a) 6   (b) 4   (c) 42.659 MeV   (d) 27.8 μW
33. 12.6 fm
35. 37.5 μCi
39. 1.93 mW
41. 1.04×10^{-20} s^{-1}, 9.0×10^8 s^{-1}
43. (a) 7.52×10^{-4} MeV   (b) 3.26×10^{-4} MeV
45. (a) 4.67×10^{-9} eV   (b) 1.95×10^{-3} eV
    (c) 0.097 mm/s

Chapter 13

1. (a) $^1_1\text{H}_0$ (c) $^{30}_{15}\text{P}_{15}$
3. 98 b
5. 8.6×10^{-4} b
7. (a) 0.5   (b) 0.75   (c) 0.9325
11. 1.58 μCi
13. (a) -10.313 MeV   (b) 2.314 MeV
15. 12.201 MeV, 11.152 MeV or
    22.706 MeV, 0.647 MeV
17. (a) 235.3 MeV   (b) 202.0 MeV
19. (a) 6.546 MeV   (b) 4.807 MeV
21. (a) 1.943 MeV, 1.199 MeV, 7.551 MeV, ...
23. 14.1 MeV
25. 7.274 MeV
31. (a) 5.594 MeV   (b) 0.56 W
33. 4.49 b, 2.20 b
35. 8.662 MeV
37. 9.043 MeV
39. (a) 3.2×10^8 J   (b) 6.8×10^8 J

Chapter 14

1. (a) Strong   (b) Electromagnetic   (c) Weak
7. (a) L_e (b) S
11. (a) 1.29 keV   (b) 0.21 MeV
13. 179.4 MeV
15. (a) 34 MeV   (b) 218 MeV
17. (a) 43 MeV, 19 MeV   (b) 30 MeV, 4 MeV
19. 387 MeV, 34.9°
21. (a) 181 MeV   (b) -605 MeV
23. (a) 2205 MeV   (b) 903 MeV
25. (a) $u\bar{u}$ annihilation, creation of 2 $s\bar{s}$ pairs
27. (a) $s \rightarrow u + W^-$ and $W^- \rightarrow d + \bar{u}$
29. $c\bar{u}$, etc.
31. 0.999635c, 40.2 GeV
33. (a) 0.360c (b) 0.637c (c) 279 MeV
35. 47 MeV

Chapter 15

1. (a) 590.0 nm   (b) 637.1 nm
3. (b) 6.64×10^{-4} eV
5. 0.089 nm
7. 1.6×10^{-2} Hz
11. (b) 128 rev/s
15. (a) 5.8×10^{12} K   (b) 6.7×10^{-6} s
17. (a) 1.0×10^9 K   (b) 225 s   (c) 6600 K, 1.6×10^5 y
19. 0.27 m^{-3}
21. 3.0 eV
23. (a) 2.9×10^5 y   (b) 1.17 eV   (c) 1.24
25. (a) 9.79 km   (b) 3.01×10^{39} J
    (c) 6.02×10^{30} J   (d) 6.98×10^{25} W
    (e) 6.21×10^{-14} W

PHOTO CREDITS

Chapter 1
Opener: Cassini Interplanetary Trajectory illustration. Courtesy of NASA.

Chapter 2
Opener: Statue of Albert Einstein at NAS in Washington, D.C. ROGER L. WOLLEN-BERG/UPI/Newscom. Page 29: © SSPL/The Image Works. Page 31: Time, Inc. UPI Photo Service/NewsCom. Figure 2.7: GIPhotoStock/Photo Researchers, Inc.

Chapter 3
Opener: Heat radiation from a building using Thermography. Ted Kinsman/Getty Images. Page 79: © Mary Evans Picture Library/The Image Works. Page 85: © SSPL/Science Museum/The Image Works. Page 88: Weber Collection/AIP/Photo Researchers. Figure 3.2: GIPhotoStock/Photo Researchers, Inc. Figure 3.7b: Omikron/Photo Researchers, Inc. Figure 3.7c: Educational Images Ltd./Custom Medical Stock Photo, Inc. Figure 3.8: Omikron/Photo Researchers, Inc.

Chapter 4
Opener: Scanning tunnelling micrograph (STM) showing interactions between cobalt and silver atoms and a copper surface. DRS A. YAZDANI & D. J. HORNBAKER/SCIENCE PHOTO LIBRARY/Photo Researchers, Inc. Page 102: Meggers Gallery/AIP/Photo Researchers, Inc. Page 114: Bettmann/Corbis Images. Figure 4.1 (top): GIPhotoStock/Photo Researchers, Inc. Figure 4.0 bottom (Courtesy C. Joensson, Institut Für Angewandte Physik der Universitat Tubingen.) Figure 4.2: Courtesy Sumio Iijima, Arizona State University. Figure 4.3: Omikron/Photo Researchers, Inc. Figure 4.16: SCIMAT/SCIENCE PHOTO LIBRARY/Photo Researchers, Inc. Figure 4.27: Reprinted with permission from Akira Tonomura, Hitachi, Ltd., T. Matsuda, and T. Kawasaki, Advanced Research Laboratory. From *American Journal of Physics* 57,117. (Copyright 1989). American Association of Physics Teachers.

Chapter 5
Opener: Schrödinger's cat. MEHAU KULYK/SCIENCE PHOTO LIBRARY/Photo Researchers, Inc. Page 140: AIP/Photo Researchers, Inc. Figure 5.18: Image originally created by IBM corporation.

Chapter 6
Opener: Structure of an atom. Michael Dunning/Photo Researchers, Inc. Page 172: © SSPL/The Image Works. Page 184: The Granger Collection, New York.

Chapter 7
Opener: Distribution drawing representing the probability to locate the electron in the $n = 8$ state of hydrogen for angular momentum quantum number $l = 2$ (top) and $l = 6$. © John Wiley & Sons, Inc. Page 219: Photo from Wikipedia: http://commons.wikimedia.org/wiki/File:ZeemanEffect.GIF.

Chapter 8

Opener: Structure of an atom of neon. KENNETH EWARD/BIOGRAFX/ SCIENCE PHOTO LIBRARY/Photo Researchers, Inc. Page 226: © Mary Evans Picture Library/Alamy. Page 243: Science Source/Photo Researchers, Inc.

Chapter 9

Opener: A Buckminsterfullerene molecule. PASIEKA/SCIENCE PHOTO LIBRARY/Photo Researchers, Inc. Page 273: Time & Life Pictures/Getty Images. Figure 9.17: Dr. Francesco Moresco FU Berlin now TU Dresden. Figure 9.19: CHRISTIAN DARKIN/SCIENCE PHOTO LIBRARY/Photo Researchers, Inc.

Chapter 10

Opener: Fractal Geometry. GREGORY SAMS/SCIENCE PHOTO LIBRARY/ Photo Researchers, Inc. Figure 10.20: Photo from Wikipedia: http://commons. wikimedia.org/wiki/File:Liquid_helium_Rollin_film.jpg. Figure 10.21: Image from Wikipedia: http://en.wikipedia.org/wiki/File:Bose_Einstein_condensate.png.

Chapter 11

Opener: Scanning electron micrograph (SEM) of tungsten crystals. OWER AND SYRED/SCIENCE PHOTO LIBRARY/Photo Researchers, Inc.

Chapter 12

Opener: Gamma scan of a patient with prostate cancer. GJLP/CNRI/SCIENCE PHOTO LIBRARY/Photo Researchers, Inc. Page 370: Bettmann Corbis Images. Page 384: Science Source/Photo Researchers, Inc.

Chapter 13

Opener: Nuclear reactor core of Advanced Test Reactor (ATR) at the Idaho National Engineering and Environmental Lab. Science Source/Photo Researchers, Inc. Page 416: © Photo Researchers, Inc./Alamy. Page 434: Dave Pickoff/AP/Wide World Photos. Figure 13.1b: Courtesy Argonne National Laboratory. Figure 13.2b: Courtesy Purdue University. Figure 13.15: Dietmar Krause/Princeton Plasma Physics Laboratory. Figure 13.18: LLNL/Jacqueline McBride. Figure 13.23: Courtesy D. Bruce Sodee, MD. Figure 13.24: Courtesy Dr. Steve Petersen.

Chapter 14

Opener: Particle tracks from lead ion collisions seen by ALICE (a large ion collider experiment) in CERN. CERN/SCIENCE PHOTO LIBRARY/Photo Researchers, Inc. Page 449: Pictorial Parade/Getty Images, Inc. Page 453: PHYSICS TODAY COLLECTION/AMERICAN INSTITUTE OF PHYSICS/ SCIENCE PHOTO LIBRARY/Photo Researchers, Inc. Figure 14.1: FermiLAB. Figure 14.3: Photo Courtesy Lawrence Berkeley National Laboratory. Figure 14.4: FermiLAB. Figure 14.5: FermiLAB. Figure 14.19: Kamioka Observatory, ICCR (Institute for Cosmic Ray Research), The University of Tokyo.

Chapter 15

Opener: Starry Night by Vincent van Gogh, oil on canvas. SuperStock/Getty Images, Inc. Page 480: EMILIO SEGRE VISUAL ARCHIVES/AMERICAN INSTITUTE OF PHYSICS/SCIENCE PHOTO LIBRARY/Photo Researchers, Inc. Page 482: Carl Iwasaki/Time Life Pictures/Getty Images, Inc. Page 486: Richard T. Nowit/Photo Researchers, Inc. Figure 15.1: Data courtesy Hale Observatories. Figure 15.7: Courtesy of The Observatories of the Carnegie Institute of Washington. Figure 15.26: Courtesy of NASA. Figure 15.27: Courtesy Hale Observatories, California Institute of Technology.

INDEX

INDEX TO TABLES